INDUSTRIAL VENTILATION

INDUSTRIAL VENTILATION:
Engineering Principles

ROBERT JENNINGS HEINSOHN PhD, PE
Professor of Mechanical Engineering
The Pennsylvania State University

A Wiley-Interscience Publication
John Wiley & Sons, Inc.
New York / Chichester / Brisbane / Toronto / Singapore

In recognition of the importance of preserving what has been
written, it is a policy of John Wiley & Sons, Inc. to have
books of enduring value published in the United States
printed on acid-free paper, and we exert our best efforts to
that end.

Library of Congress Cataloging in Publication Data:
Heinsohn, Robert Jennings.
   Industrial ventilation: engineering principles / Robert Jennings
Heinsohn.
       p.   cm.
   Includes bibliographical references (p.
   Includes index.
   ISBN 0-471-63703-3

   1. Factories--Heating and ventilation.   2. Industrial hygiene.
I. Title.
TH7684.F2H38   1991
697.9'2--dc20   90-12378
                                                   CIP

Printed in the United States of America

10 9 8 7 6 5 4 3 2 1

To my alma mater Rensselaer Polytechnic
Institute, a venerable institution that taught a young
man he could do things he didn't think he was
capable of

Industrial ventilation is the field of applied science concerned with controlling airborne contaminants to produce healthy conditions for workers and a clean environment for the manufacture of products. To claim that industrial ventilation will prevent contaminants from entering the workplace is naive and unachievable. More to the point, and within the realm of achievement, is the goal of controlling contaminant exposure within prescribed limits. To accomplish this goal, we must be able to describe the movement of particle and gaseous contaminants in quantitative terms that take into account

(a) The spatial and temporal rate at which contaminants are generated.
(b) The velocity field of the air in the workplace.
(c) The spatial relationship between source, workers, and openings through which air is withdrawn or added.
(d) Exposure limits (time-concentration relationships) that define unhealthy conditions.

The body of knowledge in industrial ventilation consists of general design guidelines that lack precision and detail. Present practices provide no way to estimate the concentration at locations where workers are apt to be stationed. Thus, whether a process produces contaminants that warrant the installation of a control system or, if one is installed, whether it reduces concentrations to safe levels, cannot be predicted. Financial prudence requires that we develop the ability to make such predictions and to include them as integral steps in the design of products, systems, or processes. We can no longer afford to examine health and safety considerations after a design has been completed and to modify it to satisfy these considerations by costly cycles of testing and retrofit.

Scholarship is the creation and dissemination of knowledge. There is a sizable body of information on industrial ventilation, but up to this time it lacked the organization, scientific foundation, and mathematical rigor needed in baccalaureate courses in engineering. The primary goal of this text is to organize existing information in ways compatible with other topics in baccalaureate engineering disciplines. A secondary goal is to contribute to the body of knowledge itself and to try and advance our understanding of controlling contaminants in the workplace. Specifically, the book seeks to instruct in

Medical, economic, regulatory, and professional aspects of industrial ventilation
Contemporary engineering practices

Bodies of knowledge available in engineering disciplines relevant to industrial ventilation

Developments in computational fluid dynamics suitable to industrial ventilation

Occupational health and safety, and industrial ventilation in particular, is of interest to three types of students:

(a) Engineering students wishing to incorporate industrial ventilation in their work.
(b) Engineering students planning to make industrial hygiene a career.
(c) students whose baccalaureate degrees are not in engineering but who are enrolled in graduate programs of industrial hygiene.

The book is written as a textbook for seniors and graduate students in mechanical, chemical, environmental, architectural, mining, agricultural, and biomedical engineering whose interests are listed above. The book is also written for students who seek a rigorous presentation of industrial ventilation based on mathematics and principles of fluid mechanics. The objective of the book is to establish fundamental quantitative relationships that govern the movement of particle and gaseous contaminants in the workplace. Once established, these relationships can become the tools with which designers predict contaminant exposure. Simply defining the relationships is not enough however, a textbook must instruct readers how to use the relationships. Computers make it possible to predict concentration fields that could not be predicted in the past. As computer hardware and software improve, so also will our ability to analyze more complex flow fields. Computer techniques are integral parts of the book. Students should copy and modify the programs, much as they would do in learning the language of a foreign country in which they were living. The programs build on each other; latter programs using earlier programs as subprograms. The programs are written in BASIC and can be run on microcomputers (personal computers). The programs require no expertise in computers other than what is acquired in conventional baccalaureate programs in engineering.

The Appendices contain data needed in the homework problems. In addition, they are organized to illustrate the technical data engineers need in professional practice. Instructors should place the following source materials on reserve in the library and create homework problems or class projects that requires students to use this material:

OSHA General Industry Standards (33, 350)
ACGIH Industrial Ventilation, A Manual of Recommended Practice (27)
ASHRAE Handbooks

It is important that students become familiar with these original materials, not merely abridgments of them. Copious Apppendices are too cozy and foster habits of lazy scholarship. Lastly, instructors should locate source books of thermodynamic data in the reference section of the library and assign homework that requires students to locate data not provided by the text.

The book is written for readers familiar with the fundamental concepts of fluid mechanics, thermodynamics, and heat transfer. The book seeks to bridge the disciplines of engineering and industrial hygiene. Engineering knowledge contains the fundamental relationships needed to describe the movement of contaminants in the workplace; industrial hygiene contains knowledge that relates time and concentration to conditions injurious to health. Both disciplines are needed to create quantitative relationships that can become criteria upon which products, systems, and processes can be designed.

The book is organized in 10 chapters, beginning with fundamental relationships in fluid mechanics, thermodynamics, and heat and mass transfer and ending with the elements of future methods that will be used to predict contaminant concentrations in the vicinity of industrial processes equipped with industrial ventilation systems. Chapters 1 and 2 establish a quantitative relationship between air contaminants and risks to health. Chapter 3 defines criteria engineers must consider in the design of industrial ventilation systems. Chapter 4 provides ways to estimate the rate at which contaminants are generated by industrial processes. Chapter 5 shows how the assumption of "well-mixed" can be used to predict contaminant concentrations. Chapter 6 summarizes contemporary practices used in industrial ventilation. Chapter 7 develops the fundamental equations to predict the velocity field upstream of air inlets. Chapter 8 presents the fundamental equations describing the movement of particles in air withdrawn by inlets. Chapters 9 and 10 describe computational methods to predict the concentration of particle and vapor contaminants in the vicinity of an industrial process and a ventilation system. Each chapter contains homework problems suitable for use by instructors. Throughout the text accessible references in the professional literature are identified that readers should become familiar with. Important terms and concepts in each chapter are noted and are also listed in the Index. Each chapter contains a table of nomenclature that defines terms in the operational units of mass (M), force (F), length (L), mols (N), time (t), temperature (T), and energy (Q). Operational units are used for convenience, rather than using their fundamental counterparts (M, L, t).

In a report to the nation, the Department of Health and Human Services proposed the objective that ". . . by 1990, at least 70% of all graduate engineers should be skilled in the design of plants and processes that incorporate occupational safety and health control technologies." In the area of industrial ventilation it could be achieved were it not for the paucity of instructional materials that present the subject with the analytical rigor expected in engineering education. There is no fundamental reason why industrial ventilation cannot be presented with rigor. Faculty and students take seriously subjects possessing rigor. Without rigor, the subject is apt to be considered intellectually inferior or inappropriate to baccalaureate engineering education. Readers may consider these attitudes snobbish; perhaps they are, but they exist nonetheless, and are impediments to accomplishing the national objective. For some reason the technology in industrial ventilation is primitive, it is "stuck" at the level practiced in the 1950s. There is no reason to remain stuck. There is considerable advanced technology in other fields of engineering that can be applied to problems in industrial ventilation.

The current shortage of instructional material is reminiscent of the situation in the field of air pollution 25–30 years ago. Not only were there few instructional materials, but little agreement about what should be taught and who should do it. In 1974 an engineering textbook was written by H.C. Perkins (180) containing a series of topics that had not been grouped together before. In the ensuing years these topics were refined and enlarged in texts by Wark and Warner (181), Seinfeld (182, 239), Crawford (19), and Flagan and Seinfeld (264). Today these topics are recognized as the backbone of the body of engineering knowledge in air pollution. I have tried to write an engineering textbook that assembles the essential topics for a body of engineering knowledge in industrial ventilation. Some topics are drawn from other fields of engineering and science, some topics are new, and some are topics long associated with industrial ventilation but are perhaps presented in a different fashion. In all cases I wish to present the material in a quantitative fashion, suitable for baccalaureate programs of engineering and graduate programs of industrial hygiene. In the final analysis, numbers are the currency of the realm for intellectual discourse in science and engineering. It is my wish to organize material so that industrial ventilation and occupational health and safety in general will become integral parts of baccalaureate engineering programs.

The text has been used for several years in my classes but errors and ambiguities are bound to have escaped detection. The majority of the problems at the end of the chapters have been used in class but errors no doubt remain. I will compile corrections as they are brought to my attention and will send them to readers if requested to do so.

Achieving the national objective is also impeded by inadequate support for research in occupational safety and health in general and industrial ventilation in particular. The relationship between research and instruction is not obvious, but it is an intimate one. Faculty speak with students about subjects on the faculty member's mind and at the major research universities, from which most engineers graduate, research is the subject on the mind of the faculty. Secondly, faculty's interest in research mirrors the interest of the agencies funding it and, currently, agencies interested in weapons, space, communications, and energy provide most of the support for university research. Industrial ventilation and occupational safety and health are as intellectually stimulating as any other fields pursued in universities. Intellectual curiosity is stimulated by a variety of things, but sustaining it requires money. If substantial and continued support are forthcoming, the number of engineering faculty committing themselves to research and instruction in occupational safety and health will increase, without it there is no reason to expect change.

Books may have a single author, but they are generally the end result of intellectual stimulation and professional collaboration stretching over years. This book is such a product and I wish to acknowedge several elements that have led to its publication. I am grateful to the students in my classes whose reaction to concepts in the book enabled me to refine these ideas. Key portions of the book are based on research suported by the National Institute of Health and conducted by the following graduate students: Donald H. Behnke, Anne Bennett Fredo, Myung-Sik Choi, David W. Gilbert, Gail M. Haberlin, Kwang-Chung Hsieh, Dale D. Johnson, Kathy A. Knouse, Rebecca D. McCall, Mohammed H. Megherhi, Vincent D. Mortimer, Sheng Tao Yu, David A. Zak, and Mourad D. Zarouri. Gerald W. Anderson provided valuable help in preparing the computer program for the design of ducts. I am grateful for advice from colleagues at other academic institutions; Charles E. Billings, William A. Burgess, Morton Corn, Michael J. Ellenbecker, Melvin W. First, and Linda M. Hanna. I also appreciate the support and encouragement of John T. Talty and Robert T. Hughes from the National Institute for Occupational Safety and Health. Lastly, I am indebted to the advice and cooperation of a faculty colleague, Dr. Charles L. Merkle, with whom I have had the pleasure to collaborate in research in industrial ventilation. His talent and insight about modeling physical systems has been invaluable. His support as a friend and collaborator epitomizes the phrase "faculty colleague."

Writing this book has been a personally satisfying experience. It took several years to write, but is really a distillation of experience and knowledge acquired over a lifetime. In this regard, my wife Anne and daughters Janet and Beth are the true sources of inspiration and solace that have made this effort possible.

ROBERT JENNINGS HEINSOHN

University Park, PA
September, 1990

◼︎ CONTENTS

PREFACE   vii

1. INTRODUCTION   1

    1.1   Occupational Health and Safety, 1
    1.2   Risk Assessment, 6
    1.3   Liability, 16
    1.4   Contaminant Control Strategy, 17
         1.4.1   Administrative Controls, 19
         1.4.2   Engineering Controls, 20
         1.4.3   Personal Protective Devices, 22
    1.5   Components of Industrial Ventilation Systems, 24
    1.6   Objective, 26
    1.7   Deficiencies in Present Knowledge, 27
    1.8   Spurious Air Currents, 29
    1.9   Selecting a Ventilation System, 29
    1.10   Professional Literature, 43
    Problems, 47

2. RESPIRATORY SYSTEM   49

    2.1   Physiology, 49
         2.1.1   Nasopharyngeal Region, 52
         2.1.2   Tracheobronchial Region, 52
         2.1.3   Pulmonary Region, 55
    2.2   Respiratory Fluid Mechanics, 59
         2.2.1   Bohr Model, 62
         2.2.2   Bulk Convection, 65
         2.2.3   Pendelluft, 66
         2.2.4   Asymmetric Velocity Profiles, 68
         2.2.5   Taylor-Type Dispersion, 69
         2.2.6   Molecular Diffusion, 71
    2.3   Analytical Models of Heat and Mass Transfer, 72
         2.3.1   Extended Bohr Model, 76
         2.3.2   Distributed Parameter Models, 81

2.4   Spirometry, 89

2.5   Toxicology, 92

    2.5.1   Airway Irritation, 96

    2.5.2   Cellular Damage and Edema, 97

    2.5.3   Pulmonary Fibrosis, 98

    2.5.4   Allergic Response, 98

    2.5.5   Radon Carcinogenicity, 99

2.6   Dose–Response Characteristics, 100

Nomenclature, 109

Problems, 112

3. DESIGN CRITERIA   116

3.1   Contaminant Concentration, 118

3.2   Contaminant Exposure Levels, 120

    3.2.1   Time-Weighted Average, 121

    3.2.2   Short-Term Exposure Limits, 121

    3.2.3   Ceiling, 121

    3.2.4   TLV For Mixtures, 121

    3.2.5   Exposure Control Limits, 125

    3.2.6   Biological Exposure Index, 126

3.3   Fire and Explosion, 127

3.4   Noise, 138

3.5   Heat Stress, 145

3.6   Odors, 151

3.7   Engineering Economics, 153

    3.7.1   Total Capital Cost, 154

    3.7.2   Total Revenue Requirements, 155

    3.7.3   Time Value of Money, 158

Nomenclature, 161

Problems, 163

4. CONTAMINANT GENERATION   171

4.1   Field Measurements, 172

4.2   Empirical Expressions, 173

    4.2.1   Liquids in Vessels, 173

    4.2.2   Industrial Spills, 174

    4.2.3   Grinding, 175

    4.2.4   Pouring Powders, 176

    4.2.5   Spray Finishing and Coating, 177

4.3 Emission Factors, 178

4.4 Puff Diffusion, 179

4.5 Evaporation and Diffusion, 182

4.6 Diffusion Through Stagnant Air, 187

4.7 Diffusion Through Moving Air, 192

4.8 Single-Film Theory for Multicomponent Liquids, 198

4.9 Two-Film Theory for Multicomponent Mixtures, 203

4.10 Evaporation of Droplets, 211

Nomenclature, 216

Problems, 220

5. GENERAL VENTILATION WELL-MIXED MODEL 228

5.1 Unventilated Enclosures, 232

5.2 Dilution with 100% Makeup, 236

5.3 Time-Varying Source or Ventilation Flow Rate, 238

5.4 Removal by Solid Surfaces, 244

5.5 Recirculation, 246

5.6 Partially Mixed Conditions, 249

5.7 Basic Equations for Well-Mixed Model, 252

5.8 Characteristics of Sources and Air-Cleaning Devices, 258

    5.8.1 Wall-Loss Coefficient, 259

    5.8.2 Source Strength, 259

    5.8.3 Efficiency of Air-Cleaning Devices, 260

5.9 Cleanrooms, 260

5.10 Infiltration and Exfiltration, 263

5.11 Sequential Box Model, 264

5.12 Effectiveness Coefficient, 271

5.13 Makeup Air Operating Costs, 276

Nomenclature, 277

Problems, 280

6. PRESENT LOCAL VENTILATION PRACTICE 294

6.1 Control of Particles, 297

6.2 Control of Gases and Vapors, 304

6.3 Control Systems for Specific Applications, 307

6.4 Bulk Materials Handling, 310

6.5 Canopy Hoods for Buoyant Sources, 314

6.6 Air Curtains for Buoyant Sources, 319

6.7 Building Air Inlets and Exhaust Stacks, 324

6.8 Unsatisfactory Performance, 328

6.9   Exhaust Duct System Design, 333

    6.9.1   Duct Friction, 336

    6.9.2   Fittings, 336

    6.9.3   Hood Entry Losses, 338

    6.9.4   Fan Inlet and Outlet Losses, 341

    6.9.5.   Equivalent Length, 341

    6.9.6   Design Procedure, 341

6.10   Duct Design Computer Program, 342

Nomenclature, 353

Problems, 356

7. IDEAL FLUIDS   **360**

7.1   Fundamental Concepts, 362

7.2   Two-Dimensional Flow, 375

    7.2.1   Line Sink, 376

    7.2.2   Uniform Streaming Flow, 377

    7.2.3   Line Sink in a Plane in Streaming Flow, 377

7.3   Axisymmetric Three-Dimensional Flow, 381

    7.3.1   Point Sink, 382

    7.3.2   Uniform Streaming Flow, 383

    7.3.3   Point Sink in a Plane in Streaming Flow, 383

7.4   Flow Around Bodies, 387

7.5   Flanged and Unflanged Rectangular Slots in Quiescent Air, 392

    7.5.1   Flanged Slot, 392

    7.5.2   Unflanged Slot, 402

7.6   Flanged and Unflanged Inlets in Streaming Flow, 403

    7.6.1   Flanged Slot in Streaming Flow, 403

    7.6.2   Flanged Circular Inlet, 407

    7.6.3   Flanged Elliptical Inlet, 408

    7.6.4   Unflanged Slot in Streaming Flow, 409

7.7   Multiple Flanged Rectangular Inlets, 412

7.8   Flanged Inlets of Arbitrary Shape, 414

Nomenclature, 416

Problems, 418

8. PARTICLE DYNAMICS   **429**

8.1   Drag, 429

8.2   Physical Properties of Aerosols, 431

    8.2.1   Flow Regime, 431

    8.2.2   Particle Shape, 433

    8.2.3   Size Distribution, 434

8.3   Overall Collection Efficiency, 444

    8.3.1   Collectors in Series, 445

    8.3.2   Collectors in Parallel, 446

8.4   Equations of Particle Motion, 447

8.5   Freely Falling Particles in Quiescent Air, 448

8.6   Horizontal Motion in Quiescent Air, 453

8.7   Particles Traveling in a Moving Gas Stream, 454

8.8   Gravimetric Settling in Chambers, 460

    8.8.1   Laminar Settling Model, 461

    8.8.2   Well-Mixed Settling Model, 461

8.9   Gravimetric Settling in Ducts, 462

    8.9.1   Laminar Settling Model, 463

    8.9.2   Well-Mixed Settling Model, 463

8.10   Stokes Number, 465

8.11   Inertial Deposition in Curved Ducts, 466

    8.11.1   Laminar Model, 469

    8.11.2   Well-Mixed Model, 472

    8.11.3   Cyclone Collectors, 474

8.12   Impaction Between Moving Particles, 478

    8.12.1   Spray Chambers, 480

    8.12.2   Venturi Scrubbers, 484

8.13   Filtration, 487

Nomenclature, 495

Problems, 498

9. CONTROL OF PARTICLE   **515**

9.1   Reach of Inlets, 515

    9.1.1   Uniform Streaming Flow of Particles, 519

    9.1.2   Line Source of Particles in Quiescent Air, 519

9.2   Particle Sampling, 519

9.3   Particle Removal by Line Sink in Streaming Flow, 525

9.4   Flanged and Unflanged Rectangular Inlets in Streaming Flow, 530

    9.4.1   Flanged Slot, 531

    9.4.2   Unflanged Slot, 531

    9.4.3   Upper Dividing Streamline for a Flanged Slot, 532

    9.4.4   Bounding Trajectory, 533

9.5   Finite-Difference Methods, 538

9.6   Grinding Booth, 544

9.7   Windbreaks, 552

Nomenclature, 558

Problems, 560

10. CONTROL OF GASES AND VAPORS   **563**

10.1   Tunnels, 565

10.2   Laminar Flow Model, 571

10.3   Numerical Solution—SIMPLER Algorithmn, 577

10.4   Rim Exhauster for Open Vessels, 581

10.5   Turbulent Flow Model, 588

10.6   Push–Pull Ventilation Systems, 594

10.7   Future Developments, 602

Nomenclature, 606

Problems, 610

REFERENCES   **611**

APPENDIX

A-1   OSHA Permissible Exposure Limits for Common Indus-
       trial Materials as of 1989   633

A-2   Emission Factors for Particles from Uncontrolled Metal-
       lurgical Processes   637

A-3   Emission Factors for Volatile Hydrocarbons from Uncon-
       trolled Sources   641

A-4   Emission Factors for Uncontrolled Mineral Processes   645

A-5   Emission Factors for Uncontrolled Chemical Processes   648

A-6   Emission Factors for Uncontrolled Food Processes   651

A-7   Emission Factors for Indoor Processes and Activities   653

A-8   Vapor Pressures and OSHA PEL for Industrial Volatile
       Liquids   655

A-9   Henry's Law Constant and Diffusion Coefficients of
       Contaminants in Air and Water   658

A-10   Critical Temperatures, Pressures, and PEL for Common
       Toxicants   660

A-11   Thermophysical Properties of Air   661

A-12   Runge–Kutta Solution to a First-Order Differential Equa-
       tions   663

A-13   Newton–Raphson Method for Simultaneous Equations   666

A-14   Newton's Method   669

A-15   Basic Iteration Methods (Bisection and Newton Method)   671

A-16   Outdoor Air Requirements for Ventilation   673

A-17   Ventilation Duct Design   674

A-18   Common Physical Constants and Conversions   676

INDEX   **679**

INDUSTRIAL VENTILATION

CHAPTER 1

Introduction

Goal: Students will be expected to learn,
  – how to anticipate and rank risks quantitatively
  – strategies to control workplace contaminants

1.1 OCCUPATIONAL HEALTH AND SAFETY

Occupational hazards have always been associated with industry, and as industry grew during the 19th century so did illness and injury related to industrial activity. Beginning in the later part of the 19th century and continuing to the present, an understanding of occupational diseases and injury grew which lead to administrative and technological efforts to reduce the number of injuries and diseases per 100,000 workers. The relationship between occupation and injury is immediate and recognition of the hazard is obvious. The relationship between occupation and illness is often delayed, complicated by nonoccupational sources of disease and far more difficult to document.

To appreciate the risks associated with human endeavors, Starr (144) suggested that it was useful to categorize risks as voluntary and involuntary. *Voluntary risks* are assumed by individuals of their own free will. Examples include risks associated with recreational sports and flying private aircraft. *Involuntary risks* are imposed on individuals because of circumstances beyond their control. Examples include risks associated with using elevators in tall buildings and commercial aircraft. The issue becomes cloudy when discussing private automobiles because individuals voluntarily drive their own automobiles yet are at risk from accidents caused by others.

Figure 1.1 shows how voluntary and involuntary *risks* vary with the *benefit* people perceive the risk provides. For a first approximation the risk is computed as the statistical probability of fatality per hour of exposure associated with the activity. The benefit derived from each activity was converted into a dollar equivalent as a measure of the integrated value to the individual. For voluntary activities the amount of money spent on the activity by the average individual was assumed to be proportional to its benefit. In the case of involuntary activities, the contribution of the activity to the individual's annual income was assumed to be proportional to its benefit. While the approximations used by Starr are crude, it is apparent that the large difference between voluntary and involuntary risks suggests that individuals accept higher risks in voluntary activities but expect lower risk in involuntary activities. Other than self-employed people and some

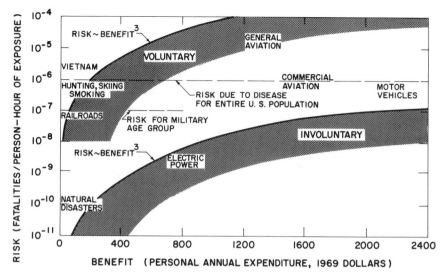

Fig. 1.1 Risk versus benefit for various kinds of voluntary and involuntary activities (adapted and redrawn from ref. 144).

exotic occupations, risks associated with most employment are considered involuntary and the public expects these risks to be no larger than those associated with naturally occurring events.

An interesting example showing the complexity of voluntary and involuntary risks concerns the political action to enact nonsmoking ordinances and the use of *wood-burning stoves*. The concentration of carbon monoxide, total suspended particles (TSP), and several polycyclic aromatic hydrocarbons (including benzo[a]pyrene) in homes using wood-burning stoves (236) is in general comparable to the concentration of these materials in public places in which smoking is allowed (24,29). Indoor pollution from wood-burning stoves is a voluntary risk to the home owner, while tobacco smoke in public buildings is an involuntary risk to the nonsmoker. While the risks to health may be comparable, many people object to the involuntary risk yet accept with equanimity the voluntary risk. Some observers would consider this hypocrisy, others would consider it the freedom of choice. The issue is even more complex because in some regions of the United States up to one half the outdoor TSP is due to wood-burning stoves. The irony is often lost that the voluntary risk to home owners using wood-burning stoves is the involuntary risk of outdoor air pollution to their neighbors.

Data on occupational injuries and deaths is sufficiently accurate and comprehensive to provide guidance for prevention. Data about occupational illness is far less accurate or comprehensive. Useful benchmarks for comparing occupational disease and injury statistics with those of the population as a whole is the fact that within the United States each year people die (332) from the following:

765,000 from heart attacks

454,000 from cancer

390,000 diseases related to smoking (approximately 1 in every 6 deaths)

100,000 from alcohol

48,000 from motor-vehicle accidents

30,000 from suicides

19,000 homicides

11,000 from falls

5,600 from drowning

4,800 from fire and burns

353 from drowning in their bathtubs

91 from lightning

The Bureau of Labor Statistics (BLS) of the U S Department of Labor compiles the most reliable data on occupationally related deaths, although it only counts about three-fourths of the nation's workforce. Adjusted to include the entire nation's workforce, the Congressional Office Of Technology Assessment (OTA) found (166) that approximately 6000 deaths occurred from occupational injuries, or about 25 deaths per each working day. In 1912 the work force was only half as large as now, yet an estimated 18,000–21,000 workers lost their lives. Table 1.1 shows the number of deaths and disabling injuries from several industries. On the basis of the number of workers, the most dangerous industry is mining, followed by construction, agriculture, and transportation. The BLS data show that about half the fatal occupational injuries involve motor vehicles, off-road industrial vehicles and falls. An ironic occupational hazard are the United States retail trades where there are approximately 1600 homicides associated with robberies! In 1986 the National Safety Council (NSC) reported (332) that there were 1.8 million injuries that resulted in "lost workdays." The NSC reported that the total cost of work injuries was approximately 34.8 billion dollars in 1986. The leading (28%) disabling, nonfatal injuries are overextensions (largely to the back).

Risks to maintenance workers are particularly high. While safe procedures are generally adopted for scheduled maintenance, workers often have to make emergency repairs following equipment breakdowns. To accomplish these repairs, safety controls and established procedures are sometime circumvented. Unless equally safe alternatives are adopted, workers are at high risk. Because of the adverse conditions under which these repairs have to be made and pressure from superiors to resume production, careless practices are apt to occur.

Health care costs have grown rapidly in the United States and industry in general has taken bold steps to reduce these costs. In the last two decades health-care expenditures, including workers' compensation, have risen from approximately 2% the gross national product to 11% in 1989. If there had been a commensurate improvement in workers' health and safety, these expenditures would be seen as worthwhile. Unfortunately, workers' health and safety have improved only marginally, and certainly not by a factor of 5. The reasons for the precipitous rise in costs are complex, but the public can ill afford to indifferent to the burden these costs are to business.

In 1982 the National Institute for Occupational Safety and Health (NIOSH) ranked occupational diseases and injuries in the following order of importance:

TABLE 1.1 Work Accidents in 1986[a]

Industry Division	Workers[a] (millions)	Deaths 1986	Deaths Change from 1985 (%)	Death Rates[b] 1986	Death Rates Change from 1985 (%)	Disabling Injuries[a] in 1986 (thousands)
All industry	108.9	10,700[c]	-6	10	-9	1800[c]
Agriculture[d]	3.2	1700	+6	52[e]	+6	170
Mining[d]	0.8	400	0	50	+25	40
Construction	6.3	2100	-9	33	-13	220
Manufacturing	19.3	1100	-15	6	-14	310
Trade[a]	25.7	1100	-15	4	-25	310
Services[d]	31.6	1500	-12	5	-17	350
Government	16.5	1300	0	8	0	240
Transportation and public utilities	5.5	1500	0	27	0	160

Source: National Safety Council estimates (rounded) based on data from the National Center for Health Statistics, state vital statistics departments, and state industrial commissions. Numbers of workers are based on Bureau of Labor Statistics data and include persons aged 14 and over.

[a] Abstracted from ref. 332. See definitions on inside back cover of ref. 332.

[b] Deaths per 100,000 workers in each group.

[c] About 3900 deaths and 200,000 of the injuries involved motor vehicles.

[d] Agriculture includes forestry and fishing (see also p. 94 of ref. 332). Mining includes quarrying and oil and gas extraction. Trade includes wholesale and retail trade. Services includes finances, insurance, and real estate.

[e] Agriculture rate excludes deaths of persons under age 14. Rates for other industry divisions do not require this adjustment. Deaths of persons under age 14 are included in the agricultural death total.

(1) Occupational lung disease (including lung cancer, pneumonoconioses, and occupational asthma).
(2) Musculoskeletal injuries (including back injury, carpal tunnel syndrome, arthritis, and vibration white finger disease).
(3) Occupational cancers (other than lung cancer).
(4) Traumatic deaths, amputations, fractures, and eye losses.
(5) Cardiovascular diseases (including myocardial infarction, stroke, and hypertension).
(6) Reproductive problems.
(7) Neurotoxic illness.
(8) Noise-induced hearing loss.
(9) Dermatologic problems (including dermatoses, burns, contusions, and lacerations).
(10) Psychological disorders.

Occupational illness is difficult to define for three reasons. First, many occupational diseases are indistinguishable from nonoccupational diseases. Second, the relationship between a disease and an occupational environment is not always recognized. Last, many diseases have long latency periods and occur after occupational exposure has ceased or the worker has changed jobs or retired. The last reason is the most troubling factor inhibiting accurate assessments of risk. Serious diseases such as respiratory and neurologic disorders and cancers are not generally captured in BLS records of work-related illness, but after considerable debate it is generally agreed (166) that approximately 5% of the national 20,000 cancer deaths are related to occupational exposures.

Data on occupational illness obscure the important fact that occupational illness is preventable, that workers in some industries have disproportionate risks and that once an occupational disease is identified controls can be adopted to reduce risks. Improvement of workplace health and safety proceeds in three steps:

(1) Identify the hazard and its causative agents.
(2) Select preventive strategies.
(3) Decide to control the hazard.

Steps 1 and 2 are technical and require specialists. Step 3 is administrative and involves managers, employers and elected officials. Depending on the severity of the hazard, step 3 can be taken before the hazards are fully identified and controls developed.

Toxic substances in the air in the workplace is one of many hazards workers to which are exposed. It is one of the most pervasive hazards since it affects workers in seemingly nonhazardous jobs merely because they are in proximity to the source of contamination. Unlike the contamination of groundwater with its large remediation period, contaminants in air can be removed almost immediately by merely eliminating their generation.

1.2 RISK ASSESSMENT

Entrepreneurs strive to compress the time between invention and commercial development to a few years rather than decades, as was the custom in the past. Governmental regulatory procedures exist for premarket testing of new materials and devices, yet the number of new products and devices is too large for an orderly testing of all. Our political system recognizes this fact and tempers its regulatory responsibility with the realization that the nation's future economic strength depends on its ability to sustain a preeminent role by developing new products and to reassume a position of leadership in process technology (253). The dilemma has been phrased aptly (161) as "Where should we spend whose money to undertake what programs to save which lives with what probability?"

A decision to develop new product is often cast in terms of *cost–benefit ratio*. If the ratio is small, conventional wisdom argues that the venture should proceed. All things being equal, it is financially prudent to pursue the venture which has the lower ratio, but a low cost–benefit ratio is not a basic moral norm that is intrinsically "good." The cost–benefit ratio is not a moral norm because it sidesteps the fundamental question "Who benefits ... who pays?" A low cost–benefit ratio is not intrinsically good because it ignores the question of which political constituency derives the benefit and which bears the cost. The majority of large societal issues such as pollution and poverty are issues where the two political constituencies are not the same (256–262). If an individual plans to buy a personal computer with his or her own funds, the person buying it also profits from its use. Thus if two machines have the same capabilities, buying the one with the lowest cost–benefit ratio has obvious virtue. On the other hand, consider citizens in eastern Ohio troubled by large pollutant concentrations of sulfur dioxide from electric utility plants. Increasing smokestack heights is cheaper than installing sulfur dioxide removal systems (scrubbers). Defining costs and benefits is a dilemma. A tall stack does not reduce the total emission of sulfur dioxide to the atmosphere, it only lowers the ground-level concentration for Ohioans, and does little to alleviate the ground level concentrations from Pennsylvania to Canada. Thus the cost–benefit ratio of stack extensions is lower than scrubbers for Ohioans, but it ignores people downwind. People downwind would prefer the installation of scrubbers, but they do not have to bear the cost!

To illustrate the difficulty in assessing risk, consider the creation of new materials. The age of metals gave way to the age of petrochemicals, which today is giving way to the age of ceramics and composite materials. The emerging materials industry is quite different from past industries in several significant ways. First, unusual chemical compositions and manufacturing processes are created before the health and safety implications can be understood fully. Second, the pace of these developments is faster than the pace at which health tests can be conducted. Third, new products are used by a diverse group of small manufacturers who are difficult to identify or monitor. Last, only small amounts of these new materials are actually used in final products. Thus new materials are manufactured in plants requiring smaller capital investments than those that produced metals or petrochemicals. The building blocks of advanced materials are metals, polymers, ceramics, semiconductors, and composites that are assembled sometimes molecule by molecule by unique processes, many of which are not automated and may indeed be labor-intensive. Some steps in the process involve exotic hazardous materials such as carcinogenic organics, highly toxic gases, and submicron particles on whose surface highly active organics have been added.

The majority of new materials are not invented by *Fortune 200* companies, but by small entrepreneurial firms that are labor-intensive. The number of the exposed people is small, but the exposure is liable to affect a large percentage of the total workforce. These firms are the least likely to have personnel whose sole occupational duty is health and safety. Thus there are no specialists to call upon, as can be called upon in *Fortune 200* companies. Last, the volatility of these small firms and the mobility of their employees prevents defining the exposed population accurately, which in turn inhibits the accuracy of epidemiological studies.

The greatest barrier to maintaining safe and healthy working conditions is the inadequacy of the data to estimate the health risk of new materials. For the most part, new materials are not obvious hazards such as pesticides, herbicides or explosives. There are approximately 60,000 chemicals and 2 million mixtures in commercial use. Each year more than 1000 new chemical compounds are synthesized. The majority of these materials do not reach the public as end products but are used as intermediary materials in the production of finished products. These intermediate chemicals are often called *"chemicals of commerce"*.

Actions presently followed by regulatory agencies are based on the following assumptions.

(a) Every perceived risk cannot be evaluated at the same time.
(b) Plans must be developed to determine which materials pose the largest potential risk.
(c) The actual risk of materials having the largest potential risk will be determined first.

Evidence of risk involves several components. Acute (short-term) hazards are the easiest to diagnose because symptoms are immediate and sometimes dramatic. The following toxins produce chronic (long-term) effects and are more difficult to assess:

(a) Carcinogens, teratogens, neuropathogens, mutagens
(b) Disorders of the pulmonary, cardiovascular, and immune systems.
(c) Disorders of the skeletal system, blood, and bone marrow.
(d) Disorders of the skin and mucous membrane.
(e) Hypersensitivity.

Assessing risk requires knowledge of the actual amounts of the material used and the percent discharged to the workplace. Merely listing materials used can create a false sense of risk. To begin with, engineers should conduct an *audit* that indicates the amounts of material used over a period of time. In addition, one must estimate the amounts remaining in the product, the amounts removed as waste, and the amounts that may escape to the workplace. Such audits are really huge mass balances and identify clearly processes and materials posing a health risk.

Responding to the *Toxic Substances Control Act* (*TSCA*), the EPA catalogued more than 56,000 manufactured or imported substances used in manufacture. The list is called the *TSCA inventory*. The TSCA inventory excludes classes of materials regulated under other federal statues: food additives (8627 items), prescription and nonprescription drugs (1815 items), cosmetic ingredients (3410 items) and pesticides (3350 items).

Identifying properties of chemicals that require special handling can be a complicated matter. Lists of chemicals are published by numerous governmental, trade, and professional agencies. The following are four familiar lists:

(a) National Institute For Occupational Safety And Health (NIOSH)—*Registry of Toxic Effects of Chemical Substances* (*RTECS*) (164).

(b) Occupational Safety and Health Administration (OSHA)—List Of Toxic And Hazardous Substances (33).

(c) National Institute For Occupations Safety and Health (NIOSH)—Pocket Guide To Chemical Hazards (160).

(d) American Conference of Governmental and Industrial Hygienists (ACGIH)—Threshold Limit Values for Chemical Substances and Physical Agents in the Workroom Environment with Intended Changes (published and updated yearly) (94).

The *Registry* is published and updated by NIOSH every few years in compliance with the 1970 Occupational Safety and Health Act. The *Registry* is the most comprehensive source to consult. It contains toxicity data extracted from the scientific and professional literature for a fraction of the approximately 65,000 chemical compounds listed in the registry. The "toxicity data should not be considered a definition of values for describing safe dose for human exposure" (164). Figure 1.2, taken from a study (165) of a representative group of chemicals, shows the limited knowledge about the toxic properties of chemicals listed in RTECS. Most alarming is the fact that there is no toxicologic information about nearly 75% of all of the chemicals in the three production categories which represent the majority (both in number and volume) of chemicals used in the United States. The TSCA act also requires the National Institutes of Health (NIH) and Environmental Protection Agency (EPA) to create a computer data base and search programs for chemicals listed in RTECS. The data base is called the *Chemical Information Service* (*CIS*) and is an iterative, on-line service that enables users to retrieve toxicologic data for chemicals identified by their RTECS numbers. In addition, CIS enables users to search for chemicals with specific toxicologic properties, structures, and doses. The American Chemical Society created a registry that uniquely identifies specific compounds by a *Chemical Abstract Service* (*CAS*) number. The CAS inventory contains many substances posing no hazard to health and is considerably larger than RTECS. Many professional journals require authors to list the CAS numbers for all chemicals included in their article. Chemicals listed in the RTECS and CIS also cite the CAS numbers.

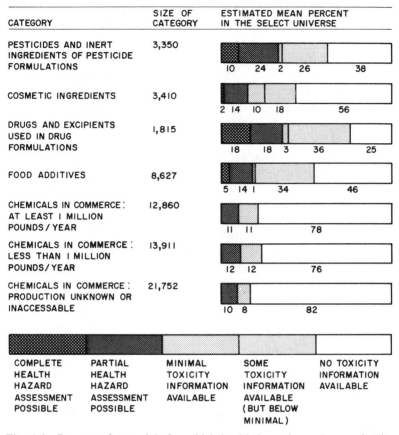

CATEGORY	SIZE OF CATEGORY	ESTIMATED MEAN PERCENT IN THE SELECT UNIVERSE
PESTICIDES AND INERT INGREDIENTS OF PESTICIDE FORMULATIONS	3,350	10 24 2 26 38
COSMETIC INGREDIENTS	3,410	2 14 10 18 56
DRUGS AND EXCIPIENTS USED IN DRUG FORMULATIONS	1,815	18 18 3 36 25
FOOD ADDITIVES	8,627	5 14 1 34 46
CHEMICALS IN COMMERCE: AT LEAST I MILLION POUNDS/YEAR	12,860	11 11 78
CHEMICALS IN COMMERCE: LESS THAN I MILLION POUNDS/YEAR	13,911	12 12 76
CHEMICALS IN COMMERCE: PRODUCTION UNKNOWN OR INACCESSABLE	21,752	10 8 82

COMPLETE HEALTH HAZARD ASSESSMENT POSSIBLE	PARTIAL HEALTH HAZARD ASSESSMENT POSSIBLE	MINIMAL TOXICITY INFORMATION AVAILABLE	SOME TOXICITY INFORMATION AVAILABLE (BUT BELOW MINIMAL)	NO TOXICITY INFORMATION AVAILABLE

Fig. 1.2 Percent of materials for which health hazards are known (165).

In addition to these readily available lists are compilations published by NFPA, IRC, CAG, and SEVESO, to name just a few associations. Commercial book publishers publish handbooks (such as 222, 237, 238), compendia and research monographs on a regular basis that contain this material. Last, states such as Massachusetts and California publish a "right-to-know" list of substances. It is not the shortage but plethora of lists that causes confusion.

Two procedures required by the federal government help to identify hazardous substances and to handle them properly. Whenever chemicals are transferred between buyer and seller OSHA requires that the transfer be accompanied by a *Material Safety Data Sheet* (*MSDS*) that specifies the following properties of the material:

(a) Manufacturer's name and chemical synonym.

(b) Hazardous ingredients (pigments, catalysts, solvents, additives, vehicle).

(c) Physical data (boiling temperature, vapor pressure, solubility evaporation rate, percent volatile material).

(d) Fire and explosion data (flash point, flammability limits, ignition temperature, fire fighting procedures).

(e) Health hazard (threshold limit, effects of overexposure, first aid procedures).

(f) Spill and leak procedures, waste disposal methods

While the MSDS's don't often have all the data an engineer wants, they at least serve to alert engineers of possible hazards. Companies are obliged to file MSDS's and to make them available to workers, thus a "paper trail" can be followed to trace materials from their creation to destruction. The second procedure guides the public about hazardous new materials. The TSCA legislation requires that the EPA evaluate new substances and methods used to manufacture them to determine potential release points, estimate potential exposures, and determine whether it will be necessary to specify specific procedures to minimize exposure. Companies planning to produce or import chemicals not on the TSCA inventory are required to notify the EPA at least 90 days prior to action. The notification is called a *Premanufacture Notice* (*PMN*). The document requires the formula, chemical structure, its use, details about the production process so that points of release can be anticipated, and estimates of the emissions and human exposure. Other data required are the physical and chemical properties of the substance (vapor pressure, solubility in water or solvents, normal melting and boiling temperature, particle size if it is a powder, Henry's law constant, pH, flammability, volatilization from water, and any toxicologic data that may be available). New substances are screened by the EPA and assigned a risk category that obliges the company to control release (through general ventilation, protective clothing, respirators, glove boxes, etc.) or to test the compound for its toxic or environmental effects. Toxicity is assessed by examining the toxicity of analogous substances (i.e., comparable molecular structure and physical properties). Both MSDS and PMN procedures were created to minimize workplace exposures, prevent chronic exposures of people living near manufacturing plants, and reduce acute exposure of people affected by transportation accidents.

Key elements in risk assessment are the fields of epidemiology and toxicology. *Epidemiology* is the branch of medicine that investigates the cause of disease; *toxicology* is the study of the adverse effects of chemical agents on biologic systems (149). Epidemiology begins with the symptoms human subjects display and relates them to certain causative factors that can be shown to be statistically significant. In a sense, epidemiology looks backward to what has already happened. Thus it is inherently limited in its ability to assess the effects of new materials and new manufacturing processes. There are several limitations to epidemiology in occupational health:

(a) Information is often lacking about the duration and concentration of the exposure.

(b) There is frequently a long time between exposure and symptoms of the disease. Induction periods of 5–50 years are common in lung disease and cancer.

(c) Workers change occupations and/or employer and it is difficult to define exposure.

(d) The number of individuals in the sample is often small, which complicates determining biologically significant elevations in risk.

(e) Multiple exposures to several chemicals in complex industrial settings makes it difficult to determine causative agents.

(f) Individuals are exposed to agents that affect health (smoking, alcohol, drugs, etc.) which are unrelated to an industrial exposure.

Toxicological studies require considerable time and involve large populations of subjects. Because the subjects are often animals with different physiologies, extrapolating the results to humans who might be subjected to low doses is fraught with debate and controversy. Occupational exposures that produce chronic effects require tests over the animal's lifetime. Occupational exposures related to pregnancy, offspring, and fertility of offspring require tests over several generations of animals. Finally, there are very few tests for neurological damage.

Because of the rapid rate at which new materials and new processes appear on the industrial scene it is obvious that detailed epidemiological and toxicological studies must be restricted to chemicals for which previous screening indicates that more detailed study is warranted. There is an urgent need for rapid, economical tests that provide accurate, albeit partial data about the health hazard of chemicals. Such tests exist and there is reason to be optimistic that they will continue to be developed in the future. The *Ames test* (requiring 24 hours) is used to show if certain types of mutations in salmonella cells occur when exposed to chemical agents. Mutagenesis is not carcinogenesis, but it often precedes it. Thus a positive Ames test is cause to suspect a potential carcinogen. From a legal position, a positive Ames test is grounds for treating mutagens as potential carcinogens. There are several short-term (1–30 day) inexpensive bioassays using hamsters, because the short-term response of the hamster is believed to be similar to the human response. One such test involves exposing hamsters to a chemical agent for prescribed periods of time and concentration, after which its lungs are rinsed with a saline solution and the biochemical and cellular material in the removed fluid is examined. The test is given the acronym BAL (*bronchoalveolar lavage fluid*) (162). By repeating the test for known toxins and benign materials, one can define a relative toxicity. Parameters that can be measured cover a large number of toxic effects: inflammation, pulmonary edema, cellular damage, cellular secretions, and the endocytic capacity of pulmonary macrophage.

Risk assessment is a formal field of study in science and technology. An area where it has received considerable attention has been the design and operation of nuclear reactors. A considerable body of knowledge exists on the theory of risk and the subject has been treated mathematically with great sophistication (e.g., reference 162). Applying the theory to occupational health and safety is hampered by the inability to define the relationships between the sources of accident and disease and to express the occurrence of each as a probability. When safety and disease can be explained in probabilistic terms, the theory of risk analysis can be tapped.

Following an assessment of risk in quantitative terms, it is also necessary to place risks in *perspective*, that is, to view risks in relation to one another (256–261) and to personal activities citizens engage in that are either both voluntarily and involuntarily. If perspective is lost, risk data acquire a false concreteness in which priorities are apt to be set foolishly.

We live in an era when *perceptions*, including those about risk, become so important they are apt to be taken as reality (289). If perceptions were based on fact and presented dispassionately, there would be little need to worry. Unfortunately, this is not the case. We also live in an era when propaganda is ubiquitous (e.g. advertising and political discussion). *Propaganda* is the dissemination of information designed to provoke a certain response by arousing our emotions rather than engaging our minds. Perceptions can be manipulated by skillful propaganda that is an impediment to formulating sound public policies. For example, terrorist bombing of commercial airliners is abominable. News of it captivates the nation's attention from time to time, yet since 1976 an average of only 61 people have died per year throughout the world. Americans lose sight of the fact that they are far more likely (332) to die from falling in their homes (5700 deaths), home fires (3900 deaths), choking on their food (2400 deaths), or drowning in their bathtubs (225 deaths).

To acquire perspectives about risk one must first understand the units with which data are presented (e.g. to ask "where are the zeros?" or if the values are small, "where is the decimal point?"). For example, if an air contaminant is described as having a concentration of 0.000001 mol fraction, or is described as being present at 1 PPM or 1000 PPB, layman are apt to respond quite differently unless they know that these concentrations are identical. The location of zeros enables one to either excite or lull without logical persuasion—0.000001 sounds small while 1000 PPB sounds large. Presenting data in this way is pernicious to logical debate, it lacks ethics, but unfortunately it is often employed. Lastly, even the units themselves tax comprehension, for example, trichloroethylene in drinking water at a concentration of 1 PPB may sound ominous, but such a ratio is equivalent to the ratio of one Chinese citizen divided by the entire population of the People's Republic of China!

Some activities have a familiarity that masks their danger, for example, driving while mildly intoxicated and using gasoline to clean paint brushes or light charcoal briquettes in a backyard grill are unfortunately practiced by many Americans. Even if the units are understood, the perception of risk (258) may produce a false consciousness or an incoherent understanding of reality. Fears about the involuntary risks of nuclear power are disproportionate with the historic record of accidents and deaths. Many elderly are in such fear of falling that they become unnecessarily sedentary. Efforts to ban smoking in public places for reasons of protecting the health of the nonsmoker, is often a disguise for matters of taste, that is, the revulsion ex-smokers have for a habit they have kicked or personal behavior nonsmokers find offensive. Table 1.2 shows the *annual risks* (probability that the event will occur per person) for several activities (256).

One must be careful in using Table 1.2 since the types of people affected are quite different. For the most part, police killed in the line of duty are healthy adults and mountaineers are healthy young people, while those killed in falls at home are elderly. Nonetheless, important differences of several orders of magnitude (a factor of 10) exist between having a motor vehicle accident, incurring cancer from eating peanut butter, and incurring cancer from drinking water containing chloroform or trichloroethylene.

An example where perspective has been debated is the *removal of asbestos* from schoolrooms. The probability of children contracting lung cancer (mesothelioma) is estimated (260) to be five per million lifetimes, less than 1/5000 the chance of death faced by other events in children's lives. The risk to a building

TABLE 1.2   **Risks of Various Activities**[a]

Action	Annual Risk	Uncertainty
All cancers	2.8×10^{-3}	10%
Cancer related to cigarettes		
(1 pack/day)	3.6×10^{-3}	Factor of 3
Electrocution	1.1×10^{-4}	5%
Police killed line of duty (total)	2.2×10^{-4}	20%
Motor vechile accident (total)	2.4×10^{-4}	10%
Mountaineering (mountaineers)	6×10^{-4}	50% alcohol
Light drinker	2×10^{-5}	Factor of 10
Motor vehicle accident (pedestrian)	4.2×10^{-5}	10%
Death due to home falls	3.5×10^{-6}	
4 tablespoons peanut butter/day	8×10^{-6}	Factor of 3
Drinking water with EPA limit		
of chloroform	6×10^{-7}	Factor of 10
Drinking water with EPA limit		
trichloroethylene	2×10^{-9}	Factor of 10

[a]Abstracted from ref. 256.

occupant for a 10-year exposure is less than one-fiftieth (1/50) the risk of a highway fatality resulting from commuting by car to and from the building (262). In addition, careless removal of asbestos poses major risks to workman, their children, and the population as a whole. Consequently, it has been decided to leave the asbestos in many buildings until the building undergoes major renovation or demolition since removal may not reduce already low concentrations and a much larger improvement in public health could be bought by spending the money on other programs.

While the techniques discussed above enable one to establish an order of magnitude to risks, other techniques can improve perspective. With regard to the carcinogenicity of certain chemical agents, such a technique has been proposed by Ames et al. (257). While animal cancer tests cannot be used to predict absolute human risk, the tests can be used to produce an index for setting priorities that reflect the carcinogenic hazard potency of certain chemical agents. A comparison of hazards from various carcinogens ingested by humans should reflect the vastly different potency carcinogens produce in humans. The usual measure of potency (TD_{50}) is the daily dose rate (in milligrams of intake per kilogram of body weight) to half the percent of tumor-free animals by the end of a standard lifetime. To arrive at an index, Ames defined a term that expressed the daily lifetime dose (in milligrams per kilogram of body weight) as a percent of the rodent TD_{50} dose (in milligrams per kilograms) for each carcinogen. Ames called the index the *Human Exposure Dose/Rodent Potency Dose* (HERP),

HERP = daily lifetime human dose (mg/kg)/rodent TD_{50} (mg/kg)

The TD_{50} values can be taken from a data base for 975 chemicals and the human exposures can be estimated from data reported in the professional literature. Table 1.3 lists several HERP values taken from Ames et al. (257).

TABLE 1.3 HERP Index[a]

Daily Human Exposure	Dose (μg) per 70-kg person	HERP (%)
Chlorinated tap water	Chloroform (83)	0.001
Contaminated well water (1 L) (Woburn, MA)	Trichloroethylene (267)	0.0004
	Chloroform (12)	0.0002
	Tetrachloroethylene (21)	0.0003
Swimming pool (1 hr, child)	Chloroform (250)	0.008
Conventional home (14 hr/day)	Formaldehyde (598)	0.6
	Benzene (155)	0.004
Mobile home air (14 hr/day)	Formaldehyde (2200)	2.1
Pesticide residue on food	PCB (0.2)[b]	0.0002
	DDE (2.2)[c]	0.0003
	Ethylene dibromide (0.42)	0.0004
	Dimethylnitrosamine (0.3)	0.003
	Diethylnitrosamine (0.1)	0.006
Cooked bacon (100 g)	Aflatoxin (0.064)	0.03
Peanut butter (32 g)	Mix of hydrazines	0.1
Raw mushroom (15 g)	Saccharin (95,000)	0.6
Diet cola (12 oz)	Ethyl alcohol (18 mL)	2.8
Beer (12 oz)	Ethyl alcohol (30 mL)	4.7
Wine (250 mL)	Formaldehyde (6100)	5.8
Worker average daily intake	Ethylene dibromide (150,000)	140.0
High-exposure farm worker		

[a]Abstracted from ref. 257.
[b]PCB—polychlorinated biphenyls.
[c]DDE—principal metabolite of DDT.

The HERP index provides an additional perspective about carcinogenic hazards. Contaminated wells pose a considerably smaller risk than diet cola, wine, beer, or several natural foods. Chlorine added to water kills bacteria but also interacts with organic matter to produce chloroform; nevertheless, the amount of chloroform in tap water results in a lower HERP index than common soft drinks and natural foods. Pesticide residues on food that once caused a great deal of anxiety are seen to result in a HERP of no particular concern. With respect to ethylene dibromide, the exposure of agricultural workers was huge compared with exposures associated to residues on foods. Plants produce their own toxins to combat a variety of insects and fungi; unfortunately, some of these possess significant HERP. The aflatoxin in peanut butter is 2 PPB, which corresponds to a HERP of 0.03% for a single peanut butter sandwich. The formaldehyde exposure for average United States workers is considerably higher than dietary intake.

It would be a mistake to use HERP data as an absolute estimate of human hazard because of the uncertainty of applying rodent cancer tests to humans. At low dose rates, human susceptibility may differ systematically from rodent susceptibility and the shape of the dose–response curves are not known. In the final analysis, the HERP index is not a scale of human risks but only a tool to help set priorities. No society is risk–free, and to believe that today is more risky than the past is doubtful and certainly unproven.

Illustrated Example 1.1—Involuntary And Voluntary Risks

It was recently announced that the drinking water in your town contained 100 PPB of chloroform (CCl_4, M = 119). (The EPA maximum allowable concentration is 100 PPB.) At a public meeting several people are alarmed at the (involuntary) risk of tumors. You assert that there is more (voluntary) risk from the saccharin in one 12-oz can of diet cola than all the tap water one could drink in a day! The audience is incredulous and hoots you down. Use Table 1.3 to estimate the number of 12-oz glasses of tap water to produce a risk equivalent to one 12-oz can of diet cola.

The concentration of CCl_4 in water, c (μg/kg water), is

$$c = (10^{-7} \text{ mols } CCl_4/\text{mols } H_2O)119/18$$
$$= 661.1/10^9 \text{ kg } CCl_4/\text{kg } H_2O = 661.1 \ \mu g \ CCl_4/\text{kg } H_2O$$

One 12-oz glass contains 340.2 g of water. Thus the mass of CCl_4 per glass of water (m) is

$$m = (340.2)(661.1)/1000 = 224.8 \ \mu g \text{ of } CCl_4/\text{glass}$$

In Table 1.3 83 μg of CCl_4 is used as the basis of comparison; thus 83 μg of CCl_4 will be ingested after the consumption of 0.37 (83/224.8) glasses of tap water. In terms of HERP, the number of glasses of water equivalent to one 12-oz can of diet cola can be found from the following,

$$\text{HERP (diet cola)}/\text{HERP (tap water)} = 0.6\%/0.001\% = 600$$
$$600 = 1 \text{ can of diet cola}/0.37 \text{ glasses of water}$$

Thus one 12-oz can of diet cola poses a risk equivalent to 222 (0.37×600) 12-oz glasses of public drinking water. The average consumption of water for an adult is 1 l (approximately three 12-oz glasses) per day. If the tap water contained only 10% of the EPA maximum allowable amount of CCl_4 (10 PPB), one can of diet cola glass produces a risk equivalent to 2,220 glasses of water!

These data illustrate the low involuntary risk associated with public drinking water compared with the larger risk associated with diet cola. Do you believe people would have become as excited if it had been reported that the CCl_4 concentration was found to be 0.1 PPM?

There has arisen the naive belief that if someone suffers harm, some agency of society must make it right and pay damages. Cigarette smokers suffering from lung disease sue cigarette companies for damage. Somehow the notion of a risk being voluntary has been forgotten. The idea that one chooses between alternatives, and that some choices turn out badly, has given way to the concept of the "deep pocket" in which any institution associated with a loss can be held responsible. It's banditry—on a par with the young boy convicted of slaying his parents throwing himself on the mercy of the court because he is an orphan!

If it were banditry alone, it could be corrected with the stroke of the pen. Unfortunately, deep-pocket attitudes reveal a nation of self-indulgent individuals preoccupied with personal health and happiness. We have spawned a generation of Americans who have been taught that any loss they believe they suffer will be corrected and someone made to pay, as if God or government were obliged to provide "no-fault" insurance to correct life's misfortunes. We are breeding a generation of pusillanimous whiners. If the country is to compete with technically advanced nations in the years to come, the fretful preoccupation with personal health and happiness will have to be put into perspective. Indeed, even the concept of competition has been distorted,—competition has become only the name for winning. Gone are the days in which winning or losing were not the issue, but rather how well you played the game.

1.3   LIABILITY

In the previous section risk was discussed in terms of injury to individuals. Engineers must be aware of another kind of risk—that which they incur in designing products whose performance or lack thereof may cause injury to others. In the event of such injury, the engineer may be liable, that is, legally bound to make good losses or damages incurred by the other party. Liability is a legal concept described by the theory of torts. A *tort* is a wrongful or injurious act for which civil action can be brought. While a full discussion of the theory of torts is beyond the scope of this text, engineers should be aware of two classes of liability—*negligence* and *strict liability*.

Literally speaking, *negligence* is the failure to use a reasonable amount of care when such failure results in injury or damage to another. The negligence standard concentrates on whether the engineer was careful, prudently trained and properly supervised. Common law negligence exists if the plaintiff can prove "the violation of a statute which is intended to protect the class of persons to which the plaintiff belongs against the risk is the type of harm which has in fact occurred" (348). An approach taken in may cases is to hold that a violation of an OSHA standard is only some evidence of negligence. Liability due to negligence requires that the plaintiff show that there is a causal relationship between the violation and the injury.

Strict liability is a phrase from the field of consumer product safety and refers to the responsibility of vendors for damages produced by products they sell that contain *design defects*.

One who sells any product in a defective condition that is unreasonably dangerous to users or consumers or to their property is subject to liability for physical harm caused to the ultimate user or consumer or to their property if

(a) the seller is engaged in the business of selling such products, and
(b) it is expected that the product reaches the user or consumer without substantial change in the condition in which it is sold.

Strict liability applies if

(a) the seller has exercised all possible care in the preparation and sale of the product, and
(b) the user or consumer has not bought the product from or entered into any contractual relation with the seller.

The principal issue under which strict liability is upheld is whether or not a defect, direct or indirect, exists. If it can be established that a defect exists, the manufacturer usually is held liable. A plaintiff need only prove the following:

(a) The product was defective.
(b) The defect was present when the product changed ownership.
(c) The defect resulted in an injury.

The courts have held that manufacturers must be as knowledgeable about their products as an expert in the field. The manufacturer must know all the requirements, standards, and codes that have been imposed by statute, issued by government agencies, and published by technical and industry associations, and even those things known as *good engineering practice* that change with the state of the art. The standards that give the necessary information are supposed to provide guidance, unfortunately, they sometimes are contradictory, omit specific details, and may appear so innocuous as to be useless. Responsible individuals do not design products that are intentionally defective, but they are nonetheless liable for technical inadequacies they are not aware of, but which are known to experts in the field, if such inadequacies cause injury to persons or property and could have been avoided.

1.4 CONTAMINANT CONTROL STRATEGY

In this text *contaminants* will be defined as airborne materials that are injurious to human health. The material may be a gas or a vapor, or it may be a suspension of solid or liquid particles. The manner in which the particles move in air will be considered, details of their pathogenicity will not be considered. Gases or vapors, or mixtures of the two, will be called a gas since their equation of state can be described by the perfect gas equation with acceptable accuracy. Particles in the atmosphere are subdivided into several categories. It is important to understand these categories and to remember their names.

An aerosol consists of airborne particles that remain in suspension for a considerably long time.

An agglomerate is a cluster of small particles which attract other small particles.

Coarse particles are those whose characteristic size is larger than 2 μm.

Dust is a suspension of solid particles formed by size-reduction processes whose characteristic size is generally greater than 1 μm.

Fibers are airborne particles which have length to breadth ratios of 10 or more.

Fine particles are those whose characteristic size is less than 2 μm.

Fog is an aerosol of liquid particles from 1 to 100 μm in diameter.

Fume is a suspension of solid particles sublimated from vapors generated by exothermic processes; the particles are generally less than 1 μm in diameter.

Inhalable particles are defined by the EPA as those whose characteristic size is less than 10 μm.

Inspirable particles are defined by the ACGIH as those which are hazardous when deposited anywhere within the bronchial system and gas exchange region.

Mist is an aerosol of liquid particles generally between 10 and 100 μm.

Nuclei are clusters of molecules that act as cites for the condensation of vapor.

Primary particles are larger than 1 μm and are emitted directly into the atmosphere by man as solids and vapors that condense.

Rain are water drops whose characteristic diameter is larger than 1000 μm.

Respirable particles are defined by the ACGIH as those which are hazardous and can be transported to the gas exchange region of the lung. In general practice, respirable particles are those whose characteristic size is less than 2 μm.

Secondary particles begin as short-lived nuclei (particles less than 0.1 μm) that grow by the processes of agglomeration and coagulation.

Smog is an aerosol produced by atmospheric photochemical reactions.

Smoke is an aerosol associated with combustion composed of solid or liquid particles generally less than 1 μm in diameter.

Spray is an aerosol of liquid particles larger than 100 μm.

Submicron particles are those whose characteristic size is less than 1 μm.

There are many ways to prevent chemical or biological toxins from contaminating air in the workplace. It is essential that each of these methods be reviewed systematically to decide which is the most economical way to achieve a healthy environment. In many instances it may not be necessary to install an industrial ventilation system. It is useful to group these methods in three main categories, each of which can be subdivided.

  (1) Administrative controls
      (a) Work practices
      (b) Labeling and warning devices
      (c) Education
      (d) Waste disposal practices
      (e) Environmental monitoring
      (f) Assignment scheduling
      (g) Medical surveillance
      (h) Housekeeping
      (i) Dust suppression
      (j) Maintenance
      (k) Sanitation
      (l) Management
  (2) Engineering controls
      (a) Elimination
      (b) Substitution
      (c) Isolation
      (d) Enclosure
      (e) Process change
      (f) Product change
      (g) Industrial ventilation
  (3) Personal protective devices

The ordering of these categories is deliberate, since administrative measures are superior and preferable to engineering controls which in turn are superior to personal protective devices. In the final analysis, combinations of methods may be the most effective for individual industries.

1.4.1 Administrative Controls

(1a) *Work practices* are decisions taken by management that specify the use of certain tools and procedures. Examples of work practices are procedures and tools used by workers in high-risk occupations, such as high-voltage electrical equipment, clean-up of toxic waste sites, radioactive materials.

(1b) *Labeling and warning systems* are written instructions attached to product containers indicating their hazards and methods to be followed in their use. Examples include labels on product containers and MSDS's accompanying the transfer of products.

(1c) *Education* involves conveying information about the hazards associated with a product or process that may not be immediately apparent and which require some degree of instruction. An example is the training and certification of workers who use certain agricultural chemicals.

(1d) *Waste disposal practices* are procedures to be followed in discarding hazardous materials that prevent them from inadvertently entering the workplace or contaminating municipal sanitation workers. An example is the need to drum certain chemicals for removal by certified waste management firms rather than discarding them into municipal sewers or solid waste sites.

(1e) *Environmental monitoring* involves the analysis of air samples taken in the workplace to insure that the concentration of certain compounds is within acceptable standards. Monitoring enables one to anticipate hazards and to prevent ill health. An example is monitoring the airborne lead concentration in plants that manufacture glass frit.

(1f) *Assignment scheduling* is regulating the time a worker is exposed to high-risk conditions so that the time-average exposure is within acceptable limits. Examples of this practice are in the glass and metals industry where workers rotate between high-temperature and moderate-temperature locations throughout the workday.

(1g) *Medical surveillance* involves screening and the routine examination of workers to detect unhealthy medical symptoms or hypersensitivity. An example is monitoring the lead in the blood of workers employed in the manufacture of glass frit (see 1e above).

(1h) *Housekeeping* involves an array of obvious procedures to clean up and tidy up. For example, removing empty containers from the workplace from which vapors could escape to the workplace, or removing inadvertent discharges (spills, drips, etc.) of materials by vacuum rather than hand brooms.

(1i) *Dust suppression* involves practices to minimize dust generated from stockpiles, bag-dumping, conveyor transfer points, or drilling. Examples include the use of wetting agents, windbreaks, or enclosures.

(1j) *Maintenance* is the scheduled inspection, repair, and replacement of components of a process to prevent failures that would result in the emission of contaminants.

(1k) *Sanitation* is the application of hygienic practices to reduce the opportunity of inhalation or ingestion hazardous materials. Sanitation involves removing work clothes before entering eating areas, properly cleaning work clothes so as not to expose those doing the cleaning, showers, chlorination or sterilization, and pasteurization.

(1m) *Management* refers to the existence of organizational structures that grant certain individuals the authority to affect change.

1.4.2 Engineering Controls

(2a) *Elimination* is the total removal of the source of contamination. For example, replacing solvent-based finishes with water-based finishes eliminates the emission of hydrocarbons as the finish dries.

(2b) *Substitution* is the replacement of one toxic substance by a substance of lesser toxicity. For example replacing benzene with toluene, replacing asbestos fibers with glass fibers, or replacing sand with steel shot for abrasive blasting of castings.

(2c) *Isolation* requires placing an impervious covering over the source of contamination or locating materials in an isolated location in a plant. For example, using glove boxes for handling radioactive materials or installing cleanrooms for operations requiring an ultrapure environment.

(2d) An *enclosure* is a physical barrier surrounding nearly the entire process and reducing substantially contaminants escaping to the workplace. For example, using hopper valves to discharge a hopper into a closed vessel equipped with a bin vent filter instead of merely dumping the hopper's contents into an open vessel and using a shake-out room where castings are removed from their molds.

(2e) A *process change* involves a new machine or process that reduces the emission of a contaminant. For example, replacing atmospheric cooking vessels with vessels that cook under vacuum so that any leakage is that of air into the vessel, or redesigning dry cleaning equipment to eliminate the need for individuals to physically transfer garments between machines.

(2f) In some instances *changing the product* can reduce workplace emissions. For example, a machine component that had to be strengthened by welding a supporting brace could be redesigned as a single casting or forging to eliminate emissions from welding. Similar advantages can be gained by eliminating gluing or soldering.

(2g) *Industrial ventilation* is the installation of a configured inlet and an air mover to withdraw air surrounding the process, capture contaminants, and prevent their transfer to the workplace.

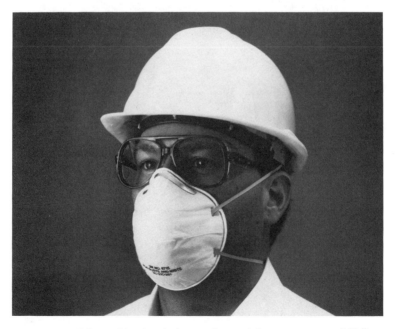

Fig. 1.3. Disposable dust/mist respirator (photo courtesy of 3M).

1.4.3 Personal Protective Devices

A variety of *respirators* covering parts of the face can remove particle, gas, and vapor contaminants from air inhaled by workers. Simple *face masks* (Fig. 1.3) can remove nuisance dust. *Half-masks* covering the mouth and nose remove gases and vapors by adsorbers located in screw-type holders mounted in the mask. *Full-face* masks covering the entire face (mouth, nose and eyes) with a tight-fitting seal remove contaminants by cartridge adsorbers mounted in the mask as well as protect the eyes (see Fig. 1.4). Lastly, *self-contained breathing apparatus* (SCBA) (Fig. 1.5) consisting of full-face respirators with independently carried air supplies are available for use in highly contaminated environments. Alternatively, air can

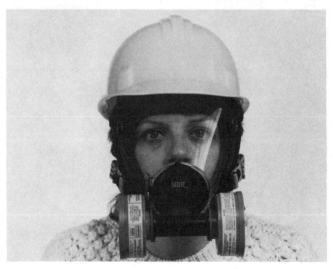

Fig. 1.4 Half-mask and full-face respirators (courtesy of Scott Aviation, Lancaster, NY).

Fig. 1.5 Grinding castings with positive pressure helmet and external air supply.

be supplied to a full-face mask through hoses connected to an external air supply. Figure 1.6 is a device in which air is blown down between the face and face shield and out around the neck. Air is moved by a fan located in the helmet, powered by a battery worn around the worker's waist or by an independent air supply.

The success of a *respirator program* depends on the effectiveness of managerial

Fig. 1.6 Helmet containing fan to move air over face. The fan and a filter can be placed on the worker's waist and filtered air ducted to the rear of the helmet (with the permission of Racal Health & Safety, Inc.).

procedures that oversee the systematic replacement of the cartridges, maintain the respirators, and insure that they fit workers properly. Securing a tight seal between the mask and the face is extremely important. The contour of men's and women's faces varies: the contour is also different for caucasian, black, and oriental face. Beards and corrective glasses require special practices. Lastly, any manager who contemplates prescribing respirators should wear one for an extended period of time before coming to a conclusion. Many people suffer from headaches induced by the tight elastic bands across the temple or skin irritation from the seal. The effectiveness of respirators is easy to appreciate, the inconvenience in wearing them is often underestimated if not ignored.

It is not essential that each of these three control strategies and their subdivisions be truly independent, but rather that individuals charged with reducing workplace contaminants consider all possible methods to eliminate the transfer of contaminants to the workplace before selecting the most economical and effective method. Selecting a control strategy is a collective act requiring participation of management, engineers, workers, and health professionals.

1.5 COMPONENTS OF INDUSTRIAL VENTILATION SYSTEMS

The function of an industrial ventilation system is to prevent contaminants from entering air in the workplace. There are several ways to accomplish this goal. Figure 1.7 shows the basic components of an industrial ventilation system. All industrial ventilation systems contain some if not all the components shown in Fig. 1.7.

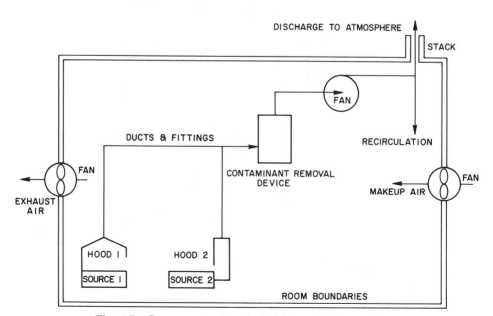

Fig. 1.7 Components of an industrial ventilation system.

(1) Source

Contaminant gases, vapors, and particles are generated by industrial activities. The rate and physical properties of the contaminants have to be described quantitatively.

(2) Hood

The configured inlet used to capture contaminants from a source.

(3) Ducts and Fittings

A system of branching ducts connects the hoods to an air moving device.

(4) Contaminant Removal Device or Air Cleaning Device

It may be necessary to remove contaminants from the collected air. If the contaminant concentrations in the collected air exceed environmental standards or if the collected air is recirculated, a removal device will be needed.

(5) Air Mover (Fan)

Air is drawn into a hood by an air mover which is generally a fan. The fan may be upstream or downstream of the contaminant removal device.

(6) Stack

A stack exhausts air to the atmosphere and prevents it from reentering the workplace.

(7) Recirculation

All or a portion of the collected air may be recirculated to the workplace. Recirculated air must be cleaned.

(8) Makeup Air

Outside air added to the workplace is called makeup air. The temperature and humidity of the makeup air may be controlled.

(9) Exhaust Air.

It may be necessary to exhaust a portion of the room air.

This book includes eight of the nine elements described above, a thorough discussion of contaminant removal devices is omitted. The selection of a contaminant removal system and a discussion of their engineering principles is a huge subject and the concern of several books on air pollution control systems. The excellent textbook by Crawford (19) and the handbook by Calvert and Englund (142) should be consulted. Computing the velocities and pressures throughout a system of ducts and fittings is an extremely important skill which engineers need to possess. Each component of the duct system must be chosen properly if the entire system is to be fabricated economically, operated cheaply, and maintained easily. The subject of duct design and fan selection has received considerable attention over many years and a great deal of material is available for engineers. A large number of books have been written on the design of duct and fan systems including the exhaust stack. Readers should consult Burgess et al (388), Alden and Kane (143), Burton (104), the ACGIH *Manual* (27), and the ASHRAE *Handbook* (97) for details. Much of the tedious, repetitive work has been eliminated by computer programs, some of which are available for personal and minicomputers.

1.6 OBJECTIVE

In the final analysis, individuals who design and operate industrial ventilation systems function as engineers irrespective of their formal education. This book is addressed to these people. The object of the book is to establish the fundamentals of industrial ventilation in a quantitative and rigorous manner. The book is written as a

textbook for university graduate and senior-level courses in engineering and industrial hygiene;

source book for professionals enrolled in continuing education courses, workshops, and symposia;

reference book for working professionals.

The book requires readers to be familiar with the fundamental equations in thermodynamics, fluid mechanics, heat transfer, and mass transfer. The book is written for individuals familiar with vector notation, calculus, and numerical computation.

The use of the word *principles* in the title is deliberate. Principles are statements revealing the source, origin, or cause of phenomena. Many books have been written on industrial ventilation and their conclusions and recommendations are very useful, but they fail to address the subject with the analytical rigor expected of today's engineers, who are asked to predict the performance of ventilation systems to insure that they satisfy stringent government standards prior to constructing and testing the entire system.

The design of industrial ventilation systems requires making decisions in several stages:

(1) Identify the contaminants and understand their affect on health.

(2) Select the maximum exposure limits to be used as design criteria and standards to judge the systems performance.

(3) Design the "hood" and select the volumetric flow rates of exhaust air, recirculated air, and makeup air.

(4) Design the duct system, select the fans, and compute the operating costs.

(5) Select the gas cleaning device to remove contaminants before discharge to the outside environment or recirculation to the workplace.

(6) Conduct laboratory experiments of the ventilation system, test the full-scale system in the field, and sample the air in the vicinity of the source and worker to insure that (2) is satisfied.

Tasks 2 and 3 are the principal concern of this book. This is not to suggest that the other issues are less important, but that at this time these issues are adequately covered elsewhere. The level of intelligible writing on these other issues far exceeds the level in task 3. The "hood" and the selection of the volumetric flow rate are the least understood of the six stages. If contaminants are not adequately captured, the ventilation system has failed to achieve its mission, irrespective of how well the duct, fan, and gas cleaning systems operate. For the most part designers are not sufficiently critical and careful in designing the "hood." The paucity of analysis describing the motion of contaminants between source and exhaust is the principal handicap designers face.

1.7 DEFICIENCIES IN PRESENT KNOWLEDGE

There are six deficiencies or inadequacies in our understanding of how to design an industrial ventilation system to satisfy standards of performance.

(1) Inability to predict the concentration.

(2) Inability to predict off − design performance of a conventional design.

(3) Effects of drafts and wakes in the workplace.

(4) Inability to generalize acceptable designs.

(5) New or unusual contaminant sources.

(6) Concept of the capture velocity.

All these deficiencies arise from a single fundamental shortcoming—the inability to describe the motion of contaminants from their source of origin to the inlet to the ventilation system.

1. Predicting The Concentration

Contemporary texts, handbooks, and manuals, etc, provide useful descriptive information but limited quantitative information about ventilation systems for specific industrial operations (i.e., booths for swing grinders, pedestal grinders, laboratory hoods, etc.). As useful as this is, there is absolutely no way designers can estimate the contaminant concentration at selected points in the vicinity of the source and worker. Thus there is no way designers can assure themselves that the design will satisfy OSHA standards. In the latter part of the 20th century such an inability is unacceptable professionally and unnecessary technically. It is prudent that performance be demonstrated prior to the commitment of funds for construction and installation. A major design change performed after the system is installed is a poor and weak substitute for a priori analysis.

2. Off-Design Performance

Seldom is a system constructed or operated exactly as planned. Practical considerations or administrative decisions often arise that require the system's dimensions or the volumetric flow rate to change. There is presently no way designers can estimate the consequences of these changes and techniques to accomplish this are urgently needed if one is to know that the system still satisfies OSHA requirements.

3. Drafts And Wakes

Spurious air currents in the workplace interact with the velocity field produced by the ventilation system and affect its performance. Similarly, wakes occur as air passes around a worker's body, work piece, tools, work table, structures, and jigs. A region of low-speed, recirculating flow is a region in which contaminant concentrations may be high. Since drafts and wakes are commonly present in industrial operations, it is naive to design a system as if they did not exist or multiply certain design parameters by safety factors.

4. Generalization

Even if a design controls contaminant concentrations successfully and satisfies OSHA requirements, there is presently no way to scale such a design geometrically so that one can be assured that a larger or smaller design will also be successful. In addition, there is no way to know how much the volumetric flow rate should be increased or decreased.

5. New or Unusual Contaminant Sources

The professional literature contains a great deal of information on designs that have been shown to be successful. Unfortunately, it is rare that one is dealing with an industrial process that is identical to one for which the ventilation system is already published. At best it might be possible to select a ventilation system for a similar, but not identical, industrial operation. Sometimes even this will not be possible and one will be forced to design a ventilation system from scratch. In either case, designers presently have no way to estimate performance.

6. Capture Velocity

For many years "hoods" have been designed using the notion of "*capture velocity.*" The ACGIH (27) defines capture velocity as the "air velocity at any point in front of the hood or at the hood opening necessary to overcome opposing air currents and to capture the contaminated air at that point by causing it to flow into the hood." The use of this concept is regrettable. There is no way to postulate values of capture velocities for vapors and particles produced by different industrial operations. To specify such values is equivalent to assuming the contaminant's trajectory. One may wish to do so qualitatively, but quantitatively it cannot be done with the precision needed in design. As will be seen in Chapter 8, the viscosity of air reduces the relative velocity between particle and air to virtually zero in a very short time. The time it takes to do so and the distance the particle travels depend on the particle's size, initial velocity, and the carrier gas velocity. For gaseous contaminants the situation is vastly different. Thus it is illogical and fundamentally inconsistent with the principles of fluid mechanics to tabulate capture velocities. The suggestion that a moving air stream somehow captures contaminants only when it has a velocity equal to or larger than a "capture velocity" is bogus.

1.8 SPURIOUS AIR CURRENTS

Theoretical consideration of ventilation systems is frequently based on the assumption that surrounding the process is a quiescent body of air. Nothing could be further from the truth. Every workplace contains equipment and workers performing tasks that produce room air currents of unique and unpredictable character, i.e., *drafts*. The way designers cope with these currents depends on whether local or general ventilation strategies are adopted. Spurious room air currents are produced from the following.

(a) Motion of workers and machinery in the vicinity of the source of contamination.

(b) Plumes of hot air rising from the surface of hot machinery or vapors rising from volatile liquids.

(c) Jets of air from cooling fans within the machinery, doors and windows, workers personal fans, or heaters and air jets from building heating or cooling systems.

Other factors affecting the motion of contaminants are walls, partitions, machinery surfaces, jigs, and fixtures. In some cases, impediments may be eliminated, but for the most part they will have to be incorporated in the design of the ventilation system from the beginning. An even more vexing circumstance is trying to correct unwanted room air currents produced by machinery or new production methods put in place after the ventilation system has been installed. Designers may have to reassess the effectiveness of the ventilation system and perhaps even change the geometry or volumetric flow rates to achieve satisfactory performance. Designers must inspect the industrial site before beginning their work; those who do not proceed at great peril.

1.9 SELECTING A VENTILATION SYSTEM

When selecting a ventilation system, one should never overlook the most obvious corrective step—modify the process to generate smaller amounts of contaminant and take steps to prevent contaminants entering the workplace. Industrial practices are not sacrosanct, operational costs and contaminant exposure can often be reduced simultaneously by modifying the process or choosing different raw materials. For example, questions such as the following should be asked before any design is begun.

Can streams of air discharged by machinery be directed elsewhere? Can the exhaust from pneumatic equipment be directed elsewhere?

Can a liquid of lower volatility be used? Can its temperature be lowered to reduce the emission of vapor?

Can the type of lubricant or its rate of flow be changed to reduce the production of mist for metal cutting and metal punching machines?

Once one has eliminated all opportunities to reduce the generation of contaminants, a ventilation system may have to be designed. To a great extent the choice of a ventilation system is dictated by how the emissions are generated. If there are many sources of emission and they are distributed throughout a process, such as dust generated in a quarry or vapors generated during metal casting (Fig 1.8) the emissions are called *fugitive* and have to be controlled one at a time. If the emissions are generated from well defined points in a process, such as shown in Figure 1.9, they are called *point-source emissions* and can be controlled more easily.

In the field, industrial ventilation *hood* is a generic phrase given to uniquely configured air inlets through which contaminated air is withdrawn. The word denotes the function to be performed rather than any particular geometrical configuration. Table 1.4 summarizes the two general classes of ventilation sys-

Fig. 1.8 Vapors generated in metal casting (photo courtesy of 3M).

Fig. 1.9 Dust generated in drilling rock.

TABLE 1.4 Classification of Ventilation Systems

General or Dilution Ventilation

Withdrawal of air from the entire workplace and its replacement with a selected mixture of make-up fresh air and a cleaned portion of the air withdrawn from the workplace.

Local Ventilation

Withdrawal of air and contaminants from a region close to the point of generation.
Enclosure—A housing that virtually encloses the source.
 Examples: laboratory fume hood, glove box, spray booth, fumigation booth, grinding booth.
Receiving Hood—A housing that collects contaminants because of their intrinsic motion. Receiving hoods move minimal amounts of air.
 Examples: canopy hood over a hot source, pedestal grinders, kitchen range hoods.
Exterior Hood—A uniquely configured air inlet placed in close proximity of a source of contamination. Exterior hoods are also called capture hoods.
 Examples: down-draft grinding benches, welding snorkels, lateral exhausters and push–pull exhausters (with and without side panels), HVLV systems, side-draft exhausters.

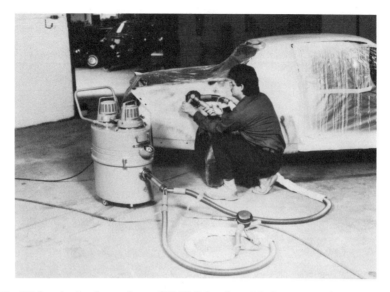

Fig. 1.10 High velocity–low volume (HVLV) local venitlation system for a surface sander (with permission of Nilfisk of America, Inc.).

tems—*general* and *local*—and subdivisions within local ventilation systems. The reader should be cautioned that the classifications are qualitative and not universally adhered to. Local ventilation systems that are inlets in close proximity of a source that withdraw small amounts of air at high velocity are called *High Velocity–Low Volume (HVLV)* systems. The phrase *close capture hood* is also used. Figures 1.10–1.12 show three examples of close capture hoods. Inlets that are farther from the source and remove large quantities of air at low velocity are

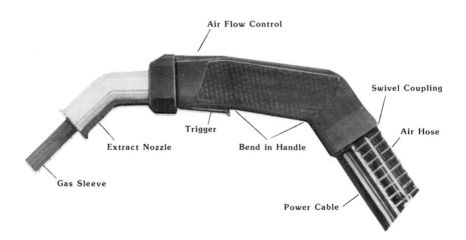

Fig. 1.11 High velocity–low-volume local (HVLV) fume extracting welding gun (with permission of the Swedish Institute of Production Engineering Research).

Fig. 1.12 High velocity–low volume (HVLV) local ventilation system for an oscillatory saw (with permission of Nilfisk of America, Inc.).

called *Low Velocity–High Volume* (*LVHV*). Examples of these are the canopy hoods shown in Fig. 1.13 and side-draft hoods (Figs. 1.14–1.16. The imprecision of the terms and the inherent confusion as to what "V" represents is obvious, but the reader should be prepared to encounter these phrases.

As a general proposition, general ventilation should be avoided as a method to control airborne toxic substances. It is more economical to address each source and install appropriate "local ventilation" systems than to allow the contaminants

Fig. 1.13 Canopy hood in a foundry (145).

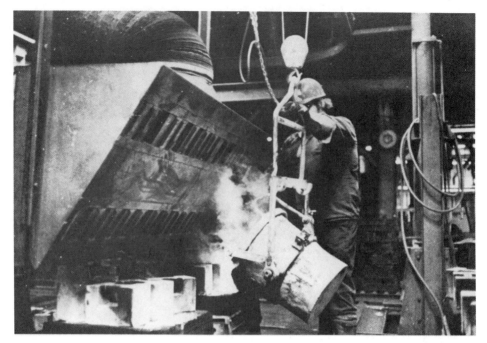

Fig. 1.14 Side-draft exterior hood for pouring molds in a pallet line (145).

to permeate the entire work space and be forced to remove and replace the entire air mass in the work space. One may wish to employ general ventilation to control temperature, humidity, and odor, but contaminant control can be achieved more economically by local rather than general ventilation. In the microelectronics and biotechnology industries there is a need for ultrapure environments (i.e., "clean-

Fig. 1.15 Side-draft hood to capture welding fume (400).

Fig. 1.16 Side-draft exterior hood for casting shake-out (145).

rooms") and general ventilation is used. Figure 1.17 shows the details of a cleanroom. While the workers health is not the primary reason the cleanrooms are needed, the movement of air and contaminants follows the same principles used in industrial ventilation.

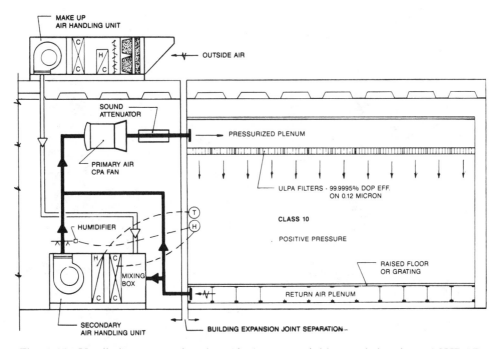

Fig. 1.17 Ventilation system for class 10 cleanroom (with permission from ASHRAE, Journal of Aug. 1986).

If it is possible to totally enclose the process, leaving the worker outside the enclosure, it should be done without hesitation. Such an *enclosure* will produce the maximum control for the minimum amount of exhaust air. A few examples where this may be possible are electric arc furnaces (Figs. 1.18 and 1.19), *laboratory fume hoods*, and machines that fill vessels with powders and granular material (Figs. 1.20 and 1.21). If the operator cannot be excluded from the enclosure, a *booth* surrounding the operator wearing personal protective apparatus (Fig. 1.22) is the next, most effective method to prevent contaminants from traveling to other parts of the workplace. Great care must be taken to insure that contaminant concentrations in the operator's breathing zone do not exceed safe values. Examples of industrial operations where booths are attractive are grinding booths for portable hand – held grinders (Fig. 1.23), paint spraying booths (Fig. 1.24), and welding booths.

Many industrial operations involve equipment and the movement of men and material that makes it impossible to enclose the source of contamination. In these cases *exterior hoods* should be used. Examples of such operations are open vessels (Fig. 1.25) for electroplating, galvanizing, pickling, etching, bag dumping stations (Fig. 1.26) "snorkle" for welding (Fig. 1.27), and long – wall automatic miners in coal mines. The air flow required for exterior hoods is larger than for enclosures and one must be careful that contaminant concentrations in the workers breathing zone do not exceed safe values. *Canopy hoods* are located directly above the process (Fig 1.13) and try to capitalize on buoyancy. *Side-draft hoods* have faces

Fig. 1.18 Close-fitting hood for an electric arc furnace (146).

Fig. 1.19 Close-fitting hood for an electric arc furnace, swung open for charging (146).

Fig. 1.20 Laboratory fume hood. (Source: Liberty Industries, Inc. Publications. Reproduced with Liberty Industries, Inc. approval.)

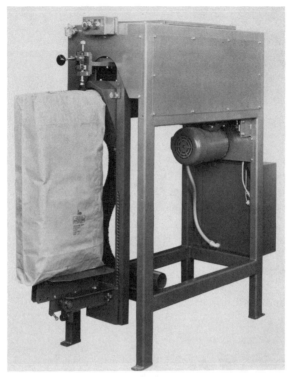

Fig. 1.21 Bag filling and weighing machine (with permission of Bemis Packaging Service Machinery Co.).

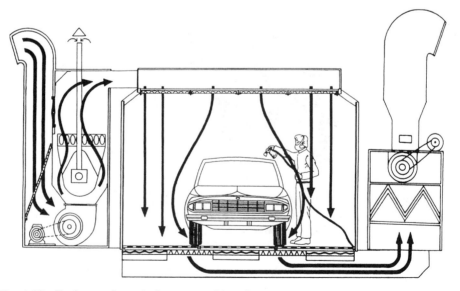

Fig. 1.22 Enclosure for painting automobiles (illustration courtesy of SAICO Paint 'n' Cure, copyright 1988).

Fig. 1.23 Down-draft grinding and chipping bench with two side panels (145).

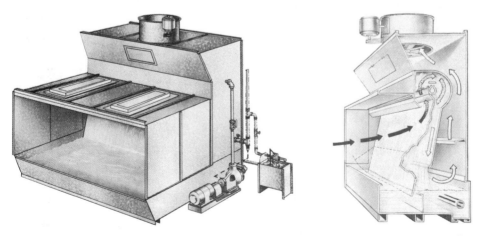

Fig. 1.24 Paint spraying booth with water curtain (with permission of the Binks Manufacturing Company).

Fig. 1.25 Open vessel with lateral exhausters (146).

oriented diagonally to the process (Figs. 1.14–1.16). *Lateral exhausters* withdraw air at right angles to the process plume (Fig. 1.25) and *down-draft benches* (Fig. 1.23) withdraw air in the downward direction.

If the contaminant is a distinct stream of heavy particles or a buoyant plume, *receiving hoods* may be applicable. Examples where receiving hoods are attractive

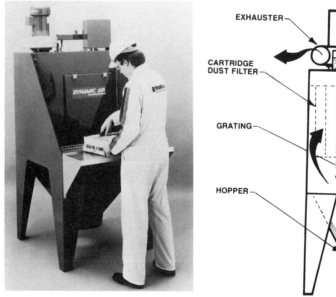

Fig. 1.26 Manual bag dumping unit, escaping dust is captured by an exhauster (reprinted with the permission of Dynamic Air, Inc.).

Fig. 1.27 Movable snorkel for welding (with permission from the Ruemelin Manufacturing Co.).

are few, but include plumes from furnaces (Fig. 1.13), wood working machines, and the swarf from pedestal grinders and swing grinders (Fig. 1.28). Even in these cases, great care must be taken to insure that the receiving hoods are located and sized to collect the contaminants since room drafts may deflect buoyant plumes or small particles. In many cases receiving hoods are used in combination with exterior hoods. Formaldehyde vapor generated in the manufacture of wood veneer can be captured by canopy hoods at the inlet and outlet of the veneer machine, Fig. 1.29.

Push–pull hoods (Chapter 10) represent a class of hoods whose full potential and limitations have not fully been assessed. The blowing jet increases the size of the region of influence in front of the inlet through which the contaminants are withdrawn. The blowing jet is a jet pump that induces room air into the jet as the composite flow sweeps over the surface of the liquid and enters the exhaust inlet. Applications where push–pull systems are attractive include large open vessels. Extreme care must be taken in the design of push–pull systems to insure that the blowing jet does not inadvertently function as a mixer and increase the dispersion of contaminant in the work place. A jet with the wrong orientation or an exhaust inlet that can not accommodate the total flow may increase the concentration in the general work space.

Fig. 1.28   Receiving hood for a swing grinder (145).

Fig. 1.29   Paten press for wood veneer panels. Close fitting canopy hoods at both ends of the press capture hut vapors generated during pressing. Note also the side draft hood for the stacked panels. Photo courtesy of NIOSH.

1.10 PROFESSIONAL LITERATURE

The literature on industrial ventilation is technically primitive. The writing lacks the precision, rigor, and application of fundamental laws of science that are required in engineering today. If professionals in the field of industrial ventilation are going to understand the motion of contaminants and are going to design collection systems skillfully, they will need to use more sophisticated analytical and experimental techniques than they have in the past. If students in mechanical, chemical, architectural, environmental and biomedical engineering are going to be attracted to the field, the level of technical discourse must be commensurate with what they have learned and use in their disciplines. If working professionals are going to be able to adopt technological developments from other fields of engineering, it will be necessary that they acquire the sophistication used by professionals in these fields. It is my belief that industrial ventilation will witness a burst of activity in the near future when engineers in the fields of fluid mechanics, heat and mass transfer awake to the interesting and important problems that need to be solved. When they discover how primitive the available technology is, I expect a stampede! The subject is ripe for picking!

Major articles are published in the leading journals (see below) on a regular basis that summarize the major publications in the field (53) or update the science and technology (54). In addition, texts on industrial hygiene (36,242,383) contain sections on industrial ventilation. It is not my intention to write an exhaustive review of all literature on industrial ventilation or the literature of fluid mechanics, numerical computation or aerosol dynamics that pertains to industrial ventilation. Rather, what follows is a brief summary of several important books and sources of information readers should be familiar with. Some of these books are old and qualitative in nature, but they are classics in the field. Anyone wishing to pursue careers in the field will find those books rich sources of practical information. Experimental and analytical techniques have changed over the years, but the knowledge that was sought and the applications to which it was to be applied have changed little. Proceeding without such information is to risk reinventing the wheel!

Hemeon (47) was published in 1955, updated, and republished in 1963. The book contains a great deal of practical qualitative information needed by individuals entering the field. Hemeon does not model ventilation systems analytically with the frequency and sophistication used in engineering today, but this eminently readable book contains many empirical equations that engineers can use to prepare initial designs of hood geometries and estimates of ventilation rates.

The third edition of Baturin (48) was published in the Soviet Union in 1965 and published in English in 1972. The book is largely qualitative, but is the most thorough technical book on the subject. The material is essential for the working professional and contains numerous analyses and experimental data not found in the United States literature of the time, for example,

Velocity fields for air curtains
Velocity fields for unusual inlets and outlets
Air changes for natural ventilation in buildings
Velocity fields in buildings containing heat sources

It is believed that with advances in the field of computational fluid mechanics (CFD), there will be decreasing need for the material in Baturin and Hemeon because analytical models can provide reliable guidance to designers more quickly and easily than in the past. Nevertheless, readers should review Baturin and Hemeon to acquire physical insight about the movement and control of contaminants. Unfortunately the English translation of Baturin contains long convoluted sentences which are often unclear. There are also many minor errors in terms and complicated nomenclature.

The *American Conference of Governmental and Industrial Hygienists* (ACGIH) is a professional organization which disseminates information on methods to maintain a healthy environment in the workplace and methods to measure the concentration of contaminants. The Committee on Industrial Ventilation within ACGIH reviews developments in industrial ventilation and every year (since 1951) publishes a new edition of *Industrial Ventilation: A Manual of Recommended Practice*. The 20th edition (27) contains sketches of specific ventilation systems for nearly 150 different industrial operations and a discussion of how to design a duct system, how to select the ventilation flow rate, the proper fans, and the appropriate gas cleaning devices, and how to test ventilation systems. The manual does not base its recommendations on rigorous engineering principles, but it is nevertheless an invaluable reference book for those responsible for the design and operation of ventilation systems. The manual provides valuable guidance to prepare the preliminary design of a ventilation system and to estimate the volumetric flow rate. The manual is the most widely used publication in the field and engineers must be conversant with it. In the field of mine ventilation, the *Manual of Mine Ventilation Design Practices* (391) provides a similar function.

The *American Society of Heating, Refrigerating and Air-Conditioning Engineers* (ASHRAE), is an engineering professional society which addresses the issue of industrial ventilation. Each year ASHRAE updates and republishes one volume of its four – volume set of handbooks. Section II of the *Systems Handbook* (28) is devoted to industrial ventilation. Of invaluable use to engineers are four ASHRAE handbooks (28). The material is written for engineers and is more quantitative and precise than what is contained in ACGIH literature. The handbooks presume familiarity with subjects included in accredited engineering Bachelors of Science degree programs.

The *Center For Chemical Process Safety* is a directorate within the *American Institute Of Chemical Engineers* and was established to create and disseminate knowledge on health, safety and loss prevention. Many educators believe that embedding health and safety within the required courses in the student's major is the most effective way to prepare engineers. Accordingly the Center developed 90 student problems and an instructor's guide (398) for undergraduate thermal sciences courses in all engineering disciplines, not just Chemical Engineering.

The text by Alden and Kane (143) has been used for many years and provides a quantitative basis to ACGIH procedures. The book by Goodfellow (167) was published in 1986 and contains the best and most complete description of industrial ventilation practices that is available and is indispensable to practicing design engineers. McDermott (49) was published in 1976, and while titled a handbook, it is largely a summary of material that can be found in the ASHRAE (28) and ACGIH (27) handbooks. Constance (50) is similar to McDermott in content and style. The text by Burgess, et al. (388) published in 1989 contains both theory and practical applications to the design of industrial ventilation systems.

The book by Hayashi et al (243) is a detailed description of design procedures adopted by the Japanese Labor Department based on several decades of experimental research by T Hayashi. The procedures are presently requirements for Japanese industry. The design procedures are similar to ACGIH procedures but incorporate ways to include cross drafts, round and rectangular sources (of different aspect ratios), buoyant plumes, and so on, in more quantitative terms than ACGIH. The procedures contain considerably more information and computational opportunities than ACGIH procedures. Hayashi's concepts are based on what is called *cooperation theory*. The theory rests on the idea that contaminant control is governed by a "pull" flow which may be aided by a "push" flow. With or without the push flow, air is induced from the surrounding environment and drawn over the source of contamination and into the pull receiver. The essential element in good design, that is, "cooperation," is to induce the minimum, essential amount of air using as little pull and push as possible. The concept is curiously described by Hayashi in terms of harmony (250), reminiscent of the way Herrigel (244) writes of Zen in the art of archery.

The textbook by Crawford (19) is technically the most thorough exposition on the design, operation and performance of air pollution control systems. It's Chapter 5 is devoted to the subject of industrial ventilation and several earlier chapters analyze of the motion of particles in moving gas streams. Crawford's approach to the subject is quantitative and his writing is lucid. Analytical developments begin with engineering principles and proceed logically based on the assumptions that are made. The book is a paradigm for what is needed in industrial ventilation. The text is an excellent reference book for engineering professionals engaged in the design and operation of industrial ventilation systems. The handbook by Calvert and Englund (142) presents (but does not develop) the essential analytical relationships describing the operation of air pollution control systems. Textbooks by Cooper and Alley (186) Licht (330), Wark and Warner (181), Flagan and Seinfeld (264) and Seinfeld (182, 239) discuss the overall issue of air pollution including control systems.

A major activity in architectural engineering is to describe the velocity field within (and between) rooms in buildings so as to achieve high entrainment of room air in jets of fresh air and simultaneously produce low air speeds in occupied zones. In Europe, Japan, and the United States there is considerable interest in using CFD to model the transport of air contaminants and water vapor to insure health and comfort. Students of industrial ventilation should acquaint themselves with this literature (360–370) because the results, arguments, modeling techniques, and experimental techniques are directly applicable to industrial ventilation.

Engineers working in occupational safety and health often encounter questions about physiology, toxicology, epidemiology and immunology. Each of these subjects is a specialized field of study in the biological sciences. Each field is based on a substantial body of knowledge, replete with excellent textbooks, and professional journals. For an initial inquiry into these fields, engineers should learn to use *Patty's Industrial Hygiene and Toxicology* (*147*) (Volume I—General

Principles; Volume II—Toxicology; Volume III—Theory and Rationale Of Industrial Hygiene Practice) (147).

The following publications of professional societies contain material relevant to industrial ventilation.

> American Industrial Hygiene Association Journal (AIHA)
>
> Applied Industrial Hygiene
>
> Annals of Occupational Hygiene
>
> Transactions of the American Society of Heating, Refrigerating, and Air Conditioning Engineering (ASHRAE)
>
> Journal of the Air & Waste Management Association
>
> Transactions of the American Society of Mechanical Engineers (ASME)
>
> Environmental Science and Technology
>
> Atmospheric Environment
>
> Aerosol Science and Technology
>
> Journal of Aerosol Science
>
> Annals of Biomedical Engineering

The following agencies of the United States government have specific missions devoted to the control of contaminants. They publish technical reports expressly related to the design of industrial ventilation systems.

> Occupational Safety and Health Administration (OSHA)
>
> National Institute of Occupational Safety and Health (NIOSH)
>
> Environmental Protection Agency (EPA)
>
> Mine Safety and Health Administration (MSHA)
>
> Department of Energy (DOE)

The following trade associations are involved in various aspects of industrial ventilation and publish standards, procedures, and other material engineers will find useful:

> Air Movement and Control Association (AMCA)
>
> American National Standards Institute (ANSI)
>
> National Electrical Manufacturers Association (NEMA)
>
> Sheet Metal and Air Conditioning Contractors National Association (SMACNA)

National and international symposia, conferences, and workshops are held virtually every year. These meetings usually prove to be more valuable than one can anticipate. Learning about research findings and new engineering practices through chance meetings with people in the hall, or at dinner is easier than trying to keep up with the professional literature. The proceedings of many conferences are published, but are apt not to appear in university or corporate libraries. International meeting are of particular importance because the Japanese and

Europeans follow practices that in many respects are superior to those used in the United States.

Lastly, engineers should subscribe to trade journals and magazines in occupational health and safety which can be received by engineers working in the field at no cost. These publications contain primarily advertisements about new equipment, products, and services and written articles about general topics in the field. Subscribing to these publications is an excellent way educators can keep abreast of new equipment and products that can be conveyed to their students. For engineers in industry the publications are essential.

PROBLEMS

1. An article in a professional journal lists the following chemicals and their CAS numbers: Pyridine CAS 110-86-1, methyl chloride CAS 74-87-3, methylene chloride CAS 75-09-2, and acrolein CAS 107-02-8. For each of these compounds find the following:

   RTECS number
   Permissible exposure limit (PEL)
   Chemical formula
   Vapor pressure at 25 C
   Phase at 25 C, 1 atm
   Lower explosion limit (LEL)
   Health hazard symptom
   Recommended personal protection equipment and sanitation

2. You plan to conduct an experiment that requires the use of the following chemicals: dimethylamine, trimethylamine, and pentaborne. Consult one the chemical registries and find the following for each compound:

   CAS and RTECS numbers
   Permissible exposure limit (PEL)
   Chemical formula
   Vapor pressure at 25 C
   Phase at 25 C, 1 atm
   Lower explosion limit (LEL)
   Health hazard symptom
   Recommended personal protection equipment and sanitation

3. Visit the supply room in your institution and ask the attendant to show you the MSDS for materials your institution uses that you know to be hazardous. (See Tables 3.1 and 3.2 in Chapter 3 for materials that OSHA considers hazardous.)

4. Visit a local auto body repair shop, automotive repair garage, construction site, commercial printer, or dry cleaning establishment, take a public tour of a company; or walk through any place where industrial activity is taking place. Identify specific industrial activities you believe are occupational hazards. Choose the most egregious example and discuss the applicability of *each* of the separate control strategies within the three main categories, administrative controls, engineering controls, and personal protective devices.

5. A smoker typically inhales one cigarette during an elapsed time of 20 seconds at a volumetric flow rate of 5 L/min. The concentration of smoke particles in the inhaled air is 10^{15} particles/m^3. Consider a nonsmoker in a room with smokers in which the smoke concentration is 0.3 mg/m^3. Assume the nonsmoker inhales air at a volumetric flow rate of 4 L/min. How long would the nonsmoker have to remain in the room to receive a dose of smoke equivalent to smoking one cigarette. Assume that smoke particles are spherical with a uniform diameter (D_p) of 0.1 μm and density 800 kg/m^3.

6. Adding chlorine to drinking water kills microscopic disease-causing organisms, but it also reacts with organic matter and produces small amounts of chloroform. The issue is more serious with treatment of surface water supplies because ground water contains negligible amounts of plant or animal matter. The United States average chloroform concentration in tap water is 83 μg/L. Chlorinating swimming pools also produces chloroform and while as a rule children don't drink the water they ingest some of it and breath air above the water surface containing chloroform and receive on the average a total dose of 250 μg/hr per hour of exposure. Assess the risk of playing in a pool compared with drinking tap water. Specifically, how many 8-oz glasses of tap water would a child have to drink to receive a chloroform dose equivalent to 1 hour's play in a pool? The average consumption of water for an adult is 1 L/day.

7. How many hours of exposure to formaldehyde in a conventional home produces risks equivalent to drinking one 12-oz can of beer?

8. How many sandwiches containing 32 g of peanut butter produce risks equivalent to formaldehyde exposure inside a conventional home for 14 hr?

9. Many students have had summer jobs in which they encountered what they believe were hazardous working conditions. Identify specific industrial activities you believe were occupational hazards. Choose the most egregious example and discuss the applicability of *each* of the separate control strategies within the three main categories: administrative controls, engineering controls, and personal protective devices.

CHAPTER 2

Respiratory System

Goal: Students will be expected to learn,
  – name and function of components of the
    respiratory system
  – how to model the motion of air in the respiratory
    system
  – how to model gas exchange in the respiratory
    system
  – primary occupational lung diseases

The large surface area of the lung, its airways, and the thin membrane separating air space and capillaries make the lung the primary organ for the adsorption of toxins. Table 2.1 lists common industrial toxicants and the lung diseases they produce. "*Toxicology* is the study of the adverse effects of chemical agents on biologic systems" (149). *Physiology* is the study of function in living matter (148). Engineers interested in air pollution and industrial hygiene should learn the fundamental concepts of these two fields of study as they pertain to the respiratory system. Physiology and toxicology are foundational fields of knowledge in the biological sciences. Both fields contain textbooks and professional journals engineers will need in the course of their work. It is recommended that engineers become familiar and learn to use Patty's (147), Guyton (148), and Casarett and Doull's (149). Several chapters (303–307) in research monographs and three physiology texts used by first-year medical students (377–379) are particularly suited for engineers since they describe the motion of gas in the lung in terms congenial to engineers. The purpose of this chapter is to describe the function of the respiratory system and how air contaminants affect its function.

Readers who have not been educated in the biological sciences will encounter many new technical phrases which will be defined in the context of the subject under discussion. These phrases and many unfamiliar phrases defined in a common dictionary should be learned since they will appear in subsequent discussion without elaboration. Since this text is written for students in the physical sciences, special care has been taken to define these terms and the reader should study these terms carefully and not take them for granted.

2.1 PHYSIOLOGY

The lungs expand and contract in two ways: (1) by the downward and upward movement of the diaphragm and (2) by the expansion and contraction of the rib

TABLE 2.1 Selected Industrial Toxicants Producing Lung Disease Through Inhalation [Abstracted (149)]

TOXICANT	CHEMICAL COMPOSITION	OCCUPATIONAL SOURCE	COMMON NAME OF DISEASE	SITE OF ACTION	ACUTE EFFECT	CHRONIC EFFECT
Asbestos	Fibrous silicates (Mg, Ca, and others)	Mining construction, shipbuilding, manufacture of asbestos-containing materials	Asbestosis	Parenchyma		Pulmonary fibrosis, pleural calcification lung cancer, pleural mesothelioma
Aluminum	Al_2O_3	Manufacture of abrasives, smelting	Aluminosis	Upper airways, alveolar interstitium	Cough, shortness of breath	Interstitial fibrosis
Ammonia	NH_3	Ammonia production, manufacture of fertilizers, chemical production, explosives		Upper airway	Immediate upper and lower respiratory tract irritation, edema	Chronic bronchitis
Arsenic	As_2O_3, AsH_3 (arsine) $Pb_3(AsO_4)_2$ $Pb_3(AsO_4)_2$	Manufacture of pesticides, pigments, glass, alloys		Upper airways	Bronchitis	Lung cancer, bronchitis, larygnitis
Berylium	Be, $Be_2Al_2(SiO_3)_6$	Ore extraction, manufacture of alloys, ceramics	Berylliosis	Alveoli	Severe pulmonary edema, pneumonia	Pulmonary fibrosis, progressive dyspnea, interstitial granulomatosis, cor pulmonale
Chlorine	Cl_2	Manufacture of pulp and paper, plastics, chlorinated chemicals		Upper airways	Cough, hemoptysis, dyspnea, tracheobronchitis, bronchopneumonia	
Chromium (IV)	Na_2CrO_4 and other chromate salts	Production of Cr compounds, paint pigments, reduction of chromite ore		Nasopharynx, upper airways	Nasal irritation bronchitis	lung tumors and cancers
Coal dust	Coal plus SiO_2 and other minerals	Coal mining	Pneumoconiosis	Lung parenchyma, lymph nodes, hilus		Pulmonary fibrosis
Coke oven emissions	Polycyclic hydrocarbons, SO_x, NO_x, and particulate mixtures of heavy metals	Coke production		Upper airways		Tracheobronchial cancers
Hydrogen fluoride	HF	Manufacture of chemicals, photographic film, solvents, plastics		Upper airways	Respiratory irritation, hemorrhagic pulmonary edema	
Iron oxides	Fe_2O_3	Welding, foundry work, steel manufacture, hematite mining, jewelry making	Siderotic lung disease: silver finisher's lung, hematite miner's lung, arc welder's lung	Silver finisher's: pulmonary vessels and alveolar walls; hematite miner's: upper lobes, bronchi and alveoli; arc welder's: bronchi	Cough	Silver finisher's: subpleural and perivascular aggregations of macrophages; hematite miner's diffuse fibrosis-like pneumoconiosis; arc welder's; bronchitis
Kaolin	$Al_4Si_4O_{10}(OH)_8$ plus crystalline SiO_2	Pottery making	Kaolinosis	Lung parenchyma, lymph nodes, hilus		Pulmonary fibrosis
Nickel	NiCO (nickel carbonyl), Ni, Ni_2S_3 (nickel subsulfide), NiO	Nickel ore extraction, nickel smelting, electronic electroplating, fossil fuel		Parenchyma (NiCO), nasal mucosa (Ni_2S_3), bronchi (NiO)	Pulmonary edema delayed by two days (NiCO)	Squamous cell carcinoma of nasal cavity and lung
Oxides of nitrogen	NO, NO_2, HNO_3	Welding, silo filling, explosive manufacture		Terminal respiratory bronchi and alveoli	Pulmonary congestion and edema	Emphysema
Ozone	O_3	Welding, bleaching flour, deodorizing		Terminal respiratory bronchi and alveoli	Pulmonary edema	Emphysema
Phosgene	$COCl_2$	Production of plastics, pesticides, chemicals		Alveoli	Edema Pulmonary edema	Bronchitis
Percloroethylene	C_2Cl_4	Dry cleaning, metal degreasing, grain fumigating			Pulmonary edema	
Silica	SiO_2	Mining, stone cutting, contruction, farming, quarrying	Silicosis, pneumoconiosis	Lung parenchyma, lymph nodes, hilus		Pulmonary fibrosis
Sulfur dioxide	SO_2	Manufacture of chemicals refrigeration, bleaching, fumigation		Upper airways	Bronchoconstriction cough, tightness in chest	

TABLE 2.1   *Continued*

TOXICANT	CHEMICAL COMPOSITION	OCCUPATIONAL SOURCE	COMMON NAME OF DISEASE	SITE OF ACTION	ACUTE EFFECT	CHRONIC EFFECT
Talc	$Mg_6(SiO_2)OH_4$	Rubber industry, cosmetics	Talcosis	Lung parenchyma, lymph nodes		Pulmonary fibrosis
Tin	SnO_2	Mining, processing of tin	Stanosis	Bronchioles and pleura		Widespread mottling of x-ray without clinical signs
Toluene 2.4-diiso-cyanate (TDI)	$CH_3C_6H_3(NCO)_2$	Manufacture of plastics		Upper airways	Acute bronchitis, bronchospasm, pulmonary edema	
Xylene	$C_6H_4(CH_3)_2$	Manufacture of resins, paints, varnishes, other chemicals, general solvent for adhesives		Lower airways	Pulmonary edema	

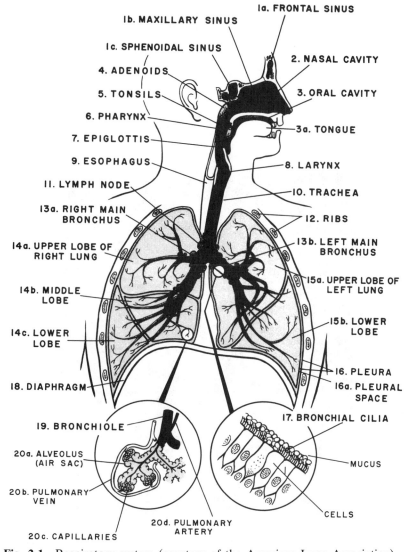

Fig. 2.1   Respiratory system (courtesy of the American Lung Association).

cage. Normal quiet breathing is accomplished with the diaphragm. The expansion and contraction of the lungs causes the gas pressure within the alveoli to become negative or positive with respect to atmospheric pressure. The pressure difference for normal breathing is approximately ±1 mm of mercury.

Figure 2.1 shows the principal elements of the respiratory system. The respiratory tract can be divided into three regions: nasopharyngeal, tracheobronchial, and pulmonary.

2.1.1 Nasopharyngeal Region

The nasopharynx region is the region between the nostrils and *larynx*. The nasopharyngeal region is also called the *upper* or *extrathoracic airways* and has a volume of approximately 50 ml. The nasal passages are lined with cellular tissue and mucous glands (vascular mucous epithelium). The nasopharynx removes large particles and is the region where the majority of heat and water vapor are exchanged with the respired air.

2.1.2 Tracheobronchial Region

The tracheobronchial region (Fig. 2.2) consists of the *trachea*, *bronchi*, and terminal *bronchiole* that act as conducting airways. The tracheobronchial region is

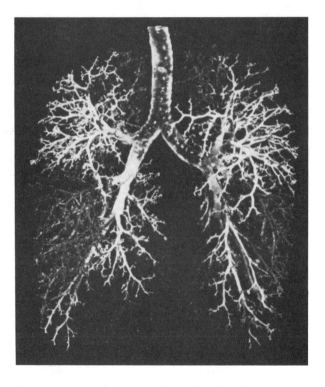

Fig. 2.2 Cast of the airways of the human lung. The alveoli have been pruned away but the conducting airways from the trachea to the terminal bronchioles can be seen. (reproduced with permission from ref. 377).

also called the *conducting airway*. All of these passageways, from the nose to the terminal bronchiole, are lined with ciliated epithelium and coated with a thin layer of *mucus* secreted by mucus-secreting cells. This surface is also called the *mucosal layer* or *mucous membrane*. Mucus production is about 10 ml per 24 hr. The thickness of the mucus layer is from 5–10 μm in humans. *Cilia* are hair-like organs protruding 3–4 μm above the surface of the cell (Figs. 2.3–2.4). As many as 200 cilia protrude from special epithelial cells lining air passages. *Mucociliary escalation* is a whip-like movement of cilia that occurs in the nasal cavity and respiratory tract. The whip-like motion causes a layer of mucus to move toward the pharynx at a rate of approximately 1 cm/min. Once at the pharynx the mucus passes through the digestive system. Cilia in the nose beat downward to the pharynx. The cilium move forward with a sudden rapid stroke 10–20 times a second, bending sharply so that they remain within the mucus layer. They move backward slowly. The whip-like movement propels mucus forward but the slow return motion does not displace the mucus. The mechanism of cilia and mucus that transfers material from the lung to the digestive tract is called *lung clearance*.

The trachea is called the *first generation respiratory passage*. The trachea divides into the *right and left bronchi* at a point called the *carina*. The main right and left bronchi are called the *second generation*. Each division of bronchi thereafter is an additional generation. There are between 20 and 25 generations before air finally reaches alveoli. The final few generations in which the passage-

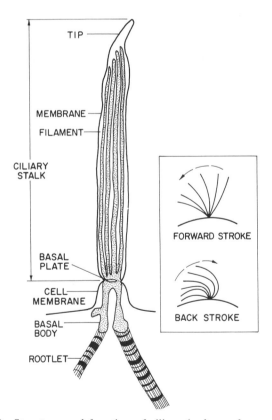

Fig. 2.3   Structure and function of cilium (redrawn from ref. 148).

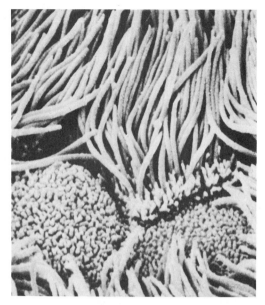

Fig. 2.4 Micrograph showing a bronchial wall with cilia to propel mucus (reproduced with permission from ref. 377).

ways are less than 1–1.5 mm in diameter are called *bronchiole* Fig. 2.5. All the intermediate passageways between the trachea and bronchiole are called *bronchi*. To keep the trachea from collapsing, multiple rings of cartilage surround 5/6's of the circumference. In the bronchi smaller amounts of cartilage provide partial rigidity. The bronchiole have no rigidity and expand and contract along with alveoli. The walls of the trachea and bronchi are composed of cartilage and smooth muscle. The walls of the bronchiole are entirely smooth muscle. The

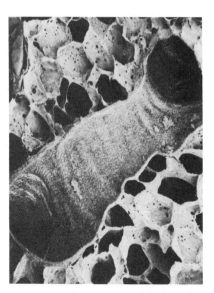

Fig. 2.5 Section of lung showing a small bronchiole and many alveoli. The pulmonary capillaries run in the walls of the alveoli. The holes in the alveolar walls are the pores of Kohn (reproduced with permission from ref. 377).

upper and conducting airways are relatively rigid conduits and comprise a volume of 150–200 ml, they are also called the *anatomic dead space* since very little gas exchange occurs.

Many obstructive lung diseases cause narrowing of the muscular bronchiole. Under *quiet breathing* (12–15 breaths/min) the pressure drop between alveoli and trachea is approximately 1 mm of mercury. The bronchi and trachea are very sensitive to touch, light, and particles or certain gases or vapors in inhaled air. The larynx and carina are particularly sensitive. Nerve cells in these passageways initiate a series of involuntary actions that produce coughing. *Coughing* begins with inspiring about 2.5 l of air followed by the spontaneous closing of the *epiglottis*, tightening the vocal cords that traps air in the lungs. The abdominal muscles then contract, forcing the diaphragm to contract the lung, collapse the bronchi, constrict the trachea and increase the air pressure in the lungs to values as high as 100 mm Hg. The epiglottis and vocal cords suddenly open and air under pressure is expired. The rapidly moving air carries upward foreign matter from the bronchi and trachea to the esophagus. *Sneezing* is another involuntary reflex similar to the cough, except that the *uvula* is depressed and large amounts of air pass rapidly through the nose and mouth.

2.1.3 Pulmonary Region

The pulmonary portion of the respiratory system consists of tiny sacs called *alveoli* clustered in groups and interconnected by openings called alveolar ducts (Fig. 2.1). The pulmonary region is also called the *respiratory airspace*. Figure 2.5 show portions of the pulmonary system. In a crude sense the lung resembles a sponge. Alveoli are thin-walled polyhedral pouches whose characteristic width is 250–350 μ and have at least one side open to either a *respiratory bronchiole* or an *alveolar duct* to another alveoli. Each terminal bronchiole supplies air for a segment of the lung called an *acinus*. The function of the alveoli is to provide a surface for the exchange of oxygen, carbon dioxide, and volatile metabolites between air and blood in the capillaries. The air side of alveoli consist of squamous epithelial cells and rounded septal cells. Mobile phagocytic cells and macrophage lie on the inner surface of the alveolus. *Macrophage* are white blood cells 7–10 μm in diameter. The function of the macrophage is to metabolize inhaled particulate material, spores, bacteria, and so on. The process is called *phagocytosis*. The metabolizing cells are also called *phagocytes*. Infectious particles are usually killed by macrophage, except some chronic bacterial and fungal infections such as tuberculosis and some viral diseases. The mean thickness of the tissue between alveoli is about 9 μm. The alveoli are served by a labyrinth of venous and arterial capillaries approximately 8 μm in diameter. Approximately 90–95% of the alveolar surface is served by capillaries. There are 300–500 million alveoli in the adult human.

The branching airways in the lung is an ordered structure in which each airway divides into two airways of identical geometry. The *Weibel symmetric model* (302) is widely used to describe the structure (Fig. 2.6). In the Weibel symmetric model the tracheobronchial region consists of generations 0–16. Generations 0-3 contain cartilage and are called *bronchi* (BR). Generations 4–16 contain no cartilage and are called *bronchiole* (BL). Generation 16 is called the *terminal bronchiole* (TBL). The *pulmonary region* consists of generations 17–23, which are subdivided

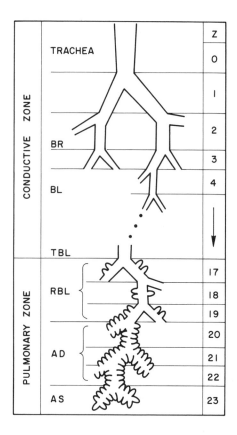

Fig. 2.6 Symmetrical Weibel model. Z is the airway generation, (BR) bronchi, (BL) bronchioles, (TBL) terminal bronchiole, (RBL) partially alveolated respiratory bronchioles, (AD) fully alveolated ducts, (AS) terminal alveolar sacs (adapted from refs. 302 and 303).

into 3 generations of *partially-alveolated respiratory bronchiole* (RBR), three generations of *fully-alveolated alveolar ducts* (AD), and the dead-ended or terminal, *alveolar sacs* (AS).

Figure 2.7 shows the airway cross sectional area, surface area, velocities, and Reynolds numbers versus distance from the nasal tip. Shown also are the Weibel airway generations. In the nose there are large convoluted surfaces called *nasal turbinates* that contain many blood vessels. The nasal turbinates are covered by a mucous membrane that transfers energy and water to fresh air during inhalation and exiting gas during exhalation. Nasal turbinates are the primary sites for the absorption of water soluble contaminants in air. Under quiet breathing most of the air passes through these nasal airways. Under vigorous exercise the mouth becomes the primary airway for inhalation and exhalation, the *switching point* being on the average a respiratory rate (also called minute volume) of 34.5 l/minute. The upper airway volume is approximately 50 ml, and the flow path between lips and glottis is approximately 17 cm long and 22 cm between the nose to glottis. Figure 2.7 shows that once inside the lung the flow decelerates and the flow path decreases rapidly. While the Reynolds numbers are low, one should not assume that the flow is fully developed since bronchial bifurcations produce regions of recirculation and unsteadiness. Fully developed flow does not occur until $L/L_e > 1$, which only occurs in the bronchiole, that is, airway generations beyond 6.

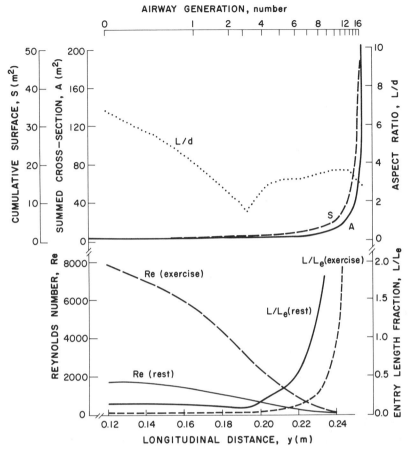

Fig. 2.7 Geometrical and aerodynamic characteristics of a symmetric Weibel model for two ventilation volumetric flow rates, rest (0.4 l/sec) and exercise (1.6 l/sec). L_e is the length needed for the flow to become fully developed (adapted from ref. 307).

The volume of the lung changes as air is inhaled and exhaled. Thus the honeycomb-alveoli structure expands and contracts. The total *alveolar surface area* in the adult human is about 35 m^2 during expiration and 100 m^2 during deep inspiration. By comparison, the area of a singles tennis court is 195.7 m^2. Great debate exists about the existence and role of a thin (0.5–1 μm) liquid layer on the air side of alveoli. The aqueous layer is called a *surfactant* and is a complex mixture of proteins, predominately dipalmitoyl lecithin. It is believed (148) that the surface tension of this liquid influences the expansion and contraction of the alveoli during breathing. In the absence of surfactant, small alveoli coalesce into larger alveoli and the area for gas exchange is reduced.

Arteries carry blood containing nutrients to tissue and veins carry blood containing waste products away from tissue. Each nutrient artery entering an organ divides six to eight times after which the remaining vessels are called *arterioles*. Arterioles are approximately 40 μm in diameter. Arterioles subdivide two to five times and become *capillaries* with diameters of 8 or 9 μm. The body contains nearly 10 billion capillaries with a total surface area of 500–700 m^2.

Nearly all cells are within 20–30 μm of a capillary. Blood flows through capillaries in an intermittent fashion. On the return leg, capillaries recombine to form *venules*, which in turn recombine to form *veins*. *Perfusion* is the phrase used to describe the flow of blood through the blood vessels in the lung. There is a large pressure drop associated with blood flow in arteries and arterioles. Thus the walls of these vessels are considerably stronger and thicker than veins and venules. Capillary walls are composed of unicellular endothelial cells surrounded by basement membrane. The thickness of the wall is about 0.5 μm. Material on the outside of capillaries is called *interstitial fluid*.

Materials are transferred through the capillary membrane by a variety of processes broadly called *diffusion*. Cells in the capillary wall are composed of different materials in which different nutrients are soluble. If the nutrient is soluble in *lipids* (materials within cells that are soluble in fat solvents but not water) such as oxygen and carbon dioxide, it will diffuse through the cells in the capillary membrane containing lipids. Water soluble but lipid-insoluble materials such as sodium and chloride ions and glucose pass through the membrane through slit-pores (6–7 nm wide). The phrase *permeability* is used to describe the diffusing capacity of different materials to pass through the capillary membrane. The concentration difference is the driving potential.

Approximately one-sixth of body tissue is the space between cells and is called *interstitium*. Fluid in this space is called generally, *interstitial fluid*. Figure 2.8 shows the elements of the interstitium. Solid material is composed of two major elements, collagen fiber bundles and proteoglycan filaments. *Collagen* fiber bundles have large tensile strength and provide tissue a tensile strength. *Proteoglycan* filaments are small coiled protein molecules that form a "mat." Interstitial fluid is a plasma derived from capillaries. The combination of proteoglycan filaments and interstitial fluid is called *tissue gel*. After passing through the capillary wall, materials diffuse through the tissue gel, molecule by molecule to cells receiving the material. Although virtually all fluid is bound in some way, small rivulets of "free fluid" are present. *Edema* is the name of a physical disorder that occurs when the amount of free fluid accumulates in pockets and the rivulets expand enormously.

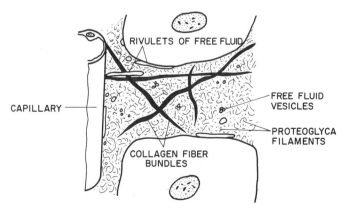

Fig. 2.8   Structure of the interstitium (redrawn from 148).

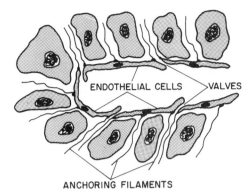

Fig. 2.9 Lymphatic capillaries showing structures that enable material of large molecular weight to enter circulation (redrawn from 148).

Not all material can be transferred to the blood by diffusion, for example, insoluble, indigestible bacteria and dust particles. The *lymphatic system* is an additional route by which these materials can be transferred from the interstitial space to the blood. Many large inhaled particles find their way into the lymphatic system where they pose a serious hazard to health. Small lymphatic vessels called capillaries carry proteins and large particles of foreign matter from interstitial spaces. About one tenth of the fluids from the arteries passes through the lymphatic system rather than returning through venous capillaries. Substances with large molecular weights that cannot diffuse may enter lymphatic capillaries. Figure 2.9 illustrates how the unique structure of lymphatic capillaries enables them to accept very large molecules and foreign particles. Anchoring filaments of the lymphatic capillary are anchored to the connective tissue between cells interstitial space. Adjacent ends of endothelial cells (of the lymphatic capillary) overlap themselves and produce a "flap". Large particles in the interstitial fluid are able to push open these flaps and flow directly into the lymphatic capillary.

Lymphatic fluid is derived from interstitial fluid but is richer in the concentration of various proteins. Lymphatic capillaries branch and merge as do veins and arteries. Lymphatic channels merge at *lymph nodes* where additional chemical processes occur. Approximately 100 ml of lymph flow through the thoracic duct of a resting individual. An estimated additional 20 ml of lymph flow through other channels. Lymph is believed to be pumped by several methods. In part pumping is achieved by movement of muscles and part by the numerous flaps that pass material to larger collecting lymphatic vessels.

2.2 RESPIRATORY FLUID MECHANICS

The elastic properties of the lung are due to elastic fibers within the lung tissue and the surface tension of the surfactant liquid lining the alveoli surface. The change in the volume of the lung with respect to pressure is called *compliance*. Compliance is comparable to the elastic coefficient or Young's modulus in solid mechanics or isothermal expansion coefficient in thermodynamics. Figure 2.10 is a graph of the volume of air input or removed from the lung (within the thorax) as a

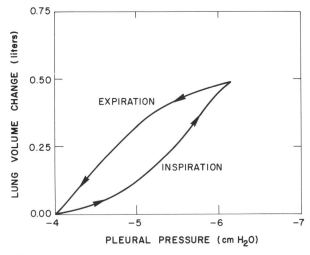

Fig. 2.10 Compliance of the lung (redrawn from 148).

function of the pressure in the space between the lung and rib cage (a space called the *pleura*). Two things are evident, (1) the slope is not constant and (2) the lung displays *hysteresis* in that expiration and inspiration do not produce coincident curves. Both effects are believed to be due to the viscoelastic properties of human tissue, the honeycomb structure of alveoli, and the surface tension of the *surfactant*. A linear relationship connecting the end points in Fig. 2.10 is approximately 0.13 l/cm water pressure drop. If the lungs were removed from the thorax, the slope would be 0.22 l/cm of water pressure. The difference is due to the elastic properties of the thoracic cage.

The lungs expand and contract due to the muscles of the diaphragm and air is drawn into and out of the lungs by the resulting positive or negative pressure in the lung cavity. *Spirometry* is the analysis of the volume and volumetric flow rate of air during respiration. Figure 2.11 shows the classic way to measure the volume of respired air corrected to STP. It is important that readers learn the names given to the various volumes of air. The *tidal volume* (V_t) is the volume of air (at STP) inhaled (and exhaled) in *normal breathing*. The normal tidal volume is approximately 500 ml. The *vital capacity* (V_{vc}) is the maximum volume of air one can expel from the lungs after first filling the lungs to their maximum extent and then expiring to the maximum extent. A typical value of the vital capacity is 4600 ml. The *residual volume* (V_r) is the volume of air remaining in the lungs following the maximum expiration one can produce. The typical residual volume is 1200 ml. The *anatomic dead space* (V_d) is the volume of the airways from the mouth to the respiratory bronchiole. The volume of the anatomic dead space increases with lung expansion varying from 144 ml in a collapsed lung to 260 ml after maximum inspiration. Approximately 50% of the anatomic dead space volume is located in the upper airways. The *functional residual capacity* (V_{fr}) is the difference between no gas in the lung and the minimum of the tidal volume (about 2300 ml). The *total lung capacity* (V) is the maximum volume to which the lungs can be expanded with the greatest possible inspiratory effort (about 5800 ml). All pulmonary volumes and capacities are about 20–25% less in women than in men. They tend also to be greater for athletic persons.

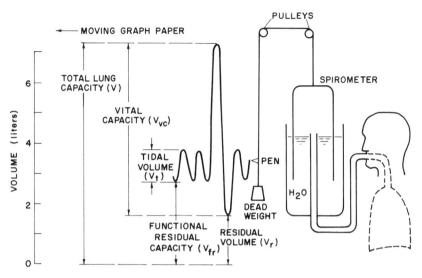

Fig. 2.11 Lung volumes and elements of spirometry. A pen records changes in air volume on graph paper that moves to the left. The residual volume and functional residual volume cannot be measured with the spirometer (redrawn from ref. 149).

The *minute respiratory rate* (or more simply the *ventilation rate*) (Q_t), also called the "minute volume," is equal to the tidal volume times the respiratory rate. The *normal tidal volume* (V_t) is 500 ml and the normal respiratory rate is 12–15 breaths/min. Using 12 breaths/min, the *normal minute respiratory rate* is approximately 6 l/min. A person can live for short periods of time at 2–4 breaths/min (1–2 l/min). Under unusual conditions the tidal volume can be as large as the vital capacity, or the respiratory rate can rise to as high as 40–50 breaths/min. At these rapid respiratory rates, a person usually cannot sustain a tidal volume greater than one half the vital capacity. The *alveolar ventilation* (Q_a) (Fig. 2.12) is the volumetric flow rate of fresh air that reaches the alveoli that is available for gas exchange. It is equal to the breathing rate times the difference between the tidal volume and the (anatomic) dead volume,

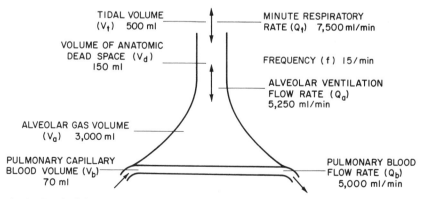

Fig. 2.12 Typical lung volumes and volumetric flow rates. There may be considerable individual differences in these values. (Redrawn from ref. 149).

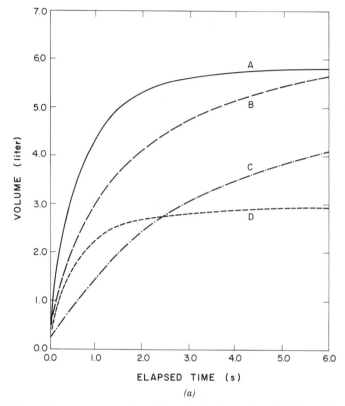

(a)

Fig. 2.13a   Spirometric measurements of forced vital capacity (FVC) and forced expiratory volume at one second (FEV$_1$) of (A) normal lung, (B) abnormal lung showing slight
obstruction, (C) advanced chronic obstructive pulmonary disease, (D) classic constricted
lungs.

$$\text{alveolar ventilation} = (500 \text{ ml}-150 \text{ ml})(12 \text{ breaths/min})$$

$$= 4200 \text{ ml/min}$$

The rate at which an individual inhales and exhales depends on the level of
exertion. A great deal of information about lung disease can be gained by
measuring the rates of inhalation and exhalation. Of particular importance is the
volume of gas exhaled in the first second and is called *forced expiratory volume at
one second* (FEV$_1$). This parameter is particularly reproducible and a sensitive
indication of obstructions in lung airways (see Fig. 2.13).

2.2.1 Bohr Model

The *Bohr model* (303), shown in Fig. 2.14, is a simple but useful analytical model
that has been used for many years to illustrate the distribution of inspired air
between the rigid conducting airways and the expandable alveolar region. The
model consists of a rigid volume called the anatomic dead space (V_d) and an
expandable volume called the alveolar region (V_a). Gas inside the alveolar region

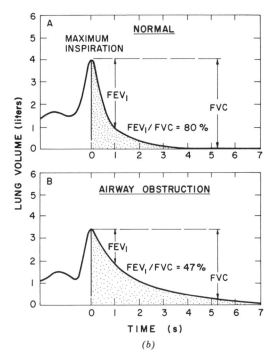

Fig. 2.13b Forced vital capacity (FVC) and forced expiratory volume at one second (FEV$_1$) (redrawn from ref. 148).

VENTILATION VOLUMETRIC FLOW RATE

$Q_t = f V_t$

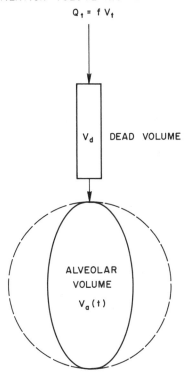

Fig. 2.14 Bohr model.

is assumed to be well mixed while the gas flowing through the dead space is assumed not to mix. During inspiration a portion of the volume of inspired air, that is, tidal volume (V_t) will occupy the dead space (V_d) and the remainder $(V_t - V_d)$ will mix with gases in the alveolar region. The product of the breathing frequency (f) and the tidal volume is a volumetric flow rate called the minute volume or more simply the ventilation rate (Q_t)

$$Q_t = fV_t \tag{2-1}$$

The variation of the alveolar volume with time (dV_a/dt) is called the *alveolar ventilation* (Q_a). Thus the volume of air entering the alveolar region during inspiration is

$$V_a = (V_t - V_d) = V_t \left[1 - \frac{V_d}{V_t} \right] \tag{2-2}$$

Differentiate with time and assume that the ratio (V_d/V_t) is constant

$$\frac{dV_a}{dt} = \frac{dV_t}{dt} \left[1 - \frac{V_d}{V_t} \right] \tag{2-3}$$

Assume that the alveolar ventilation (Q_a) is given by

$$Q_a = \frac{dV_a}{dt} \tag{2-4}$$

and that dV_t/dt is the ventilation rate (fV_t). Thus

$$Q_a = fV_t \left[1 - \frac{V_d}{V_t} \right] = Q_t \left[1 - \frac{V_d}{V_t} \right] \tag{2-5}$$

The Bohr model is useful in understanding the role of pulmonary and cardiac function under varying degrees of exercise. Table 2.2 shows values of ventilation rate (Q_t) and blood flow rate (Q_b) during exercise. Alveolar ventilation can be

TABLE 2.2   Ventilation, Blood Flow, and the Ventilation Perfusion Ratio (R_{vp}) During Exercise

	Rest	Light	Moderate	Heavy
Ventilation rate Q_t (l/min)	11.6	32.2	50.0	80.4
Frequency (l/min)	13.6	23.3	27.7	41.1
Tidal volume V_t(l)	0.85	1.38	1.81	1.96
V_d/V_t	0.34	0.20	0.16	0.16
Blood flow Q_b (l/min)	6.5	13.8	18.4	21.7
$R_{vp} = Q_a/Q_b$	1.18	1.87	2.28	3.11

Source: Abstracted from refs. 307 and 319.

computed from Eq. 2-5. The ratio of the alveolar ventilation rate (Q_a) to the blood volumetric flow rate (Q_b) is called the *ventilation–perfusion ratio* $(R_{vp} = Q_a/Q_b)$. By measuring this ratio and monitoring the concentration of specific gases in the inspired air and in the blood, one can obtain quantitative estimates of the effectiveness of the lung to transfer the gas to the blood. This transfer is commonly called *gas uptake*. In the analysis described above it was assumed that the blood absorbs all the material diffusing through the alveolar barrier. If only a fraction of the material is really absorbed, the analysis can be modified to include the solubility constant of the material in blood (see the extended Bohr model in Section 2.3)

The elementary Bohr model shows that exercise increases the ventilation volumetric flow rate by a factor of 6.9, while the blood flow rate increases by only a factor of slightly over 3. The fraction of the inspired air actually reaching the alveolar region increases only slightly (0.64–0.84) and the ventilation–perfusion ratio increases by a factor of 3.3. Thus the uptake of gases by the blood is limited more by the supply of blood than it is of air reaching the alveolar region.

The greater the ratio of dead space to tidal volume (V_d/V_t), the smaller the fraction of inspired air reaching the alveoli. Thus effective alveolar ventilation only occurs if the tidal volume (V_t) exceeds the anatomical dead volume (V_d). If the respiratory frequency is sufficiently high, there is evidence (304) that effective gas exchange occurs even when the tidal volume is smaller than the dead space. The phenomena is called *high frequency ventilation*. An example is panting in dogs.

A question that has concerned physiologists for many years is how gas exchange in the lung is maintained when the tidal volume (500 ml) is only a fraction of the vital capacity (4600 ml). The transport of gases between the mouth or nose and the alveoli can (152,303–306) be divided into five modes, each of which occurs in different Weibel airway generations:

(1) Direct alveolar ventilation by bulk convection.
(2) Convection by high-frequency "pendelluft".
(3) Convective dispersion due to asymmetric inspiration–expiratory velocity profiles
(4) Taylor-type dispersion.
(5) Molecular diffusion.

2.2.2 Bulk Convection

Bulk convection is the conventional flow of air in a passageway. If boundary layers are negligible, the volumetric flow rate is the cross-sectional area times the average (plug flow) velocity. The flow of the tidal volume in the trachea is bulk convection. The development of boundary layers and the resulting modification of the velocity profile outside the boundary layer can be analyzed by conventional boundary layer theory. One must be cautious. The flow is pulsatile, the airways contain constrictions (such as the vocal cords) and bifurcations and have a moving layer of mucus on the walls. Thus laminar, steady flow is only rarely encountered.

2.2.3 Pendelluft

There is considerable *asymmetry* in the human bronchial tree. Because of the location of the heart and other organs, the right and left lungs are not identical. The right lung is cleft by horizontal and oblique fissures into three lobes; the left lung is cleft into two lobes. When a tidal volume of air is inhaled, the fresh air may reach upper alveoli but not lower alveoli, as shown in Fig. 2.15. The path of the inspired air and the number of alveoli immediately reached is a function of the level of physical activity, number of breaths per minute, and the size of the tidal volume. There is asynchronous filling and emptying of the lungs that leads to an exchange of air between parallel lung units (Fig. 2.16).

Inhalation and exhalation can be modeled crudely as two, resistance–capacitance circuits in parallel driven by a common square-wave power supply (Fig. 2.17). The square-wave power supply corresponds to the positive and negative pressure difference produced by the diaphragm that fills and empties the lungs. The pressure drop (δP) to force air through a series of branching bronchiole can be thought to be a product of a resistance (R) and the volumetric flow rate (Q). Since the lung volume (V) changes, the pressure drop (δP) needed to fill the lung can be thought to be proportional to the lung volume (V) and inversely proportional to a lung capacitance (C). Thus for a single lung,

$$\delta P = RQ + \frac{V}{C} \tag{2-6}$$

It is recognized that the lumped-parameter circuit in Fig. 2.17 is simplistic since the lung resistance (R) and capacitance (C) are not truly constant. Nevertheless, assuming that R and C constant affords an opportunity to analyze gross behavioral characteristics of pulmonary ventilation. Applying the law of the conservation of mass for air entering an expandable lung unit, one finds that if the air density is constant, the rate of change of the lung volume is equal to the volumetric flow rate of air into the lung,

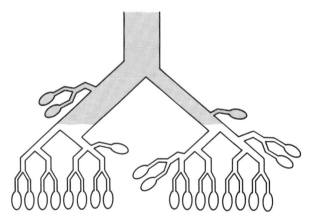

Fig. 2.15 Only a fraction of the alveoli are filled in direct ventilation even though the inspired volume is smaller than the total anatomical dead space (redrawn from ref. 151).

A — END OF EXPIRATION   B — END OF INSPIRATION

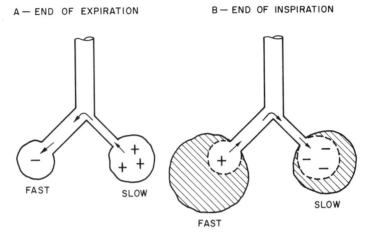

FAST   SLOW   SLOW

FAST

Fig. 2.16 When the breathing frequency is large, the lung resistance dominates the rate of filling and emptying for parallel lung units having different time constants ($\tau = RC$). Case A—expired air from the slow unit is transferred to the fast unit. Case B—at the end of inspiration, the fast unit will transfer air to the slow unit. The plus (+) and minus (−) signs indicate the pressure relative to atmospheric pressure (adapted from ref. 151).

$$\frac{dV}{dt} = Q \tag{2-7}$$

Combining these equations,

$$\delta P = R \frac{dV}{dt} + \frac{V}{C} \tag{2-8}$$

Integrate Eq. 2-8.

$$\int_{V_0}^{V(t)} \frac{dV}{\left[\dfrac{\delta P}{R} - \dfrac{V}{RC}\right]} = \int_0^t dt$$

$$\frac{V_f - V(t)}{V_f - V_0} = \exp\left(-\frac{t}{RC}\right) \tag{2-9}$$

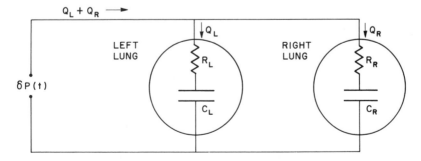

Fig. 2.17 Schematic diagram of pendelluft (source 151).

where V_0 is the initial lung volume and V_f is the final volume of the inflated lung ($V_f = C\delta P$). The product (RC) is called the time constant (τ) and is equal to the time the lung volume is 64% of it final volume, that is, $V(\tau)/V_f = 0.64$. The volumetric flow rate of air at any instant entering the lung unit can be obtained from Eq. 2-6,

$$Q = \delta P - \frac{V}{RC} \qquad (2\text{-}10)$$

and by removing V_f by using Eq. 2-9.

   If two lung units having different final volumes (V_f) and time constants (τ) are linked in a parallel, as shown in Fig. 2.16 and subjected to a common pressure difference (P_d), one can compute the volume and volumetric flow rates for each lung unit. It is clear that the response characteristics of each lung will be different. At the end of a rapid expiration of air, the unit with the smaller time constant (the fast unit) on the left is ready to fill while the slower unit on the right is still emptying. Thus there is a flow of gas from the slower to faster unit where the breathing frequency is large. At the end of a rapid inspiration, air will flow from the fast unit to the slow unit which is still filling. This "sloshing" between lung units is known as *pendelluft*.

2.2.4 Asymmetric Velocity Profiles

Experiments have shown that the velocity profiles in expiratory flow are flatter than those in inspiratory flow, which tend to resemble fully developed profiles. Figure 2.18 illustrates a simple explanation (153). Imagine that a flow passes a reference point as plug flow to the right. After a certain rightward displacement, the flow becomes parabolic if the Reynolds numbers are *laminar*. Thus it moves rightward as *fully established*. Upon exhalation, flow emerges from the bronchial tree system as individual fully established profiles from each bronchiole. Upon merging, the velocity profiles flatten such that the net profile is *uniform plug flow* to the left. Consider a parcel of air at three radial locations in an airway:

   (1) along the centerline, $r = 0$;
   (2) at the value r', where $U(r') = U_{ave}$, and $U_{ave} = Q/A$;
   (3) in the annular region $r > r'$

Consider the net displacement of these air masses as a result of several cycles of inspiration and expiration. The displacement of the air mass is equal to the air velocity (U) integrated over the elapsed time. Figure 2.18 shows the result of this integration. Air in the center of the airway moves to the right, that is, toward the alveoli, while air along the airway walls moves to the left, that is, toward the pharynx. At a radius r', the air has no net displacement. While the total volumetric flow rate through the airway as a whole during inspiration and expiration is zero, there is net flow to the right through the central portion of the airway and a net flow to the left in the outer annular portion of the airway. Consequently, the flow is both *pulsatile and countercurrent*.

INSPIRATION : PARABOLIC VELOCITY PROFILE, U > 0

$$U(r)/U_0 = \{1 - (r/R)^2\}$$

$$U(0) = U_0$$

$$U_{ave} = Q/A = U_0/2$$

$$r_{ave} = 0.707 R$$

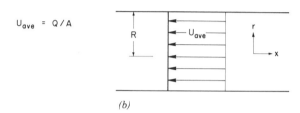

(a)

EXPIRATION : PLUG FLOW, U < 0

$$U_{ave} = Q/A$$

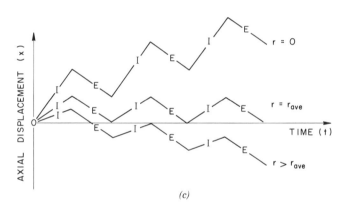

(b)

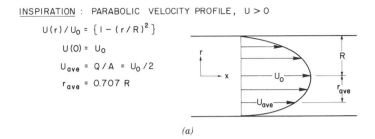

(c)

Fig. 2.18 (*a*) Inspiration: parabolic velocity profile, $U > 0$. $U_r/U_0 = \{1 - (r/R)^2\}$; $u(0) = U_0$; $U_{ave} = Q/A = U_0/2$; $r_{ave} = 0.707R$. (*b*) Expiration: plug flow, $U < 0$. $U_{ave} = Q/A$. (*c*) Displacement of a fluid element of three radial locations during inspiration (I) and expiration (E).

2.2.5 Taylor-Type Dispersion

Within alveoli, mass transport is by molecular diffusion. Within bronchiole, mass transport is a combination of convective transport and diffusion in an oscillating fluid, that is, a fluid that reverses its direction of motion owing to inspiration and expiration. Taylor-type dispersion is the name given to this composite flow. *Dispersion* is a phrase borrowed from meteorology and the mixing of plumes in the atmosphere in which the entire transport process is described by

$$\frac{\partial c}{\partial x} = K\left[\frac{\partial^2 c}{\partial y^2} + \frac{\partial c^2}{\partial z^2}\right] \tag{2-11}$$

where x is the direction of flow and y and z are in the transverse directions. The coefficient K is called the dispersion coefficient and is proportional to the molecular diffusion coefficient and inversely proportional to the convective velocity U. The transport of gases during breathing is a function of complex oscillating flow and a highly complex compliant (expanding/contracting) system of bronchiole. To begin understanding Taylor-type dispersion, consider a fully established, laminar flow in a duct of constant radius a. The average (plug flow) velocity is one-half the maximum centerline velocity. At time zero imagine that a diffusible material is injected (continuously) into the flow at a velocity equal to the local liquid velocity. The material will be convected downstream (longitudinally) but will also diffuse radially. Relative to a stationary observer the steady-state concentrations are given by Eq. 2-11, where variations in z are neglected. Now imagine a reference system moving in the direction of flow at a velocity U (plug flow velocity). Taylor showed that the transport of mass can be described as *virtual diffusion* governed by Fick's law in which the mass transport of material $\dot{m}$ relative to a moving frame of reference is described by

$$\dot{m} = -D'A\,\frac{dc}{dx'} \tag{2-12}$$

where $x' = x - Ut$, and the virtual diffusion coefficient D' is

$$D' = \begin{bmatrix} (\text{lam}) = \dfrac{(aU)^2}{\mathscr{D}} \\[2mm] (\text{turb}) = \text{proportional }(Ua) \end{bmatrix} \tag{2-13}$$

and $\mathscr{D}$ is the molecular diffusion coefficient. In reality, air moves in the bronchioles in an oscillatory fashion and one must also cope with inertial and viscous effects. The parameter used to characterize oscillatory flow is the *Womersley Number* (Wo), (382)

$$\text{Wo} = a\sqrt{\frac{2\pi f}{\nu}} \tag{2-14}$$

where f is the frequency and ν is the kinematic viscosity of air. If Wo is less than unity, the flow can be analyzed as quasi-steady viscous flow; if Wo is considerably larger than unity the flow must be analyzed as unsteady viscous flow. For laminar oscillatory flow (Wo $\gg$ 1) the dispersion coefficient is

$$K = \text{proportional to }\frac{\text{Re}^2}{\text{Wo}^n} \tag{2-15}$$

where n is a constant and the Reynolds number is

$$\text{Re} = \frac{2aU(\text{rms})}{\nu} \tag{2-16}$$

where U(rms) is the root mean square axial velocity and n is approximately three. Dispersion in the bronchiole is very complex. The flow is often unsteady and it is clear that mass transport solely on the mechanism of Taylor-type dispersion is inadequate.

2.2.6   Molecular Diffusion

Molecular diffusion is the mechanism for the transfer of gas across the alveolar membrane to the capillaries and is also a function of gas-phase diffusion of oxygen and carbon dioxide near the membrane (Figs. 2.19 and 2.20)

The overall efficiency of gas exchange is a function of the five modes of gas transport. They are not mutually exclusive and certainly interact; however, for a given set of physical conditions, one mode may be dominant in a certain airway generation. Figure 2.21 is a summary of the dominant modes of transport in the lung. In the trachea and main-stem bronchi, turbulent Taylor-type dispersion should be important. If the tidal volume is large, convective flow can clear this portion of the dead space and ventilate some alveoli directly. In the medium-sized airways, large phase lags and oscillatory convective flow can occur. Mixing in the conducting zone of the lung occurs either by convective dispersion resulting from asymmetry of the lungs or out-of phase bulk flow. In the small peripheral airways in the respiratory zone, out-of-phase oscillatory motion may be responsible for the ventilation of some lung units. Finally, in the alveoli and near the gas-exchange surface, molecular diffusion is the dominant mode of gas exchange.

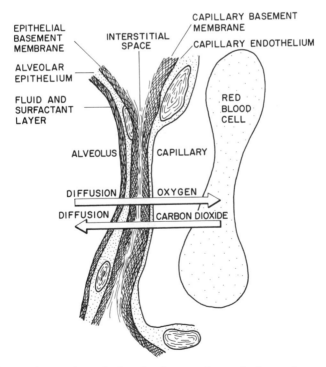

Fig. 2.19   Gas exchange through the alveolar membrane (redrawn from ref. 148).

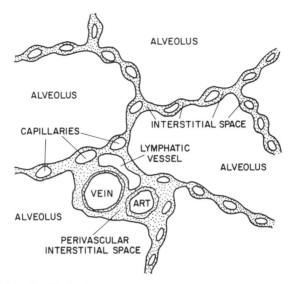

Fig. 2.20 Capillaries in the alveolar walls (redrawn from ref. 148).

2.3 ANALYTICAL MODELS OF HEAT AND MASS TRANSFER

Lung *permeability* is the rate at which materials penetrate the epithelial lining of the bronchial tree and lung respiratory surfaces. Air contaminants migrate to different body organs and affect them differently. Sulfur dioxide is soluble in water and is primarily removed by mucus in the upper respiratory system. Ozone has low solubility in water and penetrates the tracheobronchial region where

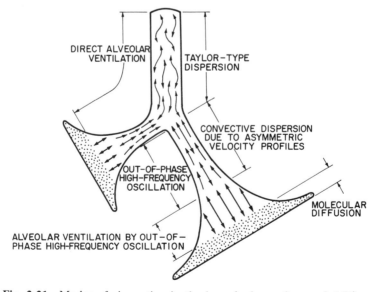

Fig. 2.21 Modes of air motion in the lung (redrawn from ref. 151).

much of it is absorbed and reacts chemically with mucus; however, a portion penetrates to attack underlying tissue. Carbon monoxide and volatile hydrocarbons have low solubility in water, do not react with mucus, and penetrate the pulmonary region where they diffuse through the alveolar–capillary barrier and are absorbed in the blood. These examples illustrate widely different effects the modeler must address in designing an analytical model. Modelers must ask themselves a number of questions such as the following:

Is the contaminant soluble in water?

Does the contaminant react with mucus or tissue?

Can the contaminant be absorbed by blood?

Do the effects of the contaminant vary with airway generation?

Can well-mixed conditions be assumed?

The functional residual capacity of the lungs (V_{fr}) is approximately 2300 ml but only 350 ml of inspired air is added to the alveolar volume and 350 ml of alveolar air is expired. Table 2.3 shows the typical composition of inspired air, gas in the alveoli, and the expired gas. Trace amounts of volatile solvents such as carbon tetrachloride and benzene produced in metabolism are also transferred to the expired gas. The large volume of alveolar air with respect to the volume of inspired gas insures slowly changing conditions within the alveoli and prevents sudden changes in the gas exchange rates to the blood. An elementary model of alveolar ventilation will be useful. Such a model, shown in Fig. 2.22 assumes that the alveolar volume is a well-mixed region where inspired air enters at 4200 ml/min, mixes with 2300 ml of alveolar air, and 4200 ml/min of alveolar air leaves on a steady-state basis. Oxygen is removed from the volume at a rate dictated by the body's metabolic rate of 250 ml/min (restful condition) and carbon dioxide enters the volume at a rate also dictated by the metabolic rate 200 ml/min. Note that 250 and 200 ml/min do not agree with the difference in oxygen and carbon dioxide entering and leaving the volume. The disparity arises because materials also enter and leave the body as solids and liquids. If the equations for the conservation of mass for oxygen and carbon dioxide are written (see the chapter on well-mixed models), one finds that the steady-state concentration of oxygen and carbon dioxide vary with the ventilation rate shown in Figs. 2.23 and 2.24. The intersection of the curves with the known values of oxygen and carbon dioxide in the alveolar volume intersect at a lung ventilation rate of 4200 ml/min.

TABLE 2.3 Mol Fraction of Respiratory Gases that Enter and Leave the Lungs at Sea Level

Species	Humid Air[a]	Alveolar Air	Expired Air
N_2	0.7413	0.749	0.745
O_2	0.1964	0.136	0.157
CO_2	0.0004	0.053	0.036
H_2O	0.0618	0.062	0.062

Source: Abstracted from ref. 148.
[a]Humid air: 25 C, 101 kPa, 50% relative humidity.

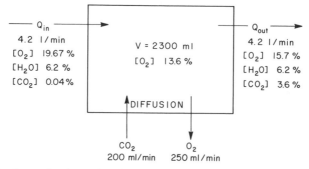

Fig. 2.22 Control volume for alveolar gas exchange (adapted from ref. 148).

If individuals exercise and produce a metabolic process requiring 1000 ml of oxygen and producing 800 ml of carbon dioxide per minute, lung ventilation rates given by the dashed lines will be needed.

Figures 2.19 and 2.20 show the composition of the alveolar walls through which gases are exchanged. The labyrinth of capillaries is so dense that the alveolar surface can be thought to be a sheet of blood contained within a thin membrane approximately 0.6 μm thick. The volume of blood in the lung is between 60 and 140 ml and is contained within capillaries about 8 μ in diameter. The movement of blood through the myriad of blood vessels in the alveolar surface is called *perfusion*. The rate of gas exchange is expressed in terms of a "diffusing capacity" times the difference in partial pressure of the diffusing species in the blood and gas within the alveoli. The diffusing capacity should not be confused with the molecular diffusion coefficient used for mass transfer within a single phase media. Under strenuous exercise (elevated metabolic rate), the pulmonary capillaries

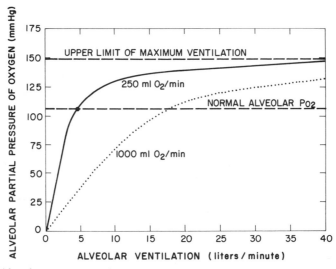

Fig. 2.23 Alveolar oxygen partial pressure versus the alveolar volumetric flow rate for two oxygen absorption rates in the blood (adapted from ref. 148).

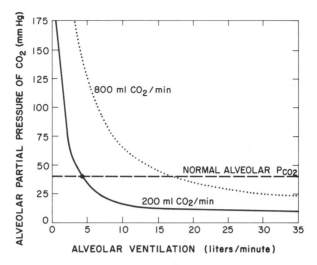

Fig. 2.24 Alveolar carbon dioxide partial pressure versus alveolar volumetric flow rate for two carbon dioxide desorption rates in the blood (adapted from ref. 148).

dilate, the blood flow rate increases, and the gas exchange rate increases. Thus the diffusing capacity may vary by a factor of 3. The diffusing capacity of carbon dioxide is considerably higher than for oxygen. Thus if the alveoli is damaged, individuals suffer from a lack of oxygen rather than excess of carbon dioxide. Under restful conditions, the diffusing capacities of oxygen and carbon dioxide are 21 ml/min per millimetre Hg and 65 ml/min per millimetre Hg, respectively.

A useful parameter to diagnosis dysfunctions of gas exchange in regions of the lung is Q_a/Q_b (the ventilation–perfusion ratio) where Q_a is the alveolar air ventilation flow rate and Q_b is the alveolar blood flow rate. Experiments can be run to measure the oxygen and carbon dioxide partial pressures in the alveolar capillaries for individuals suffering different lung disorders. Table 2.3 shows the ratio of oxygen to carbon dioxide for the normal lung. Ratios of O_2/CO_2 less than this value indicate that oxygen is not reaching the blood and that the gas exchange is insufficient.

Measurements of the ratio of the dead space to tidal volume (V_d/V_t) show that it decreases slightly as a result of exercise. From Eq. (2-5), it can be concluded that the alveolar ventilation (Q_a) increases. The volumetric flow rate of blood (Q_b) also increases even more rapidly with exercise. Consequently, the ventilation–perfusion ratio Q_a/Q_b decreases with exercise and suggests that the transfer of oxygen and carbon dioxide in the lung is more limited by cardiac function than the availability of fresh air to the alveolar region.

The ratio Q_a/Q_b is not uniform within the lung. In the upright lung it is larger at the bottom of the lung than at the top. The smaller ventilation–perfusion ratio near the top of the lung enables the concentration of contaminants accumulating in the alveolar tissue to be larger in the top part of the lung than in the lower part of the lung. The higher the solubility of the contaminant, the greater the disparity in tissue concentration between top and bottom of the lung.

Once oxygen enters pulmonary blood it is transferred to the cells. The solubility of oxygen in blood is low and the principal mechanism to transport oxygen to the cell is by hemoglobin in red blood cells. Following use in cells, carbon dioxide is absorbed in the blood and carried to the lung. The diffusion coefficient of carbon dioxide is 20 times larger than oxygen so that ordinary diffusion is sufficient to carry it to the lungs. Typical concentrations (as partial pressure) for oxygen and carbon dioxide in venous and arterial blood are as follows:

Oxygen: 40 mm Hg (venous) and 104 mm Hg (arterial)
Carbon dioxide: 45 mm Hg (venous) and 40 mm Hg (arterial)

When arterial blood reaches tissue cells the partial pressure of oxygen is approximately 95 mm Hg.

To model the heat and mass transfer processes in the respiratory system one has to apply the laws that conserve mass, energy, and momentum. What is less obvious is how to define the region within which the laws apply. A *control volume* is the region defined by the user within which they will apply the conservation equations. A *control surface* is the (closed) surface or boundary surrounding the control volume. Users define a control volume because it encompasses the region of interest and because they are capable of defining mass or energy transferred across the control surface. Thus there are really no right or wrong control volumes, but only useful and not useful ones. Analytical models can be divided into two broad categories, *compartmental models* and *distributed models*.

Compartment models may contain one or several compartments connected in series and/or parallel. Within each compartment *well-mixed* conditions are assumed. Consequently, within each compartment the gas composition is spatially uniform but may vary with time. Along the boundaries of the compartment a variety of processes may occur that remove or add material to the compartment. Compartmental models enable engineers to cope with mass and energy transfer processes that may vary with time in the respiratory system. Spatial variations can be accommodated only crudely by judiciously defining control volumes that correspond to Weibel generations of interest. The Bohr model (Fig. 2.14) is a compartmental model that can be extended to describe gas uptake by the blood.

2.3.1 Extended Bohr Model

OSHA standards proscribe exposure to vapors of common organic solvents such as trichlorethylene, acetone, methyl chloride, styrene, and carbon tetrachloride that may be present in the workplace atmosphere. They may also be present in indoor air. Many of these volatile organic compounds (VOC) are only mildly soluble and mildly reactive in human tissue. Thus they are not removed in the upper branches of the airway generation system and reach the alveolar region where they diffuse to the blood stream and are transported to organs. Under increased physical activity the ventilation rate increases and one may presume that the rate of absorption by the blood (i.e., dose rate) is directly proportional to the ventilation rate. The Bohr model can be extended to model the absorption of hydrocarbons by the blood, *hydrocarbon uptake*. Define an *uptake absorption efficiency* as the ratio of contaminant actually absorbed divided by the amount inhaled. Figure 2.25 shows the elements of an extended Bohr model. The alveolar

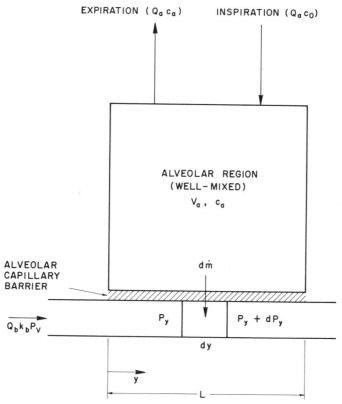

Fig. 2.25 Extended Bohr model that illustrates mass transfer of a nonreacting gas through the alveolar–capillary barrier.

region is assumed to be well-mixed such that the contaminant concentration c_a is uniform. Contaminant with a partial pressure P_0 in inspired air is mixed with residual air in the alveolar region and the partial pressure is reduced to P_a. The contaminant diffuses across an alveolar–capillary barrier to a venous blood stream in which the partial pressure approaching the alveolar region is P_y. Hydrocarbon is absorbed to the blood and the partial pressure in the arterial blood is increased to P_r. Blood with a volumetric flow rate Q_b and a solubility k_b for the hydrocarbon flows through the capillaries.

The three transfer processes in series dictate the overall rate at which hydrocarbons enter the bloodstream. The overall rate of transfer of a contaminant ($\dot{m}$, kg/sec) can be written as the product of an overall mass transfer coefficient (K, m/s) times the capillary surface area (S, m^2) times the difference in the partial pressure in the alveoli (P_a) and the partial pressure in the blood (P_r),

$$\dot{m} = \frac{MKS(P_a - P_r)}{R_u T} \tag{2-17}$$

The partial pressure in air (P_a) and the concentration (c_a, mols/volume of air) are related by the ideal gas law,

$$c_{\text{a}} = \frac{P_{\text{a}}}{R_{\text{u}} T}$$ (2-18)

By analogy to Ohms law, the reciprocal of the overall mass transfer coefficient $(1/K)$ can be thought of as a mass transfer resistance and equal to the sum of the individual resistances comprising the alveolar–capillary barrier.

In chemical and biomedical engineering the general expression for the resistance through a layer is $f/k_{\text{s}} k_{\text{m}}$ where k_{s} is the *solubility* of the contaminant in the layer, k_{m} is the *mass transfer coefficient* for the layer, and f is the fraction of the contaminant that remains after depletion because the contaminant may react chemically with materials in the layer. The solubility k_{s} is a thermodynamic property of the contaminant and materials in the layer and is independent of the rate of transfer. The mass transfer coefficient k_{m} is a transport property and is dependent on the rate of transfer, velocity field of the blood and/or air, and geometry and diffusivity of the layer. The fraction of the contaminant that diffuses without depletion depends on the chemical kinetics of the contaminant and materials in the layer. If the contaminant is transferred without reaction the fraction is unity. For a *four-layer barrier* shown in Fig. 2.26 the overall resistance to transfer is

$$\frac{1}{K} = \sum_{i} \left(\frac{f}{k_{\text{s}} k_{\text{m}}} \right)_{i}$$ (2-19)

where the subscript i pertains to each of the four layers. For contaminants that do not react with materials in the four layers, the fraction f is unity and the overall resistance is governed by the layer with the smallest product $(k_{\text{s}} k_{\text{m}})$. The alveolar–capillary barrier consists of four layers, but the mucus layer would be replaced by the pulmonary surfactant.

Consider an element of the capillary and the mass transferred into it. Transfer across the alveolar–capillary barrier can be written as

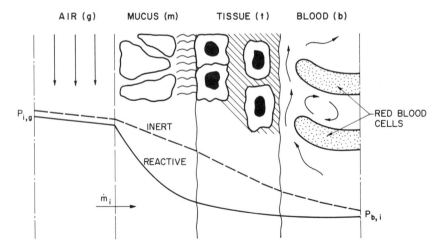

Fig. 2.26 Schematic diagram of four-layer diffusion through the bronchial wall. In the alveoli, mucus and cilia are replaced by surfactant (adapted from ref. 307).

$$dm = \left[\frac{MK(P_a - P_y)}{R_u T} P' \right] dy \qquad (2\text{-}20)$$

where P' is the perimeter of the capillary into which mass is transferred, P_y is the contaminant partial pressure at a variable point somewhere in the capillary, and K is the overall mass transfer coefficient for the alveolar–capillary barrier.

Write a mass balance for the element of the capillary absorbing the material,

$$\frac{dm}{M} + \frac{Q_b k_b P_y}{R_u T} = \frac{Q_b k_b (P_y + dP_y)}{R_u T} \qquad (2\text{-}21)$$

$$\frac{dm}{M} = \frac{Q_b k_b dP_y}{R_u T}$$

where k_b is the solubility of the transferred material in blood (volume of contaminant absorbed per volume of blood) since it is assumed that the alveolar tissue does not absorb the hydrocarbon. Eliminate dm and solve the resulting differential equation to find P_y.

$$\int_0^y \frac{K P' dy}{k_b Q_b} = \int_{P_v}^{P_y} \frac{dP_y}{(P_a - P_y)} \qquad (2\text{-}22)$$

After simplification

$$P_a - P_y = (P_a - P_v) \exp\left[-\frac{y K P'}{k_b Q_b} \right] \qquad (2\text{-}23)$$

Return to Eq. 2-20, replace P_y by Eq. 2-23 and integrate.

$$\frac{R_u T}{M K P'(P_a - P_v)} \int_0^m dm = \int_0^L \exp\left[-\frac{y K P'}{k_b Q_b} \right] dy$$

$$\dot{m} = \left(\frac{M(P_a - P_v) Q_b k_b}{R_u T} \right) \left(1 - \exp\left(-\frac{L K P'}{k_b Q_b} \right) \right) \qquad (2\text{-}24)$$

Using the alveolar region as a control volume write a mass balance

$$V_a \frac{dc_a}{dt} = Q_a c_0 - Q_a c_a - \left(\frac{\dot{m}}{M} \right) \qquad (2\text{-}25)$$

where V_a is the volume of the alveolar region and c_a is the molar contaminant concentration in the alveolar region. Assuming steady-state conditions ($dc_a/dt = 0$) obtain

$$\frac{c_a}{c_0} = 1 - \frac{\dot{m}}{M Q_a c_0} \qquad (2\text{-}26)$$

Assume that $P_a \ggg P_v$, divide both sides of Eq. 2-24 by $Q_a c_0$ and simplify. One obtains the ratio of the hydrocarbon absorbed by the blood to rate of inhaling the hydrocarbon, that is, uptake *absorption efficiency* ($\dot{m}/M Q_a c_0$)

$$\frac{\dot{m}}{MQ_a c_0} = \left[1 + \left\{\frac{R_{vp}}{1 - \exp(-N_D)}\right\}\right]^{-1} \qquad (2\text{-}27)$$

where

$$N_D \text{ (difussion parameter)} = \frac{KS}{Q_b k_b} \qquad (2\text{-}28)$$

$$S \text{ (area over which mass is transferred)} = P'L \qquad (2\text{-}29)$$

$$R_{vp} \text{ (ventilation–perfusion ratio)} = \frac{Q_a}{Q_b k_b} \qquad (2\text{-}30)$$

Figure 2.27 is a graph of the uptake absorption efficiency versus the ventilation–perfusion ratio for several values of the diffusion parameter. The following general conclusions can be drawn regarding the influence of exercise on the absorption of hydrocarbons.

(1)   Large values of N_D imply good diffusion of gas through the alveolar-capillary barrier and the uptake absorption efficiency is uniformly high. The location of N_D in the exponential term of Eq. 2-27 insures that at large values of N_D uptake absorption efficiency is only a function of the R_{vp}. When R_{vp} becomes large, the efficiencies decrease and approach approximately uniform values irrespective of the diffusion parameter (N_D) since absorption is now limited by blood flow.

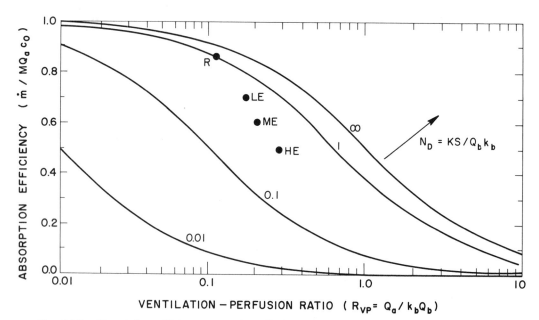

Fig. 2.27   Absorption efficiency versus ventilation–perfusion ratio for different values of the diffusion parameter corresponding to rest (R), light (LE), moderate exercise, (ME) and heavy exercise (HE) (adapted from ref. 319).

(2) Consider for the moment a rest state that corresponds to

$$\text{Rest: } N_D = 1.0 \quad \text{and} \quad R_{vp} = 0.1$$

and assume that the overall diffusion coefficient K and surface area S remain constant. Using the data in Table 2.2 to illustrate light, moderate, and heavy exercise, one finds that N_D decreases since it is inversely proportional to the blood flow rate Q_b and R_{vp} increases due to increases in the ventilation–perfusion ratio. As the rate of exercise increases, the uptake absorption efficiency decreases. From a health perspective, exercise does not increase the uptake absorption efficiency although the uptake itself, that is, the dose (kmols/sec), increases since the alveolar ventilation flow rate and blood flow rate increase. The original presumption that the bodily dose is proportional to the ventilation rate is overly pessimistic, and while the body dose increases it is not strictly proportional to the ventilation rate.

(3) Consider the transfer of oxygen through the alveolar–capillary barrier. Individuals suffering from *emphysema* experience a reduction in the surface area for gas exchange (S) and hence a reduction in the diffusion parameter N_D. Thus they experience a reduction in absorption efficiency nearly proportional to the reduction in surface area irrespective of the ventilation–perfusion ratio.

2.3.2 Distributed Parameter Models

There is considerable experimental evidence that ozone causes short-term bio-chemical functional changes in the lung. Specifically there is reduction in the one-second forced expiratory volume (FEV_1) (309). The distributed parameter model enables one to model these effects.

If one is interested in studying the effects of contaminants on portions of the bronchial tree as gas flows in the longitudinal direction, compartment models are of little help and one must turn to a distributed parameter model. A single differential equation can be written for the transport of contaminants in the direction of flow. The bifurcations of the bronchial tree are included assuming the airway is a single conduit resembling a *trumpet* (Fig. 2.28) whose cross-sectional area and perimeter varies with longitudinal distance in accord with the Weibel model. Geometrical data for the airway is provided by Fig. 2.7. Mass transfer between air passage and blood vessels is modeled as mass transfer through several layers in series, as shown in Figs. 2.26 and 2.28. The parameters describing the transfer across each layer may vary widely depending on the contaminant gas. Some gases such as carbon monoxide diffuse to the blood and are not absorbed or do not react chemically with materials in the layers. Gases such as ozone react chemically with the mucosal layer, while others such sulfur dioxide are mildly soluble in the mucosal layer.

The differential equation describing the longitudinal (y-direction) transport of contaminant species i through the airway is

$$\left(\frac{A}{R_u T}\right)\frac{\partial P_i}{\partial t} = -\left(\frac{Q}{R_u T}\right)\frac{\partial P_i}{\partial y} + \frac{1}{R_u T}\frac{\partial\left[\mathscr{D}A\,\dfrac{\partial P_i}{\partial y} - \dfrac{P'\dot{m}_i}{S}\right]}{\partial y} \qquad (2\text{-}31)$$

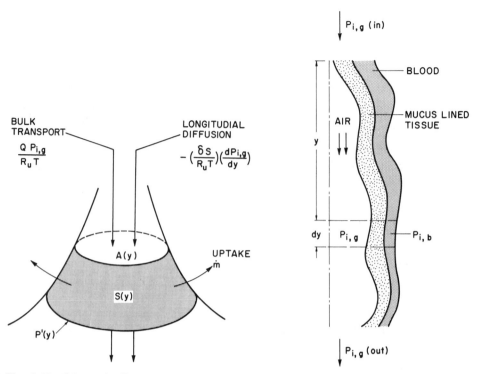

BULK
TRANSPORT

$$\frac{Q\,P_{i,g}}{R_u T}$$

LONGITUDIAL
DIFFUSION

$$-\left(\frac{\delta\,S}{R_u T}\right)\left(\frac{dP_{i,g}}{dy}\right)$$

A(y)

UPTAKE
$\dot{m}$

S(y)

P'(y)

$P_{i,g}$ (in)

BLOOD

AIR

MUCUS LINED
TISSUE

y

dy $P_{i,g}$

$P_{i,b}$

$P_{i,g}$ (out)

Fig. 2.28 Schematic diagram of a distributed parameter model (redrawn from ref. 307).

where,

$A,\ P'$ = cross sectional area and perimeter of the airway, both vary with y
Q = volumetric flow rate
$\mathscr{D}$ = diffusion coefficient
P_i = partial pressure of contaminant i
S = total surface area over which mass is transferred to the blood
$\dot{m}$ = rate of contaminant mass transferred to the blood, that is, uptake

The left-hand term in Eq. 2-31 represents the rate of accumulation and can be set to zero for steady-state solutions. The terms on the right-hand side correspond respectively to bulk transport, longitudinal diffusion, and uptake through the walls of the air passage. A differential equation for uptake is Eq. 2-17, where the overall mass transport coefficient K depends on properties of each of the four layers in Fig. 2.26.

Equation 2-31 can be solved numerically. Miller et al. (310) used such a model to describe the effects of ozone inhalation and predicted the total dose received in branches of the bronchial tree and the dosage penetrating to the tissue. Figure 2.29 shows the relative dose (dose/tracheal ozone concentration) versus location at the inside surface of the airway (mucus–air interface for airway generations up to 16 and surfactant–air interface in the pulmonary region) for several different first-order rate constants (k_r) for the ozone–mucus reaction. Figure 2.30 shows

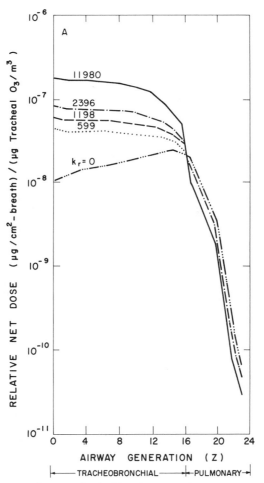

Fig. 2.29 Relative ozone dose to the inside airway surface versus airway generation for several rate first-order constants (k_r) from the ozone mucus reaction (redrawn from ref. 310).

the relative dose that penetrates the tissue for four levels of physical activity (1, rest to 4, heavy exercise) similar but not identical to those in Table 2.2. The difference between the two relative dosages is the ozone that reacts in the mucus layer or surfactant layers. While the rate constants for the chemical reactions are not known with precision, Figure 2.29 shows clearly that the dose at the mucus–air interface decreases with airway generation, but the net dose to tissue increases in the tracheobronchial region (TB) and decreases in the pulmonary region (P). Thus the tracheobronchial tissue is protected by *ozone absorption and chemical reaction with mucus* and the protection this provides tissue is superior to that provided by pulmonary surfactant. The tracheobronchial removal efficiency increases with the rate constant. The highest relative dose experienced by tissue occurs at the 16th airway generation, a conclusion borne out by experiment. With exercise, tissue dose in the tracheobronchial region is affected only slightly, but the point of maximum tissue dosage penetrates into the pulmonary region.

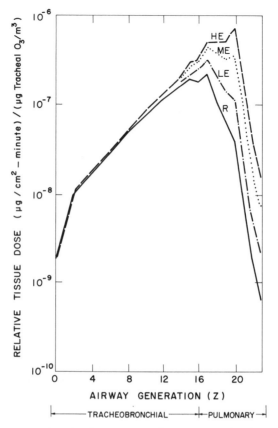

Fig. 2.30 Relative ozone dose to tissue versus airway generation for rest (R), light (LE), moderate exercise (ME), and heavy exercise (HE) (adapted from ref. 310).

The distributed parameter model has also been used by Hanna (153–155) to model the transfer of energy and water vapor during inhalation and exhalation. The upper respiratory tract, the region between the nasal tip to the midtrachea, is very susceptible to inflammation and infection. The upper respiratory tract is also the region within which the majority of energy is transferred to heat the incoming air to body temperature and to add water vapor to the incoming air, that is, *conditioning inhaled air.* Some energy and water is recovered upon on expiration. The amount of water and energy transferred to the inspired air depends on the temperature and humidity of the incoming air; the amount transferred during expiration is affected by the blood temperature and blood flow rate in the nasal and oral cavity. The bronchi may constrict because of the loss of energy and water from the mucosal surface, thus it is important to be able to describe quantitatively the transfer of energy and water from the tracheobronchial passages.

To preserve the functioning of the alveoli, inspired air must be heated to the body temperature and contain its maximum amount of water vapor. This conditioning is performed upstream in the trachea and bronchi. The mucus membrane conditions incoming air and collects pathogenic organisms and particles. The particles are removed by ciliary motion. The mucus secretion rate is a function of air temperature and humidity. The rheological properties of the mucus are also a function of net loss of energy and water.

The majority of conditioning occurs in the nasal cavity where during normal breathing the temperature is raised to 70% of body core temperature (37 C) and the air becomes saturated with water vapor. The temperature of the mucosal surface is 32-33 C. After the first third of the bronchial tree (Fig. 2.6), the temperature of the air and mucosal surface are equal to the body core temperature. The energy associated with water evaporation in the respiratory system is approximately 85% of the entire body's heat loss. The difference in the humidity of the expired and inspired air is the largest source of water loss by the body. During expiration 20–25% of the energy lost by evaporation during inhalation is regained by condensation. For quiet breathing, the net heat transfer to heat the inspired air is about 350 kcal or 15% of the basal metabolic rate and the net transfer of water is 250–400 ml for each breathing cycle.

Respiratory air flow is oscillatory, but for purposes of analysis the flow can be assumed to be quasi-static since the Womersley numbers are less than unity. Experimental measurements of air temperature in the nose and trachea show that steady-state temperatures are reached after 20% of the total time for inspiration and expiration. Measurements of relative humidity in air expired through the mouth show steady-state values after 20% of the total time. Measurements of the temperature at the interface between mucus layer and surface air passage show that the temperature is constant throughout the respiratory cycle. Beyond the second or third division of the bronchial tree the temperature and humidity of the air do not vary over the respiratory cycle. This point in the bronchial tree is called the *isothermal saturation boundary* (ISB). The precise location of ISB varies with the ambient temperature tidal volume and breathing rate.

Within the mucus layer and tissue steady-state conditions are achieved within characteristic times equal to the layer thickness divided by the thermal conductivity or mass diffusivity. Assuming the thickness of the layer is 50 μm, these characteristic times are of the order of a millisecond, whereas the duration of inspiration and expiration is 2.5 sec if the breathing rate is 12 breaths/min. In summary, assuming quasi-static conditions for inspiration and another set for expiration is a reasonable assumption for flow in the trachea and first two generations of the bronchial system.

Figures 2.31–2.33 show the results of an analysis based on an elementary quasi-static model of the transport of mass and heat in the respiratory tract. Air is assumed to pass uniformly through a duct whose walls are a composite of the mucous membrane, bronchial wall, and a layer composed of the tissue containing capillaries. The energy transferred from the blood through the mucus–air interface is assumed to be equal to the latent heat of evaporation of the evaporated water plus energy transferred to the air by convection. The blood is assumed to have constant temperature.

$$\frac{k_t}{y} [T_b - T_a] = h_c[T_m - T_a] + \dot{n}_w M_w h_{fg} \tag{2-32}$$

where

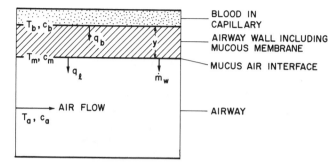

Fig. 2.31 Transverse heat and mass transfer across layers of an airway passage during inspiration. Quantities are functions of the longitudinal distance along the airway (redrawn from refs. 153–156).

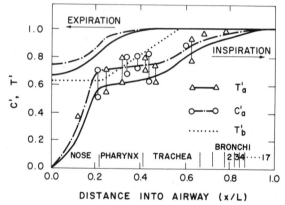

Fig. 2.32 Dimensionless air temperature (T_a'), blood temperature (T_b'), and water vapor concentration (c_a') as a function of the longitudinal distance in the bronchial system during inspiration and expiration in rest condition (redrawn from refs. 153–156).

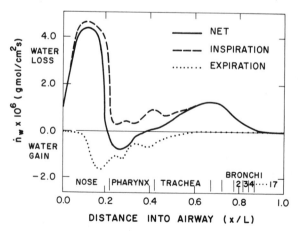

Fig. 2.33 Molar transfer rate ($\dot{n}_w$) of water to air as a function of the longitudinal distance in the bronchial system during inspiration and expiration in rest condition. Positive values of $\dot{n}_w$ denote the body's loss of water (redrawn from refs. 153–156).

x = direction of flow

T_b = blood temperature (constant)

$T_m(x)$ = mucus–air interface temperature

$T_a(x)$ = air temperature

M_w, h_{fg} = molecular weight and enthalpy of vaporization of water

$\dot{n}_w(x)$ = number of mols of water per unit time and area

y = thickness of airway wall including mucus membrane (constant)

$k_t(x)$, $h_c(x)$ = thermal conductivity of airway wall, convection heat transfer coefficient at mucus–air interface

The number of mols of water that evaporate per unit time and area ($\dot{n}_w$) is equal to the mass transfer of water vapor to air,

$$\dot{n}_w = k_c[c_m - c_a] \tag{2-33}$$

The parameter $k_c(x)$ is the convective mass transfer coefficient. The concentration of water vapor in air $c_a(x)$ is a variable, but the concentration of water vapor at the mucus–air interface is equal to its saturation value at temperature $T_m(x)$ and can be given by the empirical equation (154)

$$c_m(x) = 22.4 \exp \frac{-4900}{T_m} \tag{2-34}$$

where $c_m(x)$ is in mols/cm^3. Alternatively the Clausius–Clapeyron equation can be used. The diffusion of water vapor in the axial direction is small compared to bulk transport and can be neglected (153). Thus the conservation of mass for water vapor is given by

$$U \frac{dc_a}{dx} = \left(\frac{P'k_c}{A} \right)(c_m - c_a) \tag{2-35}$$

The conservation of energy is given by

$$U \frac{dT_a}{dx} = \left(\frac{P'}{A\rho c_{p,a}} \right) \{ h_c(T_m - T_a) + c_{p,w}M_w\dot{n}_w(T_m - T_a) \} \tag{2-36}$$

where

U = velocity (assumed to be plug flow)

$A(x)$, $P'(x)$ = airway cross sectional area and perimeter

ρ = air density

$c_{p,w}$, $c_{p,a}$ = specific heat of water vapor and air

The thermal conductivity of the airway wall and mucus membrane, k_t, can be taken as a constant and the transport parameters $h_c(x)$ and $k_c(x)$ are variables that can be evaluated (153) by dimensionless relationships from the literature of heat and mass transfer.

Equations 2-32 to 2-36 constitute a set of coupled ordinary differential equations that can be solved numerically. The values of the blood temperature T_b, airway cross-sectional area $A(x)$, and perimeter $P'(x)$ can be taken as constants at the appropriate portion of the bronchial tree. The air velocity U will also be taken as a constant and computed from the tidal volumetric flow rate divided by $A(x)$ that exists at the appropriate point in the bronchial tree.

The computations can be performed for both inspiration and expiration associated with restful room air breathing. Figure 2.32 shows the predicted air temperatures and water vapor concentration which are normalized with respect to the temperature of the inspired air $[T_a(\text{insp})]$,

$$T(x)' = \frac{T_a(x) - T_a(\text{insp})}{T_c - T_a(\text{insp})} \tag{2-38}$$

where T_c is the body-core temperature. The water vapor concentration is normalized in a similar fashion. The values of the transport coefficients were varied by 50% and shown to have only moderate effect on the results for inspiration and little effect for expiration. Of all the parameters in the analysis, the blood temperature and volume of the nasal cavity were found to have the most importance. The evaporation rate of water is shown in Fig. 2.33 for both inspiration and expiration.

Figures 2.32 and 2.32 show vividly that the nasal cavity is the primary organ that *conditions inspired air*. Within the nasal cavity the water vapor concentration in air nearly reaches equilibrium with the blood during both inspiration and expiration. During expiration, water vapor condenses on to the cooler nasal mucosa. The mucosa of the trachea and larynx lose water upon inspiration, but regain some of it upon expiration. For the cycle of inspiration and expiration the nasal cavity, lower trachea, and bronchial tree have a net loss of water, but the pharynx has a net gain of water. Air is predicted to be expired at nearly the nasal blood temperature and nearly saturated. These results are confirmed by experiment. Downstream of the nasal cavity, the airstream reaches 60–70% of body core temperature and is fully saturated during inspiration. Relatively little conditioning of the inspired air occurs within the pharynx and upper trachea. The upper portions of the bronchial tree fully condition inspired air in a relatively short distance. Conditioning in the tracheobronchial tree is the result of a large surface area of the numerous bifurcations. As the temperature of the inspired air decreases or the tidal volumetric flow rate increases, less conditioning is accomplished in the nasal cavity and more is accomplished in the tracheobronchial tree.

The health implications of these studies are important. The pharynx contains no mucus secreting cells and cannot retain water effectively. Water must be transported to the pharynx since it lacks cilia and mucus secreting glands and cells. The pharynx is thus particularly vulnerable to drying, disease, organisms, bacterial infiltration, irritation, and assault by air pollutants. These facts help confirm the validity of the analytical model that predicts that the pharynx has a net gain of water during the respiratory cycle. The conclusion is also reinforced by exercise experiments involving sulfur dioxide (240), in which the dose resulting from mouth breathing is larger than from nasal breathing owing to the efficient scrubbing that occurs in the nasal cavity.

In the outdoor environment sulfur dioxide forms sulfuric acid, which may be partially neutralized by atmospheric ammonia NH_3 to form ammonium bisulfate NH_4HSO_4 and ammonium sulfate $(NH_4)_2SO_4$. It is reported (247) that sulfuric acid particles are 10 times more potent than $(NH_4)_2SO_4$ and 33 times more potent than NH_4HSO_4 when inhaled small particles impact various surfaces in the tracheobronchial region. Because of the buffering capacity and volume of mucus, it is believed (247) that the pH of the tracheobronchial mucus does not change appreciably. However, the acid concentration in very small individual particles may be sufficient to produce localized *irritant signals* in the lower airways to increase mucus secretion and contribute to processes involved in chronic bronchitis. Depending on the exact pH depression to produce a signal and the effect of neutralization by NH_3 in the upper respiratory tract, the minimum size acid particle required to produce a signal is believed to be between 0.4 and 0.7 μm.

In order to predict the deposition of aerosols or the absorption of gases and vapors in the various parts of the respiratory system, it will be necessary to develop accurate fluid mechanics mathematical models. These models must account for the asymmetry of the respiratory system, the expansion and contraction of the air passages, and the mucus ciliary motion. Attempts to secure models that explain observed phenomena have gone on for several decades and have improved steadily. The work of Nixon and Egan (286), Yu and Xu (287,288) Miller (310), and Ultman (303,307) should be consulted.

2.4 SPIROMETRY

Spirometry is the technique used to diagnosis pulmonary disorders from measurements of the time variation of pressure and volume of respired air. To dispel air, muscles compress the lung cage, increase the internal air pressure, and force air from the alveoli and bronchiole through the bronchial tree and out the nose and mouth. Like any other fluid system, the volumetric flow rate depends on the pressure difference (δP) between lung and atmosphere and the frictional forces within the air passageways. The frictional forces can be lumped together and called the *airway resistance R* and the volumetric flow rate Q and expressed as

$$Q = \delta PR \qquad (2\text{-}39)$$

The airway resistance may vary throughout the respiratory cycle and the value for inspiration may be different than that for expiration (Fig. 2.10).

If the instantaneous volumetric flow rate versus the volume of expired air are measured by the equipment shown in Fig. 2.11, a graph similar to Fig. 2.34 can be obtained. Note that the normal lung a vital capacity of approximately 4.6 l and a residual volume of 1.2 l and that the maximum volumetric flow rate is about 400 l/min and is achieved rather quickly after one begins to exhale.

Constricted lungs (sometimes called restricted) cannot be fully expanded because they contain lesions or because abnormalities in the chest cage prevent individuals from inflating their lungs fully. The constricted lung is similar to the normal lung

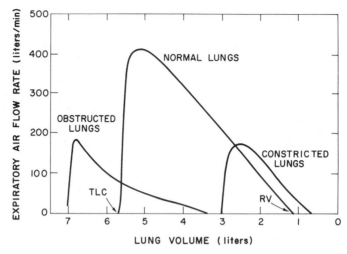

Fig. 2.34 Expiratory flow rate versus lung volume during maximum expiratory effort for obstructed and constricted lungs; TLC, total lung capacity, RV, residual volume (redrawn from ref. 148).

except that the vital capacity and maximum volumetric flow rate are smaller. The constricted lung occurs in people suffering from fibrosis of the lungs, thoracic deformities, major tumors, or other problems that restrict the amount of air that can be put into the lungs.

Individuals with *obstructed lungs* have to exert a larger pressure to expel air because airways are narrower and can not be emptied as easily as normal lungs. Thus the residual volume is also larger than the normal lung. Obstructed lungs exist in people suffering from chronic asthma, asthmatic bronchitis, emphysema, mucosal, edema and inflammation of the conducting air passages.

A test for pulmonary abnormalities is to expel as much air as one can from maximally filled lungs and to do so as quickly as possible. The volume so measured is called the *forced vital capacity* (FVC). Figure 2.13a shows the trace of normal (A) lungs, partially obstructed lungs (B), chronic obstructive pulmonary disease (C), and constricted lungs (D). Individuals with obstructed lungs will not be able to inspire quite as much as unobstructed lungs, but most importantly it will take a longer time to expel the air they have inhaled. A common way to diagnose lung disorders is to calculate two parameters, the forced vital capacity (FVC), the *forced vital capacity during the initial second* (FEV1), and the ratio, that is, FEV1/FVC.

Normal lung: $0.7 < [FEV1/FVC] < 0.75$

Obstructed lung: $[FEV1/FVC] < 0.7$

Constricted lung: $[FEV1/FVC] > 0.75$, but $FVC \ll FVC_{normal}$

An additional technique to quantify these lung abnormalities involves the concept of *moments* (158,159) since it provides insight into the pathophysiology of disease and allows one to have quantitative means to discriminate between curves such as Fig. 2.13a and b. The concept of moments is similar to what will be discussed in Section 5.11. Define the *arithmetic mean transient time* (t_m) as the arithmetic average time it takes to expire an amount of air equal to the vital capacity. If the instantaneous volume of the lung is V, the conservation of mass for air during the process of expiration (assuming the density remains constant) is

$$Q = -\frac{dV}{dt} \tag{2-40}$$

where a positive value of Q is the expiration volumetric flow rate. Rearrange, multiply both sides by time t, and integrate over the time t_{vc}, the time it takes to expire a volume of air equal to the vital capacity (V_{vc}).

$$\int_0^{t_{vc}} tQ \, dt = -\int_{V_{vc}}^0 t \, dV \tag{2-41}$$

Define the arithmetic mean transient time

$$t_m = \int_0^{V_{vc}} t \, \frac{dV}{V_{vc}} = \frac{1}{V_{vc}} \int_0^{V_{vc}} t \, dV \tag{2.42}$$

Combining Eqs. 2-41 and 2-42, the arithmetic mean transient time becomes

$$t_m = \int_0^{t_{vc}} t \, \frac{Q}{V_{vc}} \, dt = \frac{1}{V_{vc}} \int_0^{t_{vc}} Qt \, dt \tag{2-43}$$

The nth moment is generally defined generally by,

$$\alpha_n = \int_0^{t_{vc}} t^n \, \frac{Q}{V_{vc}} \, dt \tag{2-44}$$

integrated over the time period t_{vc}. The first moment ($n = 1$ or α_1) is the arithmetic mean transient time described in Eq. 2-42.

$$\alpha_1 = \int_0^{t_{vc}} t \, \frac{Q}{V_{vc}} \, dt = \frac{1}{V_{vc}} \int_0^{T_{vc}} tQ \, dt \tag{2-45}$$

The second moment ($n = 2$ or α_2)

$$\alpha_2 = \int_0^{t_{vc}} t^2 \, \frac{Q}{V_{vc}} \, dt = \frac{1}{V_{vv}} \int_0^{t_{vc}} t^2 Q \, dt \tag{2-46}$$

The square root of α_2 is the root mean square of the transient time. Because of the nature of the tails in the curves in Fig. 2.13, it can be seen that higher moments weigh the importance of the smaller expired volumes of air. The arithmetic mean transient time (α_1) provides an index of the average time it takes for increments of volume to leave the lung. With obstruction this value increases. With obstruction, the value of α_2 also increases. If α_1 and α_2 are normalized by the integral of (Q/V_{vc}), it can be shown that

$$\sqrt{\left(\frac{\alpha_2}{2}\right)} = \alpha_1 \qquad (2\text{-}47)$$

when (Q/V_{vc}) is an exponential function. If on the other hand, an obstruction is present, the relationship between α_1 and $\sqrt{\alpha_2}$ changes. Where there is an obstruction in the upper airway, Fig. 2.13 shows that the lung volume does not decrease rapidly and the $\sqrt{\alpha_2}$ does not fall relative to α_1. In cases of airway obstruction because of emphysema and age, $\sqrt{\alpha_2}$ rises relative to α_1. Permutt and Menkes (159) report that analysis of 57 nonsmoking males shows that α_1 and α_2 and age (in years) can be related by

$$\sqrt{\alpha_2} = 0.229 + 1.165\,\alpha_1 + 0.00347\,\text{age} \qquad (2\text{-}48)$$

with a standard deviation of 0.078. The companion equation for a group of smokers is

$$\sqrt{\alpha_2} = 0.371 + 0.676\,\alpha_1 + 0.00981\,\text{age} \qquad (2\text{-}49)$$

with a standard deviation of 0.206.

2.5 TOXICOLOGY

Respiratory diseases and disorders produce symptoms that are grouped in several categories—hypoxia, cyanosis, dyspnea, and hypercapnia. *Hypoxia* is an inadequate supply of oxygen to support bodily functions. Hypoxia can be subdivided as follows:

(a) **Hypoxic Hypoxia.** Sufficient oxygen does not reach the alveoli; hypoxic hypoxia can be caused by environmental factors.
(b) **Anemic Hypoxia.** Inadequate hemoglobin prevents sufficient oxygen from reaching the cells.
(c) **Circulatory Hypoxia.** Blood flow rate carrying oxygen to cells is insufficient.
(d) **Histotoxic hypoxia**. Tissues cannot use oxygen properly.

Cyanosis refers to a blue hue skin acquires because of excessive amounts of deoxygenated hemoglobin and may be a symptom of respiratory insufficiency. *Hypercapnia* is a condition of excess carbon dioxide in the blood. If the ventilation rate is abnormally high, both the oxygen and carbon dioxide concentrations become excessive. Excess carbon dioxide results in dyspnea. *Dyspnea* is a state of mind, that is, anxiety, related to the inability to provide the body with sufficient air. It is clear that the symptom may have several causes, such as hypoxia, hypercapnia, or purely emotional factors.

Unlike the exchange of oxygen and carbon dioxide, the transport of toxic gases occurs throughout all parts of the respiratory tract. Throughout this section the term "gas" includes both gases and vapors. The rate at which gases are taken up and distributed to body organs varies considerably. Anesthetics produce their effect rapidly. Toxic gases such as hydrogen cyanide and hydrogen sulfide are lethal within minutes. Hallucinogens and carbon monoxide take longer to produce physiological effects.

Toxic gases either react directly on portions of the respiratory tract or are transported to other organs before their effect is registered (Table 2.1). Examples of the first type are ozone and sulfur dioxide; examples of the second type are carbon monoxide and hydrogen cyanide. Exposure to ozone for 2 hours reduces FVC and FEV1 in healthy adults by small but statistically significant amounts (4%) and its effects persist for approximately 18 hr (373). Highly reactive agents soluble in water are apt to damage tissue (e.g., anhydrous acids and strong oxidants), while less reactive gases such as nickel carbonyl diffuse through tissue to react with endothelial cells. Other gases may damage capillaries.

In Section 2.3 it was shown that the nasal cavity is the principal organ that transfers water and energy to and from respired air. Because the surface of the nasal cavity contains a great deal of water, any toxic gas soluble in water will be removed. Anhydrous acids and sulfur dioxide are more apt to be removed than ozone owing to its lower solubility in water. The nasal cavity is the body's most efficient wet scrubber. In the tracheobronchial region toxic gases encounter mucus lining the airways. Gases penetrating the mucus lining contact goblet or ciliated cells. Ciliated cells are generally more sensitive than goblet cells and reducing the number of cilia per unit area of passageway impairs the cilial flushing mechanism.

Reduction of the cross sectional area of the bronchi is a common tracheobronchial disorder. Nerve cells in the bronchi trigger automatic mechanisms that cause muscles surrounding bronchi to contract, reduce the cross sectional area of the passageway, reduce the tidal volume, and/or increase the effort to breathe. For individuals with impaired cardiovascular capacity, such constriction can be life-threatening.

Once gases reach the pulmonary region, their uptake can be described by the same principles as oxygen uptake. Diffusion dominates the physical processes. Mass transfer within the lung is a function of the axial and radial gradients within the bronchiole and solubility and reactivity of the gas in the blood. The first few generations of the bronchial tree and upper alveoli are more susceptible to ozone and the nitrogen oxides.

The body's response to particles contained in inspired air is entirely different than its response to gases. *Clearance* is the process by which particles are removed from the lung. *Mucociliary clearance* is the process by which the lung conducting airways remove depositing particles and carry them to the larynx on surface mucus propelled by cilia. *Alveolar clearance* is the process by which particles are removed by nonciliated surfaces in the gas-exchange region of the lung. Clearance mechanisms include ingestion by macrophage followed by the migration from the lung and the gradual dissolution of the particle.

Figures 2.35 and 2.36 show that particles deposit themselves throughout all regions of the lung through a variety of processes (168–170). In the *nasopharyngeal* region *inertial impaction* is the dominant mechanism and relatively large particles are removed. Particles removed in the *trachea* and *bronchial tree* are removed by a combination of *inertial* and gravitational settling (*sedimentation*) processes. Submicron particles penetrate the *alveolar* region and are removed by

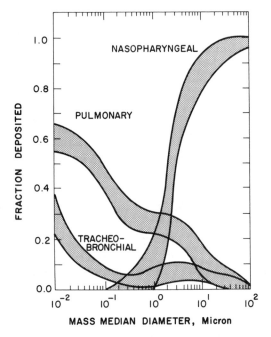

Fig. 2.35 Predicted regional deposition of particles in the respiratory system for a tidal volumetric flow rate of 21 l/min. Shaded area indicates the variation resulting from two geometric standard deviations, 1.2 and 4.5. (Redrawn from ref. 169).

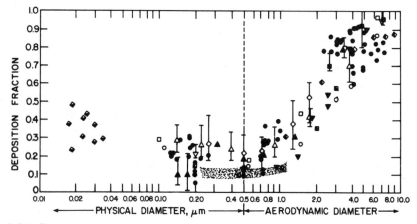

Fig. 2.36 Deposition of monodisperse aerosols in the total respiratory tract for mouth breathing as a function of particle diameter. Above 0.5 μm, the diameter refers to the aerodynamic diameter and below 0.5 μm to the actual physical diameter (reprinted with permission from ref. 239).

diffusion. The terms *inhalable* and *respirable* are used to categorize particles that pose a hazard to health. Particles removed in the nasopharyngeal region enter the digestive tract and pass through the body in a short time. Unless these materials enter the bloodstream through the digestive system, they are of little importance. Particles entering the trachea are removed by cilia and mucus and transported to the digestive tract. Particles penetrating the bronchial tree are flushed more slowly. Exactly which particles penetrate which region of the respiratory tract depends on the breathing rate and tidal volume. Thus a population of energetic workers in a dusty industrial environment may deposit more particles of a given size to the bronchial tree and alveoli than individuals in a nonindustrial environment under restful breathing. For slender particles of fibers like asbestos and cotton, deposition processes must take into account their unusual shape (171,172). The EPA defines *inhalable* as particles having an aerodynamic diameter less than 10 μm, while ACGIH defines *respirable* as particles having an aerodynamic diameter less than 2 μm. The *aerodynamic diameter* is the diameter (D_p) of a spherical water drop that has the same settling velocity as the actual particle. To a first approximation, the actual and aerodynamic diameters are related by

$$D_p(\text{aero}) = D_p(\text{act})\sqrt{\frac{\rho_p}{\rho_w}} \tag{2-50}$$

The definitions of respirable and inhalable may change or new terms may be coined as our understanding of the health effects of inhaled particles improves.

Particles deposited in the *alveoli* are acted upon by one or a combination of four processes.

(1) Particles may be phagocytized and pass up the tracheobronchial tree by the mucociliary escalator.

(2) Particles may be phagocytized and transferred to the lymphatic drainage system.

(3) Particle surface material may be dissolved and transferred to the bloodstream or lymphatics.

(4) Particles and some dissolved material may be retained in the alveoli permanently.

The process of *phagocytosis* is one of the body's essential protective mechanisms. Figure 2.37 illustrates the principal steps in the process for a small particle. Organs of the body, including the alveoli, contain special cells called *macrophage* that attack foreign matter by a variety of processes. *Phagocytosis* begins when the macrophage membrane attaches itself to the surface of the particle. Receptors on the surface of the macrophage bond with the particle. The edges of the membrane spread outward rapidly and attempt to engulf the particle (mechanism of *pinocytosis*). The macrophage membrane contracts and pulls the particle into its interior where cell lysosomes attach themselves to the particle. *Lysosomes* are special digestive organelles within cells that contain enzymes called *hydrolase* that digest foreign matter. The engulfed particle is called a *vesicle*. Following digestion, the vesicle containing indigestible material is excreted through the mac-

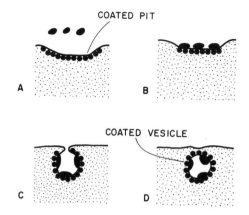

Fig. 2.37 Pinocytosis (redrawn from ref. 148).

rophage membrane. Macrophage are generally thought to originate in bone marrow. How they migrate to different parts of the body including alveoli is not fully understood except that they are very flexible and capable of amoeba-like expansion and contraction that enable them to pass through minute openings in tissue.

Many toxic material enter the body through the respiratory system but only some affect the lung directly, the others enter the blood system or lymphatic system and harm other body organs. Toxins that cause lung disease directly will be divided into five categories:

(1) **Irritation.** Air passageways become irritated leading to constriction and perhaps edema and secondary infection.

(2) **Cell Damage.** Cells lining the air passageways are damaged resulting in necrosis, increased permeability, and edema within the airway.

(3) **Allergies.** Materials in pollen, cat hair, and so on excite nerve cells that cause muscles surrounding bronchiole to contract, which constricts airways and taxes the cardiovascular system.

(4) **Fibrosis.** Lesions consisting of stiff protein structures appear on lung tissue, which inhibits lung function. Fibrosis of the pleura may also occur, which restricts movement of the lung and produces pain.

(5) **Oncogenesis.** Tumors are formed in parts of the respiratory system.

Table 2.1 lists the principal occupational diseases of the lung and describes the effects of exposure to several occupational toxic substances.

2.5.1 Airway Irritation

The cross-sectional area of elements in the bronchial tree is affected by many industrial chemicals. Most chemicals reduce the cross-sectional area of the airway, but a selected few enlarge the airway. *Dyspnea* is the anxious feeling that one cannot breath deeply and rapidly enough to satisfy respiratory demand and may be caused by a narrowing of the airways. Gases such as ammonia, chlorine, and formaldehyde vapor (344) are highly soluble in water and produce dyspnea. Exposure does not produce chronic (residual) respiratory damage, but high concentrations can result in death. *Asthma* is a disorder brought on by spastic contractions of the bronchiole that can increase the lung resistance 20-fold. Allergic hypersensitivity of the bronchiole to foreign substances such as plant pollens and atmospheric pollutants triggers involuntary nervous response. The functional residual capacity and residual volume of the lung increase because of the difficulty to expire air. *Hay Fever* is a similar disorder induced by hypersensitivity that affects the upper respiratory tract.

2.5.2 Cellular Damage and Edema

A variety of industrial substances can damage cells of the bronchial tree and alveoli, which in turn release fluid into the these passageways. The cite of the damage depends on the solubility of the material in water, the more soluble materials damage the nasal cavity and less soluble materials damage upper elements of the respirator tree. The production of fluids (edema) may take considerable time to evidence itself. Phosgene irritates the nasal cavity and upper respiratory passages because it reacts with the abundant supply of water to form carbon dioxide and hydrochloric acid. Ozone and nitrogen dioxide on the other hand are less soluble and penetrate the bronchiole and alveoli. Cadmium oxide fume is a submicron particle that travels to the alveoli and produces edema. Sustained exposure results in a irreversible destruction of alveoli and reduction of oxygen uptake by the blood. Nickel oxide, nickel sulfide fumes and vapors of nickel carbonyl damage the cells of whatever surface they reside on. Hydrocarbon vapors such as xylene and perchlorethylene have low solubility in water and travel to alveoli where they diffuse to capillaries and are transported to the liver and other organs where toxic symptoms occur. Oxygenated intermediaries produce pulmonary edema. *Emphysema* is a debilitating disease that reduces vital capacity of the lung and taxes the cardiovascular system. The disease is the result of three events.

(1) Chronic infection of alveoli or bronchial tissue produced by tobacco smoke or other irritants increases mucus excretion and paralyzes cilia.
(2) Infected lung tissue, entrapped air, and obstructing fluids in alveoli and alveolar ducts destroy alveolar walls and their web of capillaries.
(3) Airway resistance increases, diffusing capacity decreases, and ventilation–perfusion ratios decrease.

Chronic emphysema progresses slowly. Insufficient oxygen produces *hypoxia* and taxes the cardiovascular system. *Tuberculosis* is the disease produced by the tubercle bacilli that infect alveoli. Macrophage attack the bacillus and fibrous lesions are formed that wall off the infected region. The walled-off lesion is called a *tubercle*. Unfortunately, the process may be repeated as tubercle bacilli spread throughout the lung. As a result, the vital capacity and breathing capacity are reduced. The diffusing capacity is reduced and the ventilation–perfusion ratios are abnormal. *Pneumonia* is a disease in which the alveoli are filled with fluid.

Bronchitis is an inflammation of the mucous lining of the bronchial tree and results in an increased production of mucus. A common pneumonia is caused by bacterial infection. Infected alveoli, lung fluid, red and white blood cells fill alveoli which may fill the entire lung. Reduction of the total respiratory membrane and decrease of the ventilation–perfusion ratio is called *hypoxemia*.

2.5.3 Pulmonary Fibrosis

Pneumoconiosis is the phrase for the general class of disease in which pulmonary fibrosis is the central feature. Silica oxide exists in several forms, two of which, *cristobalite* and *tridymite*, induce fibrosis, quartz does not. *Silicosis* is formation of silicotic nodules of concentric fibers of collagen 1–10 mm in diameter that appear in lymphatics around blood vessels beneath the pleura in the lungs and sometimes in mediastinal lymph nodes. The complete explanation of how pulmonary lesions are formed is lacking, even though silicosis has been recognized for centuries. Macrophage attach themselves to silica particles and for reasons not fully understood go awry. It is believed that the lysosomal membrane of the macrophage ruptures and releases lysosomal enzymes that "digest" the macrophage. New macrophage repeat the process and the cycle is repeated. The damaged macrophage release some unknown material that stimulates the formation of *collagen* (a protein and major constituent in the intercellular connective tissue of meats that is not readily digested by most enzymes). The nodules may fuse and block blood vessels and reduce the flow of blood. Alveolar walls may be destroyed. The size of the alveolar sacs and ducts may enlarge, gas exchange can be reduced, and symptoms of emphysema occur. *Asbestos* is the generic name for a group of hydrated silicates existing in the form of clustered fibers. The length, diameter, and number of fibrils (fibers of smaller diameter attached to the larger fiber) on each fiber vary. *Asbestosis* is the name of the chronic disease resulting from exposure to asbestos. Asbestosis has a latency period of several decades, it has many forms including diffuse interstitial fibrosis, calcification, fibrosis of the pleura, and bronchogenic carcinoma and mesothelial tumors.

As with silicosis, a complete explanation describing how fibrosis occurs does not exist, although a great deal is known about some aspects of the process. The length and diameter of the fiber and the character of the fibrils comprising the total fiber are believed to influence fibrosis. Similar to silicosis, the incidence of disease is strongly enhanced by smoking. Bronchogenic carcinoma can be found in all portions of the bronchial tree. Interstitial fibrosis is most commonly found in the lower lobes of the lung. Asbestos itself is chemically inactive, but it is believed that chemical carcinogens on the fiber surface may also initiate cancer within the lung.

2.5.4 Allergic Response

Over 35 million Americans suffer from some kind of allergy. Fifteen million suffer from hay fever and 9 million have asthma. Approximately 40 people die each year from an acute allergic response to stings from insects. Asthma claims the lives of over 4000 people annually. Allergic responses are initiated by materials called allergens, which to most individuals are harmless materials found in the environment such as pollens, molds, and animal hair. Certain types of white blood cells in allergic persons produce antibodies, which attach themselves to mast cells. *Mast cells* are usually found in the respiratory system, the gastrointestinal tract, and the skin. When these antibodies detect an allergen, a large number of chemicals are produced, the best known of which is *histamine* which produces watery eyes, runny nose, itching and sneezing. The type of allergic response (e.g., asthma, hay fever, or hives) depends on which part of the body interacts with the activated mast cells.

Asthma attacks may be induced by viral infections or allergic reactions. Mast cells that induce the reaction are located in the bronchial tubes and lungs. Symptoms include swelling of the bronchial tubes and contraction of the muscles surrounding the bronchial tubes, both actions reduce the internal cross-sectional area of the bronchial tube. Simultaneously, lung sensory nerve endings release chemicals which stimulate the secretion of excessive amounts of mucus. Asthma attacks are associated with wheezing and shortness of breath.

Farmer's Lung results from the inhalation of certain spores that inflame alveoli and produce fever and dyspnea. *Bagassosis* is a disease having similar symptoms inflicting workers handling dried and partially fermented sugar cane. *Byssinosis* is constriction of the bronchial tree producing dyspnea in workers handling cotton, flax, and hemp. *Black lung* is a similar disease associated with coal mining. Numerous industrial chemicals such as toluene diisocyanate (TDI) used in the manufacture of polyurethane plastic and methylisocyanate (MCI) used in the manufacture of insecticide initiate allergic-like symptoms and more serious symptoms if the exposure and concentration are sufficiently high.

2.5.5 Radon Carcinogenicity

Radon is a gas generated by the slow radioacitive decay of uranium-238 that is distributed throughout the earth's crust. Thus radon gas is ubiquitous, its concentration in air is large where the source of uranimum-238 is rich and a buildings ventilation is poor. Of the several radon isotopes produced by the decay of uranium-238, radon-222 has the most important impact on health. Radon-222 decays with a half-life of 3.82 days into a seris of short-lived radioactive isotopes called the *"daughters of radon"*. The first four short half-life isotopes emit alpha, beta and gamma particles and pose the highest cancer risk. These isotopes (in order of appearance) are,

- polonium-218, alpha emitter with a half-life 27 minutes,
- lead-214, beta and gamma emitter with a half-life of 27 minutes,
- bismuth-214, beta and gamma emitter with a half-life of 20 minutes,
- polonium-214, alpha and gamma emitter with a half life of 0.000164 seconds.

These radioactive atoms can attach themselves to inert respirable airborne particles which when inhaled may deposit themselves on the mucus lining of the bronchi or on alveoli membrane. Because of their small size, these particles are not removed by the lung clearance mechanism before they decay and emit damaging radiation. The ciliated epithelial lining of the bronchi and cells in the lung are exposed to highly ionizing alpha radiation in sufficient dose to damage cells that may cause lung cancer.

Radon and its daughters are transferred to dwellings from the underlying soil, groundwater and possibly utility gas. Measured radon concentrations in buildings vary widely. For any particular building it will also vary widely from room to room. The radiation dose from the inhaled daughters of radon-222 is a significant part of the natural background radiation, constituting about half of the total effective dose equivalent the general population receives from natural radiation (390). Radon concentrations are expressed in terms of their radioactivity rather than in conventional terms of concentration, i.e. mg/m^3 or PPM. A variety of studies suggest that radon-222 concentrations in residences average about 1 pCi per liter of air (a picocurie, pCi, is 3.7×10^{-2} radioactive disintegrations per second). The incidence of lung cancer associated with this level of radon-222 activity, is estimated to result in thousands of cases per year among the US population. For individuals who spend 75% of their time in a home for 70 years, exposure to 1 pCi/L can be thought of as the risk of a non-smoker to lung cancer. The EPA recommended guideline exposure for indoor radon exosure is 4 pCi/L which corresponds to an increased lifetime risk of 13 to 50 excess lung cancers per 1000 for people who spend 75% of their time in the home for 70 years. Such exposure is comparable to 200 chest X-rays per year.

2.6   DOSE-RESPONSE CHARACTERISTICS

The relationship between disease and the duration of exposure and contaminant concentration is the object of study in the fields of toxicology, occupational medicine, and epidemiology. Safe exposure limits are derived from industrial experience and experimental studies on humans and animals. Individuals in the workforce comprise a population that is narrow compared to the general population. Individuals in the workforce are on the whole, healthy adults in the prime of life who are exposed to contaminants for only several hours of the day and can return home to rest and breath another atmosphere for the remainder of the day. Thus standards of occupational exposure should not be confused or equated to general environmental standards applicable for the general population. In setting standards for the outside environment, infants, the aged, and the ill and infirmed have to be included in the population. Since there is no escape from the outside environment, the standards are more conservative.

Each individual responds to the exposure to toxins in a unique way. Luckily groups of individuals respond in ways that can be analyzed statistically and parameters can be defined that have statistical significance. Compilation of this data constitutes the basis for health standards and public policies prescribing the maximum exposure individuals may be subject to in the workplace or the outside environment. Underlying these standards are the "dose–response" characteristics of individuals for specific toxins. The definitions of "dose" and "response" depend on the specific toxin but have the following general properties;

Dose refers to the total amount of toxin to which the body is subjected. The dose is a function of both the concentration of the toxin, duration of exposure and the rate at which it is introduced to the body.

Response refers to measurable physiological changes produced by the toxin.

Toxicity may be acute or chronic. An *acute response* is of short duration while *chronic response* is of long duration. Toxicity may also be local or systemic. *Systemic toxicity* is distributed throughout the body while *local toxicity* is concentrated in particular organs. Entry to the body can be through the skin, by ingestion, or by inhalation. Figure 2.38 is a schematic diagram showing the different pathways a toxin may take.

Toxins can be categorized by the response they produce. Shown below are these categories and some example of toxins (95).

Asphyxiant (insufficient oxygen in blood)—carbon monoxide

Irritant (inflames tissue upon contact)—ammonia

Corrosive (damages tissue upon contact)—chromic acid

Allergen (material that affects the immune system)

Site–specific toxins

    Pulmonary toxin (lung)—asbestos, beryllium, chromium

    Hepatotoxin (liver)—carbon tetrachloride, nitriles

    Nephrotoxin (kidney)—kepone, lead, allyl chloride

    Brain toxin—narcotics, ketones

    Skin toxins—benzyl chloride

    Cardiotoxin (heart)—chloroethane

    Neurotoxin (nervous system)—mercury, malathion

    Ocular toxin (eyes)—methyl chloride, methanol, phenol

    Hematopoietic toxin (blood stream)—lead, benzene

    Bone toxin—inorganic fluorides

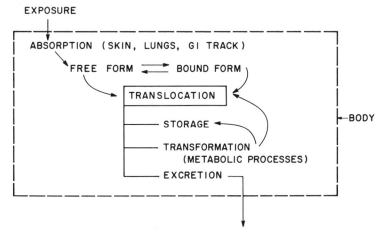

Fig. 2.38 pathways for the movement of toxic material.

Carcinogens (materials that produce cancer)—nickel carbonyl

Reproductive toxins—materials that produce low sperm production, genetic damage to egg or sperm cells, menstrual disorders, for example, polychlorinated biphenyls

Mutagen—materials that alter the character of genetic material in cells

Gametoxin—material that damages sperm or ova

Teratogens—material that interferes in the normal development of the fetus after conception and may result in miscarriage, visible birth defects, or defects not noticeable at birth (e.g., thalidomide)

Transplacental carcinogen—cancer-producing substance that crosses the placenta and reaches the fetus

Occupational exposure is traditionally assessed by comparing the concentration of the contaminant in inhaled air with concentration standards promulgated by OSHA or other professional health organizations. One could logically ask why measure the concentration in air presuming that it produces certain concentrations is the blood, urine, soft tissue or bone, why not measure the concentration of the material or its byproducts in these parts of the body. Certainly measuring the concentration in blood and urine poses no more difficulty than measuring the concentration in air. Such *biological monitoring* may replace air monitoring in the future, but before it does a series of *Biological Exposure Indices (BEI)* (354) should be developed. The biological exposure index (BEI) is the mass of contaminant per volume of the body component in which the contaminant is stored, i.e blood, soft tissue, urine, etc. Biological monitoring actually measures the concentration, the BEI infers the value based on mesurement of external parameters.

Steps to develop such standards is currently in progress and rests on the subject of *pharmacokinetics*, the science that relates the rate processes of absorption, distribution, metabolism and excretion of chemical substances in a biological system.

To establish a BEI, begin by establishing a relationship between the mass of contaminant in the body called the *body burden* and the concentration in inhaled air. For a first approximation, the relationship can be expressed as first-order kinetics. The mass rate of accumulation of a contaminant in the body is equal to its rate of absorption from inspired air minus its rate of removal (by all forms),

$$\frac{dm_{bb}}{dt} = FQc_a - k_r m_{bb} \qquad (2\text{-}51)$$

where

m_{bb} = mass of contaminant in the body, referred to as the *body burden*

Q = volumetric flow rate of inspired air

c_a = contaminant mass concentration in air

k_r–first-order rate constant describing the overall process that transforms the contaminant and, or removes the contaminant from the body

F = *bioavailability*, ratio of the mass of contaminant absorbed by the body to the mass of contaminant in inspired air

Integrating, one obtains

$$m_{bb} = m_{bb,ss}[1 - \exp(-k_r t)] \tag{2-52}$$

where the steady-state value of the body burden is given by

$$m_{bb,ss} = \frac{FQc_a}{k_r} \tag{2-53}$$

The concentration of the material in the blood is related to the body burden by

$$c_b = \frac{m_{bb}}{V_b} \tag{2-54}$$

where V_b is the volume of body in which the material is stored. The blood may be the major component but it is not the only component, indeed it may not even be the major component. Thus for any contaminant, V_b represents an unknown that must be determined.

Independent experimental measurements can be conducted to reveal the values of F and k_r. The bioavailability is determined by comparing the area under the concentration–time curves following equal doses of a chemical by intravenous injection and by inhalation. The first-order rate constant is the half-life of the contaminant in the body,

$$k_r = \frac{0.693}{t_{1/2}} \tag{2-55}$$

that is, the time it takes the body burden to decrease by a factor of two after exposure to the contaminant in inhaled air has ceased.

Illustrated Example 2.1—BEI of Acetone
Consider exposure to acetone vapor in air. Suppose independent measurements show that the bioavailability (F) is 45%, that the half-life is 4 hr, and that the average volumetric flow rate of inspired air is 1.25 m^3/hr. Suppose also that measurements of acetone in workers blood is 10 mg/l following their exposure to acetone vapor in air of 545 PPM (1293 mg/m^3) for a period of 2 hr.

To find V_b, substitute the data in the equations above,

$$V_b = FQc_a t_{1/2}/(0.693 \, c_b)[1 - \exp(0.693)t/t_{1/2})]$$
$$= (0.45)(4)(1.25)(1293)/(10)(0.693)[1 - \exp(-(0.693)(2)/4)]$$
$$= 123 \, l$$

The BEI is equal to the blood concentration corresponding to the 8-hr exposure to the limiting value of 750 PPM. Substituting this value of V_b into the equations above, the BEI corresponding to the 8-hr standard of 750 PPM can be found to be 35 mg/l of blood.

Similar correlations can be obtained for other air contaminants that are absorbed primarily in the blood. For contaminants that are absorbed in the urine, soft tissue or bone, similar correlations can be obtained but the task becomes progressively more difficult.

The body's response to a contaminant depends on how it affects various organs. In the simplest case, the contaminant is a mild irritant that metabolizes rapidly and the response is directly proportional to the ambient concentration (c_a),

$$R_p = kc_a \qquad (2\text{-}56)$$

where R_p is some measurable response and k is a constant. At higher concentrations, the initial response may be superceded by a secondary acute or chronic response that may be cumulative. If the material accumulates in the body, the response can be expressed in a form called *Haber's law*,

$$R_p = kc_a t^n \qquad (2\text{-}57)$$

The coefficient n is not necessarily unity nor is it necessarily constant and have higher values for progressive stages of the disease.

The body burden is highest for fat-soluble organics, organic bases and weak acids. Asbestos, chlorcarbons and radioactive fluorides may be stored for life, while lead and DDT are retained for years. In contrast, the body burden for water-soluble compounds is low.

At the cellular level, the response is a function of the effective local concentration (c) and its rate of change with time,

$$R_p = k\,\frac{dc}{dt} \qquad (2\text{-}58)$$

The effective local concentration is the concentration within tissue in contrast to c_a which is the dose concentration (concentration in ambient air). For soluble material, the effective local concentration is smaller than the dose concentration, i.e. $c < c_a$. The effective concentration at the site (c) depends on the dose concentration (c_a), rate of metabolism, and rate of absorption and is often written as

$$\frac{dc}{dt} = k_{ab}c_a - k_r c \qquad (2\text{-}59)$$

where k_{ab} is the rate of absorption and k_r is the metabolic removal rate. For drugs a value of k_r is sought that is close to unity so that all of the therapeutic agent reaches its target and the dose can be kept low. The respiratory system is an efficient transfer system and many contaminants can be readily absorbed and transmitted to certain organs. Carbon monoxide is a classic example.

Biotransformation can render a contaminant harmless or toxic. Formaldehyde is converted to formic acid by the liver which is a normal metabolic by-product and harmless at modest concentrations. On the other hand, aromatic compounds such as benzene and benzo-pyrene are transformed (metabolized) into highly reactive compounds called epoxides that initiate tumors.

The response of an individual to a given dose can not be predicted accurately owing to the unique characteristics of the individual. Groups of individuals on the other hand display dose-response characteristics that can be described statistically. Thus the statistics of normal distributions can be used. Of great use are cumulative distribution functions, median value and standard deviations. Three ranges are easily determined by experiment.

(a) A dose from zero to a threshold value (c_t) at which a response is observed
(b) A dose at which 100% of the subjects manifest a certain response
(c) A selected mid-range between (a) and (b), for example a median value of 50%

It is common to express dose-response data as a cumulative distribution graph and to identify the median 50% response dosage and the standard deviation.

Whether a substance is toxic depends on its dose. Almost any material is toxic if taken to excess. Most chemicals have more than one effect, depending on concentration. For example, arsenic and selenium are important trace metals needed by the body. If the daily dose is below some threshold, a deficiency occurs which may cause disease. As the dose increases above the necessary level, the body will eliminate it. If the dose exceeds the body's ability to metabolize or eliminate the material, it will accumulate and cause another type of disease. In pharmacology an *Ehrlich index*, shown in Fig. 2.39, is used to elucidate the safety of substances. The Ehrlich index is the difference in concentration between the 95% point of the curative dose–response curve and the 5% point of the toxic dose–response curve. The larger the Ehrlich index, the safer the material. In the case of mercury, which was once used to treat syphilis, the index is small. For anesthetics the index is larger, but not large enough to be indifferent to their effects.

Experimental data is often acquired at large doses well in excess of normal

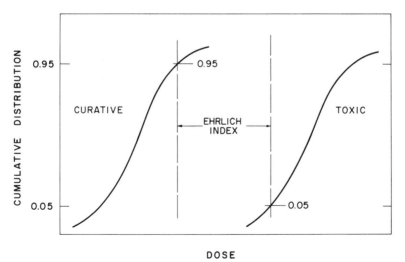

Fig. 2.39   Ehrlich index.

exposures and the data extrapolated to estimate the response at lower dosage. The way extrapolations are made is extremely important and the cause for much controversy. Figure 2.40 shows three ways extrapolations can be made. Naturally occurring materials such as CO, CO_2, NH_3, and formaldehyde fit curve III because there is always a certain amount of these materials in the lungs and airways in excess of what is present in the atmosphere. Certain highly toxic bacteria such as the tubercle bacillus and some carcinogens fit curve I. Other air contaminants are quite harmless at low concentrations and possess a certain threshold such that a straight-line extrapolation through the threshold (curve II) is accurate.

Since toxins produce different effects it is difficult to generalize dose–response characteristics for all toxins. For illustration purposes, the dose–response characteristics of sulfur dioxide (157) on exercising asthmatics will be presented to show how data are analyzed. What follows is not meant to be a comprehensive description of sulfur dioxide toxicity, but rather one study that illustrates how researchers and officials responsible for public policy operate.

Sulfur dioxide constricts the tracheobronchial system. Thus the airway resistance is higher than normal. The percentage airway resistance is the increase in airway resistance above the control (preexposure) value normalized by control values for both individual and groups of individuals serving as their own control. Two dose parameters can be considered:

$$\text{total dose} = D_t = \int_0^t Qc \, dt \tag{2-60}$$

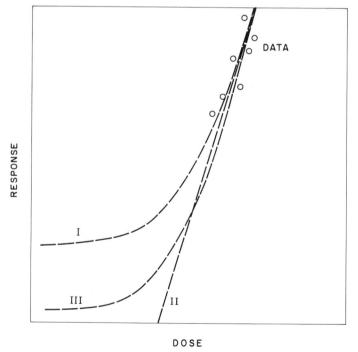

Fig. 2.40 Methods to extrapolate dose–resonse data for three classes of toxins.

$$\text{dose rate} = D_{\min} = Qc \tag{2-61}$$

The total dose is the cumulative (or time-averaged) amount of sulfur dioxide introduced to the body, but it ignores and perhaps blurs any possible distinction between slow breathing, length of exposure period, and concentration that may exist. The dose rate is the instantaneous assault of sulfur dioxide on the body. To study the effect of sulfur dioxide on exercise in which the inhaled volumetric flow rates vary considerably, the dose rate is the more attractive dose parameter. Previous studies failed to establish a close correlation between response and total dose. Secondly, sulfur dioxide irritates and inflames bronchial tissue and the instantaneous assault is expected to correlate better with response than the total dose. If the toxin is cumulative and time to be transported to body organs significant, the total dose might be more attractive.

The use of dose rate is logical for toxins like sulfur dioxide, ozone, PAN, and other irritants that produce an immediate response. *Carbon monoxide* is a well-studied toxin (399). It is toxic because *hemoglobin* absorbs carbon monoxide more readily than oxygen. The brain fails to receive sufficient oxygen and produces effects such as reduced visual acuity, psychomotor skill, pulmonary function and eventually death. For sedentary individuals, breathing at approximately the same volumetric flow rate, the total dose is an attractive dose parameter. The response parameter could be altered by behavioral responses such as loss of psychomotor skills, but the direct measurement of *carboxyhemoglobin* in the blood is a more accurate measure of response. Figure 2.41 is a dose–response curve for carbon monoxide.

Ozone is a highly reactant gas and strong oxidant. It reacts with body fluids and tissue to impair lung function in the short term (Figs 2.29–2.30). In simple terms, ozone limits the ability to take a deep breath. A demonstrative effect of ozone is a reduction in the forced expiratory volume at 1 sec (FEV$_1$). Figure 2.42

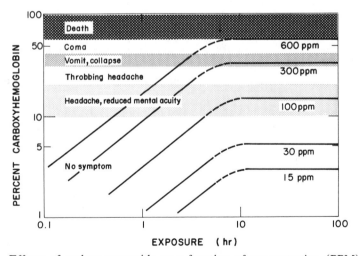

Fig. 2.41 Effects of carbon monoxide as a function of concentration (PPM) and time. OSHA 8-hr PEL is 35 PPM; EPA Primary Air Quality Standard is 9 PPM (redrawn from ref. 239).

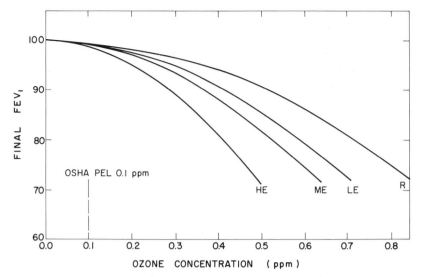

Fig. 2.42   Final FEV_1 for healthy young adult males exposed to ozone for 2 hr consisting of alternating 15-min periods of exercise and rest at four levels of exercise (redrawn from ref. 343). Rest (R), $Q_t < 23$ l/min. Light exercise (LE), 24 l/min $< Q_t < 43$ l/min. Moderate exercise (ME), 44 l/min $< Q_t < 63$ l/min. Heavy exercise (HE), 64 l/min $< Q_t$.

is a compilation (343) of several studies and shows that the forced expiratory volume in 1 sec is reduced for healthy males. Following exposure, lung function is partially restored in 7 days and complete restoration in 2 weeks. In the long term, the effects on health are twofold. There is a transient reduction in the resistance to infection and it is believed that there may be chronic damage to the gas-exchange region in the lung.

The dose–response relationships for cancer are quite different than those for ozone, carbon monoxide, and sulfur dioxide. *Cancer* can be thought to progress through several stages—initiation, promotion, and progression (326). Initiation is an irreversible lesion in the DNA that leads to cancer if further attack occurs. The attack can occur through exposure to chemicals or other agents such as viruses. Promotion is a biochemical process that accelerates progression of the uninitiated cell to cancer. If a promoter attacks an uninitiated cell, the damage is thought to be reversible.

Cancer can be caused by a "hit" of a toxicant on a DNA molecule in the target organ. The resulting *point-mutation* may or may not be reversible. The carcinogen–DNA complex is called an *adduct* and may be removed from the DNA molecule as new unaffected DNA comes into being. If the DNA adduct is retained, it is likely to alter cellular control and initiate the cancer process. Adduct formation is an index of exposure, but not necessarily an index of initiation since continued attack on the DNA does not mean that cancer necessarily develops.

Some chemicals are transformed into carcinogens through human metabolism. For example, benzo[a]pyrene must be metabolized to an epoxide which reacts with DNA in the cell to induce cancer. Other chemicals, such as bis(chloromethly)ether, do not appear to need metabolic conversion to be reactive.

The existence of a threshold for a cancer is a point of contention among professionals who assess risk. The argument against a threshold assumes that a single hit will lead to the uncontrolled growth of a somatic cell and eventually produce cancer. Arguments for a threshold are based on the existence of gene repair mechanisms and immune defenses.

The *one-hit model* assumes that a single hit causes irreversible damage to DNA and leads to cancer. In the *multi-stage model*, a cell line must pass through N stages before a tumor is initiated irreversibly. The rate at which cell lines pass through one or more stages is a function of the dose rate. In the *multi-hit model*, N' dose-related hits to sensitive tissue are required to initiate a cancer. The *Weibull model* assumes that these hits occur in a single cell line and that different cell lines compete independently in producing a tumor. In all these models, it is assumed that the rate at which dose-related hits occur is a linear function of the dose rate. The most important difference between the multi-stage mode and either the Weibull or multi-hit models is that in the Weibull and multi-hit models, all hits must result from the dose, whereas in the multi-stage model, passage through some of the stages can occur spontaneously.

NOMENCLATURE

Symbol	Description	Dimensions*
a	radius of airway	L
$A(y)$	total airway cross sectional area as a function of longitudinal distance y	L^2
C	lung capacitance	L^5/F
c_a, c_b, c_m	molar concentration in air, blood and at mucus-air interface	M/L^3
c	concentration	M/L^3
c_a	concentration in the alveolar region	M/L^3
c_y	concentration at an arbitrary location (y) in the capillary	M/L^3
c_0	inlet concentration	M/L^3
$c_{p,w}, c_{p,a}$	specific heat of water vapor and air	Q/MT
C_L, C_R	capacitance of left and right lung	L^5/F
d	airway diameter	L
D'	virtual diffusion coefficient	L^2/t
$\mathcal{D}$	molecular diffusion coefficient	L^2/t
D_t	total dose	M
D_{min}	dose rate	M/t
D_{act}, D_{aero}	actual and aerodynamic diameter of a particle	L
f	breathing frequency	t^{-1}
f	fraction of contaminant remaining after depletion by chemical reaction during mass transfer (Bohr model)	

*Q – energy, F – force, L – length, M – mass, N – mols, t – time, T – temperature

f_i	fraction of species i depleted by chemical reaction	
F	bioavailability	
h_c	convective heat transfer coefficient of mucus-air interface	Q/TtL
h_{fg}	enthalpy of vaporization	Q/M
k	constant of proportionality	
k_{ab}	absorption coefficient	L^3/t
$k_{b,i}$	solubility of molecular species i in the blood	
k_c	convective mass transfer	L/t
k_r	first-order rate constant for the removal of any contaminant	L^3/t
$k_{s,i}$	solubility constant of molecular species i in a layer	
k_t	thermal conductivity of airway wall	Q/TLt
$k_{m,i}$	mass transfer coefficient of molecular species i in the layer	L/t
K	dispersion coefficient	L
K	overall mass transfer coefficient	L/t
L	longitudinal length of airway	L
L/d	airway aspect ratio	
L_e	length of airway needed for fully developed flow	L
m	mass	M
m_{bb}	body burden	M
$\dot{m}_i$	mass transport of molecular species i	M/t
M_i	molecular weight of molecular species i	M/N
n	constant	
$\dot{n}_w$	number mols of water per unit area and time	N/tL^2
N_D	diffusion parameter	
P	pressure	F/L^2
P_0	inlet partial pressure	F/L^2
P_a	partial pressure in alveolar region	F/L^2
P_r	partial pressure in arterial blood	F/L^2
P_v	partial pressure in venous blood	F/L^2
P_y	partial pressure at an arbitrary location (y) in the capillary	F/L^2
$P_{i,g}, P_{i,b}$	partial pressure of molecular species i in gas or blood	F/L^2
P'	airway perimeter	L
δP	pressure drop	F/L^2
Q_L, Q_R	volumetric flow rates in right and left lung	L^3/t
q_1, q_b	total heat transfer from the liquid and blood	Q/t
Q	volumetric flow rate	L^3/t
Q_a	alveolar ventilation, volumetric flow rate into (or out of) alveolar region (Bohr model)	L^3/t
Q_b	blood volumetric flow rate	L^3/t
Q_t	minute respiratory rate	L^3/t
r	radius	L
r_{ave}	radius at which the local velocity is equal to U_{ave}	L
R_L, R_R	resistance of left and right lung	Ft/L^5

R_p	response	M/L^3
Re	Reynolds number	
R_u	universal gas constant	Q/NT
R_{vp}	ventilation-perfusion ratio	
S	capillary surface area, cumulative airway surface area	L^2
t	time	t
t_m	arithmetic mean transient time	t
t_{vc}	time to expire volume of air equal to vital capacity	t
$T_a(x)'$	dimensionless temperature defined by equation	
T_b, T_a, T_m	temperature of blood, air and mucous	T
T_c	body core temperature	T
$T_a(x)'$	normalized air temperature, defined by equation	
U	velocity	L/t
U_{ave}	average (plug flow) velocity	L/t
U_0	centerline velocity	L/t
U_{rms}	root mean square velocity	L/t
$V, V(t)$	lung volume	L^3
$V_a(t)$	alveolar volume	L^3
V_b	pulmonary capillary blood volume	L^3
V_{vc}	vital capacity volume	L^3
V_t	tidal volume	L^3
V_d	anatomic dead space volume	L^3
V_0	initial value of lung volume	L^3
V_f	final volume of inflated lung	L^3
V_{fr}	lung functional residual capacity	L^3
V_r	lung residual volume	L^3
W_0	Womersley number defined by equation	
x, y, z	spatial coordinates	L
Z	airway generation number	

Greek

$\alpha_1, \alpha_2, \alpha_n$	constants defined by equation	
π	3.14159	
μ	dynamic viscosity	Ft/L^2
ν	(μ/ρ) kinematic viscosity	L^2/t
ρ	density	M/L^3
τ	lung time constant ($\tau = RC$)	t

Subscripts

$(\ )_a$	air property
$(\ )_b$	blood property
$(\ )_i$	molecular species i
$(\ )_m$	mucus property
$(\ )_p$	particle
$(\ )_{ss}$	steady-state
$(\ )_w$	water property
$(\ )_0$	initial value

Abbreviations

AD	fully alveolated alveolar ducts
AGCIH	American Conference of Governmental Industrial Hygienists
AS	alveolar sacs
BEI	biological exposure index
BL	bronchiole
BR	bronchi
EPA	Environmental Protection Agency
FEV1	forced expiratory volume in one second
FVC	forced vital capacity
ISB	isothermal saturation boundary
MCI	methyisocyanate
OSHA	Occupational Safety and Health Administration
P	pulmonary region
RBR	partially alveolated respiratory bronchiole
Re	Reynolds number
STP	standard temperature and pressure
SR_{aw}	airway resistance
TB	tracheobronchial region
TDI	toluene diisocyanate
TL	terminal bronchiole
VOC	volatile organic compound

PROBLEMS

1. Unlike engineering, where knowledge is based on writing and using quantitative relationships, knowledge in physiology and toxicology is based on naming things and explaining their function. The following are *diseases*, *parts of the body*, *or physiological parameters* that you should be able to recall from memory. If the word is a *disease*, state its symptoms and explain the physiological impairment. If the word is *part of the body*, define it succinctly and explain its physiological function. If the word is a *physiological parameter*, describe what it represents.

Aerodynamic particle diameter

Alveoli

Allergic responses

Asthma, bronchitis, and emphysema

Bronchi, bronchiole, and alveolar ducts

Capillary permeability

Cilia

Collagen

Convective flow, pendelluft, Taylor-type diffusion, molecular diffusion

Dose and dose rate

Dyspnea

Edema

Epithelium and endothelium cells

Hypoxia

Interstitium

Lipids

Lung compliance

Lymphatics

Lysosome and hydrolase

Mucus and surfactant

Phagocytosis, macrophage

Pharynx, trachea, esophagus and epiglottis

Pneumonia and tuberculosis

Respirable and inhalable particles

Silicosis and asbestosis

Spirometer, tidal volume, vital capacity, residual volume, anatomical dead space, functional residual capacity, forced expiratory volume in one second

Veins, arteries, arterioles, venules, capillaries

Ventilation perfusion ratio

2. Write and solve the differential equations that predict the volumetric flow rate in the right and left lung (Q_R and Q_L) arranged in parallel as shown in Figs. 2.15–2.17 Assume that during exhalation (when the pressure is positive) that the pressure (centimeters water) varies as follows:

$$P = 6t/0.1t_c \qquad\qquad 0 < t < 0.1t_c$$
$$P = 6 \qquad\qquad 0.1t_c < t < 0.4t_c$$
$$P = 6[1 - (t - 0.4t_c)/0.1t_c] \qquad 0.4t_c < t < 0.5t_c$$

where t_c is the length of the cycle (inhalation followed by exhalation) in seconds. During inhalation, the cycle is repeated, except the pressure pulse is negative. The frequency of oscillation is to be varied over a range of 5 to 20 times the normal breathing frequency. The frequency for normal breathing is 13.6 cycles per minute ($t_c = 4.41$ sec). For computation purposes assume that the capacitance (C_L) and resistance (R_L) of the left lung are 0.083 l/cm water and 0.8 cm of water per LPM (where LPM is the abbreviation for l/min). Assume that the right lung has a capacitance and resistance 30% larger that the left lung. Vary the breathing frequency from 13.6 to 50 cycles per minute and show when Pendelluft occurs by plotting the volumetric flow rates in the right and left lungs.

3. Prove that Eq. 2-47 is correct.

$$\alpha_1 = \sqrt{\alpha_2/2}$$

4. If (Q/V_{vc}) can be characterized by an exponential function, show that for higher moments (normalized by the integral of (Q/V_{vc}))

$$\frac{\left(\alpha_r\right)^{1/r}}{\alpha_1} = \left(r!\right)^{1/r}$$

5. Describe the acute and chronic effects the following toxicants have on the pulmonary system. Using Appendix A-1, list the OSHA PEL for each of the substances.

ammonia
chlorine
hydrofluoric acid vapor
perchloroethylene vapor
silica
sulfur dioxide
toluene vapor

6. The following are characteristic sizes of different airborne particles.

Tobacco smoke: 0.01–0.10 μm
Pollen: 10–100 μm
Sea salt: 0.1–1.0 μm
Bacteria: 0.5–50 μm
Insecticide dusts: 1–10 μm

On the basis of information in Figs. 2.35 and 2.36, estimate the amount (in percentage of inlet) you'd expect to be deposited in different parts of the respiratory system. If the ventilation rate (volumetric flow rate of inhaled air, m^3/min) tripled, would you expect these values to change?

7. Derive fully the equations for the extended Bohr model in Section 2.3 to establish the validity of Eq. 2-27.

8. Emphysema is a debilitating chronic disease in which the alveolar surface area (S) decreases with time, thus reducing the exchange of oxygen and carbon dioxide in the lung. Using the extended Bohr model, evaluate the following features of the disease.

(a) Derive an equation that expresses the change in gas uptake efficiency $E_u = \dot{m}/MQ_a c_0$ with alveolar surface area (S); that is, find dE_u/dS assuming all other parameters remain constant.

(b) Assume that the diffusion parameter (N_D) for a healthy 18-year old adult at rest is 1.0 and that the data from Table 2.2 can be used.

Activity	Q_a/Q_b	Q_b (L/min)
Rest (R)	1.18	6.5
Light exercise (LE)	1.87	13.8
Moderate exercise (ME)	2.28	18.4
Heavy exercise (HE)	3.11	21.7

The solubility of gas in blood (k_b) is 10.7, and the alveolar surface area (S) decreases linearly with time such that at any age (t) after 18 the surface area can be expressed as

$$S(t)/S(18) = 1 - [(t - 18)/120] \qquad \text{for } t > 18$$

Thus by age 78, one-half the alveolar surface will have disappeared. Plot the uptake efficiency (E_u) versus time (t), that is $18 <= t <= 78$ for four levels of activity $(R, LE, ME$ and $HE)$. If you wish, assume that the total surface area for gas exchange at age 18 is 100 m^2.

9. Assume people with impaired cardiovascular function suffer a 10% reduction in blood flow rate (Q_b) (i.e. multiply Q_b in the table in the preceding problem by 0.9). Using the extended Bohr model compute the percentage reduction in gas uptake efficiency (E_u) between individuals with healthy and impaired hearts for four levels of activity assuming all other parameters remain the same.

10. Many riding lawn tractors are equipped with exhaust pipes that discharge engine exhaust gases a few feet upwind of the operators face. Operators inhale carbon monoxide at a concentration of 200 PPM (228.6 mg/m^3) over a period of time and may display the effects of carbon monoxide poisoning. Using the extended Bohr model estimate the dose (kmols) of carbon monoxide an operator receives in 2 hours,

$$\text{dose} = \int_0^2 E_u Q_a c_0 \, dt$$

where E_u and Q_a are the appropriate values for light exercise (LE).

CHAPTER 3

Design Criteria

Goal: Students will be expected to learn
- safe contaminant concentrations
- to anticipate fire and explosion
- safe levels of noise and heat stress
- to rank odors quantitatively
- estimate the cost of controlling contaminants

Preventing contaminants from entering the workplace should be a criterion in the design of all new industrial processes. In many cases it is and control is adequate. Unfortunately, the history of industrial development is replete with examples where health and safety were either ignored or underestimated and engineering controls were designed after the process equipment was in place.

The design of an industrial ventilation system is the creation of a geometrical configuration that satisfies certain performance specifications and must be accomplished within a prescribed amount of time, and money. The performance specifications, time and money are not chosen by engineers, rather they are imposed on engineers by the agency who retains their service. Time, cost and performance specifications are the criteria by which one judges the success or failure of the venture.

Engineering Design is the method by which devices or systems that perform a desired function are created within the constraints of budget and time. Design is the central activity in engineering. It is often assumed that science and engineering are either ends of a continuum. Alas such is not the case. To paraphrase the famous aerodynamicist Theodore von Karman, science discovers what is— engineering invents what never was. The goal of science is discovery and explanation, while the goal of engineering turns on invention and making things. While science is essential to engineering, so also is finance and marketing. In the final analysis, engineering is as much business as it is science. A second difference between science and engineering is that they have ends and means that are different. The end of science is an explanation of natural phenomena; the end in engineering is the creation of a device or system believed to useful by the agency financing the endeavor. The means of science is the *scientific method*; the means of engineering is the *design method*. The scientific method consists of observation, hypothesis, prediction and verification. In the context of industrial ventilation, the essential steps in the design method are

Selecting a geometrical configuration.

Predicting its performance based on contaminant generation rate and mass transport.

Constructing the ventilation system.

Verifying its performance through experiment.

In the final analysis the scientific method establishes truth and certainty while the design method establishes usefulness and verifies that the device performs as planned.

For the most part industrial ventilation systems are designed to *design standard* sometimes called *specification requirements* design standards while compliance is determined on the basis of *performance standards* also called performance criteria. It is very important that engineers understand the difference between specification requirements and performance standards and are able to explain it to others. *Specification requirements* prescribe certain critical dimensions and volumetric flow rates while *performance standards* prescribe that a worker's exposure should not exceed certain government standards (or more stringent standards an industry may choose to adopt). It is recommended that in the future ventilation systems should be designed to satisfy performance standards rather than specification requirements. The two types of standards are not inimical, indeed they should complement each other. The point is that performance standards make the individual undertaking the design primarily responsible for the satisfactory performance, rather than side-stepping the responsibility and hoping performance will be satisfactory or relying on modifying the system after its installation. Embracing performance standards is also the tradition in engineering and essential for coping with in the competitive environment of the future.

Design criteria are factors upon which decisions are made in the design of an industrial ventilation system. Sometimes decisions are made in a deliberate and formal manner involving numerous people having different interests in the outcome of the design. At other times, a designer advances through the steps of the design process quickly with less formality. The formality of the design process depends on administrative practices within the company and the cost of the project. There is a great deal of professional literature (31,32) on decision making in design. In Chapter 1 a great deal was written about reducing exposures by modifying the process and substituting materials. It will be presumed that these steps have been taken and that the decision has been made to design an industrial ventilation system. The objective of this chapter is to discuss several important criteria designers must address:

Contaminant exposure levels

Fire and explosion

Noise

Heat stress

Odors

Engineering economics

An apocryphal practice used by an enterprising engineering consultant may provide some insight to designers who are asked to correct problems in safety and health. The consultant first speaks with the highest ranking official in the company to ascertain what management believes the problem to be and how receptive management will be to accepting changes. Second, the consultant speaks with the workers directly associated with the process to find out what they perceive the problem to be and what remedies they believe are appropriate. Armed with such information the consultant is aware of remedies acceptable to the workers whose health and safety the consultant was retained to protect and how serious management is about implementing the advice given. If the workers ideas are sound and can be incorporated in the design of the control system, the consultant has facilitated an exchange of information that should have already taken place. If the worker's ideas cannot be incorporated in the design, the conversations will help designers understand the institutional environment in which they are asked to make recommendations.

An industrial ventilation system should not impede the worker's movement, visibility, or mental concentration with the process. If the design is for a process that has been in operation for quite some time the designer should learn where workers stand to observe the process and how they move to add or remove material from the process. It is not that such movement is necessarily the best, but workers will operate in their best interest and if the design interferes with these activities the designer can anticipate difficulty. If the worker's movement must change substantially, management must be prepared for a major reeducation program. If the process is new, the designer should become familiar with the expected movement of the worker, indeed it is advisable to involve the worker's opinion in the design process.

3.1 CONTAMINANT CONCENTRATION

Air contaminants may be either particles or as gases. *Particles* may be either solid or liquid. *Gaseous contaminants* may be truly gases or they may be vapors that obey the perfect gas law and Dalton's Law of partial pressures. Thus both contaminant gases and vapors will be called simply "gas" throughout the text. *Fume* is the generic name given to contaminants generated by exothermic processes, often metallurgical processes. Fumes are mixtures of gases, vapors, and particles in varying amounts. The particles are tiny (characteristic dimensions are generally considerably less than a micrometer), which for the most part are condensates of vapors produced by the exothermic process. Fumes may also contain small solid particles that never changed phase.

The mass of particles per unit volume of carrier gas is called the *mass concentration* (c). Often just the term *concentration* is used and readers must be careful to understand what units are implied since some users will be thinking of the number of particles per unit volume of carrier gas. There is no consensus on the phrase concentration. To clarify the issue, users should demand that the units of the concentration be given. A variety of units are used for the mass concentration, but milligram per cubic meters is common.

The mass concentration of gaseous contaminants is defined the same way as it is for particles. If the molecular weight of the contaminant is known, the concentration is often given as a mol fraction. The relationship between mol fraction and mass concentration can be developed from the *perfect gas law* and *Dalton's law of partial pressure*.

A single species of a perfect gas satisfies

$$PV = Z\left(\frac{m}{M}\right) R_u T = Z n_t R_u T \tag{3-1}$$

where n_t is the total number of mols, R_u is the *universal gas constant* and Z is the *compressibility factor*. The *compressibility factor* for an ideal gas is unity. The density of dry air (ρ) is

$$\rho = \left(\frac{m}{V}\right) = \frac{PM}{(ZTR_u)} \tag{3-2}$$

If the air contains water vapor, the *density of moist air* can be computed from

$$\rho(\text{moist}) = \frac{PM_a}{(R_u TZ)}\left[1 - \left(\frac{M_v}{M_a}\right)\phi f \frac{P_v(T)}{P}\right] \tag{3-3}$$

where M_a and M_v are the molecular weights of water and air, $P_v(T)$ is the saturation pressure of water at temperature T and ϕ is the relative humidity (as a fraction). Jennings (274) claims that the constant f is equal to 1.004 for air over a wide range of temperatures and pressures. If the pressure is well below the critical pressure ($P <<< P_c$), the compressibility factor Z is unity. For a *mixture of perfect gases*, the ith molecular species satisfies the law of *partial pressure*.

$$P_i V = \left(\frac{m_i}{M_i}\right) R_u T = n_i R_u T \tag{3-4}$$

The mol fraction of the gaseous species i can be written as

$$y_i = \frac{n_i}{n_t} = \frac{P_i}{P} \tag{3-5}$$

In this text the phrase PPM will NEVER be used to represent a mass fraction. A mol fraction (y_i) of 10^{-6} is equivalent to saying one molecule of species i is contained in a mixture consisting of one million molecules. Thus,

$$1\,\text{PPM} = y_i = 10^{-6} \tag{3-6}$$

It is suggested that the reader visualize PPM as the ratio of the number of molecules of species i per million molecules of carrier gas.

The mass of gaseous species i per total volume (the mass concentration) and the mol fraction of species i and mass fraction of species i are related as follows

$$c_i = \frac{m_i}{V} = \frac{P_i M_i}{R_u T} = \frac{M_i y_i P}{R_u T} \tag{3-7}$$

At *standard temperature and pressure* (STP),

P(standard) = 1 atm = 101.325 kPa

T(standard) = 298.15 K

the concentration (c_i) in milligram per cubic meters is

$$c_i \left(\frac{mg}{m^3} \right)_{STP} = \frac{y_i M_i}{24.5}$$ (3-8)

where the mole fraction (y_i) in Eq. 3-8 *must be expressed* in PPM. At arbitrary pressure P and temperature T, the mass concentration is

$$c_i \left(\frac{mg}{m^3} \right)_{STP} = \left(\frac{y_i (PPM) M_i}{24.5} \right) \left(\frac{P}{P(STP)} \right) \left(\frac{T(STP)}{T} \right)$$ (3-9)

where the absolute pressure and absolute temperature are expressed in compatible units. The reader must keep in mind that since the mol fraction is defined by Eq. 3-5 above, the mol fraction is independent of pressure and temperature.

3.2   CONTAMINANT EXPOSURE LEVELS

Throughout the world, governmental agencies, professional organizations, and individual corporations establish industrial exposure limits (217). Only those in the United States will be discussed, but individuals are urged to be cautious and to seek guidance when working in foreign nations which may have different standards. In the United States three agencies promulgate standards for occupational exposure that the student is likely to encounter.

(a) The American Conference of Governmental and Industrial Hygienists (ACGIH) publish (94) a group of standards called *threshold limit values* (TLV). These standards are reviewed yearly and revised when necessary.

(b) The Occupational Safety and Health Administration (OSHA) in the Department of Labor publishes standards (33,350) known as *permissible exposure limits* (PEL). The standards are not as detailed as the TLV's and in some instances are not as stringent. Nevertheless PEL's are backed by the power of the law.

(c) The American National Standards Institute (ANSI) publishes exposure standards called *maximum acceptable concentrations* (MAC) which are in general compatible with TLV's and PEL's. In many foreign nations MAC's are used.

Agencies that promulgate standards review them regularly and sometimes revise them. Appendices A-1, A-8 and A-10 contain an abbreviated listing of the new and revised OSHA PEL's (350) that will be used in this text. Engineers should always ask OSHA for current PEL values. Readers should be careful when using TLV or PEL and to use only the most up to date values. Since PEL and TLV are reviewed regularly and sometimes changed, readers should always use currently approved values in the most recent editions (94) and (350). The Mine Safety and

Health Administration (MSHA) imposes current ACGIH standards on United States coal mining operations.

There are several categories of threshold limit values.

3.2.1 Time-Weighted Average

The time weighted average is the concentration to which nearly all workers can be exposed repeatedly during a normal 8-hr day, 5 days per week without adverse effect.

$$TWA(8\text{-hr}) = \left(\frac{1}{8}\right) \int_0^8 c(t)\, dt \tag{3-10}$$

$$TWA(40\text{-hr}) = \left(\frac{1}{40}\right) \int_0^{40} c(t)\, dt \tag{3-11}$$

Appendices A-1, A-8 and A-10 list PEL-TWA for several common toxicants.

3.2.2 Short-Term Exposure Limits

The short-term exposure limit is the maximum concentration to which workers can be exposed for a continuous period up to 15 min without suffering from (1) irritation, (2) chronic or irreversible tissue change, or (3) necrosis of sufficient degree to increase accident proneness, impair self − rescue, or materially reduce work efficiency. In addition, there can be no more than four exposures per day, with at least 60 min between exposure periods. Lastly, the daily TLV–TWA can not be exceeded. The STEL should not be considered a maximal allowable concentration or absolute ceiling during a 15 min excursion.

$$STEL = \left(\frac{1}{15}\right) \int_0^{15} c(t)\, dt \tag{3-12}$$

3.2.3 Ceiling

The ceiling concentration should never be exceeded even instantaneously.

$$c(t) < TLV–C$$

3.2.4 TLV For Mixtures

If a source is a liquid mixture and the effects of its components are believed to be additive and it is assumed that the composition of the vapor is composed of components in similar proportion to the original liquid, it is useful to define a TLV for the liquid mixture. When a *liquid mixture* containing N toxic substances and there is no synergistic relationship between the substances, and that the atmospheric composition produced by the mixture is similar to that in the liquid phase the ACGIH recommends (94) that an equivalent TLV can be computed,

$$\text{TLV–Mixture}\left(\frac{\text{mg}}{\text{m}^3}\right) = \frac{1}{\sum\limits_{i}\left(\frac{f}{\text{TLV}}\right)_i} \tag{3-13}$$

$$f_i = x_i \frac{M_i}{M_{avg}}$$

$$M_{avg} = \sum_{i} x_i M_i \tag{3-14}$$

where Σ denotes the mathematical process of summing over i, f_i is the mass fraction of species *i* in the liquid phase, and TLV is expressed as mg/m³. M_i is the molecular weight and M_{avg} is the average molecular weight (see Section 4.8 for details). Alternatively, if a variety of contaminants are present in the gas phase, OSHA assesses hazards by an *exposure parameter* (En) defined as

$$\text{En} = \sum_{i} \left(\frac{y}{\text{PEL}}\right)_i \tag{3-15}$$

where y_i and PEL are expressed as mol fractions. If En is greater than unity, exposure is considered to be beyond acceptable limits and within acceptable limits if En is less than unity. The OSHA standard for a mixture of gaseous contaminants is similar to Eq. 3–15 but is written in terms of PEL's.

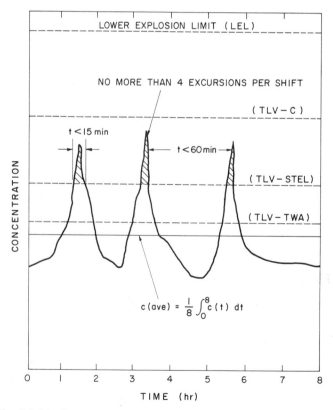

Fig. 3.1 Varying concentrations that satisfy ACGIH standards.

$$En = \sum_i \left(\frac{c}{L}\right)_i \qquad (3\text{-}16)$$

where c_i is the measured concentration and L_i is the PEL in comparable units of concentration. If En is less than unity, exposure is considered safe, and unsafe if En is greater than unity. Shown in Fig. 3.1 is a record of concentration versus time for an 8-hr day, illustrating that the ACGIH TLV standards would be met.

Illustrated Example 3.1   Exposure Limits of Mixtures

A mixture has been prepared of the following liquids.

Material	% by weight	TWA-TLV in PPM (mg/m^3)
Trichlorethylene	35	50 (270)
Benzene	25	10 (30)
Methylene chloride	40	50 (175)

Since the mixture composition is known, the TWA-TLV for the mixture can be computed from Eqs. 3-13 and 3-14

$$TWA\text{-}TLV = [(0.40/175) + (0.35/270) + (0.25/30)]^{-1} = 83.9 \, mg/m^3$$

In another part of the plant waste liquid hydrocarbons are stored in a vessel. Vapors from the mixture escape into the workplace atmosphere and concentration measurements show that the air contains the following.

Material	Concentration (PPM)
Methylene chloride	40
Trichlorethylene	30
Benzene	8

To determine whether such an atmosphere is a violation of OSHA standards, compute the parameter En (Eq. 3-15)

$$En = [(40/50) + (30/50) + (8/10)] = 2.2$$

Thus the workplace atmosphere is in violation of OSHA standards.

The Occupational Safety and Health Act of 1970 charged the Department of Labor with the responsibility to enforce standards that maintain safe and healthy conditions in the workplace. The act also established OSHA as the agency to enforce these standards. Appendix A-1 and A-8 list an assortment of PEL's for a variety of industrial materials. Readers should take the time to review the complete set of OSHA PEL values (33,350) as well as the complete set of ACGIH TLV's (27) since they will need this data in industrial practice. The designation "skin" in Appendix A-1 indicates that exposure also involves transfer through the skin, mucous membrane or eye.

The EPA uses criteria called *ambient air quality standards* to establish safe environmental conditions. A comparison of these standards and PEL's shows that for the most part PEL's are higher. The reason is that PEL's apply to workers who for the most part are healthy adults in the prime of life who are exposed to contaminants for only a short period (typically 8 hr) of the day. Ambient air quality standards pertain to exposure for 24 hr and therefore must be more stringent. Also, these standards must safeguard the health of infants, the elderly, and the infirm.

Engineering control systems are designed to satisfy performance standards established at the beginning of the project. Since compliance with OSHA regulations requires that the PEL's not be exceeded, it is logical to use them, or reduced values of them (PEL/K, $K > 1$) as the performance standards. Alternatively, TLV's could be used instead of PEL's. The designers's task is then to devise a control system or to modify a process such that the predicted contaminant concentrations in the worker's breathing zone do not exceed PEL's. It must be emphasized that PEL's and TLV's are not precise, indisputable criteria distinguishing "safe" from "unsafe" working conditions; consequently, dividing the PEL's and TLV's by the factor K is wise. The value of K is an issue to be decided by the designer in consultation with the company at the beginning of the project. In the absence of other agreed upon standards to characterize safe and unsafe conditions, PEL's and TLV's divided by K are useful parameters to judge the effectiveness of control systems. Management, insurance carriers, and state or local governmental agencies may devise more stringent standards, but in lieu of such standards PEL's and TLV's divided by K can be used as indicators to judge the effectiveness of engineering control systems. It is vitally important that engineers understand by what standards and by which agency will the performance of the systems they design be judged.

TABLE 3.1–OSHA Cancer-Suspect Agents

Asbestos fibers longer than 5 μm
(chrysotile, amosite, crocidolite, tremolite,
anthophyllite, actinolite)
4-Nitrobiphenyl (92933)[a]
α-Naphtylamine (134327)
4,4' Methylene bis(2-chloroaniline) (101144)
Methyl chloromethyl ether (107302)
3,3' Dichlorobenzidine (and its salts) (91941)
Bis-chloromethyl ether (542881)
β-napthylamine (91598)
Benzidine (92875)
4-Aminodiphenyl (92671)
Ethylemeimine (151564)
β-Propiolactone (57578)
2-Acetylaminofluorene (53963)
4-Dimethylaminoazobenzene (60117)
n-Nitrosodimethylamine (62759)
Vinyl chloride (75014)

[a] CAS number.

Concentrations used to estimate a worker's exposure must be measured close to the face. Such a region is called the *breathing zone*. Compliance with health standards must be based on concentrations that exist near equipment or processes where workers are expected to be stationed, not in obscure locations where they do not stand or sit.

Because of variations in individual susceptibility, there will be isolated cases of individuals in a population of workers which may experience discomfort at or below TLV or PEL exposures, for example, alcoholics and heavy smokers, the chronically ill, hypersensitive or allergic individuals, elderly individuals, pregnant women, diabetics, and individuals on medication.

Certain materials are known or suspected to be carcinogens. OSHA is obligated to identify these materials and recommend how they are to be manufactured, processed, repackaged, released, handled, and stored and what protective equipment workers using them must wear. Shows Table 3.1 materials (and their Chemical Abstract Service Registry Numbers) identified by OSHA as cancer-suspect agents.

3.2.5 Exposure Control Limits

In the pharmaceutical industry workers may be exposed to pharmacological compounds at concentrations and dosages that may be dangerous. Workers are protected by *exposure control limits* (ECL). An ECL is a "time-weighted average (air) concentration for a normal 8-hr workday and 40-hr workweek to which nearly all workers may be repeatedly exposed day after day without adverse effect" (308). In developing these limits the pharmaceutical industry follows procedures similar to those used by the ACGIH to develop TLV's. Data needed to compute ECL's are generated as part of the normal investigations required by the Food and Drug Administration in applications for new drugs. To compute an ECL from pharmacological oral dosage data the following is used:

$$\text{ECL} = \frac{(\text{NOEL})m_b F_s \alpha}{Q t_s} \tag{3-17}$$

where the terms and their common units are as follows:

$$\text{ECL}\left(\frac{\text{mg}}{\text{m}^3}\right) = \text{exposure control limit}$$

NOEL (mg day/kg) = *no observable effect* limit; milligrams per killigram of body mass per day of exposure

m_b (kg) = mass of body; 70 kg for males, 50 kg for females

F_s = safety factor; selected by the industry, values of 10–1000 depending on the response the drug produces

α = fraction of the compound absorbed by the body; determined by laboratory experiment, generally unity

Q (m³/day) = inhalation volumetric flow rate; 10 m³/day

t_s (days) = time to achieve a plasma steady-state concentration

Every drug produces a unique response depending on its pharmacokinetics. For chronic, intermittent exposures such as occur in the workplace, the steady-state level is directly proportional to the amount of exposure and the biological half-life of the drug as long as the elimination processes are first order. For most drugs the rate of elimination is rapid, and the half-life is of the order of several hours and there is minimal accumulation in the body. In these cases a value of 1 day would be used for t_s. If the half-life is large, the time to achieve a plasma steady-state (t_s) is determined experimentally by administering an oral dose of the drug once a day and observing the blood plasma concentration as a function of time. Upon administration, the plasma concentration increases and then falls to a detectable level. After the next administration the concentration rises to a higher level before falling. The pattern is repeated over a 5-day period. In many cases the peak plasma concentrations are reached in 1–2 days and hence a value of 2 days is used for t_s. Literally "steady state" plasma concentrations do not occur. Rather, t_s represents the elapsed time required by the body to exhibit a repeatable maximum concentration produced by a daily dose of the drug.

3.2.6 Biological Exposure Index

Biological monitoring assesses a workers exposure to chemicals by measuring the concentration of certain materials in the body. *Air monitoring* assesses a workers exposure to chemicals by measuring the concentration of the chemicals in the workplace atmosphere. TLV's and PEL's serve as reference values for air monitoring. Biological Exposure Indices (BEI) serve as reference values in biological monitoring to indicate potential health hazards. The BEI for a chemical is the concentration of the chemical (or its metabolite) in the blood, urine, exhaled air, etc that is likely to be observed in a healthy worker who has been exposed to the chemical at an inhalation exposure equal to the TLV–TWA (94).

BEI's do not define a sharp distinction between hazardous and nonhazardous exposures. Due to biological varability, an individual's measured concentrations may be in excess of BEI without incurring increased health risk. Exposure to air contaminants and their (or their metabolites) appearance in the body of certain materials is influenced by,

Physiological and health status: age, body build, diet, enzymatic activity, sex, pregnancy, medication, disease states,

Exposure factors: intensity of physical work, exposure intensity and variation with time, temperature, humidity, skin exposure, exposure to other chemicals

Environmental factors: community and home pollutants in air, water and food

Individual habits: personal hygiene, social activities, working and eating habits, smoking, alcohol and drug consumption,

Methodological factors: specimine collection, storage and analysis

Action on unexpected concentrations should not be taken based on a single isolated measurement but on measurement of multiple samplings. If however, numerous measurements for the individual over a period of time are above BEI, or the majority of measurements from a group of individuals at a specific time are above BEI, the cause of the excessive values should be investigated and corrective actions taken.

Shown below are BEI, the indices, timing and PEL-TWA for several workplace chemicals (CAS number).

Carbon disulfide (75-15-0)
    TWA-PEL: 4 PPM
    indices: 2-thiothiazolidine-4-carboxylic acid in urine
    BEI: 5 mg/g creatinine at end of shift
Methyl ethyl ketone (MEK) (78-93-3)
    TWA-PEL: 10 PPM
    indices: MEK in urine
    BEI: 2 mg/l at end of shift
Xylene (1330-20-7)
    TWA-PEL: 100 PPM
    indices: methylhippuric acids in urine
    BEI: 1.5 g/g creatinine at end of shift
    BEI: urination rate of 2 mg/min during the last 4 hours of shift

Individuals that wish to design a biological monitoring program or interpret BEI data should consult the most recent version of (94) and related literature.

3.3 FIRE AND EXPLOSION

A fuel-oxidant mixture that generates sufficient energy upon ignition to cause a flame to propagate throughout the unburned mixture will be defined as a *flammable* mixture. A *non-flammable* mixture is one that can not (of itself) sustain combustion. While the phrase *inflammable* is a synonym of flammable, it is often interpreted to be just the opposite and will not be used. Occasionally a nongaseous explosive material contains both a fuel and oxygen: also, some gaseous materials are in themselves explosive, such as azomethane and hydrazine. These types of material will not be discussed. *Ignition* is a source of energy that is transferred to the fuel and oxidant at such a rate within a minimal volume to initiate reaction. A *combustion zone* is a thin region within which a series of reactions occur involving the transfer of heat from the burned to unburned regions and simultaneously the diffusion of unique chemical species (chain carriers, free radicals, etc.) from the burned to unburned region. What one sees and calls a *flame* is radiation in the visible spectrum that accompanies the chemical reactions. However, not all reactants (e.g., hydrogen) produce visible radiation.

Combustion zones are of two types—detonation waves and deflagration waves. A *deflagration wave* (also called a combustion wave or a laminar flame) propagates by the diffusion of heat and chemical species to the unburned gas. A *detonation wave* is a shock wave with a combustion wave on its backside sustained by the energy released by combustion and the increased pressure and temperature associated with a shock wave. The pressure difference across a flame is negative and very small, whereas the pressure difference across a detonation wave is positive and very large. Combustion waves possess a subsonic definable speed called the *flame speed* while detonation waves travel at supersonic speeds. A combustible mixture may support either a combustion wave or detonation wave depending on the fluid mechanics and heat transfer associated with the chemical reaction, which in turn depends on the geometry of the container enclosing the mixture.

Fire is associated with deflagration waves, while explosions are associated with detonations. As a practical matter, it is often difficult to determine if the source of a fire was a flame or an explosion. A delineation between fire and explosion may be made by analyzing a steady-state, one-dimensional flow involving an exothermic chemical reaction. The analysis yields a set of classical equations called the *Rankine–Hugoniot* relations. A full discussion can be found in texts on combustion (265–269,331). The thin combustion zone will be modeled as a one dimensional discontinuity that propagates into a motionless combustible gas mixture (Fig. 3.2a). The upstream state (unburned state) will be denoted by the subscript 1 and downstream state (burned state) will be denoted by the subscript 2. Other assumptions are the following:

Inviscid, adiabatic, steady flow

Ideal gases with constant specific heats

All gradients are zero upstream and downstream of the combustion zone

Define a control volume (Fig. 3.2b) relative to the stationary combustion zone and apply the equations for the conservation of mass, momentum, and energy.

Continuity: $\nabla \cdot (\rho U) = 0$

$$U_1 \rho_1 = U_2 \rho_2 = \dot{m} \tag{3-18}$$

Momentum: $\dfrac{U dU}{dx} = -\left(\dfrac{1}{\rho}\right)\dfrac{dP}{dx}$

$$\rho_1 U_1^2 + P_1 = \rho_2 U_2^2 + P_2 \tag{3-19}$$

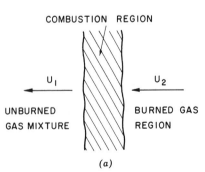

(a)

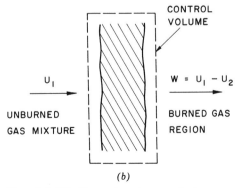

(b)

Fig. 3.2 (a) Combustion region propagating into unburned gas.

Fig. 3.2 (b) Control volume relative to the combustion region.

Energy:
$$Q - W = \dot{m}\left(h_2 - h_1 + \frac{U_2^2}{2} - \frac{U_1^2}{2}\right)$$

$$h_1 + \frac{U_2^1}{2} = h_2 + \frac{U_2^2}{2} \tag{3-20}$$

Using the continuity equation the following identity can be created:

$$\rho_2 U_2^2 = \rho_1 U_1^2 = (\dot{m})^2\left[\left(\frac{1}{\rho_2}\right) - \left(\frac{1}{\rho_1}\right)\right] \tag{3-21}$$

enabling the momentum equation to be written as

$$\frac{P_2 - P_1}{\left(\dfrac{1}{\rho_2}\right) - \left(\dfrac{1}{\rho_1}\right)} = -(\dot{m})^2 \tag{3-22}$$

Equation 3-22 is called the Rankine equation and its graph is called the *Rankine curve*. The energy equation can also be rearranged,

$$h_2 - h_1 = -\frac{(\dot{m})^2}{2}\left[\left(\frac{1}{\rho_2}\right)^2 - \left(\frac{1}{\rho_1}\right)^2\right] \tag{3-23}$$

Eliminating the mass flux ($\dot{m}$) one obtains,

$$h_2 - h_1 = \left(\frac{1}{2}\right)[P_2 - P_1]\left[\left(\frac{1}{\rho_2}\right) + \left(\frac{1}{\rho_1}\right)\right] \tag{3-24}$$

Equation 3-24 is called the *Hugoniot equation*. The velocity with which the combustion zone propagates into the unburned mixture, U_1, can be found from the material discussed above,

$$U_1 = \left(\frac{1}{\rho_1}\right)\left\{(P_2 - P_1)\Big/\left[\left(\frac{1}{\rho_1}\right) - \left(\frac{1}{\rho_2}\right)\right]\right\}^{1/2} \tag{3-25}$$

The gas mixture can be described by the perfect gas equation

$$P_1 V_1 = n_1 R_u T_1$$
$$P_2 V_2 = n_2 R_u T_2 \tag{3-26}$$

and one must be careful to account for any change in molecular species or number of moles resulting from the chemical reaction. The enthalpy difference can be represented as

$$h_2 - h_1 = c_p(T_2 - T_1) - h_{RP} \tag{3-27}$$

where h_{RP} is the enthalpy change associated with combustion.

For a given fuel–oxidant composition and values of P_1 and ρ_1, all the burned states P_2, ρ_2 lie at the intersections of the Rankine and Hugoniot curves as shown in Fig. 3.3. Define point A as the upstream state P_1, ρ_1 and construct a number of Rankine curves (i.e., straight lines from point A) that intersect the Hugoniot curve at points B, C, F, G, J, and K. Define angle alpha by the points A–P_1 and one of the points (B, C, F, G, J, or K). Thus

$$\tan \alpha = \frac{P_2 - P_1}{\left(\dfrac{1}{\rho_1}\right) - \left(\dfrac{1}{\rho_2}\right)} \tag{3-28}$$

From Eq. 3-22 it can be seen that the tangent of angle α corresponds to the square of the mass flux ($\dot{m}$). Thus the propagation velocity U_1 is

$$U_1 = \left(\frac{1}{\rho_1}\right)\sqrt{\tan \alpha} \tag{3-29}$$

Define a velocity W

$$W = U_1 - U_2 \tag{3-30}$$

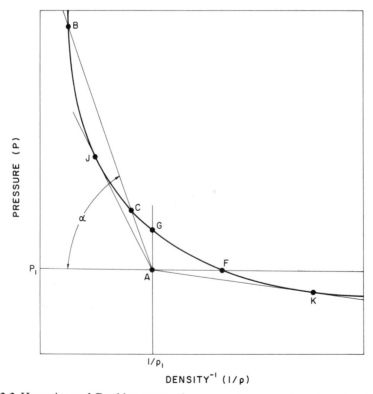

Fig. 3.3 Hugoniot and Rankine curves for a one-dimensional combustional region.

The gas velocity W is the velocity of the burned gas velocity relative to the control volume.

$$W = U_1 - U_2 = \left[\left(\frac{1}{\rho_1} \right) - \left(\frac{1}{\rho_2} \right) \right] \sqrt{\tan \alpha} \qquad (3\text{-}31)$$

A negative value of W implies that the burned gas leaves the combustion zone faster than the gas enters the combustion zone.

With respect to the quadrants defined by point A in Fig. 3.3, it can be seen than only the upper left and lower right quadrants correspond to physically significant results since the positive value of the mass flux ($\dot{m}$) in the Rankine equation requires angle α to satisfy

$$0 < \alpha < \frac{\pi}{2} \quad \text{and} \quad \pi < \alpha < \frac{3\pi}{2} \qquad (3\text{-}32)$$

The upper left quadrant is called the *detonation branch* of the Hugoniot curve and the lower right quadrant is called the *deflagration branch*. Various regions of Fig. 3.3 can be analyzed for physical significance.

Point G ($\rho_1 = \rho_2$) corresponds to combustion in a constant volume container that produces a pressure rise.

Point F ($P_1 = P_2$) corresponds to constant pressure combustion with an increase in volume corresponding to the work done due to the energy addition.

The deflagration branch (region FK) represents a region in which the density decreases and the pressure only slightly decreases. The burned gases expand and increase in velocity after passing through the zone. The value of W is negative and the burned gas moves in a direction opposite to the movement of the combustion zone. This region corresponds to an ordinary combustion wave (ordinary flame) and the equation for the propagation velocity becomes

$$U_1^2 = \frac{P_1 - P_2}{\rho_1 \left\{ \left(\frac{\rho_1}{\rho_2} \right) - 1 \right\}} \qquad (3\text{-}33)$$

In the region between G and F, the square root of $\tan \alpha$ is imaginary and therefore the region has no physical significance.

The region from G upward is the detonation branch of the Hugoniot curve in which the burned gases are compressed, that is, the pressure and density increase. On the burned side of the combustion zone the flow is in the same direction as the propagating combustion zone.

The solution of both the Rankine and Hugoniot equations is the intersection of the Rankine equation (straight line Eq. 3-22) and the Hugoniot curve (Eq. 3-24) and can be satisfied by either the Rankine line intersecting the upper branch of the Hugoniot curve at points B and C or being tangent to the curve at point J. A similar argument can be made for the deflagration branch. It can be shown that for a stable combustion zone, only the tangent points J and K satisfy both the requirements of thermodynamic and mechanical stability. Points J and K are called the upper and lower *Chapman–Jouguet* points respectively. It can be shown (265) that point J corresponds to a detonation wave with a supersonic velocity (U_1) while point K corresponds to subsonic combustion zone called the flame whose velocity is called the *flame speed or burning velocity*. A thorough presentation of the theory of Rankine–Hugoniot flow can be found in texts on combustion (267,269 and 331).

From this discussion it can be seen that a combustion zone traveling thorough a combustible mixture can exist either as a deflagration wave (flame) or a detonation wave. Which of the two states occurs depends on the presence of a heat absorbing body which enables the combustion zone to exist as a stable flame. In the absence of a heat absorbing body, the deflagration wave (flame) undergoes a transition, becomes a detonation wave, and one experiences an explosion.

Flames are categorized as premixed or diffusion flames. *Diffusion flames* are ones in which the fuel and oxygen diffuse toward the combustion region from opposite directions, mix, and burn. A match flame or flame surrounding a burning fuel drop are diffusion flames. *Premixed flames* are ones in which the fuel and oxygen are initially mixed and the flame propagates into the unburned mixture. A flame on a bunsen burner is a premixed flame. The flame speed for premixed flames is a function of the fuel–oxidant composition, temperature, and pressure. There are certain fuel–oxidant compositions at which flames cannot be produced.

Engineers must insure that the concentration of flammable materials in air never achieve limits that will enable the materials to burn. The *limit of flammability* is defined as the lowest percentage of one of the reactive components in a gaseous mixture that can sustain combustion. There are both fuel-lean and fuel-rich limits of flammability for both flames or detonation waves. The fuel-lean limit is called the *lower explosion limit* (LEL) and the rich-rich limit is called the *upper explosion limit*. Figure 3.4 shows the flammability limits of mixtures of

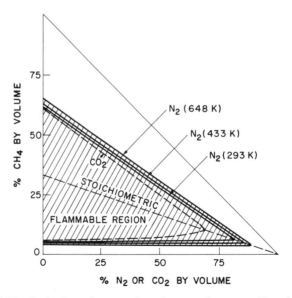

Fig. 3.4 Flammability limits for mixtures of methane and oxygen diluted with nitrogen or carbon dioxide and different temperatures (redrawn from ref. 266).

methane and oxygen at different temperatures diluted with various amounts of nitrogen or carbon dioxide. These curves show that adding a diluent narrows the limits of flammability; increasing the temperature widens the limits, but changing the diluent from nitrogen to carbon dioxide has only a minimal effect on the limits. Figure 3.5 shows that the lean flammability limit is virtually unaffected by pressure, whereas the rich flammability limit increases rapidly with pressure. The upper explosion limit is seldom encountered in industrial ventilation, indeed even the lower explosion limit is considerably larger than TLV or PEL. Unusual cases are materials that are not toxic and where care must be taken to insure that LEL are never exceeded. As a rule of thumb insurance carriers require that the concentration of an inflammable material should never exceed 10% of the LEL. Table 3.2 is an abbreviated table of LEL for a variety of common industrial materials. Figure 3.6 shows that a correlation exists between the LEL and heat of combustion for many hydrocarbons in air.

Anticipating fire and explosion requires one to recognize

Sources of ignition

Fuel–oxidant concentrations sufficient to support combustion

Unusual combinations of oxygen enrichment and high temperatures and
pressures

In order to initiate combustion, a high temperature must be produced in a small volume whose dimensions are comparable to the thickness of a flame. Minimum ignition energy is inversely proportional to pressure and temperature, thus at elevated pressure and temperature sparks and short circuits will initiate combustion where they do not at STP.

It is often thought that the ignition temperature is the temperature to which a combustible mixture has to be raised before combustion can occur. While raising

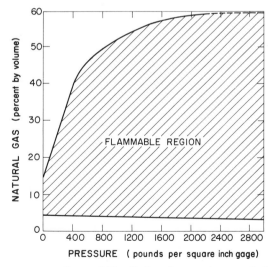

Fig. 3.5 Effect of pressure on flammability limits of natural gas and air (redrawn from ref. 265).

TABLE 3.2 Flammability Limits and Flash Temperatures of Common Industrial Materials (abstracted from 27 and 265)

Material	Molecular Weight	Flash Point (F)	Explosive Limit* Lower	upper
Acetic acid	60.05	104	5.4	
Acetic anhydride	102.09	121	2.67	10.13
Acetone	58.08	0	2.55	12.8
Benzene	78.11	12	1.4	7.1
n-Butanol	74.12	84	1.45	11.25
Butyl acetate	116.16	72	1.39	7.55
Carbon tetrachloride	153.84	Nonflammable		
Chloroform	119.39	Nonflammable		
Cyclohexane	84.16	1	1.26	7.75
Ethane	30.0	Gas	3.0	12.5
Ethylene	28.0	?	2.75	28.6
Ethylene oxide	44.05	—	3.0	80.0
Ethyl acetate	88.10	24	2.18	11.4
Ethyl alcohol	46.07	55	3.28	18.95
Ethyl chloride	64.52	−58	3.6	14.80
Gasoline	86	−50	1.3	6.0
Heptane	100.2	25	1.1	6.7
Hexane	86.17	−7	1.18	7.4
Hydrogen sulfide	34.08	Gas	4.3	45.5
Methane	16.0	Gas	5.0	15.0
Methanol	32.04	54	6.72	36.5
Methyl acetate	74.08	15	3.15	15.6
Methyl formate	60.05	−2	4.5	20
Octane	114.22	56	0.95	3.2
Isopropanol	60.09	53	2.02	11.8
Propyl acetate	102.13	43	1.77	8.0
Propane	44.09	Gas	2.12	9.35
Toluene	92.13	40	1.27	6.75
Trichloroethylene	131.4	Nonflammable		
Xylene	106.16	63	1.0	6.0

Source: Abstracted from refs. 27 and 265.
[a]Flash point measured by closed cup experiments
[b]Flammability limits in % by volume

the temperature of a combustible mixture is not a safe practice, temperature alone is not the determining factor producing ignition. The *ignition temperature* is defined as the temperature at which there is an inflection point in the flame temperature profile, $(\partial^2 T/\partial x^2) = 0$. Figure 3.7 is a schematic diagram of a typical temperature profile. The temperature gradients are very large considering that at STP a flame is approximately 1 mm thick and the burned temperature is over a thousand degrees centigrade above the unburned temperature. On the unburned side of the flame energy is absorbed by the gas mixture and on the burned side of the flame energy is transferred to the unburned gas mixture. Thus if one examines the temperature profile across a flame there is a point where the second derivative is zero, corresponding to the point separating the region that absorbs energy from

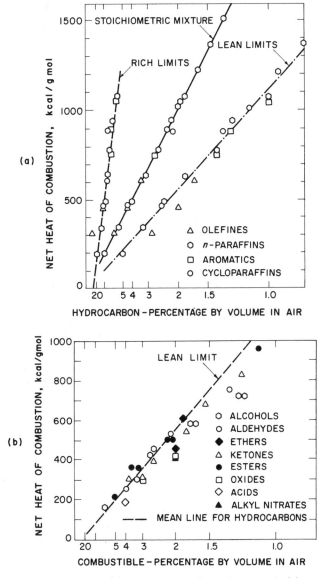

Fig. 3.6 Flammability limits of (*a*) common hydrocarbons and (*a*) common organic substances (redrawn from ref. 267).

the region that liberates energy. Historically (265) such a point was called the ignition temperature and represents the minimum temperature below which the chemical reaction does not occur.

Ignition can also occur due to spontaneous combustion or from heated surfaces. Often cited in the professional literature is the flash-point temperature. The *flash-point* temperature is defined as the minimum temperature at which a flammable liquid will generate enough vapor and produces a flame. The flash-point temperature is not an intrinsic property of the fuel and is a function of the

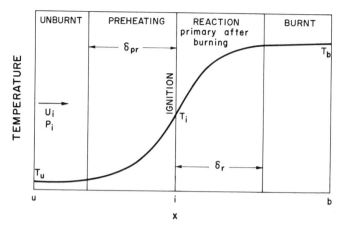

Fig. 3.7 Structure of a premixed flame front.

experimental apparatus used to measure it. Nonetheless, it provides engineers some way to anticipate when a hot surface can initiate ignition.

Aerosols of combustible solids can support flames and detonation waves, and the same precautions must be taken as with flammable gases and vapors. History records fires and explosions of coal dust, sawdust, grain dust, and aerosols of metals such as AL, Zn, Ti, Mn, Sn, and pure Fe. Fires and explosions are a particular hazard in equipment used to collect particles from the indoor work environment (e.g., filters, cyclones) or in powder-handling equipment (e.g., transfer points of conveyors and pulverizers). Sparks formed by fan motors, fan blades that scrape on fan housing, or electrical discharges produced by static electrification are often sufficient to ignite these materials.

It is more difficult to describe the combustion of *dust–air mixtures* than gas–air mixtures because they consist of two phases (suspended solids in air) while gas–air mixtures are of a single phase. The *explosion of dust–air mixtures* depends on the existence of a flammable mixture and the presence of an ignition source (electrical discharge, chemical exothermic reaction, heated surface, etc.) of sufficient magnitude to initiate combustion. Flammability describes the state at which given a sufficiently high temperature, the particles will burn in air. Explosion describes the state at which, in addition to flammability, a source of energy exists to ignite the particles. Thus there are aerosol compositions that can burn but can not be ignited and explode.

Hertzberg et al. (273) propose that three sequential processes occur in the propagation of a flame through a dust-air mixture:

(1) Heating and devolatization of dust particles.
(2) Mixing of these volatiles with air.
(3) Combustion of the volatile–air mixture.

Explosions of dust–air mixtures are predictable in terms of a minimum mass concentration, maximum particle size, and minimum ignition energy. There is no fuel-rich explosion limit. There are three reasons why it is more difficult to describe the explosion limits of heterogeneous dust-air mixtures:

Composition

It is not possible to describe the mixture composition in terms of a fuel–air ratio because a dust–air mixture is heterogenous—it is a two-phase mixture in which compositional properties are discontinuous. In addition, an aerosol may be explosive at one instant of time but owing to gravimetric settling, the particle mass concentration can decrease or the mixture can segregate.

Particle Size

There is a unique particle size below which the explosion limit is independent of ignition energy. (A 1-kg lump of coal suspended in 1-m^3 container of air will not explode but a monodisperse suspension of the same amount of coal with a particle size of 100 μm diameter is explosive.)

Minimum Ignition Energy

The ignition energy must be sufficiently large so as to produce a flame front that propagates freely once ignition occurs.

The dependence of the lean explosion limit on the minimum ignition energy for coal (Pittsburgh coal seam), polyethylene dust and oil shale is shown in Fig. 3.8. Explosion limits reach their asymptotic limits at effective energies in the kJ level while shale oil still shows a sensitivity to ignition energy even at this high energy level. Hertzberg et al. (272) conclude that the product of the lean limit mass dust

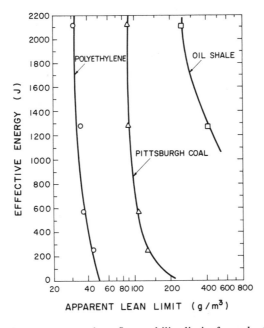

Fig. 3.8 Effect of ignition energy on lean flammability limits for polyethylene, coal, and oil shale dusts. Mass mean diameters 37, 50, and 60 μm, respectively (redrawn from ref. 272).

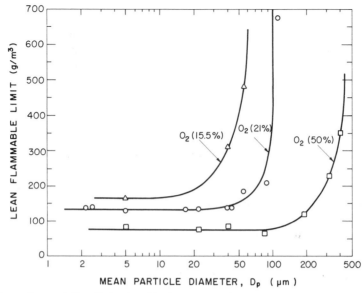

Fig. 3.9 Lean flammability limits for Pittsburgh bituminous coal as a function of particle size and three oxygen concentrations (redrawn from ref. 273).

concentration and the combustible volatile percentage is constant. Below a critical particle size, Fig. 3.9 shows that the lean explosion limit is independent of particle size. They conclude that below this diameter that the rate of devolatization is so rapid that combustion is controlled by the diffusion and combustion of volatiles. For the Pittsburgh coal seam, the critical particle diameter is approximately 50 μm, while for particles of polyethylene (which are composed entirely of volatile materials) the critical particle diameter is approximately 100 μm.

A comparison between lean explosion limits and TLV's or PEL's shows that there is a considerable difference. The ACGIH (27) TLV is 2 mg/m^3 for coal dust containing a respirable fraction less than 5% quartz. The lean explosion limit is approximately 100 g/m^3, a difference of a factor of 5×10^4.

3.4 NOISE

Longitudinal pressure waves from 20 to 20,000 Hz are called sound waves. A vast number of superimposed sound waves in which there is no predetermined relationship between the frequencies and amplitudes of the waves is *noise*. Hearing can be impaired permanently if individuals are exposed to sound or noise above certain amplitudes and for a certain duration. It is important that designers of industrial ventilation systems be able to predict the noise associated with their designs and to take steps to reduce the generation of noise, add material to absorb it, or know when to recommend personal protective devices.

The square of the speed of sound (a) is defined as the change of pressure with respect to density at constant entropy.

$$a^2 = \left(\frac{\partial P}{\partial \rho}\right)_s \tag{3-34}$$

For sound traveling in a gaseous medium which can be approximated as a perfect gas, the speed of sound can be found from

$$a = \sqrt{\frac{kR_uT}{M}} \tag{3-35}$$

k is the ratio of specific heats (c_p/c_v), R_u is the universal gas constant, M is the molecular weight of the gaseous medium, and T is the absolute temperature. The speed of sound in air at STP is 344.5 m/s. Sound travels much faster in solids and liquids. The speed of sound in solids and liquids can be derived from Eq. 3-34 and is equal to

$$a = [\beta\rho]^{-0.5} \tag{3-36}$$

where β is the isothermal expansion coefficient,

$$\beta = \frac{1}{\rho}\left(\frac{\partial \rho}{\partial P}\right)_T \tag{3-37}$$

Consider a sound source located in free space. At a distance r from the source, the sound can be characterized by three parameters—*pressure* (P in force/area), *intensity*(I in power/area), and *power* (W in energy/time). Assuming that pressure waves travel through air without loss (which is consistent with the assumption of constant entropy), the power distributed over a spherical surface is equal to the acoustical power of the source. The power associated with the sound wave per unit area is called the intensity and is related to the pressure by

$$I = \frac{P^2}{\rho a} \tag{3-38}$$

Thus sound power (W) is related to sound intensity by

$$W = I[4\pi r^2] = (4\pi r^2)\left(\frac{P^2}{\rho a}\right) \tag{3-39}$$

If the source of sound is not located in free space, and reflecting surfaces direct the sound in preferential directions, it is convenient to modify the equation above as follows:

$$W = \frac{I[4\pi r^2]}{Q} \tag{3-40}$$

where Q is called the *directivity factor*. Sound directivity is defined (36) as the ratio of the sound power of a small omnidirectional hypothetical source to the sound power of an actual source that produces the same sound–pressure level at a point of measurement. If the sound source is bounded by acoustically reflecting surfaces,

$Q = 1$, sound source in free space

$Q = 2$, sound source located in an infinite reflecting plane

$Q = 8$, sound source located at the intersection of three mutually perpendicular intersecting reflecting planes

It is convenient to define a sound–pressure level L_P related to the sound pressure P,

$$L_P = 20 \log_{10}\left(\frac{P}{P_0}\right) \qquad (3\text{-}41)$$

The term P_0 is a reference value,

$$P_0 = 2 \times 10^{-5} \ \frac{N}{m^2} \qquad (3\text{-}42)$$

which at one time was thought to be an approximate normal threshold value for a young person at 1000 Hz. In a similar fashion it is useful to define a sound intensity level L_I

$$L_I = 10 \log_{10}\left(\frac{I}{I_0}\right) \qquad (3\text{-}43)$$

where the reference value I_0 corresponds roughly to the reference pressure level (P_0)

$$I_0 = 10^{-12} \ \frac{W}{m^2} \qquad (3\text{-}44)$$

Sound power (W) can also be expressed as a sound power level (L_W),

$$L_W = 10 \log_{10}\left(\frac{W}{W_0}\right) \qquad (3\text{-}45)$$

where

$$W_0 = 10^{-12} \ W \qquad (3\text{-}46)$$

The units used to express the sound-pressure level, sound-intensity level and sound-power level are called *decibels* (dB).

It is of advantage to manufacturers to reduce the acoustic power of equipment they manufacture. Consequently, the acoustic power of equipment is information that is often available to the purchaser. The acoustic power can be used to predict the sound-pressure level (L_P) at arbitrary locations from the equipment. Substituting Eqs. 3-38 and 3-39 into 3-40 and evaluating the constant terms, it can be shown that

$$L_P = L_W - 20 \log_{10} r(m) + 10 \log_{10} Q - 11 \qquad (3\text{-}47)$$

Illustrated Example 3.2—Sound-Pressure Level of a Punch Press
A punch press is rated as having a free − field sound level of 85 dB at 10 m. What is the acoustic power of the equipment and what is the sound-pressure level at work stations 0.5 and 2.0 m from the press if the press is located in the corner of a building.

The acoustic-power level of the source (L_W) is found by substituting $L_p = 85$ dB, $r = 10$ and $Q = 1$ in Eq. 3-47. One finds that the acoustic-power level of the source (L_W) is 116 dB. The acoustic power is

$$L_W = 116 = 10 \log_{10}\left(\frac{W}{10^{-12} \text{ W}}\right)$$

$$11.6 - 12 = -0.4 = \log_{10}(W) = \frac{\ln(W)}{2.303}$$

$$W = 0.398 \text{ W}$$

The sound-pressure level at $r = 0.5$ and 2.0 m are found from the same equation using $L_W = 116$ dB and $Q = 8$.

$$L_p(r = 0.5 \text{ m}) = 120 \text{ dB}$$

$$L_p(r = 2.0 \text{ m}) = 108 \text{ dB}$$

Since noise consists of several simultaneous sound waves of different frequencies and intensities, the ear is subjected to a composite wave in which the total power is the sum of the power of each individual sound wave. Thus the sound intensity at a point in space subject to *several sources* of noise is

$$\frac{I}{I_0} = \sum_j \left(\frac{I_j}{I_0}\right) \tag{3-48}$$

where j refers to the jth source of noise. The right hand side of the equation above can be evaluated by taking antilogs of Eq. 3-43

$$\frac{L_{I,j}}{10} = \log_{10}\left[\frac{I_j}{I_0}\right] \tag{3-49}$$

The total sound-intensity level of several sources of noise can be expressed as

$$L_I(\text{dB}) = 10 \log_{10}\left[\sum_j 10^{n_j}\right] \tag{3-50}$$

$$n_j = \frac{L_{I,j}}{10}$$

Illustrated Example 3.3—Sound-Intensity Level in a Grinding Room
A company wishes to place six pedestal grinders in a room. What is the total sound-intensity level produced by all six grinders at a certain point (A) in the room if at point A each grinder separately produces the following sound-intensity levels.

Grinder	Sound-Intensity Level (dBA) at Point A
1	85
2	92
3	90
4	84
5	93
6	87

Substituting into Eq. 3-50 shows that the total sound level is 97 dBA.

$$L_I = 10 \log_{10}[10^{8.5} + 10^{9.2} + 10^{9.0} + 10^{8.4} + 10^{9.3} + 10^{8.7}] = 97 \text{ dB}$$

The human function of hearing depends on both frequency and sound level, as shown in Fig. 3.10. To simulate the ear it is convenient to define three sound

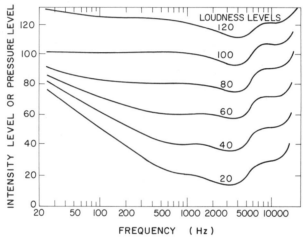

Fig. 3.10 Free-field equal loudness countours of pure tones (adapted from ref. 96).

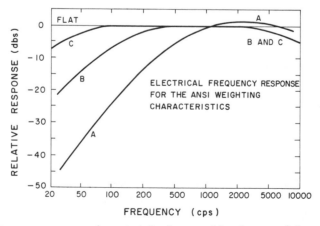

Fig. 3.11 Frequency-response characteristics for sound-level meters (adapted from ref. 95).

levels (A, B, and C scales) in which there are different weighing factors for selected bands of frequencies, as shown in Fig. 3.11. The OSHA Permissible Noise Exposures shown in Table 3.3 are defined in terms of the A scale. While the 8-hr standard is 85 dBA, higher sound levels are allowed for shorter periods of time. When the daily exposure is composed of different periods of exposure of different level sound levels, the hazard is evaluated by the following:

$$En = \sum_i \left(\frac{c}{t}\right)_i \tag{3-51}$$

where the subscript i refers to the sound level, c_i is the duration of period of exposure at one sound level, t_i is the total duration of exposure permitted at that level (Table 3.3), and the summation is over the number of periods of exposure. If the En exceeds unity, the mixed exposure is considered to have exceeded the OSHA limit for safe conditions. Figure 3.12 shows the correspondence between sound-pressure level and commonplace sources of noise. Noise produced by impact should not exceed 140 dB. *Impact noise* is noise produced by a time-varying source in which maxima occur at intervals greater than 1 sec. If maxima occur at intervals less than 1 sec, the noise is defined as continuous.

The punch press in Illustrated Problem 3.2 is so noisy that it should not be used in its present condition because while workers using it may wear personal protective devices, other individuals in its proximity will be affected. The sound level in the grinding room clearly exceeds the 8-hr OSHA standard and an individual located at the point in question should not remain there for more than 3 hr unless a personal hearing protective device is worn. *Hearing impairment* develops long after exposure and is irreversible. The susceptibility of individuals to hearing impairment is illustrated in Fig. 3.13 and shows that even at 90 dB, a significant portion of a working population will suffer impairment. Individuals that design industrial ventilation systems must be sensitive to the noise produced by fans, air jets, or indeed any vortices, recirculation, or flow separation that is apt to occur. Not only must OSHA standards be met, but one must minimize noise in general. Worker productivity and morale decrease if annoying noise is present. The noise characteristics of fans can be provided by manufacturers and

TABLE 3.3 OSHA Threshold Limit Values for Noise

Duration per day (hr)	Sound Level (dBA)
16	80
8	85
4	90
2	95
1	100
1/2	105
1/4	110
1/8	115

Source: Abstracted from ref. 33.

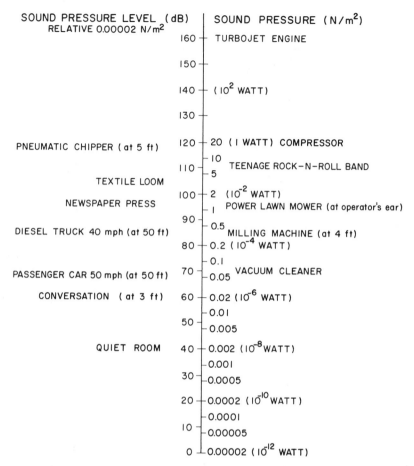

Fig. 3.12 Relationship between decibels, sound energy, and sound pressure level (adapted from ref. 36).

information is available from the Air Moving and Conditioning Association (AMCA) to compute the sound level for a particular installation.

Industrial ventilation involves devices that both remove contaminated air (exhausters) and return makeup air (blowers). Exhausters may have large openings producing low face velocities such as side-draft hoods (Figs 1.14 and 1.15) or small openings and produce large face velocities such as low volume–high velocity local exhausters (Figs. 1.10 to 1.12). The *face velocity* is the air velocity that exists at the surface of the air inlet. Blowers such as makeup air units produce low face velocities. On the other hand, the face velocities for *air curtains* and *fresh-air islands* (regions of the workplace bathed by a stream of fresh air) are considerably larger. Noise is generated by all these devices. The noise may be merely distracting, fatiguing, or a nuisance. If however, the face velocities are large, the noise might even exceed OSHA standards. Figure 3.14 shows experimental data (173,174) illustrating the relationship between sound level, frequency, face velocity, and distance from the inlet plane for exhausters of different geometry. The two most important parameters affecting noise are the *face velocity* and distance

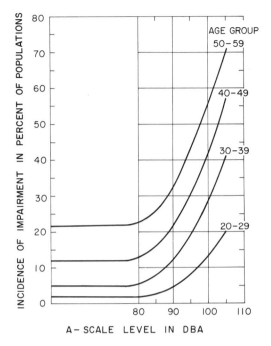

Fig. 3.13 Prevalence of impaired hearing and sound-levels for different age groups (adapted from ref. 95).

from the inlet plane of the fan. Flanged surfaces act as reflecting surfaces and intensify the noise potential. Whether the opening is round, circular, or rectangular and of different aspect ratios is of minor significance.

A systematic exposition of how to estimate noise generated by the components of an industrial ventilation system is beyond the scope of this text. Chapter 35 in the ASHRAE Systems Volume (255) describes how to estimate noise generated by fans and duct fittings, for example, and how to attenuate the noise. Much of the discussion concerns hot air heating and air-conditioning systems for public buildings, and homes where noise standards are more stringent than in the workplace. Nonetheless, the procedures are applicable to industrial ventilation systems.

3.5 HEAT STRESS

When working in a hot environment, the heart beats faster to increase the blood circulation rate in an attempt to cool the body. In addition to general discomfort, workers may experience dizziness, cramps and nausea, which are key indicators that they should cease working. At other times the symptoms may be less obvious and workers may not appreciate the stress to which they have been subjected. In severe cases, workers may lose consciousness and lapse into a coma.

Internal body temperatures are regulated within narrow limits by the flow of blood from sites of energy production in muscles and deep tissues to the skin where energy is transferred to the air by radiation, convection, and evaporation.

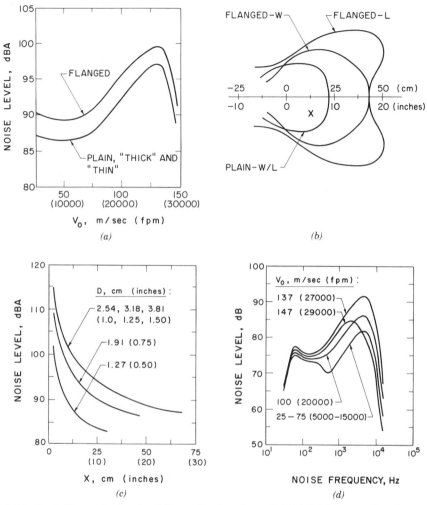

Fig. 3.14 Centerline noise levels of flanged and unflanged (plain) inlets as a function of (*a*) face velocity, flanged and plain circular inlets; (*b*) 90-dBA contours, flanged and unflanged rectangular inlet ($W/L = 0.1$); (*c*) upstream distance, unflanged inlets of different diameters; and (*d*) frequency distribution, unflanged circular inlets at different face velocities. (Adapted from refs. 173 and 174).

When energy production is equal to energy transfer, internal temperatures are controlled within narrow limits and a state of *thermal homeostasis* is said to exist. When heat transfer from the skin is not equal to energy production in the body, internal temperatures do not remain within narrow limits, the heart is stressed and the condition is referred to as *heat stress*. Heat stress taxes the heart and beyond certain limits and is unsafe and unhealthy. It is important for designers to quantify heat stress so that deleterious effects can be anticipated. By virtue of age, physical fitness, health status, living habits, and ability to acclimatize, an individual's ability to tolerate heat stress varies.

Consider the body to be a closed system of constant mass. Define the rate of heat transfer into the body as the sum of the rates by radiation, convection, and evaporation. Application of the *first law of thermodynamics* requires that the rate at which work is done by the body (W) plus the rate at which energy is stored in the body (dU/dt) is equal to the metabolic rate (M) plus the energy transferred by radiation, evaporation, and convection. Work is interpreted in the literal thermodynamic sense of the body exerting a force on an external body and displacing it through a distance, i.e. lifting a weight.

$$\frac{dU}{dt} = M - W + R + E + C \tag{3-52}$$

R, C – rate of radiant and convective heat transfer into the body

E – rate of energy transferred into the body by evaporation

M – metabolic rate of energy production

The signs on E, R, and C will be positive if energy is transferred into the body from the environment (note that this is the usual thermodynamic convention in engineering).

Table 3.4 shows the rate of oxygen uptake and the metabolic rate for different human activities. Energy needed by the body to perform physical work is derived from the enzymatically controlled oxidation of fats, proteins, and carbohydrates to produce carbon dioxide, water, and nitrogen wastes. These exothermic reactions constitute the metabolic rate M. The *metabolic rate* per square meter of body surface can be expressed as (241)

$$M\left(\frac{W}{m^2}\right) = 5.87\left[0.23(RQ) + 0.77\right]Q_{oxy}\left(\frac{60}{A_s}\right) \tag{3-53}$$

where

RQ = respiration quotient, ratio of exhaled carbon dioxide to inhaled oxygen; typical values are 0.83 (resting) $< RQ < 1.0$ (heavy exercise)

A_s = body surface area, typically 1.8 m^2 for an adult

Q_{oxy} = oxygen consumption rate in liters per minute at STP

The amount of physical work performed by the body is sometimes difficult to compute and thus the metabolic rate shown in Table 3.4 is the rate of energy

TABLE 3.4 Activity, Oxygen Uptake and Metabolic Rate

Activity	Oxygen Uptake (l/min)	Metabolic Rate (M) (kcal/hr)
Rest (seated)	0.3	90
Light machine work	0.66	200
Level walking (3.5 MPH)	1.0	300
Forging	1.3	390
Shoveling (depends on rate and load)	1.5–2.0	450–600
Slag removal	2.3	700

Source: Abstracted from ref. 36.

required to sustain human activity and to a first approximation will be interpreted as $(M - W)$.

Energy transferred into the body by *convection* depends on the velocity of air passing over the body and the temperature difference between skin and air. Wadden and Scheff (242) suggest an expression of the following form:

$$C\left(\frac{\text{kcal}}{\text{min}}\right) = KA_s(0.0325 + 0.1066U_a^{0.67})[T_a - T_s] \qquad (3\text{-}54)$$

where
  K = fraction of skin exposed to atmosphere (a value of 0.6 should be used for fully clothed)
  U_a = ambient air velocity (m/s)
  T_a, T_s = ambient and skin temperatures; for a first approximation use
    $T_s = 35$ C (308 K)

Energy transferred into the body by *radiation* depends on the skin and radiating surface temperatures, the emissivity of the radiating surfaces, and various shape factors (335). In most industrial situations the identity, location, and temperature of the radiating surfaces is difficult to ascertain. A useful parameter to account for these factors is the *globe temperature*. The globe temperature is the air temperature inside an enclosure subjected to both convection with ambient air and radiation from nearby hot or cold surfaces. In the presence of hot radiating surfaces, the globe temperature will be larger than the ambient temperature. The globe temperature depends on the location of the radiating surfaces. Thus the globe temperature is a physical measurement that must be made where the workers perform their work. The globe temperature integrates surrounding radiating wall temperatures into a mean effective radiation temperature. To simplify the computation, the conventional difference between temperatures raised to the fourth power will be replaced by a first-order difference between a *wall temperature* and skin temperature. Wadden and Scheff (242) report that the following expression can be used to relate the *wall temperature* and the measured globe and ambient temperatures:

$$T_w(K) = [T_G^4 + (0.248 \times 10^9 U_a^{0.5})(T_G - T_a)]^{0.25} \qquad (3\text{-}55)$$

where all temperatures are given in Kelvin and U_a is the air speed in m/s. Wadden and Scheff (242) report that the rate of energy transferred to the body by *radiation* can be expressed in the following form,

$$R\left(\frac{\text{kcal}}{\text{min}}\right) = 0.0728A_s K(T_w - T_s) \qquad (3\text{-}56)$$

Under steady-state conditions there is no storage or depletion of energy within the body (i.e., $dU/dt = 0$). Thus the required heat transfer by evaporation can be found from Eq. 3-52,

$$E_{\text{req}} = -(M - W) - R - C \qquad (3\text{-}57)$$

The reader must be careful with signs and remember that positive heat transfer implies energy transferred into the body. The equations for R and C above presume input heat transfer and as such the signs in Eq. 3-57 will take care of themselves and the reader should not impose another sign convention.

A widely used method to quantify heat stress is the *heat stress index* (HSI),

$$\text{HSI} = 100\left(\frac{E_{\text{req}}}{E_{\text{max}}}\right) \tag{3-58}$$

where E_{max} is the maximum amount of evaporation the body can achieve assuming the entire surface of exposed skin is covered with sweat. When $\text{HSI} > 100$, body heating occurs since the body transfers less energy by evaporation than is generated by metabolism convection and radiation and an unhealthy state exists. Wadden and Scheff (242) report that E_{max} can be computed from,

$$E_{\text{max}}\left(\frac{\text{kcal}}{\text{min}}\right) = 0.198 K A_s U_a^{0.63}[\phi P_a - P_s] \tag{3-59}$$

where ϕ is the *relative humidity* of the air, P_a the *saturation pressure* (mm Hg) of water based on T_a and P_s is the *saturation pressure* (mm Hg) of water based on T_s. The ratio of the required to maximum evaporative cooling reflects whether the body is capable of maintaining equilibrium by evaporation and prevent energy from accumulating (or being depleted) within the body. Shown in Table 3.5 are the physiological implications of different values of the heat stress index which will enable engineers to anticipate hazardous conditions for 8-hr exposures.

Heat stress can be controlled in a number of ways. Appropriate clothing should be worn that does not impede cooling by evaporation. The length of time individuals work under stressful conditions should be regulated. Lastly, workers may need a specially designed stream of moving air to cool them. The air stream can be permanent and incorporated into the building's return air system (fresh-air islands) or, as is often the case, it is provided by portable fans called *spot coolers*

TABLE 3.5 Heat Stress Index

HSI (%)	Consequence of 8-hr Exposure
0	No thermal strain.
10-30	Mild to moderate heat stress. Manual dexterity and mental alertness may suffer but there is little impairment to perform heavy physical work.
40-60	Severe heat stress. Health may be threatened unless physically fit. This condition should be avoided by people with cardiovascular or respiratory impairment or chronic dermatitis.
70-90	Very severe heat stress. Only specially selected people are capable of sustaining these conditions for 8-hrs. Special care must be taken to replace water and salt.
100	Maximum heat stress. Only acclimated, physically fit young people can withstand this for 8-hrs.

Adapted from ref. 241.

or *man coolers*. Spot cooling that directs a stream of fast-moving air over the worker may interfere with the operation of a nearby local ventilation system that captures contaminants unless spot cooling was incorporated into the design of the ventilation system in the first place.

A more detailed discussion of heat stress can be found in the ASHRAE Fundamentals Handbook (241). In many instances workers are clothed in special wearing apparel that produces unique heat stress. Examples include workers engaged in rescue operations (e.g., fire-fighters), workers engaged in hazardous chemical clean-up, individuals exposed to hot environments, individuals working in confined spaces, i.e. welding, spray painting, etc. Under these conditions readers should consult specialty texts on the subject. The control of body temperature can also be modeled as an engineering control system called a proportional controller with negative feedback (36). A procedure proposed by McKarns and Brief in 1966 and described in reference 36 can be followed. The procedure is unsophisticated and enables one to quickly estimate a maximum exposure to heat stress.

Illustrated Example 3.4—Heat Stress Index in a Hot Environment

Estimate the HSI for kitchen workers in the kitchen in a small restaurant. The air in the work space has a velocity (U_a) of 1.5 m/s. The wet bulb temperature (T_{wb}) is 27 C and the dry bulb (ambient) temperature (T_{db}) is 30 C, which correspond to a relative humidity of 80%. The globe temperature (T_G) is measured and found to be 60 C. It is estimated that the tasks the workers perform require a net metabolic rate (M-W) may be as high as 5 kcal/min or as low as 1.5 kcal/min. Assume the worker is clothed with 60% of the skin area available for heat transfer.

The wall temperature can be found from Eq. 3-55,

$$T_w = [333^4 + (0.248 \times 10^9 1.5^{0.5})(333 - 303)]^{0.25} = 382.5 \text{ K } (109.5 \text{ C})$$

The radiation heat transfer can be found from Eq. 3-56,

$$R = (0.0728)(1.8)(0.6)[382.5 - 308] = 5.86 \text{ kcal/min}$$

The convection can be found from Eq. 3-54,

$$C = (0.6)(1.8)\left[0.0325 + (0.1066)(1.5^{0.67})\right][303 - 308]$$
$$= -0.93 \text{ kcal/min}$$

The required energy can be found from Eq. 3-57,

$$E_{req} = -(M - W) - R - C = -5 - 5.86 - (-0.93) = -9.93 \text{ kcal/min}$$

The maximum amount of evaporative cooling can be found from Eq. 3-59,

$$E_{max} = (0.198)(1.8)(0.6)(1.5^{0.63})[(0.8)(31.95) - 42.35]$$
$$= -4.64 \text{ kcal/min}$$

The HSI is

$$HSI = 100(E_{req}/E_{max}) = 214$$

If the workers shed most of their clothing, the air speed increased to 5 m/s, and the work load decreased such that the metabolic level was 3 kcal/min, the following heat stress values can be expected.

Heat Stress Index Under Different Working Conditions

Metabolic Rate	$U_a = 1.5$ m/s	$U_a = 5.0$ m/s
$M = 5$ kcal/min		
clothed (K = 0.6)	214	114
unclothed (K = 1.0)	171	94
M = 3 kcal/min		
clothed (K = 0.6)	171	94
unclothed (K = 1.0)	145	82
M = 1.5 kcal/min		
clothed (K = 0.6)	138	79
unclothed (K = 1.0)	126	73

It is clear that of $U_a = 1.5$ m/s workers should not work continuously for 8 hr. If the air speed can be increased to 5 m/s (rather unlikely in a kitchen) stressful conditions will be reduced substantially. Alternatively. The work schedule could be divided by periods of rest or steps taken to reduce thermal radiation, that is, reduce the globe temperature.

3.6 ODORS

By far the majority of complaints about air pollution or inadequate ventilation concern odors. An *odor* is the physiological response from the interaction of particular airborne molecular species and the olfactory sense organs located in the upper part of the nose. Although there is no satisfactory comprehensive theory of odors, chemicals that elicit odors have low vapor pressures and certain molecular structures. Odors may be harmless, indeed even pleasant (such as food) but if odors persist, are intense, or inundate individuals, they constitute an involuntary intrusion and are a legitimate basis for complaint. The resolution of such complaints is another matter. If the odor is associated with toxic materials, the mere existence of the odor is to be taken seriously since odor threshold levels are generally not vastly different from TLV and PEL values. Table 3.6 is a summary of *odor thresholds* taken from several sources (192–194,222) and PEL values for some common toxicants. While many odor threshold values are larger than PEL or TLV, one should not generalize, for example,

Material	PEL (350), TLV (27) (PPM)(ref)	Threshold Odor Values (PPM)(ref)
Methylene chloride	100 (27)	214
Ethylene oxide	1 (350)	500
Butyl acetate	150 (27)	0.037
Trimethylamine	10 (27)	0.00021

TABLE 3.6 Odor Threshold Values for some Common Toxicants

Chemical	Threshold Value[a] (PPM)(ref.)	TWA–TLV(350) (PPM)	Description
Acetone	100(142)	750	Sweet, pungent
Acetic acid	1(192)	10	Sour
Acrolein	0.21(142)	0.1	Burnt sweet, pungent
Acrylonitrile	21.4(192)	2(160)	Onion, garlic, pungent
Ammonia	47.0(142)	35(STEL)	Pungent
Aniline	1.0(192)	2	Pungent
Benzene	4.7(142)	10(160)	Solvent
Benzyl chloride	0.047(192)	1	Solvent
Butyl acetate	0.037(193)	150	Sweet, ester
Carbon tetrachloride	21.4(192)	2	Sweet, pungent
Chlorine	0.314(192)	0.5	Bleach, pungent
Diethylamine	0.06(193)	10	Musty, fishy, amine
Ethyl acetate	13.2(193)	400	Sweet, ester
Ethyl acrylate	0.00047(142)	5	Hot plastic, earthy
Formaldehyde	1(192)	3(160)	Hay, pungent
Methyl ethylketone	10(142)	200(95)	Sweet
Methyl mercaptan	0.0021(192)	0.5	Sulfidy, pungent
Nitrobenzene	0.0047(142)	1	Shoe polish, pungent
Phenol	0.047(192)	5(160)	Medicinal
Phosgene	1(142)	0.1	Hay-like
Pyridine	0.021(142)	5	Burnt, pungent
Toluene	2.14(192)	100	Moth balls
Toluene diisocyanate	2.14(192)	0.005	Medical bandage, pungent

[a]Threshold Recognition level based on 100% recognition

As a general proposition, it is safe to assume that if a toxic material produces an odor, its concentration is probably within the range of its PEL and certainly warrants attention. *Threshold odor values* (along with TLV and PEL) are quantitative parameters that should be included in design performance standards. The ASTM compilation (207) is the most systematic tabulation of odors.

Olfaction is a sensitive physiological system that responds to contaminants at low concentrations, sometimes less than parts per billion. The olfactory system is capable of discriminating one odor from a background of different odors. Unfortunately, the olfaction system exhibits *fatigue* and individuals may become insensitive to an odor until the concentration changes. Odors are hedonistic in that they elicit an involuntary response that pleases or offends individuals. While no two individuals are alike, groups of individuals exhibit responses that can be quantified with statistical significance. There is a strong correlation between the senses of smell and taste since the neurological response involves the same organs. Food flavors are generally associated with aromas, which are airborne mixtures of molecular species that stimulate the sensory organs. Practices followed in the foods and flavors industry in which the sensory response is described are also used to describe environmental odors. To *quantify* odors, at least three independent properties are needed.

Quality (*or Character*). A description of the odor using familiar functional groupings.

Acceptability (*or Hedonic Tone*). Sensual pleasure, annoyance or offense the odor evokes.

Intensity. A quantitative response proportional to the concentration.

The intensity or strength of a response can be described in terms of the *Weber–Fechner* law:

$$S_i = K \log(c_i) \tag{3-60}$$

where S_i is the magnitude (intensity) of the response, c_i is the concentration of the material, and K is a constant. Experiments to establish threshold values involve trained observers that ascribe a numerical value to diluted samples of the odor sometimes spanning 6–8 orders of magnitude. Minimal detectable levels are established by diluting samples until the odor is no longer detected. Statistical methods are used to establish minimal detectable levels at which 50 and 100% of trained observers detect the odor. The most reliable parameter is the *recognition threshold* which is defined as the lowest concentration at which 100% of trained observers can positively identify an odor and which is consistent with the response at all higher concentrations. The *detection threshold* is defined as the concentration which produces the first evidence of the odor above the background. Unfortunately, there are different detection thresholds depending on the background. An additional way to categorize odors is the *odor index*, which is the quotient of the actual vapor pressure to its value at the odor threshold value.

$$\text{Odor index} = \text{vapor pressure/threshold level (partial pressure)} \tag{3-61}$$

Another way to express threshold values is called the *dilution factor* (194) and is defined as the total number of volumes to which one volume of air saturated with the vapor or odorant must be diluted to reach the odor threshold. The odor index and dilution factor are the same, and because the values are large, are often reported in terms of their logarithms. Both terms are useful because they take into account the escaping tendency of the material (its vapor pressure) and the ability for individuals to recognize its odor.

3.7 ENGINEERING ECONOMICS

Engineers must be cognizant of the economic ramifications of their design recommendations. There are two classes of costs or expenses that must be clearly distinguished:

Total Capital Cost (TCC)

Total capital cost (TCC) is the initial or *first cost*. Total capital costs consist of all money spent to design, build, and install the ventilation system. The costs are incurred at the beginning of the project and money must be taken from savings or borrowed to put a ventilation system in place.

Total Revenue Requirements (TRR)

Total revenue requirements (TRR) are monies that must be built into the price of the product to recoup the money spent in (TCC) and provide the annual cost to operate the system.

The TCC represents all initial expenditures needed to establish the system in operation and theTRR represent expenditures needed to operate the system over a period of time. The TCC is an "up front" cost borne by the company making the product and the TRR is a cost passed on to the customer. The TRR is the yearly cost to capitalize the initial cost of the system plus the cost to operate the system. The TRR is an additional expense the manufacturer must recover and is reflected in the purchase price of the product for which the ventilation system is required. The objective of this section is to show how these costs can be computed.

The section follows procedures described by Molburg and Rubin (37) and Neveril et al. (39). The reader should also consult established texts on engineering economics, such as Grant and Ireson (38), for details on capitalize cost, depreciation, and capital recovery cost.

3.7.1 Total Capital Cost

The TCC, or first cost, consists of total direct costs (TDC) plus total indirect cost (TIC). Table 3.7 shows items to be included in these categories and the amounts (%) to be expected.

TABLE 3.7 Total Capital Cost and Total Revenue Requirements

Total capital cost (TCC = TDC + TIC)	
Total direct cost (TDC)	Total indirect cost (TIC) in % TDC
Equipment	Construction expense (10–15%)
Labor	Contingencies (5–30%)
Materials	Contractors fees (4–5%)
Structures	Engineering (4–6%)
Engineering consulting fees	Interest during construction (10–25%)
	Start-up costs (10–15%)
	Working capital (2–4%)
	Total ICF (45–100%)
Total revenue requirements (TRR = TVC + TFC)	
Total variable cost (TVC)	Total fixed cost (TFC) in % TCC
Administration	Capital recovery cost (11–23%)
Electric, gas, water	Taxes (3–7%)
Maintenance labor	Insurance (1–3%)
Maintenance material	Interim replacement (1–7%)
Operating labor	Tax credits (0 − 5%)
Supervision	
Raw materials	Total FCF (16–40%)

Source: Abstracted from refs. 37 and 39.

$$TCC = TDC + TIC \qquad (3\text{-}62)$$

The TDC consists of the predictable cost to purchase and transport equipment, cost of labor to install the equipment, cost of major materials, engineering consulting fees, and the cost of new permanent structures. The TIC consists of construction expenses, contractors fees, engineering costs, start-up costs, and interest during construction. The TDC is arrived at by negotiations with vendors supplying major pieces of equipment. The cost to install an industrial ventilation system depends on the size of the system, whether the system is being installed in a facility under construction or an established facility and whether an air pollution control device is needed (39). The TIC may include internal transfers of funds between divisions of the user's company if assembly and installation of the ventilation system is to be done by the user's personnel. The TIC needs to be established by the user. If much of the assembly and installation is done by outside personnel, the associated costs should be included under TDC.

Often TIC is represented as a fraction of TDC. The fraction is called the *Indirect Cost Factor* (ICF).

$$TCC = TDC + TIC = TDC(1 + ICF) \qquad (3\text{-}63)$$

3.7.2 Total Revenue Requirements

The TRR refer to yearly expenditures required to operate the ventilation system over a period of time. The TRR is composed of *total variable costs* (TVC), often called *annual operating cost*, and *total fixed costs* (TFC). Examples of the fixed and variable costs in TRR are shown in Table 3.7.

$$TRR = TVC + TFC \qquad (3\text{-}64)$$

Total variable costs include the expense to operate the ventilation system, and to provide and heat makeup air, money spent on material and labor to maintain the system, plus associated *overhead expenses*. Overhead costs depend on how the company is organized and includes the costs to administer and supervise personnel operating the ventilation system. In small companies these costs may be small, but in large companies where permanent staff are hired to maintain ventilation systems these costs may be substantial. Total variable costs depend on the utilization of the ventilation system, the type of contaminant being removed (gas, vapor, or particles) and include the cost of electricity, water, labor, and raw materials used to capture contaminants. In general, the greater the automation the lower the TVC of labor, supervision, and administration. The total fixed costs are constant and include taxes and capital recovery.

The category called taxes includes federal, state. and local taxes and is included in the TFC. Unfortunately, tax laws change rapidly and taxes are difficult to estimate. The installation of a ventilation system may increase the company's taxable base and hence increase its taxes. On the other hand, the purchase of new equipment and the installation of air pollution control devices often results in tax credits that reduce the company's taxes. The tax status of the company changes rapidly and the design engineer needs to seek the advice of the company tax accountant or tax attorney. The installation of an industrial ventilation system may also change the cost of insurance. Once again, the design engineer needs to seek advice.

Table 3.7 shows that the major item in TFC is the *capital recovery cost* (CRC). It is obvious that interest charged by a bank to a company which borrowed money to cover TCC is a cost for which revenue must be generated. Suppose however, that TCC was paid entirely from the company's savings. One might assume that an interest charge is now inappropriate. Wrong! By spending its savings, the company has lost the income that money would have provided had it remained an investment. Thus *interest* is a proper annual cost. Lastly, the ventilation system wears out and an annual depreciation cost should be included in TFC. There are many accounting procedures used to compute the cost of interest and depreciation. The CRC (38,39) can be determined from the *fixed cost factor* (FCF),

$$CRC = (TCC)(FCF)$$

$$FCF = \frac{[i(1 + i)^t]}{[(1 + i)^t - 1]} \tag{3-65}$$

where i is the annual interest rate and t is the capital recovery period (system life time in years). The fixed cost factor is also called the *capital recovery factor* (38). For average interest rates over 10% over a recovery period of 10 yr, the fixed cost factor is equal to 0.16275. For a 20 − yr period, the fixed cost factor reduces to 0.11746.

The capital recovery costs are often used instead of the total fixed costs, thus

$$TRR = TVC + TFC = TVC + CRC$$

$$= TVC + (TCC)(FCF) \tag{3-66}$$

Illustrated Example 3.5—Annual Operating Cost of a Ventilation System

Estimate the annual operating cost of a new ventilation system for a firm that manufactures electronic circuit boards. The company manufactures 25,000 circuit boards per year. What additional revenue per board will be needed to cover these annual operating costs? The circuit boards are prepared by chemical etching processes involving open vessels containing volatile toxic liquids. The open vessels are presently not equipped with any kind of local ventilation system, the general ventilation of the work area is woefully ineffective and OSHA standards are not being met. An engineering firm has recommended the installation of several lateral exhaust systems operated by independent fans. Three lateral exhausts can be purchased ready-made, others will be constructed from PVC sheet by company personnel. Shown below is a list of the materials and equipment to be bought, and the prices negotiated with the vendor. Shown also are TIC estimated by the manufacturer to construct and assemble the system with company personnel. Total variable costs to operate the system have also been estimated. The lifetime of the system is to be 10 years, during which time the average interest rate is expected to be 10%.

Total capital cost
  Total direct cost
    Equipment

Three ready-built PVC lateral exhausters	3,000
Two 4 HP, 3000 SCFM PVC fans	4,000
Two 5 HP, 4500 SCFM PVC fans	4,500

    Materials

160 ft^2, 3/8 − inch PVC sheet	1,000
40 ft × 24 − inch Diameter, PVC duct	1,500
45 ft × 12 − inch Diameter, PVC duct	500
PVC cement	50
Elbows, tees, other fittings	1,500
Electrical controls	600
TDC subtotal	**$16,650**

Total Indirect Cost

Construction expense	2,500
Contingencies	3,500
Engineering	1,000
TIC subtotal	**$7,000**

TCC = TDC + TIC = $23,650

The TIC is only 42% of the TDC, somewhat below what is suggested in Table 3.7. Because the system is small and company personnel will undertake much of the construction and installation, it is realistic.

Total revenue requirements
  Total variable cost

Electricity	6,000
Maintenance labor	400
Maintenance material	500
Makeup air heating	5,000
TVC subtotal	**$11,900**

Total fixed cost

Capital recovery cost (for 10 yr, at 10%)	3,848
Insurance	100
Taxes	300
Interim replacement	100
TFC subtotal	**$4,348**

TRR = TVC + TFC = $16,248

The TFC is 18.4% of the TCC and within the range suggested by Table 3.7. The TRR or annual operating cost is equal to $16,248. If 25,000 circuit boards are manufactured each year, $0.65 should be added to the price each board to provide the $16,248 of needed revenue.

3.7.3 Time Value of Money

Engineers recommend actions that involve the time value of money. The ultimate cost of an undertaking must include the obvious initial cost, lifetime (n) of the device, taxes, yearly operating and maintenance costs, but also the cost of borrowing money at an interest rate i_b and the return which would be gained if the money was invested at a rate i_i. To compute the ultimate cost and to compare the ultimate costs of several alternative proposals on an equivalent basis, the following parameters can be used:

Present Worth P—Cost, value, or payment at the present time.

Annual cost A—Cost or payment distributed equally over a number of years.

Future Worth F—Cost, value, or payment realized at a prescribed time in the future.

If a technical recommendation involves considering alternatives in which the lifetime (t) is the same, comparisons can be made using either A, P or F. If however, the time periods (t) are different, an easily understood basis to establish the equivalency of several alternatives is the annualized cost (A). The relationship between A, P and F can be found in handbooks (18) and texts on engineering economics (38). For an interest rate i and lifetime (t), the results are summarized below:

Compound Interest   future worth/present worth, (F/P)

$$F/P = (1 + i)^t \tag{3-67}$$

Present value of an annuity—Present worth/annualized cost, (P/A)

$$P/A = \frac{(1 + i)^t - 1}{i(1 + i)^t} \tag{3-68}$$

From these two equations all other relationships between A, P and F can be developed: for example future worth of an annuity.

Future worth/annualized cost, (F/A)

$$F/A = (F/P)(P/A) = \frac{(1 + i)^t - 1}{i} \tag{3-69}$$

The use of these equations to perform engineering economic analyses can be seen in the following illustrated problems.

Illustrated Example 3.6—Buy or Lease?

An electronics company needs to install an exotic electronic air cleaner to purify air for a clean room. The air cleaner can either be bought or leased from the vendor. If it is leased, the company pays the vendor an annual leasing fee for 5 yr and the vendor provides yearly maintenance and attends to unexpected malfunctions in the equipment for this period. At the end of 5 ys, the vendor removes the equipment. If purchased, the company provides it's own maintenance and service but owns the equipment which has a salvage or trade – in value at the end of 5 ys. A summary of the terms for these arrangements are:

Lease agreement:
(A) Yearly rental fee = $3500
(A) Yearly operating expenses = $1000
(t)   Period of use = 5 yr

Purchase agreement:
(P) Initial purchase price = $10,000
(A) Yearly operating expense + maintenance = $2000
(F) Salvage or trade-in = $2000
(t)   Period of use = 5 yr

If the interest rates (i_b, i_i) during the period are 10%, which agreement is the cheapest?

One basis of equivalence is to compare the future cost (F) at the end of 5 years.

$$F(\text{lease}) = \$3500\,(F/A) + \$1000\,(F/A) = \$4500\,(F/A)$$

$$F(\text{purchase}) = (\$10,000)\,(F/P) + \$2000\,(F/A) - \$2000$$

From Eqs. 3-67 and 3-69,

$$(F/A) \text{ for 5 yr and 10\% interest} = 6.105$$

$$(F/P) \text{ for 5 yr and 10\% interest} = 1.6105$$

Thus, $F(\text{lease}) = \$27{,}472$, $F(\text{purchase}) = \$26{,}315$,   and   purchasing   is   4.3% cheaper.

Alternatively, a comparison can be made on the basis of annualized costs(A).

$$A(\text{lease}) = \$3500 + \$1000 = \$4500$$

$$A(\text{purchase}) = (\$10,000 - \$2000)(A/P) + \$2000$$

From Eq. 3-68,

$$(A/P) \text{ for 5 yr and 10\% interest} = 0.2638$$

Thus $A(\text{lease}) = \$4500$, $A(\text{purchase}) = \$4100$, and purchasing is cheaper.

The two analyses illustrate how to compare two different offers on an equivalent basis. The limitations in the analysis lie with omissions and inadequacies in the assumptions, not the analysis. For example, the analysis shows that $F(\text{purchasing})$ is cheaper than $F(\text{leasing})$ by $1157, providing maintenance and repairs cost no more than $1000 per year. If for any reason serious repairs have to be made, or there is a lack of trained personnel within the company to provide them, the savings could disappear. Leasing removes many these difficulties since the terms of the contract require the vendor to attend to the company's needs upon request. In many small firms where there are no skilled personnel to perform maintenance, the leasing arrangement is clearly more attractive since the plant manager knows that skilled help is always available. Lastly, the issue of taxes has been ignored. Leasing provides a yearly business expense that reduces the company's taxable gross income while owning the equipment increases its capital investment and property taxes but also affords a deduction for depreciation. Since it is highly probable that the company's tax status may change in 5 yr, taxes will have a bearing on the ultimate cost of operation not revealed by the given analysis above. Thus while it is relatively easy to compute which alternative is cheapest, many more factors (often nonquantifiable ones) have to be considered to determine which alternative is the best.

Illustrated Example 3.7—Value of Quality Equipment

A firm discharges noxious odors to the environment. The state environmental pollution control agency requires the company to stop. Incineration appears to be the only feasible solution and the company has applied for state permit. After consulting with vendors, the company learns that incineration produces exhaust products that corrode the incinerator. One vendor proposes an incinerator made of ordinary steel costing $10,000 that lasts 25 yr while another vendor recommends a stainless-steel and ceramic incinerator that costs $15,000, lasts 50 yr and has yearly maintenance and operating costs $200 less than the incinerator made of ordinary steel. The company president likes firstrate equipment and prefers the stainless steel incinerator. Which incinerator is cheaper?

Since the time periods are different, a useful way to determine which alternative is cheaper is to compare annual costs (A). Assume also that the interest rate is 10% for both proposals.

A(ordinary steel) = $10,000 ($A/P$, 25 yr, 10%)

$\qquad$ + yearly maintenance and operating cost

A(stainless steel) = $15,000 ($A/P$, 50 yr, 10%)

$\qquad$ + yearly maintenance and operating cost − $200

From Eq. 3-68

$$A/P \text{ for 25 yr at } 10\% = 0.11017$$
$$A/P \text{ for 50 yr at } 10\% = 0.10086$$

and

$$A\text{(stainless steel)} - A\text{(ordinary steel)} = \$221.20$$

From the point of view of expenditure only, the president's preference for high quality costs $221.20 more a year. The amount is not large and, as in the previous example, the best incinerator to buy should be determined by considering factors not included in the analysis such as taxes or whether an incinerator will even be needed beyond 25 yr.

NOMENCLATURE

Symbol	Description	Dimensions*
a	Speed of sound	L/t
A_s	Body surface area	L^2
A/P	Ratio of annualized cost to present worth	
c, c_i	Mass concentration	M/L^3
C	Heat transfer from body by convection	Q/t
c_p	Specific heat	Q/MT
D_p	Particle diameter	L
E	Rate of energy transfer from body by evaporation	Q/t
En	Parameter defined by equation	
E_{req}	Required evaporation rate	Q/t
E_{max}	Maximum evaporation rate	Q/t
f_i	Mass fraction	
F_s	Safety factor	
h	Enthalpy	Q/M
h_{rp}	Enthalpy change associated with combustion	Q/M
i	Annual rate of interest	
i_b, i_i	Interest rate on borrowed or invested money	
I	Sound intensity	Q/tL^2
I_0	Reference value of sound intensity	Q/tL^2
k	Ratio of specific heats (c_p/c_v)	
K	Constant defined by equation	
L	Flange width, long dimension	L
L_i	Exposure limit for molecular species i	M/L_3
L_p, L_I, Lw	Sound level defined in terms of pressure, Intensity, and power	
m	Mass	M
m_b	Mass of body	M
$\dot{m}$	mass flow rate	M/t
$\dot{m}_i$	mass flow rate of species i	M/t
M	Metabolic rate of energy production	Q/t
M, M_i	Molecular weight, molecular weight of species i	M/N
n, n_i	Number of mols, number of mols of species i	N
n_t	Total number of mols	N
O_{oxy}	Oxygen consumption rate	L^3/t
P, P_i	Pressure, pressure of species i	F/L^2
P_0	Sound reference pressure	F/L^2
P_a, P_s	Saturation pressures based on T_a and T_s	F/L^2
Q	Sound directivity	
Q	Inhalation rate	L^3/t
Q	Heat transferred into body	Q/t
r	Radius	L

*Q – energy, F – force, L – length, M – mass, N – mols, t – time, T – temperature

R_u, R_i	Universal gas constant, gas constant of species i	Q/MT Q/t
R	Radiant heat transfer from body	Q/MT
s	Entropy	
S_i	Odor strength during period i	t
t	Time	t
t_s	Time to achieve steady state concentration	T
T	Temperature	
T_{wb}, T_{db}, T_G, T_0	Wet bulb, dry bulb, globe, and ambient temperature	T T
T_w	Wall temperature	T
T_a, T_s	Air and skin temperatures	
T_u, T_i, T_b	Unburned, ignition and burned gas temperatures	T L/t
U_1, U_2	Gas velocities	L/t
U, U_a	Air velocity	L/t
U_f	Face velocity	
U_u	Unburned gas velocity approaching a premixed flame	L/t Q
U	Stored energy in body	L^3
V	Volume	Q/t
W	Acoustical power	L
W	Flange width, short dimension	Q/t
W_0	Reference value of acoustical power	Q/t
W	Rate at which work is done by body	L/t
W	Velocity defined by equation	L
x	Distance	
y_i	Mol fraction of molecular species i	
Z	Compressibility factor	

Greek	Description	Dimensions*
A	Angle defined by equation, fraction of material absorbed by the body	
β	Isothermal expansion coefficient	T^{-1}
δ_{pr}, δ_r	Thickness of the preheat and reacting zones in a premixed flame front	L
π	3.14159	
ρ	Density	M/L^3
ϕ	Relative humidity	

Subscripts	Description
$(\ )_1$, $(\ )_2$	locations 1 and 2
$(\ )_i$	molecular species i
$(\ )_t$	total

Abbreviations

A	Annual cost
ACFM	Volumetric flow rate (ft^3/min) measured at the actual temperature and pressure

ACGIH	American Conference of Governmental Industrial Hygienists
A/P	Ratio of annualized cost to present worth
BEI	Biological exposure index
C	Ceiling value
CAS	Chemical Abstract Service
dB	Decibel
ECL	Exposure control limit
F	Future worth
F/A	Ratio of future worth to annualized cost
FCF	Fixed cost factor
F/P	Ratio of future worth to present worth
HSI	Heat stress index
ICF	Indirect cost factor
NOEL	No observable effect limit
OSHA	Occupational Safety and Health Administration
P	Present worth
PEL	Permissible exposure limit
PPM	Parts per million (mol fraction)
SCFM	Volumetric flow rate (ft^3/min) that would exist at STP
STEL	Short term exposure limit
STP	Standard temperature (25 C) and pressure (101 kPa)
TCC	Total capital cost
TDC	Total direct cost
TFC	Total fixed cost
TIC	Total indirect cost
TLV	Threshold limit value
TRR	Total revenue requirements
TVC	Total variable cost
TWA	Time weighted average

PROBLEMS

1. Individuals engaged in occupational safety and health engineering must become familiar with the OSHA General Industry Standard (33). There is no better way to do this than by locating specific passages in the document to determine whether certain industrial situations are or are not in compliance with the OSHA standard. Study the standard, state the page and identify the part, subpart, section, paragraph, subparagraph, division, and subdivision of the OSHA standard that applies to the situation described below. Determine whether a violation exists.

   (a) Approximately 1 gal of a flammable liquid used by a company is kept in an uncovered container on a shelf in the equipment repair room, on the first floor of the building. The flash point of the fluid is 20 F.

   (b) A motor–generator is mounted on an open-sided platform 10 ft above the main floor of a factory. There is no guard railing (or its equivalent), nor is there a toeboard to protect workers passing beneath the platform.

(c)   Respirators worn by workers in a plant that inadvertently discharges ethylene oxide into the workplace are found to contain cartridges that only capture particles.

(d) An X-ray technician not wearing a personal monitoring device is found in an area where she receives ionizing radiation 25% above the allowable exposure defined by Table G − 18.

(e) A metal stamping firm has not maintained noise exposure records for a period of 2 yr.

(f) Full-term sampling is conducted of workers operating a dip tank containing soluble salts of chromium. It is found that the 8-hr TWA is $0.9 \, mg/m^3$ at STP.

(g) A worker removing asbestos from a public building is exposed to asbestos particles. A personal monitor shows that the 8-hr TWA for particles greater than 5 μm is 16 fibers/cm^3.

(h) A table-top circular saw in the university shop does not have a guard that completely encloses the blade when it is operating.

2. Workers use isopropyl alcohol (normal household rubbing alcohol) to clean parts of electric motors prior to their assembly. What is the OSHA PEL for isopropyl alcohol (in PPM)? What is the OSHA PEL in the units of milligrams per cubic meters on a day in which the workplace temperature is 5 C?

3. Workers in a leather finishing firm spray solvents on leather to produce various colors, textures, water proofing, and so forth. What is the 8-hr TWA–TLV (mg/m^3) of the solvent if it is composed of the following materials (if TLV data is not available, use PEL data, Appendix A-1, Table 3.6)?

Material	Mass Fraction (%)
Ethyl acetate	15
Isopropyl alcohol	25
Toluene	25
Isobutyl acetate	15
Isobutyl alcohol	10
Nontoxic materials	10

4. Cotton swabs are used to apply carbon tetrachloride to clean small metal parts prior to applying a finish. The process is such that the concentration of carbon tetrachloride in the workers breathing zone varies with time during an hour-long event as a ramp function, that is,

$$c \, (PPM) = \begin{bmatrix} 5 & 0 < t < 0.5 \, h \\ 5 + 95(t - 0.5)/0.5 & 0.5 < t < 1.0 \\ 5 & t = 1.0 \end{bmatrix}$$

These hour-long events are repeated throughout the day. Compute an 8-hr TWA. Are OSHA standards violated?

5. [This problem requires the use of ref. (33.)] What is the maximum allowable dust concentration (mg/m^3) in a limestone quarry? The particle size distribution is log-normal. Eighty percent (80%) of the particles (by mass) are respirable (i.e., diameter equal to or less than 15 μm). The composition of the respirable portion of the particle sample and the composition of the entire particle sample are as follows:

Respirable particles
>   Quartz, 8%
>   Other silicates, 5%
>   Inert material, 80%

Total particle sample
>   Quartz 4%
>   Other silicates, 6%
>   Inert material, 80%

6. Individuals engaged in occupational safety and health engineering encounter numerous acronyms and medical terms. It is necessary to understand the phrases shown below. Many of these terms can be found in the *Merck Manual* (334).

(6.1) What do the following acronyms mean?
>   PEL
>   TLV
>   8-hr TWA–TLV
>   STEL–TV

Identify the organization using the acronym and discuss the similarities and differences in the concepts represented by the acronym.

(6.2) What is homeostasis?

(6.3) What are alveoli?

(6.4) With regard to pulmonary ventilation, what is the tidal volume?

(6.5) Define metabolic rate. Estimate its value for walking.

(6.6) Estimate the volume of blood in the body of an adult.

(6.7) Define the following:
>   Teratogens
>   Carcinogens
>   Mutagens
>   Allergens

(6.8) Contrast acute and chronic toxicity. Contrast local and systemic toxicity.

(6.9) Contrast hepatoxins and nephrotoxins.

(6.10) Identify the following acronyms for organizations and contrast the functions the organizations perform:
>   OSHA
>   NIOSH
>   EPA
>   MSHA

(6.11) Contrast sound levels scales A, B, and C.

(6.12) What is the difference between the sound intensity level of a free − field source and the sound pressure level at a distance r meters away?

(6.13) Discuss the major components in the ear canal, middle ear, and inner ear.

(6.14) Discuss the ways in which the respiratory system reacts to airborne particles and noxious gases.

(6.15) What is the OSHA 8-hr PEL for trichloroethylene?

(6.16) Discuss the differences between ionizing and nonionizing radiation.

(6.17) What do the acronyms RAD and REM denote? Discuss similarities and differences.

(6.18) Contrast asthma and bronchitis.

(6.19) What is heat stress and why is it a hazard to health?

(6.20) What is isometric work? Contrast it with conventional work associated with raising a weight.

(6.21) Which of the following occupational disorders is most common?
Lower back pain
Bursitis
Tendonitis
Raynaud syndrome
Carpal tunnel syndrome

(6.22) What is the difference between a syndrome and a disease?

(6.23) Discuss the role of mucus, cilia, macrophage, and lymphatic fluid in the respiratory system.

(6.24) Contrast emphysema and pulmonary fibrosis.

(6.25) Contrast carcinomas, sarcomas, lymphomas, and mesotheliomas.

(6.26) What is meant by a dose − response curve?

(6.27) Which of the following occupational diseases are the top three. List in order with the most serious disease first.
Traumatic death, amputations, eye loss
Musculoskeletal injury (back, arthritis, etc.)
Lung disease (cancer, pneumoconiosis, asthma)
Noise − induced hearing impairment
Occupational cancers (other than lung)
Cardiovascular disease

(6.28) Contrast viruses and bacteria.

(6.29) Contrast the scientific fields toxicology, physiology, and epidemiology.

(6.30) What is the physiological response to heat stress?

(6.31) Contrast hypothalamus with hypothermia.

(6.32) Discuss the pathways by which toxic material travels through the body.

(6.33) Discuss individuals who may not be protected by TLV's and PEL's.

7. In a factory manufacturing wall paper, a sample of the air shows the presence of the following hydrocarbons:

Material	Concentration (PPM)
Ethyl acetate	80
Isopropyl alcohol	100
Toluene	10
Isobutyl acetate	75
Isobutyl alcohol	20

Is this environment in compliance with OSHA standards? Use Table 3.6 and Appendix A-1.

8. An underground tank of methyl isocyanate (molecular weight = 61.25) begins to leak and discharge vapor into the interior of an unheated storage building in which the temperature is −5 C. The initial concentration inside the building is 0.001 PPM and the concentration increases linearly with time according to the equation

$$c(t) = c_0 + At$$
$$A = 0.0033 \text{ PPM/hr}$$
$$c_0 = 0.001 \text{ PPM}$$

If a worker is inside the building for 8 hr each day, will OSHA standards be met? What is the concentration (mg/m^3) at the end of 8 hr?

9. A worker in a refrigerated enclosure (−20 C, 95 kPa) is repairing a pump used to transport ethylene oxide (M = 44). The initial concentration of ethylene oxide in the enclosure is negligible, but a leak in the pump seal allows the gas to enter the enclosure and cause the concentration (measured at −20 C, 95 kPa) to increase linearly with time a rate of 5 mg/m^3 per hour. What is the concentration at the end of an 8-hr day? Does the worker's exposure over an 8-hr day constitute a violation of OSHA standards?

10. Ethylene oxide (molecular weight 44) stored in a holding vessel begins to leak (as a vapor) into an unheated storage building in which the temperature is −5 C. The initial concentration inside the building is 0.1 PPM and the concentration increases linearly:

$$c(t) \text{ (PPM)} = 0.1 \text{ (PPM)} + 0.6 \text{ (PPM/hr)} \ t(\text{hr})$$

(a) If workers are present in this building for 8 hr a day, what is the 8-hr time-weighted average? Will they be within OSHA standards?
(b) What is the concentration in mg/m^3 inside the building at the end of 8 hr?
(c) Can the odor of ethylene oxide be detected after 8 hr? (See table in Section 3.6)

11. Derive Eqs. 3–18 to 3–25.

12. To illustrate the difference in value between TLV and LEL, compute the ratio of TLV/LEL for the materials listed in Table 3.2.

13. The maximum allowable concentration for inert dust (OSHA standards) is 5 mg/m^3 for respirable dust and 15 mg/m^3 for total dust. To illustrate the difference between TLV and LEL for inert dusts, compute the ratio of TLV/LEL for coal dust, polyethylene dust, and oil shale dust shown in Figs. 3.8 and 3.9. If OSHA standards are satisfied, what are the chances that fire or explosion will occur?

14. The dust concentration in many dusty industrial environments is often measured in the number of particles per unit volume of air. Review Fig. 3.9. Assuming that the oxygen concentration is 21% by volume and that the density of coal particles is 1200 kg/m^3, replot the LEL of coal dust in the units of number of particles per cubic centimeter versus the particle size D_p in micrometers for particles between 2 and 100 μm.

15. A stamping machine is rated as having a free-field sound-pressure level of 85 dB at 10 m. What is the output acoustic power (watts)? What is the sound-pressure level (L_p) at a distance r (meters) from the machine when the machine is located in the corner of a building (a) $r = 2$ m; (b) $r = 0.5$ m?

16. Six pedestal grinders will be located in a foundry. The noise (sound-pressure level) at a point P in the room for each machine is found to be, 85, 92, 90, 84, 93, and 87 dBA. What is the sound-pressure level at the point P when all six machines are working at the same time?

17. A plant manager buys 5 swing grinders, each of which is rated as having an acoustic power output of 0.01 W. The five machines will be located 5 m apart midway along the long wall of a building. The building dimensions are 30 m by 100 m by 20 m high. Estimate the sound pressure level of the operator at the center machine working 0.2 m from his machine.

18. Three identical stamping machines are located in the center of a large building. The layout of the machines is in the form of an equilateral triangle, 2 m on a side. The free-field sound-pressure level of each machine is 85 dB at a distance of 0.1 m from the machine. When all three machines are in use, what sound pressure level dB can be anticipated for an operator?

19. Estimate the sound-pressure level experienced by an aircraft ground crew chief who guides a 4-engine jet aircraft into it's berth at an airport. Each engine produces 10,000 W of acoustical power. At the closest point to the aircraft, the crew chief stands directly in front of the aircraft's nose, 10 m from the inboard engines and 20 m from the outboard engines.

20. Derive Eq. 3-47.

21. Derive Eq. 3-50.

22. Workers are asked to conduct an EPA Method − 5 particle sampling test on a tall stack from a power plant. The work is not strenuous but it is tedious and one must be attentive. The workers stand on a narrow platform surrounding

the stack over a hundred feet above the ground. They place a long sampling probe at various positions inside the stack and operate equipment that withdraws a gas sample from the stack. The wind speed is 60 FPM and the air temperature is 96 F. The wet bulb temperature (T_{wb}) is 80 F and the dry bulb temperature (T_{db}) is 96 F. The globe temperature is 100 F. Stack sampling requires a metabolic rate of 500 BTU/hr. It is estimated that the test requires the workers to be on the platform continuously for 4 hr. Can this be done without unhealthy heat stress? Is the work any safer if the air speed is 10 ft/sec, air temperature is 90 F and, wet bulb temperature is 65 F?

23. A ventilation system is run by a fan that exposes a worker to 80 dBA when standing 5 m from the fan. The plant manager wishes to install an identical fan 3 m on the other side of the worker. What sound level (dBA) will the worker be exposed to?

24. It is necessary to replace the ties on a railroad line. Conventional ties, costing $5.00 in place, have a life of 6 yr. What expenditure per tie is warranted for using a tie containing an improved preservative, if the life of the tie is extended to 9 yr? Assume interest rates are 10%.

25. Consider Illustrated Example 3.6. What is the yearly maintenance cost at which leasing and purchasing are equivalent?

26. A floor surface in a food processing plant has to be replaced every 5 yr at a cost of $1500 per hundred square feet. How long should a new floor surface last that costs twice as much to justify the larger expenditure (interest is 10%)?

27. Two 25-HP electrical motors are being considered for purchase. The first costs $200 and has an efficiency of 85%. The second costs $150 and has an efficiency of 82%. If all annual expenses, such as depreciation, insurance, and maintenance, amount annually to 15% of the original cost and electricity costs $0.051 kW-hr, how many hours of full-load operation per year are necessary to justify purchasing the more expensive motor?

28. An incinerator is to be installed in a plant that "coats" (applies color, texture, etc.) leather used in wearing apparel, book binding, upholstery, and so on. The process emits hydrocarbons and gummy particles that can be removed only by thermal (direct flame) incineration. The following costs are contained in a quotation from the incinerator manufacturer.

Base unit price—$1,055,000
Incinerator and heat exchanger, fans, ducts, fans, motors and drives, flow control valves, burner controls, hydraulics, service platforms, control cabinet, painting, drawings, operating instructions

Operating costs
    Operation   2500 hr/yr
    Exhaust volume   26,500 SCFM

Inlet exhaust temperature	70 F
Contaminant rate	120 lbm/hr
Incineration temperature	1500 F
Incineration residence time	0.5 sec
Percent thermal energy rcovery	95%
Natural Gas consumption rate	1.383 million BTU/hr

Energy costs
Fuel cost (estimated)	$7.00/million BTU
Fan electric motor (80% efficiency)	300 HP
Other electrical drives	10 HP
Yearly maintenance	$2000
Electricity	$0.051/kW-hr

Other Initial Costs Borne By Company
Labor to erect	$ 5,000
Concrete foundation	$35,000
Consulting engineering fees	$10,000
Legal fees	$10,000
Permits	$ 5,000
Connections for utilities	$10,000
Temporary shelter for materials	$ 8,000

Yearly taxes and fees (state and local)	$ 7,500

Assume that the incinerator will operate for 25 yr. Estimate the additional cost per square foot of leather that will be passed on the customers if the plant's production rate is a million square feet of leather per year. Assume an interest rate of 10%.

29. An electrical motor to operate a ventilation system is needed. The features of two available motors are listed below. The total variable cost for maintenance is 15%, the total capital cost and the cost of electricity is $0.05 per kW-hr. If the indirect cost factor is 45%, interest is 10%, what is the total revenue requirement to operate the motors for 2500 hr/yr?

Motor Number 1	Motor Number 2
Efficiency = 85%	Efficiency = 80%
Life time = 10 yrs	Lifetime = 5 yrs
Initial cost = $300	Initial cost = $150
Salvage = 0	Salvage = 0

CHAPTER 4

Contaminant Generation

Goal: Students will be expected to learn to,
  – estimate contaminant generation rate using
    empirical equations
  – predict contaminant generation rate from
    evaporation

The precision with which a ventilation system is designed will be no better than the precision with which the designer estimates the generation of contaminants. Specifically, designers need to know the following information to design a ventilation system by any of the methods to be presented in later chapters of the book:

Particles

Generation rate or source strength (mass/area–time) and how it varies with location and time.

Particle shape, density, and size distribution.

Initial particle velocities, and how they vary with particle size, location, and time.

Gases and Vapors

Generation rate or source strength (mass/area–time) and how it varies with location and time.

Chemical composition (molecular species that comprise the contaminant mixture and the mole fraction of each species).

Local contaminant concentration at the interface of air and source and how it varies with location and time.

There are four ways contaminant generation data can be expressed quantitatively,

Field measurements.
Empirical expressions.
Emission factors.
Prediction from physical principles.

4.1 FIELD MEASUREMENTS

The most relevant information is what can be measured at the site. Unfortunately, these measurements may be difficult and expensive and may require equipment not always available to the designer. One should keep in mind that the purpose of these measurements is to obtain input data for design computations and thus one should use techniques that provide accuracy commensurate with this objective. Determining particle size, gas composition, interface mol fraction, and so on involves sophisticated equipment and sampling techniques that are difficult to master and that have to be adapted to each source. A useful summary of air sampling instruments and measuring techniques is published by the ACGIH (51). New editions of the book are published every few years.

Measurement of the actual mass generation rate is difficult. For example, if one measures the mass lost by evaporation or material removed by grinding over a period of time

$$\frac{m(t_2) - m(t_1)}{t_2 - t_1} \qquad (4\text{-}1)$$

the accuracy of the value is apt to be poor since one may be subtracting two quantities of nearly equal value. Equation 4-1 has been used with success (282) to estimate the emission of methylene chloride (CH_2Cl_2) occurring with the use of aerosol spray cans for paint stripping and removing and applications of metallic aerosol finishes and clear polyurethane finishes.

In the case of evaporation of pools of volatile liquid, one must also realize that the measurement may not be the same when a ventilation system is employed, for example, the evaporation rate of a volatile liquid in quiescent air will be different if the air is moving. Lastly, Eq. 4-1 yields the average mass generation rate, the actual mass generation rate might very well vary with time.

Over the last one or two decades, the public has become concerned about contaminants generated by *household products and appliances*. Several books have been written on indoor air pollution and indoor air quality (98,225,226,242). In addition, countless articles have been published in the professional literature. These books and articles describe the rate at which the contaminants are emitted and the environmental factors affecting the emission rate. This data is useful to engineers to estimate indoor contaminant concentrations and to design ventilation systems. The articles may be exhaustive survey papers, research monographs, or merely timely reports of important discoveries. The following is sample of such literature for several common household activities.

Organic vapors emitted by consumer products (108,117,227,245,251,283–285)
Kerosene space heaters (30,113,116,118,275,358)
Wood and coal stoves (120,218,236,281)
Gas kitchen ranges (116,119,275)
Tobacco (24,29,46,112,223,224,276,278–280,349,353,374,375)

Shown below is an example of the range of indoor average source strengths of carbon monoxide (CO), total suspended particulate (TSP), and benzo[a]pyrene (BaP) for several "airtight" and "non-airtight" *wood-burning stoves* operating steadily in a home (236).

Stove	CO (cm^3/hr)	TSP (mg/hr)	BaP (μg/hr)
Airtight	10–140	2.5–8.7	0.02–0.76
Non-airtight	220–1800	16–320	2.2–57

It is impractical to provide a complete and current summary of such literature for every household or industrial activity and designers must keep abreast of the field themselves. Computer-aided literature searches can simplify the search considerably. Appendix A-7 summarizes the emission for several household devices.

Large manufacturing companies assemble data about emissions associated with specific manufacturing processes. Such data is used to design industrial ventilation systems to maintain health and safety standards. In some cases companies belonging to trade associations pool this data for their members and sell the data to the public. An example of this information (175) is the publications of the American Welding Society on the characteristics of fume and gases produced in welding.

4.2 EMPIRICAL EXPRESSIONS

While not possessing the sophistication researchers may desire or the detail designers desire, empirical expressions enjoy widespread industrial use and offer an expedient way to characterize the source. Empirical equations express contaminant generation in terms of easily measurable quantities. The expressions often contain constants that require the use of certain units and designers must be careful. The acceptance of a design will be enhanced if it is based on data and empirical expressions describing contaminant generation data that are used by industry.

4.2.1 Liquids in Vessels

The EPA has the task of assessing the hazard from the countless number of new chemical substances developed in the United States each year. They use several elementary expressions (184) to estimate the rate at which vapors enter the atmosphere due to the following:

Filling containers
Obtaining samples from vessels containing liquids
Cleaning and maintaining containers
Open vessels

The rate at which vapor is generated by filling operations can be expressed as

$$S_f = \frac{fVP_vM_iL_r}{R_uT} \tag{4-2}$$

where

S_f = evaporation rate due to filling (mass/time)
f = dimensionless filling factor
V = volume of the container (length cubed)

P_v = vapor pressure (atm) see Appendix A-8
M_i = molecular weight of molecular species i
R_u = universal gas constant
 T = liquid temperature (absolute)
L_r = loading rate, number of containers per unit time

If the liquid is poured into the vessel (*splash filling*), the filling factor is 1.0. If the liquid enters the vessel through a pipe beneath the liquid surface (*submerged or bottom filling*), f is equal to 0.5.

When a liquid fills a vessel that formerly held a volatile liquid, it *displaces* air containing vapor of that former liquid. The rate at which this vapor is displaced and enters the atmosphere can be expressed as

$$S_d = \frac{M_i V P_v L_r}{R_u T} \tag{4-3}$$

where S_d is the evaporation rate because of vapor in the displaced air. The total rate at which vapor enters the atmosphere is $(S_f + S_d)$.

Vapor enters the atmosphere when workers open a lid, hatch, or port of a vessel to obtain a sample of a volatile liquid for analysis. Vapor also enters the atmosphere when a volatile liquid is stored in an open vessel or when a vessel is cleaned. In all these cases the EPA estimates the *evaporation rate* by the following:

$$S_e = \frac{M_i K A P_v}{R_u T} \tag{4-4}$$

where K is the mass transfer coefficient, A is the area over which evaporation occurs, and the remaining terms are the same as in the expressions above.

The mass transfer coefficient K is assumed to be proportional to the diffusion coefficient to the power 2/3, and the diffusion coefficient is assumed to be inversely proportional to the square root of the molecular weight M_i. Thus, using water as a reference value, the mass transfer coefficient can be expressed as

$$K(\text{cm/s}) = 0.83(18/M_i)^{0.33} \tag{4-5}$$

4.2.2 Industrial Spills

Evaporation of a volatile liquid from an industrial spill may be quite different from evaporation from an open vessel. Evaporation of a highly volatile material requires heat to be transferred to the evaporating liquid. If the liquid is stored in a large vessel, there may be sufficient heat transfer within the liquid to maintain isothermal conditions at the liquid–air interface. In an industrial spill the pool is shallow, the surface to volume ratio is large, and heat transferred to the liquid pool may be low, causing the pool temperature to decrease. Since the partial pressure is strongly dependent on temperature, the rate of evaporation may be affected significantly. An expression used by the US Army (59) for cases where heat transfer is negligible is

$$S_s = 0.3 U^{0.8} A^{0.9} T^{-0.8} M_i P_v \left\{ \frac{[(3.1 + \rho^{-0.33})^2]}{K} \right\}^{-0.67} \tag{4-6}$$

where

$$K = T^{0.5}[(1/29) + (1/M)]^{0.5}$$

S_s = evaporation rate (kg/hr)
U = air speed (m/s)
A = spill area (m^2)
M_i = molecular weight of volatile liquid
T = temperature (K)
ρ = molar density of volatile liquid (g-mol/cm^3)
P_v = liquid vapor pressure at interface (mm/Hg)

While developed for the outside environment, it is nevertheless suitable for evaporation from spills within a building where the wind speed is low and the length of the Reynolds number is less than 20,000. Kunkel (59) discusses several analytical models that predict the evaporation rate for spills in which heat transfer is of varying importance.

4.2.3 Grinding

Hahn and Lindsay (11) suggest that the total rate at which metal is removed by grinding (V_g) can be expressed as

$$V_g \text{ (vol/time)} = \Lambda F_n \qquad (4\text{-}7)$$

where F_n is the normal force (lbf) between the workpiece and grinding wheel and Λ is the metal removal parameter (vol/min lbf) which is a function of the wheel speed and the physical properties of the workpiece. Properties of the grinding wheel have only a slight affect on the metal removal rate. Grinding particles are primarily metal particles, only a small amount of the wheel becomes airborne. Values of Λ between 0.0001 and 0.025 (in^3/min-lbf) can be expected (11) for the majority of industrial grinding operations. The *metal removal parameter* can be measured experimentally or estimated analytically. Figure 4.1 (11) shows values

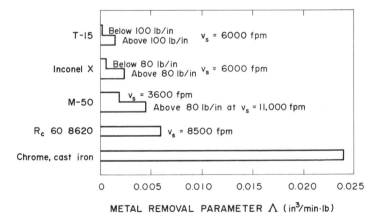

METAL REMOVAL PARAMETER Λ (in^3/min·lb)

Fig. 4.1 Metal removal parameter for five metals for different normal forces and grinding wheel speeds (v_s). T-15: typical difficult to grind steel alloy. Inconel X: a difficult to grind steel alloy. M-50: typical easy to grind steel alloy. 8640, Rockwell hardness (R_c) 60: easy to grind steel alloy. Chrome, cast iron: easy to grind material (abstracted from ref. 11).

of Λ for several alloys, wheel speeds, and normal forces. The dimensionless particle generation rate at points around the wheel's periphery can be expressed (12) as

$$g(\alpha) = 0.17 - 3.53 \times 10^{-3}\alpha + 0.23 \times 10^{-4}\alpha^2 - 0.4 \times 10^{-7}\alpha^3 \qquad (4\text{-}8)$$

where α is the angular displacement (in degrees, $0 < \alpha < 90$) measured from the point of contact with the workpiece. The function $g(\alpha)$ is a dimensionless generation rate equal to the ratio of the local particle mass generation rate at angle α to the total *metal removal rate* ($\dot{m}_t$). Thus the particle mass emission rate at the point on the grinding wheel S_g is

$$S_g(\alpha) = \int_m V_g g(\alpha) \qquad (4\text{-}9)$$

The particles can be assumed to leave the wheel with a velocity equal to the tangential velocity of the wheel (13). Figure 4.2 shows the *particle size distribution* for different metal removal rates.

4.2.4 Pouring Powders

There are countless industrial operations in which workers pour small quantities of powdered materials into vessels. Such operations require skill to minimize generating dust, but nonetheless the operations are notoriously dusty. Several laboratory techniques are available to estimate the fraction of dust that is

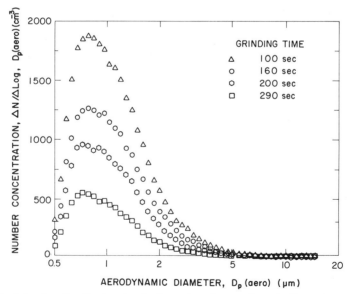

Fig. 4.2 Particle size distribution in the boundary layer of a grinding wheel 270 degrees from the point of contact after grinding has begun. See Chapter 8 for details about the cumulative number distribution function (N) and the aerodynamic particle diameter (redrawn from ref. 203).

generated (345). For representative drop heights of 25 cm, Cowherd et al. (345) suggest the following empirical equation to predict the mass of airborne dust (mg) per kilogram of material poured.

$$L = 16.6 f_w^{-0.75} \sigma_g^{3.9} \rho_b^{-1.2} D_{p,50}(\text{mass})^{-0.45} \tag{4-10}$$

where

L = mg of airborne dust per kg poured powder
f_w = powder moisture content (%)
σ_g = geometric standard deviation of particle size distribution
ρ_b = powder bulk density (g/cm^3)
$D_{p,50}(\text{mass})$ = mass median particle diameter of powder (μm)

The units shown above must be used in Eq. 4-10.

4.2.5 Spray Finishing and Coating

Surface coatings are applied to countless products by spray nozzles (361). Workplace contaminants are generated by the evaporating solvents and from overspray. *Overspray* are atomized finish particles that do not land on the intended surfaces. The total mass of solvent *volatile organic compound* (VOC) entering the workplace environment is equal to the VOC content in the finish (a value listed on the MSDS) times the application rate (gallons/m$_2$ of surface coated). The VOC generation rate is divided between the VOC generated during spraying and VOC generated during drying. The former depends on the type of sprayer and spraying rate (gallons/min) and the latter depends on the drying time, a value that can be provided by the coating supplier.

Examples of spray finishing include automobile refinishing and coatings for wood and metal furniture, machinery, and appliances. The major constituents in finishes are binders, pigments, solvents, and additives. Liquid coatings are classified broadly as those that use volatile organic compounds as solvents for the binders, pigments, and additives and those that use water. Coatings can also be categorized as either thermoplastic or convertible. *Thermoplastic coatings* are soft when heated and harden upon cooling. A *convertible coating* undergoes an irreversible chemical change as the film forms. *Binders* are non-volatile materials that cement the pigments to the film to the material to be finished. Synthetic and natural resins and drying oils are used as binders. *Pigments* are fine insoluble powders dispersed in a liquid medium. Pigments are used to impart color, inhibit corrosion, and produce a certain opacity or absortivity. *Additives* are used in relatively small amounts to increase the finish shelf-life and to impart certain features of the finish not obtained by pigments.

Industrial finishes are applied by sprayers that use compressed air or high liquid pressure. Compressed air is the most widely used method to atomize the finish and transport it to the intended surface. Unfortunately, a great many of the particles fail to land on the intended surface (i.e., overspray). Finer control of the particle size and a higher percentage of particles can be applied by *airless spraying* in which atomization is achieved as the liquid is forced through the nozzle at high liquid pressure. *Electrostatic spraying* consists of applying an electrical charge to the particle and an opposite charge to the surface to be painted.

The mass generation rate associated with overspray cannot be calculated accurately, but only estimated from manufacturers' literature. The amount of solvent that evaporates can be estimated on the basis of finished used and the percent solvent, but the rate at which the solvent evaporates is not easily calculated. The particle size distribution generated by spray nozzles has to be provided by the vendors of the spray equipment.

4.3 EMISSION FACTORS

The phrase *emission factor* is perhaps more appropriate to pollution in the outdoor environment than in the workplace, but there are a number of industrial processes relevant to industrial ventilation for which emission factors can be used. The concept has been used with great success by the EPA and there is no reason why industrial ventilation cannot benefit equally as well.

Emission factors express the amount of contaminant typically released to the environment per unit quantity of material being processed. The quantity of material being processed is chosen to be a quantity the source operator is keenly aware of (i.e., fuel firing rate, production rate, mass of product being stored). It must be stressed that emission factors only typify the amount of contaminant generated by a class of industrial operations. Individual sources may generate quite different amounts of contaminant. Nevertheless, in lieu of actual experimental data for a particular process, empirical expressions, or analytical expressions that can be derived from first principles, emission factors may be the only data available to the designer. The EPA has compiled emission factors for a large variety of industrial activities (6 – 10). Emission factors for a variety of industrial processes are shown in Appendices A-2 to A-6. The EPA updates and expands these tables on a regular basis. Contained in these tables is data taken from other publications which, while not strictly emission factors, are presented in similar fashion and can be used for the same purpose. Emission factors can also be found in the separate chapters of reference 142 that describe air pollution control systems for specific industrial processes and in several handbooks (320,321). Emission factors are also available on microcomputer databases available commercially and reviewed in professional journals (270,271).   Emission factors for a variety of indoor processes and activities are contained in Appendix A-7 and are drawn from a variety of sources. For materials such as formaldehyde that is emitted from home-building materials and furnishings, only average values are given as emission factors. The actual emission rate decreases with age as formaldehyde escapes from the material. The actual emission rate also depends on the temperature and relative humidity (227).

The following is a summary of emission factors in the Appendix:

A-2   Emission Factors For Particles From Uncontrolled Metallurgical Processes

A-3   Emission Factors For Volatile Hydrocarbons From Uncontrolled Sources
A-4   Emission Factors For Uncontrolled Mineral Processes
A-5   Emission Factors For Uncontrolled Chemical Processes
A-6   Emission Factors For Uncontrolled Food Processes
A-7   Emission Factors Of Indoor Processes and Activities

4.4 PUFF DIFFUSION

Contaminants may enter the atmosphere in the form of bursts. *Bursts*, or *instantaneous sources* as they are called formally, occur when a vessel containing gas or vapor ruptures, when a safety valve pops open, or when a mass of powder falls on the floor. The movement of particles or gas through the surrounding air is called *dispersion*. The overall phenomena of the burst followed by dispersion is called *puff diffusion*. During its early stages, puff diffusion is distinctly different than the dispersion of contaminants from continuous sources. Dispersion from continuous point sources concerns the plumes from smoke stacks, volcanic plumes, forest fires, and plumes from other large-scale sources that transport contaminants hundreds of kilometers for periods of hundreds of hours. Nevertheless, a great deal of the theory of *atmospheric dispersion* can be applied to puff diffusion in the workplace.

Once in the air, contaminants are transported by the bulk motion of the carrier gas (convection) and diffusion. The concentration downwind of the source is given by the equation

$$\frac{\partial c}{\partial t} = -\mathbf{Div}(\mathbf{U}c) + \nabla^2(\mathscr{D}_{12}c) \tag{4-11}$$

where $\partial c/\partial t$ is the partial derivative of concentration with respect to time and $\mathbf{U}$ is the velocity of the carrier gas. If there is no bulk motion ($\mathbf{U} = 0$) the coefficient $\mathscr{D}_{12}$ is the bimolecular molecular diffusion coefficient, which from this point on will often be shortened to $\mathscr{D}$ for simplicity. Values of the bimolecular diffusion coefficient for various gases and vapors in an air contained in Appendix A-9. If there is convection, the coefficient $\mathscr{D}$ must reflect turbulence in the carrier gas. In quiescent air, the solution of Eq. 4-11 is the Gaussian formula (41,105):

$$c(r, t) = \left[\frac{M}{\{(2\pi)^{1.5}\sigma^3\}}\right] \exp\left[\frac{-(r/\sigma)^2}{2}\right] \tag{4-12}$$

where M is the mass released instantaneously for each puff, r is the distance from the center of the puff to a point in space and σ is called the standard deviation. From the field of atmospheric dispersion, σ is also called the *dispersion coefficient*.

Values of the dispersion coefficient depend on time. Hanna et al. (105) report that when the elapsed time is short and the puff dimensions are small with respect to the size of eddies in the air, σ is linearly proportional to time. When the puff has grown and is larger than the eddies, σ is proportional to the square root of time. Cooper and Horowitz (106) assumed that the *dispersion coefficient* can be given by

$$\sigma = \sqrt{2\mathcal{D}t} \tag{4-13}$$

where $\mathcal{D}$ is the diffusion coefficient of particles of diameter D_p can be estimated from the following expression from Fuchs (41):

$$\mathcal{D} = \frac{kTC}{3\pi\mu\mathcal{D}_p} \tag{4-14}$$

where k is the Boltzmann constant, μ is the viscosity of the carrier gas, and C is the unitless Cunningham correction factor (see Chapter 8). For particles larger than 2 μm, C is essentially unity. Values of the diffusion coefficient for particles in air at STP are given in Table 4.1

The concentrations at various radii from the origin of the puff were studied by Cooper and Horowitz (106) assuming σ is given by Eq. 4-13. Values predicted from Eq. 4-12 are shown in Fig. 4.3 as functions of $\mathcal{D}t$. The graph predicts two interesting features about puff diffusion.

TABLE 4.1 Particle Diffusion Coefficients in air at STP

$D_p(\mu m)$	$\mathcal{D}\ (cm^2/s)$
0.01	1.35×10^{-4}
0.05	6.82×10^{-6}
0.10	2.21×10^{-6}
0.50	2.74×10^{-7}
1.00	1.27×10^{-7}
5.00	2.38×10^{-8}
10.00	1.38×10^{-8}

Source: Abstracted from ref. 41.

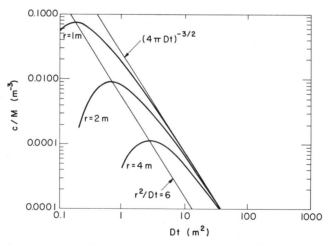

Fig. 4.3 Ratio of concentration to released mass (c/M, m^{-3}) versus Dt (m^2) for puff diffusion (redrawn from ref. 106).

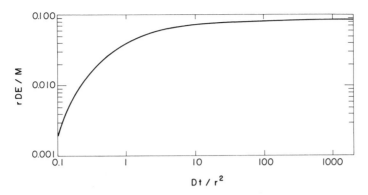

Fig. 4.4 Exposure versus Dt/r^2 for puff diffusion (redrawn from ref. 106).

(1) As the puff moves radially outward, the concentration (at any radius) increases with time, reaches a maximum, and then decreases. For larger radii the pattern remains the same, except the peak concentration occurs later and is less than what it was closer to the source. Both results are consistent with the conservation of mass of a spherically propagating wave. For large values of $\mathcal{D}t$, the exponential function approaches unity and the concentration varies as $(4\pi\mathcal{D}t)$ to the negative 3/2 power at any radii from the source.

(2) The second interesting feature predicted by Fig. 4.3 concerns the maximum values of the concentration. Drawing a line through the maximum values, one finds that they occur for values of radius and time such that $(r_2/\mathcal{D}t) = 6$.

The contaminant *exposure* of individuals affected by a puff can be computed from

$$E(t) = \int_0^t c(t)\, dt \qquad (4\text{-}15)$$

The integration has been evaluated numerically (106) and is shown in Fig. 4.4. At large values of $\mathcal{D}t/r^2$, the exposure approaches the value $(0.080M/r\mathcal{D})$.

Illustrated Example 4.1   *Momentary Release of Gas in Laboratory*

A pressurized vessel of methyl chloride located in the center of a small laboratory develops a leak and 200 g of methyl chloride vapor suddenly enter the air. The vessel is 2 from a laboratory technician. Assess the hazard to the worker. The TLV for methyl chloride is 100 PPM (210 mg/m^3) and the molecular diffusion coefficient is 1.3×10^{-5} m^2/s. There is no perceptible motion to the air in the laboratory.

The concentration at 2 m can be found by direct substitution. The following results are obtained:

Time (hr)	Concentration at 2 m (mg/m^3)
3.7	40
9.3	100
18.6	180
37	100
186	20
370	4

The predicted times do not jibe with what one would expect. Intuition or experience suggest that the times should be 2 or 3 orders of magnitude smaller. The anomaly lies with the assumption that the atmosphere is quiescent, that is, that there is no bulk motion. Such an environment does not exist in the workplace. Air currents are always present and are more important than diffusion to transport contaminants.

Using Eq. 4-13 in Eq. 4-12 suffer from two serious deficiencies:

(a) A momentary release of material produces a jet that disturbs the surrounding air. Assuming quiescent conditions during the first few moments following discharge is unrealistic.

(b) The workplace environment is not strictly quiescent. There will be movement of air whether one is aware of it or not. Thus diffusion is small in comparison with bulk motion.

To account for these two realities, the dispersion coefficient must be related to the turbulent properties of the carrier gas in addition to time. It is believed that the general form of Figs. 4.3 and 4.4 will remain the same but the dispersion coefficients will have to be computed in another fashion. Hanna et al. (105) report that in experiments involving a wide variety of instantaneous point sources in the atmosphere, σ correlates well with time (even short durations of several seconds) by the following:

$$\frac{\sigma^2}{2Kt} = \left(\frac{t}{t_{Lv}}\right) - \left[1 - \exp\left(\frac{-t}{t_{Lv}}\right)\right]$$

$$-0.5\left[1 - \left(\frac{v_0^2 t_{Lv}}{K}\right)\right]\left[1 - \exp\left(\frac{-t}{t_{Lv}}\right)\right]^2$$

(4-16)

where K is the eddy diffusivity, v_0 is an initial velocity of the puff, and t_{Lv} is the Lagrangian turbulence time scale and have the following values:

$$K = 50,000 \text{ m}^2/\text{s}, \qquad v_0 = 0.15 \text{ m/s}, \qquad t_{Lv} = 10,000 \text{ s} \qquad (4\text{-}17)$$

4.5 EVAPORATION AND DIFFUSION

A variety of volatile liquids are used in industrial operations. Unless precautions are taken, vapors from these liquids will produce workplace concentrations that exceed TLV. To design ventilation systems to control these vapors, it will be necessary to know the rate at which the vapors are generated. In certain cases one can estimate these rates accurately from the principles of mass transfer and obtain more accurate rates than predicted by Eq. 4-2 or emission factors. Texts on the theory of mass transfer (17, 20–22) should be consulted to acquire a thorough understanding of the physical concepts and principles of mass transfer.

In order to model the generation of contaminants, one must know the thermodynamic and physical properties of air and the contaminant. The properties of air are readily available, but the properties of contaminants are less abundant. Appendix A-8 lists the partial pressures of several organic liquids for a variety of temperatures. Primary references and handbooks (16,18) should be consulted for data for specific contaminants or for more accurate data at different temperatures.

Values of partial pressure (P_2) at untabulated temperatures (T_2) can be estimated from their values at P_1 and T_1 using the *Clausius Clapeyron* equation

$$\ln\left(\frac{P_2}{P_1}\right) = \left(\frac{h_{fg}}{R_u}\right)\left(\frac{T_2 - T_1}{T_1 T_2}\right). \tag{4-18}$$

where R_u is the universal gas constant and h_{fg} is the *enthalpy (heat) of vaporization* at T_1 or T_2. The reference values P_1 and T_1 are often 1 atmosphere and the saturation temperature at atmospheric pressure. The saturation temperatrue at atmospheric pressure is called the *normal boiling temperature*. The accuracy of this calculation is high if T_1 and T_2 are not vastly different.

Illustrated Example 4.2   Computation Of The Partial Pressure At An Arbitrary Temperature

To illustrate the usefulness of Eq. 4-18, compute the partial pressure (in atmospheres) of toluene at 136.5 and 31.8 C if the only information that is available is that the normal boiling temperature is 110.6 C and the heat of vaporization (at STP) is 8580.5 cal/g-mol.

The "normal boiling temperature" is the saturation temperature at 1 atm pressure. Likewise, the "heat of vaporization" is also the value of (h_{fg}) at 1 atm pressure. Substituting a temperature 409.65 K (136.5 C) and introducing constants to convert units,

$$\ln(P/1) = (8580.5)(4.186)(409.65 - 383.75)/(409.65)(383.75)(8.314)$$

$$P = 2.037 \text{ atm}$$

Substituting a temperature of 304.95 K (31.8 C),

$$\ln(P/1) = (8580.5)(4.186)(304.95 - 383.75)/(8.314)(304.95)(383.75)$$

$$P = 0.05452 \text{ atm (41.4 mm Hg pressure)}$$

Tabulated values of saturation pressures (P_{sat}) shows that

$P_{sat}(136.5 \text{ C}) = 2.0 \text{ atm}, \quad \text{or } 1.8\% \text{ below the computed value}$

$P_{sat}(31.8 \text{ C}) = 40 \text{ mm Hg pressure}, \quad \text{or } 3.5\% \text{ below the computed value}$

Considering the approximations in the Clausius Clapeyron equation, these errors are small.

Appendix A-9 contains a listing of the *molecular diffusion coefficients* or *diffusivities* for a variety of hydrocarbons in air and water at 0–25 C. Handbooks or source books should be consulted to find diffusion coefficients for other materials. Gas phase *binary diffusion coefficients* ($\mathscr{D}_{12}$ in cm^2/s) can also be estimated by an equation proposed by Chen and Othmer and reported by Vargaftik (16).

$$\mathscr{D}_{12}(cm^2/s) = \left(\frac{0.43}{K}\right)\left(\frac{T}{100}\right)^{1.81} \sqrt{\left[\left(\frac{1}{M_1}\right) + \left(\frac{1}{M_2}\right)\right]} \qquad (4\text{-}19)$$

where

$$K = P\left(\frac{T_{c1} T_{c2}}{10000}\right)^{0.1405} E$$

$$E = \left[\left(\frac{v_{c1}}{100}\right)^{0.4} + \left(\frac{v_{c2}}{100}\right)^{0.4}\right]^2$$

where v_c and T_c are *critical* volume (cm^3/g-mol) and temperature (K), M_i is the molecular weight, and pressure P is in atmospheres. More accurate equations to predict the diffusion coefficient can be found in refs. (17, 18, 21 and 22). Appendix A-10 lists values of critical temperature and values of critical volume can be computed from

$$v_c = \frac{Z_c R_u T_c}{P_c} \qquad (4\text{-}20)$$

where Z_c is the *compressibility factor* at the critical point. Values of Z_c vary from about 0.23 to 0.33 for different substances, but a value of 0.27 is a good average value for many substances encountered in industrial ventilation.

The diffusion coefficient $\mathscr{D}$ depends on the temperature and pressure. At temperatures and pressures other than STP, one can estimate the diffusion coefficient from

$$\mathscr{D}(P, T) = \mathscr{D}(STP)\left(\frac{1}{P}\right)\left(\frac{T}{298}\right)^{1.81} \qquad (4\text{-}21)$$

where P is in atmospheres and T is in degrees Kelvin.

A review of tabulated diffusion coefficients for a variety of materials reveals that the range of values is small. For design purposes, inaccuracies in the diffusivity of 10% or so are not significant because actual flow fields for industrial ventilation systems are virtually always turbulent and the *turbulent diffusivity* is considerably larger than the molecular diffusivity. Computing turbulent diffusivities is difficult in itself, and one should be content with using molecular diffusivities of modest accuracy.

Prediction of liquid phase diffusion coefficients is more difficult than predictions of gas phase diffusion coefficients. Determining diffusivities in concentrated solutions are different than diffusivities in dilute solutions. Diffusivities of electrolytes is particularly complex. Primary reference sources should be consulted for liquid phase diffusivities.

Illustrated Example 4.3 Computation Of Diffusion Coefficients

To illustrate the value of Eq. 4-19, compute the diffusion coefficient of toluene in air at STP. From handbooks and Appendix A-10 the following data can be found:

Air (use nitrogen)	Toluene
$T_c = 126.2 \text{ K}$	$T_c = 593.75 \text{ K}$
$v_c = 89.9 \text{ cm}^3/\text{g-mol}$	$v_c = 315.5 \text{ cm}^3/\text{g-mol}$
$M = 28.02 \text{ g/g-mol}$	$M = 92.13 \text{ g/g-mol}$

Substituting these values in Eq. 4-19 results in

$$\mathscr{D}_{12} = 0.0766 \text{ cm}^2/\text{s}, 0.77 \times 10^{-5} \text{ m}^2/\text{s}$$

The value of diffusivity tabulated in Appendix A-9 is $0.71 \times 10^{-5} \text{ m}^2/\text{s}$

Figure 4.5 depicts a column of gas containing two diffusing molecular species A and C. The motion of molecular species C will be expressed in terms of the mols of C crossing a unit area (normal to it) per unit time. For a nonuniform solution

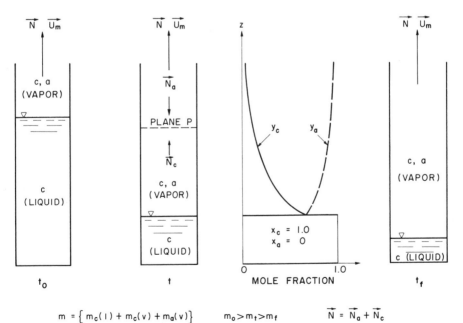

$$m = \{ m_c(1) + m_c(v) + m_a(v) \} \qquad m_o > m_t > m_f \qquad \vec{N} = \vec{N}_a + \vec{N}_c$$

Fig. 4.5 Evaporation of a volatile liquid through stagnant air.

of constituents, the motion of molecular species C can be described in terms of two fluxes,

N_c (kmols/m^2 s)—molar flux rate of C relative to a fixed observer
J_c (kmols/m^2 s)—molar flux of C relative to the average molar velocity of all the constituents

The molar flux rate $\mathbf{N}_c$ is important for the purpose of design and performance of equipment. The second reflects the nature of the diffusing species and the medium through which it moves and is expressed by *Ficks law*,

$$\mathbf{J}_c = -\mathscr{D}_{ac}\,\mathbf{Grad}\,c_c \tag{4-22}$$

where **Grad** is the vector operator representing the gradient. Consider the element of fluid in Fig. 4.5, of unit cross sectional area in which high concentrations of air (A) and a contaminant (C) exist at either ends of the element. At a plane somewhere between the ends of the element, air (A) and contaminant (C) diffuse in opposite directions, each moving from a region of high concentration to low concentration. The velocity (actually speed) of species A and species C are related by $U_a = -U_c$, since the total volume of material on either side of the plane remains the same. The rate at which moles of A pass the stationary observer per unit area is $\mathbf{N}_a$,

$$\mathbf{N}_a = \mathbf{U}_a c_a \tag{4-23}$$

where c_a is the molar concentration of A. The rate at which moles of C pass the same observer per unit area is $\mathbf{N}_c$,

$$\mathbf{N}_c = \mathbf{U}_c c_c \tag{4-24}$$

The net molar transfer rate per unit area past the observer is $(\mathbf{N}_a + \mathbf{N}_c)$. The *molar average velocity* $(\mathbf{U}_m)$ is defined as

$$\mathbf{U}_m c = \mathbf{N}_a + \mathbf{N}_c = \mathbf{U}_a c_a + \mathbf{U}_c c_c$$
$$\mathbf{U}_m = \frac{(\mathbf{N}_a + \mathbf{N}_c)}{c} = \frac{(\mathbf{U}_a c_a + \mathbf{U}_c c_c)}{c} \tag{4-25}$$

where c is the total molar concentration, $c = c_a + c_c$. The molar flux of species A with respect to a stationary observer must be larger than the diffusional flux rate $(\mathbf{J}_a)$ by the amount of A in the volume flow rate per unit area $\mathbf{U}_m$:

$$\mathbf{N}_a = \mathbf{U}_m c_a + \mathbf{J}_a = (\mathbf{N}_a + \mathbf{N}_c)\frac{c_a}{c} - \mathscr{D}_{ac}\,\mathbf{Grad}\,c_a \tag{4-26}$$

Similarly for contaminant C,

$$\mathbf{N}_c = \frac{(\mathbf{N}_c + \mathbf{N}_a)c_c}{c} - \mathscr{D}_{ac}\,\mathbf{Grad}\,c_c \tag{4-27}$$

Equations 4-26 and 4-27 relate concentration to molar flow rate. To obtain an equation involving just concentration, consider a volume element in space through which a fluid composed of several molecular species flows. A mass balance results in the following expression, hereafter called the *species continuity equation*,

$$\frac{\partial c_i}{\partial t} + \text{Div}(\mathbf{U}c_i) = \mathcal{D}_{ai}\nabla^2 c_i + S_i \tag{4-28}$$

where **Div** is the vector operation of taking the divergence of the vector $(\mathbf{U}c_i)$, S_i represents the rate of production of molecular species i by a chemical reaction, that is, a chemical source of i. In a nonreacting gas mixture, S_i is zero.

Consider steady-state diffusion in the z-direction in which both N_a and N_c and $\mathcal{D}_{ac}$ are constant. Since there is only one spatial coordinate (z), vector quantities in the z direction such as $N_{c,z}$ and $J_{c,z}$ will be written simply as N_c and J_c for the sake of brevity. Equation 4-27 becomes

$$-\int_{c_{c,1}}^{c_{c,2}} \frac{dc_c}{[cN_c - c_c(N_a + N_c)]} = \frac{1}{(\mathcal{D}_{ac}c)} \int_{z_1}^{z_2} dz \tag{4-29}$$

which when integrated and rearranged becomes,

$$N_c = \frac{N_c}{(N_a + N_c)} \left[\frac{c\mathcal{D}_{ac}}{(z_1 - z_2)} \right] \ln\left(\frac{E}{F}\right) \tag{4-30}$$

$$E = \left[\frac{N_c}{(N_c + N_a)} \right] - \left[\frac{c_{c,2}}{c} \right]$$

$$F = \left[\frac{N_c}{(N_c + N_a)} \right] - \left[\frac{c_{c,1}}{c} \right]$$

The general expression given above can be used to predict the generation of contaminants for two common processes of interest to industrial ventilation

(a) Evaporation of a pure liquid through stagnant air.
(b) Evaporation of a volatile species from a homogeneous liquid mixture through stagnant air.

4.6 DIFFUSION THROUGH STAGNANT AIR

Consider first the case in which the vapors of a volatile liquid (denoted by C) diffuse through a stagnant layer of air (denoted by A) before entering the workplace. An example of such diffusion is the evaporation from a partially filled barrel of liquid that is open at the top (Fig 4.6). Assume that room air currents sweep the vapors away from the opening of the barrel (point 2) but do not induce any motion of the air and vapor inside the barrel. Assume also that air is not absorbed by the volatile liquid and that air currents carrying away vapors establish

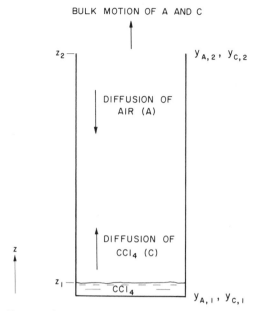

BULK MOTION OF A AND C

z_2 —

$y_{A,2}, y_{C,2}$

DIFFUSION OF
AIR (A)

DIFFUSION OF
CCl_4 (C)

z

z_1 —

CCl_4

$y_{A,1}, y_{C,1}$

Fig. 4.6 Evaporation from a partially filled barrel open at the top.

a vapor concentration at the mouth of the barrel that is negligibly small. The temperature and pressure will be assumed to be constant. Assume that radial gradients are zero and that there is only one spatial coordinate (z). Thus vector quantities will be understood to be in the z direction.

Since air is not absorbed by the liquid, $N_a \cong 0$ and N_c is essentially constant and $N_c/(N_c + N_a) \cong 1$, and

$$N_c = \left(\frac{c\mathscr{D}_{ac}}{(z_2 - z_1)}\right) \ln\left[1 - \left(\frac{c_{c,2}}{c}\right)\right] \Big/ \left[1 - \left(\frac{c_{c,1}}{c}\right)\right] \qquad (4\text{-}31)$$

Assuming that air and vapor can be described by the ideal gas equation the mol fraction (y_c) is related to concentration and partial pressure

$$P = cR_u T \qquad (4\text{-}32)$$

$$P_c = c_c R_u T \qquad (4\text{-}33)$$

$$y_c = \frac{c_c}{c} = \frac{P_c}{P}. \qquad (4\text{-}34)$$

The molar flux becomes,

$$N_c = \left[\frac{\mathscr{D}_{ac}P}{R_u T(z_2 - z_1)}\right] \ln\left[\frac{(1 - y_{c,2})}{(1 - y_{c,1})}\right] \qquad (4\text{-}35)$$

But

$$1 = y_{c,1} + y_{a,1}$$
$$1 = y_{c,2} + y_{a,2} \qquad (4\text{-}36)$$

and
$$y_{c,2} - y_{c,1} = y_{a,1} - y_{a,2}$$

$$1 = \frac{(y_{c,2} - y_{c,1})}{(y_{a,1} - y_{a,2})} \tag{4-37}$$

Using these relationships Eq. 4-35 can be rearranged,

$$N_c = \left(\frac{\mathscr{D}_{ac}P}{R_u T y_{cm}(z_2 - z_1)}\right)(y_{c,1} - y_{c,2}) \tag{4-38}$$

where y_{cm} is the *mol fraction log mean ratio*,

$$y_{cm} = \frac{(y_{c,1} - y_{c,2})}{\ln\left(\dfrac{y_{c,1}}{y_{c,2}}\right)} \tag{4-39}$$

Since $y_{c,2}$ is negligibly small,

$$y_{a,2} = 1 ; \qquad y_{a,1} = 1 - y_{c,1} \tag{4-40}$$

Thus,

$$y_{c,1} \cong y_{c,1} - y_{c,2} \tag{4-41}$$

$$y_{cm} = \frac{1 - (1 - y_{c,1})}{\ln(1 - y_{c,1})}$$

$$y_{cm} = -\frac{y_{c,1}}{\ln(1 - y_{c,1})} \tag{4-42}$$

The molar flux of C per unit area can now be written as

$$N_c = -\left\{\frac{P\mathscr{D}_{ac}}{R_u T(z_2 - z_1)}\right\} \ln(1 - y_{c,1}) \tag{4-43}$$

Equation 4-38 can also be written in terms of partial pressures and rearranged in a form encorporating a mass transfer coefficient k_G.

$$N_c = \left[\frac{\mathscr{D}_{ac}P}{R_u T P_{cm}(z_2 - z_1)}\right](P_{c,1} - P_{c,2})$$

$$N_c = k_G(P_{c,1} - P_{c,2}) \tag{4-44}$$

where P_{cm} is the partial pressure log mean difference and defined similarly to Eq 4-39.

The concentration profile that accompanies diffusion through a stagnant air layer can be found by substituting, $N_a = 0$, into Eq 4-27 and obtaining

$$N_c = -\frac{c\mathscr{D}}{1 - y_c}\frac{dy_c}{dz} \tag{4-45}$$

Since N_c is constant,

$$\frac{dN_c}{dz} = \frac{d}{dz}\left[-\frac{c\mathscr{D}}{1-y_c}\frac{dy_c}{dz}\right] = 0 \tag{4-46}$$

Integrating twice,

$$-\ln(1-y_c) = C_1(z) + C_2 \tag{4-47}$$

and evaluating the two constants from knowledge of the concentration at z_1 and z_2,

$$y_c(z_1) = y_{c,1} \qquad y_c(z_2) = y_{c,2} \tag{4-48}$$

Upon rearrangement the concentration profile becomes

$$\left(\frac{1-y_c}{1-y_{c,1}}\right) = \left(\frac{1-y_{c,2}}{1-y_{c,1}}\right)^E \qquad E = \frac{z-z_1}{z_2-z_1} \tag{4-49}$$

Illustrated Example 4.4—Evaporation From An Empty Barrel
Consider an open 55 – gal drum containing a thin layer of carbon tetrachloride at the bottom. The drum is left uncovered in the workplace. Estimate the rate at which carbon tetrachloride evaporates and enters the workplace.

Drum cross sectional area, $A = 0.25\,\text{m}^2$
Drum height, $z_2 - z_1 = 0.813\,m$
Room temperature, 80 F (26.7 C)
Diffusion coefficient of carbon tetrachloride in air, $\mathscr{D} = 0.62 \times 10^{-5}\,\text{m}^2/\text{s}$
Carbon tetrachloride partial pressure is 15.6 kPa

$$N_c = \frac{-(100\,\text{kPa})(0.62 \times 10^{-5}\,\text{m}^2/\text{s})\ln(1-15.6/100)}{(8.314\,\text{kJ/kmol K})(299.7\,\text{K})(0.813\,\text{m})}$$

$$N_c = 5.19 \times 10^{-8}\,\text{kmol/m}^2\text{s}$$

The rate of evaporation from the open drum is thus

$$\dot{m}_c = (5.19 \times 10^{-8}\,\text{kmol/m}^2\text{s})(0.25\,\text{m}^2)(154\,\text{kg/kmol})$$

$$\dot{m}_c = 7.164\,\text{g/hr}$$

If the barrel contained other volatile liquids, the evaporation rate could be found in a similar fashion. A compilation of evaporation rates for other volatile liquids is shown below. It should be noticed that the rates of evaporation related to carbon tetrachloride are in good agreement with the data shown in the emission factors for volatile hydrocarbons (Appendix A-3).

Hydrocarbon	Evaporation rate (g/hr)	Evaporation Relative to CCl_4
Toluene	1.21	0.17
Benzene	3.90	0.54
Acetone	7.95	1.11
Acetic acid	0.75	0.10
Chlorobenzene	0.23	0.03

Illustrated Example 4.5—Emissions Of HCl From Aqueous HCl Solutions

Consider the emission of HCl from drums of aqueous HCl solutions inadvertently left open in the workplace. Assume the drums are half full and that the distance $(z_2 - z_1)$ is 50 cm. Assuming room air sweeps the acid fumes away from the opening such that $y_{c,2}$ is very small with respect to unity, the emission of HCl can be estimated from Eq. 4-43. In attempting to perform the calculations two questions present themselves,

What is the diffusion coefficient of HCl in air?

Can the aqueous solution of HCl be considered to be homogeneous, that is, is the acid concentration at the air–liquid interface equal to the overall concentration?

The diffusion coefficient of polar compounds in air is difficult to predict, but for a first approximation Eq. 4-19 will be used. The computation results in

$$\mathscr{D} \text{ HCl, air} = 1.864 \times 10^{-5} \text{ m}^2/\text{s}$$

Uncertainty about the value of the acid concentration at the liquid–air interface is more difficult to resolve. If a steep concentration gradient exists in the liquid phase, a "two − film theory" will have to be used. In the case of strong electrolytes dissolved in water, the diffusion rates are those of the individual ions which move rapidly [21]. Thus a good estimate of the maximum emission rate is to neglect the liquid film and assume that the liquid HCl concentration at the liquid–air interface is equal to the overall liquid concentration. The partial pressure of HCl over an aqueous HCl solution can be obtained from ref. 18. The table below shows the result of using Eq 4-43 to estimate the HCl emission rate at 25 C for a vartiety of HCL concentrations

% HCl (by mass)	$P_{c,1}$ (mm Hg)	N_c (kmols HCl/m^{2}s)
20	0.32	6.30×10^{-10}
24	1.49	2.95×10^{-9}
28	7.05	1.40×10^{-8}
32	32.5	6.57×10^{-8}
36	142	3.11×10^{-7}
40	515	1.70×10^{-6}

To illustrate the sensitivity of HCl emissions to temperature, the computations shown above can be repeated for a 30% acid concentration with varying temperature.

T (C)	$P_{c,1}$ (mm Hg)	N_c (10^{-8} kmols/m^{2}s)
15	7.6	1.51
20	10.6	2.11
25	15.1	3.00
30	21.0	4.21
35	28.6	5.77

From these calculations it should be apparent that the vapor pressure of the volatile component is the dominant factor controlling the emission rate of the contaminant. Thus whether it be the concentration in the liquid phase, or temperature, anything that increases the vapor pressure increases the rate of evaporation.

4.7   DIFFUSION THROUGH MOVING AIR

When evaporation occurs in conjunction with the bulk motion of air over the liquid surface, the molar flux rate of C normal to a unit area is expressed as

$$N_c = k_G(P_{c,i} - P_{c,g}) \tag{4-50}$$

where (typical units)

N_c = molar flux rate $(kmols/m^2 s)$
k_G = *gas phase mass transfer coefficient* $(kmols/m^2 s\ kPa)$

$$k_G = \frac{\mathcal{D}_{ac}P}{[R_u T P_{cm}(z_{a,g} - z_{a,i})]} \tag{4-51}$$

R_u = universal gas constant $(kJ/kmol\ K)$
P = overall pressure (kPa)
P_{cm} = *log mean partial pressure difference* (kPa)

$$P_{cm} = \frac{(P_{c,g} - P_{c,i})}{\ln\left(\dfrac{P_{c,g}}{P_{c,i}}\right)} \tag{4-52}$$

$P_{c,i}$ and $P_{c,g}$ = partial pressure of the vapor at the liquid–gas interface and in the bulk gas (kPa)
$(z_{a,g} - z_{a,i})$ = effective boundary layer thickness for mass transfer (m)

Mechanical engineers (335,336) often express the molar flux rate (N_c) as a function of the difference in molar concentrations,

$$N_c = h_m(c_{c,i} - c_{c,0}) \tag{4-53}$$

where h_m is called the average *convection mass transfer coefficient* and has the units (m/s) and is related to k_G by $h_m = k_G R_u T$. This form of the equation is preferred because of the similarity of the rate of mass transfer to the average rate of heat transfer by convection $(q/A, kJ/s\ m^2)$ and its representation by the *Newton convection heat transfer* equation,

$$q/A = h_L(T_s - T_0) \tag{4-54}$$

where h_L is called the *average convection heat transfer coefficient* $(kJ/s\,K\,m^2)$ and T_s and T_0 are the average surface and far-field temperatures. If one is interested in the local rate of convection heat transfer $(q/A)_x$, that occurs at a point a distance x downstream of the leading edge and can be expressed as

$$(q/A)_x = h_x(T(x)_s - T_0) \tag{4-55}$$

where h_x is called the *local convection heat transfer coefficient*. Likewise, if one is interested in the local convection mass transfer rate $(N_{c,x})$ it can be expressed as

$$N_{c,x} = h_{x,m}[c(x)_{c,i} - c_{c,0}] \tag{4-56}$$

In the field of heat transfer, h_L and h_x are often called simply *film coefficients* where the specific meaning is implicit in the situation being analyzed.

Evaporation is an endothermic process, but it will be assumed that there is sufficient heat transfer to maintain a constant temperature. In the event the volatile liquid is a mixture, it will be assumed that mass transfer within the liquid is sufficiently rapid so as to maintain a constant concentration of each species within the liquid mixture.

For mass transfer by bulk flow it is useful to express the gas-phase mass transfer coefficient in dimensionless form,

$$Sh = C_1(Re)^{a_1}(Sc)^{b_1} \tag{4-57}$$

where C_1, a_1, and b_1 are dimensionless constants that depend on the geometry of the evaporating surface and the Reynolds number

$$Sh\ (\text{Sherwood number}) = (k_G R_u T P_{cm} L / P \mathscr{D}_{ac})$$
$$= \left(\frac{h_m L}{\mathscr{D}_{ac}}\right)\left(\frac{P_{cm}}{P}\right) \tag{4-58}$$

$$Re\ (\text{Reynolds number}) = \rho\,\frac{LU}{\mu} \tag{4-59}$$

$$Sc\ (\text{Schmidt number}) = \frac{\mu}{\rho\,\mathscr{D}_{ac}} \tag{4-60}$$

It is the practice in heat transfer to express the film coefficient in dimensionless form,

$$Nu_L = C_2\,(Re)^{a_2}\,(Pr)^{b_2} \tag{4-61}$$

where C_2, a_2, and b_2 are dimensionless constants that depend on the geometry of the surface and the Reynolds number. The other parameters are dimensionless and defined as

$$Nu_L\,(\text{Nusselt number}) = (h_L)L/k \tag{4-62}$$
$$Pr\,(\text{Prandtl number}) = \mu c_p/k \tag{4-63}$$

In all the equations above, the viscosity (μ), density (ρ), and thermal conductivity (k) pertain to air. Table 4.2 is a compilation of familiar expressions from heat transfer that will be useful in industrial ventilation.

From the theory of heat and mass transfer (see refs. 335 and 336 for details) it can be shown that the Prandtl number (Pr) is proportional to the ratio of the local velocity boundary layer thickness (δ_v) to the local thermal boundary layer thickness (δ_t),

$$\frac{\delta_v}{\delta_t} = \text{Pr}^n \qquad (4\text{-}64)$$

where n is a positive number. Similarly the Schmidt number (Sc) is proportional to the ratio of the velocity boundary layer thickness to the concentration boundary layer thickness (δ_m),

$$\delta_v/\delta_m = \text{Sc}^n \qquad (4\text{-}65)$$

The ratio of the Schmidt and Prandtl numbers is called the *Lewis number* (Le),

$$\frac{\delta_t}{\delta_m} = \left(\frac{\text{Sc}}{\text{Pr}}\right)^n = \text{Le}^n \qquad (4\text{-}66)$$

Heat and mass transfer processes are considered to be analogous since both can be described by fundamental equations containing convection and diffusion terms of similar form. When these equations are made dimensionless (355) one finds that each equation is related to the velocity field through the Reynolds number (Re_L) and the temperature and concentration are related through the Prandtl and Schmidt numbers, respectively. Because the dimensionless equations are similar it is concluded that heat and mass transfer all analogous processes. If the respective dimensionless parameters are equal, the solution of one equation is interchangeable with the solution of the other equation providing the boundary conditions are the same.

The *Reynolds analogy* is the name given to the condition if both the Prandtl and Schmidt numbers are unity and the solutions of both equations are identical. In actuality the Schmidt and Prandtl numbers are not unity, nor even equal to each other. Nevertheless, the Reynolds analogy is so useful that it is often assumed to be valid. As a result, Eqs. 4-57 and 4-61 can be set equal to one another and,

$$C_1 = C_2 : b_1 = b_2 : a_1 = a_2$$
$$\text{Sh} = \text{Nu}(\text{Sc}/\text{Pr})^{b_1} \qquad (4\text{-}67)$$

Thus the gas phase mass transfer coefficient k_G can be written as

$$k_G = \text{Nu}\left(\frac{\mathscr{D}_{ac}}{L}\right)\left(\frac{\text{Sc}}{\text{Pr}}\right)^{b_1}\left(\frac{P}{P_{cm}}\right)\frac{1}{R_u T} \qquad (4\text{-}68)$$

TABLE 4.2 Summary of Convection Heat Transfer Correlations

1. *Parameters*

$$\text{Nu}_x = xh_x/k \qquad \text{Nu}_L = Lh_L/k \qquad \text{Pr} = \mu c_p/k$$

$$\text{Re}_x = x\rho U_0/\mu \qquad \text{Re}_L = L\rho U_0/\mu \qquad \text{Re}_D = D\rho U_0/\mu$$

2. *Flow Over a Horizontal Flat Plate*

Laminar ($\text{Pr} > 0.6$, $\text{Re}_x < 5 \times 10^5$)

Local:	$\text{Nu}_x = 0.332\,\text{Re}_x^{1/2}\,\text{Pr}^{1/3}$
Average:	$\text{Nu}_L = 0.664\,\text{Re}_L^{1/2}\,\text{Pr}^{1/3}$

Turbulent ($0.6 < \text{Pr} < 60$, $5 \times 10^5 < \text{Re}_x < 10^8$, $\text{Re}_L < 10^8$)

Local:	$\text{Nu}_x = 0.0296\,\text{Re}_x^{4/5}\,\text{Pr}^{1/3}$
Average:	$\text{Nu}_L = (0.037\,\text{Re}_L^{4/5} - 871)\text{Pr}^{1/3}$

3. *Fully Developed Flow in Duct of Diameter D*

Laminar: (constant wall temperature, $\text{Pr} > 0.6$): $\text{Nu}_D = 3.66$

Turbulent: ($L/D > 60$, $\text{Re}_D > 10,000$, $0.6 < \text{Pr} < 160$): $\text{Nu}_D = 0.023\,\text{Re}_D^{4/5}\,\text{Pr}^{1/3}$

4. *External Flow*

(a) Flow Across Bodies of Various Cross Sectional Areas

$$\text{Nu}_L = C\,\text{Re}_L^m\,\text{Pr}^{1/3} \qquad (\text{Pr} > 0.7)$$

Body	Re_L	C	m
Horizontal cylinder: (L is D)	0.4–4	0.989	0.330
	4–40	0.911	0.385
	40–4000	0.683	0.466
	4000–40,000	0.193	0.618
	40,000–400,000	0.027	0.805
Square (L by L): (U_0, 90-degrees to side)	5000–100,000	0.246	0.588
Square (L by L): (U_0, 45-degrees to side)	5000–100,000	0.102	0.675
Vertical plate (height L): (U_0, 90-degrees to plate)	4000–15,000	0.288	0.731

Stationary Sphere (L is D)
($0.71 < \text{Pr} < 380$, $3.5 < \text{Re}_L < 7.6 \times 10^4$ and $1.0 < \mu_0/\mu_s < 3.2$)

$$\text{Nu}_D = 2 + (0.4\,\text{Re}_L^{1/2} + 0.06\,\text{Re}_L^{2/3})\,\text{Pr}^{0.4}\,(\mu_0/\mu_s)^{1/4}$$

Source: Abstracted from refs. 61, 335, and 336.

and the convection mass transfer coefficient h_m can be written as

$$h_m = \text{Nu}\left(\frac{\mathscr{D}_{ac}}{L}\right)\left(\frac{\text{Sc}}{\text{Pr}}\right)^{b_1}\left(\frac{P}{P_{cm}}\right) \tag{4-69}$$

Illustrated Example 4.6—Evaporation from an Open Vessel

Consider evaporation from a horizontal layer of carbon tetrachloride at room temperature over which air passes at speeds typically encountered in the workplace. The layer of liquid could be a "spill" on a workroom floor on a large open vessel filled to the top. Assume the following physical properties.

L (width of the spill) = 2 m
U (room air current) = 2 m/s
T_c (carbon tetrachloride) = T_a (air) = 25 C
μ/ρ (of air at 25 C, 1 atm) = 1.5×10^{-5} m²/s
Pr (air) = 0.708
ρ (air) = 1.2 kg/m³
$\mathscr{D}_{ac}$ (diffusion coefficient of C and A) = 0.62×10^{-5} m²/s
Sc (air) = $(\mu/\rho)/\mathscr{D}_{ac}$ = 2.419
k (air) = 26.24×10^{-6} kJ/s m K
$c_{p,a}$ (air) = 1.0057 kJ/kg K

The Reynolds number based on length (L) is

$$\text{Re}_L = UL/(\mu/\rho) = (2)(2)/1.5 \times 10^{-5} = 2.66 \times 10^5$$

The flow can be characterized as near the upper end of laminar flow. The average Nusselt number for laminar flow over a horizontal surface can be found from the equations in Table 4.2.

$$\text{Nu}_L = 0.664 \, \text{Pr}^{0.33} \, (\text{Re}_L)^{0.5} = (0.664)(0.708)^{0.33}(2.66 \times 10^5)^{0.5} = 305$$

The log mean partial pressure difference

$$P_{cm} = \frac{(P_{c,g} - P_{c,i})}{\ln(P_{c,g}/P_{c,i})}$$

can be simplified by expanding the log term as a Taylor series and retaining only the first term,

$$\ln\left(\frac{P_{c,g}}{P_{c,i}}\right) = \frac{(P_{c,g} - P_{c,i})}{P_{c,i}}$$

The vapor concentration in the bulk gas will be neglected ($P_{c,g} = 0$) and the vapor partial pressure at the interface will be assumed to be equal to the carbon tetrachloride saturation pressure at 23 C, see Appendix A-8. After substitution, the log mean partial pressure difference reduces to

$$P_{cm} = P_{c,i} = 100 \text{ mm Hg} \times (101 \text{ kPa}/760 \text{ mm Hg}) = 13.3 \text{ kPa}$$

Setting $b_1 = 0.33$ in Eq. 4-68, the average gas-phase mass transfer coefficient becomes,

$$k_G = \text{Nu}(\text{Sc}/\text{Pr})^{b_1}P\mathcal{D}_{ac}/(R_u TLP_{cm})$$

$$= \frac{(305)(2.419/0.708)^{0.33}(101 \text{ kPa})(0.62 \times 10^{-5} \text{ m}^2/\text{s})}{(8.314 \text{ kJ}/\text{kmol K})(298 \text{ K})(13.3 \text{ kPa})(2 \text{ m})}$$

$$k_G = 4.36 \times 10^{-6} \text{ (kmol m/ kJ s)(kJ/kPa m}^3)$$

$$k_G = 4.36 \times 10^{-6} \text{ kmol/m}^2 \text{ kPa s}$$

The molar evaporation rate per unit area can be found from Eq. 4–56,

$$N_c = k_G(P_{c,i} - P_{c,g}) = k_G P_{c,i} = 5.803 \times 10^{-5} \text{ kg mol/m}^2 \text{ s}$$

The evaporation rate per unit area for a spill is over 10,000 times larger than evaporation from the partially empty barrel (Illustrated Example 4.4). If the spill is 2 m in diameter, the total evaporation rate of carbon tetrachloride is

$$\dot{m}_c = (5.803\text{x}10^{-5} \text{ kg mol/m}^2 \text{ s})(3.14 \text{ m}^2)(153.6 \text{ kg/kg mol}) = 28 \text{ g/s}$$

Illustrated Example 4.7   Evaporation From A suspended "Wet" Workpiece
Consider evaporation from the surface of a spherical body 0.5 m in diameter that was just removed from a tank of carbon tetrachloride and suspended in air to allow the excess liquid to drip off. Assume the same flow conditions as in Illustrated Example 4.5. Compare the molar flux rate with that in Illustrated Example 4.6.

$$\text{Re} = UD/(\mu/\rho) = (0.5)(2)/(1.5 \times 10^{-5}) = 66{,}667$$

Table 4.2 indicates that for this situation the Nusselt number can be calculated from

$$\text{Nu} = 2 + (0.4 \text{ Re}^{0.5} + 0.06 \text{ Re}^{0.67}) \text{ Pr}^{0.4}$$

$$\text{Nu} = 181$$

The Reynolds analogy will be used to compute the evaporation rate. Since the Nusselt number depends on the Prandtl number to the power (b = 0.4), it will be assumed that an overall relationship of the form Eqs. 4-57 and 4-61 will also have $b_1 = 0.4$. The mass transfer coefficient can be found from

$$k_G = \text{Nu } (\text{Sc}/\text{Pr})^{0.4}PD_{ac}/R_u TDP_{cm}$$
$$= (181)(2.419)^{0.4}(101)(0.62)/(0.708)^{0.4}(8.314)(298)(0.5)(13.3)(10^{-5})$$
$$= 1.124 \times 10^{-5} \text{ (kmol m/kJ s)(kJ/kPa m}^3)$$
$$= 1.124 \times 10^{-5} \text{ kmol/m}^2 \text{ s kPa}$$

The molar flux rate is

$$N_c = k_G(P_{c,i} - P_{c,g}) = (1.124)(13.3)/100,000 = 1.5 \times 10^{-4}4 \text{ kmols/m}^2 \text{ s}$$

The evaporation rate is

$$\dot{m}_c = N_c \pi D^2 = (1.5)(3.14)(.25)/10^4 = 1.178 \times 10^{-4} \text{ kmols/s} = 18.09 \text{ g/s}$$

4.8 SINGLE FILM THEORY FOR MULTICOMPONENT LIQUIDS

Since mixtures of liquids will be discussed in the next few sections, it will be helpful to summarize relationships describing the concentration of individual species for mixtures. For simplicity, it will be assumed that the liquids are ideal. An *ideal solution* is one in which no energy is generated (or consumed) when the individual species are mixed. Similarly, there is no change in volume accompanying the mixing. Thus the total volume (V_t) of the mixture containing N distinct molecular species is equal to the sum of the volumes of the individual components,

$$V_t = \sum_i V_i \tag{4-70}$$

The total mass (m_t) of the mixture and the total number of moles (n_t) in the mixture can be expressed as

$$m_t = \sum_i m_i$$

$$n_t = \sum_i n_i = \sum_i \left(\frac{m}{M}\right)_i \tag{4-71}$$

From Eq 4-71 define the *average molecular weight* M_{avg} such that

$$m_t = n_t M_{avg} = \sum_i n_i M_i$$

$$M_{avg} = \sum_i \left(\frac{n_i}{n_t}\right) M_i = \sum_i x_i M_i \tag{4-72}$$

Divide m_t by the total volume (V_t), and an *average density* (ρ_{avg}) can be defined as

$$\rho_{avg} = \frac{m_t}{V_t} = \sum_i \left(\frac{m_i}{V_t}\right) = \sum_i \rho_i \tag{4-73}$$

where ρ_i is the *partial density* which can be interpreted as the density species i would have if it alone occupied the entire volume. Dividing the total number of moles (n_t) by the total volume (V_t) an *average molar concentration* (c_{avg}) or *total molar concentration* (c_t) can be defined,

$$c_{avg} = c_t = \frac{n_t}{V_t} = \sum_{i=1} \left(\frac{n_i}{V_t} \right) = \sum_{i=1} \left(\frac{m_i}{M_i V_t} \right) = \sum_{i=1} \left(\frac{m_i}{V_t} \right) \frac{1}{M_i}$$

$$c_{avg} = c_t = \sum_i \left(\frac{\rho_i}{M_i} \right) = \sum_i c_i \qquad (4\text{-}74)$$

where c_i is the *partial molar concentration* which can be interpreted as the concentration species i would have if it alone occupied the entire volume.

For a single species, the molar concentration is related to the density by

$$c = \frac{n}{V} = \frac{(m/M)}{V} = \frac{(m/V)}{M} = \frac{\rho}{M} \qquad (4\text{-}75)$$

$$M = \rho/c$$

For a mixture, the *average molecular weight* (M_{avg}) can also be defined as,

$$M_{avg} = \frac{\rho_t}{c_{avg}} = \frac{\rho_t}{c_t} = \left(\frac{m_t}{V_t} \right) \bigg/ \sum_i \frac{(m/M)_i}{V_t}$$

$$\frac{1}{M_{avg}} = \sum_i \frac{(m_i/m_t)}{M_i} = \sum_i \left(\frac{f}{M} \right)_i \qquad (4\text{-}76)$$

where f_i is the *mass fraction* defined as

$$f_i = \frac{m_i}{m_t} = \frac{m_i}{\sum m_i} = \frac{(m_i/V_t)}{\sum (m_i/V_t)_i}$$

$$f_i = \frac{\rho_i}{\rho_t} \qquad (4\text{-}77)$$

The *mol fraction* is defined as

$$x_i = \frac{n_i}{n_t} = \frac{n_i}{\sum n_i} = \frac{(n_i/V_t)}{\sum (n_i/V_t)} \qquad (4\text{-}78)$$

$$x_i = \frac{c_i}{c_t}$$

Mol fraction and mass fraction are related by

$$f_i = \frac{m_i}{m_t} = \frac{\left(\dfrac{m}{M} \right)_i}{\left[\dfrac{m_t}{M_{avg}} \right] \left[\dfrac{M_{avg}}{M_i} \right]}$$

$$f_i = \left(\frac{n_i}{n_t} \right) \left(\frac{M_i}{M_{avg}} \right) = x_i \left(\frac{M_i}{M_{avg}} \right) \qquad (4\text{-}79)$$

A mixture of liquids in which components are separate molecular species is called a *multicomponent liquid*. If any of the species are volatile, they will evaporate at a unique rate. The composition of the remaining liquid changes because of such preferential evaporation. In order to assess the toxicity of evaporating multicomponent liquids, it will be necessary to estimate the rates at which individual species enter the atmosphere. The task is difficult because it is difficult to express the liquid's vapor pressure as a function of its changing composition. As the multicomponent liquid evaporates, the more volatile liquids are lost first and the mixture's total vapor pressure falls. If the liquid is a single molecular species, the resistance to mass transfer is entirely in the air phase because there is no necessity for material to diffuse to the liquid–air interface. Evaporation is entirely controlled by the transfer of vapor through the single air film, that is, *single film*, air resistance. If the liquid consists of several liquid species, each species must diffuse to the air–liquid interface before it can evaporate. Unfortunately, the species diffuse toward the interface at different rates and the composition on the liquid side of the interface will not be equal to the composition at lower depths. Once at the surface, each evaporating species must diffuse through a layer of air and a mixture of vapors. Mass transfer of this sort is analyzed by *two-film theory*. There are situations when the liquid film resistance is small compared with the air phase resistance and the mass transfer can be approximated by *single-film theory*.

The following analysis illustrates the fundamental aspects of evaporating multicomponent liquids, and is similar to analyses suggested by Drivas (110) and Stiver et al. (337). Consider a circular pool of liquid over which air passes at a velocity U. The diameter D_0 and depth h of the pool may vary slowly with time, but will be neglected at this time. The liquid in the pool is composed of a mixture of liquid species having the following properties:

(a) *Ideal solution*—molar properties of each species are equal to their partial molal properties, for example, the total volume of the mixture is equal to the sum of the volume of the components (no change in volume upon mixing).

(b) At the interface between a multicomponent liquid and a multicomponent vapor, the partial pressure of molecular species i in the gas phase (P_i) is,

$$P_i = y_i P = x_i P_{i(sat)}$$

where $P_{i(sat)}$ is the saturation pressure of pure species i at the temperature of the liquid

(c) Negligible liquid film resistance, well-mixed conditions within the liquid (uniform concentrations of species within the liquid phase).

(d) Negligible concentration of vapor in the air far above the liquid surface, Fig. 4.7.

(e) The total number of mols per volume of liquid is constant; the temperature of the liquid is equal to the temperature of the air (T_a) and the temperatures are constant.

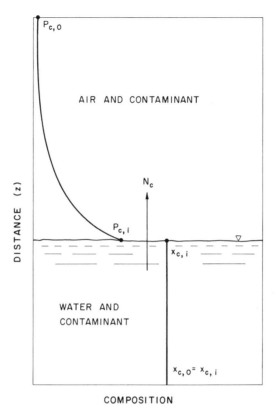

Fig. 4.7 Concentration profile for one-film resistance.

Liquid mixtures satisfying these assumptions require some sort of internal mixing mechanism to eliminate concentration gradients within the liquid. The liquid mixture may also be one in which the concentration of solute (evaporating species) is small with respect to the solvent (the remaining liquid that undergoes little evaporation). These assumptions may also be satisfied during the early stages of evaporation when the total number of mols of solution is large (i.e., the relative change in the total number of mols is small). These assumptions are restrictive, but the evaporation of many industrial liquids satisfy them.

The evaporation rate (kmols/s m^2) of molecular species i can be found from Eq. 4-50. Neglecting the concentration of species in the far-field ($P_{c,0}$), Eq. 4-50 becomes

$$N_i = k_G\left(\frac{c_i}{c_t}\right)P_i = k_G x_i P_i \qquad (4\text{-}80)$$

where x_i is the mol fraction of species i in the liquid and P_i is the saturation pressure of pure species i at the temperature of the liquid. Providing the mol fraction of species i (x_i) does not change, the evaporation rate (N_i) is constant. However, in a multicomponent liquid the most volatile species will evaporate faster than less volatile species and the ratio of volatile to less volatile will decrease with time. Consequently, while the total number of mols of liquid may decrease slowly with time, the mol fraction of the least volatile will increase slowly with time simultaneously with a decrease in the mol fraction of the most volatile species. The following analysis shows how to estimate the instantaneous evaporation rate of any molecular species (N_i).

Define a control volume in a layer of liquid of cross sectional area A and thickness of h of the layer. From the conservation of mass,

$$\frac{dc_i}{dt} = -\frac{k_G c_i P_i}{(hc_t)} \tag{4-81}$$

where (typical units)

P_i = saturation pressure of pure species i evaluated at temperature T_a (atm)

c_i = number of mols of species i per volume of liquid

k_G = mass transfer coefficient, assumed to be constant (mols/m^2 atm s)

c_t = total number of mols per volume of liquid

h = thickness of liquid pool (assumed to decrease very slowly with time)

Integrating Eq 4-81 yields the concentration of i at any instant

$$c_i(t) = c_{i,0} \exp\left[\frac{-k_G P_i t}{(hc_t)}\right] \tag{4-82}$$

where $c_{i,0}$ is the initial molar concentration of species i in the liquid. The total mass of liquid per unit area (m_t, kg/m^2) can be expressed as

$$m_t = \sum_i (hc_i M_i) \tag{4-83}$$

where the summation is taken over the total number of molecular species and M_i is the molecular weight of species i. The total rate of evaporation in which several species evaporate at individual rates is obtained by differentiating Eq. 4-83 with respect to time to obtain the total emission rate. The overall evaporation rate ($\dot{m}_t$, kg/s m$_2$) is equal to $-dm_t/dt$

$$\frac{dm_t}{dt} = -\dot{m}_t = \sum_i hM_i \frac{dc_i}{dt} \tag{4-84}$$

Substituting Eq. 4-81,

$$\dot{m}_t = \sum_i \left(\frac{k_G P_i c_i M_i}{c_t}\right) \tag{4-85}$$

Replace c_i by Eq. 4-82

$$\dot{m}_t = \sum_i M_i k_G P_i \left(\frac{c_{i,0}}{c_t}\right) \exp\left[-\frac{k_G P_i t}{(hc_t)}\right] \tag{4-86}$$

Multiply Eq 4-86 by the initial mass per unit area of the liquid $m_{t,0}$ where

$$m_{t,0} = hc_t M_{avg,0} \qquad (4\text{-}87)$$

$$M_{avg,0} = \sum_i x_{i,0} M_i \qquad (4\text{-}88)$$

where $x_{i,0}$ is the initial mol fraction of species i. Rearrange, simplify, and obtain,

$$\dot{m}_t = E \sum_i \left\{ x_{i,0} M_i \exp\left[-\frac{k_G P_i t}{(hc_t)} \right] \right\}$$

$$E = \frac{m_{t,0} k_G P_i}{c_t h M_{avg,0}} \qquad (4\text{-}89)$$

For *nonideal solutions*, Raoult's law can modified by including the liquid *activity coefficient* (Γ_i) so that Eq. 4-81 can be rewritten as

$$-\frac{dc_i}{dt} = \frac{k_G c_i \Gamma_i P_i}{(hc_t)} \qquad (4\text{-}90)$$

The remainder of the derivation proceeds in the same fashion, with the activity coefficient carried through as a known parameter.

The mass transfer coefficient (k_G) can be found by the methods described in the previous section. For large spills of petroleum on bodies of water, spills of the order of hundreds of meters in diameter, Drivas (110) uses the expression

$$k_G = \frac{0.0292 U^{0.78}}{R_u T D_0^{0.11} Sc^{0.67}} \qquad (4\text{-}91)$$

for all the evaporating species. The parameters must use the following units,

k_G = gmol/m^2 atm hr

U = air velocity (m/hr)

D_0 = diameter of the spill (m)

Sc = gas phase Schmidt number (use a mass weighted average for a liquid mixture)

R_u = universal gas constant (8.206×10^{-5} atm m^3/gmol K)

T = temperature (K)

4.9   TWO-FILM THEORY FOR MULTICOMPONENT MIXTURES

The evaporation of contaminants from a multicomponent liquid mixture is complicated by the fact that each species must be transported from the interior of the liquid through a boundary layer inside the liquid as well as through a boundary layer on the air side of the liquid–air interface (Fig. 4.8). Thus mass is transferred through two "*films or resistances.*" One film may offer more resistance than the other, but until this is known to be true, both films have to be dealt with as if they both were equally as important. If the mass of liquid and air are

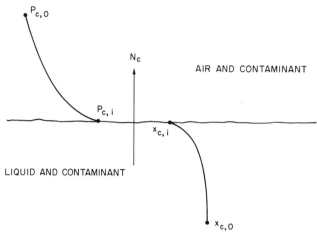

Fig. 4.8 Concentration profiles for two-film resistance.

large, such as evaporating a volatile organic compound from a lagoon of industrial wastes, a steady state mass transfer rate can be achieved. If on the other hand the mass of liquid and/or air is small, the analysis is complicated by the depletion of material in the liquid phase and the accumulation in the gas phase. Under these conditions, the concentration profile varies with time as shown in Fig 4.9. Examples of this phenomena are evaporating volatile organic compounds from liquid drops in a spray booth, bathroom shower (252), and car wash. Only steady-state mass transfer through two films shown in Fig. 4.8 will be considered at this time. Readers should consult texts (17,20,21) and professional papers on the more general subject (14,109,111)

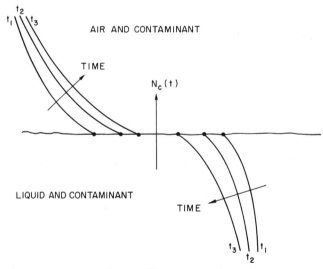

Fig. 4.9 Instantaneous concentration profiles for two-film resistance showing depletion in the liquid and accumulation in the gaseous phase.

Before postulating equations to describe the mass transfer, it is essential to understand the equilibrium of a liquid–air-contaminant system. Shown in Fig 4.10 is a cylinder equipped with a frictionless piston. The cylinder contains water and air, each of which contains a certain amount of dissolved contaminant C. The vessel is immersed in a constant temperature bath (T_0). Water containing contaminant C is placed in the vessel along with (pure) air. Some of contaminant evaporates from the water and enters the air. Eventually the system comes to equilibrium. At equilibrium the mol fraction of C in water (x_c) and the partial pressure of C (P_c) in the air are measured. The experiment is repeated for a variety of mol fractions of C in water and temperatures (T_0). A typical graph of the partial pressure versus the liquid mol fraction at equilibrium is shown in Fig 4.11. If the temperature is varied and the experiment repeated, a series of equilibrium lines can be found in which at any given partial pressure, the equilibrium liquid mol fraction is inversely proportional to the temperature.

For purposes of relevance to air pollution and industrial ventilation, only the portion of the equilibrium curve in the vicinity of the origin will be considered. In this range the equilibrium isotherms satisfy *Henry's law*, which can be expressed in several forms depending on the units chosen to represent the solute concentration.

$$P_c = Hc_c = Hx_c c_L = H'x_c \qquad (4\text{-}92)$$

$$Hc_L = H'$$

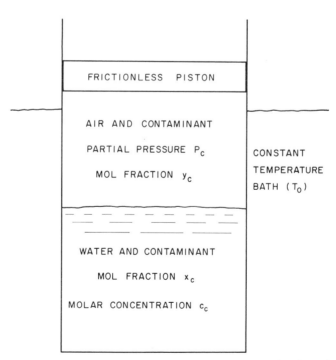

Fig. 4.10 Multicomponent equilibrium for contaminant in air and water.

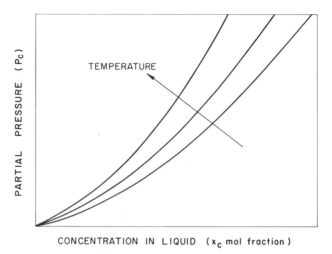

PARTIAL PRESSURE (P_c)

TEMPERATURE

CONCENTRATION IN LIQUID $(x_c$ mol fraction$)$

Fig. 4.11   Absorption equilibrium isotherm for contaminant in air and water.

The parameters H and H' are the slopes of the equilibrium curve and are called the *Henry's law constants*. The units of H are (atm m^3/kmol) and the units of H' are atm. Alternatively, the units of pressure could also be N/m^2. The parameters c_c and c_L are the solute and liquid molar concentrations. Henry's constant is a physical property that reflects how a molecular species partitions itself between air and water. Chemicals with low values of Henry's constant are those that partition themselves more in the liquid phase than in the air phase, while higher values of the Henry's constant suggest the opposite. Henry's constant can be predicted from thermodynamic data (338) or computed from experimental measurements of vapor pressure in air and solubility in water (334). Appendix A-9 tabulates Herny's law constants (H') and diffusion coefficients for many contaminants of importance to industrial ventilation.

If a liquid containing a contaminant is brought into contact with air containing the contaminant, there will be evaporation if the coordinates (P_c, x_c) lie below the equilibrium line on Fig. 4.11. In terms of the variables shown in Fig. 4.11, the mass transfer rate can be expressed as

$$N_c = k_L(x_{c,0} - x_{c,i})c_L = \frac{k_G(P_{c,i} - P_{c,0})}{R_u T} \tag{4-93}$$

unfortunately both the properties at the interface ($x_{c,i}$ and $P_{c,i}$) and the mass transfer coefficients (k_L and k_G, are unknown. (Note also that k_G in Eq. 4-93 differs from k_G in Eq. 4-50 by the amount $R_u T$). An alternative expression for the mass transfer rate can be postulated that is more useful,

$$N_c = K_L(x_{c,0} - x_c^*)c_L = \frac{K_G(P_c^* - P_{c,0})}{R_u T} \tag{4-94}$$

where K_L and K_G are called the *overall mass transfer coefficients* in the liquid and gas phases, and the *star states* (x_c^* and P_c^*) are hypothetical values defined by the equilibrium diagram, Fig. 4.12:

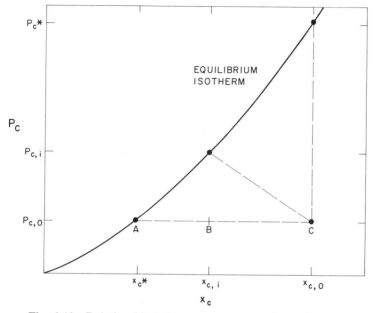

Fig. 4.12 Relationship between star state and actual state.

x_c^* = the hypothetical mol fraction in the liquid phase corresponding to the actual far-field gas phase partial pressure $P_{c,0}$

P_c^* = the hypothetical partial pressure in the gas phase corresponding to the actual bulk liquid mol fraction $x_{c,0}$

It is important to recognize that the star states are hypothetical states which can only be defined by the graphical relationships shown in Fig. 4.12. The overall mass transfer coefficient now can be expressed by using the following construction:

$$\left(x_{c,0} - x_c^*\right) = \left(x_{c,0} - x_{c,i}\right) + \left(x_{c,i} - x_c^*\right) \tag{4-95}$$

From Fig. 4.12 it can be seen that $x_{c,i}$ is related to $P_{c,i}$ by Henry's law, as is x_c^* related to $P_{c,0}$,

$$P_{c,i} = H' x_{c,i} \tag{4-96}$$

$$P_{c,0} = H' x_c^* \tag{4-97}$$

Thus

$$\left(x_{c,0} - x_c^*\right) = \left(x_{c,0} - x_{c,i}\right) + \frac{\left(P_{c,i} - P_{c,0}\right)}{H'} \tag{4-98}$$

The differences contained in the three bracketed terms above can be replaced by expressions involving the molar flux N_c by using Eqs. 4-82 to 4-94. In so doing one obtains

$$\frac{N_c}{c_L K_L} = \frac{N_c}{k_L c_L} + \frac{P_{c,i} - P_{c,0}}{H c_L}$$

$$\frac{N_c}{K_L} = \frac{N_c}{k_L} + \frac{R_u T N_c}{H k_G} \tag{4-99}$$

Thus the overall mass transfer coefficient (K_L) is related to the specific mass transfer coefficients by the following

$$\left(\frac{1}{K_L}\right) = \left(\frac{1}{k_L}\right) + \left(\frac{R_u T}{H k_G}\right) \tag{4-100}$$

It is useful to treat each one of the terms in Eq. 4-100 as a resistance. The overall resistance to mass transfer ($1/K_L$) is equal to the resistance of the liquid film ($1/k_L$) plus the resistance of the air film ($R_u T / H k_G$). The mass transfer rate can now be written as

$$N_c = K_L \left(c_{c,0} - \frac{P_{c,0}}{H}\right) \tag{4-101}$$

Equation 4-101 has received considerable attention in the field of air pollution where it has been used to estimate the rate of evaporation of volatile organic hydrocarbons from ocean spills and industrial waste lagoons. Mackay and Yeun (115) recommend the following expressions to estimate the liquid and gas mass transfer coefficients k_G and k_L in the units of m/s,

$$k_G = 1.0 \times 10^{-3} + 46.2 \times 10^{-3} U^* \text{Sc}_G^{-0.67} \tag{4-102}$$

$$k_L = 1.0 \times 10^{-6} + 34.1 \times 10^{-4} U^* \text{Sc}_L^{-0.5}; \qquad U^* > 0.3 \tag{4-103}$$

$$k_L = 1.0 \times 10^{-6} + 144 \times 10^{-4} U^{*2.2} \text{Sc}_L^{-0.5}; \qquad U^* < 0.3 \tag{4-104}$$

The parameter U^*(m/s) is called the *friction velocity* and is defined as

$$U^* = \sqrt{\frac{\tau_a}{\rho_a}} \tag{4-105}$$

where τ_a is the shear stress of air passing over the liquid surface. In the outdoor environment where the liquid surface is large and evaporation is driven by the wind, Mackay and Yeun (115) recommend that friction velocity (U^* in the units of m/s) can be computed from

$$U^* = U_{10} \sqrt{6.1 + 0.63 U_0} \tag{4-106}$$

where (U_{10}) is the wind speed (m/s) 10 m above the liquid surface. In the laboratory (or indoor environment) they recommend that U^* (in the units of m/s) can be computed from

$$U^* = 0.02 U_0^{1.5} \tag{4-107}$$

where U_0 (m/s) is the room air speed.

From laboratory experiments with aqueous solutions containing mixtures of volatile hydrocarbons, Mackay and Yeun (115) found that each molecular species evaporates at rates independent of the other species. If however, the wind speeds are large and produce surface waves or if the liquid contains adsorbing suspended solids the expressions used above may prove to be inappropriate.

Illustrated Example 4.8—Volatilization of Benzene from Industrial Cooling Water

Chemicals are added to the cooling water in an industrial process to prevent biological activity from occurring. Benzene, ($M = 78$) is among the chemicals added and there is concern that unsafe amounts of benzene may evaporate from the cooling water and enter the air breathed by workers. You have been asked by your supervisor to estimate the benzene concentration in an air stream passing over an open channel containing the cooling water that is virtually motionless. The water and air are at 25 C.

The open channel is 1 m wide and 5 m long and is enclosed by a hood to prevent material from falling into the open channel. The hood is 10 cm above the free surface (air–liquid interface) and air passes through the open space (between hood and liquid surface) in the long direction at an average velocity (U_0) of 5 m/s. The average upstream benzene concentration in the water is 1 g/m^3 of water. For a first approximation assume that the benzene concentration below the water surface is 1 g/m^3 and decreases as one approaches the water–air interface. If a worker stands near the downstream end of the open channel and breathes the air passing between the hood and the upper surface of the cooling water, is a worker exposed to benzene vapor at concentrations above OSHA PEL values?

It will be assumed that the partial pressure of benzene in air ($P_{c,0}$) is negligibly small (an assumption whose validity must be checked at the end of the problem). The overall mass transfer coefficient (K_L) will be obtained from Eq. 4-100. The value of Henry's law constant (H') for benzene in air is 3.055×10^7 N/m^2. The Schmidt numbers of benzene in air and water are

$$\mathrm{Sc_G} = [\mu/\rho\mathscr{D}]_G = (1.5 \times 10^{-5}\ \mathrm{m^2/s})/(0.77 \times 10^{-5}\ \mathrm{m^2/s}) = 1.95$$

$$\mathrm{Sc_L} = [(\mu/\rho\mathscr{D}]_L$$

$$= (10^{-3}\ \mathrm{kg/m\ s}/1000\ \mathrm{kg/m^3})/(1.02 \times 10^{-9}\ \mathrm{m^2/s}) = 980$$

The friction velocity (U^*) is

$$U^* = 0.02(U_0)^{1.5} = (0.02)(5)^{1.5} = 0.2236\ \mathrm{m/s}$$

From Eqs. 4-102 and 4-104,

$$k_L = 10^{-6} + (144/10^4)(0.2236)^{2.2}/(980)^{0.5} = 18/10^6\ \mathrm{m/s}$$

$$k_G = 10^{-3} + (46.2/10^3)(0.2236)/(1.95)^{0.67} = \frac{7.68}{10^3}\ \mathrm{m/s}$$

The value of the Henry's law constant (H) in Eq. 4-100 is

$$H = H'/c_L$$

$$= (3.055 \times 10^7 \, N/m^2)/[(1000 \, kg/m^3)(kmol/18 \, kg)]$$

$$= 549,900 \, N \, m/kmol$$

$$= (549,000 \, N \, m/kmol)(atm/100 \, kPa)(kPa/1000 \, N/m^2)$$

$$= 5.5 \, atm \, m^3/kmol$$

The overall mass transfer coefficient K_L is

$$\frac{1}{K_L} = \frac{1}{k_L} + \frac{R_u T}{H k_G}$$

$$= \frac{10^6}{18} \, (s/m)$$

$$+ (0.082 \, m^3 \, atm/kmol \, K)(298 \, K)/(5.5 \, atm \, m^3/kmol)(7.68/10^3 \, m/s)$$

$$= 55,555 + 578 = 56,133 \, s/m$$

$$K_L = 1.78 \times 10^{-5} \, m/s$$

Compare the values of the mass transfer resistances of the liquid and gas films,

$$\text{Gas film: } R_u T/k_G H = 578 \, s/m$$

$$\text{Liquid film: } 1/k_L = 55,555 \, s/m$$

It can be seen than that the resistance in the liquid phase is large compared to the resistance in the gas phase. Thus the liquid phase film governs the rate of mass transfer more than the gas phase film. If one neglects the liquid phase film resistance, the mass transfer can be analyzed on the basis of the single-film theory in Section 4.8

To find the evaporation rate from Eq. 4-101, the average benzene concentration in the liquid ($c_{c,0}$) has to be found.

$$c_{c,0} = (1 \, gm/m^3)(1 \, kmol/78000 \, g) = 1.282 \times 10^{-5} \, kmol/m^3$$

The total surface area (A_s) over which evaporation occurs is $5 \, m^2$. The evaporation rate (kg/s) from the free surface is

$$\dot{m}_c = A_s M K_L c_{c,0} = (5 \, m^2)(1.78/10^5) \, (m/s)(1.282/10^5) \, (kmol/m^3)(78 \, kg/kmol)$$

$$= 8.9 \times 10^{-8} \, kg/s$$

If the benzene concentration is uniformly distributed in the air leaving the hood,

$$c_c(avg) = \dot{m}_c/(U_0 A)$$
$$= (8.9 \times 10^{-8} \, kg/s)/[(5 \, m/s)(0.1 \, m^2)$$
$$= 1.78/10^4 \, g/m^3$$
$$= 0.178 \, mg/m^3$$

In terms of PPM this concentration is equivalent to

$$PPM = c_c(avg, mg/m^3)24.5/78 = 0.056 \ PPM = 56 \ PPB$$

Thus the assumption that $P_{c,0}$ is negligibly small is valid. Since the evaporation rate is small, the assumption that the average benzene concentration in the water is constant is also valid.

4.10 EVAPORATION OF DROPLETS

Drops of liquid are formed in countless industrial operations, for example

Surface coating
Transferring liquids between vessels, in open channels
Scrubbing, washing, and cleaning
Formulating liquid mixtures, chemical reactions
Electroplating
Aeration

Drops may be formed directly such as in spraying, atomizing or aerating. Liquid drops are also generated when gas bubbles traveling upward through a liquid collapse as they break through the surface of a liquid. Every drop has a liquid–air interface through which the liquid itself evaporates or liquid species dissolved in the drop evaporate. Per unit volume of liquid, the drop surface area is inversely proportional to the drop diameter, D_p. Thus the smaller the drop, the larger the surface area, the larger the evaporation rate and the shorter the lifetime of the drop.

Consider evaporating droplets of a pure substance (single molecular species) suspended in air. In the analysis that follows it will be assumed that the number of drops per unit volume of air is low enough such that one drop does not influence the motion of another drop. Thus the total evaporation rate from a *cloud of drops* is the evaporation rate for each drop summed over all the drops. A thorough analysis of the issue can be found in texts on the microphysics of clouds (341) in the field of meteorology; nonetheless, an analytical model can be developed using fundamental concepts taught to all engineering undergraduates.

Before modeling the process mathematically, consider the physical processes that occur. Vapor escapes from the surface of the drop shown in Fig. 4.13 because the vapor pressure of the saturated liquid (based on the drop surface temperature) exceeds the partial pressure of the vapor in the far-field. As liquid evaporates, the drop diameter D_p decreases, which in turn influences the rate of evaporation. Simultaneously, the evaporating liquid extracts energy from the drop (a process called *evaporative cooling*), which lowers the drop temperature below the local air temperature, which in turn lowers the saturation liquid vapor pressure at the drop-air interface. Because evaporation tends to lower the drop temperature below the air temperature, energy is transferred to the drop from the surrounding air by convection. Thus both mass and heat transfer are coupled and control the rate at which the drop evaporates.

In terms of mass per unit time the evaporation rate from a drop a/diameter D_p is

$$\dot{m} = N\pi D_p^2 M \tag{4-108}$$

where the molar flux rate N is

$$N = h_m(c_i - c_0) \tag{4-109}$$

Since the drop contains only one species of molecular weight M, the nomenclature can be simplified so that the concentration subscripts pertain to the liquid–air interface (subscript i) and to the far-field concentration (subscript 0). The molar concentrations can be expressed as a partial pressure using the perfect gas equation if it is assumed that the mol fractions in air are small. Thus

$$\dot{m} = \left(\frac{\pi M D_p^2 h_m}{R_u}\right)\left(\frac{P_{v,i}}{T_p} - \frac{P_{v,0}}{T_0}\right) \tag{4-110}$$

where $P_{v,0}$ is the partial pressure of the contaminant in the far-field and $P_{v,i}$ is the liquid saturation pressure based on a drop temperature (T_p). The convection mass transfer coefficient h_m can be related to other flow variables using the *Reynolds analogy*. Combining Eqs. 4-69 in which the characteristic length L is the drop diameter (D_p) and 4-110,

$$\dot{m} = \pi M D_p \, \mathrm{Nu} \left(\frac{\mathscr{D}_{ac}}{R_u}\right)\left(\frac{\mathrm{Sc}}{\mathrm{Pr}}\right)^{b1}\left(\frac{P}{P_{cm}}\right)\left\{\left[\frac{P_{v,i}}{T_p}\right] - \left[\frac{P_{v,0}}{T_0}\right]\right\} \tag{4-111}$$

The Nusselt number (Nu) for a sphere can be expressed as a function of the Reynolds and Prandtl numbers (Eq. 4-61),

$$\dot{m} = C_1 \pi M D_p \left(\frac{\mathscr{D}_{ac}}{R_u}\right)\mathrm{Sc}^{b1}(\mathrm{Re})^{a1}\left\{\left[\frac{P_{r,i}}{T_p}\right] - \left[\frac{P_{v,0}}{T_0}\right]\right\} \tag{4-112}$$

where C_1 is a constant and the Reynolds number is

$$\mathrm{Re} = \frac{\rho D_p U_0}{\mu} \tag{4-113}$$

and U_0 is the velocity of the drop relative to air.

As the drop evaporates, its diameter decreases. To compute the diameter at any instant of time $D_p(t)$, write the conservation of mass (m) for the drop,

$$\dot{m} = -\frac{dm}{dt} = -\frac{d(\pi\rho D_p^3/6)}{dt} = -\frac{\rho\pi D^2}{2}\frac{dD_p}{dt} \tag{4-114}$$

If the drop temperature at the liquid–air interface is known, the diameter can be found as a function of time by equating Eqs. 4-112 and 4-114. Unfortunately, one does not know the liquid drop temperature (T_p) nor can one assume that the drop surface temperature is equal to the air temperature. A separate calculation is needed to compute the drop temperature.

From the field of heat transfer (335,336) it can be shown that if the quantity $h_L D_p/2k$, *where* h_L is the convection heat transfer film coefficient and k is the thermal conductivity of the liquid, is less than approximately 0.1, then the drop temperature varies little with radius, and

$$T(r, t) = T(t) \qquad (4\text{-}115)$$

The quantity $h_L D_p/2k$ is called the *Biot number* (Bi). Note that while the Biot and Nusselt numbers are composed of the same variables, the thermal conductivity (k) in the Biot number refers to the liquid while the value of k in the Nusselt number it refers to air. Assuming the drop has a spatially uniform temperature, it is called the *lumped heat capacity* assumption. Physically it means that the resistance to heat transfer by conduction within the drop is less than the resistance to heat transfer by convection with air. Thus if the total heat transfer is conceived as governed by two "resistances," the resistance by convection exceeds that due to conduction. For small drops that constitute an evaporating cloud, the lumped heat capacity assumption is generally valid and will be adopted in this analysis.

To compute the drop temperature as a function of time, write the conservation of energy for the evaporating drop shown in Fig 4.13. Define the volume of the drop at time t as the control volume. The integral form of the equation for the conservation of energy written for the control volume contains an unsteady term that becomes important because the mass and internal energy within the control volume change with time. Neglecting shear work and the kinetic energy of the evaporating vapor, one finds that

$$q = \frac{d(mu_f)}{dt} + \dot{m}h_g(T)$$

$$q = \frac{mc_v dT_p}{dt} + u_f \frac{dm}{mt} + \dot{m}h_g(T) \qquad (4\text{-}116)$$

where the change in internal energy of the saturated liquid (u_f) is expressed as the liquid specific heat (c_v) times the change in temperature, that is, $c_v\, dT_p$. The mass of the drop at any instant (m) is

$$m = \frac{\rho\pi D_p^3}{6} \qquad (4\text{-}117)$$

and its change with time is given by Eq. 4-114. The rate of heat transfer q is given by Eq. 4-54.

Upon substitution, Eq. 4-116 becomes

$$h_L \pi D_p^2(T_0 - T_p) = \left(\frac{\rho\pi D_p^3}{6}\right)c_v \frac{dT_p}{dt} + \dot{m}[h_g(T_p) - u_f(T)] \qquad (4\text{-}118)$$

The quantity $[h_g(T_p) - u_f(T_p)]$ is approximately equal to the enthalpy of vaporization $h_{fg}(T_p)$. Replacing the variable T_p with the difference $(T_0 - T_p)$, Eq. 4-118 can be rearranged as follows:

$$\frac{d(T_0 - T_p)}{dt} = -\left(\frac{6h_L}{\rho D_p c_v}\right)(T_0 - T_p) + \frac{6\dot{m}h_{fg}}{\rho\pi D_p^3 c_v} \qquad (4\text{-}119)$$

Equations 4-112, 4-114 and 4-119 constitute a set of simultaneous differential equations whose solution yields the evaporation rate, temperature, and diameter of a drop as functions of time. Since the equations are coupled, numerical techniques should be used. A significant simplification can be achieved if the product $\rho D_p c_v / 6h_L$ is small. The product can be thought of as a time constant (τ) for a thermal R–C circuit. The time constant is the product of a thermal resistance $(h_L D_p / k)$ and a thermal capacitance $(k / \rho c_v D_p^2)$. If the time constant is small, the system responds quickly to changes in the thermal environment and achieves equilibrium quickly. If equilibrium is achieved quickly, the left-hand side of Eq 4-119 can be assumed to be zero and the drop temperature given by

$$T(\text{steady-state}) = T_{p,ss} = T_0 - \frac{\dot{m}h_{fg}}{\pi D_p^2 h_L} \qquad (4\text{-}120)$$

Illustrated Example 4.9—Droplet Evaporation

Consider a cloud of water droplets in quiescent air at STP. Estimate the evaporation from water droplets of diameters 5, 10, 50, 100, 500, and 1000 μm assuming that the only relative velocity between the particles and air is due to gravimetric settling. The settling velocities (v_t) can be found from Fig. 8.13.

The relevant properties are:

Air: $T_0 = 25$ C $P_0 = 101$ kPa, $\mu/\rho = 1.5 \times 10^{-5}$ m^2/s
       $k = 0.0258$ W/mK   Pr $= 0.707$ $\mathscr{D}_{ac} = 0.26 \times 10^{-4}$ m^2/s
       Sc $= 0.5769$ Relative humidity $= 50\%$

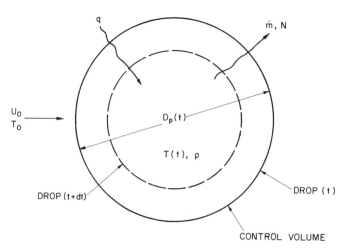

Fig. 4.13   Control volume for evaporating drop.

Water: $k = 0.613$ W/mK $\rho = 1000$ kg/m^3 $c_v = 4.18$ kJ/kg/K
P(sat, 25 C) = 3.169 kPa $P_{v,0} = (0.8)(3.169$ kPa)
h_{fg}(25 C) = 2442.3 kJ/kg

The Reynolds number is calculated on the basis of the settling velocity and is thus very small. Using the appropriate equation from Table 4-2, the Nusselt number can be computed for a sphere from which the convection heat transfer coefficient h_L can be computed. The Biot numbers can be computed and are seen to be small for particles less than 100 μm. For these particles the lumped heat capacity assumption is reasonable. The time constant $\tau = \rho D_p c_v / 6h_L$ can be computed directly. The time constant is less than 0.1 second for particles smaller than 100 μm and one would expect that the drop temperature is equal to its equilibrium value predicted by Eq. 4-120.

D_p (μm)	v_t (m/s)	Re	Nu	h_L (W/m^{2}K)	$h_L D_p/k$	$\tau = \rho D_p c_v / 6h_L$ (s)
5	0.0008	0.001	2	10,320	0.042	0.0003
10	0.003	0.002	2	5160	0.042	0.001
50	0.078	0.258	2.18	1125	0.046	0.031
100	0.255	1.7	2.54	655	0.054	0.106
500	2.0	66.7	5.81	300	0.122	1.162
1000	4.0	266.7	10.07	250	0.204	2.788

For purposes of reference one can categorize the above particles as follows,

Fog: $1\,(\mu m) < D_p < 10\,(\mu m)$
Mist: $10\,(\mu m) < D_p < 100\,(\mu m)$
Drizzle: $100\,(\mu m) < D_p < 1000\,(\mu m)$
Rain: $D_p > 1000\,(\mu m) = 1$ mm

For particles less than 100 μm, one concludes that heat transfer within the drop is rapid so that the temperature within the drop is uniform and that particles have their equilibrium temperatures throughout the evaporation process. However, for particles larger than 100 μm, one will have to solve the coupled equations of heat and mass transfer.

The evaporation rate can be computed from Eq. 4-112. Because of the form of the relationship Nu = const $Re^{a_1} Pr^{b_1}$, used to determine the heat transfer from a sphere, it is not obvious what value the coefficient a_1 should be. Because $0.4\,Re^{0.5} > 0.06\,Re^{0.67}$, a value of $a_1 = 0.5$ was used in Eq. 4-112. The reader must realize that as the diameter decreases, the evaporation rate decreases also. Shown below are the mass of the drop and the evaporation rates for particles from 5 to 100 μm. Knowing the evaporation rate one can find the steady-state temperature of the particles from Eq. 4-120. Shown also are the temperature differences brought about by evaporative cooling. It can be seen that the particle temperature is virtually the same as the ambient temperature for particles less than 10 μm. Larger particles experience greater temperature depression, which in turn reduces the water vapor pressure at the air–liquid interface and reduces the rate of evaporation.

The evaporation rate is proportional to the particle diameter (D_p) raised to the power $1 + a_1$,

$$\dot{m} = \text{const } D_p^{(1+a_1)}$$

$$\frac{d\dot{m}}{dD_p} = \text{const } D_p^{a_1}$$

where a_1 is a number less than unity. Thus the evaporation rate increases with the diameter. The lifetime of the particle can be found by equating Eqs. 4-112 and 4-114 and integrating over the time the particle diameter is reduced to zero. An approximate measure of the lifetime is to divide the particle mass by the evaporation rate.

D_p (μm)	m (kg)	$\dot{m}$ (kg/s)	$T_0 - T_p$ (C)	$(m/\dot{m})_{\text{approx}}$ lifetime (s)
5	6.542×10^{-14}	1.08×10^{-10}	0.45	0.0006
10	5.230×10^{-13}	2.16×10^{-10}	0.63	0.0024
50	6.542×10^{-11}	1.18×10^{-9}	6.58	0.0554
100	5.230×10^{-10}	2.74×10^{-9}	14.48	0.1908

Compare the lifetimes above to the lifetime of water droplets that comprise steam plumes. Water vapor (invisible) from a process condenses as it leaves an exhaust stack and is cooled with ambient air. Thus a "steam" plume is really a plume of small liquid water drops. After a short period the drops evaporate and the plume disappears. A distinctive feature of a steam plume, in contrast to particle plumes, that is, flyash, is that steam plumes have distinct boundaries and end abruptly. Particle plumes on the other hand have diffuse boundaries and persist for a considerable distance downwind of the exhaust stack.

Using intuition gained from observing steam plumes, it seems that the above lifetimes are too short, even taking into account the crudeness of the Reynolds analogy and the approximate manner in which the lifetime was computed.

NOMENCLATURE

Symbol	Description	Dimensions*
a_1, a_2	Exponents	
A	Area	L^2
b_1, b_2, b_3	Exponents	
c_i	Molar concentration, mass concentration of molecular species i	$N/L^3, M/L^3$
c_f	Average skin friction coefficient	
c_p	Specific heat at constant pressure	Q/MT
c_v	Specific heat at constant volume	Q/MT
C_1, C_2	Constants defined by equation	

*Q – energy, F – force, L – length, M – mass, N – mols, t – time, T – temperature

$\mathscr{D}$	Diffusion coefficient	L^2/t
D_0	Diameter of spill (m) defined by equation	L
$\mathscr{D}_{12}$	Binary diffusion coefficient abbreviated as $\mathscr{D}$	L^2/t
D_p	Particle diameter	L
$D_p(\text{aero})$	Aerodynamic particle diameter	L
$D_{p,50}(\text{mass})$	Powder mass median particle size	L
E	Constant defined by equation	
$E(t)$	Contaminant exposure defined by equation	Mt/L^3
F	Constant defined by equation	
F_n	Grinding normal force	F
f	Dimensionless filling factor	
f_w	Powder moisture content	
g	Acceleration of gravity	L/t^2
$g(a)$	Function defined by equation	
H	Henry's law constant	FL/N
H'	Henry's law constant	F/L^2
h	Thickness of volatile liquid pool	L
h_m	Average convection mass transfer coefficient	L/t
$h_{x,m}$	Local convection mass transfer coefficient	L/t
h_x	Local convection heat transfer coefficient	Q/tTL^2
h_g	Enthalpy of a saturated vapor	Q/M
h_{fg}	Enthalpy of vaporization	Q/M
h_L	Average convection heat transfer coefficient	Q/tTL^2
J_a	Molar flux of A relative to average molar velocity of all constituents	N/tL^2
k	Thermal conductivity	QL/T
k	Boltzmann constant	Q/T
K	Overall mass transfer coefficient	L/t
K	Constant defined by equations	
k_G, k_L	Gas phase and liquid phase mass transfer coefficients	L/t
K_G, K_L	Overall gas and liquid phase mass transfer coefficients	L/t
L	Characteristic length airborne dust mg/kg of powder	L
L_r	Loading rate	t^{-1}
$m(t)$	Mass at time t	M
M	Mass released per puff	M
M_i	Molecular weight of species i	M/N
$m(t)_i$	Mass of molecular species i at any instant of time	M
$\dot{m}_c$	Mass transfer rate of species C	M/tL^2
$\dot{m}_t$	Total mass emission rate metal removal rate	M/t
N	Cumulative number distribution	
$\mathbf{N}_a$	Molar flux rate of A relative to fixed observer	N/tL^2
P_v	Vapor pressure	F/L^2
P	Pressure	F/L^2
P^*	Star state pressure, defined by equation	F/L^2
P_a	Partial pressure of molecular species A	F/L^2

P_c	Critical pressure, partial pressure of contaminant species	F/L^2
$P_{c,i}, P_{c,g}$	Partial pressure of molecular species C at liquid-air interface and in the far-field gas phase	F/L^2
P_{cm}	Log mean partial pressure difference defined by equation	F/L^2
P_i	Saturation pressure of pure species i	F/L^2
$P_{(sat)}$	Saturation pressure	F/L^2
$P_{v,i}$	Partial pressure at liquid drop–air interface	F/L^2
$P_{v,0}$	Partial pressure in the far-field	F/L^2
q/A	Convection heat transfer flux	Q/L^2t
R_u	Universal gas constant	Q/NT
r	Radial distance	L
S_a	Rate of production of species A	M/t
S_e	Rate vapor enters air when vessel is opened	M/t
S_f	Evaporation rate when vessel is filled	M/t
S_d	Evaporation rate due to vapor in displaced air	M/t
S_g	Mass generation rate due to grinding	M/t
S_i	Rate of generation of species i by chemical reaction	N/tL^3
S_s	Source evaporation rate due to an industrial spill	M/t
t	Time	
t_{Lv}	Lagrangian turbulence scale	t
T	Temperature	T
T_c, T_p	Critical temperature, particle temperature	T
T_i, T_g	Temperature at interface, in gas phase	T
u_f	Internal energy of a saturated liquid	Q/M
U	Air speed	L/t
U_0	Far field air speed	L/t
U^*	Friction velocity, defined by equation	L/t
U_a	Velocity of species A	L/t
U_m	Molar average velocity defined by equation	L/t
V	Volume	L^3
v_c	Critical specific volume	L^3/M
v_0	Initial velocity of puff	L/t
v_s	Tangential velocity of grinding wheel	L/t
V_g	Metal removal rate due to grinding	L^3/t
x^*	Star state mol fraction defined by equation	
x_i	mol fraction of molecular species i in liquid phase	
y_i	mole fraction of molecular species i in gas phase	
y_{cm}	mol fraction log mean difference defined by equation	
z_1, z_2	spatial coordinates	L

Z	constant defined by equation	
Z_c	critical point compressibility factor	

Greek

α	Displacement angle	
β	Thermal expansion coefficient	T^{-1}
Γ_i	Activity coefficient of species i	
δ	Boundary layer thickness	L
Λ	Metal removal parameter	L^3/tF
μ	Viscosity	Ft/L^2
ρ_b	powder bulk density	M/L^3
ρ_m	metal density	M/L^3
ρ	molar density (g-mol/cm^3)	N/L^3
σ	dispersion coefficient	L
σ_g	geometric standard deviation of powder particle size distribution	
τ	time constant	t
τ_a	shear stress of air	F/L^2
ν	kinematic viscosity (μ/ρ)	L^2/t

Subscripts

$()_a$	Air
$()_{avg}$	Average value
$()_c$	Characteristic or molecular species C
$()_{cm}$	Log mean difference
$()_g$	Gas phase
$()_i$	Interface, molecular species i
$()_L$	Liquid, or for length
$()^*$	Hypothetical star state defined by graph
$()_0$	Far field properties
$()_m$	Mass transfer boundary layer
$()_p$	Particle
$()_s$	Surface condition
$()_{sat}$	Saturation conditions
$()_t$	Thermal boundary layer
$()_x$	Local value
$()_v$	Velocity boundary layer
$()_w, ()_{wall}$	Wall

Abbreviations	*Description*
ACFM	actual volumetric flow rate (ft^3/min) at actual temperature and pressure
ACGIH	American Conference of Governmental Industrial Hygienists
Bi	Biot number
CFM	ft^3/min
Nu_x, Nu_L	Local and average Nusselt numbers
PPM	Parts per million

Pr   Prandtl number
Ra   Rayleigh number
Re_x Reynolds number based on local distance x
Re_L Reynolds number for length L
Sc   Schmidt number
SCFM   Volumetric flow rate (ft^3/min) that would exist at STP
Sh_x Local Sherwood number
Sh_L Average Sherwood number over length L
STP   standard temperature (25 C) and pressure (101 kPa)
TSP   total suspended particle matter
VOC   volatile organic compound

PROBLEMS

1. Small horsepower gasoline engines are tested in a laboratory 8 m × 8 m × 3 m. Each test consists of filling the tanks (splash loading—normal service) with 250 ml of gasoline and allowing the engines to run until they are out of fuel. The volume of each fuel tank is 2 gal. The engine exhausts are discharged outside the building. Properties of gasoline can be found in Appendix A-8.

   (a) How much gasoline vapor is emitted to the room each time the tank is filled with 250 ml of gasoline ?

   (b) An engine is tested every 10 min. The room is not ventilated and the gasoline vapor accumulates. Estimate how long it will take before the gasoline vapor concentration is equal to the lower explosion limit (LEL), 1.3% by volume (mol fraction = 0.013)?

2. Glass fiber is formed into a "wool" which in turn is compressed into a mat used for thermal insulation. The "wool" is formed by a rotary spinning process. The entire process is conducted inside a large building. Estimate the rate (kg/hr) at which particles are generated if the production rate is 2 tons per hour.

3. In an automotive body shop, autos are "refinished" at a net rate that corresponds to $5\,m^2$/hr. Refinishing consists of applying all the following materials,

   Primer—100 g/m^2 per coat, one coat applied
   Enamel—200 g/m^2 per coat, two coats applied
   Lacquer—150 g/m^2, one coat applied

   Using emission factors, estimate the rate (kg/hr) at which hydrocarbon vapor enters the workplace environment.

4. Using the empirical equations in Section 4-2, estimate the rate at which trichlorethylene vapor will be generated (kg/hr) when a conventional closed 55-gal drum is filled at a rate of 2 GPM under conditions called "splash loading."

5. If trichlorethylene is spilled on the floor of a building, estimate the rate $(kg/hr\,m^2)$ at which vapor enters the environment using the empirical equations in Section 4.2. Assume the following conditions,

    Air speed = 2 m/s   Liquid density = 12.3 lbm/gal
    Temperature = 25 C   Spill area = 2 m × 2 m

6. An open 55-gal drum is filled with jet fuel (kerosene) at a rate of 5 GPM. The temperature and pressure are 25 C, 1 atm. Estimate the vapor emission rate in kilogram per second.

7. A group of workers using hand-held grinders and 1-in grinding wheels smooth welds in the hold of a ship under construction. The metal removal parameter (Λ) for steel is 0.005 in^3/min-lbf. If there are 9 grinders in a hold (1250 m^3),
    (a) Estimate the total particle generation rate (lbm/sec).
    (b) Estimate the particle emission rate (kg/s) between the point of contact and 30 degrees.

8. Crushed limestone is transferred from one conveyor to another. To minimize the generation of fugitive dust, an enclosure is built to surround the operation. Using the emission factors for screening, conveying, and handling in stone quarries, estimate the mass of respirable dust generated per ton of stone. For calculation purposes define respirable dust as particles 10 μm or less. Assume that the density of limestone is 105 lbm/ft^3 and that the respirable mass fraction is 15%.

9. Using emission factors, estimate the contaminant generation rate in the units of kg/s.
    (a) Your firm makes office furniture and paints the metal surfaces with an enamel paint (8.0 lbm/gal). The consumption of paint is 5 gal/min. What is the hydrocarbon emission rate.
    (b) You work for a slaughterhouse and in the Fall of the year local hunters bring venison for smoking. Estimate the emission of particles (smoke) if meat is processed at the rate of 50 lbm/hr.
    (c) You use an unvented kerosene space heater in your hunting cabin. Estimate the rate at which carbon monoxide is emitted while you sleep. Kerosene (density 7.5 lbm/gal, heating value 40,000 kJ/kg) is burned at the rate of 0.1 gal/hr.
    (d) Six people smoke simultaneously during a meeting in a small conference room. Estimate the emission rate of carbon monoxide assuming only sidestream smoke.
    (e) You are in charge of loading concrete mix trucks. The space where this is done is small. Estimate the particle emission rate if trucks are loaded at a rate of one every 10 min. Each truck carries 10 tons of concrete mix.
    (f) Estimate the emission rate of gasoline fumes from a filling station (having no fume controls) if on the average 10 gal of gasoline are added to each

auto and if tanks are filled ("splash filling") at the rate of one every 5 minutes.

10. A worker is stationed 1 m away from on a production line in a plant that manufactures a commercially available pesticide. Each time a 80-lb bag is placed on the conveyor belt, it generates a "puff" of fine dust that drifts toward the worker.

(a) If 10 mg of dust is made airborne for each puff of which only 10% by mass is respirable, estimate the maximum concentration of respirable dust the worker is subjected to for each puff. Because there are drafts and random air currents in the workplace, the effective diffusion coefficient $\mathscr{D}$ has a value 0.1 m^2/s.

(b) If the workers remain at their station for 8 hrs each day, compute the total exposure for a production rate of 5 bags per minute and compare it with the OSHA PEL value.

11. Women working at a health food store fill small bags of flour with a machine that pours 5 lb of flour into the bag. Unfortunately, at the beginning of each filling a "puff" of flour dust is expelled from the empty bag and travels outward into their faces. The women's patience is at its end and they complain. They are also apprehensive about effects to their health. Estimate the maximum concentration (mg/m^3) they experience and the total exposure per filling if the following parameters pertain to the situation.

Distance between face and machine: $r = 0.3$ m

Mass of respirable dust released per puff: $M = 10$ g

Diffusion coefficient: $\mathscr{D} = 2 \times 10^{-7}$ cm^2/s

12. Figure 4.3 shows how the dust mass concentration c varies with radius r and time t when a burst of dust (mass M) is released in the form of a "puff".

(a) Show that at any radius r, the ratio c/M achieves its maximum value at a time t_m approximately equal to $r^2/6\mathscr{D}$.

(b) Show that well after t_m, that the concentration decreases with time such that c/M is essentially equal to $(4\pi\mathscr{D}t)^{-1.5}$

(c) Show that at any radius r the maximum concentration is equal to 0.0736 M/r^2.

13. Compute the TWA for puff diffusion for an individual a distance r from the source.

$$\text{TWA}(r) = (1/t) \int_0^t c(r, t) \, dt$$

14. Show that as time approaches infinity, the exposure $E(t)$ defined by Eq. 4-15 is equal to 0.08 $M/(r\mathscr{D})$.

15. At what temperature (C) will water boil if the barometric pressure is 85 kPa?

16. Methylene chloride (CH_2Cl_2, molecular weight 84.94) has a normal boiling temperature of 40 C. What is the minimum pressure (psia) at which it can be stored as a liquid? A handbook reports that the "heat of vaporization" of methylene chloride at 80 F is 144.8 Btu/lbm R. (The odd mixture of units typifies what engineers will encounter when they use several handbooks to obtain data.)

17. Trichlorethylene is a common nonflammable hydrocarbon used to clean metal surfaces. The chemical formula is C_2HCl_3 and the molecular weight is 131. The enthalpy of vaporization at 85.7 C is 57.24 cal/g. Shown below are saturation temperatures and pressures:

P (mm Hg)	1	5	0	20	40	60	100	200	400	760
T (C)	−43.8	−22.8	−12.4	−1.0	11.9	20.0	31.4	48.0	67.0	86.7

The enthalpy of vaporization at 85.7 C is 57.24 cal/gm. Estimate the vapor pressure at 25 C.

18. Using Equation 4-19 compute the diffusion coefficient of trichlorethylene in air in square meters per second. The critical temperature and pressure for trichlorethylene are 571 K and 4.91 MPa.

19. Derive Eq 4-30.

20. An open 55-gal drum contains a thin layer of toluene at the bottom. The drum is left uncovered in a storage room. Estimate the rate (g/hr) at which toluene evaporates and enters the storage room atmosphere. Use first principles and compare your answer with Illustrated Problem 4.4.

21. An open 55-gal drum contains a thin layer of trichlorethylene at the bottom. The drum is left uncovered in a storage room.
    (a) Using first principles, estimate the rate (kg/hr) at which vapor enters the environment.
    (b) Using emission factors (Appendix A-3) and Illustrated Problem 4.3, estimate the rate (kg/hr) at which vapor enters the atmosphere.

22. A young male worker is assigned the task of cutting sidewalk pavement with a gasoline-powered chain saw equipped with a concrete cutting blade. The saw produces a free-field noise level of 100 dBA at 1 m. The saw has no dust control equipment. It is estimated that cutting concrete produces a respirable dust concentration of 4 grains/ft^3 (7000 grains = 1 lbm) in the worker's breathing zone. The weather is hot and dry (90 F, RH 20%) and the wind speed is 3 m/s. The globe temperature is measured to be 100 F. The worker is wearing a T-shirt, long pants, hard hat, and goggles, but no protective ear plugs or respirator. Assume that the properties of concrete dust are the same as in Problem 5, Chapter 3. If the work assignment takes 1 hr, assess the health risks associated with silica, heat stress, and noise.

23. Benzene (M = 78) at 28 C (83 F) lays in a shallow pool in a flat bottomed cylindrical reactor 3 m in diameter and 5 m high. The normal boiling temperature is 80 C, the heat of vaporization is 433.5 kJ/kg at 25 C and the diffusion coefficient of benzene in air is 0.77×10^{-5} m²/s. The reactor has a circular opening (diameter 0.5 m) at the top through which materials are added. The lid on the opening has been removed and benzene vapor escapes to the workplace atmosphere (25 C). Estimate the rate (kg/hr) at which benzene vapor escapes through the opening.

24. If trichlorethylene completely fills a square pan 2 m × 2 m over which air passes at 2 m/s, estimate the rate (kg/hr m²) at which vapor enters the environment using the relationships in Section 4.7 and compare the results with Problem 21 above.

25. An artist is casual about handling chemicals in his small unvented hot studio. To save money the artist cleans his brushes with used allyl alcohol brought to him by a friend who obtains it from his place of work and is unaware of its toxicity. (Unfortunately neither the friend or the artist are aware of MSDS literature!) Uncapped 1-gal cans (4 in × 6.5 in base, 9.5 in high, 1-in diameter cylindrical opening, 0.5 in high) with pools of alcohol at the bottom are stashed in corners of the room. Near his easel he keeps a 6-in cup filled to the rim with alcohol for cleaning his brushes. The artist uses a room fan to cool himself while he works. Air passes over the open cup at a velocity of 3 m/s. Assume the diffusion coefficient for allyl alcohol in air is 1×10^{-5} m²/s. See Appendix A-8 for other properties of allyl alcohol.
    (a) Estimate the vapor pressure (kPa) and enthalpy of vaporization (kJ/kmol K) at 25 C.
    (b) Using first principles, estimate the emission rate of fumes (kg/hr) from an open gallon can assuming,
        (1) That evaporation is governed by diffusion through a stagnant air column 1-in in diameter, 9.5 in high,
        (2) That the concentration of alcohol in the air inside the can is dictated by its vapor pressure and that diffusion through the opening can be approximated as diffusion through a stagnant air column 1/2 in high.
    (c) Using first principles, estimate the emission rate of fumes (kg/hr) from the cup.

26. Automobile parts are forged from heated steel. After forging the parts have a surface temperature of 150 C and are placed on an overhead conveyor where they are transported to several machining operations. While on the conveyor they cool by forced convection (air velocity 10 m/s) with room air at 25 C. After being cooled to 25 C and after several machining operations the parts are dipped in trichloroethylene (TCE) to remove oils, cutting fluids and soon. After being dipped in TCE the parts have a 1-mm film of TCE over the entire surface. The parts are placed on an overhead conveyor. While being conveyed, TCE vapor evaporates. You have been asked to estimate the initial heat transfer rate (kJ/s) after forging and the evaporation rate (kg/s)

immediately after being removed from the TCE cleaning tank. The part can be assumed to resemble a sphere of diameter 0.5 m.

27. A pie plate is filled with water. The upper diameter of the pie plate is 10 in. Water evaporates at a rate of 20 mg/s when the air and water are 25 C and the air is absolutely dry (relative humidity is zero).

    (a) What is the overall convection mass transfer coefficient?

    (b) Estimate the evaporation rate if the relative humidity is 70%

    (c) Assuming that the convection mass transfer coefficient in part a remains the same, estimate the mass evaporation rate (kg/s) if the relative humidity is 50% and the air and water temperatures are 50 C.

28. Cylindrical brass stock 1/4 in in diameter leaves a machine with a thin (1.0 mm) layer of water coating its surface. The speed with which the brass stock moves is 0.1 m/s. With room air at 25 C, relative humidity 20% passes over the brass perpendicular to the axis at a velocity of 10 m/s and evaporates the water. The temperature of the water film remains constant (30 C) due to the large heat capacity and thermal conductivity of the brass. Estimate the

    (a) Overall convection mass transfer coefficient (h_m).

    (b) Local evaporation rate (kg/s m^2) at a point 5 sec after it has left the machine.

    (c) Water film thickness at the point in b above.

29. Moth repellant (paradichlorobenzene) in the form of a long cylinder of circular cross sectional area is hung in an air stream (at STP) moving with a uniform velocity U. The moth repellant has uniform chemical composition, density (ρ) and sublimates.

    (a) Show that the sublimation rate ($\dot{m}$, kg/s) varies as $\dot{m} = $ constant U^m, where m is a constant given in Table 4.2.

    (b) Show that the rate of change in diameter (dD/dt) is $dD/dt = -$const. D^m.

30. Basketballs ($D = 14$ in) are manufactured by a process that causes the balls to emit hydrogen cyanide (HCN, M $= 27$) after they are taken from molds. Air blows past the balls at 3 m/s. If the vapor pressure of HCN at the surface of the ball is 10 mm HG, estimate the rate of generation of HCN (kg/s).

31. Ethyl alcohol (C_2H_5OH) is accidentally spilled on the floor of a store room in which the temperature and pressure are 23 C, 1 atm. The average air velocity is 2 m/s and the spill is 2 m in diameter. Estimate the rate of evaporation in kilogram per square meter per second.

32. A square vessel (1.5 m $\times$ 1.5 m) is filled to the rim with benzene at room temperature (70 F, 21 C). Room air blows slowly over the vessel at 1 m/sec. Estimate the total evaporation rate (kg/sec) of benzene into the workplace.

33. A large open vessel contains an aqueous mixture of volatile wastes. The

vessel is 10 m in diameter, 3 m high, and filled to the rim. The vessel is stored in a remote part of an industrial building at 25 C over which air passes at a velocity of 3 m/s parallel to the surface. Initially the waste liquid consists of the following:

Hydrocarbon	Mol fraction
Benzene	0.05
Methyl alcohol	0.05
Carbon tetrachloride	0.05
Toluene	0.05

The remainder of the liquid is water.

(a) Following procedures in Chapter 3, estimate the TWA–TLV for the mixture.

(b) Following procedures in Section 4.8, estimate the evaporation rate (kg/hr) of volatile materials after 10, 100 and 1000 hr, etc assuming that the wastes are agitated to make the concentration within the vessel uniform. Does the evaporation rate change with time? Does the composition of the remaining liquid remain constant?

34. A liquid hydrocarbon mixture consisting initially of equal molar amounts of acetic acid, carbon tetrachloride, and nitrobenzene (see Appendices A-8 and A-9 for properties) is spilled on a body of water and floats as a pool 10 m in diameter and 1 mm thick. The concentration of species in the floating pool are uniform (well mixed) and remain so such that the liquid film resistance in the liquid pool can be neglected. Assume that the room air is at STP and that the velocity is 3 m/s. The density of acetic acid and nitrobenzene and carbon tetrachloride are 1049 kg/m^3, 1205 kg/m^3 and 1595 kg/m^3 respectively. Plot the following until the pool disappears.

(a) Overall evaporation rate (kg/s m^2)

(b) Individual species evaporation rates (kg/s m^2)

(c) Mol fraction of each species

Note that the average molecular weight, total molar concentration, and the mass and thickness of the pool change slowly with time.

35. Questions are sometimes raised (252) about the exposure humans experience due to volatile organic compounds (VOC's) that evaporate from household tap water. If the concentration of trichloroethylene (TCE) in household tap water is 1 mg/l of water, estimate the rate of evaporation of TCE from a bathtub of water at 25 C using two-film theory. For a first approximation, neglect any TCE in the air.

36. Estimate the evaporation rates (kg/hr) of benzene and chloroform from a waste lagoon containing 100 mg/l of benzene (M = 78) and 100 mg/l of chloroform (M = 119). The lagoon is 25 m × 40 m × 3.5 m deep. The air and liquid temperatures are 25 C and the wind speed is 3.0 m/s. The Henry's law

constants are, benzene, 5.5×10^{-3} atm-m^3/gmol, and chloroform, 3.39×10^{-3} atm-m^3/gmol. The Schmidt numbers are benzene and water, 1000; benzene and air, 1.76; chloroform and water, 1100; chloroform and air: 2.14

37. You work in a research facility that models rivers and flood control projects, tests the design of ship hulls and so on. Topographically scaled models are laid out on the floor of the building to simulate a river, bay, lake, and impounded water behind a dam. Thus a large water surface is exposed to room air and evaporation occurs. At other times a tank of water is used and scale models of ship hulls are towed through the water so that drag and wave patterns can be studied. In all these experiments certain values of the Reynolds, Prandtl, and Froude numbers are needed so that researchers can achieve proper similitude. To control the water's viscosity, unique hydrocarbons are added to the water. Unfortunately researchers select volatile toxic materials (unfortunately they do not read MSDS literature) and toxic vapors escape into the room. Using two-film theory, estimate the evaporation rate (kmols/sm^2) of toxic vapor from the water under the following conditions:

The characteristic length of water surface is 100 m.

Room air has an average velocity of 2 m/s.

Negligible water movement.

Henry's law constant, diffusivities, and other physical properties of the additive are similar to perchloroethylene.

Initial bulk mol fraction $(x_{c,0})$ in water is 0.01.

CHAPTER 5

General Ventilation Well-Mixed Model

Goal: Students will be expected to learn to
- estimate maximum contaminant concentrations in enclosed spaces
- model ventilated well-mixed spaces containing sources and sinks
- use the sequential box model
- quantify and rank mixing within a room

The phrase *general ventilation* suffers from ambiguity. As a control strategy it denotes the practice of removing air from the entire workplace and replacing it with cleaned or outside air. *Dilution ventilation* and *displacement ventilation* are other names for this strategy. As an analytical concept, general ventilation denotes the assumption that the concentration is uniformly distributed throughout the enclosure. The concentration may vary with time (t) but it does not vary with location, that is, *spatial uniformity*

$$\text{General ventilation: } c(x, y, z, t) = c(t) \qquad (5\text{-}1)$$

If the phrase is used, individuals must be careful and indicate whether they are referring to the strategy or the analysis, because adopting the strategy does not insure that the analysis will be valid. One can always remove and add air to a room but only experiment will verify that Eq. 5-1 is valid. It is suggested that "general ventilation" be used only to refer to the strategy of adding and removing air from the entire workplace and that another phrase be used to refer to the analysis wherein Eq. 5-1 is valid. Equation 5-1 is not valid physically unless air is added and removed in a unique fashion such that mixing removes any spatial variations of the contaminant concentration. In other fields of engineering, enclosures in which the concentration is spatially uniform are called *well-mixed*, *perfectly stirred*, or *well-stirred*. It is suggested that industrial ventilation adopt the same language. The phrase "well-mixed" will be used in this text to describe enclosures in which Eq. 5-1 is valid. In air pollution control the well-mixed model is called the *box model*.

The ASHRAE does not use general and dilution (ventilation) as synonyms. For heating and ventilation purposes, the objective of general ventilation is to satisfy personal comfort and the objective of dilution ventilation is to reduce contaminant concentrations to acceptable levels. For gaseous contaminants, ASHRAE (325) recommends a dilution air flow rate (Q_d),

$$Q_d = \frac{Q_g K 10^6}{\text{TLV}} \tag{5-2}$$

where Q_g is the gaseous contaminant generation rate and TLV is given in parts per million. For volatile organic solvents, ASHRAE (325) recommends a dilution air volume (V_d in m^3)

$$\frac{V_d}{m_s} = 24.45 \times \frac{10^6 K}{M_s(\text{TLV})} \tag{5-3}$$

where m_s and M_s are the mass (in kilogram) and molecular weight of the organic solvent and TLV is given in parts per million. The parameter K in both equations is a safety or mixing factor whose value is between 3 and 10 depending on the toxicity, type and effectiveness of the local ventilation system and the quality of the dilution air. Both the equations above lack precision and should be used only when it is known that well-mixed conditions exist.

Throughout the analyses of industrial ventilation systems, the phrases *control volume*, *control surface*, and *open and closed systems* will be used often.

A *control volume* is any volume in space users wish to define through which they can assess the mass flow rates and transfer of energy and work. Control volumes are also called *open systems*. The mass of material within the control volume may change but the boundaries of the control volume must be definable. The surface of a control volume is called a *control surface* or *open system boundary*.

A *closed system* is a fixed mass of material users wish to define. Heat may be transferred or work performed on the mass within the closed system. The surface of a control mass is also called the *system boundaries*. A *control mass* is another name for a closed system.

Another way to describe general ventilation is to consider it a *global* , *lumped parameter* or *control volume* perspective. Such an analytical approach ignores details of the velocity field, assumes a general flow path, and concentrates on mass entering and leaving a control volume. Only global information about the flow field is required and the finite-sized control volume becomes a lumped mass of air containing uniformly distributed contaminant. Lumped parameter models involve only ordinary differential equations for the conservation of mass. Such mathematical models are called *macro-models* by some because details of the flow field are obscured and because the size of the control volume is considerably larger that what would be called a *differential volume*. Macro-models are easy to analyze mathematically and predict time-varying and steady-state concentrations.

In contrast to the macro-model is the *micro-model* that will be the subject of study in Chapter 10. Micro-models are mathematical models based on a very large number of elemental volumes (i.e., *distributed parameters*) and partial differential equations describing the conservation of mass, momentum, and energy. Details of the flow field are not assumed as they are in macro-models, but are unknowns to be computed along with temperatures and concentrations throughout the space under study. Micro-models may be either two or three-dimensional, depending on the physical phenomena being modeled. Micro-models are conceptually easy to analyze but involve extensive computer programs and inventive numerical (computer) solution techniques.

A common ventilation practice is to select the volumetric flow rate (Q) of makeup air in terms of the *number of room air changes per unit time*. The contaminant concentration is directly related to the number of room air changes only if well-mixed conditions exist. If there are spatial variations in concentration, the well-mixed model should not be used since it is of no value whatsoever to insure that PEL or TLV values are satisfied. If the time scale is large, the assumption of well-mixed may be valid because diffusion and slow mixing within the enclosure distributes the contaminant (albeit slowly) to insure the uniformity of concentration. Unfortunately one does not always know the mixing time scale to know whether several hours or several days is the appropriate time scale. The number of room air changes is an anachronism and it is unfortunate that it is used in building codes with such aplomb. Its general use should be abandoned, but its specific use in isolated cases in which one can show conclusively that Eq. 5-1 is valid should continue.

A variation to the number of room air changes is a requirement of certain quantities of fresh outside air based on the number of occupants in a room. For example, the revised ASHRAE standard 62–1989, *Ventilation for Acceptable Indoor Air Quality*, recommends

20 CFM/person in offices

30 CFM/person in bars, cocktail lounges

60 CFM/person in office smoking lounges

The standard is an improvement because it addresses the wide variation in the number of people per cubic foot of room and the activity within the room, for example, cramped quarters for low-paid clerical workers to the spacious conditions in executive offices. Nonetheless, the underlying premise of well-mixed conditions is retained. Appendix A-16 lists other ventilation standards in which the volumetric flow rate is expressed in terms of floor area (m^2) or room volume (m^3).

An obvious situation in which well-mixed conditions occur is a room in which the inlet and outlet ventilation ducts are located uniformly around the room and there is a great deal of activity within the room that mixes the air thoroughly. The contaminant generation rate and ventilation flow rate can be constant or they may vary with time. If there is a circulation fan or if individuals are moving about, such activity can mix the contaminants rapidly so that at any instant Eq. 5-1 is valid. Under these conditions, the "well-mixed" assumption is physically reasonable. The well-mixed model may also be appropriate in enclosures where mixing occurs more rapidly than the time scale of the problem. For example, if one were concerned about the hydrocarbon concentration at points within a new mobile home whose furnishings liberate hydrocarbons in a decreasing rate over a period of months, the well-mixed model would be appropriate if measurements were concerned with the concentration at weekly intervals for a period of many months.

The well-mixed model represents the upper limit of a real process because it predicts what would happen if mixing was instantaneous and helps engineers to place the results from more accurate models in perspective. If the designer is expressly concerned with spatial variations, the well-mixed model should never be used since the model assumes explicitly that there are no spatial variations. For example consider contaminant concentrations related to an open vessel containing a volatile liquid that is located in a poorly ventilated enclosed space. A designer and OSHA are particularly interested in the concentration at the edge of the tank where workers stand rather than directly above the tank where the concentration should be larger. For such applications the well-mixed model is inappropriate and the designer will have to use some other analytical model to predict the concentration at the edge of the tank. If, however, one wished to predict the concentration of tobacco smoke in a small, well-ventilated conference room in which people were smoking and moving about, the well-mixed model would be a reasonable physical model to choose. In general, when the source is concentrated or small with respect to the size of the enclosure, when there is not vigorous mixing, or when the time scale of the problem is short, the well-mixed model should not be used.

A *floor sweep* is an inlet register placed at floor-level in storage rooms to capture contaminants that could produce fires. It is often alleged that contaminants heavier than air settle so that the concentration near the floor of an enclosure will be larger than near the ceiling. Under certain physical conditions the observation may be correct, but never because the contaminant gas is heavier than air. To understand why the general statement is false, recall the second law of thermodynamics which requires that molecular "disorder" increases in a spontaneous process. Disorder refers to the freedom of the contaminant molecules to distribute themselves throughout the entire enclosure rather than restricting themselves to any subsection of the enclosure. Thus if an initially homogeneous gas mixture in an isolated system segregated itself spontaneously into layers of equal molecular weight, the entropy of the system would increase. This is an outcome prohibited by the second law of thermodynamics.

Consider two cases: (a) contaminant added to an enclosure containing only quiescent, uncontaminated air and (b), the equilibrium state of an initially uniform mixture of air and contaminant.

(a) *Contaminant Added to Quiescent Air* If a volatile liquid leaks onto the floor of a room, vapors will rise as the liquid evaporates. The process is dynamic and the rate of evaporation and concentration vary with time. Mass transport is upward, and the concentration will be highest near the spill and decrease with height. The situation is true irrespective of whether the molecular weight of the liquid is less or greater than that of air, although the molecular weight of the liquid affects the rate of evaporation and diffusion upward. Once evaporation ceases, the contaminant distributes itself until the concentration gradients become zero and the concentration is uniform throughout the room.

If a jet of vapor or gas enters an enclosure near the floor, diffusion and bulk motion produced by the jet gradually distribute the contaminant throughout the enclosure. Prior to achieving equilibrium, the concentration near the floor might be larger than near the ceiling but once the jet stops the material distributes itself uniformly throughout the enclosure.

The mixing described in these two situations concerns material of the same phase, that is, a gas. (A mixture of a gas or vapor in air is still a single phase.) The mixing of gas molecules can be described in a simple fashion by kinetic theory or if one wishes more detail, by more sophisticated theories. In any case, the molecular weight of each molecular species influences the rate of mixing but not the equilibrium state of uniform concentration.

If the contaminant consists of liquid or solid particles the situation is entirely different because two phases are involved (i.e., solid particles suspended in a gas, an aerosol). Recall that particles are considerably larger than gas molecules. For example, smoke particles are typically $0.01–1.0$ μm in diameter, while an oxygen or nitrogen molecule is approximately 0.0003 μm (i.e., smoke is $30–3000$ times larger). In addition, the density of solid particles is nearly 1000 times larger than air. The motion of particles is governed by gravity, viscosity, and motion of the gas through which the particles move. Thus if particles enter a quiescent enclosure, gravity will cause them to settle and produce regions of high concentration near the floor and low concentration near the ceiling. In a quiescent enclosure, there will be a region near the ceiling virtually devoid of particles. If, however, the enclosure is well-mixed, the particle concentration will be uniform but decrease with time.

(b) *Initially Uniform Conditions* Now consider the opposite condition to (a) above in which the initial concentration inside the enclosure is uniform. What is the final equilibrium state? Is it logical to expect gaseous constituents to segregate themselves spontaneously into layers of equal molecular weight? My, wouldn't the chemical industry love to use inexpensive gravity to separate gaseous mixtures into their constituent parts! No, spontaneous separation does not occur. Gases and vapors remain uniformly distributed while only particles (solid or liquid) settle.

The reason why storage rooms have exhaust registers near the floor (floor sweeps) is because a spill initially produces high concentrations near the floor. It is the momentary condition after a spill that is of concern and an exhauster near the floor removes vapor and lessens the chance of fire. As a general proposition, hazardous conditions occur because air and contaminants are not uniformly mixed in the first place, not because they separate after mixing. The fallacy of gravimetric separation is the fallacy of presuming that conditions are mixed in the first place. Rather, the hazard arises because contaminants are not mixed to begin with.

5.1 UNVENTILATED ENCLOSURES

Unventilated enclosures are unlikely to be encountered in industry because it is virtually impossible to prevent air from entering or leaving an enclosure and also because OSHA and fire codes prohibit them. Nonetheless, it is instructive to consider unventilated enclosures in which volatile liquids are stored and determine whether there is an upper limit to the amount of a volatile liquid that can evaporate since such a value represents the upper limit to the concentration in a ventilated enclosure. The rate at which the maximum concentration is approached depends on the particularities of the problem, but the maximum concentration is a simple matter of thermodynamics. In an unventilated enclosure one might assume a volatile liquid will completely evaporate when there is sufficient time to do so. The assumption is wrong because thermodynamics is the limiting condition. Thermodynamic principles dictate that the maximum partial pressure a vapor may have when a vapor is mixed with other gases is equal to its *saturation pressure*, $P(\text{sat}, T)$.

To understand the issue, compare room temperature and pressure (T_r, P_r) to the *critical temperature and pressure* (T_c, P_c) of various contaminants. Appendix A-10 is a table of common workplace contaminants and shows that all the materials have critical pressures in excess of atmospheric pressure and most have critical temperatures greater than room temperature. Thus contaminants fall into two categories;

$$\text{Category 1:} \quad P_c > P_r \quad \text{and} \quad T_c > T_r$$
$$\text{Category 2:} \quad P_c > P_r \quad \text{and} \quad T_c < T_r$$

Figure 5.1 shows these two categories on T-s and P-v phase diagrams.

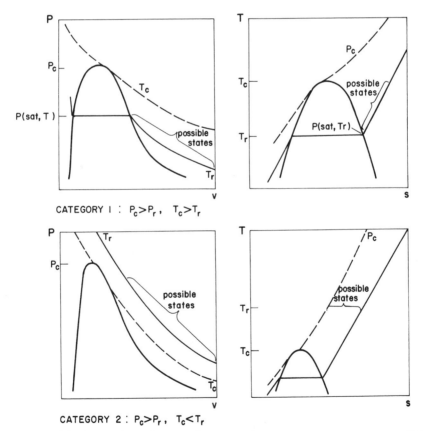

Fig. 5.1 Possible thermodynamic states for contaminant vapor at a temperature (T_r) and pressure (P_r).

Category 1: $P_c > P_r$ *and* $T_c > T_r$. Figure 5.1 shows the possible states of such an air–vapor mixture. The maximum amount of the contaminant that exists as a vapor has a partial pressure equal to its saturation pressure based on room temperature. Thus, providing there is sufficient liquid to evaporate, the maximum vapor mol fraction (y) is the ratio of the contaminant saturation pressure $P(\text{sat}, T_r)$ divided by the total pressure $P(\text{total})$.

$$y(\text{maximum}) = \frac{P(\text{sat}, T_r)}{P(\text{total})} \tag{5-4}$$

Whether this value is ever achieved depends on the initial amount of volatile liquid and air. If the number of mols of volatile liquid contaminant divided by the number of mols of air is less than the value predicted by Eq. 5-4, then the actual maximum concentration will be determined by the amount of contaminant available for evaporation, for example,

$$y(\text{maximum}) = \frac{\text{mols of liquid}}{(\text{mols of liquid} + \text{mols of air})} \tag{5-5}$$

In summary, for Category 1 the actual maximum concentration is the smaller of Eqs. 5-4 and 5-5.

Category 2: $P_c > P_r$ and $T_c < T_r$. Figure 5.1 shows the possible states of such an air–vapor mixture. In this case thermodynamics does not establish an upper limit and the maximum concentration is dictated only by the Eq. 5-5.

The following example illustrates the use of this rationale.

Illustrated Example 5.1—Maximum Concentration in an Unventilated Enclosure   Chemicals are stored in an unventilated outdoor shed. The shed volume is $5\,\text{m}^3$ and ambient temperature and pressure are 18.4 C and 1 atm. What is the maximum contaminant mol fraction if the following accidents occur to materials left in the shed?

(a) An open vessel containing 1.2 kg of toluene is left for several days.

(b) A 1-L cylinder containing nitric oxide with an initial pressure of 2 atm develops a leak

(a) Toluene   From Appendix A-10 toluene is found to have a critical temperature and pressure of 320.8 C, 41.6 atm. Thus, $P_c > P_r$ and $T_c > T_r$. From Fig. 5.1 it can be seen that while there is a wide range of possible toluene concentrations, there is a maximum value dictated by the thermodynamic consideration that vaporous toluene cannot exist in the vapor state with a partial pressure greater than its saturation value based on room temperature. From Appendix A-8 the saturation pressure of toluene at 18.4 C is 20 mm Hg. Thus if the total pressure of toluene and air is 1 atm (760 mm Hg), the maximum mol fraction of toluene allowed by thermodynamic considerations is

$$y(\text{max}) = \frac{P(\text{sat}, 18.4\,\text{C})}{P(\text{total})} = \frac{20}{760} = 0.0263$$

If all the toluene evaporates, the mol fraction in air would be

$$y(\text{complete evap}) = \frac{\text{mols toluene}}{(\text{mol air} + \text{mols toluene})}$$

$$\text{mols toluene} = \frac{1.2\,\text{kg}}{92\,\text{kg/kmol}} = 0.0130\,\text{kmols}$$

$$\text{mols air} = \frac{PV}{R_u T} = \left[\frac{(760-20)}{760}\right]\frac{(101)(5)}{(8.314)(273+18.4)} = 0.2029\,\text{kmols}$$

$$y(\text{complete evap}) = \frac{0.013}{(0.013+0.2029)} = 0.0602$$

Thus while complete evaporation would yields a mol fraction of 0.0602, thermodynamics only allows evaporation until the mol fraction is 0.0263. Thus there will be some liquid remaining in the open vessel. Thermodynamics prevents the complete evaporation of all the liquid toluene and the final mol fraction is 0.0263 (26,300 PPM). Such a value is cause for alarm because not only does it exceed TLV (100 PPM) by a considerable amount but it also exceeds the lower explosion limit (LEL). Review of Table 3.2 (27) shows that the mol fraction at the lower explosion limit is 0.0127. Thus there is a strong possibility that an explosion could occur, caused by a spark in a light switch or any other source of ignition.

(b) *Leaking Cylinder of Nitric Oxide* Appendix A-10 shows that the critical temperature and pressure of nitric oxide are -93 C and 64 atm. Under these conditions Fig. 5.1 shows that thermodynamics does not limit the amount of nitric oxide that can be mixed with air and that the final mol fraction depends only on the amount of nitric oxide leaking into the air. The nitric oxide stops leaking when the pressure inside the cylinder is 1 atm. Heat transfer between the cylinder and air maintains the final temperature equal to the room temperature.

$$y(\text{max}) = \frac{\text{mols NO}}{(\text{mols NO} + \text{mols air})}$$

Since the critical pressure is so much larger than the cylinder pressure, one may assume that the nitric oxide is an ideal gas. The mols of NO escaping to the atmosphere is

$$\text{escaping mols NO} = (\text{initial mols NO} - \text{final mols NO})$$

$$= \frac{[P(\text{initial}) - P(\text{final})]V}{R_u T}$$

$$= \frac{(200-100)(0.001)}{(8.314)(273+18.4)}$$

$$= 4.18 \times 10^{-5}\,\text{kmols}$$

The final nitric oxide mol fraction is

$$y(\text{final}) = \frac{(4.18 \times 10^{-5})}{(0.00004 + 0.208)} = 200 \times 10^{-4}$$

$$= 200 \text{ PPM}$$

This value is in excess of the PEL (25 PPM). At room temperature nitric oxide and oxygen readily react to form nitrogen dioxide and hence one is apt to find 200 PPM of nitrogen dioxide.

5.2 DILUTION WITH 100% MAKEUP

Consider the enclosed space of volume V in which the concentration is spatially uniform (Fig. 5.2). Initially ($t = 0$) the mass concentration in the enclosed space is c_0 and a source begins to generate contaminants at a constant rate S (mass/time). Outside air containing contaminant at a concentration of c_a is added to the enclosed space at a constant volumetric flow rate Q and contaminated air is removed from the enclosed space at the same rate. The density of the air entering and leaving the enclosure are assumed to be constant and equal to the density in the enclosed space.

Using the enclosed space as a control volume and applying the equation for the conservation of mass for the contaminant,

$$\frac{d(Vc)}{dt} = Qc_a - Qc + S$$

$$\frac{Vdc}{dt} = (Qc_a + S) - Qc \tag{5-6}$$

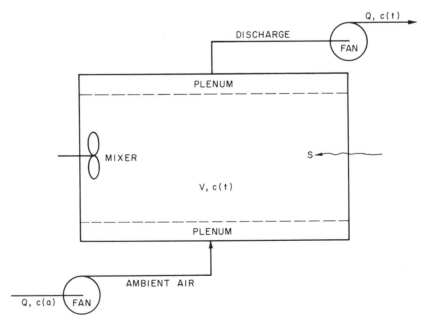

Fig. 5.2 General ventilation with 100% makeup.

There is no need to write an equation expressing the conservation of mass for the air since the inlet flow rate is equal to the outlet flow rate and the density is the same. Integrating Eq. 5-6 between $t = 0$ and a time when the concentration achieves the value $c(t)$,

$$\int_{c_0}^{c(t)} \frac{dc}{[(Qc_a + S) - Qc]} = \left(\frac{1}{V}\right) \int_0^t dt \qquad (5\text{-}7)$$

one obtains after rearrangement

$$\frac{[c_{ss} - c(t)]}{[c_{ss} - c_0]} = \exp\left(-\frac{Qt}{V}\right) \qquad (5\text{-}8)$$

where (c_{ss}) is the *steady state concentration* obtained by setting the left-hand side of Eq. 5-6 equal to zero,

$$c_{ss} = c_a + \frac{S}{Q} \qquad (5\text{-}9)$$

If both the initial and ambient concentrations are zero, the above reduces to

$$\frac{c(t)}{c_{ss}} = 1 - \exp\left(-\frac{Qt}{V}\right) \qquad (5\text{-}10)$$

A great deal of industrial ventilation literature expresses the ventilation rate Q in terms of the *number of room air changes per unit time N*

$$N = \frac{Q}{V} \qquad (5\text{-}11)$$

Thus for the case of zero inlet and initial concentration, Eq. 5-10 can be written as

$$\frac{c(t)}{c_{ss}} = 1 - \exp(-Nt) \qquad (5\text{-}12)$$

For example, if the ventilation rate was 10 room air changes per hour ($N = 10/$ hr), Eq. 5-12 predicts that the concentration will be 64% of its steady state value in 6 min.

It is easy to see how the phrase *"number of room air changes per unit time"* came into being. The practice of assessing the worth of general ventilation or of setting building standards in terms of room air changes per unit time is unwise and should be abandoned. The general practice is poor because it is unwise to assume that enclosed spaces are well-mixed until some independent means shows that they are. Rarely is the activity within an enclosed space so robust and the inlet and outlet ducts so judiciously located as to enable one to assume that the air is well-mixed.

Illustrated Example 5.2—Safe Entry into Room Containing HCN A fire has occurred in a motel room ($85\,\mathrm{m}^3$) in which an upholstered sofa containing polyurethane foam has burned and filled the room with smoke containing HCN.

Immediately after the fire has been extinguished, the smoke concentration is $100 \, g/m^3$, of which 10% by mass is HCN. Fireman evacuate the smoke with an exhaust fan and blow outside air into the room at the same rate, 1000 CFM ($28.33 \, m^3/min$). Because of the fire, the outside air contains HCN at a concentration (c_a) of $1 \, mg/m^3$. On the basis of the well-mixed model, estimate the length of time that must elaspe before individuals can enter the room after the fire has been extinguished. The fire company uses a Short Time Exposure Limit (STEL) of $5 \, mg/m^3$ as the HCN concentration that must be achieved before individuals can enter the room.

The initial HCN concentration is

$$c_0 = 100 \times 1000 \times 0.1 = 10{,}000 \, mg/m^3$$

Figure 5.2 depicts the flow of HCN in and out of the room. A mass balance for HCN is

$$\frac{V \, dc}{dt} = c_a Q - Qc = -Q(c - c_a)$$

This can be integrated directly and one obtains

$$\frac{(5-1)}{(10{,}000-1)} = \exp\left(\frac{-t \, 28.33}{85}\right)$$

$$t = 23.5 \, min$$

Considering the toxicity of HCN and the fact that the room may not be well-mixed and there may be regions where the HCN concentration is large, it is clear that respirators should be worn by the fireman.

The quantity V/Q (the reciprocal of the number of room air changes per unit time) has the units of time and is visualized by some as the time it takes a mass of make-up air equal to the volume of the room to traverse the room and displace all the contaminated air. Of course, make-up air does not traverse a room as a continuous mass, i.e. as a piston displaces gas in a cylinder, but the visualization is often used in industrial ventilation. Visualizing make-up air displacing contaminated air, is the reason the concept general ventilation is sometimes called *displacement ventilation*.

5.3 TIME-VARYING SOURCE OR VENTILATION FLOW RATE

Often the source strength S or the volumetric flow rate Q is not constant and Eq. 5-7 can not be integrated in closed form, for example,

Variable number of smokers in a conference room
Leaky valve, faulty kerosene space heater
Erratic ventilation controller
Poorly designed duct system to supply makeup air

Generation of volatile organic compounds (VOC) within the home from waxes and cleaners for floors and furniture, moth balls and moth flakes, dry-cleaned cloths, drapes, new carpets, furniture, wallpaper

Under these conditions Eq. 5-6 can be solved numerically. A first-order *Runge Kutta* method (see Appendix A-12) will be used. Rewrite Eq. 5-6 as

$$\frac{dc}{dt} = -\frac{Qc}{V} + \frac{[S + Qc_a]}{V}$$

$$= Ac + B \qquad (5\text{-}13)$$

where

$$A = -\frac{Q}{V} \qquad (5\text{-}14)$$

$$B = \frac{[S + Qc_a]}{V} \qquad (5\text{-}15)$$

Using Runge–Kutta, the concentration at the end of a time step δt can be written as

$$c(i+1) = \frac{\left(1 + \dfrac{A\,\delta t}{2}\right)c(i) + B\,\delta t}{1 - \dfrac{A\,\delta t}{2}} \qquad (5\text{-}16)$$

The solution begins at $t = 0$, $i = 0$, when the initial concentration is c_0. The coefficients A and B on the right hand side of Eq. 5-16 are the values at the beginning of the time step. The concentration at the end of the first time step, $c(1)$ is then calculated. Using the newly found concentration $c(1)$ as a new initial value, the coefficients A and B are recalculated and the concentration is then calculated at the end of the next time step, $c(2)$. The calculations are repeated until the elapsed time has been spanned.

Written in this form the concentration can be computed as functions of time for a variety of variable source strengths (S) and volumetric flow rates (Q). One only needs to know the values of S and Q during each time step. Thus S and Q can be any of the following:

Continuous (but difficult to integrate) functions
Discontinuous functions
Highly irregular tabulated functions

Illustrated Example 5.3—Conference Room with 100% Makeup Air Analyze the smoke concentration in a conference room (33.31 m^3) assuming well-mixed conditions. The conference room contains six people that smoke at irregular times as shown in Fig. 5.3. Assume that the smoke particles are generated at a rate of 1100 μg/min for each cigarette (Repace and Lowery, 24).

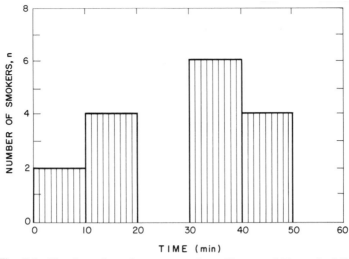

Fig. 5.3 Number of smokers versus time, Illustrated Example 5.3.

A ventilation system removes air from the conference room at a rate of 6.3 m^3/min and replaces it with ambient air at the same rate. Such a condition corresponds to 100% makeup air and 0% recirculated air. If the smoke concentration in the ambient air and the initial smoke concentration in the conference room are 20 $\mu g/m^3$, compute $c(t)$ produced by Fig. 5.3. Compare the concentration with the EPA 24-hr Primary Air Quality Particulate Standard of 150 $\mu g/m^3$. The computer program below is based on Eq. 5-16. The irregular generation rate is expressed by the series of If–Then statements. The results shown in Fig. 5.4 are obtained for $\delta t = 0.1$ min.

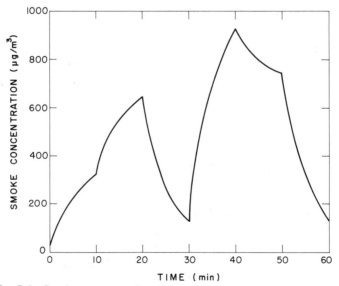

Fig. 5.4 Smoke concentration versus time, Illustrated Example 5.3.

```
1 rem filename-general ventilation with time varying
source

2 rem this program computes c(t) for a time varying
source

3 rem variables:
   q=6.3 (cum/min)        v=33.31 (cum)
   dt=0.1 min, time step   s=1.1 mg/min per cigarette

   c and c(a)=smoke concentration (micrograms/cum)

4 rem standards:
   epa 24-hr ambient primary air quality standard is 260
   micrograms/cum
   national ambient air quality 24-hr standard for sig-
   nificant harmful effects 1 mg/cum

5 print ''time (min)  number of smokers  concentration''
6 print ''_____    _____    _____''

10 ca=20
15 q=6.3
20 v=33.31
25 dt=0.1
30 a=-q/v
35 c=ca
40 j=0
45 for t=0 to 60 step 0.1
50 if t>0 and t<=10 then n=2
55 if t>10 and t<=20 then n=4
60 if t>20 and t<=30 then n=0
65 if t>30 and t<=40 then n=6
70 if t>40 and t<=50 then n=4
75 if t>50 and t<=60 then n=0
80 s=n*1100
85 b=(s/v)+(q*ca/v)
90 c=((1+a*dt/2)*c+b*dt)/(1-a*dt/2)
95 if t=0 then print t,n,c
100 if j=20 then print t,n,c: j=0
105 j=j+1
115 next t
120 end
```

A ventilation rate of 6.3 m^3/min corresponds to 37 CFM/person, which is similar to the old standards of 35 CFM/person for public facilities containing smokers. The ASHRAE 62-1989 standard calls for 60 CFM per person for an office smoking lounge. Cain et al. (52) report that 75% of a sample (mixed smokers and nonsmokers) found that 35 CFM per person reduced tobacco odor to an acceptable level. Leaderer et al. (46) found that this ventilation rate was adequate to maintain safe levels of CO. Figure 5.4 shows that 37 CFM/person results in particulate concentrations that exceed the EPA 24-hr Primary Air Quality Standard (24). Similar results were obtained by Leaderer et al. (46). The concentration of 150 $\mu g/m^3$ pertains to a 24-hour, continuous exposure in the out of doors. A short-time standard for indoor conditions can be expected to be a great deal higher.

Illustrated Example 5.4—Computation of Minimum Ventilation Flow Rate For Safe Conditions Estimate the minimum volumetric flow rate for the conference room in Illustrated Problem 5.3 for a "worst case" condition in which six people smoke continuously for an entire hour. Your supervisor asserts that "safe conditions" consist of a maximum smoke concentration [$c(\max)$] of 520 $\mu g/m^3$.

For reference, it will be useful to calculate Q_{ss}, the steady-state ventilation flow rate corresponding to $c(\max)$. From Eq. 5-9,

$$c_{ss} = c(\max) = 520 \ \mu g/m^3$$

$$c_{ss} = c_a + \frac{S}{Q}$$

$$Q_{ss} = \frac{[(6)(1100) \ \mu g/min]}{(520 - 20) \ \mu g/m^3}$$

$$Q_{ss} = 13.2 \ m^3/min \ (465.8 \ CFM)$$

It is unwise to assume that the answer to the problem is the value above since only the value of Q that results in $c(60) = c(\max)$ is desired and it is not known how rapidly the concentration varies with time for the enclosure in question.

Since S, V are known and Q is constant, Eq. 5-8 can be used directly. Unfortunately Eq. 5-8 will not allow Q to be calculated explicitly and an iterative solution will be needed. Shown below is an algorithm to compute the value of Q to achieve $c(\max)$ in 60 min. The algorithm uses the program developed for Illustrated Example 5.3. One guesses a value of Q and computes $c(60)$. If the value of $c(60)$ is more than $c(\max)$ a higher value of Q is used and the process is repeated until convergence is obtained. The reader should refer to Appendix A-15 for basic iterative computer programs that will be used elsewhere in this book. The algorithms are efficient and converge rapidly.

The results of the computation are shown below. A ventilation flow rate of 13.31 m^3/min is obtained. The value is essentially equal to the steady-state value calculated above. The fact that the two are virtually equal is coincidence. There is no way of predicting such an outcome and it would be imprudent to assume that such an outcome is going to be true for all problems like this. For example suppose the problem was repeated for an elapsed time of 30 min, would the value of the minimum flow rate be the same?

1 rem filename—computation of optimum ventilation flow rate

2 rem this iterative program computes the minimum flow rate to achieve c < c(max) in 60 minutes

3 rem variables:
v=33.31 cum   ql, qh, qg=vol flow rate (cum/min)
dt=0.5 min   s=1.1 mg/min per cigarette
c(a) and c=smoke concentration (micrograms/cum)

```
 5 print ''time (min)  q (cu m/min)  c(60)
   (microgram/cum)''
 6 print ''_____      _____      _____''
10 cmax=520
15 ca=20
20 v=33.31
25 dt=0.5
30 c=20
35 s=6*1100
40 ql=0.1*s/(cmax-ca)
45 qh=10*s/(cmax-ca)
55 for t=0 to 60 step 0.5
60 qg=(ql+qh)/2
65 a=-qg/v
70 b=(s/v)+(qg*ca/v)
75 c=((1+a*dt/2)*c+b*dt)/(1-a*dt/2)
80 next t
85 r=c/520 : print t, qg, c
90 if abs(r-1)<=0.01 then goto 110
95 if abs(r)>1 then ql=qg
100 if abs(r)<then qh=qg
105 goto 55
110 print t, qg, c
115 end

run
```

time (min)	$q(m^3/min)$	$c(60)$ $(\mu g/m^3)$
60.5	66.66	119.01
60.5	33.99	214.17
60.5	17.66	393.83
60.5	9.49	715.65
60.5	13.57	506.32
60.5	11.53	592.45
60.5	12.55	545.88
60.5	13.06	525.33
60.5	13.32	515.64
60.5	13.32	515.64

5.4   REMOVAL BY SOLID SURFACES

The preceding sections presumed that the exhaust stream was the only way contaminants were removed from the enclosed space, that is, that there were no *wall losses*. In the case of tobacco smoke anyone who has cleaned house knows that various surfaces in a home adsorb tobacco smoke and odors and furthermore, when smoking ceases, desorption occurs so that the room reacquires an odor of tobacco. Adsorption such as this occurs for many combinations of contaminant and surface material. The adsorption of contaminants on walls is also called *plate-out*. As a crude first approximation, one may assume that the mass removal rate (mass/time) is equal to the product of the room's surface area (A_s), the concentration (c) and an *adsorption rate constant* (k_{ad}). The statement may also be phrased that plate-out is *first-order* with respect to the concentration. If the contaminant is a gas or vapor, plate-out occurs by the physical process called adsorption and k_{ad} is truly an adsorption coefficient. If, however, the contaminant is an aerosol, removal occurs because the particle travels to a solid surface with a unique deposition velocity, strikes it, and adheres to it. The magnitude of the deposition velocity depends on

  Orientation of the surface, for example, vertical, upward-facing horizontal, downward-facing horizontal

  Particle size

  Natural convection, that is, air currents produced by surfaces that have a different temperature than the gas

  Bulk motion and homogeneous turbulence in the gas

Nazaroff and Cass (342) provide equations for the deposition velocity that include these factors. For particles above the respirable range ($D_p > 2$ μm), gravimetric settling (sedimentation) on upward-facing horizontal surfaces is the dominant mechanism. In this case, k_{ad} can be set equal to the gravimetric settling velocity (Chpt 8). Particles smaller than this may be removed by surfaces of all orientation since the processes of thermophoresis, Brownian motion, eddy diffusion and bulk motion may be as important, or more important than gravity.

The conservation of mass for the contaminant can be written (342,357) as

$$V \frac{dc}{dt} = Qc_a + S - c(Q + A_s k_{ad})$$

(5-17)

The adsorption rate constant is a parameter that can be expressed as functions of the gas transport properties or it may be determined experimentally. The units of k_{ad} are length/time. The units are those of velocity, but it should not be confused with a velocity unless the contaminant is an aerosol. For constant values of Q, S and (k_{ad}), Eq. 5-17 can be solved in closed form for a variety of initial conditions (c_0),

$$\frac{[c_{ss} - c(t)]}{[c_{ss} - c_0]} = \exp\left[-(Q + A_s k_{ad})\left(\frac{t}{V}\right)\right]$$

(5-18)

where the *steady-state concentration* (c_{ss}) is obtained by setting $dc/dt = 0$ in Eq. 5-17

$$c_{ss} = \frac{(Qc_a + S)}{(Q + A_s k_{ad})} \tag{5-19}$$

Adsorption is seen to be a removal mechanism working in parallel with, but independently of, the exhaust ventilation rate. Both removal mechanisms produce a smaller steady state concentration and both enable a room to achieve its steady state concentration faster.

To illustrate the importance the adsorption of tobacco smoke consider the results of experiments of Repace and Lowery (24), Fig. 5.5. A single cigarette was burned steadily and then extinguished in an isolated, small room ($22\,\mathrm{m}^3$). The total mass of suspended particle matter was measured during the entire period. In one experiment, well-mixed conditions were produced by fans within the room. In another experiment, the only mixing was by natural air currents within the isolated room. The generation of smoke was expressed by

$$S = \frac{nM_p}{t_b} \tag{5-20}$$

where n is the number of cigarettes, (t_b) the length of time burning occurred, and (M_p) the total mass of particle matter produced per cigarette, determined by tests conducted by the Federal Trade Commission.

The effect of mixing is clearly seen in Fig. 5.5. Following the extinction of the cigarette, the concentration in the well-mixed experiment fell rapidly. The slope of the curve enables one to estimate the removal of contaminants by adsorption

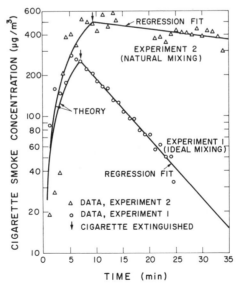

Fig. 5.5 Smoke concentration from one smoldering cigarette in an isolated room ($22\,\mathrm{m}^3$) with and without internal mixing (redrawn from ref. 24).

on solid surfaces. Rewriting Eq. 5-17 for the case in which Q and S are zero, and beginning the integration at the maximum concentration, one obtains

$$\frac{c(t)}{c_{max}} = \exp\left(-\frac{tAk_{ad}}{V}\right) \qquad (5\text{-}21)$$

Repace and Lowery (24) found the quantity V/Ak_{ad} had a value of 10 min, which corresponds to an adsorption rate constant k_{ad} of 0.00078 m/s. If the surface area of the room is 30 m^2, the rate of adsorption is equivalent to an exhaust ventilation flow rate of 1.4 m^3/min (50 CFM). Thus adsorption in this small room lowered the concentration in a fashion equivalent to a ventilation flow rate of 50 CFM or 4.2 room air changes per hour.

Contaminants may *desorb* from a surface. If this occurs, desorption becomes a source term in Eq. 5-17. The rate at which contaminants *desorb* (mass/time area) from a surface is not known. It is expected that it is proportional to the concentration of the contaminant on the surface of the material (mass/area). In such a case, the desorption rate constant (k'_{ad}) will have the units of (length/time) and will have to be determined from experiment.

Up to this point the discussion presumed that the contaminant was removed in a first-order process described by a single differential equation, Eq. 5-17. If the contaminant consists of a range of submicron particles, separate equations have to be written for each particle size range since the deposition velocities depend on the particle diameter. In addition, particles coagulate to form a single particle of a new size. Thus the physical process *coagulation* is now a significant removal process that must be accounted for. The coagulation process involves a removal (and production) term proportional to the product of particle concentrations, that is, $c(D_{p,1})c(D_{p,2})$. Thus instead of a single equation of the form Eq. 5-17 one now needs to write a simultaneous set of equations, one equation for each particle size, and solve a coupled set of nonlinear differential equations.

5.5 RECIRCULATION

Consider an enclosed space of volume V in which only a fraction (f) of the return air flow is fresh air, that is, makeup air (Q_m) and the remainder is a portion of the exhausted air in which the contaminant concentration has been reduced by an air cleaning device of efficiency η. The cleaning efficiency of an air cleaning device is defined as the ratio of the removed material to the incoming material.

Figure 5.6 is a schematic diagram of a ventilation system using recirculation. If the volumetric flow rate of air into and out of the enclosure is the same ($Q_i = Q$), the conservation of mass applied to air at a point upstream of the air cleaner is

$$Q_r = Q - Q_m = Q(1 - f) \qquad (5\text{-}22)$$

The factor f is the *makeup air fraction*,

$$f = \frac{\text{makeup fresh air}}{\text{input air}} = \frac{Q_m}{Q} \qquad (5\text{-}23)$$

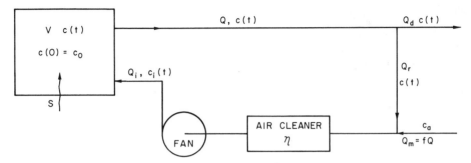

Fig. 5.6   recirculation with makeup air.

Applying the equation for the conservation of mass to the contaminant in the enclosed space,

$$V \frac{dc}{dt} = Qc_i + S - Qc \tag{5-24}$$

where c_i is the concentration in the air entering the enclosure. The value of c_i can be found by applying the equation for conservation of mass to the collector,

$$\eta = 1 - \left[\frac{Qc_i}{(cQ_r + fQc_a)} \right]$$

$$c_i = (1 - \eta)[fc_a + c(1 - f)] \tag{5-25}$$

Substituting Eq. 5-25 into Eq. 5-24,

$$V \frac{dc}{dt} = -cQ[1 - (1 - f)(1 - \eta)] + S + Qfc_a(1 - \eta) \tag{5-26}$$

The *steady state* concentration (c_{ss}) can be found by setting $dc/dt = 0$

$$c_{ss} = \frac{[S + Qf(1 - \eta)c_a]}{[Q\{1 - (1 - \eta)(1 - f)\}]} \tag{5-27}$$

If the ambient concentration is negligible ($c_a = 0$), the steady state concentration is

$$c_{ss} = \frac{S}{Q[1 - (1 - \eta)(1 - f)]} \tag{5-28}$$

To find the concentration at any time, Eq. 5-26 can be integrated from an initial condition in which the concentration in the enclosed space is (c_0) to a time when the concentration is $c(t)$. One obtains

$$\frac{[c_{ss} - c(t)]}{[c_{ss} - c_0]} = \exp\left[-\left(\frac{Qt}{V}\right)\{1 - (1 - \eta)(1 - f)\} \right] \tag{5-29}$$

If the source or volumetric flow rate vary with time, Eq. 5-26 is valid although its solution must now be obtained numerically.

Illustrated Example 5.5—Hospital Operating Room   The ventilation system for a hospital operating room is shown in Fig. 5.7. The maintenance department inadvertently installs the exhaust from a chemical laboratory hood within a few feet of the makeup air inlet. A laboratory technician evaporates ethyl alcohol in the laboratory hood on a day in which an operation is in progress. Compute the steady state concentration of ethyl alcohol that will exist in the operating room and the length of time before the operating personnel begin to smell the alcohol. Assume the threshold odor limit for ethyl alcohol is $40 \, mg/m^3$. The following parameters characterize the situation.

V = volume of operating room ($50 \, m^3$)

A_s = total area of adsorbing surfaces in operating room ($85 \, m^2$)

k_{ad} = adsorption rate constant ($0.001 \, m/s$)

S = rate at which ethyl alcohol is vaporized inside operating room ($1 \, g/min$)

c_0 = initial alcohol concentration inside operating room ($10 \, mg/m^3$)

c_a = concentration of ethyl alcohol entering the make up air duct ($100 \, mg/m^3$)

Q_e, Q_r, Q_a, Q_s = volumetric flow rate of exhausted air, recirculated air, make up air, and supply air ($Q_s = 20 \, m^3/min$)

$f = Q_a/Q_s$ = make up air fraction ($f = 0.9$)

η_1, η_2 = efficiencies of the activated charcoal filter (0.95)

Assuming the operating room to be well-mixed, a mass balance for the alcohol in the operating room results in the following:

$$V \frac{dc}{dt} = S + Q_s c_s - c Q_e - c A_s k_{ad}$$

A mass balance for the air results in the following

$$Q_s = Q_a + Q_r$$
$$Q_e = Q_s$$

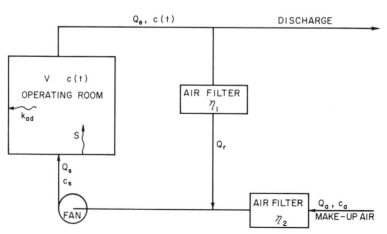

Fig. 5.7   Ventilation system for hospital operating room.

At the inlet to the fan, the following mass balance can be written for alcohol:

$$c_a Q_a (1 - \eta_2) + c Q_r (1 - \eta_1) = c_s Q_s$$

From the definition of f, the concentration c_s can be written as

$$c_s = c_a f (1 - \eta_2) + c (1 - \eta_1)(1 - f)$$

Upon substitution, the differential equation can be arranged as

$$\frac{dc}{dt} = Ac + B$$

where, after the substitution of numerical values, A and B become

$$A = \frac{\{-Q_s - A_s k_{ad} + Q_s (1 - f)(1 - \eta_1)\}}{V}$$

$$A = -\frac{25 \, m^3/min}{50 \, m^3} = -0.5 \, min^{-1}$$

$$B = \frac{[S + f Q_s c_a (1 - \eta_2)]}{V}$$

$$B = 0.0218 \, g/min \, m^3$$

The steady state concentration can be found by setting $dc/dt = 0$. Thus

$$c_{ss} = -\frac{B}{A} = 43.6 \, mg/m^3$$

The differential equation can be integrated in closed form since both A and B are known constants.

$$\int_{10}^{40} \frac{dc}{[Ac + B]} = \int_0^t dt$$

$$t = 4.47 \, min$$

5.6 PARTIALLY MIXED CONDITIONS

Sections 5.2–5.5 are analyses in which the concentration is uniform throughout the enclosed space although it may vary in time, i.e., spatial uniformity but not temporal uniformity. If the ventilation volumetric flow rate (Q), source strength (S) and adsorption rate (k_{ad}) are constant, Eqs. 5-17 and 5-26 can be integrated in closed form. If these parameters vary with time, the equations can be integrated numerically. It must be emphasized that the notion of spatial uniformity is critical to the validity of the well-mixed model and the solutions that follow from it.

Unfortunately, in many industrial applications, both spatial and temporal variations in concentration occur. Analyses of these situations is difficult since both the equations of mass and momentum transfer have to be solved simultaneously. Analytical procedures are available for this and, will be discussed in Chapter 10.

Over the years an alternative computational technique has arisen that many workers in industrial ventilation find useful. The technique employs a scaler constant called *mixing factor* (*m*) to modify the equations of the well-mixed model to account for spatial variations in concentration, that is, to account for nonuniform concentrations brought on by poor mixing.

Consider the ventilated enclosed space with 100% recirculation shown in Fig. 5.8. Other geometric configurations can be modeled in a comparable fashion. Assuming well-mixed conditions and neglecting adsorption on the walls, the following expression for the contaminant can be written,

$$V \frac{dc}{dt} = S + (Qc_a) - (Qc) - (c\eta Q_r) \tag{5-30}$$

To account for nonuniform mixing, a mixing factor *m* can be adopted, and Eq. 5-30 rewritten as,

$$V \frac{dc}{dt} = S + (mQc_a) - (mQc) - (mc\eta Q_r) \tag{5-31}$$

When integrated, Eq. 5-31 becomes

$$\frac{[c_{ss} - c(t)]}{[c_{ss} - c_0]} = \exp\left[-(mQ + mQ_r\eta)\left(\frac{t}{V}\right)\right] \tag{5-32}$$

where

$$c_{ss} = \frac{(S + mQc_a)}{(mQ + mQ_r\eta)} \tag{5-33}$$

Esmen (26) states that values of *m* are normally 1/3 to 1/10 for small rooms and possibly less for large spaces. Table 5.1 contains values of *m* referenced by

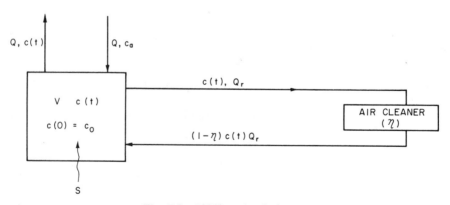

Fig. 5.8 100% recirculation.

TABLE 5.1 Mixing Factors For Various Enclosed Spaces

Enclosed Space	m
Perforated ceiling	1/2
Trunk system with anemostats	1/3
Trunk system with diffusers	1/4
Natural draft and ceiling exhaust fans	1/6
Infiltration and natural draft	1/10

Repace and Lowery (24). If m is less than unity, the concept of mixing factor suggests that a fraction of the flow (mQ and mQ_r) is well-mixed while another fraction [$(1-m)Q$ and $(1-m)Q_r$] bypass the enclosure. Consequently,

$$m = 1 \text{ implies a well-mixed model and concentration that is spatially uniform}$$
(5-34)

$$m < 1 \text{ implies nonuniform mixing and spatial variations in concentration}$$
(5-35)

The parameter "m" is a discount rate or handicap factor. It implies that the enclosed space is a well-mixed region in which the effective ventilation rate is a fraction m times the actual volumetric flow rate. Conversely, the reciprocal of m could be called a safety factor, that is, the actual flow rate is equal to the well-mixed value times the safety factor.

The difficulty in selecting the proper value of m can be seen in Fig. 5.9, taken from Ishizu (25). Six cigarettes were allowed to smolder in the center of a room ($70 \, m^3$). Ventilation consisted of $32 \, m^3/min$ of ambient air and $8 \, m^3/min$ of cleaned recirculated air. No information was given on the location of the inlet and outlet ducts. The concentration of smoke was measured in the center of the room before, during and after the cigarettes were burned. During the burning phase, a steady-state concentration was not reached even though the well-mixed model predicted adequate time for one to occur. The maximum concentration exceeded the steady state, well-mixed value by a factor of 3, clearly indicating nonuniform conditions within the enclosed space. After extinction the concentration decreased but a single value of m could not explain the data. The value varied between 0.4 and 0.3. Figure 5.9 shows how sensitive the calculations of concentration are on the choice of m.

Uniform mixing is synonymous with the well-mixed model. It is not possible to insert a constant, scaler multiplier into the equations for the well mixed-model and expect to acquire equations appropriate for nonuniform concentrations. There are several fundamental flaws in the concept of mixing factor.

(a) The principles of science governing the motion of air and contaminants do not justify the use of a scaler multiplier m.

(b) Experimental values of m are unique to the volumetric flow rates, geometry of the enclosed space, location of inlet and outlet duct openings, and the location of the point where the contaminant is measured.

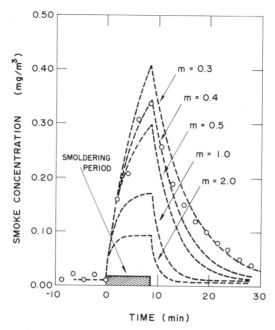

Fig. 5.9 Comparison of measured smoke concentrations in a ventilated room ($V = 71\,\text{m}^3$) with analytical predictions for different mixing factors (m); $Q = 32\,\text{m}^3/\text{min}$, $Q_r = 8\,\text{m}^3/\text{min}$ (redrawn from ref. 25).

(c) The value of m can not be predicted with any precision. Once it is found experimentally for a particular enclosure it can't be generalized for other enclosures.

(d) The range of values used for m is so large as to make it an ineffective parameter for design and economic analysis.

In the final analysis, modifying an equation based on the well-mixed model to account for nonuniform concentrations is a contradiction in terms. Either the concentration is uniform in space or it is not, and if it is not, no amount of fudging will yield meaningful answers. Nevertheless, arcane practices that have been used for a considerable time have a sizeable following and are not going to be changed simply because they are illogical. Uniform mixing and well-mixed implies something concrete, that is, $c(x, y, z, t) = c(t)$. Nonuniform mixing and the concept of "mixing factor" mean that the equality does not hold, but the concept does not predict how, where, or in what way the concentration varies. The concept of mixing factor lacks concreteness and should be abandoned.

5.7 BASIC EQUATIONS FOR WELL-MIXED MODEL

Readers may have detected common features in the previous equations for concentration, $c(t)$. Indeed they should. The equations are similar because all are solutions to the differential equation for the conservation of mass which has the general form of Eq. 5-13, i.e.

$$\frac{dc}{dt} = Ac + B \qquad (5\text{-}36)$$

The values of A and B depend on the enclosure volume, volumetric flow rates of exhausted air, makeup air and recirculated air, and the strength of sources and sinks and are often constant. It must be emphasized that once the volume is known, details about the geometry of the enclosure and location of inlets and outlets are irrelevant because the well-mixed model assumes the existence of mixing mechanisms that distribute contaminants uniformly throughout the enclosure. Furthermore, arguments about how well or how poorly contaminants are transported to certain regions in the enclosure are also irrelevant since the basic tenets of "well-mixed" presupposes that mixing is instantaneous.

The *steady-state* concentration (c_{ss}) can always be found by setting $dc/dt = 0$, from which follows,

$$c_{ss} = -\frac{B}{A} \qquad (5\text{-}37)$$

The general solution for the concentration at any time, $c(t)$, is

$$\frac{[c_{ss} - c(t)]}{[c_{ss} - c_0]} = \exp(At) \qquad (5\text{-}38)$$

For any geometrical configuration, the reader must draw a sketch depicting the flow of air in and out of the enclosure and write the differential equation for the conservation of contaminant within the enclosed space. Once done, the equation should be rearranged in the form of Eq. 5-36 and A and B defined.

The parameter A is always negative and $(-1/A)$ may be thought of as a *time constant* for the well-mixed model. When the elapsed time is equal to the time constant, the left-hand side of Eq. 5-38 will decrease to 0.367 and when the elapsed time is seven times the time constant, the left-hand side is essentially zero. If there is 100% makeup air (see Eq. 5-12), the time constant is the reciprocal of the number of room air changes per unit time.

Illustrated Example 5.6—Clever Outdoorsman

Illustrated Example 5.6—Clever Outdoorsman Hemoglobin in the blood has a stronger affinity for CO than for oxygen. Carbon monoxide absorbed by hemoglobin produces carboxyhemoglobin (COHb), while oxygen absorbed by hemoglobin produces oxyhemoglobin (O2Hb). Both quantities can be measured. When the ratio of (COHb) to (O2Hb) in the blood exceeds 10%, a person is apt to lose consciousness. If exposure continues, a person is apt to die of asphyxiation. See also Fig. 2.41. For the purpose of analysis assume that at any instant the ratio (COHb)/(O2Hb) in the blood is linearly related to the concentration of CO in the air,

$$R = \frac{(COHb)}{(O2Hb)} = CO\,\frac{(PPM)}{1000}$$

An outdoorsman with a bent toward engineering lives in a small cabin 12 ft × 12 ft × 8 ft. The cabin is heated by a single kerosene space heater that is not vented to the outside. Outside air leaks into the cabin (infiltration) at the rate of 0.1 room air changes per hour. Air inside the cabin escapes to the atmosphere at the same rate. The concentration of CO in the outside air is 10 PPM (11.42 mg/m^3). Carbon monoxide is adsorbed by materials on the walls, floor and ceiling of the cabin. The adsorption rate constant (k_{ad}) is 0.25 m/hr.

The outdoorsman considers himself to be clever and designs a thermostat for the kerosene heater. Without knowing it, he inadvertently alters the combustion process so that CO is generated in the following fashion:

$$S\frac{(\text{mg CO})}{\text{hr}} = 4500[1 + \sin(2\pi t)]$$

where t is measured in hours.

Compute the concentration of CO in the air and the (COHb)/(O2Hb) in the blood as functions of time. Will the clever outdoorsman suffer from CO poisoning? Will he survive the night?

Figure 5.10 is a schematic diagram of the cabin showing how CO enters and leaves the air inside the cabin. Assuming well-mixed conditions, the conservation of CO becomes

$$\frac{dc}{dt} = Ac + B$$

$$\frac{dc}{dt} = \frac{c(-Q - k_{ad}A_s)}{V} + \frac{(S + Qc_a)}{V}$$

where A_s is the total area of the adsorbing surfaces 62.46 m^2 (672 ft^2). Since S varies with time, the above equation will be integrated numerically using the program developed for Illustrated Example 5.2. Shown below is the program for this problem.

```
1 rem filename-clever outdoorsman

2 rem this program computes the ratio of carboxyhemo-
  globin to oxyhemoglobin in the blood
```

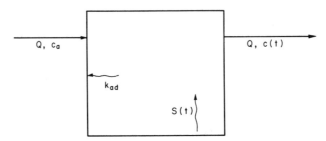

Fig. 5.10   Carbon monoxide entering and leaving air in cabin, Illustrated Example 5.6.

```
3 rem variables:
  v=volume (cu m)           c,a and c=co conc (mg/cu m)
  a,s=surface area (sq m)   dt=0.05 hr time step
  Q=infiltration (cu m/hr)
  k,ad=adsorb rate const (m/hr)

4 rem CO poisoning occurs when ratio exceeds 0.1
5 print    ''time (hr)    CO (mg/cu m)    ratio''
6 print    ''_____    _____    _____,,
7 dt=0.05
10 v=32.64
15 ca=11.42
20 as=62.46
25 kad=0.25
30 q=3.264
35 pi=3.14159
40 c=ca
45 for t=0 to 8 step 0.1
50 s=4500*(1+sin(pi*t*2))
55 a=(-q-kad*as )/v
60 b=(s/v)+(q*ca/v )
65 c=((1+a*dt/2)*c+b*dt)/(1-a*dt/2)
70 r=c/(ca*100)
75 print t,c,r
80 next t
85 end
```

The results of the program are shown in Fig. 5.11. This unfortunate out-doorsman will display symptoms of CO poisoning. Indeed if the CO emission rate were somewhat higher, the thermostat would be the last device he ever designs!

It may be argued that the way in which wall losses and the source strength are modeled is too simplistic and that more refined expressions are needed. If refined expressions exist, they can be included in the differential equations for the well-mixed model without fundamentally changing the methods used to solve them.

Consider the case of *formaldehyde emissions* from *building materials* and *home products* such as

Urea-formaldehyde foam insulation

Fiberglass, sealants, and adhesives

Gypsum wallboard and pressed wood products such as wood paneling and particle board

Carpeting, wall coverings, and upholstery

The net emission of formaldehyde depends on formaldehyde emitted by the material minus wall losses due to adsorption. Unfortunately adsorption depends

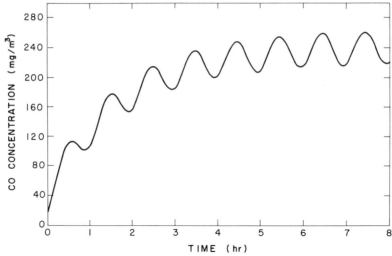

Fig. 5.11 CO concentration versus time, Illustrated Example 5.6. NIOSH 8-hr PEL = 35 mg/m³. EPA Primary Air Quality Standard: 1-hr avg = 40 mg/m³; 8-hr ave = 10 mg/m³.

on the temperature and the concentration of formaldehyde in the material (called a *bulk concentration*, c_b), the concentration of formaldehyde in air and the temperature and relative humidity (ϕ) of the air. There are even atmospheric conditions in which formaldehyde will desorb from the material.

Rather than deal with a separate source strength and wall loss, Matthews and Hawthorne (283–285, 340) suggest using a *net source strength S'*.

$$S' = k_0(T, \phi)[r_1(t)r_2(T)r_3(\phi)][c_b - c] \qquad (5\text{-}39)$$

where $k_0(T, \phi)$ is the air transport property that reflects dependence on temperature and humidity. Hawthorne and Matthews (283) suggest that the term can be expressed as the product

$$k_0(T, \phi) = k_0'[1 - a_1(T - T_0)][1 - a_2(\phi - \phi_0)] \qquad (5\text{-}40)$$

where the subscript 0 refers to a reference state and a_1, a_2 are model constants unique to the application. The functions r_1, r_2, and r_3 are functional relationships to account for the fact that the rate of formaldehyde generation depends on the age of the material, and the temperature and relative humidity respectively.

Age dependence:

$$r_1(t) = \exp - \left[\frac{(t - t_0)}{t'} \right] \qquad (5\text{-}41)$$

where $(t - t_0)$ is the age or the material since measurement of the emission rate and t' is a characteristic time of the order of 1–5 yr.

Temperature dependence:

$$r_2(T) = \exp\left\{-B\left[\left(\frac{1}{T}\right) - \left(\frac{1}{T_0}\right)\right]\right\}$$ (5-42)

where B is a coefficient to be determined and T_0 is a reference temperature.

Relative humidity dependence:

$$r_3(\phi) = \left(\frac{\phi}{\phi_0}\right)^{a_3}$$ (5-43)

where the exponent a_3 in Eq. 5-43 is a coefficient to be determined and ϕ_0 is a reference value.

Before one can use these expressions, the model constants (a_1, a_2, a_3, B) and reference values (k_0, T_0, ϕ_0, t_0) have to be determined from either data obtained in the literature or from experiment. Silberstein et al. (339) suggest an alternative to Eq. 5-39 that includes relative humidity, age, and temperature. But like Eq. 5-39, it also includes a number of parameters and reference states that have to be determined experimentally.

Consider the enclosed space shown in Fig. 5-12 containing an internal air cleaner of efficiency η_r that cleans and recirculates internal air at a volumetric flow rate Q_r. Room air is discharged at a volumetric flow rate Q_s and ambient fresh air added at a volumetric flow rate Q_s after first passing through an air cleaner which has an efficiency of η_m. If the formaldehyde source is expressed as Eq. 5-39. Differential equation for the concentration is,

$$V \frac{dc}{dt} = S' - cQ_s - c\eta_r Q_r + Q_s c_a(1 - \eta_m)$$ (5-44)

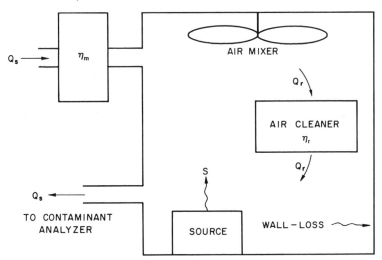

Fig. 5.12 Test apparatus to measure source strength, chamber wall loss (plate-out), and performance of air-cleaning device.

The (bulk) formaldehyde concentration in and on the surface of the building material (c_b) is not a constant. But since it decreases slowly, one can integrate the differential equation by assuming a representative constant value during the period of integration. If the bulk concentration (c_b) decreases rapidly, or if the period of integration is long with respect to t', or if one wishes to be truly rigorous, another differential equation must be written based on a formaldehyde mass balance for the material. In such a case initial conditions must be specified for the building material and the pair of ordinary differential equations solved simultaneously.

For the most part, the bulk concentration c_b decreases slowly and only a single differential equation arranged in the form of Eq. 5-36 has to be solved. Solution of the equation can be obtained by the Runge-Kutta techniques used previously, but additional data concerning the temperature and relative humidity along with the model constants have to be provided. The equation can even be solved if the temperature and relative humidity vary with time. All one has to do is to insert updated values of temperature and relative humidity into the Runge–Kutta equation (Eq. 5-16) for each time step.

5.8   CHARACTERISTICS OF SOURCES AND AIR CLEANING DEVICES

While general ventilation is of narrow and limited use to control contaminants in the workplace, the well-mixed model is ideally suited as a laboratory technique (248,249) to measure

Wall-loss coefficient (k_{ad})
Source characteristics (S)
Efficiency of room air cleaners (η)

Consider a test apparatus shown schematically in Fig. 5.12. Air inside the chamber is sampled at a rate Q_s and the concentration of the contaminant (particle or gas) is measured by a suitable analyzer. Air at the same volumetric flow rate (Q_s) is added to the chamber after passing through an "absolute filter" ($\eta_m = 100\%$) (HEPA or ULPA filter) or adsorber, that removes all but a negligible amount of contaminant. The value of Q_s may be very small, in which case it can be neglected from the analysis. It will be assumed in what follows that no contaminant enters the chamber. It will also be assumed that there is no air infiltration or exfiltration. Inside the chamber a mixing fan insures well-mixed conditions. The concentration inside the chamber satisfies the following:

$$V \frac{dc}{dt} = S - c(A_s k_{ad} + Q_s + Q_r \eta_r) \qquad (5\text{-}45)$$

where

V = chamber volume
t = elapsed time
S = source strength

k_{ad} = wall-loss coefficient

A_s = chamber internal surface area

Q_s = volumetric flow rate through instrument sampling

Q_r = volumetric flow rate through air cleaning device

η_r = efficiency of air cleaning device (contaminant mass removal rate divided by the contaminant mass flow rate entering the device)

In the analysis that follows it will be assumed that the concentration and time are the only variables and that all the other parameters have constant (but perhaps unknown) values.

5.8.1 Wall-Loss Coefficient

To use the test chamber to measure source characteristics or room air cleaner performance, the *wall-loss coefficient* (k_{ad}) must be known for each contaminant to be studied. Alternatively, the researcher's explicit goal may be to study the adsorption characteristics of wall hangings, furniture, and so on. The wall-loss coefficient is also called *plate-out coefficient* or *adsorption rate constant* (k_{ad}). To measure the wall-loss coefficient, shut off the air cleaning device ($Q_r = 0$) and let the source fill the chamber with contaminant. Once a satisfactory concentration is achieved (which need not be the steady-state value), shut off the source and measure the decreasing concentration over a period of time. Under these conditions Eq. 5-45 becomes

$$\left(\frac{1}{c}\right) \frac{dc}{dt} = \frac{d \ln c}{dt} = -\frac{[A_s k_{ad} + Q_s]}{V} \tag{5-46}$$

Assuming the sampling rate Q_s is a known constant, the slope of $\ln c$ versus time enables the one to determine the wall-loss coefficient (k_{ad}).

5.8.2 Source Strength

The *source emission rate* (*source strength*) (S) can be found from several measurements which can be used to confirm one another. Begin by running the air cleaning device over a long period of time without the source ($S = 0$). When a steady minimum concentration is obtained, turn off the air cleaner ($Q_r = 0$), turn on the source, and measure the rise in concentration. Equation 5-45 becomes

$$V \frac{dc}{dt} = S - c(A_s k_{ad} + Q_s) \tag{5-47}$$

Immediately after the source is activated and while the concentration is still small the second term on the right-hand side is small with respect to S. Thus

$$S = V \frac{dc}{dt} \tag{5-48}$$

and the source strength can be found from the slope of concentration versus time. Measuring S by Eq. 5-48 is inherently inaccurate owing to the difficulty of computing a derivative from a few concentration values obtained over a short period of time. If the concentration rises slowly, the accuracy improves. There are two ways to check the accuracy of the value of S obtained from Eq. 5-48: (a) computing S from values of concentration obtained over a period of time after the initial period and (b) computing S from steady-state concentrations.

After the initial period, the source strength can be found by integrating Eq. 5-47.

$$S = -(A_s k_{ad} + Q_s)\left[c_1 \exp\left(\frac{-(Ak_{ad} + Q_s)(t_2 - t_1)}{V} \right) - c_2 \right] \Big/ G$$

$$G = 1 - \exp\left[\frac{-(A_s k_{ad} + Q_s)(t_2 - t_1)}{V} \right] \qquad (5\text{-}49)$$

where c_1 and c_2 are the concentrations after the elapsed times t_1 and t_2. Lastly, one can wait until equilibrium conditions occur and the left-hand side of Eq. 5-45 is zero. Under these conditions the source strength and steady-state concentration c_{ss} are related by

$$S = c_{ss}(k_{ad} A_s + Q_s) \qquad (5\text{-}50)$$

5.8.3 Efficiency of Air-Cleaning Devices

To find the efficiency (η_r) of an air-cleaning device, run the source and air-cleaning device at steady rates for a long period of time until a steady-state concentration (c_{ss}) is obtained, i.e. $c = c_{ss}$ and $dc/dt = 0$. Under these conditions Eq. 5-45 reduces to

$$\eta_r = \frac{\left[\left(\dfrac{S}{c_{ss}} \right) - Q_s - A_s k_{ad} \right]}{Q_r} \qquad (5\text{-}51)$$

Alternatively, allow the source to produce a significant concentration (although not necessarily its steady state value), remove the source ($S = 0$) and turn on the air cleaning device. The concentration will begin to fall and the efficiency can be found from,

$$\eta_r = \frac{-V\left(\dfrac{d \ln c}{dt} \right) - A_s k_{ad} - Q_s}{Q_r} \qquad (5\text{-}52)$$

5.9 CLEANROOMS

Cleanrooms (see Fig. 1.17) are large enclosures in which individuals work and in which the following atmospheric properties are controlled within stringent limits:

Concentration of particles
Temperature
Humidity
Concentration of contaminant gases and vapors

The geometry and operation of cleanrooms varies, but all are designed to provide an environment that will protect a manufactured product from contamination. Unfortunately, many materials used in cleanrooms are very toxic.

In the manufacture of semiconductors, the principal concern is to remove airborne small particles which can short circuit the minute integrated circuits on silicon wafers. Often overlooked, however, are emissions of vapors of corrosive, reactive, and toxic materials used to fabricate wafers. The amounts of these materials are small, but there may be a great deal of hand labor that can produce spills, splash, and airborne emissions.

The fabrication process begins by applying photoresist to the wafer. *Photoresist* is an ultraviolet-sensitive, polymeric material dispersed in a mixed solvent carrier. Following curing in ovens, the wafer is exposed to ultraviolet light and placed in an alkaline developer. The exposed resist dissolves in the developer, leaving the open surface for subsequent processing. A *wet etching* process using corrosive acids such as hydrofluoric acid or bases may be used to remove unwanted material. Alternatively a *dry etching* process involving an RF plasma can be used to remove unwanted material, but in so doing a vast variety of gaseous chemical compounds may be formed that must be controlled. Thin films of material such as silicon nitride and silicon dioxide are then deposited on the wafer by liquid and gaseous processes involving silane, tetraethylorthosilicate, phosphine, diborane, and ammonia. Next, highly toxic or reactive materials called *dopants* (arsenic, phosphorous, arsine, phosphine, or boron trifluoride) having unique electrical properties are imbedded into the surface of the silicon wafer. Arsenine, phosphine, and diborane are gases and are supplied by gas cylinders. Liquid solutions of dopants may be passed into high temperature furnaces by bubblers using inert gases, whereupon the dopant atoms diffuse to the silicon surface. Because of the acute toxicity of dopant materials, sophisticated controls must be used and explicit safety procedures adhered to strictly. Layers of noble or common metals such as gold, aluminum, titanium, tantalum, or tungsten are next deposited as thin films by evaporative or sputtering processes. In between all steps in the process wafers are cleaned by a variety of solvents such as carbon tetrachloride, methylene chloride, and trichloroethylene having long-term toxicity.

Cleanrooms should not be confused with laboratory *fume hoods* (*Fig* 1.20), *biological cabinets, or gloveboxes*. The objective of cleanrooms is to protect a product that is being manufactured as distinct from protecting the worker. The standards for the purity of the air in cleanrooms are considerably more stringent than those to insure the health and safety of workers. Air entering cleanrooms is cleaned and conditioned continuously. Well-mixed conditions are achieved because of the unique ways air enters and leaves the cleanroom rather than because there is a vigorous mixing mechanism within the room.

The cleanliness of a cleanroom is defined by federal standards and is determined by counting the total number of particles greater than 0.5μm (in diameter) per cubic foot of air. Figure 5.13 defines the size distribution for different class cleanrooms in the United States, Japan and West Germany. For example, a class 100 cleanroom is one in which the number of particles greater than 0.5μm per cubic foot of air is less than 100.

Workers in cleanrooms are clothed in garments designed to prevent particles from being generated from clothing and the body. The humidity is set to values appropriate for the product being manufactured and equipment being used. The

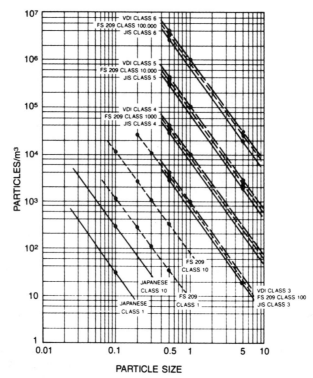

Fig. 5.13 Comparison of West German, US, and Japanese cleaniness classes (Japanese class 1 and class 10 were tentative when this figure was prepared) (reprinted with permission from *Microcontamination*, August 1987, Copyright 1987, Canon Communications, Inc.).

temperature is normally set at 68 F. Floor, ceiling and wall surfaces are designed so as not to generate particles. In addition, floor coverings and garments are designed so as not to generate static electricity.

In vertical (Fig 1.17) laminar flow cleanrooms, the entire ceiling is a high efficiency (HEPA or ULPA) filter and the floor is the receiving plenum. Typical air velocities entering a vertical laminar flow cleanroom are 60-100 FPM. Temperature and humidity control is achieved by a separate air handling system. Class 100 conditions can be achieved by such designs. The performance of laminar-air flow rooms is hampered by wake regions downstream of equipment and personnel. Such wakes become dead regions that tend to collect particles and prevent their removal.

It must be emphasized that while air may enter a laminar-flow room in a laminar fashion, the existence of wakes and dead air spaces produce limited degrees of turbulence that are unavoidable. In addition, the Reynolds numbers for the rooms themselves or the obstacles around which the air passes are large such that the notion of "laminar" is incorrect.

Illustrated Example* 5.7   *Time To Achieve Cleanroom Conditions   Consider a cleanroom such as Fig. 1.17 in which there is 100% recirculation two inlet filters in series and no make up air. The floor area of the cleanroom is $100\,m^2$ and the height is 3 m. Cleaned air enters the room through the ceiling with an average velocity of 15.24 m/min (50 FPM). Assume that there is no adsorption of particles on solid surfaces. Maintenance is performed on the cleanroom in which ambient air containing 50,000 particles $/cm^3$ (particles larger than 0.5 μm) fills the cleanroom. After maintenance is performed, how long will it take to achieve class 100,000 to class 10 cleanroom conditions?

The differential equation describing the conservation of mass for particles in a well mixed environment is

$$V \frac{dc}{dt} = -Qc[1 - (1 - \eta_1)(1 - \eta_2)]$$

where η_1 and η_2 are the removal efficiencies of the prefilter and HEPA filter, 98% and 99.98% respectively. The equation can be integrated directly. A summary of the time required to achieve various class environments is shown below.

Class Cleanroom	Time (min)
100,000	1.88
10,000	2.33
1,000	2.79
100	3.24
10	3.70

While these time periods are small, one should realize that a face velocity of 50 FPM distributed over a ceiling of area $100\,m^2$ requires a hefty air-handling system to produce $1524\,m^3/min$ (53,792 SCFM) and approximately 304.8 room-air-changes per hour!

5.10 INFILTRATION AND EXFILTRATION

The transfer of air into and out of an enclosure is equal to the deliberate input and removal of air (ventilation) plus the uncontrolled air leakage. The uncontrolled flow of air into a building is called *infiltration* and the uncontrolled removal of air is called *exfiltration*. Infiltration and exfiltration are produced by the pressure difference between the building interior and atmosphere due to the aerodynamics of air flowing around and over the building. To a lesser extent it is also due to the temperature difference between the building interior and the atmosphere. To a first approximation one may assume that the volumetric for rates of infiltration and exfiltration are equal. The relative air leakage of a building is distributed as follows (97):

Walls (top and bottom joints, plumbing and electrical penetrations), 35%
Ceiling, 18%
Heating system, 15%
Fireplaces, 12%

TABLE 5.2 Infiltration and Exfiltration[a]

Kind of Room	Single Glass—No Weatherstripping	Storm Sash or Weatherstripping
No windows or exterior doors	0.5	0.3
Windows or exterior doors on one side	1.0	0.7
Windows or exterior doors on two sides	1.5	1.0
Windows or exterior doors on three sides	2.0	1.3
Entrance halls	2.0	1.3

[a]Air changes per hour occurring under average conditions in residences exclusive of air provided for ventilation

Vents in conditioned spaces, 5%

Diffusion (conduction) through walls, <1%

For purposes of industrial ventilation, the forced flow of air into and out of a building, that is, ventilation, is generally much larger than infiltration and exfiltration. Infiltration can be estimated in three ways:

Air changes per hour

Empirical equations

Equations based on pressure drop and construction details

Table 5.2 is a condensation of the ASHRAE (97) recommendations for air leakage in terms of number of air changes per hour (infiltration plus exfiltration) for average buildings under average conditions. Infiltration can also be estimated in a detailed fashion as functions of building construction details such as wall construction, types of shingles, window and door specifications, wall construction of elevator and stair shafts, and so on. These relationships will be omitted, and the reader is advised to consult ASHRAE's literature on "ventilation and infiltration" (97).

Wadden and Scheff (98) report that the following empirical equation for infiltration volumetric flow rate divided by the building volume can be used:

$$\frac{Q}{V} = 0.315 + 0.0273U + 0.0105(T_a - T_0) \qquad (5\text{-}53)$$

where Q/V is in 1/hr, U is the wind speed in miles per hour, and T_a and T_0 are the ambient and room temperatures in degrees Fahrenheit.

5.11 SEQUENTIAL BOX MODEL

If it is known or suspected that the flow field within an enclosure distributes contaminants nonuniformly, as shown in Fig. 5.14, it would be inaccurate to model the entire enclosure as a well-mixed region (i.e. Eq. 5-36). One could

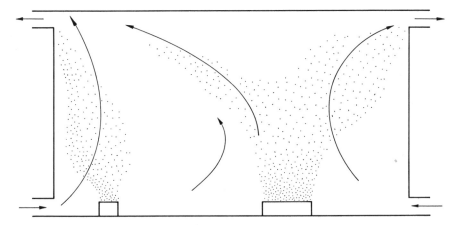

Fig. 5.14 Partially mixed enclosure containing two sources, two makeup air vents, and two exhaust vents.

assume partially mixed conditions and use mixing factors (Section 5.6) with the attendant uncertainty in selecting the value of the mixing factor (m), but in the final analysis the partially mixed model would still predict spatially uniform concentrations and if one needed a model to predict concentrations that had some dependence on location as well as time, the partially mixed model would be of no use. An alternative is the *sequential box model*, also called the *multi-cell well-mixed model* (99,100,223,224,327).

Consider a building containing well-defined rooms of measurable volume, measurable volumetric flow rates from one room to another and well-mixed conditions within each room. The sequential box model is ideally suited to predict the instantaneous concentration in each room. In addition to the obvious application to rooms within a building, the sequential box model can also be applied to situations where there are not physical boundaries separating cells and the volumetric flow rates between cells is not known.

The sequential box model divides the enclosure into a number of cells, within each, well-mixed conditions occur, and between each, contaminant is transferred. Figure 5.15 is a schematic diagram of a two-cell, sequential box model. The concept can be expanded to as many cells as the user wishes. For convenience in analysis, the fractions x_1, x_2 and y_1 and y_2 will be used to denote the fraction of the total air volumetric flow rate (Q) entering and leaving cells across their exterior boundaries. Thus into the enclosure, the volumetric flow rates are

$$Q_{1,i} = x_1 Q \tag{5-54}$$

$$Q_{2,i} = x_2 Q \tag{5-55}$$

$$Q = Q_{1,i} + Q_{2,i}$$

$$Q = Q(x_1 + x_2)$$

$$1 = x_1 + x_2 \tag{5-56}$$

and similarly for y for flow out of the enclosure

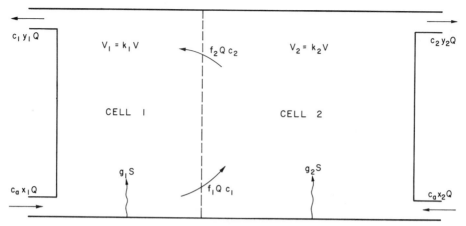

Fig. 5.15   Schematic diagram of a two-cell sequential box model.

$$Q_{1,o} = y_1 Q \qquad (5\text{-}57)$$

$$Q_{2,o} = y_2 Q \qquad (5\text{-}58)$$

$$1 = y_1 + y_2 \qquad (5\text{-}59)$$

Values of these fractions are computed from the known amounts of make-up air, recirculation, exhaust, and infiltration. Contaminant generated in each cell will be designed by g_1 and g_2 times the total contaminant generating rate S,

$$S = S(g_1 + g_2)$$

$$1 = g_1 + g_2 \qquad (5\text{-}60)$$

The volume of cells 1 and 2 are expressed as fractions k_1 and k_2 of the total volume V,

$$V = V(k_1 + k_2)$$

$$1 = k_1 + k_2 \qquad (5\text{-}61)$$

Fractions x_1, x_2, y_1, y_2, g_1, g_2, k_1, and k_2 must be specified by the user as input parameters to the analysis.

Of critical importance in the sequential box model are parameters f_1 and f_2, called *exchange coefficients*, that account for the volumetric flow rate transferred across the internal boundary between cells 1 and 2. Values of f_1 and f_2 may be *less or greater than unity*. Values of f_1 and f_2 are selected by the designer based on some idea that well-mixed conditions exist within certain portions of the enclosure. The parameters f_1 and f_2 are difficult to select since the designer may have little information about the flow field within the enclosure or how much air is transferred between cells. For each cell the conservation of mass for air reduces to

$$\text{Cell 1: } x_1 Q + f_2 Q - y_1 Q - f_1 Q = 0$$

$$x_1 - y_1 + f_2 - f_1 = 0 \qquad (5\text{-}62)$$

$$\text{Cell 2: } x_2Q + f_1Q - y_2Q - f_2Q = 0$$

$$x_2 - y_2 + f_1 - f_2 = 0 \tag{5-63}$$

For cells 1 and 2, the equations for the conservation of mass for the contaminant are

$$k_1 V \frac{dc_1}{dt} = g_1 S + x_1 Q c_a + f_2 Q c_2 - f_1 Q c_1 - y_1 Q c_1 \tag{5-64}$$

$$k_2 V \frac{dc_2}{dt} = g_2 S + x_2 Q c_a + f_1 Q c_1 - f_2 Q c_2 - y_2 Q c_2 \tag{5-65}$$

The concentration in the air entering cells 1 and 2 from the outside is assumed to be constant and equal to c_a.

The number of air changes N (Eq. 5-11) will be used

$$N = \frac{Q}{V}$$

Simplify the nomenclature by letting, $k = k_1$, $g = g_1$ and $f = f_1$. Thus from Eq. 5-62, $f_2 = f + y - x$. Equations 5-64 and 5-65 reduce to the following pair of simultaneous, first-order differential equations with constant coefficients

$$\frac{dc_1}{dt} = A + Bc_1 + Dc_2 \tag{5-66}$$

$$\frac{dc_2}{dt} = E + Fc_2 + Gc_1 \tag{5-67}$$

where the constant coefficients are

$$A = \left(\frac{N}{k}\right)\left[x_1 c_a + \left(\frac{gS}{Q}\right)\right] \tag{5-68}$$

$$B = -\left(\frac{N}{k}\right)[f + y_1] \tag{5-69}$$

$$D = \left(\frac{N}{k}\right)[x_2 - y_2 + f] \tag{5-70}$$

$$E = \left[\frac{N}{(1-k)}\right]\left[x_2 c_a + \left\{(1-g)\frac{S}{Q}\right\}\right] \tag{5-71}$$

$$F = -\left[\frac{N}{(1-k)}\right](x_2 + f) \tag{5-72}$$

$$G = \frac{Nf}{(1-k)} \tag{5-73}$$

Skaret (99) gives the following general solution for the concentration in cells 1 and 2 as a function of time

$$c_1(t) = K_1 \exp(Nts_1) + K_2 \exp(Nts_2) + c_{1,ss} \tag{5-74}$$

$$c_2(t) = MK_1 \exp(Nts_1) + LK_2 \exp(Nts_2) + c_{2,ss} \tag{5-75}$$

where $c_{1,ss}$ and $c_{2,ss}$ are the final (steady state) cell concentrations given by

$$c_{1,ss} = \frac{(AF - ED)}{(DG - BF)} \tag{5-76}$$

$$c_{2,ss} = \frac{(EB - AG)}{(DG - BF)} \tag{5-77}$$

The remaining constants in Eqs. 5-74 and 5-75 are

$$s_1 = \left(\frac{1}{2N}\right)\left[(B + F) + \sqrt{(B - F)^2 + 4DG}\right] \tag{5-78}$$

$$s_2 = \left(\frac{1}{2N}\right)\left[(B + F) - \sqrt{(B - F)^2 + 4DG}\right] \tag{5-79}$$

$$M = \frac{(Ns_1 - B)}{D} \tag{5-80}$$

$$L = \frac{(Ns_2 - B)}{D} \tag{5-81}$$

$$I_1 = c_1(0) - c_{1,ss} \tag{5-82}$$

$$I_2 = c_2(0) - c_{2,ss} \tag{5-83}$$

$$K_1 = \left[L - \left(\frac{I_2}{I_1}\right)\right]\left[\frac{I_1}{(L - M)}\right] \tag{5-84}$$

$$K_2 = \left[\left(\frac{I_2}{I_1}\right) - M\right]\left[\frac{I_1}{(L - M)}\right] \tag{5-85}$$

where $c_1(0)$ and $c_2(0)$ are the initial cell concentrations.

Applying Eqs. 5-74 and 5-75 is straightforward except for the selection of f. All the other constants in the solution are parameters are known to the designer. The selection of k depends on what portions of the enclosure can be assumed to be wellmixed. The value of g depends on how contaminants are generated within the enclosure and will be known with some certainty. The value of f is the most difficult to forecast. The best way of proceeding is to solve the problem for several values of f that the designer believes bracket the actual situation. In this way upper and lower bounds of the solution can be found which then may suggest measurements or the means to bracket the value of f more accurately. At one extreme, a value of zero makes the coefficient G zero, decouples the variable c_1 from Eq. 5-67, and isolates cell 2 from contaminant generated in cell 1. As the value of f increases, the exchange of material between each cell increases. Note, values of f may exceed unity. When f is approximately 15 (100), the concentration in both cells approaches what would be predicted if the entire enclosure was analyzed as a single well-mixed cell. Thus, if the volumetric flow rate between cells is approximately 15 times the total volumetric flow rate into the entire enclosure, well-mixed conditions can be presumed to exist throughout the entire enclosure.

The use of only two cells is crude, but better than assuming the entire enclosure is well-mixed. The next logical step is to divide the enclosure into several more cells and repeat the technique. The complexity of analyzing multiple cells increases as approximately the square of the number of the cells (327). Even assuming values of x_n, y_n, g_n and k_n are known for each cell (n), a differential equation of the form Eq. 5-62 must be written for each cell and a pair of equations of the form Eqs. 5-62 and 5-63 must be written to describe the exchange of air between each set of adjacent cells. Thus selecting the value of f_n for each cell becomes critical to the accuracy of the solution. The solution of several simultaneous, first-order differential equations is not overly complicated and several computer programs exist to perform the task. Once again, finding the values of the parameters $f_1, f_2, \ldots f_n$ is the most difficult task. The complexity of the problem escalates quickly and the usefulness of the solution may decrease accordingly. In the final analysis, if hundreds of cells are used, the solution resembles finite difference techniques that will be discussed in Chapter 10.

The sequential box model is useful for predicting concentrations that vary with both time and location. The accuracy of the model is limited by the inability to describe the exchange of air between cells in sufficient detail. Setting the derivative dc_i/dt (where i is the box number) equal to zero yields expressions from which the steady-state concentration ($c_{i,ss}$) can be found. If transient ccncentrations are sought, the ordinary differential equations (similar to 5-64 and 5-65) can be solved. If the volumetric flow rate and, or source strength vary with time, the ordinary differential equations can be solved by Runge Kutta methods. Sequential box models have been used to analyze indoor radon concentrations (359), tobacco smoke in commercial aircraft cabins (327, 342, 223), contaminant concentrations associated with painting and welding in confined spaces (360) and oxygen deficiencies in confined spaces (394, 395).

Illustrated Example 5.8—Transfer of HCL Vapor Between Rooms   A small electronics firm plans to manufacture circuit boards. The process consists of dipping circuit boards in open vessels containing various aqueous HCl solutions in room number 1 followed by assembly operations in room number 2. The liquid surface area is 75 ft^2 and HCl vapor is emitted at a rate of 0.02 g/s m^2. The room containing the open vessels is 30 ft × 30 ft × 15 ft high and is connected by a doorway 10 ft wide and 15 ft high to the assembly room that is 50 ft × 30 ft × 15 ft high. No HCl is generated in the assembly room. The plant manager wishes to place an air curtain in the doorway to prevent HCl vapor from entering the assembly room: however, he suspects that the air curtain may not provide isolation, and worse yet, may act as a mixing device and transfer air from one room more vigorously than without the air curtain. Each room will have its own heater and air conditioner. Each device moves only air within the respective room so that assuming well-mixed conditions within each room is a reasonable assumption, at least for a first approximation. Within each room infiltration and exfiltration are equal to one change per hour. A ventilation system delivers 600 CFM of outside air to the assembly room and 600 CFM of contaminated air is removed from the room containing the open vessels. You have been asked by the plant manager to estimate the steady-state concentrations in each room, how fast the concentrations will increase with time, and whether PEL (5 PPM, 7 mg/m^3) will be exceeded in either room. Assume that adsorption can be neglected, that the initial HCl concentration in both rooms is zero and that the concentration of HCl in the outside air is zero.

Values of the parameters used in the equations above are computed to be

$$x_1 = 0.1875 \quad x_2 = 0.8125 \quad y_1 = 0.6875 \quad y_2 = 0.3125$$

$$k = 0.375 \quad g = 1.0 \quad Q = 72{,}000 \text{ CFM} \quad S = 502 \text{ g/hr} \quad N = 2/\text{hr}$$

$$A = 0.0372 \text{ g/hr m}^3$$

$$B = -5.33(f + 0.6875)\text{per hr}$$

$$D = 5.33(f + 0.5)\text{per hr}$$

$$E = 0$$

$$F = -3.2(f + 0.8125)\text{per hr}$$

$$G = 3.2(f)\text{per hr}$$

A useful first step is to compute the steady state concentration and time constant if the plant is a single enclosure, a calculation that occurs if f exceeds values of 15. Such a calculation is very crude, but it helps place bounds on later calculations. Using Eqs. 5-74 and 5-75, one finds that a steady-state concentration is 175 PPM and the time constant is 0.5 hr. On the basis of this alone, the plant manager can expect to exceed PEL before much of the working day has passed.

Figure 5.16 shows how the concentration in cells 1 and 2 vary with time and several values of f. When f is zero, cell 1 becomes well-mixed and cell 2 receives no HCl vapor. As the value of f increases, mixing increases and the steady state concentrations in both room approach the value that would be predicted if the

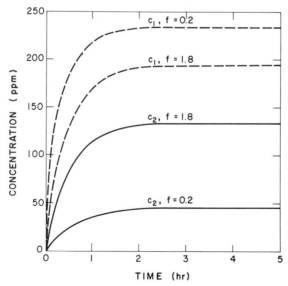

Fig. 5.16 Concentration in cells 1 and 2 as functions of time and exchange coefficient—results of Illustrated Problem 5.8.

entire plant was well-mixed. Values of f also affect the rate at which the concentration increases with time.

Selecting appropriate values of f remains an elusive task. If tracer experiments could be conducted, the change of concentration with time could be determined, and from such knowledge, designers could select the appropriate value of f.

5.12 EFFECTIVENESS COEFFICIENT

Several ventilation design strategies are in common use throughout the world. In one strategy deliberate decisions are made about the locations of air inlets and outlets in order to achieve well-mixed conditions. In another strategy deliberate steps are taken to produce an air velocity that is uniform across the enclosure such as a vertical-laminar flow cleanroom, Fig. 1.17. A ventilation system in which this is achieved is called *displacement ventilation*. In ordinary engineering parlance, such flow fields are called *plug flow*. In between these two unique flow fields, there exist ventilation systems in which there are regions of relatively stagnant air and regions in which air "short circuits" the majority of the enclosure.

If the circulation of air in an enclosure is known to produce nonuniform concentrations, that is, partially mixed conditions, it would be useful to know which parts of the enclosure receive too little of the incoming air and which points receive too much. For example, point P in Fig. 5.17 is located in a low-velocity eddy region in which the concentration changes slowly, whereas point Q is located in a high velocity region in which the concentration changes rapidly. However, just because the concentration changes slowly does not mean the concentration is large or small, but only that it changes slowly. While the mixing factor (see Section 5.6) for the entire room is less than unity, it would be useful to characterize the mixing at points P and Q in a quantitative way. Two enclosures may have the same overall mixing factors, but the mixing at distinct points in either room may be quite different.

Sandberg (114) has developed a comprehensive theory to analyze nonuniform conditions within enclosures based on measuring the concentration over period of time (t_1) and computing a *dose* (or zero moment):

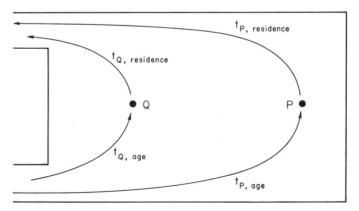

Fig. 5.17 Age and residence time in an enclosure.

$$\text{dose:} \int_0^{t'} c(t)\, dt \tag{5-86}$$

and a first moment of the concentration,

$$\text{The (first) moment:} \int_0^{t'} c(t)t\, dt \tag{5-87}$$

where t' is the time period of interest. Partially mixed conditions can be characterized (100–102) by an *effectiveness coefficient e*, defined as follows:

$$e = \frac{t_N}{t_i} \tag{5-88}$$

where t_i is called the *local mean age*,

$$t_i = \frac{\displaystyle\int_0^{\infty} c(t)t\, dt}{\displaystyle\int_0^{\infty} c(t)\, dt} \tag{5-89}$$

and t_N is the reciprocal of the number of room air changes per unit time,

$$t_N = \frac{1}{N} = \frac{V}{Q} \tag{5-90}$$

For well-mixed conditions, one can show mathematically that the local mean age t_i is equal to t_N. Thus if at a point in an enclosure, the effectiveness coefficient is unity ($e = 1$), mixing at that point is similar to what would be predicted if the entire space were well-mixed. If $t_i > t_N$, the effectiveness coefficient e is less than unity and it means that the point in space is located in a stagnant region where mixing is poor. Finally, if the effectiveness coefficient e is greater than unity, mixing at that point in space is very vigorous and one may visualize that makeup air passes through the point and "short circuits" the enclosure as a whole.

At this point the reader must be careful to understand the difference between three rather overlapping terms:

m = mixing factor (Eq. 5-31)
f_1, and f_2 = exchange coefficients (see Fig. 5-15), Eq. (5-62, 5-63)
e = effectiveness coefficient (Eq. 5-88)

The mixing factor m is an empirical constant included in the differential equation for a well-mixed model to account for the fact that only a portion of the inlet air mixes with air inside the enclosure. Even when a mixing factor is included in the differential equation, the concentration remains spatially uniform, that is, $c(x, y, z, t) = c(t)$. The exchange coefficient is a scaler quantity which when multiplied by the total volumetric flow rate Q expresses the total volumetric flow rate transported between adjacent cells in the sequential box model. The effectiveness coefficient is a scalar quantity evaluated at a specific point in space that reflects the level of mixing occurring at that point.

In summary,

$$e > 1 \text{ requires } t_i < t_N \ldots \text{good mixing}$$

$$e < 1 \text{ requires } t_i > t_N \ldots \text{poor mixing}$$

(5-91)

From tracer experiments the exchange coefficients can be measured and since the effectiveness coefficients are proportional to the exchange coefficients f used in the sequential box model, the designer can select values for calculation with better precision.

Suppose the flow field in an enclosure is as shown in Fig. 5.17. If the velocity field was known, the *age of the makeup air* reaching points P and Q could be computed from the integral,

$$t(\text{age}) = \int^{\text{Pathline from inlet to point}} \frac{dl}{s}$$

(5-92)

where l is the length of the fluid pathline from the inlet to point P or Q and s is the speed of the air along the pathline. In a similar fashion, the *residence time* of contaminants generated at points P and Q could be computed from

$$t(\text{residence}) = \int^{\text{Pathline from point to outlet}} \frac{dl}{s}$$

(5-93)

If point Q receives makeup air faster than another point P, the age of the makeup air at point P is larger than at point Q and the effectiveness coefficient at point P is less than it is at point Q. The greater the disparity of effectiveness coefficients (e) throughout the enclosure, the more uneven the distribution of makeup air and the less valid an assumption of well-mixed becomes. By determining the effectiveness coefficients (e) at various points, one can evaluate partially mixed conditions on a quantitative basis and characterize the enclosure in a more detailed fashion than merely using a mixing factor (m).

If the flow through an enclosure is in one direction and spatially uniform, plug flow, the integral of dl/s taken over the pathline from inlet to outlet is the total elapsed time it takes the makeup air to pass through the enclosure. The elapsed time can be found from the integral,

$$\text{Elapsed time} = \frac{dl}{s} = \frac{dl}{(Q/A)} = A\frac{dl}{Q} = \frac{V}{Q}$$

(5-94)

Thus the conventional time constant for well-mixed conditions is equal to the physical time air resides in an enclosure if plug flow conditions exist. The reader should be careful not to conclude that well-mixed means plug flow because it certainly does not.

The mean age of air (t_i) can be determined experimentally by *step-up*, *step-down*, or *pulse injection* tracer experiments. In step-up experiments, a tracer gas is added to the makeup air at a constant rate, the (rising) concentration at a point in the enclosure is measured over a period of time, and t_i computed from Eq. 5-89. In step-down experiments, the tracer gas is added to the enclosure and uniformly distributed. At time zero, the introduction of tracer gas is stopped and only fresh air is added to the enclosure. The (decreasing) concentration is measured at a point in the enclosure over a period of time, and the mean age, t_i, is computed,

Step-up:

$$t_i = \frac{\int_0^\infty t \left[1 - \frac{c(t)}{c(\infty)} \right] dt}{\int_0^\infty \left[1 - \frac{c(t)}{c(\infty)} \right] dt} \tag{5-95}$$

Step-down:

$$t_i = \frac{\int_0^\infty t \left[\frac{c(t)}{c(0)} \right] dt}{\int_0^\infty \left[\frac{c(t)}{c(0)} \right] dt} \tag{5-96}$$

where $c(\infty)$ is the tracer concentration at time equal to infinity and $c(0)$ is the initial tracer concentration. The experiments described above can be conducted without much difficulty because, for most rooms and buildings, the concentration changes slowly over a period of hours and rapid response instruments are not necessary. In addition, measurements can be made at several points and effectiveness coefficients computed for several points in a single experiment. The technique has been found to be very useful in characterizing ventilation in buildings and rooms within buildings (333).

Illustrated Example 5.9—Effectiveness Coefficient   Let points P, Q, and R correspond to three locations in an enclosure of volume V. The enclosure receives fresh air at a volumetric flow rate Q and discharges room air at a similar rate. A step-down tracer experiment has been run and the concentration at points P, Q, and R are found to vary in the following way.

Point P:

$$\frac{c_P(t)}{c_0} = \begin{bmatrix} 1 & \text{for} & t < t_1 \\ 0 & \text{for} & t > t_1 \end{bmatrix}$$

Point Q:

$$\frac{c_Q(t)}{c_0} = \begin{bmatrix} [1 - (t/t_2)] & \text{for} & t < t_2 \\ 0 & \text{for} & t > t_2 \end{bmatrix}$$

Point R:

$$\frac{c_R(t)}{c_0} = \exp\left(-t\frac{Q}{V}\right) \qquad \text{for all } t$$

where c_0 is the initial concentration in the room.

If $t_1 = 0.2V/Q$ and $t_2 = 5V/Q$, show that the effectiveness coefficients for the three points are equal to

$$e_P = 10.0 \qquad e_Q = 0.6 \qquad e_R = 1.0$$

Point P:

$$t_{i,P} = \frac{\displaystyle\int_0^\infty t\,\frac{c(t)}{c(0)}\,dt}{\displaystyle\int_0^\infty \frac{c(t)}{c(0)}\,dt}$$

$$t_{i,P} = \frac{\displaystyle\int_0^{t_1} t\,dt}{\displaystyle\int_0^{t_1} dt} = \frac{t_1}{2}$$

$$e_P = \frac{\left(\dfrac{2V}{Q}\right)}{t_1} = \frac{\left(\dfrac{2V}{Q}\right)}{\left(\dfrac{0.2V}{Q}\right)} = 10$$

Point Q:

$$\int_0^{t_2}\left(\frac{c(t)}{c_0}\right)dt = \int_0^{t_2}\left[1-\left(\frac{t}{t_2}\right)\right]dt = \frac{t_2}{2}$$

$$\int_0^{t_2} t\,\frac{c(t)}{c_0}\,dt = \int_0^{t_2}\left[1-\left(\frac{t}{t_2}\right)\right]t\,dt = \frac{t_2^2}{6}$$

$$t_{i,Q} = \frac{\left[\dfrac{t_2^2}{6}\right]}{\left[\dfrac{t_2}{2}\right]} = \frac{t_2}{3}$$

$$e_Q = \frac{\left(\dfrac{3V}{Q}\right)}{t_2} = \frac{\left(\dfrac{3V}{Q}\right)}{\left(\dfrac{5V}{Q}\right)} = 0.6$$

Point R:

$$\int_0^\infty \frac{c(t)}{c_0}\, dt = \int_0^\infty \exp\left(-\frac{tQ}{V}\right) dt = \frac{V}{Q}$$

$$\int_0^\infty t\, \frac{c(t)}{c_0}\, dt = \int_0^\infty t \exp\left(-\frac{tQ}{V}\right) dt = \left(\frac{V}{Q}\right)^2$$

$$t_{i,R} = \frac{\left[\left(\frac{V}{Q}\right)^2\right]}{\left[\frac{V}{Q}\right]} = \left(\frac{V}{Q}\right)$$

$$e_R = \frac{\left(\frac{V}{Q}\right)}{t_{i,R}} = 1.0$$

Point P is a location where the mean age t_i is considerably smaller than the time constant and consequently e_P is greater than unity. Such a location has very good mixing. Point Q on the other hand, is a location where the average age t_i is considerably larger than the time constant, e_Q is less than unity and indicates that mixing is very poor. Point R is a location where mixing is what one would predict if the entire volume was well-mixed.

5.13 MAKEUP AIR OPERATING COSTS

Makeup air must be heated before it enters the workplace. In some cases it may also have to be cooled and cleaned and have water added or removed. Costs associated with conditioning makeup air are significant but unfortunately are apt to be ignored. Such an omission is serious and care must be taken to avoid it.

For purposes of illustration, consider only the yearly cost to heat makeup air. The rate at which energy is required to heat outside air from an ambient temperature (T_a) to say 65 F is equal to the makeup air mass flow rate (ρQ) times the enthalpy change $c_p(65 - Ta)$. The energy added throughout the year is this rate integrated over the time energy is added. Since the ambient temperature (T_a) varies throughout the year, such an integral would be tedious to compute. The value of this integral is called *number of heating degree-days* and average values for specific communities have been measured by the US Weather Bureau for the period 1931–1960. Tables of values for communities throughout the United States can be found in the ASHRAE Handbook (97). Shown below is a sample summary:

Wash DC	NYC	Chicago	Denver	Boston	Pitt	St Louis	Atlanta
4224	4871	6155	5524	5634	5987	4900	2961

If energy is acquired from a heater, the yearly fuel cost is equal to the energy gained times the number of hours heating is required times the cost per unit of fuel (C_{fu}) divided by the efficiency of conversion. The ACGIH manual (27) recommends the following equation:

$$\text{Annual cost} = 0.154Q \; dg \; \frac{tC_{fu}}{q} \qquad (5\text{-}97)$$

where

Q = volumetric flow rate (CFM)
dg = annual heating degree-days
t = operating time (hr/wk)
C_{fu} = fuel cost (\$/unit)
q = available energy per unit of fuel

Typical values (27) of the available energy per unit of fuel are

Fuel	Efficiency	Available energy per unit
Coal	50%	6000 BTU/lbm
Oil	75%	106,500 BTU/gal
Gas (heat exchanger)	80%	800 BTU/ft^3
Gas (direct fire)	90%	900 BTU/ft^3

Illustrated Example 5.10—Cost to Heat Makeup Air Assume that the firm manufacturing circuit boards (see Illustrated Example 3.5) is located in Pittsburgh, PA. Estimate the annual cost to heat makeup air to 65 F assuming a direct-fired modular heater is used. The following data pertain:

Q = 34,500 CFM
t = 40 hr/wk
dg = 5987
C_{fu} = \$0.005/ft^3
q = 900 BTU/ft^3

$$\text{Annual cost} = \frac{0.154(34,500)(40)(5987)(0.005)}{(900)} = \$7069/\text{yr}$$

The estimated makeup air heating cost used in Illustrated Example 3.5 was \$5000 and below what should have been used.

NOMENCLATURE

Symbol	Description	Dimensions*
a_1, a_2, a_3	Constants defined by equation	
A	Constant defined by equation	

* Q – Energy, F – Force, L – Length, M – Mass, N – Mols t – Time, T – Temperature

A	Area	L^2
A_s	Surface area of enclosure over which Adsorption occurs	L^2
B	Constant defined by equation	
$c(i+1)$	Concentration determined in iteration number $(i+1)$	M/L^3
$c(t)$	Concentration	M/L^3
c_i	Concentration of molecular species i, Inlet concentration	M/L^3
$c(0), c(\infty)$	Initial and final concentration	M/L^3
c_a	Ambient concentration	M/L^3
c_b	Bulk concentration	M/L^3
c_p	Specific heat at constant pressure	Q/MT
c_s	Concentration in supply air	M/L^3
c_{ss}	Steady-state concentration	M/L^3
C_{fu}	Cost per unit of fuel	
D	Constant defined by equation	
D_p	Particle diameter	L
dg	Annual heating degree-days	t
e	Effectiveness coefficient defined by equation	
E	Constant defined by equation	
F	Constant defined by equation	
f	Makeup air fraction defined by equation $(f = Q_m/Q)$	
$f_1, f_2,$	Exchange coefficient (multiple of total volumetric flow rate transferred between cells 1 and 2)	
G	Constant defined by equation	
g_1, g_2	Fraction of contaminant generated in cells 1 and 2	
$I_1, I_2,$	Constants defined by equation	
k_1, k_2	Fraction of total volume comprised by cells 1 and 2	
$k_0(T,\phi)$	Mass transfer coefficient	L^3/t
k_{ad}	Adsorption rate constant or wall-loss coefficient	L/t
K, K_1, K_2	Constants defined by equation	
L	Constant defined by equation	
M	Constant defined by equation	
M_i	Molecular weight of species i	M/N
M_p	Average amount of particle matter exhaled per cigarette	M
m	Mixing factor	
m_s	Mass of organic solvent	M
n	Number of cigarettes per room	
N	Number of room air changes per unit time $(N = Q/V)$	t^{-1}
P	Pressure	F/L^2
P_c	Critical pressure	F/L^2

P_0, P_r	Far field pressure, room pressure	F/L^2
$P(sat, T_r)$	Saturation pressure based on temperature T_r	F/L^2
q	Available energy per unit of fuel	Q/M
Q	Volumetric flow rate	L^3/t
$Q_e, Q_m, Q_r,$	Volumetric flow rate of air exhausted from	
Q_a, Q_s, Q_d	room, makeup air, recirculated air,	
	ambient air, supply air, dilution air	L^3/t
$r_1(t), r_2(T), r_3(\phi)$	Functions defined by equation	
R_u	Universal gas constant	Q/NT
s_1, s_2	Constants defined by equation	
S	Source, contaminant generation rate	M/t
S'	Net formaldehyde mass emission rate	M/t
t	Time	t
t'	Characteristic time	t
T	Temperature	T
t_b	Average length of time a cigarette burns	t
t_i	Local mean age defined by equation	t
t_N	Average residence time, $t_N = V/Q$	t
T_c	Critical temperature	T
T_r	Room temperature	T
U	Wind speed	L/t
V	Volume	L^3
V_d	Volume of dilution air	L^3
x_1, x_2	Fraction of total volumetric flow rate	
	entering cell 1 and 2	
y_i	Gas phase mol fraction of molecular species i	
y_1, y_2	Fraction of total volumetric flow rate	
	leaving cell 1 and 2	

Greek

δt	Increment of time	t
η_n	Contaminant removal efficiency	
	of device number n	
ϕ	Relative humidity	
ρ	Air density	M/L^3

Subscripts

$(\ )_a$	Properties of ambient air
$(\ )_{ad}$	Adsorption of contaminants on walls
$(\ )_c$	Property at (thermodynamic) critical point
$(\ )_m$	Make up air
$(\ )_{max}$	Maximum value
$(\ )_p$	Particle
$(\ )_P, (\ )_Q, (\ )_R$	Properties at point P, Q and R
$(\ )_r$	Room, recirculated
$(\ )_s$	Supply, surface condition
$(\ )_{ss}$	steady state
$(\ )_0$	Far-field properties
$(\ )_1, (\ )_2$	Properties at points 1 and 2

Abbreviations

ACFM	Volumetric flow rate (ft^3/min) that exists at the actual temperature and pressure
ACGIH	American Conference of Governmental Industrial Hygienists
ASHRAE	American Society of Heating, Refrigerating and Air-Conditioning Engineers
BTU	British thermal unit
CFM	Ft^3/min
EPA	Environmental Protection Agency
HEPA	high efficiency particle air filter
LEL	Lower explosion limit
OSHA	Occupational Safety and Health Administration
PPM	Parts per million
TLV	Threshold limit value
SCFM	Volumetric flow rate (ft^3/min) that would exist at STP
STP	Standard temperature (25 C) and pressure (101 kPa)
TSP	Total suspended particulates
ULPA	Ultra efficiency particulate air filter
VOC	Volatile organic compound

PROBLEMS

1. A storage closet ($5\,m^3$) is used to store hydrocarbons. If a leak occurs in one of the vessels containing 1 kg of liquid, and there is no transfer of air into or out of the room, compute the maximum mole fraction if the leaked hydrocarbon is acetone. How does this concentration compare with the TLV and LEL for acetone?   (a) Repeat for benzene, carbon tetrachloride and methyl alcohol.(b) repeat (a) if the initial amount is 5 kg.

2. General ventilation may be a viable control strategy when the contaminant is not toxic, the generation rate (S) is low and the room volume (V) and ventilation rate (Q) are large. If a room is ventilated as per Fig. 5.10, show that if V and Q are constant, and the outside concentration and adsorption are negligible, the concentration is linearly proportional to the source strength (S) during the first few moments of the process.

3. One kilogram of paint remover composed of 30% methyl alcohol and 70% methylene chloride are stored in an unventilated closed closet ($4\,m^3$) in a home. If the can is left open for a long period of time, find the maximum concentration of alcohol and methylene chloride in the air in the closet.

4. An underground conduit carries waste liquids to a central disposal process. The conduit contains a trap to remove solids that can accumulate and stop the flow of liquid. Every few months workers clean the trap by entering an underground chamber ($8\,ft \times 8\,ft$ and 10 ft deep) normally covered by a manhole cover. Unfortunately, gases and vapors dissolved in the waste water escape to the air in the chamber. For a period of time the conduit has carried an aqueous mixture containing dissolved carbon monoxide, hydrogen

cyanide, and benzene. The concentration of these materials in the waste water are as follows,

C_6H_6: 0.18 mg per 100 g of water
CO: 4.0 μg per 100 g of water
HCN: 12.0 mg per 100 g of water

Estimate the equilibrium partial pressure of these materials in the air using Henry's law. If there is no exchange of air between air in the underground chamber and the atmosphere, estimate the maximum concentration of carbon monoxide, hydrogen cyanide, and benzene in the air in the underground chamber. Discuss these values in terms of OSHA and ACGIH standards.

5. An underground access pit 3 m × 3 m × 4 m in a petrochemical plant contains 5 valves (with flanges on either ends) for pipes carrying phosgene gas (carbonyl chloride, Cl_2CO, molecular weight 98.92) under high pressure. The access pit has a manhole at the top and small amounts of air pass between the pit and the outside at a rate of 0.1 m^3/hr. The average temperature of the air inside the pit is 12 C. Using emission factors for fugitive emissions for pipeline valves and flanges for gas streams in petroleum refineries, estimate the steady-state phosgene concentration using the well mixed model. Is the concentration below the PEL for phosgene? Is it safe to send a workman into the pit?

6. A degreaser with an open top 1 m × 2 m containing trichlorethylene (see Chapter 4, Problems 17 and 18 for properties) is located inside a special room (2 m × 3 m × 2.5 m) in an automotive repair company. Outside air is drawn into the room and air from within the room is withdrawn by a small fan. Your supervisor wishes to know if a 300 CFM fan will be adequate. When the facility is built, the initial concentration of trichlorethylene is 1% of PEL. Using emission factors and the well mixed model, determine whether the steady state concentration is below PEL. The vessel is closed every evening (5 PM), estimate the concentration at 12 noon, 5 PM, and 8 AM.

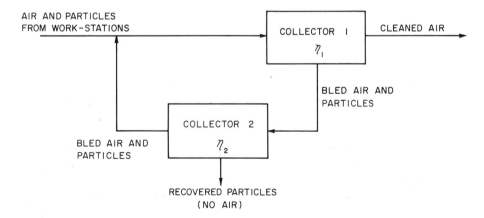

Fig. 5.18 Particle recovery system.

7. A collection system is used to recover particles generated by a buffing wheel used in making gold jewelry. Because the collected particles are very valuable, you suggest that a feedback system consisting of two collectors of efficiency η_1 and η_2 shown in Fig. 5.18 be installed. Derive an expression for the overall collection efficiency.

8. Derive an equation that expresses the time-varying particle concentration for the clean room shown in Fig. 5.19. Write an expression that gives the class of clean room as a function of the operating parameters of the problem (Q_1, Q_2, V, S, f, η_1, η_2). Assume that the particles are $1.00 \, \mu\text{m}$ in diameter.

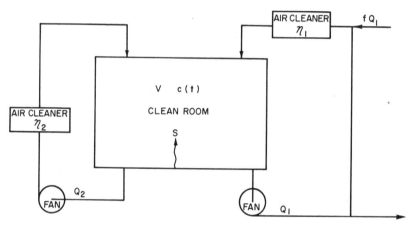

Fig. 5.19   Dual-flow clean room.

Neglect gravimetric settling and assume the particles are removed by the ventilation system. If initially there are 10^8 particles/m^3, draw a graph of the class versus time.

$Q_1/V = 0.5$ room air changes per hour

$Q_2/V = 2.0$ room air changes per hour

$V = 600 \, \text{m}^3$

$S = 2 \, \text{gm/sec}$

$f = 0.1$

$\eta_1 = \eta_2 = 0.95$

9. Consider the paint spray booth shown in Figs. 5.20 and 1.24. Paint with a density of 12 lbm/gal containing 90% hydrocarbons (assume toluene) by weight is used at a rate of 0.514 gal/min. Paint leaves the booth on the surface of painted surfaces at a rate of 0.36 gal/min. A water spray in the rear of the booth (see Fig. 1.24) removes the bulk of the unused paint particles but the hydrocarbons are discharged outside the building. Hydrocarbon-laden air is withdrawn from the booth at a volumetric flow rate of 7000 SCFM and an equal amount of room containing 0.02 PPM of toluene is supposed to enter the booth from the room. Your supervisor suggests that 1650 ACFM, $T = 300 \, \text{C}$, from a nearby dryer containing 5000 PPM of toluene (molecular weight = 92.1) can be discharged into the booth to reduce the amount of

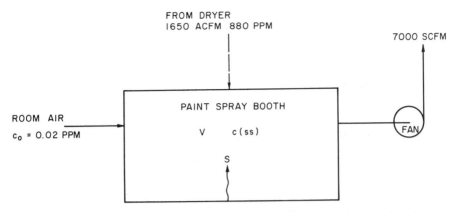

Fig. 5.20 Paint spray booth with and without dryer discharging into the booth.

room air. Such a step reduces the amount of makeup air for the building and lowers heating costs. The fire insurance carrier requires that the hydrocarbon concentration in the booth never exceed 1% of the lower explosion limit (LEL = 1.27% by volume, i.e., LEL mol fraction = 0.0127). What is the toluene concentration in the booth under normal circumstances? What is the toluene concentration if your supervisor's suggestion is adopted?

10. New homes and new automobiles contain many materials that emit hydrocarbons for a period of time (i.e., the "new car smell"). Solvents are contained in adhesives, dyes and finishes in carpets, furniture, wallpaper, paint, drapes, laminated wood paneling, thermal insulation, and so on. People normally cope with these hydrocarbons but some people are sensitive to very low concentrations and suffer serious discomfort. Assume the hydrocarbon is formaldehyde (HCOH), which has an 8-hr PEL of 3 PPM (160) and an odor threshold of approximately 1 PPM. Compute and plot the formaldehyde concentration (PPM) as a function of time in a home based on the well-mixed model assuming the following conditions.

   (a) After construction, the home is sealed shut for 10 days, whereupon it is opened and inside air is exchanged for outside air at the rate of 60 m^3/hr. The concentration of formaldehyde in the ambient air is 10 PPB.

   (b) Formaldehyde generating surfaces $A_g = 464$ m^2, home volume $V = 284$ m^3 and formaldehyde adsorbing (and desorbing surfaces) $A_s = 93$ m^2.

   (c) Formaldehyde generation rate S (mg/hr) $S = K_1 A_g \exp(-t/K_2)$; $K_1 = 75$ mg/day m^2; $K_2 = 500$ hr

   (d) For simplicity, assume that formaldehyde is adsorbed by surfaces (or desorbed by the same surfaces) at a rate S_a (mg/s) given by $S_a = K_3 A_s (c - c_0)$; $K_3 = 0.0005$ m/s; $c =$ formaldehyde concentration at any instant; and $c_0 = 10$ PPB. Thus adsorption occurs whenever $c > c_0$ and desorption occurs whenever $c < c_0$.

11. Toluene 2,4-diisocyanate (M = 171.5) called TDI is a highly toxic material. A sample of a process gas stream containing TDI has been taken and shipped to a laboratory for analysis. The sample was placed in a rigid, spherical, stainless

steel container 0.5 m in diameter. Adsorption of TDI on the walls of the container occurs at a rate given by

$$\text{Adsorption (mg/s)} = kA_s c$$

where $k = 0.00277$ m/hr; $A_s =$ container internal surface area; and $c =$ TDI concentration (mg/m^3). After a period of 30 hr, the laboratory measures a TDI concentration of 2 mg/m^3 at STP. If the pressure and temperature of the original sample as 1.5 atm, 80 C, what was the original TDI concentration? Is this value below the PEL for TDI?

12. Oil mist is used to lubricate the dies in a plant that produces metal stampings. Unfortunately, some of the mist enters the workplace atmosphere. Activity in the workplace produces a uniform (but not constant) mist concentration. Airborne oil mist is generated at a rate of 0.01 kg/sec. The workplace volume is 10^5 m^3. An oil mist eliminator withdraws 1 m^3/sec of room air and removes 95% of the oil mist from the withdrawn air and recirculates the cleaned air back into the room. A fan in the roof withdraws 1 m^3/sec of room air and discharges it outdoors. Another fan adds 1 m^3/sec of tempered outside makeup air (containing no oil mist). Adsorption and settling processes within the workplace remove mist at a rate equal to the concentration times the product of the surface area and adsorption rate constant. The quantity $(A_s k_{ad})$ is equal to 5 m^3/s. If the initial oil mist concentration is 0.01 g/m^3, what is the concentration at the end of an 8-hr day?

13. An enclosure of volume V contains processes that generate toxic vapors at a rate S (mass/time). Air is removed from the room at a rate of 10 room air change per hour and cleaned with an efficiency η. A portion of the cleaned air is returned to the room and a portion (equal to the volumetric flow rate of makeup air) is discharged. Fresh makeup air containing no contaminant is added at a rate of one room air change per hour. Neglecting wall losses and settling, write an expression that predicts the concentration as a function of time. What is the steady-state concentration?

14. Consider the general ventilation configuration in Fig. 5.7. Suppose there has been a great deal of smoking in the enclosure such that, initially, the smoke concentration is a high value $c(0)$. Everyone leaves the room but the ventilation system continues to run. In terms of the parameters in the figure, write an expression for the time it takes to reduce the concentration by a factor of 2. Neglect adsorption and desorption of smoke by the walls of the room.

15. The ventilation system for a conference room is shown in Fig. 5.7. Initially the room is empty and the concentration of total suspended particulate matter (TSP) is 10 μg/m^3. A meeting is called and a number of people begin to smoke such that throughout the meeting, smoke is generated at a constant rate 0.5 g/hr. The physical properties of the problem are:

$V = 100$ m^3 $Q_e = 25$ m^3/min

$S = 0.5$ g/hr $c_a = 0$

f (makeup air fraction) = 10%
efficiency of each collector = 80%

How long will it take before the TSP concentration reaches the Primary National Ambient Air Quality Standard of 150 $\mu g/m^3$? Is there a steady state concentration, and if so, what is it?

16. Welding fume is produced in a metal fabrication facility. Activity in the workplace produces uniform, but not constant, fume concentration throughout the room. Welding fume is generated at a rate of 1 g/sec. The room volume is 1000 m^3. A fume eliminator is placed in the room that withdraws 1.0 m^3/s and removes 85% of the fume and recirculates the cleaned air back into the room. If the initial fume concentration is negligible, what is the concentration after 1 hr?

17. The rooms of a "smoke-free" building are served by a single air conditioning system. Examine the smoke concentration that will occur if one person smokes in an office. For simplicity, assume the building is divided into two rooms of volume V_1 and V_2. Air is withdrawn from each room at volumetric flow rates Q_1 and Q_2. The withdrawn air enters a common heater that leads to a recirculating fan. Upstream of the fan a volume of air [f($Q_1 + Q_2$), where $0 < f < 1$] is discharged to the atmosphere and the same amount of outside "makeup air" is added to the remaining air. The fan returns air at volumetric flow rates Q_1 and Q_2 to rooms V_1 and V_2. Initially, the smoke concentration in rooms V_1 and V_2 is the same as the outside air (c_0). At $t = 0$, someone begins to smoke at a steady rate S (g/sec) in room V_1. Because of the design of the air conditioning system it is believed that some of the smoke will travel to room V_2. Write an expression that shows how the concentration in room V_2 varies with time and the other parameters of the design. Will the concentration in V_2 reach a steady-state value? What is it?

18. Assume that a two-man office has the general ventilation configuration shown in Fig. 5.6. An individual who smokes wishes to install a small portable air cleaner in the room so as not to annoy the other person in the office. The air cleaner discharges the cleaned air back into the room. It is claimed that the air cleaner has an efficiency of η_p and cleans air at a rate equal to N room air changes per hour. Write an expression for the instantaneous concentration inside the office.

19. A volatile liquid is stored in an open vessel inside a small, unventilated, well-mixed room. Liquid evaporates and enters the air where it remains since there is no mechanism to remove it. Show that the vapor concentration (in air) increases in an exponential fashion,

$$c = A[1 - \exp(-Bt)]$$

where A and B are constants that do not depend on the concentration.

20. Contaminants are generated at a rate S inside a well-mixed enclosure equiped

with a ventilation system similar to Fig. 5.6. There is no adsorption on the walls of the enclosure. Air from the enclosure is withdrawn at a rate Q and a fraction discharged to the atmosphere. Makeup fresh air from the outside Q_m (contaminant concentration c_a) is mixed with the recirculated air and the mixture is cleaned by an air cleaner with a removal efficiency η_3. The mixture is returned to the enclosure. A second supply of make-up air (Q_m) from the outside is sent through an air cleaner of removal efficiency η_1 and discharged directly into the enclosure. Lastly, an air cleaner is placed inside the enclosure to withdraw air at a rate Q_i from inside the enclosure, clean it with an efficiency η_2 and return all of it to the enclosure. The concentration inside the enclosure varies with time given by the differential equation

$$\frac{dc}{dt} = Ac + B$$

where A and B are constants. Evaluate the constants A and B in terms of the parameters in the problem.

21. A house painter uses an indoor air cleaner to remove paint odors in homes where he works. The efficiency of the indoor air cleaner is 95% and the volumetric flow rate is Q'. A schematic diagram of the ventilation system for the room without the indoor air cleaner is similar to Fig. 5.7. Assuming well-mixed conditions, calculate the volumetric flow rate Q' that produces a steady-state concentration one-half the value that would occur if the indoor air cleaner was not used.

$$V = 50 \text{ m}^3 \qquad S = 100 \text{ g/min}$$
$$c(0) = 10 \text{ mg/m}^3 \qquad c_a = 10 \text{ mg/m}^3$$
$$k_{ad} = 0.001 \text{ m/s} \qquad A_s = 85 \text{ m}^2$$
$$f = 0.9 \qquad \eta_1 = \eta_2 = 75\%$$
$$Q = 20 \text{ m}^3/\text{min}$$

22. You have been asked to evaluate the ventilation system for a new clean room (Fig. 5.12) used to manufacture integrated circuits. One of the operations is the removal of unwanted material from the circuits by a dry etching process using a RF plasma. Unfortunately, gaseous products are generated. The ventilation system is designed to achieve well-mixed conditions, but diffusion and adsorption of these materials to the wall (wall losses), the rate at which these materials are generated, and the air cleaner efficiency are unknown. A series of experiments are run to evaluate these quantities during which time the discharge and makeup air flow rates are set to zero and the concentration is measured. The volumetric flow rate through the sampling device (Q_s) is negligible. Properties of the ventilation system are as follows.

Volume $= 60 \text{ m}^3$

Surface area over which adsorption occurs $= 150 \text{ m}^2$

Discharge and makeup air volumetric flow rates $= 5 \text{ m}^3/\text{min}$

Volumetric flow rate through air cleaner $= 10 \text{ m}^3/\text{min}$

Efficiency of air cleaner = unknown

Wall loss coefficient = unknown

Contaminant generation rate = unknown

(a) The air cleaner, discharge, and makeup air flow are shut off and a large amount of contaminant is generated until the concentration is $0.3 \, \text{mg/m}^3$. Generation is halted, and the concentration is measured experimentally. What is the wall-loss coefficient?

concentration (mg/m^3)	Time (min)
0.300	0
0.148	5
0.060	15
0.030	21
0.020	26
0.007	35

(b) The makeup air and discharge are shut off and the air cleaner is run to reduce the contaminant concentration to a relatively low value. The air cleaner is then shut off. The etching process is begun at the desired production level and maintained at a constant value. What is the contaminant generation rate if experimental measurements show that the concentration varies as follows?

concentration (mg/m^3)	time (min)
0.010	0
0.020	1
0.063	5
0.082	10
0.089	15

(c) The makeup air and discharge are shut off, contaminant generation is maintained at the constant rate given in (b) above, and the air cleaner is turned on. If the concentration decreases and achieves a steady state value of $0.053 \, \text{mg/m}^3$, what is the efficiency of the air cleaner.

23. A smoldering cigarette emits particles in an unventilated well-mixed chamber of volume $22 \, \text{m}^3$ and surface area $47 \, \text{m}^2$. At different times particle concentrations are measured. Figure 5.5 shows that the concentration rises during the first 8 min in which the smoldering cigarette emits particles, after 8 min the cigarette ceases to emit particles and the concentration decreases as particles plate-out on the chamber walls.

(a) Estimate the source strength (mg/min) for the smoldering cigarette.
(b) Estimate the wall loss coefficient in the chamber (m/hr).

(c) In another experiment an air cleaning device is placed in the chamber and tested. The volumetric flow rate of the device is 300 ACFM. Estimate the removal efficiency of the room air cleaner. After the cigarette has ceased emitting particles, the device is turned on and the concentration is measured.

Time (min)	Concentration (mg/m^3)
0	0.240
2	0.122
4	0.067
6	0.020
8	0.018
10	0.010

24. An enclosure is not well-mixed. Step-down tracer experiments are conducted. Initially, the concentration at a point P is c_0. Fresh air is added at a volumetric flow rate Q and air containing the tracer is removed at the same rate. The concentration at point P decreases linearly with time and becomes negligible at a time t_1. Show that the exchange effectiveness is equal to $Q(t_1)/3V$.

25. An enclosure is not well-mixed. Initially, the concentration of a tracer is a low value, c_0. Tracer is added to the incoming air at a constant rate and the concentration is measured at point P. The concentration is found to rise linearly until a time t_1 when it is equal to the value in the incoming air. Compute the exchange effectiveness at point P.

26. A company sprays colored lacquer on leather. It uses many colors but all contain the solvent butyl acetate (molecular weight 116, vapor pressure 15 mm Hg at 25 C, PEL 150 PPM, $710\ mg/m_3$). Workers are inclined to leave empty 55-gall drums open during the workday, but they are closed at the end of the day. The empty drums nevertheless contain a thin puddle of solvent in the bottom. The drums are stored in a special storage-mixing room equipped with an overhead exhaust fan that removes 3000 CFM and a "floor sweep" exhaust that removes 1000 CFM from the lower corner of the room opposite the doorway. Air containing 10 PPM of butyl acetate enters the room through an open doorway ($5\ ft \times 8\ ft$) in the middle of one wall that connects the storage-mixing room to the remainder of the workplace. The room is $30\ ft \times 30\ ft$ and is 15 ft high. There are approximately 30 uncovered drums in the room.

(a) Estimate the total evaporation rate (g/min) of solvent from these drums. Use emission factors, empirical equations, and first principles.

(b) Using an evaporation rate of $4\ g/hr\ m^2$ per drum, estimate the time-varying and steady-state solvent concentrations by a two-cell, well-mixed model. Assume the initial concentration in each cell is uniform and equal to 10 PPM. Divide the room into horizontal cells with the lower cell extending from the floor to a height of 5 ft and the other cell extending to the ceiling. Draws graphs of the concentration in each cell as a function of time for several values of the exchange coefficients (f) for the upper cell over a range of values (i.e., $f = 0.001$, 0.5, 1.0, 5.0, and 10.0) and discuss the physical significance of the results.

27. Farm workers attend to chickens kept in a structure (22 m long and 11 m wide) shown in Fig. 5.21. The birds are kept in cages and the workers move through a center isle and two side isles along the outside walls. The birds generate particles of dried manure, feed, feathers at a constant rate of 0.1 mg/s per square meter of floor space in the caged areas. During the winter months fresh air (heated to 20 C) enters the building through a long plenum along the center of the building. Air is exhausted along the outside walls. The volumetric flow rates are uniform along the length of the building. The inlet air flow rate is 5 SCFM/ft^2 of floor space. Air also enters and leaves the building uniformly over the building surface by infiltration and exfiltration at rates equal to 0.02 building changes per minute (0.02 CFM/ft^3 of space). Compute and plot the dust concentration (mg/m^3) as a function of time ($0 \le t \le 8$ hr). Assume that the initial dust concentration is the same as the ambient concentration and equal to 0.1 mg/m^3. Owing to gravity, particles settle on the floor with a "plate-out" coefficient (k_{ad}) of 0.3 cm/s.

(a) Assume symmetry with respect to a plane through the center of the building. Use a sequential box model consisting of two cells defined as follows:

    Cell A: half the center aisle
    Cell B: cages and outside aisle

and an array of exchange coefficients (f) such as 0.001, 0.5, 1.0, 5.0, and 10.0.
(b) Use a sequential box model consisting of five cells defined as follows:

    Cell A: Center aisle and the ceiling.
    Cells B and C: spaced defined by the poultry cages and ceiling.

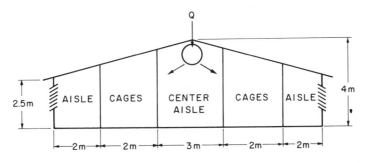

Fig. 5.21 Ventilated poultry building.

Cell D: space defined by the outside aisle and ceiling with 25% of the air exhausted through the outside wall.

Cell E: same as Cell D with 75% of the air exhausted through the outside wall.

Use a range of exchange coefficients shown in part (a).

(c) To improve conditions you suggest installing an internal air recirculation system that withdraws 3000 SCFM from above each set of cages, removes the particles with an air cleaning system with a particle removal efficiency of 75%, and returns the cleaned air to the center aisle. If all conditions are the same as in (a), replot the concentration as a function of time.

(d) "Step-up" experiments are conducted in which a tracer gas is added to the inlet air and the tracer concentration measured at different points in the structure. Calculate the mixing effectiveness coefficients (e) for Cells A, B, and D in part (b) and discuss their significance with respect to selecting values of the exchange coefficients (f) that you believe should be used in the analysis.

$$\text{Cell A: } c_a/c(\infty) = [1 - \exp(-t/t_A)] \text{ for all } t \text{ and } t_A = 15 \text{ min}$$

$$\text{Cell B: } c_B/c(\infty) = \begin{bmatrix} t/t_B & \text{for} & t < t_B = 60 \text{ min} \\ 1.0 & \text{for} & t > t_B = 60 \text{ min} \end{bmatrix}$$

$$\text{Cell D: } c_D/c(\infty) = \begin{bmatrix} 0 & \text{for} & t < t_D = 45 \text{ min} \\ 1.0 & \text{for} & t > t_D = 45 \text{ min} \end{bmatrix}$$

(e) Your supervisor questions the value of the source strength and plate-out coefficient used above. Using the well-mixed model devise an experimental way to calculate the plate-out coefficient (k_{ad}). With this same apparatus devise an experiment that will enable you to calculate the particle source strength during feeding and for other periods of the day when the birds are less active.

28. An enclosure has been found to be well-mixed. At time zero, the initial concentration of a tracer is c_0. Fresh air is blown into the enclosure and air containing the tracer is removed at an equal volumetric flow rate. Prove that the exchange coefficient is unity.

29. A new office complex ($V = 600 \text{ m}^3$) has been designed with uniquely located inlet and outlet ventilation registers. The designer claims that cigarette smoke from one work station will not travel to another work station. The total volumetric flow rate for the entire volume is $10 \text{ m}^3/\text{min}$. Step-up tracer experiments are conducted to verify the designer's claim. In these experiments the entire work space (without workers) is ventilated with fresh air and suddenly a tracer gas is added to the inlet air supply for every register and a constant tracer gas concentration is maintained. The concentration is measured experimentally at three work stations.

Station A:

$$c_A = \begin{bmatrix} 0 & \text{for} \quad t < t_A \\ c = c(\infty) & \text{for} \quad t > t_A \end{bmatrix}$$

Station B:

$$c_B = \begin{bmatrix} [tc(\infty)/t_B] & \text{for} \quad t < t_B \\ c(\infty) & \text{for} \quad t > t_B \end{bmatrix}$$

Station C:

$$c_C = c(\infty)[1 - \exp(-t/t_C)] \qquad \text{for all } t$$

From these measurements discuss the validity of the designers claim if values for t_A, t_B and t_C are as follows:

(a) $t_A = 0.2$ hr $t_B = 5.0$ hr $t_C = 1.0$ hr
(b) $t_A = 0.5$ hr $t_B = 0.5$ hr $t_C = 2.0$ hr
(c) $t_A = 2.0$ hr $t_B = 1.0$ hr $t_C = 0.5$ hr
(d) $t_A = 2.0$ hr $t_B = 0.2$ hr $t_C = 0.2$ hr

30. Your automobile is stuck in a snow drift. You start the car but open the windows a crack to allow fresh air into the car and air inside the car interior to escape to the atmosphere. You then turn on the hot air heater and sit inside the car to warm up before attempting to dig yourself out of the snow drift. Unknown to you, the snow surrounds the bottom of the car such that the auto exhaust is drawn directly into the air inlet for the heating system.

V (car interior) $= 2\,\text{m}^3$
Q_H (hot air heater) $= 0.5\,\text{m}^3/\text{min}$
Q_I (infiltration) $= 0.5\,\text{m}^3/\text{min}$
Q_E (exfiltration) $= 1.0\,\text{m}^3/\text{min}$
c_H (CO in heater air inlet) $= 1000\,\text{mg}/\text{m}^3$
c_A (CO in ambient air) $= 10\,\text{mg}/\text{m}^3$

Assume that these conditions are constant and that well-mixed conditions exist inside the car.

(a) What is the steady-state CO concentration inside the car?
(b) How long will it take for the CO concentration to reach $400\,\text{mg}/\text{m}^3$
(c) Using Fig. 2.41, how long could you remain in the car before you'd collapse?

31. Ammonia is generated by a chemical etching process at a rate of $10\,\text{mg/min}$ and escapes into a room ($10\,\text{m} \times 10\,\text{m} \times 4\,\text{m}$). The temperature and pressure are 300 K and 1 atm. Fresh air leaks into the room (infiltration) and room air

leaks out (exfiltration), and a steady-state ammonia concentration is established. The room is equipped with a large circulating fan that distributes the ammonia fumes uniformly throughout the room and a steady-state ammonia concentration of 100 PPM is established. Workers complain about the odor. Your supervisor suggests that fresh air be ducted into the room and a similar amount of room air be discharged to the outside. What volumetric flow rate (CFM) of new fresh air is needed (infiltration and exfiltration remaining the same) to reduce the concentration to one-half the odor threshold.

32. A room of volume V is ventilated by adding air at a volumetric flow rate Q at one location and withdrawing an equal amount of air at another location. Initially, the concentration of a tracer gas in the room is zero. For $t > 0$, a tracer gas is added to the incoming air. The concentration in the inlet air supply is c_0. Such an experimental technique is called a "step-up tracer input." If the instantaneous tracer concentration is found to be given by

$$\frac{c(t)}{c_0} = 1 - \exp(-at)$$

where a is a constant, show that the mixing factor m is equal to (aV/Q).

33. Air is added to a room of volume V at one location at a volumetric flow rate Q and an equal amount air is withdrawn at another location. Initially the concentration of a tracer gas in the incoming air and in the room is equal to c_0. The tracer gas in the inlet air is suddenly eliminated. Such an experimental technique is called a "step-down tracer input." If the instantaneous tracer concentration is found to be given by

$$\frac{c(t)}{c_0} = \exp(-at)$$

where a is a constant, show that the mixing factor m is equal to (aV/Q).

34. Consider fully developed flow in a tunnel of radius R. The velocity $u(r)$ at any radius is,

$$\frac{u(r)}{U_0} = 1 - \left(\frac{r}{R}\right)^2$$

where U_0 is the centerline velocity that is twice the average velocity (Q/A). Show that in a step-up tracer experiment in which a tracer gas is injected a distance L upstream of the sampling station, that the effectiveness coefficient e one would measure at any radius satisfies the following:

$$e = 4\left[1 - \left(\frac{r}{R}\right)^2\right]$$

35. Redo Illustrated Example 5.3 using the 4th order Runge Kutta method (see Appendix A-12).

36. Redo problem 5-10 using the 4th order Runge Kutta method (see Appendix A-12).

37. Describe an experiment to measure the source strength related to the generation of volatile hydrocarbons that occurs in spray painting. Address both vapor generated during spraying and during the time the paint drys.

38. Radon gas seeps into your basement. The daughters of radon gas generate radioactivity of 50 pCi per liter of air in your $300\,m^3$ basement. The radioactivity is directly proportional to the concentration of radon gas in your basement. You wish to remove some basement air and dilute the remainder with outside air (containing backgound radioactivity of 1 pCi per liter) such that the average steady-state radioacitivity in your basement is 4 pCi per liter of air. Using the well-mixed model estimate the volumetric flow rate of ouside air necessary to accomplish this goal? Using the degree-days for Pittsburgh, estimate the cost to heat the make-up air.

CHAPTER 6

Present Local Ventilation Practice

Goal: Students will be expected to learn
    – contemporary design standards and practices
    – to design a system of interconnecting ducts

A local ventilation system, often called a "hood," is designed to remove contaminants at the point(s) of generation. A *hood* is a shaped inlet designed to capture contaminated air and conduct it into the exhaust duct system (27). The options available to the designer are the geometry of the shaped inlet, its location with respect to the source, and the exhaust volumetric flow rate. *Local ventilation* systems are classified in the following broad categories:

    – Plane openings (Fig. 1.27)
    – Flanged openings (Figs. 1.14 and 1.15)
    – Lateral exhausts (Figs. 1.25 and 6.1)
    – Receiving hood (Fig. 1.28 and 6.4)
    – Full enclosures (Figs. 1.18, 1.19, 1.21, 1.22 and 6.2)
    – Partial enclosure (Figs. 1.24 and 6.3)
    – Downdraft hoods (Figs 1.23 and 1.26)
    – Canopy hoods (Figs. 1.13, 1.29 and 6.5)
    – High velocity–low volume hoods (Figs. 1.10, 1.11 and 1.12)

These photographs and other photos and sketches in Chapters 1 and 6 illustrate features of each category. Features of one category can often be combined with features from another category.

At present, the design of local ventilation systems follows practices recommended by the ACGIH and published in Industrial Ventilation, A Manual of Recommended Practice (27). Hereafter this publication will be referred to as the *ventilation manual*. The manual is not analytically rigorous or pedantic. Rather, the manual is intended to be a guide to commonly followed methods for designing and testing industrial ventilation systems. The manual is widely used throughout industry. While the practices do not enable designers to predict contaminant concentrations to insure that mandatory health standards are met, the manual provides reasonable assurance that the ventilation system will control the transfer of contaminants to the workplace environment. There is a deliberate effort to write for a diverse class of users rather than specialists. The manual is written in

Fig. 6.1 Lateral rim exhauster for crucible.

Fig. 6.2 Lids to draw-off dust generated in filling barrels.

Fig. 6.3   Canopy hood and curtains for mixer.

Fig. 6.4   Receiving hood and partial enclosure for cut-off saw.

Fig. 6.5 Close-fitting canopy and receiving hood for a reheat furnace.

uncomplicated language and contains principles that do not require the reader to possess a baccalaureate degree in engineering. The ACGIH conducts training sessions several times each year at which the ACGIH methods are taught. A helpful self-instruction manual (104) can also be used.

The Systems Edition, ASHRAE Handbook and Products Directory (103) (updated every four years) describes procedures to design local ventilation systems. Design practices recommended by ASHRAE are written in language used in the engineering profession and in four-year baccalaureate engineering education programs. The practices recommended by ASHRAE are in basic agreement with those recommended by ACGIH.

The ACGIH ventilation manual contains dozens of *design plates* of ventilation systems for specific industrial processes. If the process for which control is sought can be found in the manual, the designer ought to use the recommended control system. At least, the recommendations should be used as a preliminary design that can later be refined as conditions warrant. If the process under study cannot be found in the manual, the designs stimulate the readers imaginations and enable them to choose design concepts that have a high degree of success.

What follows is a summary of the practices recommended by both ACGIH and ASHRAE. If individuals wish to use these design practices they should consult the original sources (27,103) for a full discussion of details they must address and reference to the original material upon which the recommendations are based.

6.1 CONTROL OF PARTICLES

The central concept in contemporary ventilation practice is the *capture velocity*. The "capture velocity is the air velocity at any point in front of the hood or at the hood opening necessary to overcome opposing air currents and to capture the contaminated air causing it to flow into the hood" (27). The velocity of air at the plane of the hood inlet is called the *face velocity*. The area of the opening is called the *face area*.

The numerical value of the capture velocity is a function of the particle to be removed. Table 6.1 shows values recommended by ASHRAE and ACGIH. Once the capture velocity is selected, the designer chooses an inlet and flow rate such that the air velocity is equal to the capture velocity at the point where the contaminant is to be removed. Such a point is called the *null point*. Whether the particles are removed at the point of generation or at another location is a decision to be made by the designer. It is desirable to capture the particle as close to the point of generation as possible.

In the absence of wakes or jets, it is accurate to model an inlet as a sink in a potential flow field. For example, consider a slot of width $2w$ and length L along the edge of perpendicular flanges above which is a quiescent air mass, Fig. 6.6. If air is withdrawn uniformly along the length of the slot at a volumetric flow rate Q, the radial velocity $U(r)$ entering normal to the surface of a quarter-cylinder of radius r above the slot can be estimated from

$$U(r) = \frac{2Q}{L\pi r} \tag{6-1}$$

where $r > w$. The slot width is unimportant at $r > w$ since the slot appears to the flow as a line sink. The distance r is the distance between the center of the slot to

TABLE 6.1   Capture Velocities[a]

Characteristics of Contaminant Generation	Examples	Capture Velocity (FPM)
1. Contaminant enters quiescent air with negligible velocity	Degreasing tank, Evaporation	50–100
2. Contaminant enters slightly moving air with a low velocity	Welding, Vessel filling	100–200
3. Contaminant actively generated and enters rapidly moving air	Spray painting, Stone crushers	200–500
4. Contaminant enters rapidly moving air at high velocity	Grinding, Abrasive blasting	500–2000

Lower values of capture velocity
   Room air movement minimal or conducive to capture
   Contaminants of low toxicity
   Intermittent use or low production rates
   Large hood and large mass of air moved

Upper values of capture velocity
   Adverse room air movement
   Contaminants of high toxicity
   Heavy use and high production rates
   Small hood and small mass of air moved

Source: Abstracted from ref. 27.

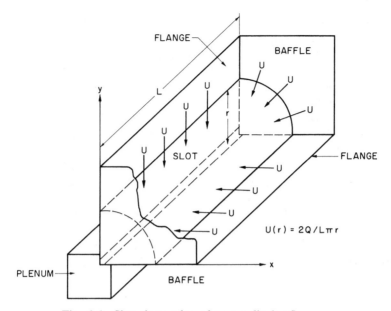

Fig. 6.6   Slot along edge of perpendicular flanges.

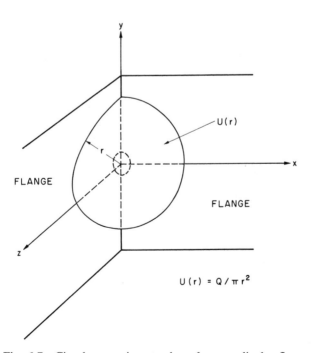

Fig. 6.7   Circular opening at edge of perpendicular flanges.

the point on the quarter-cylinder surface. Figure 6.7 gives the magnitude of the radial velocity $U(r)$ at a radial distance r from the center of an exhaust port in the corner of right-angle walls. When the point in space is closer to an entry of diameter D or width w, such that $r < D$, or $r < w$ etc, the velocities are no longer equal to Eq. 6-1 or the equations in Figs. 6.6 and 6.7, but the flow field can still be modeled as potential flow (Chapter 7).

For several decades designers have used empirical equations to predict the centerline velocity in front of flanged and unflanged circular (and rectangular) inlets that withdraw air from a quiescent environment. During this time nearly two dozen equations have been proposed for each case. Braconnier (290,297) reviewed these equations, compared their results with experimental measurements, and suggested that the following expressions are the most accurate statements for the velocity $U(r)$ along the centerline a distance r from the inlet plane for inlets with no transition regions, that is, straight-sided inlets. The average inlet velocity. (i.e. face velocity) designated as U_a and is equal to volumetric flow rate Q divided by the face area A.

Circular, unflanged, free-standing inlet (diameter D)

$$\frac{U(r)}{U_a} = \frac{1}{\left[1 + 10\left(\frac{r^2}{A}\right)\right]} \tag{6-2}$$

Rectangular, unflanged, free-standing inlet (a by b, where $a > b$)

$$\frac{U(r)}{U_a} = \frac{1}{[0.93 + 8.58p^2]}$$

$$p = \left(\frac{r}{\sqrt{A}}\right)\left(\frac{a}{b}\right)^q$$

$$q = 0.2\left[\frac{r}{\sqrt{A}}\right]^{-0.33} \tag{6-3}$$

where

$$1 < \frac{a}{b} < 16 \text{ and } 0.05 < \left(\frac{r}{\sqrt{A}}\right) < 3.$$

Circular, flanged, free-standing inlet (diameter D)
  (flange width equal to the inlet diameter)

$$\frac{U(r)}{U_a} = 1.1 \, (0.07)^{r/D} \qquad 0 < \frac{r}{D} < 0.5$$

$$\frac{U(r)}{U_a} = 0.1\left(\frac{r}{D}\right)^{-1.6} \qquad 0.5 < \frac{r}{D} < 1.5 \tag{6-4}$$

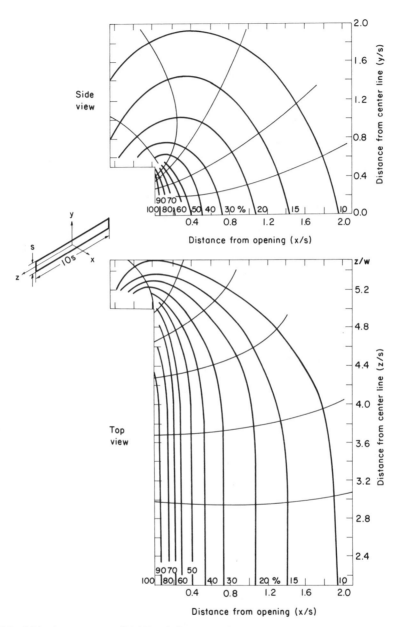

Fig. 6.8 Velocity contours (U_a/U_{face}) for an unflanged rectangular opening, aspect ratio 1:10 (adapted from ref. (48)).

Rectangular (infinite) flanged free standing inlet (a by b, $a > b$)

$$\frac{U(r)}{U_a} = 1 - \left(\frac{2}{\pi}\right) \arctan\left\{\left(\frac{2r}{a}\right)\sqrt{\left(\frac{2r}{b}\right)^2 + \left(\frac{a}{b}\right)^2 + 1}\right\} \qquad (6\text{-}5)$$

Unfortunately, the equations given above are of only limited value since seldom is an industrial environment quiescent and for the most part designers want to know the velocity at arbitrary points in front of an inlet, not at points just along the centerline. The professional literature contains numerous graphs such as Figs. 6.8 and 6.9 that show *contours of constant air speed* at arbitrary points in front of flanged and unflanged inlets. Analytical expressions can be generated by the computer to express these contours for design purposes.

A *flange* is a flat metal plate around the outside of the inlet in the form of a "collar". The flange lies in the plane of the inlet area. The width of a flange varies. It may be large and give an inlet the appearance of a hole in a plate, but more often the flange is narrow and approximately equal to the width of the inlet.

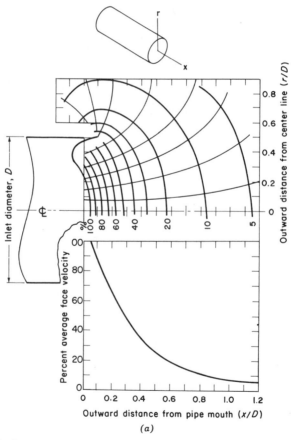

Fig. 6.9 Velocity contours (U_a/U_{face}) for (a) plain and (b) flanged circular openings (adapted from ref. 103).

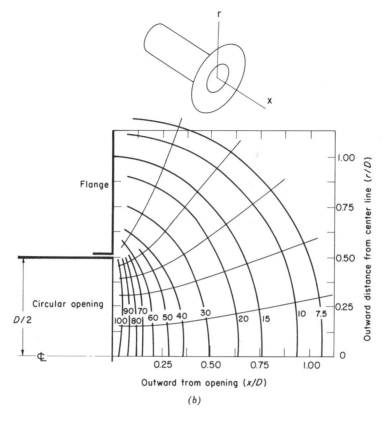

Fig. 6.9 (*Continued*)

A flange serves two purposes, it provides structural strength to the inlet and it increases the size of the region in front of the inlet from which air is withdrawn. An *unflanged inlet* is also called a *plain inlet*. Velocity gradients are very large near the lip of an unflanged inlet and room air is drawn into the opening from along the outside surface of the inlet duct which contains little contaminant. Figure 1.27 is an example of an unflanged inlet and Figs 1.14 and 1.15 are examples of flanged inlets.

Comparing the velocity contours for flanged and unflanged (Figs 6.8 and 6.9) it is clear that flanged openings have greater reach than unflanged inlets. The phrase *reach* is used to characterize the size of the region in front of the inlet within which the incoming air is capable of drawing contaminants into the inlet. At any point in front of the opening, a flanged inlet produces higher air speeds than unflanged inlets. Unflanged inlets draw air into the inlet from near the inlet rim, including regions behind the inlet plane which are of little importance if one wishes to capture contaminants emitted in front of the inlet plane. These conclusions are supported by experiments conducted by Garrison and Erig (293) who measured velocity gradients along an inlet's centerline for flanged and unflanged openings that faced an infinite solid surface one to four inlet diameters (or slot widths) away. The centerline velocity gradients were only slightly affected by the solid surface for unflanged inlets which drew much of their air from regions close to and behind the plane of the inlet rim. The centerline velocity gradients were affected significantly for flanged inlets which drew all of their air from regions in front of the inlet.

6.2 CONTROL OF GASES AND VAPORS

Vapors and gases generated by industrial processes are more difficult to control than particles which owing to their large density, have inertia that affects their motion. The ACGIH uses the concept of *control velocity* to design control systems to prevent vapors and gases from entering the workplace.

If the process can be enclosed and the evolving gases and vapors removed before they enter the workplace environment, it is obviously wise to do so. The enclosure must truly isolate the source of contamination. Provision must be made to introduce room air (makeup air) at a rate equal to the rate at which contaminated air is removed from the enclosure. Designers must be absolutely sure that the velocity at the inlet is directed into the enclosure at every point across the face of the inlet.

A control velocity is not the physical velocity of air at a specific point in space, it is a design parameter. Processes are rated in terms of a control velocity and the designer matches the needed control velocity with the control velocity a local ventilation system is capable of producing. The generation of gases and vapors by a process is rated in terms of a control velocity which depends on the hazard potential of the gas or vapor and the rate of evolution. A local ventilation system is rated in terms of the control velocity it can produce. The designer chooses a ventilation system that is capable of producing the control velocity equal to that possessed by the escaping vapor or gas.

Shown in Table 6.2 is a letter (A–D) that corresponds to the hazard potential of a contaminant and a number (1–4) that indicates the rate of evolution of the contaminant from a process. The combination of a letter and a number defines

TABLE 6.2 Hazard Potential And Rate Of Contaminant Evolution[a]

Hazard Potential (F)	Health Standard		Flash Point (F)
	Gas or Vapor	mist	
A	0–10 PPM	0–0.1 mg/m^3	–
B	11–100 PPM	0.11–1.0 mg/m^3	Under 100 F
C	101–500 PPM	1.1–10 mg/m^3	100–200 F
D	over 500 PPM	over 10 mg/m^3	over 200 F

Rate	Liquid (F)	Degrees below Boiling (F)	Evaporation time[a] (hr)	Gassing[b]
1	Over 200	0–20	0–3	High
2	150–120	21–50	3–12	Medium
3	94–149	51–100	12–50	Low
4	Under 94	Over 100	Over 50	Nil

Source: Derived from ref. 27.
[a]Time for 100% evaporation.
[b]Extent to which gas or vapor will be generated, rate depends on the physical process and the solution concentration and temperature.

TABLE 6.3 Minimum Control Velocities (FPM) for Undisturbed Locations

Class	Enclosing Hood Sides Open (1)	(2)	Lateral Hood[a]	Canopy Hood[d] Sides Open (3)	(4)
A1[b], A2[b]	100	150	150	Do not use	
A3[b], B1 B2, C1	75	100	100	125	175
B3[c], C2[c] D1[c]	65	90	75	100	150
A4[b], C3[c] D2[c]	50	75	50	75	125
B4, C4, D3[c], D4—Adequate general ventilation required					

[a]Use Table 6.4 to compute the volumetric flow rate.
[b]Do not use a canopy hood for class A processes.
[c]Where complete control of hot water is desired, design as next highest class.
[d]Use $Q = 1.4(PD)$ control velocity, where P is hood perimeter and D is distance between vessel and hood face (27).

the class of an industrial process. Table 6.3 shows the control velocity needed to cope with different classes of contaminant for enclosing, lateral, and canopy ventilation systems. Table 6.4 shows the exhaust volumetric flow rate needed to produce the control velocities for the lateral hoods listed in Table 6.3. Table 6.5

TABLE 6.4 Minimum Volumetric Flow Rates Per Unit Surface Area (CFM/ft²) For Lateral Exhaust System

Control Velocity (FPM)	Tank Width/Tank Length (W/L)[a] 0–0.09	0.1–0.24	0.25–0.49	0.5–0.99	1.0–2.0
Tank against wall or baffled[b]					
50	50	60	75	90	100
75	75	90	110	130	150
100	100	125	150	175	200
150	150	190	225	250[c]	250[c]
Free-Standing Tank[b]					
50	75	90	100	110	125
75	110	130	150	170	190
100	150	175	200	225	250
150	225	250[c]	250[c]	250[c]	250[c]

Source: Derived from ref. 27.
[a]Inlet slot along the long side (L), if $6 < L < 10$ ft, multiple takeoffs are desirable, if $L > 10$ ft, multiple in plenum are necessary if,
 $W = 20$ in, slot on one side is suitable
 $20 < W < 36$ in, slots on both sides are desirable
 $36 < W < 48$ in, slots on both sides are necessary unless all other conditions are optimum
 $W > 48$ in, lateral exhausts are not usually practical, use push–pull or enclosures
 it is undesirable to use lateral exhaust when $W/L > 1$ and not practical when $W/L > 2$.
[b]Use half width to compute W/L for inlet along tank centerline or two parallel sides of tank.
[c]While control velocities of 150 FPM may not be achieved, 250 CFM/ft² is considered adequate for control.

TABLE 6.5 Metal Surface Treatment Processes

Process	Bath	Emission[a]	T (F)	Class
Anodizing				
Aluminum	H_2SO_4	H_2, acid mist	60–80	B1
Etching				
Aluminium	NaOH, NA_2CO_3 Na_3PO_4	Alkaline mist	160-180	C1
Copper	HCl	HCl	70–90	A2
Pickling				
Aluminum	HNO_3	Oxides of nitrogen	70–90	A2
Copper	H_2SO_4	Acid mist, steam	125–175	B3, B2
Monel & nickel	HCl	Acid mist, steam	180	A2
Stainless steel	H_2SO_4	acid mist, steam	180	B1
Cleaning				
Alkaline	Sodium salts	Alkaline mist	160–210	C2, C1
Degreasing	trichloroethylene	Vapor	188–250	B
	perchloroethylene	Vapor	188–250	B
Electroplating				
Platinum	NH_4PO_4 and NH_3(g)	NH_3(g)	158–203	B2
Copper	NaOH, cyanide salts	Cyanide, alkaline	110–160	C2
Chromium	Chromic acid	Chromic acid mist	90–140	A1
Nickel	HF, NH_4F	Acid mist	102	A3
Stripping				
Gold	H_2SO_4	Acid mist	70–100	B3, B2
Nickel	HCl	Acid mist	70–90	A3
Silver	HNO_3	Oxides of nitrogen	70–90	A1

Source: Adapted from ref. 27.
[a]At high temperature, an alkaline or acid bath will produce a mist of similar composition.

shows the classifications for a variety of metal surface treatment processes obtained by this method.

Illustrated Example 6.1—Pickling Copper in Sulfuric Acid

Copper plate is to be immersed in a bath of water and sulfuric acid prior to coating the metal. You are asked to estimate the ventilation requirements for a bath 10 ft long and 3 ft wide (W/L = 0.3) for an installation that is either to be placed against a wall with a lateral exhaust along one side or a free-standing installation in which lateral exhausts will be placed along both long sides of the tank. The bath temperature is 175 F and acid fume is apt to exist in the form of a mist. The acid concentration is such that the liquid mixture will boil at 225 F. Sulfuric acid has a PEL and TLV of $1\,mg/m^3$. Using Tables 6.2 and 6.3 you conclude that the bath can be classed as either B-2 or B-3, which agrees with Table 6.5. Table 6.3 indicates that the control velocities are,

Class B2: control velocity = 100 FPM

Class B3: control velocity = 75 FPM

Table 6.4 indicates that the ventilation volumetric flow rates per area of liquid surface (Q/A in $ACFM/ft^2$) are the following,

Class	Against wall	Free-standing
B2	150	200
B3	110	150

which require an exhaust fan and plenum for the following ventilation volumetric flow rates (ACFM):

Class	Against wall	Free-standing
B2	4500	6000
B3	3300	4500

6.3 CONTROL SYSTEMS FOR SPECIFIC APPLICATIONS

The ACGIH Ventilation Manual (27) contains over 100 design plates for local ventilation systems for specific applications that have been used to control emissions. The design plates are intended as guidelines and do not take into account special conditions such as cross drafts, motion of workers or machinery, and strong buoyant plumes. Even if the ventilation manual does not contain a design plate for every industrial process, engineers will find design plates for sources similar to the one they wish to control. Each design plate contains guidelines about the following:

(a) Drawing of the system with instructions for the construction of key elements.

(b) Recommended volumetric flow rates.

(c) Recommended minimum duct transport velocities.

(d) Entry pressure losses.

(e) Other information such as slot velocities, slot dimensions and face velocities, baffles, clearance dimensions between source and inlets.

Unless stated explicitly, the design data should not be applied to materials of high toxicity, radioactive materials, or any unusual materials in which higher volumetric flow rates will be required. Details about specific control systems are also published in many professional journals, government reports, and trade publications and provide designers advice and inspiration. Figure 6.10 is similar to the design plates in the ventilation manual and describes a booth for portable grinding, polishing, or buffing operations. The OSHA General Industry Standards (33) does not as a rule provide design information and operating parameters about specific control systems. An exception is information on pedestal grinders and swing grinders.

A class of control systems that has received a great deal of attention recently is laboratory fume hoods, biological safety cabinets and biohazard cabinets.

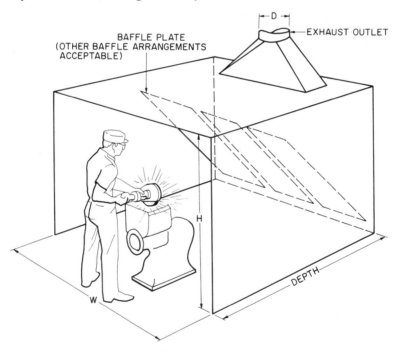

Fig. 6.10 Booth for portable grinding, polishing, and buffing adapted from ref. (76). $Q = WHV$. $W =$ equipment width + 6 ft. $H =$ equipment height + 3 ft, $H(\text{min}) = 7$ ft. $D =$ equipment depth + 6 ft. Baffle area = 0.4 WH. Face velocity $(V) = 100$ FPM for class 1 contaminants, 200 FPM for class 2 contaminants, and 400 FPM for class 3 contaminants. Entry loss for tapered outlet plus baffles = $(1.78$ slot $VP)$ + (entry loss factor for a tapered outlet) times duct VP. Entry loss for plain outlet plus baffles = $(1.78$ slot $VP)$ + $(0.5$ duct $VP)$.

Laboratory fume hoods are five-sided partial enclosures used to handle toxic, corrosive, flammable, or odorous materials in the laboratory. The devices are sometimes called *chemical fume hoods* or *fume cupboards*. A typical example is shown in Fig. 1.20. The purpose of the device is to prevent materials used or produced in experiments from entering the workplace. The performance of these devices is affected by the face velocity, drafts in the workplace, or obstructions (chemical apparatus etc.) inside the hood. The face velocity should not be skewed beyond 20% of the rated values. If the face velocity is too high, regions of high velocity or recirculation regions may be produced inside the hood that may affect the chemical experiments and processes being conducted inside the hood. OSHA requires a face velocity of 150 FPM for handling 13 carcinogens. Standard laboratory fume hoods use 100% air drawn from the laboratory. Auxiliary-air laboratory fume hoods (also called *supply-air hoods*) recirculate up to 70% of the total air flow air and as low as 30% air from the laboratory. Doors on laboratory fume hoods may open horizontally or vertically and produce eddies inside the fume hood that may affect experiments. For particularly corrosive materials, one should use a laboratory fume hood composed of stainless steel and smooth (untaped) seams. Such hoods may also contain a water-wash system.

Biological safety cabinets are similar to laboratory fume hoods but are used to provide sterile air for experiments with cells and tissue cultures. At the same time, the cabinet protects the worker and prevents any material from being transferred to the workplace. Two currents of air are used. There is a low-velocity, steady downward current of air directed over the material to be protected. Particles are removed by HEPA filters. When the movable doors are open, a second current of room air is to flow inward at such a velocity that no biologic material enters the workplace.

Bio-Hazard Cabinets are enclosures for handling very hazardous biological materials. Class II cabinets are total enclosures under negative pressure and have at least one double-door lock.

An *enclosure* is a volume in which there is no opening through which air can pass into or out of the volume other than openings explicitly constructed for this purpose. People enter or leave enclosures through doors that are closed after them. Examples of enclosures would be industrial clean rooms, motel rooms, elevators, and rest rooms in public buildings. A *confined space* is defined by OSHA (33) as a "relatively small or restricted space such as a tank, boiler, pressure vessel, or small compartment of a ship." Thus confined spaces may resemble enclosed spaces but they contain unobstructed openings through which an unknown amount of air can pass in addition to the openings and air supplies explicitly constructed for this propose. The OSHA definition also suggests that confined spaces are not intended for normal human uses and that fresh air must be supplied by auxilary means to people working in confined spaces.

The velocity and concentration fields inside enclosed and confined spaces are not readily apparent but can be predicted by the sequential box model (360) and procedures described in later chapters. With such predictions engineers can anticipate whether hazardous conditions will or will not occur for personnel performing certain jobs at different locations in these spaces. Present design practices do not specify the needed ventilation flow rates with much precision. The 1989 ASHRAE fresh air requirements for conventional rooms are shown in Appendix A-16. For a variety of industrial activities that generate vapors, gases, and particles, no such generalizations should be made. The issue is serious for people asked to perform work such as cleaning, welding, grinding and painting inside tanks on railroad cars and trucks, elevator shafts, farm silos, ship compartments and so on. In these cases contaminants are generated and more stringent ventilation standards need to be met. The OSHA standards for *welding in confined spaces* provides guidance.

In spaces less that than 10,000 ft^3/welder or rooms with ceilings less than 16 ft, general ventilation must be provided at a rate of 2000 CFM/welder except if local hoods are provided that produce velocities of 100 FPM in the welding zone or approved airline respirators are worn.

General ventilation is questionable since 2000 CFM of fresh air may never reach a welder located in a remote part of an unusually shaped confined space. The "well-mixed" presumption underlying the statement may not be valid and the air duct should be located in the close proximity of the welders breathing zone. If possible welding snorkles (Fig 1.27) should be used.

6.4 BULK MATERIALS HANDLING

Transferring bulk solids (powders, granular material, etc.) generates particles which, if not controlled, enter the workplace atmosphere. Particles can be generated from the following:

Crushing. grinding, milling, screening, classifying, size reduction, cleaning.
Conveyor belts, chutes, vibratory feeders, elevators, augers.

If processes can be enclosed, it is wise to do so. Figure 6.2 shows covers that are placed over drums when they are filled. The large flexible duct directs material from a screener to the drum and another flexible hose (not shown) withdraws air through the short exit pipe to a vacuum shown to the left. Figures 6.3 and 6.4 show a canopy hood and curtains placed around a rolling mill to provide a partial enclosure. Figure 6.11 shows an enclosure for a transfer point on a conveyor.

Emissions from vaguely defined transfer points are called *fugitive emissions* or *area emissions* because they enter the atmosphere at vaguely defined points. If emissions are contained within a duct they are called *point source emissions* because their origin is well defined. Because the momentum of the stream of moving bulk solids is large, these sources of particulate emissions are often referred to as *inertial sources*.

Enclosures in which the face velocity across the inlets and outlets are controlled within prescribed values are the only viable method known to control inertial sources. There has been scant attention paid to inertial sources. Analytical models proposed by Hemeon (47) have been modified and improved upon (228,

229) and are presently used. Transferring bulk materials generates particles by four mechanisms:

Induced air
Splash
Displaced air
Secondary air

Induced air is the name of the process whereby moving bulk material transfers momentum to surrounding air causing the air to move in the same direction as the bulk materials. When the bulk materials enter a hopper, mixer or merely an enclosure surrounding a transfer point, this air is *entrained* and drawn into the vessel with the bulk materials. Unless precautions are taken, this incoming air and the particles that are there will be made airborne by "splash" and generate a substantial amount of particulate matter. As an illustration, recall the amount of air set into motion by the stream of water from the bathroom shower (228). A commonly used equation to estimate the volumetric flow rate of induced air (Q_i) surrounding a falling stream of bulk materials is reported (228) to be

$$Q_i(\text{m}^3/\text{s}) = 0.631 \left[\frac{\dot{m}_b H^2 A_b^2}{\rho_b D_{pm}} \right]^{0.33} \tag{6-6}$$

where

$\dot{m}_b$ = mass flow rate of bulk materials (kg/s)
H = drop height (m)
A_b = cross sectional area of falling stream (m^2)
ρ_b = bulk solids density (kg/m^3)
$D_{p,m}$ = particle mass median diameter (m)

Splash is the phrase used to describe the generation of particles as the stream of bulk materials strikes a solid surface. The initial velocity of the particles may be modest and their range may be small, but if the velocity of the surrounding air is large, the particles may travel a considerable distance. At this time the only way to estimate the particle generation rate by splash is by means of emission factors.

Displaced air is the air displaced within the vessel by the incoming bulk material. The volumetric flow rate of the displaced air (Q_d) can be found by defining the enclosure as a control volume and applying the conservation of mass for the enclosure in which the bulk solids are being stored

$$\dot{m}_b - Q_d \rho_a = \frac{dM_b}{dt} \tag{6-7}$$

where ρ_a is the density of the displaced air and M_b is the instantaneous mass of the bulk solids inside the enclosure which in turn may be expressed as the rate at which the top surface of the bulk solids rises with time.

Secondary air is moving air (Q_s) entering the enclosure produced by independent means that causes particles made airborne by splash to be carried throughout the workplace. Sources of secondary air include spurious air currents (cross drafts) present in the workplace and air discharged from pneumatic equipment. Secondary air is sometimes called entrained air (228), although such a phrase tends to be confused with induced air. A distinction can be made that induced air refers to air entrained by the stream of bulk solids before splash while secondary air refers to air entrained after splash.

The volume of the enclosure (V_e) affects the splash and the amount of displaced air while the volumetric flow rate of secondary air, induced air, and air withdrawn from the enclosure (Q_w) are related. To prevent particles from escaping the enclosure, it is commonly believed that the atmospheric air entering the enclosure should have a face velocity (V_{face}) of 0.5–1.0 m/s. Thus if the enclosure is defined as the control volume, the volumetric flow rate of air withdrawn from the enclosure (Q_w) can be estimated from

$$\frac{dM_a}{dt} = \rho_a(Q_i + Q_s + Q_f - Q_w) \tag{6-8}$$

$$Q_f = V_{face}A_{face} \tag{6-9}$$

$$M_a = \rho_a V_a \tag{6-10}$$

$$V_e = V_a + V_b = V_a + M_b/\rho_b \tag{6-11}$$

where M_a, M_b are the mass of air and bulk solids in the enclosure at any instant and A_{face} is the area of the face through which atmospheric air enters the enclosure.

Figure 6.11 shows an enclosure around a transfer point between two conveyors; Eqs. 6-8 to 6-11 reduce to

$$Q_w = Q_i + Q_s + Q_f \tag{6-12}$$

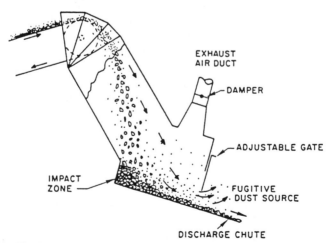

HEAD PULLEY ENCLOSURE

EXHAUST AIR DUCT

DAMPER

ADJUSTABLE GATE

IMPACT ZONE

FUGITIVE DUST SOURCE

DISCHARGE CHUTE

Fig. 6.11 Enclosure at a transfer point on a conveyor (adapted from ref. 229).

since the bulk flow rate is constant, and there is no accumulation of bulk material inside the enclosure and the process is steady. The induced air flow (Q_i) contains a large amount of particles and if the fan is not sized to accommodate this additional flow of air and particles, it will spill over at the inlet and allow particles to escape to the workplace atmosphere.

Figure 6.12 shows a bin that is being filled. Defining the vessel as the control volume and writing the continuity equations for the mass of air and bulk material results in the following:

$$\frac{dM_a}{dt} = \frac{d(\rho_a V_e)}{dt} = \rho_a \frac{dV_a}{dt} \qquad (6\text{-}13)$$

The volume of air (V_a) is

$$V_a = V_e - V_b = V_e - \frac{M_b}{\rho_b} \qquad (6\text{-}14)$$

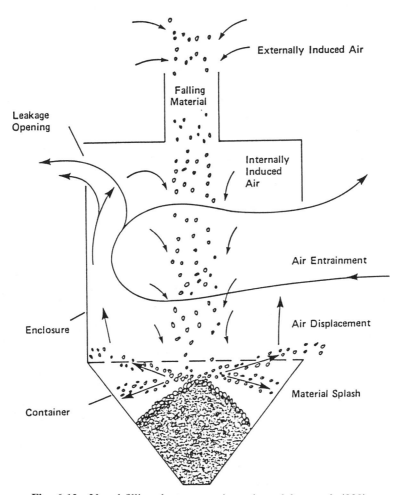

Fig. 6.12 Vessel-filling dust generation adapted from ref. (228).

Substituting Eq. 6-14 into Eq. 6-13 results in the following:

$$\frac{dM_a}{dt} = \left(\frac{\rho_a}{\rho_b}\right) \frac{dM_b}{dt}$$

$$= -\left(\frac{\rho_a}{\rho_b}\right) \dot{m}_b \tag{6-15}$$

which reduces to

$$Q_w = Q_i + Q_s + Q_f + \frac{\dot{m}_b}{\rho_b} \tag{6-16}$$

Equations 6-12 and 6-16 are alike, except that owing to the unsteady nature of the bin filling process the fan has to accommodate the additional volumetric flow rate of the displaced air.

6.5 CANOPY HOODS FOR BUOYANT SOURCES

Exothermic industrial processes that produce buoyant plumes are called *buoyant sources*. Steady processes such as the reheat furnace shown in Fig. 6.5 can be equipped with a close-fitting canopy hood that functions as a receiving hood to capture the hot plume. Batch processes are more difficult to control. Vapors emitted when oven doors are opened may be captured by a canopy hood, but unfortunately workers stand under these hoods when they open the oven doors. In the metallurgical industry, the volumetric flow rates of buoyant plumes are huge and vary with time owing to the nature of intermittent batch processes. When metal scrap is added to the furnace, or the furnace is skimmed or tapped, large amounts of fume (vapor and particles) are emitted at volumetric flow rates that are several times the values that exist for the rest of the cycle. The issue is doubly difficult because state and federal agencies will not allow these fumes to be discharged to the atmosphere and OSHA requires that concentrations within the workplace remain within safe values.

Plumes from hot sources rise because of buoyancy. In most industrial hot processes, the initial momentum is negligible in contrast to process gas streams that are discharged to the atmosphere through tall stacks external to the building. The velocity field of the buoyant plume (with or without a control device) can now be predicted by the techniques in the field of computational fluid mechanics. In lieu of such solutions empirical expressions can be used to guide engineers in the design of devices to control buoyant plumes. Figure 6.13 shows two examples of empirical equations from ACGIH (27) and Hayashi (243). There are also methods to model experimentally buoyant plumes and analytical techniques to study full-scale plumes within plants (167).

Based on work done by Morton et al. (231, 233) and Turner (232), Kashdan et al. (228) suggest that the following expression can be used to predict the steady (plume lasting more than 30 sec) volumetric flow rate of a buoyant plume a distance Z above the virtual origin of the plume.

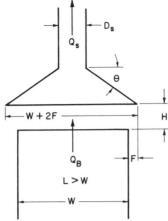

Fig. 6.13 Open, rectangular canopy hood for rectangular sources (adapted from refs. 27 and 243). $D_s/W \leq 0.3$; $H/W \leq 0.7$; $1 \leq \left(\dfrac{W+2F}{W}\right) \leq 1.5$; $0.2 \leq W/L \leq 1.0$; $\theta =$ unimportant; $Q_s = Q_B[1 + m(K_L + 3\Delta T/2500)]$; $1 \leq m$ (safety factor) ≤ 1.5; $\Delta T(C) = [T(\text{source}) - T(\text{ambient})]$; $K_L = \{1.4(H/W)^{1.5} + 0.3\}\{0.4(1+2F/W)^{-3.4} + 0.1\}\{1+W/L\}$; (from ref. 243).

$$Q_B = \begin{cases} \text{given} \\ \text{buoyant (m}^3/\text{min)} = 1.61[(L/W)\Delta T^{4/9}W^{1/3}] \end{cases}$$

(from ref. 27).

$\theta = 45$ degrees; $F = 0.4H$; $Q_s = 1.4[v_cH2(W+L)]$; 50 FPM $\leq v_c$ (capture velocity) ≤ 500 FPM.

$$Q_H\left(\frac{m^3}{s}\right) = 0.166Z^{1.67}F^{0.33} \tag{6-17}$$

where Z (virtual source height) – distance between hood and actual source plus one-half the source diameter (m),

$$Z = d + \frac{D}{2} \tag{6-18}$$

where

d = vertical distance between actual source and hood inlet

F = buoyant flux parameter (m^4/s^3)

$$F = \frac{(gq)}{(c_pT_0\rho_0)} \tag{6-19}$$

g = acceleration of gravity $(9.8\ m/s^2)$

c_p = specific heat of air $(cal/g\ K)$

ρ_0, T_0 = density (kg/m^3) and absolute temperature (K) of ambient air

q = (convection) heat transfer from source (kcal/s)

$$q = hA_s(T_s - T_0) \tag{6-20}$$

where

h = convection film coefficient (kcal/s m^2 K)

A_s, T_s = source surface area (m^2), and temperature (K)

Hemeon (47) neglected heat transfer by radiation, which seems appropriate unless it can be shown that radiation truly contributes to the buoyant movement of air.

Figure 6.14a and 6.14b shows schematic diagrams of canopy hoods for steady and unsteady buoyant sources. To capture buoyant plumes, analysis and experiment have shown (228) that the volumetric flow rate to withdraw gas from such hoods should be larger than Q_H,

$$Q_w = 1.21 Q_H \qquad (6\text{-}21)$$

Spillage is the name used to describe portions of the buoyant plume that are not captured by the hood and spill outward to other parts of the workplace. Spillage is due to cross drafts, obstructions from machinery (overhead cranes, ducts, building structural elements, etc.) that block the path of the hot plume to the canopy

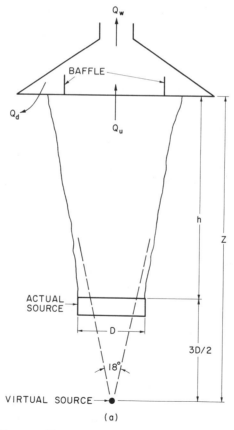

(a)

Fig. 6.14 Pool (*a*) and hopper (*b*) hoods showing displaced air (Q_d) during initial portion of buoyant plume surge (Q_u).

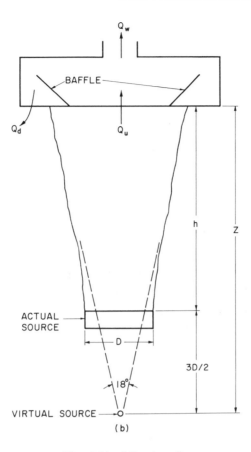

Fig. 6.14 (*Continued*)

hood, turbulent eddies within the hood causing gas to leave the hood before being withdrawn.

The dimensions of the hood are critically important, particularly canopy hoods for large-scale metallurgical processes that are intermittent in nature. For steady flow, the face velocity at the hood inlet should be 1.5 m/s or else the plume may overturn and spill from the hood. Canopy hoods for steady processes should have a diameter equal to one-half the effective height (Z). Alternatively, if there are no impediments or cross drafts, the diameter can be computed, assuming that the included angle of the hot plume is 18 degrees (228,230).

Canopy plumes for intermittent processes pose unique design problems. As a first step, the designer should estimate the plume volumetric flow rate (Q_u) throughout the cycle of the process. If the maximum and minimum values are close to each other and the economics allow it, a single canopy hood run at a constant withdrawal volumetric flow rate (Q_w) may suffice. If however, the variations (i.e., surges) in Q_u are large, such as shown in Fig. 6.15, a single hood designed for the maximum value may be very costly. A common practice

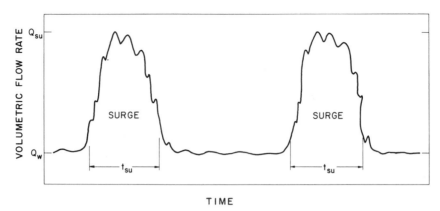

Fig. 6.15 Plume volumetric flow rate for intermittent buoyant source.

(228,230) in these cases is to design a canopy hood that captures and retains the gas for a short period of time until the surge has ended. Such canopy hoods are called *pool hoods* and *hopper hoods*. Figure 1.13 is an example of a pool hood. Internal baffles (230) are recommended to insure that only room air is displaced as the plume fills the hood and that the plume does not overturn and spill from the hood. The volume of the pool hood or hopper hood (V_h) should be no smaller than the integral of the surge in Fig. 6.15,

$$V_h = t_{su}(Q_{su} - Q_w) \qquad (6-22)$$

where t_{su} is the duration of the surge and Q_{su} is the maximum volumetric flow rate associated with the surge. The steady volumetric flow rate that gas is withdrawn from the hood (Q_w) must be sufficient that the entire volume of surge gas stored in the hood can be completely withdrawn before the next surge occurs. Thus

$$\frac{V_h}{Q_w} < \frac{1}{f_{su}} \qquad (6-23)$$

where f_{su} is the frequency of the surges. If the hood is not completely emptied, the new surge will displace gas in the hood containing remnants of the previous surge which will then enter the workplace. Often a pool and hopper hood will be formed by partitioning off a deep section between building roof trusses. With deep pool hoods, a face velocity of 0.5 m/s is adequate (228).

If cross drafts are present within the building, the volumetric flow rate of the withdrawn gas must be larger than predicted by Eq. 6-21. The following is suggested by Bender (230)

$$Q_w = Q_H\left[1 + 4.7\left(\frac{V_{cross}}{U_{max}}\right)\right] \qquad (6-24)$$

where V_{cross} is the cross-draft velocity and U_{max} is the plume centerline velocity. A cross draft also deflects the plume, so the hood has to be either off center or made large enough to accommodate the deflection.

6.6 AIR CURTAINS FOR BUOYANT SOURCES

Canopy hoods are located some distance above a source and the amount of air to be moved would be less if the contaminants were captured closer to the source. Air curtains offer this opportunity and there has been significant success in using them to control fumes in the metallurgical industry (228) and rubber milling (295). An air curtain (jet) in combination with an oppositely located exhaust is called a push–pull control system. The phrases *push–pull and air curtain* are used interchangeably.

Figure 6.16 illustrates details of the concept. In lieu of applying the full set of conservation equations to the interaction of the plume and air curtain, an elementary model proposed by Bender (230) has been used (228). The model assumes that the upward momentum of the buoyant plume (M_u) interacts with the jet momentum (M_j) inclined downward at an angle of θ such that the resultant momentum is horizontal and the combined flow enters the exhaust opening.

This concept presumes that while viscosity entrains surrounding air, the viscous stresses are negligible, and that there is no viscous dissipation. Consequently, the momentum of the push jet is conserved between the jet exit and exhaust opening,

$$M_j = \rho_j V_j^2 A_j \tag{6-25}$$

where A_j and V_j are the area and velocity at the exit of the push jet.

The upward moving buoyant stream also entrains air as it rises, and the model

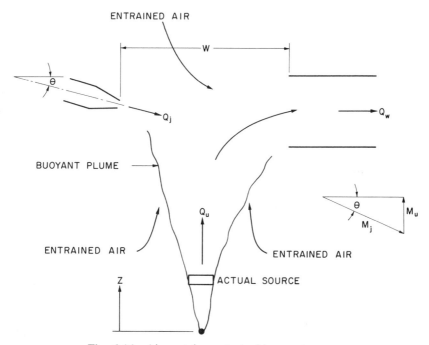

Fig. 6.16 Air curtain control of buoyant source.

assumes that at the plane between the push jet and exhaust opening, the plume has momentum given by

$$M_u = \rho_u V_u^2 A_u \tag{6-26}$$

where A_u and V_u are the plume cross sectional area and the (mass) average velocity at the plane where the plume and push jet interact, that is, the plane between the push jet and exhaust opening. Thus

$$\rho_u V_u^2 A_u = \rho_j V_j^2 A_j \sin\theta \tag{6-27}$$

which can be rearranged to express the jet velocity

$$V_j = V_u \sqrt{\left(\frac{\rho_u}{\rho_j}\right)\left(\frac{A_u}{A_j}\right)\frac{1}{\sin\theta}} \tag{6-28}$$

The model also assumes that the volumetric flow rate of the rising plume (Q_u) is given by Eq. 6-17 and that the average velocity is the plug flow velocity given by dividing the Q_u by the cross-sectional area of the plume at the plane between the push jet and exhaust opening. Air from the push jet entrains air. Kashdan et al. (228) report that for rectangular push–pull systems, the volumetric flow rate of the withdrawn gas (Q_w) can be expressed by

$$\frac{Q_w}{L}\left(\frac{\text{ACFM}}{\text{ft}}\right) = 0.88\sqrt{\left(\frac{Q_j}{L}\right)V_j W} \tag{6-29}$$

where

Q_j = jet volumetric flow rate (ACFM)
V_j = jet velocity (FPM)
W = distance separating push jet and inlet opening (ft)
L = length of system (ft)

The model assumes that the expression is valid for the interaction of rising plume and push jet.

Illustrated Example 6.2—Capture of Secondary Emissions from Copper Converter (228,234)

Copper converting is the process of transforming copper matter (produced in a smelter) into blister copper. A copper converter is a refractory-lined cylindrical vessel with an opening in the center. The vessel is rotated to various positions during charging, skimming and blowing, as shown in Figs. 6.17 and 6.18a, 6.18b. For most of the cycle (12 hr) the vessel is in the blowing phase in which oxygen-enriched air is blown through the melt and the emissions are drawn off through a close-fitting primary hood and sent to a sulfur recovery plant. During the charging and skimming phases (approximately 30 min total, in periods of 4 min each) the converter is rotated away from the primary hood and controls are needed to prevent fume from entering the workplace.

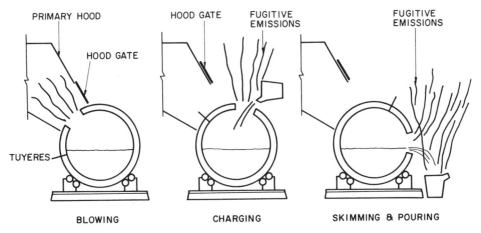

Fig. 6.17 Control of fume from copper converter (*a*) receiving hood and (*b*) air curtain (adapted from ref. 228).

The analysis begins by selecting the air curtain flow conditions and computing the necessary exhaust volumetric flow rate (Q_w). The design can be optimized by repeating the process until minimum volumetric flow rates (Q_j and Q_w) are obtained.

For the converter shown in Fig. 6.18b the following data were measured, calculated, and selected.

Jet slot width = 0.05 m (2 in)
Jet slot length (L) = 3.96 m (13 ft)

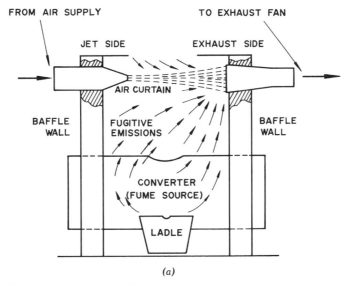

(*a*)

Fig. 6.18 Air curtain control for copper converter (adapted from ref. 228).

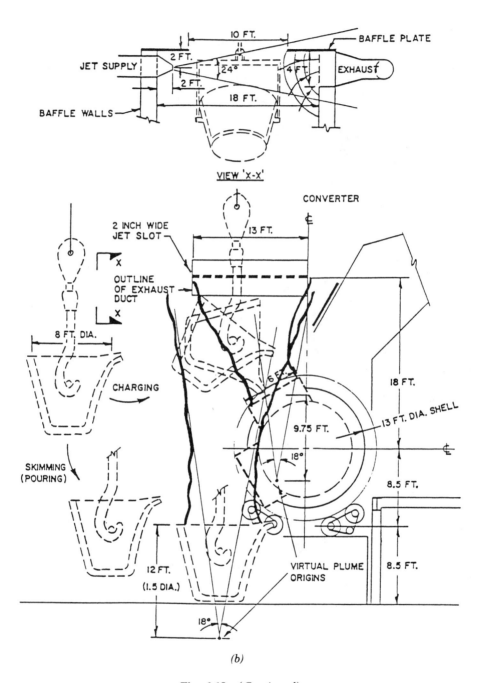

(b)

Fig. 6.18   (*Continued*)

Separation of jet and exhaust $(W) = 3.66$ m (12 ft)
Jet air temperature $= 21$ C (294 K)
Jet velocity $(V_j) = 0.7$ m/s (137 FPM)
Jet volumetric flow rate $(Q_j) = 8.4$ m^3/s (17,800 ACFM)
Angle $\theta = 15$ degrees
Ambient temperature $(T_0) = 21$ C (294 K)
Ambient density $(\rho_0) = 1.2$ kg/m^3
Plume temperature during skimming $(T_u) = 85$ C (358 K)
Plume temperature during charging $= 249$ C (522 K)
Heat transfer $(q) = 150,000$ BTU/min

From Eq. 6-19, the buoyancy flux (F) is

$$
\begin{aligned}
F &= \frac{(qg)}{(c_p T_0 \rho_0)} \\
&= \frac{(150,000)(32.2)(3600)}{(0.24)(530)(0.075)} \\
&= 1.8 \times 10^9 \text{ ft}^4/\text{min}^3 \quad \text{or} \quad 72 \text{ m}^4/\text{s}^3
\end{aligned}
$$

During charging, the plume spreads around the charging ladle and the cross sectional area of the plume cannot be based on the idealized 18-degree plume angle. Instead the 10-ft diameter of the ladle will be used as the area of the source. The height of the virtual source (Z) will be assumed to be 21 ft during charging. Consequently,

Charging:
$Q_u = 15.1$ m^3/s at 522 K (32,000 ACFM at 480 F)
$A_u = (10)^2 \pi/4 = 78$ ft^2 (7.25 m^2)
$V_u = 32,000/78 = 410$ FPM (2.08 m/s)

During skimming the plume will rise from the skimming ladle (overall diameter assumed to be 15 ft) and spread under the influence of buoyancy. From Fig. 6.18 the height of the virtual source (Z) is found to be 38.5 ft. Consequently,

Skimming:
$Q_u = 42.5$ m^3/s at 358 K (90,000 ACFM at 185 F)
$A_u = (15)^2 \pi/4 = 176$ ft^2 (16.4 m^2)
$V_u = 90,000/176 = 510$ FPM (2.6 m/s)

Since skimming produces the largest plume, the air curtain will be designed for 90,000 ACFM at 185 F. The gas entering the exhaust opening can be computed from

$$Q_w = 0.88L\sqrt{\left(\frac{Q_j}{L}\right)V_jW}$$

$$= (0.88)(13)\sqrt{\left(\frac{17,800}{13}\right)(8220)(12)}$$

$$= 133,000 \text{ ACFM } (62.8 \text{ m}^3/\text{s})$$

The process can be repeated for other values of Q_j, slot dimensions, and angle θ until a minimum Q_j and Q_w are obtained. Measurements made in the plant confirm (228) the accuracy of the values given above.

The analysis employs several untested assumptions that users must be cognizant of.

(1) The model uses a push–pull equation that neglects viscous stresses to describe the interaction of a jet directed perpendicularly into a hot, low-speed plume.

(2) The engineer cannot select an exhaust's dimensions (4 ft wide) arbitrarily and be assured that the desired volumetric flow rate ($Q_w = 133,900$ CFM) will be withdrawn. The engineer has an obligation to prove or measure that the desired volumetric flow rate is achieved. It is foolhardy to presume that spillage at the exhaust opening does not exist!

(3) The model assumes incorrectly that the mass-average velocity is the same as the momentum-average velocity.

6.7 BUILDING AIR INLETS AND EXHAUST STACKS

One of the commonest errors made in industrial ventilation is to place a makeup air inlet too close to an exhaust stack. The error arises generally because the locations of the inlet and exhaust are decided at different times by different people unaware of each other's actions. Secondly, errors are made because individuals are unaware of the size and consequences of aerodynamic *wakes* associated with buildings. Figures 6.19 and 6.20 illustrate the wakes and vortices one can expect from block-like buildings on level terrain immersed in a deep terrestrial boundary layer. If the building lies in the wake of other buildings or if the building lies on terrain that is not level, the aerodynamic wakes are somewhat different, but exist nonetheless. An aerodynamic wake is a region in which the air velocities are not equal to free stream values. Wakes can be subdivided into a recirculation cavity or eddy or a turbulent shear region. A *recirculation eddy*, is a region in which a relatively fixed amount of air moves in a circular fashion and there is little air transported across the eddy boundaries. A *turbulent shear region* is one in which there is a net convective flow but the turbulent shear stresses are larger than free stream values.

Relationships predicting the location and velocity field of wakes are not readily available, but alternatively a series of empirical equations defining the boundaries of the wake regions are widely used (183). In the discussion that follows it will be assumed that the building is a block-like structure of height H, cross-wind width W, and length in the direction of the wind L.

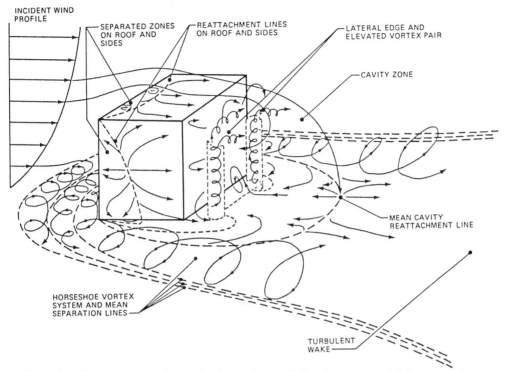

Fig. 6.19 Wakes and vortices of a block-like building in a terrestrial boundary layer (adapted from ref. 183).

In selecting the location and dimensions of an air inlet or exhaust stack, two general principals should be followed:

Do not locate air inlets in recirculation regions or at other locations susceptible to contamination from exhaust gases.

The effective stack height above the building roof should be above the roof wake boundary.

As good engineering practice the EPA recommends a *stack height* (h_s, height above ground level) of

$$h_s = H + 1.5z' \tag{6-30}$$

where z' is the smaller of W or H. The stack height (h_s) is a geometric dimension. The *effective stack height* is equal to the geometric stack height plus the plume rise due to the buoyancy and the momentum of the exhaust gas (105, 181). If zoning laws or other considerations prevent Eq. 6-30 from being followed, it will be necessary to estimate the location of the wakes. There are two regions of concern (a) the recirculation eddy and wake produced by the leading edge of the building roof and (b) large recirculation eddy directly downwind of the building.

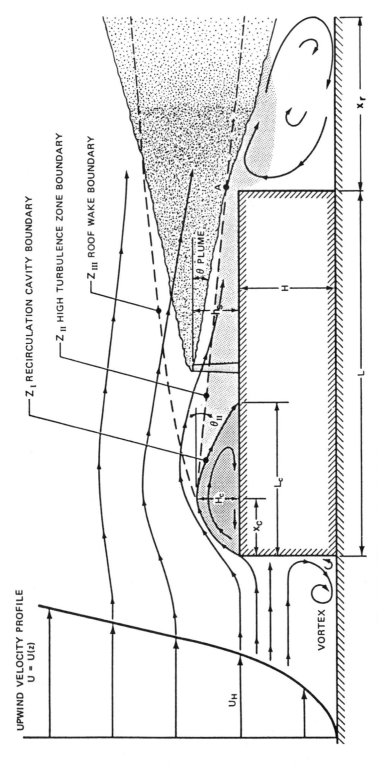

Fig. 6.20 Wakes and recirculation eddies of a two-dimensional flat building in a terrestrial boundary layer (adapted from ref. 183).

Key dimensions of the *roof eddy* are shown in Fig. 6.20 and can be expressed in terms of length parameters L_L, L_S and R', where

L_L larger of H or W

L_S smaller of H or W $\qquad\qquad$ (6-31)

$R' = L_S^{0.67} L_L^{0.33}$

The height (H_c), length (L_c) and center (x_c) of the roof eddy can be estimated from the following:

$$H_c = 0.22 \ R'; \qquad L_c = 0.9 \ R'; \qquad x_c = 0.5 \ R' \qquad (6\text{-}32)$$

The roof eddy is bounded above by a turbulent shear region (zone II) and above that by a wake region (zone III). The height of the turbulent shear zone (Z_{II}) can be estimated from

$$\frac{Z_{II}}{R'} = 0.27 - 0.1 \ \frac{x}{R'} \qquad (6\text{-}33)$$

where x is measured in the downwind direction from the roof lip. The height of the roof wake (Z_{III}) can be estimated from

$$\frac{Z_{III}}{R'} = 0.28 \left(\frac{x}{R'}\right)^{0.33} \qquad (6\text{-}34)$$

The safest design is one in which the expanding plume remains above the roof wake boundary (Z_{III}). The manner in which the plume expands depends on the atmospheric stability which is a function of wind speed and the solar radiation. If the plume cannot be kept above Z_{III}, a plume that is kept in the turbulent shear region is the next best choice.

Directly downwind of the building a very large *building eddy* is formed. Contaminants trapped in this region are apt to enter building windows and affect people and vehicles on the ground. The size of the this eddy can be seen in Fig. 6.19. The width and height of the eddy seldom exceed the building dimensions W and H by more than 50% and the downwind length of the eddy (x_r) can be estimated from the following:

$$\frac{x_r}{H} = \frac{A}{(1 + B)} \qquad (6\text{-}35)$$

For buildings in which $L/H < 1$,

$$A = -2.0 + 3.7 \left(\frac{L}{H}\right)^{-0.33} \qquad (6\text{-}36)$$

$$B = -0.15 + 0.305 \left(\frac{L}{H}\right)^{-0.33} \qquad (6\text{-}37)$$

For buildings in which $L/H > 1$, $A = 1.75$ and $B = 0.25$.

The previous material describes building wake regions and provides guidance about where to place exhaust stacks or fresh air inlets. If for some reason a vertical stack is placed on a large horizontal surface and the wind speed is constant, it is possible to be more definitive and actually predict the plume gas concentration at arbitrary downwind points, including points on the ground. *Dispersion modeling* is the name given to this body of knowledge and excellent, concise presentations suitable for engineers can be found in the literature (105,181,239).

Gases leave the stack and rise owing to momentum and buoyancy. Once the exit momentum has been dissipated and the gases have cooled to the local atmospheric temperature (*neutrally-buoyant*), the plume is carried downwind by the prevailing wind. In the lateral direction, mixing is governed by molecular and turbulent diffusion. In the vertical direction, mixing is governed in addition by the *lapse rate* (the negative value of the atmospheric temperature gradient), which in turn is governed by the solar intensity and the wind speed. Studies of how plumes are transported in the atmosphere have been of considerable interest for many years. Since the solution of the governing transport equation is the Gaussian distribution function, the plumes are called *Gaussian plumes*. There are several expressions (105,181,239) to predict the height to which a plume rises above the physical stack (i.e., *plume rise*), before being transported downwind.

Predictions of ground-level concentrations from Gaussian plumes are valid for regions no closer than approximately 100 m of the stack. There is a paucity of information for regions closer to the stack where the plume is rising owing to buoyancy and momentum. For neutrally buoyant plumes, Halitsky (380) provides expressions to predict downwind concentrations for regions downwind of the stack where the jet velocities have decayed and are essentially equal to the free stream values.

In the wake of a stack the pressure is lower than the free stream value and under some conditions may cause the exiting stack gases to fall rather than rise. Such a fall is called *downwash*. Downwash may be a serious consideration in the design of short stacks. Downwash can be prevented by keeping the ratio of the duct exit velocity to the wind speed greater than 1.5, that is, $V_j/U_0 > 1.5$. If the ratio is less than 1.5 it is suggested (105) that the downwash distance (H_{dw}) can be estimated from

$$\frac{H_{dw}}{D} = 2\left[\left(\frac{V_j}{U_0}\right) - 1.5\right], \qquad \text{if } \frac{V_j}{U_0} < 1.5 \tag{6-38}$$

where D is the stack exit diameter. The negative value computed by Eq. 6-38 implies that the downwash causes the plume centerline to fall rather than rise above the stack exit plane.

6.8 UNSATISFACTORY PERFORMANCE

It may be troublesome to the reader that design data in previous sections recommend a range of volumetric flow rates or a range of capture velocities rather than specific values. Furthermore, the range is often very large. A range acknowledges variabilities in the workplace (drafty or still air, dirty or clean environment, toxicity of the contaminant, etc.) and deliberately forces engineers to decide whether the upper or lower end of the range is to be selected. One of the principal variables is drafts, or more formally, spurious air currents. Seldom do engineers know the ambient air velocity at points throughout the workplace unless measurements are made to acquire them. Individuals are generally unaware of drafts, for example, a velocity of 100 FPM (0.5 m/s) is similar to blowing on the hand, and most people would be unaware of such a draft unless their attention was called to it. While small, such a velocity influences the capture of contaminants. Later in the text procedures are described that incorporate the velocity of the ambient air in the design process in a quantitative way. The following physical conditions produce spurious air currents:

(a) Air issuing from internal room air heaters, conditioned makeup air entering through registers.
(b) Portable fans used by workers for comfort.
(c) Windows or doors workers choose to leave open for comfort.
(d) Wakes produced by workers and machinery.
(e) Drafts produced by moving workers or vehicles.

A well-designed industrial ventilation system may perform badly because the designer is unaware of actions taken after the system is installed. The following is a partial list of actions that adversely affect performance:

(a) The volumetric flow rate may decrease because someone has blocked off part of the inlet area of the capture device or the inlet to the exhaust fan.
(b) Someone may tap into the exhaust duct for another control device.
(c) Large amounts of material collect in the exhaust duct and reduce the volumetric flow rate.
(d) Holes may be cut into a hood or enclosure to improve a worker's visibility.
(e) Workers may stand between the inlet and the source.
(f) More volatile or toxic material maybe substituted.

The list of such blunders is long and sometimes embarrassing. Many conditions can be avoided if designers observe the movement of workers and talk to them about their work before the design is begun. After the system is installed it behooves the designer to inspect the site and observe if any of the conditions above occur or if any unanticipated changes have occurred that adversely affect performance.

Inadequate makeup air is another common reason why ventilation systems fail to perform adequately. The following are symptoms of inadequate makeup air.

(a) Exhaust gases from ovens, furnaces, water heaters, cookers, and dryers, (any device having a natural draft flue) have insufficient draft and exhaust gases enter the workplace; this event is often preceded by complaints of odors.

(b) Pilot flames on ovens fail to remain lit.

(c) face velocities on ventilation systems and volumetric flow rates are well below their design values even though fan horsepower and RPM may have their correct value.

(d) Doors opening into the workplace will not stay shut or doors opening away from the workplace take a great deal of effort to open.

(e) Large overhead doors collapse inward.

These symptoms often occur because equipment such as ovens, exhaust fans, and ventilation systems are installed in the workplace in a piecemeal, uncoordinated fashion, generally over a period of time, without regard for makeup air. As a rule, each time a device is installed that exhausts air from a building, a comparable amount of makeup air must be provided. The symptoms are most severe in the winter when doors and windows are closed.

Present design practices are primitive. They lack the precision brought to engineering by advances in analytical modeling and numerical computation. These advances have greatly benefited other fields of engineering such as heat transfer, aerodynamics, and mass transfer. Nevertheless while sophistication is lacking, present design practices are not wrong nor do they lead to unsafe designs. Present practices should continue to be followed, but used as initial estimates. Analytical techniques and modeling concepts can then be used to optimize the design for the particularities of an industrial site, determine of whether mandatory health standards will be met, and estimate performance for a variety of operating conditions. Irrespective of how a ventilation system is conceived, its performance must be verified by measurements at the site. The objective of improving analysis is to

(a) Make design decisions authoritatively with a minimum amount of time.

(b) Withdraw a minimum amount of air and use a minimum amount of electrical power.

(c) increase the probability that the design will require few design changes in the field.

Illustrated Example 6.3—Control of Fume from Robotic Welder

An industrial robot performs arc welding on an assembly line for automotive chasses. You have been asked to recommend a local ventilation system of circular cross section to control welding fume. There are several design constraints:

The inlet can be no closer than 6 in (0.15 m) from the welding zone,

The cross-sectional area of the inlet must be less than 1 ft^2 (0.092 m^2).

OSHA stipulates a workplace TLV for welding fume of 5 mg/m^3 although it does not stipulate an exhaust volumetric flow rate. Table 6.1 recommends a capture velocity of 100–200 FPM (0.51–1.02 m/s). The upper value will be chosen because of the toxicity of welding fume and air currents that are known to exist in the workplace.

A local ventilation system consisting of a flanged or unflanged opening will be studied with welding performed along the centerline at $r = 6$ in. The air velocity $U(r)$ at $r = 6$ in will be set equal to 200 FPM. Under production line conditions, there will always be an obstruction at the point of welding—it is unavoidable. The way in which the obstruction affects the velocity field is hard to predict, it can lower the velocities as well as raise them, depending on the geometry of the surfaces to be welded.

The required volumetric flow rate for an unflanged opening of different diameters (D) was computed as follows:

(a) Compute r/D for $r = 6$ in and several D.
(b) Use Fig. 6.9 and compute $U(r)/V(\text{face})$ for each x/D along the duct centerline,
(c) for $U(r) = 200$ FPM, compute $V(\text{face})$.
(d) compute the volumetric flow rate (Q)

UNFLANGED OPENINGS

Diameter (in)	A (ft^2)	Q (CFM)	Entry face velocity (FPM)
2.0	0.022	509	23,113
4.0	0.087	516	5,931
6.0	0.196	538	2,744
8.0	0.349	569	1,630
10.0	0.545	608	1,116
12.0	0.785	656	836

Shown below are the volumetric flow rates for several flanged entry diameters predicted by the same procedure.

FLANGED OPENINGS

Diameter (in)	A (ft^2)	Q (CFM)	Entry face velocity (FPM)
2.0	0.022	344	15,636
4.0	0.087	357	4,103
6.0	0.196	379	1,193
8.0	0.349	410	1,173

It can be seen that a flanged entry affords the best control since it requires approximately 40% less air for the same control velocity.

On the basis of the rate given above, an 8-in entry with a 2-in flange or 6-in entry with a 3-in flange are sound recommendations. While smaller entries require less air, they have higher face velocities and produce more noise.

The ACGIH ventilation manual (27) recommends exhaust volumetric flow rates for a flanged and unflanged opening for welding. At $r = 6$ in, the manual recommends 250 CFM and a flange of 3 in and 335 CFM for a plane opening. In both cases the recommended face velocity is 1500 FPM. The agreement between this and the recommendations above is reasonably good considering that a control velocity of 200 FPM, rather than 100 FPM, was chosen for the computation.

While the design constraints require the entry to be no closer than 6 in from the point of welding, the designer should consider a low volume–high velocity ventilation system (Figs. 1.10 to 1.12). The ventilation manual (27) provides some details on the design of such systems. A *low volume-high velocity system* consists of a small entry located very close to the point of welding that produces a very large face velocity but requires considerably less exhaust air. The systems may produce considerable noise because of the high face velocity.

Human workers need an unobstructed view of the welding region, but robots do not. Furthermore, the robot performs prescribed welding operations that may tolerate a small entry located close to the point of welding. The designer should examine the constraint of 6 in and investigate the possibility of using such a system. If a clear advantage can be realized, the designer should recommend this alternative design approach.

Illustrated Example 6.4—Control of Vapor From Open Vessel

You have been asked to study lateral ventilation systems to control the vapor entering the workplace from a rectangular, free-standing open vessel (1.4 m wide by 2.5 m long) containing trichlorethylene at an elevated temperature. The width of the vessel is the smaller of the two dimensions and normal to the lateral exhaust. The open vessel is used for degreasing operations in which workers must have access to the liquid, thus it is not possible to enclose the vessel. Two lateral ventilation systems will be compared.

(a) Suction system (Fig. 6.1 is a circular version)—lateral exhaust slots along both (long) sides of the vessel following ACGIH procedures (Tables 6.1–6.4).

(b) Push-pull system—an exhaust slot along one side of the vessel and a blowing slot along the opposite side following procedures shown in Fig. 6.21.

Suction System. The TLV for trichlorethylene is 100 PPM and the normal boiling temperature is 87 C. Table 6.2 shows that the liquid can be categorized as class B1. Table 6.3 recommends a minimum control velocity of 100 FPM (0.5 m/s). Since the ratio of the tank width to length is 0.56, Table 6.4 indicates that if the vessel is free-standing the minimum exhaust volumetric flow a rate should be 225 CFM/ft^2 (1.14 m^3/s per m^2) of liquid surface. For a vessel 1.4×2.5 m, the total exhaust volumetric flow rate is 8472 CFM (4 m^3/s).

ACGIH Push-Pull System Select a continuous 1/4-inch (S_B) slot that runs the length ($L = 2.5$ m) of the vessel. ACGIH recommendations in Fig. 6.21 suggest a blowing volumetric flow rate (Q_B),

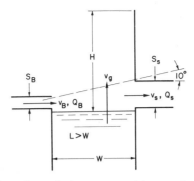

Fig. 6.21 Rectangular push–pull ventilation system with cross draft (V_0) (adapted from refs. (243) and (271)). $H/W \geq 0.2$; $W/S_B \leq 30$; $2 \leq H/S_B \leq 50$; $0.5 \leq S_S/S_B \leq 10$; $0.75(W/S_B)$ min $\leq$ (W/S_B) $\leq$ 1.5 (W/S_B) min; $Q_s = Q_B(1 + mK)$; $1 \leq m$ (safety factor) ≤ 1.5; $Q_B = (v_B S_B L)$; $K = 5.8\{(v_0/v_B)^{1.4} (W/S_B)^{0.25} + 1\}K_0$; $(W/S_B)_{min} = (W/H)$ $\{3.2 + [130(W/H)^{-1.1} + 46]^{1/2}\}^{0.91}$; $v_0/v_B = \left\{ \dfrac{K_0 + 1}{2.3 (W/S_B)^{0.25}K_0} \right\}^{0.71}$ for buoyant source replace v_0 by v_g, $v_0 \cong v_g \cong 0.3$ m/s; $K_0 = 0.55 (W/S_B)^{1.1}\{0.46(H/S_B)^{-1.1} + 0.13\}$ (from ref. 243).

Blowing jet area, A_B (ft^2); blowing plenum cross sectional area $>3A_B$ slot: $1/8'' \leq S_B \leq 1/4''$; holes: $1/4''$ Dia spaced $3/4''$ to $2''$ appart Q_B/L (CFM/ft) $= 243 (A_B/L)^{0.5}$ where A_B/L (ft^2/ft). Suction opening A_S (ft^2) $= LS_S$; $S_S = 0.14$ W, Q_S/LW (CFM/ft^2) $= 75$ (from ref. 27).

$$\frac{Q_B}{L} = 243\left(\frac{A_B}{L}\right)^{0.5} = 243\left(\frac{LS_B}{L}\right)^{0.5} = 243\left(\frac{1}{48}\right)^{0.5}$$

$$\frac{Q_B}{L} = 35.1 \text{ CFM/ft}$$

$$Q_B = 287.6 \text{ CFM } (8.15 \text{ m}^3/\text{min})$$

and a suction volumetric flow rate (Q_S/A) of 75 CFM/ft^2. Thus,

$$Q_S = (75)(2.5)(1.4)(3.28)^2 = 2824.1 \text{ CFM } (80 \text{ m}^3/\text{min})$$

The suction system requires three times more air than the ACGIH push–pull system. Thus the push–pull system is the preferred system unless it is absolutely necessary that workers have access to both sides of the vessel.

6.9 EXHAUST DUCT SYSTEM DESIGN

Collection hoods are connected to an exhaust fan (or fans) by a network of ducts whose dimensions must be selected by the designer. Which hoods are to be connected to a single exhaust fan and which hoods are to be equipped with *blast gates* (damper valves to isolate the hood when it is not in use) are important matters that must be decided at the beginning of the design process in close consultation with the users of the system. Factors entering the decision are:

Space available to locate ducts, fans, and air cleaning equipment (if needed).

Minimum cost (capital and operating costs).

Ease of operation, that is, how will opening and closing blast gates affect other hoods.

Maintenance, repair, and duct cleaning.

Once the decision has been made about which hoods are going to be connected to a particular exhaust fan, the designer should prepare a line sketch that identifies each hood, length of each duct segment, contaminant removal device, blast gate, fan, and number and type of duct fittings (elbow, tee, expansion or contraction fitting, etc). The volumetric flow rates for each collection hood are determined by the designer, based on the nature of the process to be controlled and type of collection hood selected. What is unknown is the diameter of each duct segment. For air cleaning purposes, it is wise to use ducts of circular cross section.

If particles are to be collected it is important to achieve duct velocities above certain minimum values to minimize gravimetric settling. ACGIH design plates specify minimum duct velocities and should be followed. Lacking any specific instructions, the *duct transport velocities* shown in Table 6.6 should be used. Later in the text there is a description of how to estimate the pressure drop in ducts and fittings, which vary as a function of the duct diameter, transport velocity, particle size distribution, density, and concentration. Operating at a low transport velocity will result in serious deposition which reduces the duct cross sectional area and the duct velocity, which in turn increases further deposition. Within a very short time the system will become plugged and immobilize all collection hoods downstream of the plugged point. Plugged ducts sag under the weight of the deposited particles and have to be removed. Duct systems should contain numerous *clean-out ports* easily accessible by workers for routine cleaning. In addition *drop-out boxes* should be placed at duct junctions, hoods, and so on so as to collect large particles before they enter the ducts.

If vapors are to be collected provision should be made for clean-out ports. Vapors will condense on cool duct surfaces. Flammable condensate will catch fire whenever there is a source of ignition, for example,

TABLE 6.6—Duct Transport Velocities

Contaminant	Velocity (FPM)
Gases, vapors, smoke	1000–2000
Fumes	2000–2500
Very fine light dust (cotton lint, wood flour)	2500–3000
Dry dust and powder (soap, rubber, Bakelite molding powder)	3000–4000
Average industrial dust (general materials handling)	3500–4000
Heavy dust (foundry shakeout, metal working)	4000–4500
Very heavy and sticky dust (quick-lime, moist cement, metal chips)	4500 and above

Discharge from statically electrified surfaces.

Sparks from welders torch.

Burning particles from dryers.

Sparks from sanding machines that encounter wood containing metal.

Since ducts are located near roofs and ceilings, initially minor duct fires often lead to serious building fires.

Unless a hood is to operate continuously, it should be equipped with a *blast gate*. If there is no industrial activity, the blast gate should be closed. If left open, air will be exhausted unnecessarily and increase the fan operating costs and quite possibly reduce the exhaust volumetric flow rate at hoods where there is industrial activity. The selection of the appropriate exhaust fan requires knowledge of the volumetric flow rate and overall pressure drop. The volumetric flow rate is the sum of the volumetric flow rates (corrected for density) of the separate collection hoods, taking into account which hoods will operate full time and which will be activated by blast gates.

There are several detailed procedures to design duct systems. The most authoritative discussion is contained in the ASHRAE *Fundamentals Handbook* (185). These procedures are general in nature and can be used for heating and air conditioning as well. Procedures recommended by the ACGIH (27) pertain to contaminant control, are widely used in the field of industrial ventilation, and are compatible with ASHRAE procedures.

Computing the pressure drop in a duct is the basis of all duct design procedures. Since the velocity and pressure drop in a duct are sufficiently low, one may assume that the air is a constant-density viscous fluid. Thus the pressure drop between two arbitrary points (1 and 2) in a duct can be expressed as

$$P_1 - P_2 = \rho \frac{(U_2^2 - U_2^2)}{2} + \rho g(z_2 - z_1) + \rho h_{LT}$$

$$P_{t,1} - P_{t,2} = \rho h_{LT} \tag{6-39}$$

where P_t is the *total or stagnation pressure*, U_2 and U_1 are the average velocities at points 1 and 2, and h_{LT} is the *total head loss* between points 1 and 2.

$$P_t = P + \frac{\rho U^2}{2} \tag{6-40}$$

Sometimes the term *SP* is used to designate the *static pressure* (*P*) and *VP* to designate the *velocity or dynamic pressure* ($\rho U^2/2$),

$$SP = P; \quad VP = \rho U^2/2 \tag{6-41}$$

$$P_t = SP + VP$$

For air at STP (25 C, 101 kPa) the velocity head (*VP*) in Pascals (Pa) is equal to

$$VP(\text{Pa}) = \left(\frac{U}{1.29}\right)^2 \tag{6-42}$$

when the average velocity (*U*) is in meters per second.

The term dealing with elevation changes can be neglected unless there is a huge vertical rise to be accounted for. For example, an elevation change of 10 m corresponds to a pressure difference of only 0.12 kPa (0.4 in of water) for air at STP. Thus the fundamental equation to compute the static or total pressure change between any two points 1 and 2 is

$$P_1 - P_2 = VP_2 - VP_1 + \rho h_{LT}$$
$$P_{t,1} - P_{t,2} = \rho h_{LT} = \rho(h_L + h_{LM}) \tag{6-43}$$

The total head loss (h_{LT}) is composed of *major losses* (h_L) due to friction losses in constant area ducts and *minor losses* (h_{LM}) due to flow separations occurring in

Duct fittings (elbows, tees, dampers, etc.)
Duct area changes (expansions and contractions)
Inlet losses at entrances to equipment

The dimensions of all these losses have the units of energy.

6.9.1 Duct Friction

It is customary to express the head loss due to friction (major loss h_L) in terms of a *friction factor f*,

$$h_L = f\left[\frac{L}{D}\right] VP \tag{6-44}$$

where the *friction factor* is a function of the height of the wall deposits or material roughness (*roughness height*) and the *Reynolds number* (see the Moody chart, refs. 18 and 185)). Alternatively, Fig. 6.22 can be used to compute the duct friction loss (Pa) per meter of duct for air at 20 C, 101 kPa flowing through round galvanized ducts having approximately 40 slip joints per 30 m. The figure can also be used for other duct materials of comparable roughness. If the duct cross section is not round, the diameter should be replaced by a circular equivalent called the *hydraulic diameter* (D_y) which is equal to 4 times the cross sectional area divided by the duct perimeter.

6.9.2 Fittings

It is customary to express the minor head loss (h_{LM}) as

$$h_{LM} = C_0 VP_0 \tag{6-45}$$

where C_0 is called the *local loss coefficient* and depends on the geometry of the fitting. The subscript 0 refers to the point in the fitting at which the velocity pressure is to be calculated. Generally conditions entering the fitting are used but for some fittings (T's, Y's, etc.) the point is specified on the table of loss coefficients for the fitting. Local loss coefficients represent losses due to flow

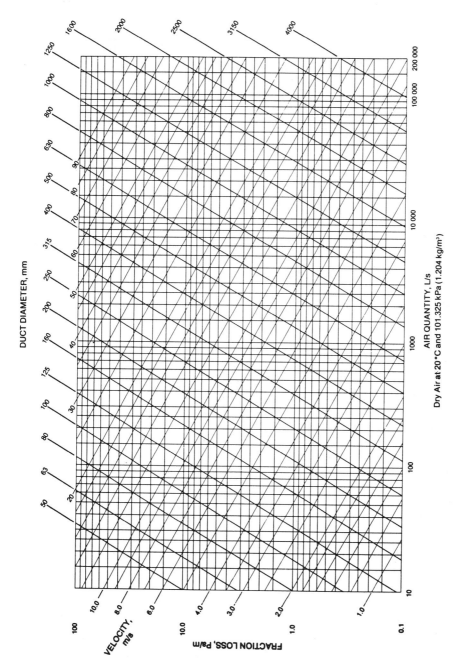

Fig. 6.22 Duct viscous friction loss (reprinted by permission from the *ASHRAE Hand-book—1985 Fundamentals*).

337

separation within the fitting, frictional losses are insignificant because of the small length of the fitting. The ASHRAE *Fundamental Handbook* (185) should be consulted for a detailed listing of local loss coefficients for a large variety of fittings. Shown in Figs. 6.23 and 6.24 are the local loss coefficients for several common fittings (105).

The total pressure loss for a converging or diverging a *branch section* (section composed of a branch joining a straight main section) is more complicated owing to the momentum exchange between the air streams. The total pressure loss through the main section is equal to

$$P_t \text{ (main section loss)} = C_{c,s} VP_c \tag{6-46}$$

and the total pressure loss through the branch section is

$$P_t \text{ (branch section loss)} = C_{c,b} VP_c \tag{6-47}$$

where VP_c is the velocity pressure at the common section (Fig. 6.24) and $C_{c,s}$ and $C_{c,b}$ are the *loss coefficients* loss for the *straight* (main) and *branch* flow paths, respectively. Converging flow junctions may have either positive or negative local loss coefficients because when two streams of different velocity converge, the momentum exchange may increase or decrease the total pressure.

6.9.3 Hood Entry Losses

In industrial ventilation, engineers use a variety of devices such as plane inlets (Fig. 1.27), flanged inlets (Fig. 6.9b), slotted inlets (Fig. 1.15), canopy hoods (Figs. 1.13 and 6.5) and partial enclosures (Figs. 1.24, 6.3, and 6.12). Head losses will occur as air passes through these devices and enters the duct. The minor loss (h_{LM}) associated with *hood entry losses* can be expressed as

$$h_{LM} \text{ (hood entry loss)} = C_0 VP_{ref} \tag{6-48}$$

where C_0 is the *local entry loss coefficient* and the subscript "ref" refers to the fact that some loss coefficients are based on the slot velocity pressure while others are based on the duct velocity pressure. ACGIH design plates and comparable design data published in the professional literature (Fig. 6.10) generally state equations similar to Eq. 6-48. In the absence of such data, Figs. 6.23 and 6.24 can be used. For example,

> Plane inlet: use Table 1-1 in Fig. 6.23 for $L/D > 1$
> Flanged inlet: use Table 5-1 in Fig. 6.24 for $\theta = 180$ and $A_1/A_0 > 10$
> Small openings (slots, orifices, etc.): use Table 4-1 in Fig. 6.24 for $\theta = 180$ and $A_1/A_0 > 16$

The static pressure difference between the atmosphere (P_0) and a reference location downstream of an entry where the velocity (U_{ref}) is known is

1-1 Duct Mounted in Wall (Hood, Non-Enclosing, Flanged and Unflanged)[1]

Rectangular: $D = 2HW/(H + W)$

			C_o				
			L/D				
t/D	0	0.002	0.01	0.05	0.2	0.5	≥1.0
≈ 0	0.50	0.57	0.68	0.80	0.92	1.0	1.0
0.02	0.50	0.51	0.52	0.55	0.66	0.72	0.72
≥ 0.05	0.50	0.50	0.50	0.50	0.50	0.50	0.50

Fitting coefficient (C_o) with screen at inlet.

Sharp Edge ($t/D \leqslant 0.05$):

$$C_o = 1 + C_s$$

Thick Edge ($t/D > 0.05$):

$$C_o = C_o' + C_s$$

where

C_o' = fitting coefficient (see Table above).
C_s = screen coefficient (see Fitting 7-8).

7-8 Obstruction, Screen in Duct, Round, and Rectangular[1]

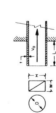

$$n = A_{or}/A_o$$

where

n = free area ratio of screen, dimensionless
A_{or} = total flow area of screen, mm²
A_o = area of duct, mm²

n	0.30	0.40	0.50	0.55	0.60	0.65	0.70	0.75	0.80	0.90	1.0
C_o	6.2	3.0	1.7	1.3	0.97	0.75	0.58	0.44	0.32	0.14	0

1-7 Hood, Tapered, Flanged or Unflanged[2]

$A_1 \geqslant 2A_o$

θ is major angle for rectangular hoods

Hood Shape: Round										
θ	0°	20°	40°	60°	80°	100°	120°	140°	160°	180°
C_o	1.0	0.11	0.06	0.09	0.14	0.18	0.27	0.32	0.43	0.50

Hood Shape: Square or Rectangular										
θ	0°	20°	40°	60°	80°	100°	120°	140°	160°	180°
C_o	1.0	0.19	0.13	0.16	0.21	0.27	0.33	0.43	0.53	0.62

3-1 Elbow, Smooth Radius (Die Stamped), Round[1,4]

$$C_o = K_\theta C_o'$$

Coefficients for 90° Elbows:

r/D	0.5	0.75	1.0	1.5	2.0	2.5
C_o'	0.71	0.33	0.22	0.15	0.13	0.12

For angle correction factors K_θ, see Note 3.

3. For elbows with angles other than 90°, multiply by the following factor (K_θ).

θ	0	20	30	45	60	75	90	110	130	150	180
K_θ	0	0.31	0.45	0.60	0.78	0.90	1.00	1.13	1.20	1.28	1.40

Fig. 6.23 Local loss coefficients for entries and elbows (reprinted by permission from the *ASHRAE Handbook—1985 Fundamentals*).

4-1 Conical Diffuser, Round[1]

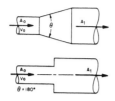

5-1 Contraction, Round & Rectangular[1]

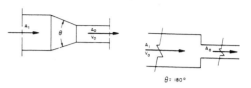

					C_o					
					θ, degrees					
A_1/A_0	8	12	16	20	30	45	60	90	120	180
2	0.11	0.11	0.14	0.19	0.32	0.33	0.33	0.32	0.31	0.30
4	0.15	0.17	0.23	0.30	0.46	0.61	0.68	0.64	0.63	0.62
6	0.17	0.20	0.27	0.33	0.48	0.66	0.77	0.74	0.73	0.72
10	0.19	0.23	0.29	0.38	0.59	0.76	0.80	0.83	0.84	0.83
≥16	0.19	0.22	0.31	0.38	0.60	0.84	0.88	0.88	0.88	0.88

			C_o				
			θ, degrees				
A_1/A_0	10	15-40	50-60	90	120	150	180
2	0.05	0.05	0.06	0.12	0.18	0.24	0.26
4	0.05	0.04	0.07	0.17	0.27	0.35	0.41
6	0.05	0.04	0.07	0.18	0.28	0.36	0.42
10	0.05	0.05	0.08	0.19	0.29	0.37	0.43

6-4 Converging Wye (30°), Conical, Round[9]

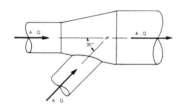

Branch, $C_{c,b}$

$\frac{A_s}{A_c}$	$\frac{A_b}{A_c}$					Q_b/Q_s					
		0.2	0.4	0.6	0.8	1.0	1.2	1.4	1.6	1.8	2.0
0.3	0.2	-2.4	-.11	1.8	3.4	4.8	6.0	7.1	8.0	8.9	9.7
	0.3	-2.8	-1.3	0.14	0.72	1.4	2.0	2.4	2.8	3.2	3.5
0.4	0.2	-1.4	0.61	2.3	3.8	5.2	6.3	7.3	8.3	9.1	9.8
	0.3	-1.8	-.54	0.42	1.2	1.8	2.3	2.7	3.1	3.4	3.7
	0.4	-1.9	-.89	-.17	0.36	0.76	1.1	1.3	1.5	1.7	1.9
0.5	0.2	-.82	0.97	2.6	4.0	5.3	6.4	7.4	8.3	9.1	9.9
	0.3	-1.2	-.15	0.71	1.4	2.0	2.5	2.9	3.3	3.6	3.9
	0.4	-1.4	-.54	0.06	0.50	0.85	1.1	1.3	1.5	1.7	1.8
	0.5	-1.4	-.66	-.15	0.21	0.48	0.68	0.84	0.97	1.1	1.2
0.6	0.2	-.52	1.2	2.7	4.1	5.3	6.4	7.4	8.3	9.1	9.9
	0.3	-.93	0.06	0.85	1.5	2.1	2.6	3.0	3.4	3.7	4.0
	0.4	-1.1	-.37	0.16	0.55	0.86	1.1	1.3	1.5	1.6	1.8
	0.5	-1.1	-.49	-.06	0.25	0.48	0.66	0.79	0.90	1.0	1.1
	0.6	-1.2	-.55	-.15	0.12	0.31	0.45	0.56	0.65	0.71	0.77
0.8	0.2	-.27	1.3	2.7	4.0	5.2	6.3	7.3	8.2	9.0	9.7
	0.3	-.67	0.18	0.90	1.5	2.0	2.5	2.9	3.3	3.6	4.0
	0.4	-.85	-.27	0.16	0.49	0.75	0.97	1.2	1.3	1.4	1.6
	0.5	-.90	-.40	-.07	0.18	0.36	0.50	0.61	0.70	0.78	0.84
	0.6	-.92	-.46	-.16	0.04	0.18	0.29	0.37	0.44	0.49	0.53
	0.7	-.93	-.49	-.21	-.03	0.10	0.19	0.25	0.30	0.34	0.37
	0.8	-.93	-.50	-.24	-.07	0.05	0.13	0.19	0.23	0.27	0.29
1.0	0.2	-.26	1.2	2.6	3.9	5.1	6.1	7.1	8.0	8.8	9.5
	0.3	-.65	0.12	0.79	1.4	1.9	2.4	2.8	3.1	3.5	3.8
	0.4	-.83	-.34	0.04	0.33	0.58	0.78	0.95	1.1	1.2	1.3
	0.5	-.89	-.48	-.20	0	0.15	0.27	0.37	0.45	0.51	0.57
	0.6	-.91	-.54	-.31	-.14	-.03	0.06	0.12	0.18	0.22	0.25
	0.8	-.93	-.59	-.38	-.25	-.16	-.10	-.06	-.03	-.01	0.01
	1.0	-.93	-.60	-.40	-.28	-.20	-.14	-.11	-.08	-.07	-.06

Main, $C_{c,s}$

$\frac{A_s}{A_c}$	$\frac{A_b}{A_c}$					Q_b/Q_s					
		0.2	0.4	0.6	0.8	1.0	1.2	1.4	1.6	1.8	2.0
0.3	0.2	4.5	2.8	1.5	0.56	-.17	-.74	-1.2	-1.6	-1.9	-2.1
	0.3	4.6	3.1	2.0	1.2	0.57	0.08	-.30	-.62	-.89	-1.1
0.4	0.2	1.6	0.85	0.16	-.43	-.92	-1.3	-1.7	-1.9	-2.2	-2.4
	0.3	1.7	1.1	0.58	0.13	-.24	-.56	-.82	-1.1	-1.3	-1.4
	0.4	1.8	1.3	0.80	0.42	0.11	-.15	-.37	-.55	-.72	-.86
0.5	0.2	0.67	0.18	-.33	-.79	-1.2	-1.5	-1.8	-2.1	-2.3	-2.5
	0.3	0.75	0.42	0.07	-.25	-.54	-.80	-1.0	-1.2	-1.4	-1.5
	0.4	0.80	0.55	0.28	0.03	-.20	-.40	-.57	-.73	-.86	-.98
	0.5	0.82	0.62	0.41	0.20	0.02	-.15	-.29	-.42	-.53	-.63
0.6	0.2	0.26	-.11	-.54	-.95	-1.3	-1.6	-1.9	-2.1	-2.4	-2.5
	0.3	0.34	0.13	-.14	-.42	-.67	-.90	-1.1	-1.3	-1.4	-1.6
	0.4	0.39	0.25	0.06	-.14	-.33	-.51	-.66	-.80	-.93	-1.0
	0.5	0.41	0.32	0.18	0.03	-.12	-.26	-.38	-.50	-.60	-.69
	0.6	0.43	0.37	0.26	0.14	0.02	-.09	-.19	-.29	-.37	-.45
0.8	0.2	-.01	-.30	-.67	-1.1	-1.4	-1.7	-2.0	-2.2	-2.4	-2.6
	0.3	0.07	-.07	-.29	-.58	-.76	-.97	-1.2	-1.3	-1.5	-1.6
	0.4	0.11	0.05	-.09	-.26	-.42	-.58	-.72	-.85	-.97	-1.1
	0.5	0.14	0.12	0.03	-.09	-.21	-.34	-.45	-.55	-.64	-.73
	0.6	0.15	0.17	0.11	0.02	-.07	-.17	-.26	-.34	-.42	-.49
	0.7	0.17	0.21	0.17	0.11	0.03	-.05	-.12	-.19	-.26	-.32
	0.8	0.17	0.23	0.22	0.17	0.11	0.05	-.02	-.07	-.13	-.18
1.0	0.2	-.05	-.33	-.70	-1.1	-1.4	-1.7	-2.0	-2.2	-2.4	-2.6
	0.3	0.03	-.10	-.31	-.55	-.78	-.98	-1.2	-1.3	-1.5	-1.6
	0.4	0.07	0.02	-.12	-.28	-.44	-.59	-.73	-.86	-.98	-1.1
	0.5	0.09	0.09	0.01	-.11	-.23	-.35	-.46	-.56	-.65	-.74
	0.6	0.11	0.14	0.09	0	-.09	-.18	-.27	-.35	-.43	-.50
	0.8	0.13	0.20	0.19	0.15	0.09	0.03	-.03	-.08	-.14	-.19
	1.0	0.14	0.24	0.25	0.24	0.20	0.16	0.12	0.08	0.04	0

Fig. 6.24 Local loss coefficients for transitions and wyes (reprinted by permission from the *ASHRAE Handbook—1985 Fundamentals*).

$$(P_0 - P_{ref}) = VP_{ref} + h_{LM} \text{ (hood entry loss)}$$

$$= \left(\frac{\rho}{2}\right)U_{ref}^2(1 + C_0)$$

$$= (1 + C_0)VP_{ref} \qquad (6\text{-}49)$$

The addition of unity to the local entry loss coefficient (C_0) is sometimes called the *acceleration factor* since the addition accounts for the pressure drop associated with accelerating still air into the hood plus the pressure drop associated with viscosity and flow separation associated through an opening.

6.9.4 Fan Inlet and Outlet Losses

If the velocity profile entering or leaving the fan is skewed because of elbows or some other fittings, *fan inlet loss coefficients* may be needed. No such losses will be used in this text and the reader should consult the ASHRAE *Fundamentals* (185) for details if such losses should be accounted for.

6.9.5 Equivalent Length

Minor head losses are sometimes expressed as

$$h_{LM} = f\left(\frac{L_e}{D}\right)VP \qquad (6\text{-}50)$$

where L_e is an *equivalent length* of straight duct. The literature on duct design is immense and several conventions are used to compute the pressure drop in duct components and in duct systems. Rather than seeking to standardize conventions, users should be flexible and accommodate themselves to these different conventions.

6.9.6 Design Procedure

1. Select the collection hood appropriate for the industrial process to be controlled. Select the exhaust volumetric flow rate for each hood.

2. Prepare a line sketch of the duct system showing the location of each component: hood, type of fitting in each duct segment, contaminant removal system (if one is needed), and exhaust fan. Determine the length of each duct segment between fittings. Identify each point where a component is to be placed and designate by number or letter each point in the duct system where junctions, fittings, and so on are to be placed.

3. Select the minimum duct velocity for the contaminant to be collected.

4. Determine the loss factor for each of the components in paragraph 2.

5. Choose a diameter for each segment of duct. Compute the static pressure drop for duct segment. Insure that whenever two or more ducts meet at a common point (P), the pressure drop between atmospheric pressure and the static pressure at point P is the same. If the pressure drop for each duct segment is not the same, select another duct diameter for one or each of the segments and

recompute the pressure drop. Repeat the process until the pressure drop in each of the duct segments is the same.

6.   Select an exhaust fan of a size, speed and HP such that the overall pressure drop and volumetric flow rate occur at the optimum point on the fan performance curve. Use the *fan laws* to estimate the HP, pressure drop, and volumetric flow rates at off-design conditions.

6.10   DUCT DESIGN COMPUTER PROGRAM

The design of a system of ducts is a tedious job requiring repetitive computations and is ideally suited for computers. The requirement that the designer make certain critical decisions about duct length, transport velocity, and so on, must be preserved, but computer programs can facilitate acquiring loss coefficients and processing numerical data. In 1990 there were nearly two dozen computer-aided duct design programs available commercially (187, 393) and many proprietary programs developed within companies for their own use. What follows is the description of how to use an uncomplicated program contained in a disk accompanying the book. The program was originally written for an Apple II computer but rewritten for IBM and IBM compatible machines that is in the public domain (188,235). The program is uncomplicated and interactive. By asking users to make certain key decisions, they will be intimately involved in the design of the duct system. More elaborate proprietary duct design programs contain loss coefficients for a larger number of fittings and perform many decisions automatically based on some optimization algorithm built into the program. Thus these programs are faster and better suited to complex duct systems. From an instructional point of view the elementary program is preferable because the deliberate involvement of users improves their understanding of what is being done.

The program contains 16 options that enable users to compute the pressure drop in a branch and tabulate, transfer, and cancel data until satisfactory performance is achieved.

Option 1—Determines pressure losses in a branch
Option 2—Balances all branches at a junction
Option 3—Determine a fan's static pressure requirements
Option 4—List of branches and their losses
Option 5—List of the branches' velocities, total losses, design volumetric flow
rates and duct diameters
Option 6—Print branch data on printer
Option 7—Determine the static pressure losses/gains through an expansion or
contraction
Option 8—Calculate the pressures at a point
Option 9—Calculate the velocity of air for a certain duct size
Option 10—Clear current data file
Option 11—Pack current data file
Option 12—Delete a branch
Option 13—Rename a branch

Option 14—Duplicate a branch

Option 15—Redetermine a branch's static pressure loss

Option 16—Load a different data file

If users become muddled progressing through any option they can always press the ESC key which returns them to the menu where they can sort things out. To clear erronous data from a data file, Option 10 should be used. To delete an actual data file, users should use the DEL function in DOS.

The program opens with Option 1. Option 1 is the basic program that computes the pressure drop in a branch. Users should compute the pressure loss for the branch furthest from the fan. The worksheets show data users provide and what the program computes. The program asks users to input data followed by striking the "carriage return, or enter key". Such steps will be noted by underlining the input data and the symbol "CR" for the carriage return. At the end of the computation the user is asked "is this satisfactory," at which time the duct diameter is either changed and the computation in Option 1 repeated or, if answered affirmatively, the list of the 16 options appears and users select what they want. The program requires users to designate segments of the duct system by symbols composed of no more than three alphanumeric symbols in which the first letter *must adhere* to the following format:

Hoods—each hood inlet must begin with H

Junctions—a junction is a fitting with two or more inlets or two or more outlets, each junction must begin with J

Fan—each fan must begin with F

Points—points in the duct system in which there is a single inlet and a single outlet must be designated by an integer, they must not start with H, J, or F

The program uses a mixture of duct design conventions. The pressure drop through a slot(s) is computed as the product of the "slot entry loss coefficient" (C_s), input by the user times the velocity pressure through the slot VP_s (computed by the program). The pressure drop through a hood is computed as the product of a "hood entry loss coefficient" (C_h), input by the user times the velocity pressure in the duct, leaving the hood VP_d (computed by the program). In the case of elbows, an *equivalent length* procedure is used. If users prefer to compute the pressure drop through a fitting as the product of a head loss coefficient times the duct velocity pressure, they should compute these losses by hand and introduce the pressure drop as a separate entity in Option 1. Other duct design computer programs may use different procedures which are more accurate, but this is not of great importance at this point. It is important that users become familiar with how to select a fitting and hood and input this knowledge to an interactive computer-aided design program.

Individuals learn how to use computer programs by reading written instructions and a great deal of "hacking." By selecting different options, a program asks users to select various items. Users learn by doing, much as one learns how to ride a bicycle, get on, fall off, get on again, . . . and so on. After a few bumps and bruises, one learns. The following illustrative problem shows how the different options of the program can be used. It is suggested that users change the input data in the problem and repeat the steps.

To begin the program, insert the disk provided with the text into Drive A and boot the disk. Follow the instructions that appear on the screen. See Appendix A-17 for details.

Illustrative Example 6.5—Duct System For A Foundry

A small foundry casts gray iron gear blanks and other small castings for an agricultural equipment manufacturer. The foundry uses scrap steel and iron from an automotive scrap operation they run adjacent to the property. The foundry has recently been cited by the state air pollution control agency and OSHA for excessive dust emissions to the environment and excessive silica concentrations in the workplace. The foundry operation has been redesigned to capture particles from key parts of the process. The electric arc melting furnace is equipped with its own air-pollution control system and is not part of the redesign. The molds are filled at a pouring station and vapors are collected by a side-draft hood similar to Figs. 1.14 and 1.15. After the castings have cooled, they are removed from their sand molds by a shake-out machine contained in an enclosure (15 ft × 15 ft × 10 ft). The flasks holding the sand molds enter and leave the enclosure on a conveyor that passes through two openings. The molding sand falls to the floor below where it is reprocessed for future use. The castings leave the shake-out enclosure to have their sprues and risers removed. The pressure drop through the shakeout enclosure is equal to the hood entry loss coefficient (C_h) 0.5 times the duct velocity pressure VP_d. All air enters the shake-out enclosure through the two openings surrounding the conveyor and leaves through an opening in the ceiling of the enclosure that is connected to the main duct leading to the fabric filter. A cyclone collector with a pressure drop of 4 in of water removes particles of sand before the air enters the main duct. The casting sprues and risers are removed by workers on two downdraft grinding benches similar to Fig. 1.23. Any surface imperfections are ground off on these downdraft benches. The total pressure drop through the pouring station is equal to the slot entry loss coefficient (C_s) of 1.78 times the slot velocity pressure VP_s plus the hood loss coefficient (C_h) times the duct velocity pressure VP_d. The pressure drop through the downdraft grinding booth is equal to the hood loss coefficient (C_h) 0.25 times the duct velocity pressure VP_d.

Ventilation air from each device (pouring station, two downdraft grinding benches, and shakeout enclosure) is collected by a common duct and the particles removed by a fabric filter before being discharged to the atmosphere. Each device is connected to the common duct by branch ducts of smaller diameter. Each branch contains a blast gate that can be closed to prevent air from being drawn through the device when it is not in use. The pressure drop through the blast gate is 1.5 in. of water when opened fully. Air is withdrawn through the entire system by a fan whose pressure drop is unknown. The fan manufacturer requires that air approaches the fan at a velocity of 1500 FPM. Thus a 30-degree expansion section must be installed between point 4 and the fan.

Figure 6.25 is a schematic diagram showing the collection devices, lengths of duct, and the fittings that will be needed.

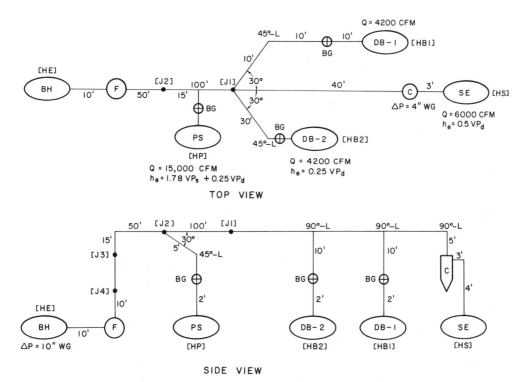

Fig. 6.25 Duct system for Illustrated Example 6.5.

Abbreviations in Fig. 6.25

BH—Baghouse	BG—Blast Gate
F—Fan	DB—Downdraft Grinding Bench
C—Cyclone	SE—Shakeout Enclosure
PS —Pouring Station	

Elbow centerline radius/duct diameter $(r/D) = 2$

Slot velocity = 2000 FPM

Maximum velocity entering fan = 1500 FPM

Selecting the individual ventilation systems (hoods) and deciding the needed volumetric flow rates and duct transport velocities has been performed using ASHRAE (185) and ACGIH (27) literature and the volumetric flow rate, slot velocity, and entry loss coefficient for each collector and these are listed in Worksheets 1 and 2.

The layout of the plant makes it necessary to connect each of the processes to a single duct of increasing diameter that carries the dusty air to the fan and fabric filter before being discharged to the atmosphere. The location of the main duct has been selected by the plant engineer, but the diameters of each of the branches, the main duct, and the fan pressure drop have yet to be computed. Between the shake-out enclosure and the duct is a cyclone to remove large particles. For the volumetric flow rate that has been selected, the pressure drop across the cyclone has been determined from manufacture's literature to be 4 in of water.

Shown in Worksheets No 1 and 2 are the key parameters for each of the devices in the duct system that will be called for in the steps that follow. The program is driven by a menu in which questions appear on the screen and users reply. The questions and answers will be denoted by <u>underlined</u> phrases. In many cases the reply is followed by striking the carriage return <u>CR</u> or enter key.

Step 1   Booting The Disk

Add the DOS System Utility to the disk in the text. Boot the computer with a DOS version 3.00 or higher, having SYS.COM. Type SYS x: <u>CR</u> where x: is the drive containing the copy of the ventilation duct design program. Finally type COPY COMMAND.COM x: <u>CR</u> to complete the process.

To run the program, place the disk in Drive A, type A: followed by <u>CR</u>, then type VENT followed by <u>CR</u>. A display will appear containing prompts at the bottom of the screen that need to be answered. At the command <u>Enter [Drive:] [Path] of data files</u> type <u>CR</u>. To the question <u>Enter name of file to read/create, OR ⟨CR⟩ to re-enter path</u> type a name for a new data file followed by <u>CR</u>. Alternatively you may read an existing file created at an earlier time by typing its name. The file extension ".dat" will be input automatically. If an existing file was chosen the screen will display the 16 options listed above. To begin select Option 1. If a new file name was entered, the program goes immediately to Option 1.

Step 2   Pressure Drop Between HS and J1

Begin by computing the pressure drop for the branch furthest from the fan. In this problem the furthest is the shake-out enclosure. The prompt <u>enter starting point</u> appears on the screen. Type <u>HS</u> followed by <u>CR</u>. The prompt <u>enter end point</u> appears, to which users answer <u>J1</u> followed by <u>CR</u>. The prompt <u>enter design Q (CFM)</u> appears, to which users enter <u>6000</u> followed by <u>CR</u>. The prompt <u>transport velocity (FPM)</u> appears, to which <u>4000</u> and <u>CR</u> is entered. The program computes the duct diameter (rounded off to commercial diameters) and recomputes the actual duct transport velocity and duct velocity pressure (<u>VP</u>). On the screen will appear,

Duct diam	16.00 in
Duct area	1.40 ft^2
Act. duct vel.	4297.18 FPM
Duct vel. pres.	1.15 in H20

The question <u>are there any hood entry losses ? [Y/N]</u> appears next. Answer <u>Y</u>, no carriage return is needed. This question pertains to losses due to slots, plenum chambers and diffusers in the hood. Worksheet 2 or 3 contain the data, if the information is not available, users will be obliged to estimate slot loss factors and hood entry loss factors from experiment or approximate them from hoods of similar design. Since there are no slots for the shake-out enclosure, the only loss to report will be hood entry loss factor (C_h) equal to 0.5. To the question <u>are there any slots?</u>, answer <u>N</u>. To the question <u>enter hood entry loss factor,</u> answer <u>0.5</u> followed by <u>CR</u>. The program will display the hood entry loss in the units of inches of water. The screen will now display the following additional information,

Add acceleration factor to Entry loss factor 1.5	1.5
Hood entry loss	1.73 in H20

To the question <u>are there any duct losses? [Y/N]</u>, users should reply with <u>Y</u>. To the question <u>length (feet)</u>, one replies <u>52</u> followed by <u>CR</u>. The screen will now display the following additional information,

Friction loss/100 ft	1.49 in H20
Duct losses	0.77 in H20

To the question <u>are there any fittings? [Y/N]</u> answer <u>Y</u>. The question <u>are there any elbows [Y/N]</u> is answered in the affirmative by striking <u>Y</u>. The program now asks the question <u>centerline radius (duct diameters) 1.5, 2, or 2.5</u> (i.e., elbow radius/duct diameter). It will be assumed that long-radius elbows will be used and the answer <u>2</u> followed by <u>CR</u> should be entered. A series of prompts appear to which the following answers are given. <u>Number of 90 elbows</u>, type <u>2</u> followed by <u>CR</u>. <u>Number of 60 elbows</u>, type <u>0</u> followed by <u>CR</u>. <u>Number of 45 elbows</u>, type <u>0</u> followed by <u>CR</u>. <u>Number of 30 elbows</u>, type <u>0</u> followed by <u>CR</u>. The screen now displays the following additional information:

Equiv. lengths	47.43 ft
Elbow losses	0.70 in H20

The question <u>are there any branching entry losses? [Y/N]</u> refers to the presence of branches shown in Fig. 6.25 that enter the main duct between HS and J1. Since the volumetric flow rate through the branch HS-J1 is 6000 CFM and larger than through the other two branches, it will be presumed that branch HS-J1 is the main duct to which branches HB1-J1 and HB2-J1 are added. Thus there are no branching entries in HS-J1, and users should type <u>N</u>. The program then prompts <u>enter other losses (i.e., evases, collectors, etc. . .) in H20</u> which refer to the pressure drop produced by any devices in the branch between the end points HS and J1. Since the sketch shows that there is a cyclone that produces a pressure drop of 4 in of water, users type the number <u>4</u> followed by <u>CR</u>. The program computes the total branch loss between HS and J1 and displays the prompt <u>total branch losses (in H2O) = 7.204</u>. The program screen also displays the question <u>Is this satisfactory?</u>. The question is important in future steps to be taken, but for now users should reply in the affirmative by striking <u>Y</u>. The menu now appears listing the 16 options and users select one of the options for whatever they wish to compute next. Select Option 1 by typing <u>1</u> followed by <u>CR</u> to compute the pressure drop in branch HB1-J1.

Step 3 Pressure Drop Between HB1 and J1

Data from Worksheets 2 and 3 are entered in a fashion similar to step 1. Figure 6.25 shows that the branch enters the larger duct by an entry that is inclined to the duct by 30 degrees. Thus to the question are there branching entry losses?, answer Y. The prompt then asks Angle of entry [A = 45, B = 30], answer B. The prompt then asks Enter other losses (ie evases, collectors, etc. . .) in H20. Enter 1.5 for the blast gate. At the end of the option the statement is displayed total branch losses (in H20) = 3.789 Is this satisfactory? to which the answer Y is given. Users are again asked to select an option. Since branch HB2-J1 is identical to HB1-J1, users should enter 14 followed by CR to calculate the initial pressure drop in the branch.

Step 4 Pressure Drop Between HB2 and J1

Option 14 begins with the prompt enter original starting point, to which users answer HB1, followed by CR. Users are next asked enter new starting point, to which they answer HB2 followed by CR. To the prompt enter original ending point, users answer J1 followed by CR. Lastly the question enter new ending point, is answered by J1, followed by CR. When the list of options is displayed on the screen, users should enter 5 to display a summary of the pressure drops from each of the three branches that join at junction J1. The following are the results:

Start	End	Vel	Losses	Q	Diam
HS	J1	4297.18	7.204	6000	16.00
HB1	J1	3928.85	3.789	4200	14.00
HB2	J1	3928.85	3.789	4200	14.00

Step 5 Tabulation of Pressure Drops at Junction J1

A balanced system is one in which the static pressure drop between the atmosphere and point J1 is the same for each of the three branches that meet at junction J1. If the computed pressure drop for each branch shown when Option 5 is selected are not the same, users must select other duct diameters, repeat steps 1 to 3, enter data as above, and compute a new branch pressure drop. The process must be repeated until the pressure drop in each of the three branches is the same within at least 20%. It must be remembered that the pressure drop is proportional to the square of the velocity and the velocity is inversely proportional to the square of the diameter. Thus the pressure drop is inversely proportional with the fourth power of the duct diameter. Thus to reduce the pressure drop, the duct diameter should be increased. Users must be careful not to increase the diameter by too large an amount or else the transport velocity will become too small and gravimetric settling will become a problem.

To reconsider a previous branch and to change its diameter so as to change its pressure drop, select Option 15 by typing 15 followed by CR. The question enter starting position and ending point appear on the screen and users enter the end points of whichever branch they wish to modify. The question duct diameter (inches), appears to which users type a value X larger or smaller value than before followed by <u>CR</u>. Other questions will be asked as per step 1 and users enter the appropriate values. Each branch that is to be reconsidered requires the use of Option 15. Even if one reconsidered branch is identical to another, each branch should be reconsidered using Option 15 rather than Option 14. At the end of the computation, Option 5 should be selected to display the pressure drop in each of the branches. Only the latest values will be shown, so if former values of diameter and pressure drop are needed by users they must tabulate them separately (on paper). The pressure drop in each of the three branches can be brought into agreement within 20% in a few iterations. Shown below are these final values.

Start	End	Vel	Losses	Q	Diam
HS	J1	3395.31	5.955	6000	18
HB1	J1	5347.61	5.852	4200	12
HB2	J1	5347.61	5.852	4200	12

Step 6   Junction Balance

Once the pressure drop for the three branches is within 20% of each other users should select Option 2 to modify the volumetric flow rates and bring about equality of the pressure drop at a junction. Type *2* followed by <u>CR</u>. The question <u>name of junction to be balanced</u> will appear to which the answer <u>J1</u> followed by <u>CR</u> should be given. The volumetric flow rate in the branch with the lowest volumetric flow rate will be increased to a value "CFM_{cor}",

$$CFM_{cor} = CFM_{design} \sqrt{\frac{\text{highest SP branch}}{\text{lowest SP branch}}} \tag{6-51}$$

If the user was lucky and selected duct diameters in which the pressure drops at the junction J1 were within 5%, Option 2 will not adjust the volumetric flow rate of the lowest value. To see the new volumetric flow rates in the three branches, type <u>5</u> followed by <u>CR</u>. No attempt should be made to correct for the larger diameter at the exit of junction J1. The equation computing the pressure drop for the entries used in Option 1 will include this effect automatically. The user will see that the three branches have also been reported as a single straight duct with approximately the same pressure drop but volumetric flow rate equal to the sum of the volumetric flow rates of the three separate branches. The program also automatically defines the starting and end points of this new effective branch with new nomenclature in which the starting point is called <u>HJ</u> and ending point is defined as <u>@1</u>. Users should use these terms in future calculations.

HS-J1	5.95 in H2O
HB1-J1	5.85 in H2O
HB2-J1	5.85 in H2O
HJ-@1	5.96 in H2O

Users may find it helpful to print the results of steps 1–6. Select Option 6 to print a summary of what has been shown on the screen plus a summary of the individual pressure drops through a slot, hood, duct, elbows, and branching entries. Select Option 8 to determine the pressure at any point, such as @1. Use Option 11 to reassemble data and eliminate data discarded as a result of using Option 15.

Step 7 Pressure Drop in a Straight Duct

The pressure drop between @1 and J2 should be computed using Option 1. The entry of branch HP-J2 should be ignored for the moment since it will be addressed in the next step. By typing @1 and indicating that the starting and end points are @1 and J2, respectively, the program will automatically use the corrected total flow rate and transport velocity from step 6. Users can override this and input another transport velocity by typing in the value followed by CR. Even though points @1 and J2 are junctions at which fittings with entries exist, users should not include them in step 7. The fittings at @1 were noted as entries in the previous steps and the entry at J2 will be included in step 9.

Step 8 Cumulative Pressure Drop

The pressure drop between the entry of the pouring station and junction J2 must be equal to the cumulative pressure drop in the duct between points HB1, HB2, HS, and junction J2. To find the cumulative pressure drop, select Option 8 by typing 8 followed by CR. Input the end point J2 followed by CR and the cumulative static pressure drop, velocity pressure and total pressure at point J2 will be shown. These values (plus or minus 20%) become the performance requirements for the branch HP and J2 which will be computed next.

Step 9 Pressure Drop In Branch HP-J2

The pressure drop in branch HP-J2 is computed using Option 1 following decisions similar to those discussed in Step 1. Since the pouring station has a hood with slots, answer Y to the question are there any slots?. Input 2000 to the question slot velocity (FPM). To the prompt enter slot entry loss factor, type 1.78. To the prompt are there any branching entries? [Y/N] answer Y and to the prompt angle of entry [A = 45, B = 30] answer B. Be sure to include the pressure drop of 1.5 in H2O for the blast gate. At the end of Option 1, the prompt appears total branch losses (in H2O) = 3.783 Is this satisfactory? Answer Y.

Some amount of iteration using Option 15 will have to be performed to bring the pressure drop to within 20% (or less) of the cumulative pressure drop in Step 8. Since the pressure drop is only 3.783 in H2O for a duct 26 inches in diameter, select a smaller duct, such as 20 inches in diameter and a transport velocity of 6876 FPM. A final balancing should be performed using Option 2 as per step 7. The following are the results of the first and last steps in the iteration. Note that point J2 will hereafter be called @2.

Start	End	Vel	Losses	Q	Diam
HJ	@1	4314.04	5.955	14,400	24.74
@1	J2	3905.61	0.951	14,400	26
HP	J2	4068.34	3.783	15,000	26
HP	J2	6876	7.346	15,000	20
HJ	@2	5523	7.346	29,400	31.24

Step 10   Pressure Drop Between Points @2 And 3

The pressure drop in the duct between points @2 and 3 is computed in the same fashion as step 7. Use the volumetric flow rate of 29,400 CFM and a transport velocity of 5523 FPM that appears on the display. A total branch loss of 1.5 in H2O is computed. To the question Is this satisfactory? answer Y. Use Option 8 to show that the static pressure at point 3 is −9.181 in H2O.

Step 11   Pressure Drop Between Point 4 and the Fan

Before computing the pressure recovery in the expansion, users must compute the pressure drop between points 4 and the fan (denoted as F) by using Option 1. It will be presumed that 29,400 CFM must enter the fan at a transport velocity of 1500 FPM, a velocity which is lower than the transport velocity at point 3.

Step 12   Pressure Rise in Expansion Fitting Between Point 3 and 4

The pressure increase from an expansion fitting is calculated by Option 7. To the question is it an (E)xpansion or (C)ontraction [E/C] type E followed by CR since it is an expansion fitting. Input the starting and end points 3 and 4 respectively. The prompt enter taper angle (degrees) appears and a total (included) angle 30 is entered followed by CR. The question is the expansion within 150 in of a disturbance? is asked because the program computes the pressure recovery lessened by an amount depending on the proximity of any upstream disturbance. The sketch should be consulted and an appropriate Y/N given followed by CR. It should be noted that to gain full value of an expansion, it is to the designers advantage to locate the expansion at least 5 duct diameters downstream of any other fitting, junction, fan discharge, hood. Since there is 15 ft of duct upstream of the expansion and 10 ft of duct downstream of the expansion answer N. Using Option 5 the following data appear:

Start	End	Vel	Losses	Q	Diam
HS	J1	3,395.31	5.955	6,000	18.0
HB1	J1	5,347.61	5.852	4,200	12.0
HB2	J1	5,347.61	5.852	4,200	12.0
HJ	@1	4,314.04	6.105	14,400	24.7
@1	J2	4,583.66	7.129	14,400	24.0
HP	J2	6,875.49	7.346	15,000	20.0
HJ	@2	5,522.94	7.681	29,400	31.2
@2	3	5,989.32	1.500	29,400	30.0
3	4	1,497.33	−0.844	29,400	60.0

Step 13 Pressure Drop Downstream Of The Fan

Computing the fan performance properties is the last calculation to be made. Prior to performing it, the pressure drop between the point F and HE must be computed. Option 1 should be selected with the 10 in of water pressure drop for the fabric filter used in answer enter the losses (evases, collectors, etc. . .) in H2O. Use the same duct diameter and transport velocity between points 4 and F as was used between points 3 and F.

Step 14 Fan Requirements

Select Option 3 and define F as the fan location. The program computes the fan static pressure requirement. Shown below are the results:

Fan static pressure difference = 18.21 in H2O
Fan design flowrate = 29,400 CFM
Intake duct diameter = 60 in
Output duct diameter = 60 in

Use Option 6 to print the results of the entire analysis including the fan requirements.

Worksheet No 1—Components

Device	CFM	Slot FPM	Slot	Hood	δP (in H2O)
			Loss Coefficients		
Shake-out	6,000	N/A	N/A	0.5	To be found
Downdraft bench	4,200	N/A	N/A	0.25	To be found
Pouring slot hood	15,000	2000	1.78	0.25	To be found
Cyclone	6,000	N/A	N/A	N/A	4.0
Blast gates	4,200	N/A	N/A	N/A	1.5
Baghouse	29,400	N/A	N/A	N/A	10.0
Expansion	29,400	N/A	N/A	N/A	To be found

Worksheet No 2 – Duct Parameters

Points	CFM	Duct FPM[a]	Length	Elbows	Entries	Other δP
HS–J1	6,000	4,000	52	2–90	0	4.0
HB1–J1	4,200	4,000	30	1–90, 1–45	30	1.5
HB2–J1	4,200	4,000	30	1–90, 1–45	30	1.5
J1–J2	14,400	4,000	100	0	0	0
HP–J2	15,000	4,000	27	1–45	30	1.5
J2–3	29,400	4,000	65	1–90	0	0
3–4	29,400	1,500	10	0	0	0
4–F	29,400	1,500	10	0	0	0
F–HE	29,400	1,500	10	0	0	10.0

[a] initial value of duct transport velocity.

NOMENCLATURE

Symbol	Description	Dimensions*
A	Parameter defined by equation	
A_B	Cross sectional area of blowing inlet	L^2
A_c	Cross sectional area of common (outlet) section of a wye	L^2
A_{face}	Face area	L^2
A_s	Cross sectional area of straight (inlet) section of a wye	L^2
A_s	Surface area of hot surface	L^2
A_S	Cross sectional area of suction inlet	L^2
A_b	Cross sectional area of branch section of a wye	L^2
A_b	Cross sectional area of flowing bulk materials	L^2
A_j	Jet cross sectional area	L^2
A_0	Cross sectional area of duct at section 0	L^2
A_u	Cross sectional area of plume at elevation (Z)	L^2
a, b	Constants defined by equation	
B	Parameter defined by equation	
c_p	Specific heat at constant pressure	Q/MT
C_s	Loss coefficient for a slot or of a screen	
C_h	Loss coefficient for hood entry	
$C_{c,s}$	Loss coefficient for main section of a wye	
$C_{c,b}$	Loss coefficient for branch section of a wye	
C_0	Local loss coefficient of fitting or inlet	
d	Actual distance between source and hood	L
D	Diameter of a circular inlet, stack exit diameter	L
D_s	Diameter of suction (exhaust) duct	L
D_s	Duct diameter in straight section of a wye	L
$D_{p,m}$	Mass median particle diameter	L
D_y	Hydraulic diameter	L
E	Parameter used in design of a push–pull ventilation system defined by equation	
f	Friction factor	
F	Buoyant flux parameter defined by equation	L^4/t^3
F	Distance canopy hood overhangs open vessel	L
f_{su}	Plume surge frequency	t^{-1}
g	Acceleration of gravity	L/t^2
h	Convection film coefficient	Q/tTL^2
h	Distance between actual source and inlet face of canopy	L
h_s	Geometric height of a stack above the height of a building	L
h_{LM}	Minor head loss	M/Lt^2
h_L	Major head loss	M/Lt^2
h_{LT}	Total head loss	M/Lt^2

* Q – energy, F – force, L – length, M – mass, N – mols, t – time, T – temperature

H	Height of a building, drop height of bulk materials	L
H	Height of suction inlet baffle for push–pull system	L
H	Distance between liquid surface and face of canopy hood	L
H_c	Height of roof leading edge recirculation cavity	L
H_{dw}	Downwash distance	L
L	Length of a building in the windward direction, length of duct, length of slot, length of open vessel	L
L_c	Length of roof leading edge recirculation cavity	L
L_e	Equivalent length of duct	L
L_L	Larger of H or W	L
L_S	Smaller of H or W	L
K, K_L, K_0	Constants defined by equation	
m	Safety factor	
$\dot{m}_b$	Mass flow rate of bulk materials	M/t
M_a	Mass of air	M
M_b	Mass of bulk materials in a vessel at an instant of time	M
M_j	Jet momentum	ML/t^2
M_u	Plume momentum	ML/t^2
n	Free area of screen, dimensionless	
P	Static pressure	F/L^2
P	Perimeter	L
p, q	Constants defined by equation	
P_t	Total pressure	F/L^2
q	Convection heat transfer from source	Q/t
Q	Volumetric flow rate	L^3/t
Q_b	Volumetric flow rate in branch	L^3/t
Q_B	Volumetric flow rate of blowing air, volumetric flow rate of buoyant plume	L^3/t
Q_i	Induced air volumetric flow rate (in)	L^3/t
Q_d	Volumetric flow rate of displaced air (out)	L^3/t
Q_j	Volumetric flow rate of air curtain jet	L^3/t
Q_s	Volumetric flow rate in straight section of a wye	L^3/t
Q_s	Volumetric flow rate of secondary air (in)	L^3/t
Q_s	volumetric flow rate of suction air	L^3/t
Q_f	Volumetric flow rate of air entering inlet face (in)	L^3/t
Q_H	Volumetric flow rate at hood inlet	L^3/t
Q_u	Volumetric flow rate of buoyant plume at elevation (Z)	L^3/t
Q_w	Volumetric flow rate of withdrawn air (out)	L^3/t
Q_1, Q_2	Volumetric flow rates for push–pull ventilation system	L^3/t

Q/L	Volumetric flow rate per unit length of slot	L^2/t
Q_{su}	Plume surge volumetric flow rate	L^3/t
r	Radius	L
R'	Constant defined by equation	
s	Slot width ($s = 2w$)	L
S_B	Width of blowing jet in push–pull system	L
S_S	Width of suction inlet in push–pull system	L
t	Duct thickness	L
t_{su}	Duration of plume surge	t
T, T_s, T_0	Temperature, source, ambient temperature	T
$U, U_i, U(r),$		
$U(z), U_a, U_H$	Air velocity, velocity at height of building	L/t
U_{max}	Maximum (centerline) velocity of plume	L/t
U_0	Wind speed	L/t
v_b	Velocity at face of blowing opening	L/t
v_B	Velocity of blowing jet	
v_c	Capture velocity	L/t
v_g	Velocity of evaporating vapor from open vessel	L/t
v_0	Velocity of room air current, i.e. draft	L/t
V_j	Jet velocity, stack exit velocity	L/t
v_S	Velocity at face of suction opening	L/t
V_u	Mass average velocity of plume at elevation (Z)	L/t
V_x	Radial velocity in front of an inlet	L/t
V_{cross}	Cross flow velocity	L/t
V_{face}	Face velocity	
V_e	Volume of an enclosure	L^3
V_a	Instantaneous volume of air in an enclosure	L^3
V_b	Instantaneous volume of bulk materials in an enclosure	L^3
V_h	Hood volume	L^3
w	Slot half-width	L
W	Width of open vessel	L
W	Width of a building in cross wind direction, width of booth	L
W	Distance between jet and exhaust opening for air curtain	L
x	Distance perpendicular to opening face	L
x	Distance downwind from roof lip	L
x_c	Distance to center of roof eddy	L
x_r	Downwind length of building eddy	L
Z	Distance between hood inlet and virtual source	L
z_1, z_2	Vertical elevation of points 1 and 2	L
z'	Smaller of W or H	L
Z_I	Height of roof leading edge recirculation cavity	L

| Z_{II} | Height of the turbulent shear zone | L |
| Z_{III} | Height of roof wake | L |

Greek

π	3.14159	
θ	Angle defined by equation or sketch	
μ	Dynamic viscosity	Ft/L^2
ρ	Gas density	M/L^3
ρ_a	Density of air	M/L^3
ρ_b	Density of bulk materials	M/L^3
ρ_0	Ambient air density at temperature T_0	M/L^3

Subscripts

$(\ )_a$	Property of air
$(\ )_b$	Bulk properties
$(\ )_B$	Blowing air
$(\ )_c$	Common section
$(\ )_{cor}$	Corrected value
$(\ )_{design}$	Design
$(\ )_e$	Equivalent
$(\ )_{face}$	Property at face of inlet
$(\ )_p$	Particle
$(\ )_{ref}$	Reference value
$(\ )_s$	Surface, suction air
$(\ )_{total}(\ )_t$	Total or stagnation property
$(\ )_0$	Far field property

Abbreviations

ACFM	Volumetric flow rate (ft³/min) under actual pressure and temperature
CFM	ft³/min
FPM	ft/min
HP	Horse power
SCFM	Hypothetical volumetric flow rate (ft³/min) that would exist under STP
SP	Static pressure
STP	Standard temperature (25 C) and pressure (101 kPa)
VP	Velocity pressure
WG	Water gauge, inches of water pressure drop

PROBLEMS

1. An acid-dipping tank contains chromic and hydrochloric acid and is used to clean the surface of steel sheet prior to dipping the metal in a molten zinc alloy. The vessel is 2 m wide and 8 m long. The tank has no side baffles and is not located near any walls. It is not possible to enclose the process and lateral

exhausts along each of the long sides is needed. Estimate the volumetric flow rate and slot dimensions for

(a) lateral suction system along both of the long sides,
(b) push–pull system along the long sides.

2. Describe a way to control dust generated when 80-lbm bags are filled with a powder containing lead. Estimate the required ventilation volumetric flow rate. If the horsepower required to run a fan is approximately 1/4 HP per 1000 CFM, estimate the cost per bag associated with dust control. Electrical power costs $0.05 per kilowatt hour.

3. In a small chemical plant several bags of different powders are added to each batch of material in a large reactor. The reactor is a 5000-gal, glass-lined, 8-ft diameter cylindrical vessel with hemispherical ends. In the center of the reactor is a helical mixer. Powders in 80-lbm bags are added by workers through a 2-ft diameter access port on the top of the vessel after the reactor is 3/4 full filled with various liquids. Presently a cloud of dust escapes through the port and into the face of the worker each time the contents of a bag are added to the reactor. Contaminants also enter air in the workplace. The company has been cited by OSHA because of unhealthy exposure of the worker and to others in the workplace. Recommend engineering controls to prevent the escape of dust. For a variety of reasons, it is not possible to eliminate the need for individual workers to add the powder directly through the access port.

4. When the reaction has ended a slurry is drained through a 6-in valve in the bottom of the reactor. The valve is 7 ft above the floor. The slurry travels over a rectangular, slightly inclined screen where liquid is extracted by applying a vacuum to the bottom side of the screen. Highly irritating vapor escapes from the top surface of the slurry burning the eyes and skin of the workers. Vapors also enter the workplace generally which irritate eyes and produce obnoxious odors. Unfortunately, the workers need to be able to see the process and scrape the material into a second vessel which is later sent to a standard dryer a few hundred feet away. The screen is wheeled under the reactor when the reactor is unloaded. Recommend an engineering control.

5. Devise a way to control dust generated from a 1/2 HP hand-held, surface sanding machine used to smooth the surface of fiberglass boats. It will be necessary to occasionally sand internal surfaces. Estimate the ventilation volumetric flow rate and discuss limitations you expect the system may possess.

6. One step in the manufacture of TV picture tubes requires workers to remove the tube from an overhead conveyor, dip it in an aqueous hydrogen fluoride bath for a few moments, remove and inspect it, and place it back on the conveyor. Recommend a ventilation system to control hydrogen fluoride vapor and estimate the volumetric flow rate. The production rate for the bath is 20 tubes per minute. The vessel is 1.1 m wide and 8 m long and workers stand along both the long sides of the vessel to perform their duties. Estimate

the cost per tube if fans require 0.75 HP per 1000 SCFM and electrical power costs $0.05 per kilowatt hour.

7. A vertical stack is to be located on the upstream side of a building 20 m high by 50 m wide (in the direction of the wind) by 100 m (cross wind length). The wind is perpendicular to the 20 m by 100 m face. What is the minimum height of the stack above the roof if there is to be reasonable assurance that the exhaust will not be trapped in the upwind recirculation eddy ? What is the EPA recommended stack height?

8. An unflanged circular inlet 6 inches in diameter is used to capture grinding particles. What should be the volumetric flow rate (CFM) if the minimum capture velocity of 200 ft/min is to be achieved over a circular region 12 inches in diameter in a plane lying 3 inches in front of the inlet?

9. An open vessel containing chromic acid is used for electroplating chromium. The tank is 3 ft wide and 10 ft long, and is placed with its long dimension against a wall. If the vessel is equipped with a lateral exhauster, what volumetric flow rate (CFM) would you recommend for the entire vessel if ACGIH design standards are to be followed?

10. You wish to control particles escaping from an open barrel being filled with powder. The barrel is 2 ft in diameter. Define a "controlled region" as one in which the air velocity along a streamline is 100 FPM or more. Management asks you to choose either a flanged round circular opening (Fig. 6.9) or a semicircular slot similar to Fig. 6.1. On the basis of the data in Figs. 6.8 and 6.9, which type of exhaust affords the maximum control per CFM?

11. It is common practice to cut concrete pavements prior to excavation so as to minimize the amount of pavement that has to be replaced. To accomplish this large and small disk cutters are used. Large machines are equipped with water for lubrication and dust control. Small hand-held machines (resembling chain saws fitted with cutting disks) often make no attempt to control dust (or noise). Recommend engineering controls to suppress the generation of dust.

12. Clay from a conveyor is discharged into a chute that leads to a blender where the clay is mixed with water and other ingredients in the first step in making vitreous china (sinks, toilet bowls, etc). A partial enclosure is built to surround the transfer point as shown in Fig. 6.11. Air is withdrawn through the exhaust duct to control fugitive emissions. Estimate the volumetric flow rate (CFM) needed to select an exhaust fan.

$\dot{m}_b = 40$ kg/s $\qquad$ H = 0.5 m
$A_b = 0.1$ m^2 $\qquad$ $\rho_b = 2000$ kg/m^3
$D_p = 100$ μm $\qquad$ Q_s = secondary air = 0.05 m^3/s
face area near adjustable gate = 0.5 m^2

13. Consider the duct system described in Illustrated Example 6.5. Management contemplates the following modifications and asks you to estimate how performance will be altered.

(a) A 6000 CFM side draft hood is to be installed to control vapors from binders used in making cores. The effluent from the hood will be joined to the main duct midway between points J2 and J3. The speed of the fan will be changed to accommodate the additional volumetric flow rate and new pressure drop. No other changes will be made. What will be the new pressure drop of the fan and will the volumetric flow rate through the existing devices suffer any reduction ?

(b) In addition to (a) above, openings in the shakeout enclosure are to be enlarged to increase the shakeout volumetric flow rate by 20%. By what amount should the volumetric flow rate of the fan be increased and if there are no other changes to the duct dimensions, what are the new volumetric flow rates through the remaining devices.

14. You have been asked to design a lateral exhauster for an open vessel 3 ft wide, 20 ft long and 3.5 ft high. The exhausters will run along both (long) sides of the vessel. The exhauster plenum should protrude above the vessel to a minimal amount. It has been decided that the volumetric flow rate per square foot of liquid surface is 175 CFM/ft^2. Use the duct-design computer program to compute the dimensions of the plenum chamber. It has been decided that air will be withdrawn through 3 inch holes located along 6 inch centers a minimal distance above the surface of the liquid. Use the program to compute the dimensions of the plenum chamber such that the pressure is constant throughout the chamber and that the volumetric flow rate through each 3 inch hole is 233 CFM. Because the holes are 3 inch in diameter, it has been decided that the width of the plenum chamber should be 6 inch. The vertical dimensions is of course unknown. You may select a plenum that withdraws air in a single direction for the full length of 20 ft or a symmetrical plenum that withdraws air in opposite directions for 10 ft.

15. An electric arc furnace is opened for a period of three minutes every 30 minutes in order to add scrap metal or minerals or for pouring. A hopper hood is located 30 feet above the furnace which collects fume which in turn is sent to a baghouse. The furnace diameter is 25 feet. Its heated surface is at a temperature is 600 K and the convection film coefficient can be computed from Table 4.2. When the furnace lid is opened the volumetric flow rate of hot gas is approximately 10^6 CFM. The baghouse is designed to handle 100,000 CFM on a steady basis. Recommend the volume and dimensions of a hopper hood.

16. A hopper-type canopy hood is to be located above an electric arc furnace used to melt scrap metal. Every 30 minutes, the lid is removed and a 10,000 CFM surge of hot gas is produced that lasts approximately 30 seconds. The lid is removed to recharge the furnace with scrap, add special minerals and to tap the molten metal. Recommend the dimensions of the hood and the volumetric flow rate of an exhaust fan if the hopper hood is to accomodate both the steady-state and surge plumes. Shown below are key parameters of the furnace.

   furnace diameter—5 meters
   furnace surface temperature—800 K
   natural convection film coefficient—10 watts/m^2 K
   distance between furnace lid and canopy face—10 m
   draft velocity in workplace—0.5 m/s
   maximum centerline velocity of buoyant plume (exclusive of the surge)—5 m/s

CHAPTER 7

Ideal Flow

Goal: Students will be expected to learn and use the
- equations of ideal flow to predict the velocity field of
- plain and flanged inlets in quiescent air
- plain and flanged inlets immersed in moving air
- multiple inlets
- flow around bodies

Chapter 6 describes contemporary practices and design standards that provide excellent guidance for designing industrial ventilation systems. Unfortunately, engineers cannot predict the effectiveness of the systems they design nor can they optimize the dimensions and volumetric flow rates prior to construction and field testing. Engineering today requires one be able to optimize designs based on analytical models of performance. There is no reason why industrial ventilation cannot and should not do the same.

The essential ingredient in optimization is accurate prediction of the velocity field upwind of inlet. The objective of this chapter and those that follow is to describe methods to predict the movement of contaminants upwind of an inlet of arbitrary geometry immersed in arbitrary moving air mass.

Figures 6.8 and 6.9 provide some detail about the velocity field upwind of rectangular and circular inlets. A larger selection of air speed contours can be found in ref. 48. While these figures show only the contours of equal air speed, engineers can generate empirical equations to fit the data and generate analytical expressions of the form $U = U(x, y)$. Curve-fitting procedures can be accurate but are tedious to generate, the tedium increasing with the level accuracy that is desired. The objective of this chapter is to describe ways to predict the velocity field based on fundamental fluid mechanics taught in engineering curricula.

Local ventilation is the capture of contaminants close to the point of generation by withdrawing air at a selected rate through a uniquely shaped inlet, opening, entry, or port. A *close capture hood*, or more generally a *hood* is the generic name given to this uniquely shaped opening located close to the source of contamination. Section 1.4 described how local ventilation hoods are classified. General ventilation describes how air from the entire work space is withdrawn. Local ventilation is preferred over general ventilation because volumetric flow rates of air are considerably smaller, capital and operating costs are smaller, and the adverse impact on heating and air conditioning is less.

The uninitiated might imagine that the velocity field near an inlet is a mirror image of the flow field near an outlet. Nothing could be further from the truth; one is not the inverse of the other, as can be seen in Fig. 7.1. A vivid illustration of the disparity is contained in the book by Hayashi et al. (243) in which two individuals try to extinguish a candle, one by exhaling and the other by inhaling. Even if both produce the same volumetric flow rate and air speed (at the mouth) and both stand the same distance from the flame, exhaling produces a flow field that extinguishes the flame while the flow field from inhaling does not. In the vicinity of an outlet, viscosity and velocity gradients profoundly affect the spread and penetration of the jet, whereas the same does not occur near an inlet. Due to viscosity, the jet entrains surrounding air and the width of the jet increases with an included angle of approximately 22 degrees. Within the core of the jet the velocity remains equal to the exit velocity, but after approximately 7 diameters the centerline velocity begins to decrease. At approximately 30 diameters downstream, the centerline velocity is approximately 10% the exit velocity. The influence of a jet (sometimes called *throw*) extends approximately 30 diameters downwind of the exit plane, whereas the influence of an intake (sometimes called *reach*) extends to approximately 1 diameter upwind of the inlet plane. At 1 diameter upstream of the inlet the velocity is approximately 10% of the inlet velocity.

Since viscosity does not affect the inlet velocity field appreciably, the viscous effects can be neglected and the flow is said to be *inviscid*. In addition, the velocities in industrial ventilation inlets are low and the density can be assumed to be constant. Fluids which are incompressible and in which the viscous effects are negligible are called *ideal fluids*. If in addition the flow is *irrotational* (zero vorticity), the flow is called *ideal flow*. If a process that generates contaminant is close to the inlet, an aerodynamic wake is apt to be produced as the air passes over and around the source. Conditions in the wake region are governed by viscosity and invalidate the assumption of ideal flow. If however the work piece is sufficiently far away, its wake may be outside the velocity field of the inlet and the assumption of ideal flow is valid. Aerodynamic wakes from any other solid body must also be outside the velocity field of the inlet if the assumption of ideal flow is valid. The size of the wake region is a matter of computation, but for a first approximation a value twice the characteristic dimension of the bluff body perpendicular to the flow field can be used.

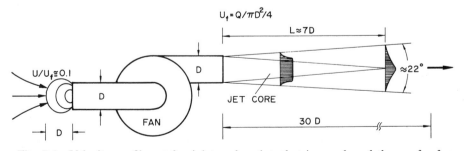

Fig. 7.1 Velocity profiles at fan inlet and outlet, that is, reach and throw of a fan.

7.1 FUNDAMENTAL CONCEPTS

If it cannot be shown that the contaminant concentration is spatially uniform, the well-mixed model described in Chapter 5 cannot be used and one must solve a series of differential equations to find the concentration. A series of differential equations describing contaminant transport includes the air velocities of the carrier gas throughout the region. There are conditions wherein the problem can be split and the air velocities computed first and the concentration computed later.

In this text an *Eulerian frame of reference* will be adopted in which attention will be focused on a point in space at a particular instant in time. Similarly, attention may be focused on a particular volume in space (control volume) through which a fluid passes. With the *Eulerian method of description*, the properties of the flow field, including contaminants it may contain, are described as functions of space and time.

It is convenient to categorize flow as *rectilinear* or *curvilinear*. There is only one vector component in rectilinear flow and two or more in curvilinear flow. A curvilinear flow in which the fluid rotates about a common center of rotation is called a *vortex*. The vorticity of a vortex may be zero, finite, or finite and constant.

Many physical phenomena in industrial ventilation can be analyzed assuming that there are only two independent spatial coordinates (e.g. flow around cylinders, flow around spheres, flow over plane surfaces containing circular openings or slots of large aspect ratio). When the flow field can be approximated as two-dimensional and ideal many simplifications can be made. A steady, two dimensional flow is *ideal* if "three I's" are satisfied:

(1) Incompressible (Div $U = 0$)
(2) Inviscid (μ is negligible)
(3) irrotational (Curl $U = 0$)

Many important flows satisfy one or two of these conditions, but for the flow to be *ideal* all three conditions must be met. If all three conditions are satisfied, the analysis can be simplified considerably.

The terms *incompressible* and *constant density* are often used interchangeably. Strictly speaking they are not synonyms, since

$$\text{Constant density: } \rho = \text{constant} \tag{7-1}$$

$$\text{Incompressible: } \frac{D\rho}{Dt} = \frac{\partial \rho}{\partial t} + U_i \frac{\partial \rho}{\partial x_i} = 0 \tag{7-2}$$

A constant density fluid is certainly an incompressible fluid, but the converse is not necessarily true. For example, if the flow is steady ($\partial \rho / \partial t = 0$) and the divergence of the velocity is zero (Div $U = 0$), the flow will be incompressible irrespective of whether the density is constant or not.

Air is viscous and compressible but there are many industrial situations in which air behaves as if its viscosity is negligible and its density and temperature are constant. In the absence of jets or wakes, the region in the vicinity upstream of the inlet (but not at the inlet face), is a region where these conditions occur. Analysis of these flow fields is not overly difficult and a vast body of literature exists (34, 35, 62) that can be employed. Many complicated analytical expressions from technical literature once considered quaint or anachronistic are important for local ventilation systems. Complicated mathematics once considered troublesome can now be handled easily with computers.

Air will be assumed to be a *perfect gas* with constant specific heats (c_v, c_p) and a Newtonian fluid. A *Newtonian fluid* is a fluid in which the *shear stress* (τ_{ij}) is equal to the dynamic viscosity (μ) times the *deformation rate* $(\partial U_i / \partial x_j)$, i.e. $\tau_{ij} = \mu \partial U_i / \partial x_j$. As a result of these assumptions the *conservation equations* for air are:

Equation of state

$$P = \rho R T \tag{7-3}$$

Continuity

$$\frac{\partial \rho}{\partial t} + \mathbf{Div}(\rho U) = 0 \tag{7-4}$$

Momentum (*Navier–Stokes equation*)

$$\frac{\partial U}{\partial t} + (U \cdot \nabla)U = -\left(\frac{1}{\rho}\right) \mathbf{Grad}\ P + g + \left(\frac{\mu}{\rho}\right)\nabla^2 U \tag{7-5}$$

Energy:

$$\rho\left[\frac{c_v \partial T}{\partial t} + c_v(U \cdot \nabla)T\right] + P(\nabla \cdot U) = \nabla \cdot (k\ \mathbf{Grad}\ T) + q_v \tag{7-6}$$

where c_v is the specific heat at constant volume, ρ is the density of air, P is the pressure, U is the velocity of air, k is the *thermal conductivity*, v is the *kinematic viscosity* and q_v is energy added by viscous dissipation, chemical reactions, and so on. *Viscous dissipation* can be expressed as

$$\mu\left[2\left(\frac{\partial U_x}{\partial x}\right)^2 + 2\left(\frac{\partial U_y}{\partial y}\right)^2 + 2\left(\frac{\partial U_z}{\partial z}\right)^2 + \left(\frac{\partial U_x}{\partial y} + \frac{\partial U_y}{\partial x}\right)^2\right.$$
$$\left. + \left(\frac{\partial U_z}{\partial x} + \frac{\partial U_x}{\partial z}\right)^2 + \left(\frac{\partial U_y}{\partial z} + \frac{\partial U_z}{\partial y}\right)^2\right] \tag{7-7}$$

The following symbols represent vector operators:

$\cdot$ = dot product
$\times$ = cross product
$\mathbf{Curl} = \nabla \times$ = curl of a vector
$\nabla = \mathbf{Grad}$ = gradient of a scalar
$\mathbf{Div} = \nabla \cdot$ = divergence of a vector
$\nabla^2 = \nabla \cdot \nabla$ = Laplacian

If the flow is frictionless ($\mu = 0$), *steady (temporally uniform,* $\partial(\ )/\partial t = 0$) and the density is constant, the equations above reduce to the following:

State

$$\rho = \text{constant} \tag{7-8}$$

Continuity

$$\boldsymbol{\nabla} \cdot \boldsymbol{U} = 0 \tag{7-9}$$

Momentum

$$(\boldsymbol{U} \cdot \boldsymbol{\nabla})\boldsymbol{U} = -\left(\frac{1}{\rho}\right) \textbf{Grad } P + \boldsymbol{g} \tag{7-10}$$

Energy

$$\rho c_{\text{v}}(\boldsymbol{U} \cdot \boldsymbol{\nabla})T = \boldsymbol{\nabla} \cdot (k \textbf{ Grad } T) \tag{7-11}$$

The form of the momentum equation in Eq. 7-10 is called *Euler's equation*. Using the vector identity

$$\boldsymbol{U} \times (\textbf{Curl } \boldsymbol{U}) = \boldsymbol{\nabla}\left(\frac{U^2}{2}\right) - (\boldsymbol{U} \cdot \boldsymbol{\nabla})\boldsymbol{U} \tag{7-12}$$

the above equations reduce to the following for the steady flow of an inviscid, constant temperature–constant density fluid.

Continuity

$$\textbf{Div } \boldsymbol{U} = 0 \tag{7-13}$$

Momentum

$$\boldsymbol{U} \times (\textbf{Curl } \boldsymbol{U}) = \boldsymbol{\nabla}\left(\frac{P}{\rho} + \frac{1}{2} U^2 + gz\right) \tag{7-14}$$

Energy

$$T = \text{constant} \tag{7-15}$$

A fluid *pathline* is the locus of points an element of fluid occupies over a period of time. If the flow is steady, all the pathlines through a point in the fluid are the same; such a pathline is called a *streamline*. Streamlines are the invisible lines in space describing the paths each element of fluid follows.

Taking the dot product of Eq. 7-14 and the velocity $\boldsymbol{U}$ one obtains,

$$(\boldsymbol{U} \times \textbf{Curl } \boldsymbol{U}) \cdot \boldsymbol{U} = \textbf{Grad}\left(\frac{P}{\rho} + \frac{U^2}{2} + gz\right) \cdot \boldsymbol{U} \tag{7-16}$$

Since $[U \times \textbf{Curl}\ U]$ is a vector perpendicular to U, its dot product with U must be zero. Thus, along a streamline,

$$\textbf{Grad}\left(\frac{P}{\rho} + \frac{U^2}{2} + gz\right) \cdot U = 0 \qquad (7\text{-}17)$$

and U must also be perpendicular to $\text{Grad}(P/\rho + U^2/2 + gz)$.

The gradient of a scalar will have its maximum value in directions perpendicular to surfaces along which the scalar is a constant. Similarly, the value of the gradient will be zero in directions tangent to surfaces along which the scalar is constant. Hence it can be concluded that $(P/\rho + U^2/2 + gz)$ is constant (called the *Bernoulli constant*) along streamlines if the flow is steady, incompressible and frictionless.

Suppose now the flow is also *irrotational*, that is, **Curl** U is everywhere zero. Hence Eq. 7-17 reduces to

$$\textbf{Grad}\left(\frac{P}{\rho} + \frac{U^2}{2} + gz\right) = 0 \qquad (7\text{-}18)$$

Now if a gradient of a quantity is zero everywhere, it can be concluded that the scalar is constant everywhere. Thus $(P/\rho + U^2/2 + gz) = $ constant everywhere if the flow is irrotational, steady, incompressible and frictionless, that is, the flow is *ideal*. Under these conditions Eq. 7-18 is called *Bernoulli's equation*.

The curl of the velocity is called the *vorticity* ($\mathbf{\Omega}$) and represents the rotation of a fluid parcel.

$$\text{Vorticity} = \mathbf{\Omega} = \frac{1}{2}\ \textbf{Curl}\ U \qquad (7\text{-}19)$$

Illustrated Example 7.1   Velocity Field Near Rotating Disk

To illustrate these fundamental principles analyze the flow field close to a hand-held surface grinder (see Fig. 1.10) of radius R rotating at an angular velocity Ω. Assume that the density is constant and that the flow is steady and inviscid. Gravitational effects will also be neglected. Assume a velocity traverse has been made and indicates that only the tangential velocity component (U_θ) is important and the velocity components in the radial and axial directions are negligible.

Velocity field

   Region I: $U_\theta = \Omega r$ $r < R$

   Region II: $U_\theta = \Omega R^2/r$ $r > R$

where Ω is a constant. A sketch of the flow field is shown in Fig. 7.2. To illustrate the previous fundamental concepts, analyze the flow field with respect to the following:

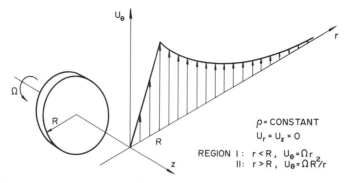

Fig. 7.2 Velocity distribution near a rotating disk, Illustrated Example 7.1.

(a) Vorticity and irrotationality.

(b) The pressure field needed to satisfy the conservation of mass and momentum.

(c) Applicability of Bernoulli's equation

(A) Vorticity

An irrotational flow field is one in which the Curl U is zero everywhere. One-half the Curl U is also the definition of vorticity. In cylindrical coordinates, the Curl U is

$$\textbf{Curl } U = \hat{r}\left[\frac{1}{r}\frac{\partial U_z}{\partial \theta} - \frac{\partial U_\theta}{\partial z}\right] + \hat{\theta}\left[\frac{\partial U_r}{\partial z} - \frac{\partial U_z}{\partial r}\right] + \hat{z}\frac{1}{r}\left[\frac{\partial(rU_\theta)}{\partial r} - \frac{\partial U_r}{\partial \theta}\right]$$

Region I: $r < R,$ $U_\theta = \Omega r$

$$\textbf{Curl } U = \hat{z}\Omega$$

Region II: $r > R,$ $U_\theta = \Omega\frac{R^2}{r}$

$$\textbf{Curl } U = 0$$

Thus the flow is irrotational in the outer region II and rotational with a constant vorticity in the inner region I. In addition, since the vorticity is constant in region I the rotational flow field is said to undergo *solid body rotation*, that is, to rotate as a solid body such as a compact disk recording. The analysis illustrates clearly that just because air rotates about a center of rotation, i.e. is a vortex, it can be either "rotational or irrotational," that is, it may or may not have vorticity. All curvilinear flow rotates about a center of rotation, but "irrotation or rotation" in the fluid dynamic sense depends on the vorticity. As a general proposition, irrotational flows have tangential velocities that vary inversely with the radius, while in rotational flows the tangential velocity is proportional to the radius raised to power (k) other than negative 1.

$$U_\theta = \Omega r^k \begin{bmatrix} k = & 1 & \text{solid body rotation, vorticity} = \Omega \\ k = -1 & \text{irrotational, vorticity is zero} \\ -1 < k < 1 & \text{rotational, vorticity is nonzero} \end{bmatrix}$$

(B) Pressure Distribution

To find the pressure distribution consistent with the continuity and momentum equations, apply the continuity and momentum equations to the velocity profile.

Continuity equation

$$\mathbf{\nabla \cdot U} = \frac{1}{r}\frac{\partial(rU_r)}{\partial r} + \frac{1}{r}\frac{\partial U_\theta}{\partial \theta} + \frac{\partial U_z}{\partial z}$$

Substituting the observed velocity distribution, it is seen that the continuity equation is satisfied in both regions I and II.

Momentum equation

$$r:\quad (\mathbf{U \cdot \nabla})U_r - \left(\frac{1}{r}\right)U_\theta^2 = -\frac{1}{\rho}\frac{\partial P}{\partial r}$$

$$\theta:\quad (\mathbf{U \cdot \nabla})U_\theta + \left(\frac{1}{r}\right)U_r U_\theta = -\frac{1}{(r\rho)}\frac{\partial P}{\partial \theta}$$

Upon substituting the observed velocity distribution, it is found that

$$r:\quad \frac{\partial P}{\partial r} = \rho\frac{U_\theta^2 r}{r}$$

$$\theta:\quad \frac{\partial P}{\partial \theta} = 0$$

in both regions I and II. Thus the pressure is only a function of the radius. The r momentum equation becomes an ordinary differential equation in both regions I and II,

$$\frac{dP}{dr} = \rho\frac{U_\theta^2}{r}$$

Begin by considering region II ($r > R$). Substitute $U_\theta = \Omega R^2/r$, and integrate the equation between r and infinity where U_θ is zero and the pressure is the ambient pressure P_0.

Region II: $r > R$

$$P_0 - P(r) = \rho\Omega^2\frac{R^4}{2r^2}$$

At the interface between regions I and II ($r = R$) the pressure must be continuous, thus $P(R)$ can be found by evaluating the equation above at $r = R$,

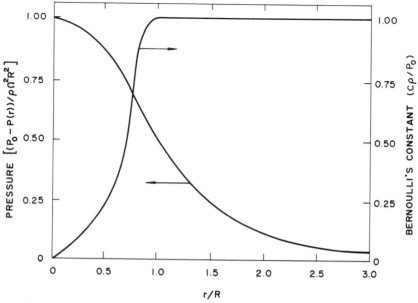

Fig. 7.3 Dimensionless pressure distribution and Bernoulli's constant near rotating disk for the case when $\Omega^2 = P_0/\rho R^2$, Illustrated Example 7.1.

$$P(R) = P_0 - \rho\Omega^2 \frac{R^2}{2}$$

Now consider region I $(r < R)$. Substitute $U_\theta = \Omega r$, and integrate between r and R,

Region I: $r < R$

$$P_0 - P(r) = \rho\Omega^2 \left[R^2 - \frac{r^2}{2} \right]$$

Figure 7.3 is a graph of the pressure distribution and shows that the pressure at the center of the vortex is one half the value at the edge of the rotational region (I).

(C) Bernoulli's Equation

Bernoulli's constant is equal to the following:

$$C = \frac{P}{\rho} + \frac{U^2}{2}$$

since gravitational effects have been neglected. Since the velocity and pressure distributions have been computed for regions I and II, the constant can be found.

Region I: $r < R$

$$C = \frac{P_0}{\rho} - \Omega^2[R^2 - r^2]$$

Region II: $r > R$

$$C = \frac{P_0}{\rho}$$

Thus Bernoulli's equation is valid everywhere in the irrotational region II, but is only valid along particular streamlines in the rotational region I. In region I the constant has a unique value for each radius. The variation of the Bernoulli constant (written in dimensionless form for the case when $\Omega^2 = P_0/\rho R^2$) with radius is shown in Fig. 7.3.

The results in A–C are consistent with the assumptions made at the beginning of the analysis. If a more detailed analysis is desired, engineers would have to consider the three dimensional and viscous features of the flow. The viscosity of air is not zero and viscous effects are significant close to the rotating disk where a boundary layer exists. Second, the predicted low-pressure region in the center of the disk will initiate an axial flow of air toward the disk which in turn produces an outwardly radial flow of air. Under these conditions, U_z and U_r are no longer zero and the full set of Navier–Stokes equations will have to be solved.

A velocity field for which the [Curl U] is zero is called a *conservative velocity field*. A property of a conservative vector field is that a scalar function $\Phi(x, y)$ exists such that

$$U = -\text{Grad } \Phi \qquad (7\text{-}20)$$

The scalar function Φ is called the *velocity potential*. In rectangular coordinates,

$$U_x = -\frac{\partial \Phi}{\partial x}; \qquad U_y = -\frac{\partial \Phi}{\partial y} \qquad (7\text{-}21)$$

(In some texts the negative sign in Eq. 7-20 is replaced by a plus sign). The terms U_x and U_y represent the components of the air velocity in the x and y directions. (Note, U_x and U_y are NOT derivatives of U with respect to x and y, which is an editorial practice used in some texts.) Similar terminology will be used for cylindrical and spherical coordinates. Substituting Eq. 7-21 into the continuity equation results in

$$\frac{\partial^2 \Phi}{\partial x^2} + \frac{\partial^2 \Phi}{\partial y^2} = \nabla \cdot \nabla \Phi = \nabla^2 \Phi = 0 \qquad (7\text{-}22)$$

Along a streamline,

$$U_x = \frac{dx}{dt}; \qquad U_y = \frac{dy}{dt} \qquad (7\text{-}23)$$

and one can write

$$\frac{U_x}{U_y} = \frac{dx}{dy} \qquad (7\text{-}24)$$

$$0 = U_y \, dx - U_x \, dy$$

Now define a continuous scalar function $\Psi(x, y)$, called the *stream function*, such that for any steady two dimensional flow field

$$U_x = -\frac{\partial \Psi}{\partial y}, \qquad U_y = \frac{\partial \Psi}{\partial x} \qquad (7\text{-}25)$$

The differential of Ψ can be expressed by the chain rule. The partial derivatives of Ψ with respect to x and y can be replaced by velocity components as expressed by Eq. 7-25. Thus,

$$d\Psi = \left(\frac{\partial \Psi}{\partial x}\right) dx + \left(\frac{\partial \Psi}{\partial y}\right) dy$$

$$d\Psi = U_y\, dx - U_x\, dy \qquad (7\text{-}26)$$

Comparing Eqs. 7-24 and 7-26 it can be concluded that along a streamline Ψ is a constant. From the definition of the streamline given earlier, it can also be concluded that between any two streamlines the mass flow rate is constant. Thus, if the density is constant, the volumetric flow rate between streamlines Ψ_1 and Ψ_2 is

$$Q = \int_1^2 d\Psi = [\Psi_2 - \Psi_1] \qquad (7\text{-}27)$$

The region bounded by two streamlines is called a *stream tube*. The volumetric flow rate in a stream tube is constant. Substituting Eq. 7-21 into the continuity equation and substituting Eq. 7-25 into the condition for irrotationality results in expressions called *Laplace's equation*,

$$\nabla^2 \Phi = 0 \qquad (7\text{-}28)$$

$$\nabla^2 \Psi = 0 \qquad (7\text{-}29)$$

Flow fields described by Eqs. 7-21, 7-25, 7-28, and 7-29 are called *potential flow fields* and the flow is called *potential flow*. Curves of constant values of Ψ define fluid streamlines and are everywhere normal to lines of constant Φ, which describe surfaces along which the speed of the fluid is constant.

Consider the unobstructed flow of air into an inlet, that is, where there are no bluff bodies producing wakes. Except for boundary layers adjacent to solid surfaces, which are of no interest at the moment, velocity gradients and viscosity are small, and the flow will be irrotational. *Irrotational flow* is a flow in which fluid elements do not rotate. Imagine that it is possible to place NESW axes on each fluid element as shown in Figs 7.4a and b. If the NESW axes retain their initial orientation, irrespective of the path (even if the fluid element travels in a circular path), the flow is called irrotational.

If a bluff body is present, the wake region will contain large velocity gradients. The combination of viscosity and large velocity gradients produce shear stresses which cause fluid elements to rotate. Such motion is called rotational flow. Air passing over a work piece, the worker's body, hand tools, and so on is apt to

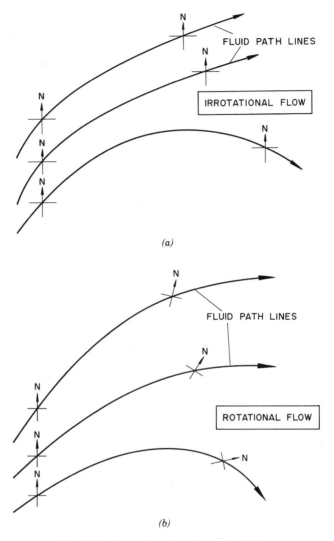

Fig. 7.4 Irrotational and rotational flow.

produce wakes. Wakes have a large effect on the motion and concentration of contaminants and cannot be neglected. Rotation can also be produced by shock waves, boundary layers, flow through air moving equipment (fans, turbines, compressors, etc.) and thermal gradients.

The rotation of a fluid element can be represented by a vector called vorticity Ω, which is defined by Eq. 7-19. By this definition, the vorticity is zero for an irrotational flow, that is,

$$\text{Irrotational flow: } \mathbf{Curl}\ U = 0 \qquad\qquad (7\text{-}30)$$

The vorticity of a rotational flow field is finite but not necessarily constant. The vorticity of a fluid element cannot change except through the action of viscosity, nonuniform heating (temperature gradients), or other nonuniform phenomena. Thus, if a flow originates in an irrotational region, it will remain irrotational until some dissipative process alters it. Flow entering an inlet from a quiescent environment is irrotational and will remain so unless a wake or nonuniform heating transforms it.

In cylindrical coordinates Eqs. 7-21 and 7-25 can be written as,

$$U_r = -\frac{\partial \Phi}{\partial r} = -\frac{1}{r}\frac{\partial \Psi}{\partial \theta} \tag{7-31}$$

$$U_\theta = -\frac{1}{r}\frac{\partial \Phi}{\partial \theta} = \frac{\partial \Psi}{\partial r} \tag{7-32}$$

where U_r and U_θ are the velocity components in the r and θ directions. In *axisymmetric three dimensional flow* in which the x-axis is an axis of symmetry, Eqs. 7-21 and 7-25 can be written as

$$U_r = -\frac{\partial \Phi}{\partial r} = -\frac{1}{r^2 \sin \theta}\frac{\partial \Psi}{\partial \theta} \tag{7-33}$$

$$U_\theta = -\frac{1}{r}\frac{\partial \Phi}{\partial \theta} = \frac{1}{r \sin \theta}\frac{\partial \Psi}{\partial r} \tag{7-34}$$

where r is the distance from the origin and angle θ is the polar angle measured from the x-axis. A summary of velocities and the derivatives of potential functions can be found in Table 7.1.

In summary, the velocity potential and stream functions have the following physical connotations, which may be useful to the reader.

(a) Lines in which the stream functions are constant are called *streamlines*.

(b) Fluid streamlines are path lines along which fluid particles move.

(c) Fluid particles do not cross streamlines.

TABLE 7.1   Velocities and Derivatives of Potential Functions.

Coordinate System	Velocities	
Two Dimensional (Rectangular)	$U_x = -\dfrac{\partial \phi}{\partial x} = -\dfrac{\partial \psi}{\partial y}$	$U_y = -\dfrac{\partial \phi}{\partial y} = \dfrac{\partial \psi}{\partial x}$
Two Dimensional (Cylindrical)	$U_r = -\dfrac{\partial \phi}{\partial r} = -\dfrac{1}{r}\dfrac{\partial \psi}{\partial \theta}$	$U_\theta = -\dfrac{1}{r}\dfrac{\partial \phi}{\partial \theta} = \dfrac{\partial \psi}{\partial r}$
Axisymmetric Three Dimensional	$U_r = -\dfrac{\partial \phi}{\partial r} = -\dfrac{1}{r^2 \sin \theta}\dfrac{\partial \psi}{\partial \theta}$	
	$U_\theta = -\dfrac{1}{r}\dfrac{\partial \Phi}{\partial \theta} = \dfrac{1}{r \sin \theta}\dfrac{\partial \psi}{\partial r}$	

(d) In two dimensions, the volumetric flow rate Q between any two stream-lines Ψ_1 and Ψ_2 is equal to

$$(\Psi_1 - \Psi_2).$$

(e) Lines in which the velocity potential is constant are called equipotential lines.

(f) Along equipotential lines, fluid particles have the same speed, that is, magnitude of the velocity vector.

Solutions to Eqs. 7-28 and 7-29 must satisfy boundary conditions dictated by the physical constraints of the particular ventilation system. The equation can be solved by analytical or numerical methods. Graphical or experimental techniques, often called *flux plotting*, can also be used. Since the accuracy of these techniques is poor, they will be omitted.

Analytical solutions to Laplace's equation are called *harmonic functions* and many have been found for a variety of phenomena encountered in engineering (56). Laplace's equation is a linear homogeneous differential equation, the combination of two or more solutions to the equation is also a solution. For example, if Φ_1 and Φ_2 are each solutions to Laplace's equation,

$$A\Phi_1; \quad (A + \Phi_1); \quad (\Phi_1 + \Phi_2); \quad (A\Phi_1 + \Phi_2) \tag{7-35}$$

where A is a constant, are also solutions. Thus, one may combine several harmonic functions and solutions to Laplace's equation. If a flow field is known to be the sum of two separate flow fields, (e.g., a sink located in a streaming flow), one can combine the harmonic functions for each to describe the combined flow field.

In order to map a particle's trajectory or to compute the transport of contaminant vapor in a moving air mass, it will be necessary to compute the air velocity (U) at any location (x, y). If analytical expressions for Φ and Ψ are known as functions of (x, y), one can differentiate and find the velocity components in the x and y directions. Expressions for Φ and Ψ have been derived (34, 35, 62) for several useful ventilation inlets:

Point sinks and line sinks in a plane
Uniform flow over a surface, (streaming flow)
Flow into a rectangular opening (flanged slot)
Flow into plain and flanged rectangular ducts
Flow into a flanged circular inlet

Table 7-2 is a summary of the potential functions for several of the above flow conditions. If one is faced with a flow field vastly different from these, it is possible to solve Laplace's equation numerically and obtain tabulated values of Ψ and Φ at any (x, y). Chapter 8 is devoted to this subject.

Equation 7-35 reveals an important feature of ideal flow fields. Suppose a flow field is produced by two independent flow fields, each of which can be treated as

TABLE 7.2 Potential Functions for Inlets.

Flow	Sketch	Potential Functions
Oblique Streaming Flow		$\phi = -U_o x \cos\alpha - U_o y \sin\alpha$ $\psi = -U_o y \cos\alpha + U_o x \sin\alpha$
Point Sink In A Plane		$\phi = -(Q/2\pi)(1/r)$ $\psi = -(Q/2\pi)\cos\theta$
Line Sink In A Plane		$\phi = (Q/L\pi)\ln r$ $\psi = (Q/L\pi)\theta$

Flow	Sketch	Potential Functions
Flanged Rectangular Inlet		$\dfrac{x}{w} = \cosh(\phi/k)\cos(\psi/k)$ $\dfrac{y}{w} = \sinh(\phi/k)\sin(\psi/k)$ $k = Q/L\Pi$
Unflanged Rectangular Inlet		$\dfrac{x}{w} = \dfrac{1}{\Pi}\left[\dfrac{2\phi}{k} + \exp\left(\dfrac{2\phi}{k}\right)\cos\left(\dfrac{2\psi}{k}\right)\right]$ $\dfrac{y}{w} = \dfrac{1}{\Pi}\left[\dfrac{2\psi}{k} + \exp\left(\dfrac{2\phi}{k}\right)\sin\left(\dfrac{2\psi}{k}\right)\right]$ $k = Q/L\Pi$
Flanged Circular Inlet		$\phi = Q/2\pi w \arcsin\left[\dfrac{2w}{a_1 + a_2}\right]$ $\psi = \dfrac{Q}{4\Pi w}\left[4w^2 - (a_1 - a_2)^2\right]^{\frac{1}{2}}$ $a_1 = \left[z^2 + (w+r)^2\right]^{\frac{1}{2}}$ $a_2 = \left[z^2 + (w-r)^2\right]^{\frac{1}{2}}$

ideal. Then the potential function of the composite field is the sum of the potential functions of the component flow fields and the velocity at any point in the composite field is the vector sum of the velocities of each one of the component flow fields. Consider an ideal flow field that is the *superposition* of two independent flow fields denoted by the subscripts 1 and 2. From Eq. 7-35

$$\Phi(\text{composite}) = \Phi_1 + \Phi_2 \qquad (7\text{-}36)$$

The x component of the velocity of the composite field is

$$
\begin{aligned}
U_x &= -\frac{\partial \Phi}{\partial x} = -\frac{\partial (\Phi_1 + \Phi_2)}{\partial x} \\
&= -\frac{\partial \Phi_1}{\partial x} - \frac{\partial \Phi_2}{\partial x} \\
&= U_{x1} + U_{x2}
\end{aligned}
\tag{7-37}
$$

A similar formulation exists for the stream function and the velocity component U_y. The concept of superposition is very useful, but readers must be careful to insure that the flow fields they wish to add vectorially are both ideal flow fields. For example, the velocity profile of a jet should never be added to the velocity profile of an inlet or streaming flow because the velocity field associated with a jet is strongly affected by viscosity, is not irrotational, and cannot be described by potential functions.

Superposition is the concept upon which "capture velocity" seems to be based. If vapor or gas are the contaminants to be removed, the velocity of the air stream in which they are contained can be added to the velocity produced by an inlet to predict the velocity of the composite flow field. Hence the underlying notion of the capture velocity, that is, velocity necessary to overcome opposing air currents, has validity. However, if the contaminant is a particle whose velocity is not equal to the air velocity or a particle whose Reynolds number is much larger than unity, the concept of superposition is not appropriate because the particle velocity and gas velocity are not the same and the particle's motion is governed by viscosity.

Many of the potential functions have *singular points* at which the function or its derivatives approach infinity. For example, point sinks and the lip of an inlet both with and without flanges, are singular points. Infinite velocities do not occur and hence users should not expect physically meaningful results in regions near singular points. Since the size of these regions is small with respect to the large region for which the solutions are valid, avoiding these regions is a small limitation on the analysis.

7.2 TWO DIMENSIONAL FLOW

Two dimensional flows will be described in x-y coordinates or r-θ coordinates depending on which pair is more useful in a particular problem. For emphasis, it will be useful to reiterate the equalities between velocity components and the derivatives of the potential functions:

$$
\begin{aligned}
U_x &= -\frac{\partial \Phi}{\partial x} = -\frac{\partial \Psi}{\partial y} \\
U_y &= -\frac{\partial \Phi}{\partial y} = \frac{\partial \Psi}{\partial x} \\
U_r &= -\frac{\partial \Phi}{\partial r} = -\left(\frac{1}{r}\right)\frac{\partial \Psi}{\partial \theta} \\
U_\theta &= -\left(\frac{1}{r}\right)\frac{\partial \Phi}{\partial \theta} = \frac{\partial \Psi}{\partial r}
\end{aligned}
\tag{7-38}
$$

7.2.1 Line Sink

Consider an infinite line in space through which an ideal fluid flows inward at a uniform volumetric flow rate Q. In the equations in this chapter Q will be positive if there is "suction," (i.e., air is withdrawn and the flow field is influenced by a sink). Through a cylindrical surface of radius r surrounding the line sink, the volumetric flow rate is

$$Q = -2\pi r L U_r$$

$$U_r = -\left[\frac{Q}{2\pi L}\right]\frac{1}{r} \tag{7-39}$$

The negative sign occurs because the velocity is in the direction of decreasing values of r. Due to symmetry,

$$\frac{1}{r}\frac{\partial \Psi}{\partial \theta} = 0, \ \frac{1}{r}\frac{\partial \Phi}{\partial \theta} = 0,$$

and consequently U_θ is zero. In terms of the potential functions Φ and Ψ,

$$U_r = -\frac{\partial \Phi}{\partial r} = -\left[\frac{Q}{2\pi L}\right]\frac{1}{r} \tag{7-40}$$

$$U_r = -\left(\frac{1}{r}\right)\frac{\partial \Psi}{\partial \theta} = -\left[\frac{Q}{2\pi L}\right]\frac{1}{r} \tag{7-41}$$

To find the velocity potential phi, integrate Eq. 7-40,

$$\Phi = \int \left[\frac{Q}{2\pi r L}\right] dr$$

$$\Phi = \left[\frac{Q}{2\pi L}\right]\ln r + g(\theta) \tag{7-42}$$

but since $\partial \Phi/\partial \theta = 0$, the function $g(\theta)$ is a constant which can be set equal to zero because constants can be added and deleted from harmonic functions. Integrate Eq. 7-41,

$$\Psi = \int \left[\frac{Q}{2\pi L}\right] d\theta$$

$$\Psi = \left[\frac{Q}{2\pi L}\right]\theta + g(r) \tag{7-43}$$

and since $\partial \Psi/\partial \theta = 0$, the arbitrary function $g(r)$ is also a constant which can be set equal to zero. Thus the potential functions for a line sink in space are

$$\Phi = \left(\frac{Q}{2\pi L}\right)\ln r \tag{7-44}$$

$$\Psi = \left(\frac{Q}{2\pi L}\right)\theta \tag{7-45}$$

7.2.2 Uniform Streaming Flow

Consider an ideal fluid moving uniformly in the negative x direction with a speed U_0. In terms of the velocity potential,

$$U_x = -\frac{\partial \Phi}{\partial x} = -U_0 \tag{7-46}$$

$$U_y = -\frac{\partial \Phi}{\partial y} = 0 \tag{7-47}$$

Integrating Eq. 7-46,

$$\Phi = xU_0 + g(y) \tag{7-48}$$

and then differentiating,

$$\frac{\partial \Phi}{\partial y} = \frac{dg(y)}{dy} = g'(y) = 0 \tag{7-49}$$

Therefore, $g(y)$ is a constant that will be set equal to zero, and

$$\Phi = xU_0 \tag{7-50}$$

Similarly, the stream function can be found from the following:

$$U_x = -\frac{\partial \Psi}{\partial y} = -U_0$$

$$U_y = \frac{\partial \Psi}{\partial x} = 0 \tag{7-51}$$

$$\Psi = yU_0 + g(x)$$

$$\frac{\partial \Psi}{\partial x} = \frac{dg(x)}{dx} = g'(x) = 0$$

and therefore $g(x) = 0$ and,

$$\Psi = yU_0 \tag{7-52}$$

Figure 7.5 shows the location of equipotential lines for *streaming flow* in the negative x-direction.

7.2.3 Line Sink in a Plane in Streaming Flow

Consider an ideal fluid passing over a horizontal surface at a uniform velocity U_0 in the negative x direction. The surface contains a line sink through which fluid is withdrawn at a volumetric flow rate Q. The potential functions for the composite flow are the sum of the potential functions for streaming flow and a line sink. Since the line sink withdraws fluid from the region above the plane only, the volumetric flow rate is twice the value for a line sink in free space.

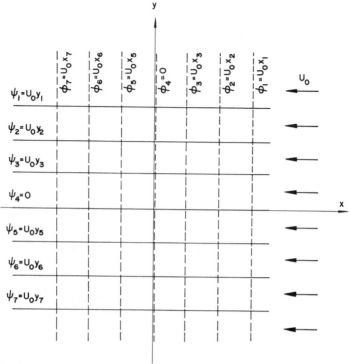

Fig. 7.5 Equipotential surfaces for streaming flow in the negative x direction.

$$\Phi = U_0 r \cos \theta + \left(\frac{Q}{L\pi}\right) \ln r \qquad (7\text{-}53)$$

$$\Psi = U_0 r \sin \theta + \left(\frac{Q}{L\pi}\right)\theta \qquad (7\text{-}54)$$

The velocity components at any point (r, θ) are

$$U_r = -\frac{\partial \Phi}{\partial r} = -\left[U_0 \cos \theta + \left(\frac{Q}{L\pi}\right)\frac{1}{r}\right] \qquad (7\text{-}55)$$

$$U_\theta = -\left(\frac{1}{r}\right)\frac{\partial \Phi}{\partial \theta} = U_0 \sin \theta \qquad (7\text{-}56)$$

$$U_x = U_r \cos \theta - U_\theta \sin \theta$$

$$U_x = -U_0 - \left(\frac{Q}{L\pi}\right)\left(\frac{x}{x^2 + y^2}\right) \qquad (7\text{-}57)$$

$$U_y = U_r \sin \theta + U_\theta \cos \theta$$

$$U_y = -\left[\frac{Q}{L\pi}\right]\left(\frac{y}{x^2 + y^2}\right) \qquad (7\text{-}58)$$

Examination of Eqs. 7-57 and 7-58, shows that the velocity of a composite ideal flow is the sum of the velocities of the component parts.

Thinking about the flow field for a minute, it can be concluded that far from the line sink, the flow may be deflected downward by the line sink as it passes from right to left, but it will be not be drawn into the line sink. Closer to the horizontal surface, fluid will be drawn into the line sink. Consequently, there must be a unique stream line that differentiates the flow that enters the line sink from that which does not. The *dividing streamline* is the name given to the streamline that divides the two regions. Figure 7.6 shows the location of the dividing streamline. The region between the x axis and the dividing streamline represents the region in which all the fluid passes through the line sink. It is useful to define the *reach* of an inlet as the region in space upwind of the inlet from which all the fluid ultimately enters the inlet. Thus the reach of a line sink is defined by the x axis and the dividing streamline.

The intersection of the dividing stream line and the x axis ($r = r_0$, $\theta = \pi$) is a *stagnation point* since U_r and U_θ are zero.

$$\text{Stagnation point: } U_r = 0, \qquad U_\theta = 0$$

Thus,

$$U_r(r_0, \pi) = 0 = -\left[U_0 \cos \pi + \left(\frac{Q}{L\pi} \right) \frac{1}{r_0} \right]$$

$$r_0 = \frac{Q}{L\pi U_0} \tag{7-59}$$

The dividing streamline passing through the *stagnation point* has the value

$$\Psi(\text{dividing}) = \Psi_d = \Psi(r_0, \pi)$$

$$= U_0 r_0 \sin \pi + \left(\frac{Q}{L\pi} \right) \theta = \frac{Q}{L} \tag{7-60}$$

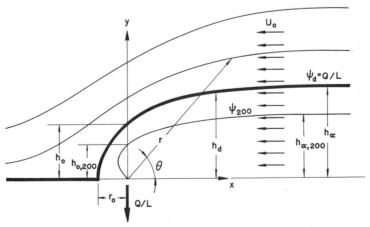

Fig. 7.6 Streamlines for Illustrated Example 7.2.

The streamline that coincides with the positive x axis has the value $\Psi = 0$. The height of the dividing streamline at $x \ggg 0$, i.e. h_∞, can be found by taking the limit of the following expression as x approaches infinity and θ approaches zero,

$$\Psi(\text{dividing}) = \Psi_d = \frac{Q}{L}$$

$$= \lim_{\theta \to 0} \left[U_0 \, r \sin \theta + \left(\frac{Q}{L\pi} \right) \theta \right] = U_0 h_\infty$$

$$h_\infty = \frac{Q}{LU_0} \tag{7-61}$$

The intersection of the dividing streamline and the y axis ($r = h_0$, $\theta = \pi/2$) can be found from

$$\Psi(\text{dividing}) = \Psi_d = \frac{Q}{L}$$

$$= U_0 h_0 \sin \frac{\pi}{2} + \left(\frac{Q}{L\pi} \right) \frac{\pi}{2}$$

$$h_0 = \frac{Q}{2LU_0} \tag{7-62}$$

In summary

$$h_\infty = \frac{Q}{LU_0} = \pi r_0 = 2h_0 \tag{7-63}$$

Illustrated Example 7.2—Reach Of A Flanged Line Sink in a Streaming Flow

Consider a line sink of strength (Q/L) equal to 350 CFM/ft ($0.542 \, \text{m}^3/\text{s m}$) lying in a horizontal plane over which air passes from right to left (Fig. 7.6). Far upwind of the line sink the air has a uniform velocity (U_0) of 100 FPM ($0.508 \, \text{m/s}$). Analyze the reach.

Using the equations given above, the following distances are computed:

$$h_\infty = 1.067 \, \text{m} \ (3.50 \, \text{ft}), \qquad r_0 = 0.339 \, \text{m} \ (1.11 \, \text{ft}), \qquad h_0 = 0.533 \, \text{m} \ (1.75 \, \text{ft})$$

The speed of the air along the dividing streamline varies from U_0 at $x \ggg 0$ to zero at the stagnation point. At the intersection of the y axis and dividing streamline ($r = h_0$, $\theta = \pi/2$), the absolute value of the air speed, $|U|$, is

$$|U| = \sqrt{U_r^2 + U_\theta^2} = 1.18 \, U_0$$

$$= 0.60 \, \text{m/s} \ (118 \, \text{FPM})$$

Suppose an industrial operation was conducted for which the capture velocity (see Chapter 6) was 200 FPM ($1.016 \, \text{m/s}$) and one wanted to locate the location of the region within which the air velocity was equal to or larger than 200 FPM. Therefore, the capture region is the locus of points (r, θ) below the dividing streamline where the air speed is never less than 200 FPM. One can find these points by solving the following equation for r at selected values of $|U| = 1.016 \, \text{m/s} = \sqrt{U_r^2 + U_\theta^2}$.

Outside the dividing streamline no air enters the sink, irrespective of the air velocity. Along the y axis, the point $(h_{0,200})$, where the air speed is $1.016 \, \text{m/s}$ (200 FPM), can be found by finding the value of y at $\theta = \pi/2$ that satisfies the expression above.

$$h_{0,200} = 0.196 \, \text{m} \quad (0.643 \, \text{ft})$$

The value of the stream function through this point (Ψ_{200}) is

$$\Psi_{200} = U_0(h_{0,200}) \sin\left(\frac{\pi}{2}\right) + \left(\frac{Q}{L\pi}\right)\left(\frac{\pi}{2}\right)$$

$$= 0.371 \, \text{m}^2/\text{s}$$

The height $(h_{\infty,200})$ of the streamline above the x axis at $x >>> 0$ can be obtained by taking the limit as x approaches infinity and θ approaches zero,

$$\Psi_{200} = 0.371$$

$$= \lim_{\theta \to 0}\left[U_0 r \sin\theta + \left(\frac{Q}{L\pi}\right)\theta\right]$$

$$= U_0 h_{\infty,200}$$

$$h_{\infty,200} = \frac{0.371}{0.580} = 0.728 \, \text{m} \quad (2.39 \, \text{ft})$$

7.3 AXISYMMETRIC THREE-DIMENSIONAL FLOW

Consider a three-dimensional flow of an ideal fluid in which there is symmetry about an axis of rotation. Table 7.1 shows the coordinates of such a flow field. The use of symmetry reduces the three dimensional problem into a two dimensional problem and enables one to use potential flow functions. The relationships between the velocity components and the derivatives of the potential functions in spherical coordinates are

$$U_r = -\frac{\partial \Phi}{\partial r}$$

$$U_r = -\frac{1}{r^2 \sin\theta}\frac{\partial \Psi}{\partial \theta} \qquad (7\text{-}64)$$

$$U_\theta = -\frac{1}{r}\frac{\partial \Phi}{\partial \theta}$$

$$= \frac{1}{r \sin\theta}\frac{\partial \Psi}{\partial r} \qquad (7\text{-}65)$$

7.3.1 Point Sink

Consider a point in space through which a constant density fluid is withdrawn at a uniform volumetric flow rate Q (a positive quantity). Applying the conservation of mass to a spherically symmetric system in which the radius is the only spatial coordinate results in the following expression for the radial velocity:

$$U_r = -\frac{Q}{4\pi r^2} \tag{7-66}$$

Since Q and r are positive quantities, the negative sign indicates that the air travels toward the sink. Using Eqs. 7-64

$$U_r = -\frac{\partial \Phi}{\partial r} = -\frac{\partial \Psi}{\partial \theta} \frac{1}{r^2 \sin \theta} \tag{7-67}$$

The velocity U_θ is zero because of symmetry. Thus

$$U_\theta = -\frac{1}{r}\frac{\partial \Phi}{\partial \theta} = \frac{\partial \Psi}{\partial r}\frac{1}{r \sin \theta} = 0 \tag{7-68}$$

To find the velocity potential Φ integrate Eq. 7-67,

$$\Phi = \int \left(\frac{Q}{4\pi r^2}\right) dr + g(\theta)$$

$$\Phi = -\left(\frac{Q}{4\pi r}\right) + g(\theta) \tag{7-69}$$

but since $\partial \Phi / \partial \theta$ is zero, $g(\theta)$ is equal to a constant which will be taken as zero. In a similar fashion, the stream function Ψ can be found by integration

$$\Psi = \int \frac{Q}{4\pi r^2} r^2 \sin \theta \, d\theta + g(r)$$

$$\Psi = -\frac{Q}{4\pi} \cos \theta + g(r) \tag{7-70}$$

Since $\partial \Psi / \partial r$ is zero, the arbitrary function $g(r)$ is a constant which will again be taken to be zero. In conclusion, the stream function and velocity potential for a point sink are,

$$\Phi = -\frac{Q}{4\pi r} \tag{7-71}$$

$$\Psi = -\left(\frac{Q}{4\pi}\right) \cos \theta \tag{7-72}$$

A scale drawing of lines of constant Φ and Ψ is shown in Fig. 7.7. If the point sink lies in a plane, Eqs. 7-71 and 7-72 can be used, but one must replace the integer 4 by 2 since the volumetric flow rate through the point sink in the plane is one half the value in free space.

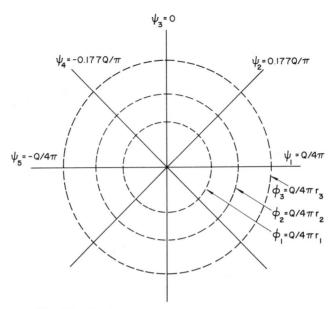

Fig. 7.7 Equipotential surfaces for a point sink.

7.3.2 Uniform Streaming Flow

Consider the flow of an ideal fluid moving uniformly in the negative x direction with a velocity U_0. The velocity potential Φ is

$$\Phi = U_0 x = U_0 r \cos \theta \tag{7-73}$$

To find the stream function Ψ, compute U_r

$$U_r = -\frac{\partial \Phi}{\partial r} = U_0 \cos \theta \tag{7-74}$$

From Eq. 7-64

$$U_r = -\frac{1}{r^2 \sin \theta} \frac{\partial \Psi}{\partial \theta} \tag{7-75}$$

When integrated,

$$\Psi = r^2 U_0 \frac{(\sin \theta)^2}{2} \tag{7-76}$$

7.3.3 Point Sink in a Plane in Streaming Flow

Consider a constant density fluid that passes over a horizontal surface containing a point sink through which fluid is withdrawn. Let U_0 be the speed of the streaming flow in the negative x direction. The potential functions for this flow can be constructed by adding the potential functions of streaming flow and a point sink

$$\Phi = \Phi_f + \Phi_w; \qquad \Psi = \Psi_f + \Psi_w \qquad (7\text{-}77)$$

The stream functions for streaming flow will be denoted by the subscript "f" and those for the opening through which air is withdrawn will be denoted by the subscript "w". Figure 7.8 shows that the x axis lies in the plane through which fluid is withdrawn. Because fluid withdrawn from above the plane never mixes with fluid withdrawn from below the plane, the sink strength used for a point sink in space is twice the volumetric rate of fluid withdrawn from above the plane. Thus the potential flow field associated with streaming flow over a point sink in a plane is

$$\Phi = U_0 r \cos \theta - \left(\frac{Q}{2\pi r}\right) \qquad (7\text{-}78)$$

$$\Psi = \left(\frac{1}{2}\right) U_0 (r \sin \theta)^2 - \left(\frac{Q}{2\pi}\right) \cos \theta \qquad (7\text{-}79)$$

The velocity components U_r and U_θ are

$$U_r = -\frac{\partial \Phi}{\partial r} = -\left[U_0 \cos \theta + \left(\frac{Q}{2\pi r^2}\right)\right] \qquad (7\text{-}80)$$

$$U_\theta = -\frac{1}{r}\frac{\partial \Phi}{\partial \theta} = U_0 \sin \theta \qquad (7\text{-}81)$$

At a point along the axis of rotation ($r = r_0$, $\theta = \pi$) there exists a stagnation point where $U_r = U_\theta = 0$. The location of the *stagnation point* can be found by setting the equations for the velocity equal to zero. The stagnation point is located a distance r_0 to the left of the origin,

$$r_0 = \sqrt{\frac{Q}{2\pi U_0}} \qquad (7\text{-}82)$$

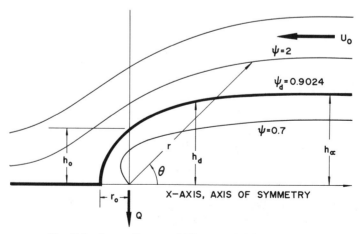

Fig. 7.8 Streamlines and Illustrated Example 7.3.

The value of the stream function that passes through the stagnation point can be found by substituting the coordinates of the stagnation point $(r = r_0, \theta = \pi)$ into Eq. 7-79,

$$\Psi(\text{stagnation point}) = \frac{Q}{2\pi} \tag{7-83}$$

The height of the dividing streamline directly above the sink (h_0) can be found from Eq. 7-79 by setting Ψ equal to $Q/(2\pi)$, $r = h_0$ and $\theta = \pi/2$,

$$\frac{Q}{2\pi} = \frac{U_0}{2}\left(h_0 \sin\frac{\pi}{2}\right)^2 - \left(\frac{Q}{2\pi}\right)\cos\frac{\pi}{2}$$

$$h_0 = \sqrt{\frac{Q}{\pi U_0}} \tag{7-84}$$

Far upwind of the sink $(x >>> 0)$, the distance between the dividing streamline and the x axis (h_∞) can be found by rewriting Eq. 7-79 as

$$\Psi = \frac{U_0 y^2}{2} - \left(\frac{Q}{2\pi}\right)\cos\theta \tag{7-85}$$

setting Ψ equal to $Q/(2\pi)$, and taking the limit as θ approaches zero. Thus

$$\frac{Q}{2\pi} = \frac{U_0 h_\infty^2}{2} - \frac{Q}{2\pi}$$

$$h_\infty = \sqrt{\frac{2Q}{\pi U_0}} \tag{7-86}$$

Figure 7.8 is a sketch of the flow field, showing values of constant Ψ and the distances r_0 and h_0.

The streamline that passes through the *stagnation point* is the *dividing streamline*. The flow below the dividing streamline enters the sink and the flow above the dividing streamline passes over it. The dividing streamline defines a three dimensional region within which gaseous contaminants will be removed by the sink.

Illustrated Example 7.3—Reach of a Flanged Point Sink in Streaming Flow

To illustrate the usefulness of the model consider an inlet in a horizontal surface over which air travels in the negative x direction with a velocity (U_0) of 1 m/s (196.8 FPM). Let the inlet be a small hole point sink through which air is withdrawn at the rate $Q = 200$ CFM (5.67 m³/min). Imagine welding, soldering, or gluing is performed upwind of the sink and your supervisor is interested in estimating the size of the region from which all the vapor, odor, or gas can be captured. Assume that diffusion and buoyancy of the contaminated stream can be neglected.

(a) Sketch the fluid streamlines.
(b) Estimate the size of region in which all the air will be withdrawn.
(c) Show how (b) varies with Q and U_0

The dividing streamline has a stagnation point located a distance r_0 from the origin.

$$r_0 = \sqrt{\frac{Q}{2\pi U_0}} = 0.123 \text{ m} \ (4.84 \text{ in})$$

The numerical value of the stream function that defines the dividing streamline can be found by substituting the equation above in Eq. 7-83,

$$\Psi(\text{dividing}) = \frac{Q}{2\pi} = 0.9024 \text{ m}^3/\text{min}$$

The equation that describes the region bounded by the dividing streamline surface can be found by substituting the values given above of r_0 and $\Psi(\text{dividing})$ into Eq. 7-79 and simplifying.

$$1 = -\cos\theta + \left(\frac{U_0 2\pi}{2Q}\right)(r\sin\theta)^2$$

$$= -\cos\theta + \left(\frac{1}{2}\right)\left(\frac{r}{r_0}\right)^2 \sin^2\theta$$

The height of the dividing streamline directly above the sink (h_0) can be found from the above by setting $h_0 = r$ and $\theta = \pi/2$. Thus,

$$h_0 = 1.414 r_0 = 0.174 \text{ m} \ (6.85 \text{ in})$$

A computer program to calculate the location of the streamlines, particularly the dividing streamline, is shown below. The program uses the Basic Iteration Program shown in the Appendix A-15.

The region bounded by the dividing streamline is the reach. The analysis shows that the x axis is an axis of symmetry, thus with respect to the x axis, the cross-sectional area of the bounded region is a semicircle. For illustrative purposes, consider the cross sectional area directly over the inlet where the radius of the semicircle is (h_0). The cross sectional area is,

$$A(\text{above inlet}) = \frac{\pi(h_0)^2}{2} = \frac{Q}{2U_0} = 0.047 \text{ m}^2$$

Using an analysis such as this, designers can estimate the size and location of the region controlled by the ventilation inlet and within which all gaseous contaminants should be captured. At a free stream velocity of 1 m/s (196.8 FPM) the reach of a 200 CFM region is indeed small. Considering the variability in direction of the streaming flow, it is safe to conclude that if the single hole is to remove gaseous contaminants, the source must be no farther than 0.123 m (4.84 in) downwind from the hole and must be contained in a region no higher than 0.174 m (6.85 in) above the inlet plane. At other ventilation flow rates or air speeds, designers can scale the results given above.

```
1 rem filename-streaming flow over point sink
2 rem this program computes the coordinates of stream-
  lines
3 rem physical conditions:
  q(actual)=5.67 cum/s
  u=1m/s in negative direction
4 print              ''psi        x(m)         y(m)''
5 print        ''_____  _____  _____''
15 u=1
20 q=5.67
25 dim si(4)
30 for j=1 to 4
35 read si(j)
40 next j
45 dim x(12)
50 for i=1 to 12
55 read x(i)
60 next i
65 for j=1 to 4
70 v=si(j)
75 for i=1 to 12
80 x=x(i)
85 yl=0.001
90 yh=100
95 y=(yh+yl)/2
100 z=sqr(x**2+y**2)
105 si=(-q*x/(2*3.14*z))+(u*y**2/2)
110 r=si/v
115 ar=abs(r)
120 if abs(ar-1) <0.001 then goto 140
125 if ar>1.0 then yh=y
130 if ar<1.0 then yl=y
135 goto 95
140 print si, x, y
145 next i
146 print
150 next j
155 data 3, 2, 1.5, 1
160 data 4, 3, 2, 1, 0, -0.25, -0.50, -0.75, -1, -2, -3, -4
165 end
```

7.4 FLOW AROUND BODIES

Consider a point source and sink of equal magnitude lying on the x axis a distance a either side of the origin (Fig. 7.9a). The potential functions at a point $P(r, \theta)$ can be found by the addition of the potential functions for a point sink and point source,

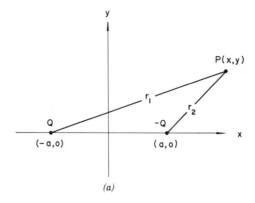

(a)

Fig. 7.9a Point sink and point source lying on the *x axis*.

$$\Phi = \frac{Q}{4\pi}\left[\left(\frac{1}{r_1}\right) - \left(\frac{1}{r_2}\right)\right] \tag{7-87}$$

$$\Psi = \frac{Q}{4\pi}\left[\cos\theta_1 - \cos\theta_2\right] \tag{7-88}$$

Now consider the limit as the distance *a* shrinks to zero but the product $[2aQ/(4\pi)]$ approaches $(4\pi M)$. Details of the mathematics in taking the limit can be found in Streeter (35) and will be omitted. The limiting process produces a *point doublet* and the parameter *M* is called the *doublet strength*. The physical nature of *M* is obscure and can be ignored for the moment. The value of Eqs. 7-87, 7-88 as *a* approaches zero and $[2aQ/(4\pi)]$ approaches $(4\pi M)$ results in the following:

$$\Phi = \left(\frac{M}{r^2}\right)\cos\theta \tag{7-89}$$

$$\Psi = -\left(\frac{M}{r}\right)\sin^2\theta \tag{7-90}$$

Figure 7.9*b* shows the stream functions for a point doublet.

Now consider a point doublet located at the origin but immersed in the flow of an ideal fluid streaming uniformly in the negative *x* direction with a velocity U_0. The potential functions for the composite flow are

$$\Phi = \left(\frac{M}{r^2}\right)\cos\theta + U_0 r\cos\theta \tag{7-91}$$

$$\Psi = -\frac{M}{r}\sin^2\theta + \frac{U_0}{2}(r\sin\theta)^2 \tag{7-92}$$

Examine the streamline $\Psi = 0$. Ψ equals zero under two conditions:

$$\text{Condition a:}\quad r = r,\quad\quad \theta = 0\text{ or }\pi \tag{7-93}$$

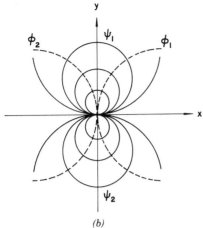

Fig. 7.9b Equipotential surfaces for a point source–sink pair (i.e., point doublet).

(b)

$$\text{Condition } b: \quad \theta = \theta, \quad r = \left(\frac{2M}{U_0}\right)^{0.33} \tag{7-94}$$

Condition a defines the positive and negative x axis. Condition b defines a sphere whose radius is equal to the cube root of $(2M/U_0)$. Physically $\Psi = 0$ is a streamline coincident with the surface of a sphere of radius a where,

$$a = \left(\frac{2M}{U_0}\right)^{0.33} \tag{7-95}$$

that intersects the sphere at its upstream and downstream *stagnation points*. Thus, if one wants to find the streamlines for flow around a sphere of radius a, one can use Eq. 7-92 with M replaced by

$$M = \frac{a^3 U_0}{2} \tag{7-96}$$

Thus for a *stationary sphere* of radius a immersed in streaming flow, the potential functions are

$$\Phi = \left(\frac{U_0 a^3}{2r^2}\right)\cos\theta + U_0 r \cos\theta$$

$$\Phi = \left(\frac{U_0}{r^2}\right)\left(r^3 + \frac{a^3}{2}\right)\cos\theta \tag{7-97}$$

$$\Psi = -\left(\frac{U_0 a^3}{2r}\right)(\sin\theta)^2 + \left(\frac{U_0}{2}\right)(r\sin\theta)^2$$

$$\Psi = U_0\left(\frac{r^3 - a^3}{2r}\right)(\sin\theta)^2 \tag{7-98}$$

Figure 7.10 shows the location of streamlines around a stationary sphere in streaming flow.

The pressure distribution on the surface of a sphere can be found from Bernoulli's equation since the fluid is ideal and Bernoulli's constant is the same for all streamlines. Thus on the sphere's surface,

$$\frac{P_0}{\rho} + \frac{U_0^2}{2} = \frac{P(a,\theta)}{\rho} + \frac{U(a,\theta)^2}{2} \tag{7-99}$$

On the surface of the sphere U_r is zero and the only velocity is U_θ.

$$U_\theta(a,\theta) = U_0\left(\frac{a^3 + 2a^3}{2a^3}\right)\sin\theta$$

$$= \left(\frac{3}{2}\right)U_0\sin\theta \tag{7-100}$$

Define the pressure coefficient C_P as

$$C_P = \frac{[P(a,\theta) - P_0]}{\rho U_0^2/2} \tag{7-101}$$

After substituting,

$$C_P = 1 - \left(\frac{9}{4}\right)(\sin\theta)^2 \tag{7-102}$$

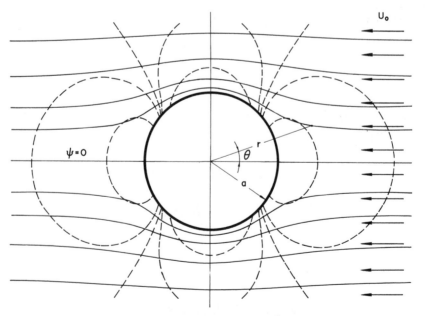

Fig. 7.10 Point doublet in streaming flow.

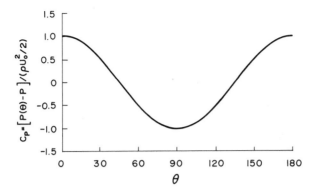

Fig. 7.11 Pressure distribution on the surface of a stationary sphere immersed in a streaming flow.

Figure 7.11 shows that the highest pressure is at the upstream and downstream stagnation points. The minimum pressure occurs at the top and bottom of the sphere where the velocity is the maximum. Since the pressure distribution is symmetrical, there is no net lift or drag on the sphere.

Tables 7.3 and 7.4 list the potential functions for several bodies relevant to industrial ventilation:

TABLE 7.3 Potential Functions for Flow Around Cylinders and Spheres.

Body	Sketch	Potential Functions
Flow Around A Stationary Cylinder		$\psi = U_o \sin\theta \left[r - \frac{a^2}{r} \right]$ $\phi = U_o \cos\theta \left[r + \frac{a^2}{r} \right]$
Flow Around A Stationary Sphere		$\psi = \frac{U_o \sin^2\theta}{2\,r}\left[r^3 - a^3 \right]$ $\phi = \frac{U_o \cos\theta}{r^2}\left[r^3 + \frac{a^3}{2} \right]$
Flow Around A Cylinder Moving Through A Stationary Fluid		$\psi = U_o \left(\frac{a^2}{r} \right) \sin\theta$ $\phi = -U_o \left(\frac{a^2}{r} \right) \cos\theta$
Flow Around A Sphere Moving Through A Stationary Fluid		$\psi = \frac{U_o a^3 \sin^2\theta}{2\,r}$ $\phi = -\frac{U_o a^3 \cos\theta}{2\,r^2}$

TABLE 7.4 Potential Functions for Flow in Corners. Distance a is the Location on the Wall Where the Velocity is the Known Value U_w

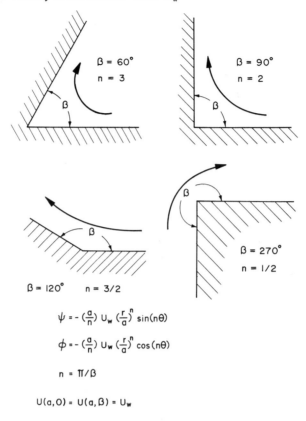

$$\psi = -\left(\frac{a}{n}\right) U_w \left(\frac{r}{a}\right)^n \sin(n\theta)$$

$$\phi = -\left(\frac{a}{n}\right) U_w \left(\frac{r}{a}\right)^n \cos(n\theta)$$

$$n = \pi/\beta$$

$$U(a,0) = U(a,\beta) = U_w$$

Stationary right circular cylinder in streaming flow
Moving right circular cylinder in a stationary fluid
Stationary sphere in streaming flow
Moving sphere in a stationary fluid
Flow into corners

Analysis of the flow around a stationary right circular cylinder is similar to the analysis for a sphere replacing a point-doublet with a line-doublet immersed in a streaming flow. Development of this and other potential functions can be found in refs. 34, 35, and 62.

7.5 FLANGED AND UNFLANGED RECTANGULAR SLOTS IN QUIESCENT AIR

7.5.1 Flanged Slot

Consider ideal flow into a *rectangular slot* of width $2w$ and length L, lying in a horizontal plane shown in Fig. 7.12. Expressions for the stream function and

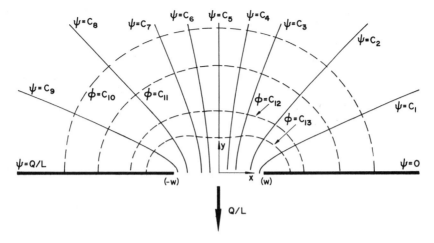

Fig. 7.12 Equipotential surfaces for flow through a rectangular slot of width $2w$.

velocity potential have been derived by Streeter (35), Milne–Thomson (34), and Kirchhoff (62),

$$\frac{x}{w} = \cosh\left(\frac{\Phi}{k}\right)\cos\left(\frac{\Psi}{k}\right) \tag{7-103}$$

$$\frac{y}{w} = \sinh\left(\frac{\Phi}{k}\right)\sin\left(\frac{\Psi}{k}\right) \tag{7-104}$$

$$k = \frac{Q}{L\pi} \tag{7-105}$$

These expressions can be rearranged to eliminate Ψ. The velocity potential satisfies

$$1 = \left(\frac{\dfrac{x}{w}}{\cosh\dfrac{\Phi}{k}}\right)^2 + \left(\frac{\dfrac{y}{w}}{\sinh\dfrac{\Phi}{k}}\right)^2 \tag{7-106}$$

Similarly, Φ can be eliminated and the stream function shown to satisfy

$$1 = \left(\frac{\dfrac{x}{w}}{\cos\dfrac{\Psi}{k}}\right)^2 - \left(\frac{\dfrac{y}{w}}{\sin\dfrac{\Psi}{k}}\right)^2 \tag{7-107}$$

Equipotential surfaces in which values of Φ are constant, are ellipses with foci at $(w, 0)$ and $(-w, 0)$. Equipotential surfaces in which values of Ψ are constant are hyperbola which have the same foci. Figure 7.12 shows these curves and certain limiting values of Ψ and Φ. The lines $\Psi = 0$, $Q/2L$, and Q/L correspond to the positive x axis, positive y axis and negative x axis respectively. The curve $\Phi = 0$

corresponds to the x axis between $x = w$ and $x = -w$. Curves of $\Phi > 0$ correspond to ellipses above it. Equations 7-103 and 7-104 can also be used to describe the flow through a slot in a surface that meets another surface at 90 degrees (i.e., baffles). Figure 7.13 illustrates configurations that can be analyzed.

The velocity components at any point (x, y) upstream of the inlet can be obtained by computing $\partial \Psi / \partial x$ and $\partial \Phi / \partial x$. Unfortunately, obtaining these derivatives from eqs. 7-106 and 7-107 is complicated. Alternatively, these derivatives can be found by *implicit differentiation*.

Equations 7-103 and 7-104 express

$$x = F(\Phi, \Psi) \tag{7-108}$$

$$y = G(\Phi, \Psi) \tag{7-109}$$

What is desired is

$$\Phi = f(x, y); \qquad \Psi = g(x, y) \tag{7-110}$$

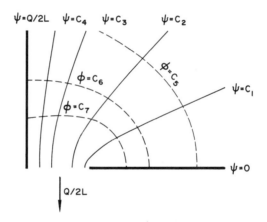

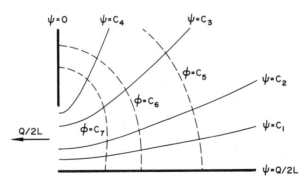

Fig. 7.13 Equipotential surfaces for flow through a rectangular slot in a 90-degree corner.

Differentiate Eqs. 7-108 and 7-109 with respect to x and remember that x and y are independent variables. One obtains

$$1 = \left(\frac{\partial F}{\partial \Phi}\right)\left(\frac{\partial \Phi}{\partial x}\right) + \left(\frac{\partial F}{\partial \Psi}\right)\left(\frac{\partial \Psi}{\partial x}\right) \tag{7-111}$$

$$0 = \left(\frac{\partial G}{\partial \Phi}\right)\left(\frac{\partial \Phi}{\partial x}\right) + \left(\frac{\partial G}{\partial \Psi}\right)\left(\frac{\partial \Psi}{\partial x}\right) \tag{7-112}$$

After evaluating the differentials, one obtains

$$1 = \frac{w}{k}\left[\sinh\left(\frac{\Phi}{k}\right)\cos\left(\frac{\Psi}{k}\right)\left(\frac{\partial \Phi}{\partial x}\right) - \cosh\left(\frac{\Phi}{k}\right)\sin\left(\frac{\Psi}{k}\right)\frac{\partial \Psi}{\partial x}\right] \tag{7-113}$$

$$0 = \left(\frac{w}{k}\right)\left[\cosh\left(\frac{\Phi}{k}\right)\sin\left(\frac{\Psi}{k}\right)\frac{\partial \Phi}{\partial x} + \sinh\left(\frac{\Phi}{k}\right)\cos\left(\frac{\Psi}{k}\right)\frac{\partial \Psi}{\partial x}\right] \tag{7-114}$$

This pair of simultaneous equations can be solved for $\partial\Phi/\partial x$ and $\partial\Psi/\partial x$ providing the *Jacobian* $J(\Phi, \Psi)$

$$J(\Phi, \Psi) = \begin{vmatrix} \dfrac{\partial F}{\partial \Phi} & \dfrac{\partial F}{\partial \Psi} \\ \dfrac{\partial G}{\partial \Phi} & \dfrac{\partial G}{\partial \Phi} \end{vmatrix} \tag{7-115}$$

is not zero. Solving Eqs. 7-113 and 7-114 one finds that

$$U_y = \frac{\partial \Psi}{\partial x}$$

$$= -\frac{k}{wA}\left[\cosh\left(\frac{\Phi}{k}\right)\sin\left(\frac{\Psi}{k}\right)\right] \tag{7-116}$$

$$U_x = -\frac{\partial \Phi}{\partial x}$$

$$= -\frac{k}{Aw}\left[\sinh\left(\frac{\Phi}{k}\right)\cos\left(\frac{\Psi}{k}\right)\right] \tag{7-117}$$

where

$$A = \left[\sinh\left(\frac{\Phi}{k}\right)\cos\left(\frac{\Psi}{k}\right)\right]^2 + \left[\cosh\left(\frac{\Phi}{k}\right)\sin\left(\frac{\Psi}{k}\right)\right]^2 \tag{7-118}$$

$$k = \frac{Q}{L\pi} \tag{7-119}$$

A check of the limits for the equations for U_x and U_y shows that along the flange surface $y = 0$, $x > w$ where Ψ is zero and along the left flange surface where $y = 0$, $x < -w$ and Ψ is equal to Q/L, that $U_y = 0$ and U_x increases as the flow approaches the slot. Along the y axis where $x = 0$, $y > 0$ and Ψ is equal to $Q/2L$,

U_x is zero and U_y increases as the flow approaches the slot. Across the inlet ($y = 0$, $-w < x < w$) U_x is zero and $U_y = -(k/w)/\sin(\Psi/k)$. Thus while one may think in terms of an average face velocity equal to $Q/(2Lw)$, the actual velocity is not constant and approaches infinity at the slot lip ($y = 0$, $x = \pm w$) which are singular points.

Illustrated Example 7.4 Map Values of Constant Φ and Ψ for a Flanged Slot in Quiescent Air

Map the flow field associated with a slot of width $2c$ in an infinite plane. Write a computer program to generate data giving the (x, y) coordinates of lines of constant Ψ and Φ Print the data in files with the following format:

$$\Psi = \text{SELECTED CONSTANT } (K) \qquad \Phi = \text{SELECTED CONSTANT } (C)$$

Φ = C	x	y		Ψ = K	x	y
C1	—	—		K1	—	—
C2	—	—		K2	—	—
.	.	.		.	.	.
.	.	.		.	.	.

To prepare such files simply turn the problem inside out and repeatedly compute x and y for selected values of Φ and Ψ using Eqs. 7-103 and 7-104 and print the data in the format above. The following program will accomplish this task.

```
1 rem intersections
2 rem this program computes locations of velocity
  potentials and stream functions for flow through a
  slot in an infinite plane
3 rem physical conditions: Q/L(actual) 3.14159 sq m/ s
  half slot width c(actual)=1 m
5 print      '' phi      psi      x(m)      y(m)''
6 print      ''____     ____     ____     ____''
9 e=2.71828
10 dim si(5)
15 read m
20 for j=1 to m
25 read si(j)
30 next j
35 dim fi(6)
40 read n
45 for i=1 to n
50 read fi(i)
55 next i
60 for i=1 to n
70 w=fi(i)
72 shw=(e**w-e**(-w))/2
```

```
 74 chw=(e**w+e**(-w))/2
 80 for j=1 to m
 85 v=si(j)
 90 x=chw*cos(v)
100 y=shw*sin(v)
110 print w,v,x,y
120 next j
140 next i
141 print·
143 print          '' psi      phi       x        y ''
145 print          ''____     ____      __      ____'',
150 for j=1 to m
160 v=si(j)
170 for i=1 to n
180 w=fi(i)
185 shw=(e**w-e**(-w))/2
187 chw=(e**w+e**(-w))/2
190 x=chw*cos(v)
195 y=shw*sin(v)
200 print v,w,x,y
210 next i
215 next j
220 data 5
230 data 0.142, 0.785, 1.571, 2.356, 3.0
240 data 6
250 data 0.2, 0.5, 1.0, 1.5, 2.0, 3.0
260 end
```

Illustrated Example 7.5 Compute Air Speed Along a Streamline for a Flanged Slot and Line Sink in Quiescent Air

Along arbitrarily selected streamlines, draw a graph that shows how the air speed varies with the distance from the origin, (i.e., the radius). Show also how the radial velocity varies with radius if the slot ($2c$) is replaced by an infinite line sink of the same volumetric flow rate.

Computing the radial velocity for a line sink is a simple matter and uses only Eq. 7-39. The absolute value of the velocity along selected streamlines is a much more difficult task since there are no equations for Φ and Ψ as explicit functions of x and y. Equations 7-108 to 7-110 can be used to compute the two velocity components. The key to the finite difference solution is to follow a particular streamline Ψ_i (see Fig. 7.14), stopping at selected values of the velocity potential Φ_i to compute the velocity components (U_x, U_y) at coordinates (x_i and y_i) in finite form,

$$U_x = \frac{(\Phi_{i+1} - \Phi_{i-1})}{(x_{i+1} - x_{i-1})}$$

$$U_y = \frac{(\Phi_{i+1} - \Phi_{i-1})}{(y_{i+1} - y_{i-1})}$$

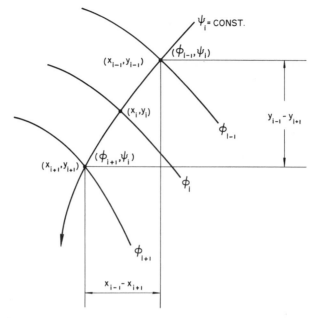

Fig. 7.14 Node description for finite difference calculations in Illustrated Example 7.5.

where Φ_{i+1} and Φ_{i-1} refer to values of Φ on either side of x_i and y_i along the streamline Ψ_i. The previous files will be needed but expanded as follows,

$$\Psi = \text{SELECTED CONSTANT } (K)$$

$\Phi = C$	x	y
C1	—	—	—
C1	—	—	—
C1	—	—	—
.	.	.	.

Once formed, U_x and U_y can be calculated using the above finite difference equations and absolute value of the velocity and radius (r) calculated at each location. At each location the finite differences must be evaluated along the streamline. Thus it is necessary to find the finite differences along the streamline keeping in mind that

$$\Psi_{i+1} = \Psi_i = \Psi_{i-1}$$

The program assumes

$$\Phi_{i+1} = 0.995\Phi_i$$

$$\Phi_{i-1} = 1.005\Phi_i$$

but other values could be used as well.

Figure 7.15 shows the results of the calculations. At radial distances approxi-

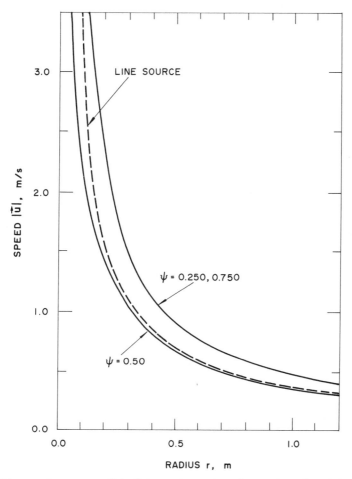

Fig. 7.15 Air speed versus radial distance to center of a rectangular slot—Illustrated Example 7.5.

mately 10 times the half slot width, the air speeds resemble each other. Closer to the slot the air speed reflects the finite width of the slot. Close to the slot, different streamlines produce different air speeds at the same radial distance from the origin. This is to be expected, remembering that the elliptical surfaces of constant Φ imply constant air speed. It must also be kept in mind that the direction of the velocity is not strictly radial since the streamlines cross the slot-plane at different distances from the origin between $-c < x < c$.

```
1 rem filename-air speed vs radius to slot center
2 rem program computes air speed along streamline, radial
    distance to slot center and line sink of
    equal flow rate
3 rem physical conditions:
  Q/L(act)=1 sq m/ s c(act)=0.1 m
4 rem variables:
  fi=phi si=psi
  v=si*pi/(Q/L) w=fi*pi/(Q/L)
```

```
 5 print  '' psi  u (slot) m/s  u (line) m/s  radius (m)''
 6 print  ''___  _____  _____  _____''
 7 Pi=3.14159
 8 e=2.71828
 9 ql=1
10 c=0.1
12 dim si(5)
15 read m
20 for j=1 to m
25 read si(j)
30 next j
35 dim fi(6)
40 read n
45 for i=1 to n
50 read fi(i)
55 next i
60 for j=1 to m
70 v=si(j)*Pi
80 for i=1 to n
90 w=fi(i)*Pi
95 wh=1.005*w
96 sh=(e**wh-e**(-wh))/2
98 ch=(e**wh+e**(-wh))/2
100 wl=0.995*w
105 sl=(e**wl-e**(-wl))/2
108 cl=(e**wl+e**(-wl))/2
110 sw=(e**w-e**(-w))/2
112 cw=(e**w+e**(-w))/2
120 xh=c*ch*cos(v)
125 xl=c*cl*cos(v)
130 yh=c*sh*sin(v)
135 yl=c*sl*sin(v)
145 uy=-(wh-wl)/(yh-yl)
146 ux=-(wh-wl)/(xh-xl)
150 us=sqr (wx**2+uy**2)
160 x=c*cw*cos(v)
170 y=c*sw*sin(v)
180 r=sqr (x**2+y**2)
190 ul=ql/3.14159*r
200 print v, us, ul, r
210 next i
220 next j
230 data 5
240 data 0.045, 0.025, 0.500, 0.750, 0.955
250 data 6
260 data 0.064, 0.159, 0.318, 0.477, 0.637, 0.955
270 end
```

An alternative approach to find the velocity potential function for a flanged slot is to assume that the flanged slot consists of an infinite number of line sinks, each of which has a strength of $Q/L2w$, see Fig. 7.16. Thus upwind of the inlet at a point P whose coordinates are (x_p, y_p) the differential of the velocity potential can be expressed as

$$d\Phi = \left(\frac{Q}{2\pi Lw}\right)\ln r \, dx \qquad (7\text{-}121)$$

where r is the radial distance from the differential line sink to point P.

$$r^2 = [(x_p - x)^2 + y_p^2] \qquad (7\text{-}122)$$

It should be noted that to compute the velocity potential at point P the user will have to intergrate Eq. 7-121 over the width of the slot $(-w$ to $-+w)$.

$$\Phi = \left(\frac{Q}{4\pi Lw}\right)\int_{-w}^{w} \ln\left[(x_p - x)^2 + y_p^2\right] dx \qquad (7\text{-}123)$$

Once integrated, the expression can be differentiated with respect to x_p and y_p to find the velocity components at point P.

This approach also lends itself to numerical computation in which the flanged slot is replaced by N line sinks spaced a distance $2w/(N+1)$ apart across the distance $(-w$ to $+w)$. Each line sink has a source strength Q/LN. The velocity potential at point P is the sum over N of,

$$\Phi = \sum_{i=1}^{N}\left(\frac{Q}{2\pi LN}\right)\ln r_i = \left(\frac{Q}{2\pi LN}\right)\sum_{i=1}^{N}\ln r_i \qquad (7\text{-}124)$$

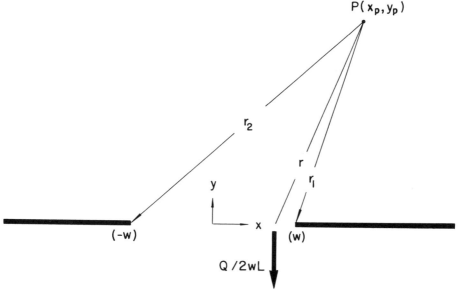

Fig. 7.16 Modeling a Flanged slot as an infinite number of line sinks.

where r_i is the radial distance from individual line sink to point P. To perform the sum, the unique value of r_i has to be used. With this method tabulated values of the velocity potential can be computed for an array of points upwind points P. The velocity components at any point P can be obtained by computing the derivative of Φ by numericaly differentiating Φ with respect to x and y.

It is expected that at distances far upwind of the inlet ($y \gg w$), that simulating a flanged slot by large number of line sinks will produce velocities reasonably close to those predicted by Eqs. 7-116 and 7-117. Along a plane close to the inlet itself, differences are expected. Equation 7-116 predicts infinite negative velocities at the slot lip and a mid-plane velocity less than the slot face velocity ($Q/2wL$) whereas the simulation is expected to yield velocities U_y closer to the actual face velocity.

7.5.2 Unflanged Slot

Consider the flow of an ideal fluid into an unflanged rectangular inlet of width $2w$, as shown in Fig. 7.17. The velocity potential and stream function satisfy the following equations:

$$\frac{x}{w} = \frac{1}{\pi}\left[\frac{2\Phi}{k} + \exp\left(\frac{2\Phi}{k}\right)\cos\frac{2\Psi}{k}\right] \tag{7-125}$$

$$\frac{y}{w} = \frac{1}{\pi}\left[\frac{2\Psi}{k} + \exp\left(\frac{2\Psi}{k}\right)\sin\frac{2\Psi}{k}\right] \tag{7-126}$$

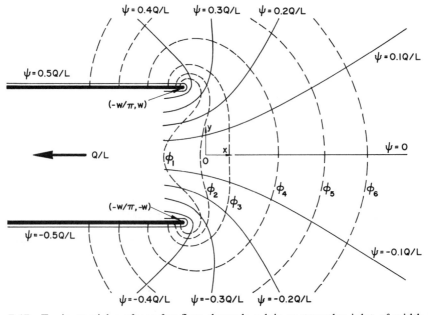

Fig. 7.17 Equipotential surfaces for flow through a lain rectangular inlet of width $2w$.

$$k = \frac{Q}{\pi L} \qquad (7\text{-}127)$$

Equipotential surfaces are shown in Fig. 7.16.

To use these equations wisely it is important to know the location of certain key stream functions. To begin, note that the origin is a point a distance w/π upwind of the inlet. Examine the positive x axis ($y = 0$, $x = x$ for $x > 0$). For Eq. 7-126 to be zero for arbitrary values of Φ, it will be necessary that Ψ be equal to zero. Thus the positive x axis corresponds to Ψ equal to zero. Now find the location where Ψ is equal to plus and minus $Q/2L$ and Φ is arbitrary. Substitution of these values into Eqs. 7-125 and 7-127 shows that $\Psi = Q/2L$ corresponds to the upper surface of the inlet while $\Psi = -Q/2L$ corresponds to the lower surface. Values of constant Φ are orthogonal to Ψ and increase as one moves in the direction of increasing x.

The velocity components U_x and U_y can be found from Eqs. 7-125 and 7-126 by the implicit differentiation process described by Eqs. 7-108 to 7-112. The results are,

$$U_x = -\frac{k\pi}{2wA} \left[1 + \exp\left(\frac{2\Phi}{k}\right) \cos\left(\frac{2\Psi}{k}\right) \right] \qquad (7\text{-}128)$$

$$U_y = -\frac{k\pi}{2wA} \left[\exp\left(\frac{2\Phi}{k}\right) \sin\left(\frac{2\Psi}{k}\right) \right] \qquad (7\text{-}129)$$

where

$$A = \left[\exp\left(\frac{2\Phi}{k}\right) \sin\left(\frac{2\Psi}{k}\right) \right]^2 + \left[1 + \exp\left(\frac{2\Phi}{k}\right) \cos\left(\frac{2\Psi}{k}\right) \right]^2 \qquad (7\text{-}130)$$

$$k = \frac{Q}{\pi L} \qquad (7\text{-}131)$$

A check of the limits on the equations for U_x and U_y shows that along the positive x axis, $x > -(w/\pi)$, $y = 0$ where Ψ is equal to zero, that U_y is zero, U_x is negative and that the magnitude of U_x increases as the flow approaches the inlet. Along the upper surface of the inlet where $x < -(w/\pi)$, $y = w$ where Ψ is equal to $Q/2L$, U_y is zero, and U_x is positive and increases as the flow approaches the inlet. Since the upper and lower lips on the inlet are singular points, the velocity approaches infinity. Along a streamline where Ψ is equal to $Q/4L$ (quadrant where both x and y are positive), both U_x and U_y are negative since the flow enters the slot from the right. The magnitude of U_x and U_y increases as the flow approaches the inlet.

7.6 FLANGED AND UNFLANGED INLETS IN STREAMING FLOW

7.6.1 Flanged Slot in Streaming Flow

Consider the withdrawal of air through an infinitely long rectangular slot of width $2w$ lying in a plane over which air passes from right to left. The volumetric flow rate per length of slot is Q/L. Air approaches the slot in the negative x direction with a uniform speed of U_0 as shown in Fig. 7.18. The potential functions for the

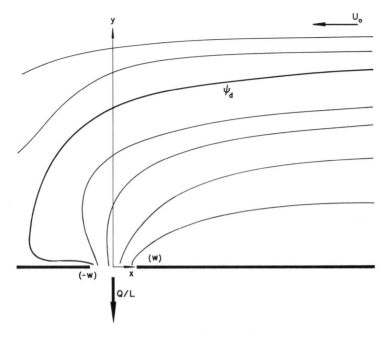

Fig. 7.18 Streamlines for flanged rectangular slot in streaming flow.

flow field are equal to the sum of the potential functions for streaming flow and flow through a flanged slot,

$$\Phi = \Phi_f + \Phi_w \tag{7-132}$$

$$\Psi = \Psi_f + \Psi_w \tag{7-133}$$

where the subscript "f" denotes streaming flow and the subscript "w" denotes the slot through which the air is withdrawn. For streaming flow the potential functions are given by Eqs. 7-50 and 7-52, but unfortunately Φ_w and Ψ_w for the flanged slot cannot be written as explicit functions of x and y. Thus procedures used earlier cannot be used in their entirety. Alternatively, the velocities U_x and U_y can be computed directly:

$$U_x = -\frac{\partial \Phi}{\partial x}$$

$$= -\frac{\partial \Phi_f}{\partial x} - \frac{\partial \Phi_w}{\partial x} \tag{7-134}$$

$$U_x = -U_0 - \left(\frac{k}{wA}\right)\left[\sinh\left(\frac{\Phi_w}{k}\right)\cos\left(\frac{\Psi_w}{k}\right)\right] \tag{7-135}$$

$$A = \left[\sinh\left(\frac{\Phi_w}{k}\right)\cos\left(\frac{\Psi_w}{k}\right)\right]^2 + \left[\cosh\left(\frac{\Phi_w}{k}\right)\sin\left(\frac{\Psi_w}{k}\right)\right]^2 \tag{7-136}$$

$$k = \frac{Q}{\pi L} \qquad (7\text{-}137)$$

Similarly, the velocity U_y can be shown to be

$$U_y = -\left(\frac{k}{wA}\right)\left[\cosh\left(\frac{\Phi_w}{k}\right)\sin\left(\frac{\Psi_w}{k}\right)\right] \qquad (7\text{-}138)$$

where A is given by Eq. 7-136

While k is equal to $(Q/L\pi)$ for the flanged slot in quiescent air, it is necessary to check that it remains the same for streaming flow over a flanged slot. Along the positive x axis for $x > w$, it can be seen that $(\Psi_w + \Psi_f)$ is zero. At the left lip of the inlet $(y = 0, x = -w)$, the dividing streamline intersects the lip and has the value Q/L. Thus at this point $(\Psi_w + \Psi_f)$ is also Q/L. Applying this condition to Eq. 7-133, it can be seen that $k = Q/(\pi L)$.

Mapping the location of the streamlines is not difficult for a flanged inlet in streaming flow. The process begins by selecting the value of the stream function one wishes to map. The values vary from zero, which is coincident with the x axis to Q/L which is the dividing streamline. Since the streamline enters the inlet the algorithm begins at a small value of y near the inlet and marches upwind computing the coordinates (x, y) of the streamline at points progressively upwind of the inlet. The following algorithm can be used.

(1) With Ψ known, select a value of y_i just above the inlet, for example, $0.01w$.
(2) Compute $\Psi_f = U_0 y_i$.
(3) Compute $\Psi_w = \Psi - U_0 y_i$.
(4) Using Eq. 7-107 solve for x_i

$$x_i = w \cos\left(\frac{L\pi\Psi_w}{Q}\right)\left[1 + \left(\frac{\dfrac{y}{w}}{\sin\dfrac{L\pi\Psi_w}{Q}}\right)^2\right]^{1/2} \qquad (7\text{-}139)$$

(5) Repeat steps 1 to 4 for the next larger y_i.

Using an infinite source of smoke traveling over flanged square inlets, Fletcher and Johnson (125) found that all *capture envelopes* could be reduced to a single curve when the variables are grouped in a dimensionless form. The dimensionless variables reduce to $y\sqrt{U_0/Q}$ and $x\sqrt{U_0/Q}$.

Flynn and Miller (371) suggest an alternative approach to predict three dimensional flow fields for plane and flanged rectangular inlets called the *boundary integral equation method*. They report that the results are in excellent agreement with accepted values.

The principal value of modeling inlets with potential functions is to enable users to predict the velocity U_x and U_y at arbitrary coordinates (x, y) upwind of the inlet. Equations 7-134 to 7-138 give the velocity components as functions of Φ_w and Ψ_w, not x and y as one would like. Phi and Ψ can be found at arbitrary

(x, y) using the *Newton–Raphson* method and the following algorithm. Readers should consult Appendix A-13 for details about the Newton–Raphson method.

(1) Define the functions F and G,

$$F = \left(\frac{x}{w}\right) - \cosh\left(\frac{\Phi_w}{k}\right)\cos\left(\frac{\Psi_w}{k}\right) \tag{7-140}$$

$$G = \left(\frac{y}{w}\right) - \sinh\left(\frac{\Phi_w}{k}\right)\sin\left(\frac{\Psi_w}{k}\right) \tag{7-141}$$

(2) Differentiate and form

$$\frac{\partial G}{\partial \Phi_w} = -\left(\frac{1}{k}\right)\cosh\left(\frac{\Phi_w}{k}\right)\sin\left(\frac{\Psi_w}{k}\right) \tag{7-142}$$

$$\frac{\partial G}{\partial \Psi_w} = -\left(\frac{1}{k}\right)\sinh\left(\frac{\Phi_w}{k}\right)\cos\left(\frac{\Psi_w}{k}\right) \tag{7-143}$$

$$\frac{\partial F}{\partial \Phi_w} = -\left(\frac{1}{k}\right)\sinh\left(\frac{\Phi_w}{k}\right)\cos\left(\frac{\Psi_w}{k}\right) \tag{7-144}$$

$$\frac{\partial F}{\partial \Psi_w} = \left(\frac{1}{k}\right)\cosh\left(\frac{\Phi_w}{k}\right)\sin\left(\frac{\Psi_w}{k}\right) \tag{7-145}$$

(3) Form the Taylor series

$$F_{i+1} - F_i = \left[\frac{\partial F}{\partial \Phi_w}\right][\Phi_{w,i+1} - \Phi_{w,i}] + \left[\frac{\partial F}{\partial \Psi_w}\right][\Psi_{w,i+1} - \Psi_{w,i}] \tag{7-146}$$

$$G_{i+1} - G_i = \left[\frac{\partial G}{\partial \Phi_w}\right][\Phi_{w,i+1} - \Phi_{w,i}] + \left[\frac{\partial F}{\partial \Psi_w}\right][\Psi_{w,i+1} - \Psi_{w,i}] \tag{7-147}$$

(4) Assuming F_{i+1} and G_{i+1} are the correct values, solve the above simultaneous equations and obtain the following,

$$(\Phi_{w,i+1} - \Phi_{w,i}) = \frac{\left[G_i\dfrac{\partial F}{\partial \Psi_w} - F_i\dfrac{\partial G}{\partial \Psi_w}\right]}{J} \tag{7-148}$$

$$(\Psi_{w,i+1} - \Psi_{w,i}) = \frac{\left[F_i\dfrac{\partial G}{\partial \Phi_w} - G_i\dfrac{\partial F}{\partial \Phi_w}\right]}{J} \tag{7-149}$$

$$J = \left(\frac{\partial F}{\partial \Phi_w}\right)\left(\frac{\partial G}{\partial \Psi_w}\right) - \left(\frac{\partial F}{\partial \Psi_w}\right)\left(\frac{\partial G}{\partial \Phi_w}\right) \tag{7-150}$$

(5) Select the values of (x, y) at which the velocity is wanted. Note these value do not change in the iterations that follow.

(6) Guess initial values of Φ_i and Ψ_i and compute the values of

$$F, G, \quad \frac{\partial F}{\partial \Phi_w}, \quad \frac{\partial F}{\partial \Psi_w}, \quad \frac{\partial G}{\partial \Phi_w}, \quad \frac{\partial G}{\partial \Psi_w}.$$

Now compute $(\Phi_{w,i+1} - \Phi_{w,i})$ and $(\Psi_{w,i+1} - \Psi_{w,i})$ using the equations in step (4) above.

(7) Repeat step (6) until the difference between successive values of Φ and Ψ are as small as one likes. The resulting values of Φ_w and Ψ_w should be interpreted as $\Psi_w(x, y)$ and $\Phi_w(x, y)$. With these values compute the velocity components U_x and U_y using Eqs. 7-134 to 7-138.

7.6.2 Flanged Circular Inlet

The velocity field produced by a *circular inlet* of radius w in a *flanged surface* in a quiescent air has been studied by Flynn, Ellenbecker et al. (121, 220, 356) who report that the potential functions are

$$\Phi_w = \left[\frac{Q}{2\pi w} \right] \arcsin\left[\frac{2w}{a_1 + a_2} \right] \tag{7-151}$$

$$\Psi_w = \left[\frac{Q}{4\pi w} \right] \sqrt{4w^2 - (a_1 - a_2)^2} \tag{7-152}$$

where the lengths a_1 and a_2 lie in the r-z plane and are equal to

$$a_1 = \sqrt{z^2 + (w + r)^2} \tag{7-153}$$

$$a_2 = \sqrt{z^2 + (w - r)^2} \tag{7-154}$$

and r and z are the radial and axial distances measured from the origin which lays in the center of the inlet plane, Fig. 7.19 The radial and axial velocity at any point upstream of the inlet are given by

$$U_z = -\frac{\partial \Phi}{\partial z} = -\frac{zQ}{(\pi a_1 a_2 a_3)} \tag{7-155}$$

$$U_r = -\frac{\partial \Phi}{\partial r} \tag{7-156}$$

$$U_r = -\left(\frac{Q}{\pi}\right)\left[\frac{(r + w)a_2 + (r - w)a_1}{(a_1 + a_2)a_1 a_2 a_3} \right]$$

$$a_3 = \sqrt{(a_1 + a_2)^2 - 4w^2} \tag{7-157}$$

Values of constant Φ and Ψ are similar to their counterparts for the rectangular flanged slot. Singular points again exist at the edge of the inlet and the velocity becomes infinite. The region influenced by the singularity is small and can be neglected. Thus for the majority of points upstream of the circular inlet Eqs. 7-155 to 7-157 can be used to predict the radial and axial velocity components.

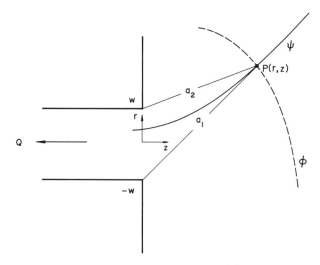

Fig. 7.19   Flanged circular inlet.

If the surrounding air passes over the inlet plane at a uniform velocity U_0, the velocity potential for streaming flow can be added to Eqs. 7-155 and 7-157 to obtain the composite velocity potential. The velocity at an arbitrary point upwind of the inlet can be found by an analysis similar to that described in Eqs. 7-140 to 7-150. Alternatively, the velocity can be found by adding the vector U_0 to U_r and U_z above. The collection efficiency of a flanged circular opening to capture gaseous contaminants released from a point source in a streaming flow was studied analytically and experimentally by Flynn and Ellenbecker (189).

7.6.3 Flanged Elliptical Inlet

The ideal flow field associated with an elliptical opening in a flanged surface (Fig. 7.20) was studied by Conroy et al. (296) who report that the velocity potential is

$$\Phi = \pm \left(\frac{Q}{4\pi}\right) \int_0^m \frac{dm}{\sqrt{m(a^2 + m)(b^2 + m)}} \tag{7-158}$$

where a and b are one half the major an minor diameters of the ellipse, Q is the volumetric flow rate and m is the positive root of the equation

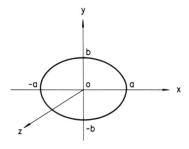

Fig. 7.20   Elliptical opening in a flanged surface lying in the x-y plane.

$$\frac{x^2}{(a^2 + m)} + \frac{y^2}{(b^2 + m)} + \frac{z^2}{m} = 1 \qquad (7\text{-}159)$$

Following considerable mathematical manipulation, the authors report that the velocity components are

$$U_x = \frac{Qx(a^2 + m)^{0.5}(b^2 + m)^{1.5}m^{1.5}}{2\pi a_4} \qquad (7\text{-}160)$$

$$U_y = \frac{Qy(a^2 + m)^{1.5}(b^2 + m)^{0.5}m^{1.5}]}{2\pi a_4} \qquad (7\text{-}161)$$

$$U_z = \frac{Qz(a^2 + m)^{1.5}(b^2 + m)^{1.5}m^{0.5}}{2\pi a_4} \qquad (7\text{-}162)$$

where

$$a_4 = (xm)^2(b^2 + m)^2 + (ym)^2(a^2 + m)^2 + z^2(a^2 + m)^2(b^2 + m)^2 \quad (7\text{-}163)$$

A useful extension of the above model lies in the ability to simulate a *flanged rectangular inlet* of various aspect ratios (length/width) by approximating the opening by the inscribed ellipse of major and minor diameters $2a$ and $2b$, that is to say a rectangular inlet of length $2a$ and width $2b$. Conroy et al. (296) measured the velocities in front of such a rectangular inlet and found that the above expressions yielded reasonably accurate results so long as one does not get close to regions near the edges of the inlet. Tim Suden, et al. (397) combined the equations for a flanged eliptical inlet with the equations from Section 6.5 for buoyant plumes to simulate the capture of arc welding fume.

There are many industrial applications in which an unflanged pipe or rectangular duct are used to capture contaminants (welding fume, local sources of obnoxious odors, etc.). On a smaller scale, the same physical principles are involved when a sample of a dust laden gas stream is withdrawn through a probe and sampling nozzle. In the first instance the engineer wants to know the reach of the inlet, especially when room air currents are present. In the second instance, one must be sure that the gas sample represents the aerosol in the gas stream from which the sample is taken, that is that the dust concentration and size distribution are the same as the sampled gas. The velocity field of the gas sample entering the inlet can be modeled by the potential functions discussed in this chapter.

7.6.4 Unflanged Slot in Streaming Flow

Figure 7.21 is a schematic diagram of an *unflanged or plain rectangular opening* of width $2w$ immersed in a gas stream of uniform velocity U_0 inclined at an angle α to the duct walls. Gas is withdrawn through the duct at a uniform volumetric flow rate Q/L per unit length of slot The average inlet velocity at the inlet plane is (U_n) and is equal to $(Q/2Lw)$. The inlet velocity (U_n) will hereafter be called the face velocity or nozzle velocity.

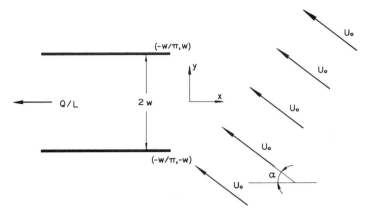

Fig. 7.21   Plain rectangular inlet in an oblique streaming flow.

The coordinate system and orientation of the streaming flow is consistent with earlier sections. The overall potential functions for the flow field are

$$\Phi = \Phi_w + \Phi_f \qquad (7\text{-}164)$$

$$\Psi = \Psi_w + \Psi_f \qquad (7\text{-}165)$$

Table 7.2 lists the analytical expressions for these functions. Since the potential functions (Φ_w and Ψ_w) for the unflanged rectangular inlet cannot be written as explicit functions of x and y, the procedures similar to those used to derive Eqs. 7-135 and 7-138 will be used. The velocity components are

$$U_x = -U_0 \cos \alpha \qquad (7\text{-}166)$$

$$-\left(\frac{k\pi}{2wA}\right)\left[1 + \exp\left(\frac{2\Phi_w}{k}\right)\cos\left(\frac{2\Psi_w}{k}\right)\right]$$

$$U_y = U_0 \sin \alpha - \left(\frac{k\pi}{2wA}\right)\left[\exp\left(\frac{2\Phi_w}{k}\right)\sin\left(\frac{2\Psi_w}{k}\right)\right] \qquad (7\text{-}167)$$

$$A = \left[\exp\left(\frac{2\Phi_w}{k}\right)\sin\left(\frac{2\Psi_w}{k}\right)\right]^2 + \left[1 + \exp\left(\frac{2\Phi_w}{k}\right)\cos\left(\frac{2\Psi_w}{k}\right)\right]^2 \qquad (7\text{-}168)$$

where $k = Q/(\pi L)$. The streamlines can be mapped by integrating the velocities and computing the displacements by the procedure in the previous section. A forward time step can be used and the *Newton–Raphson* method used to compute Φ_w and Ψ_w at arbitrary values of x and y. The dividing streamlines can be computed by the same trial-and-error procedure in Section 7.5 but since there is no symmetry with respect to the x axis both the upper and lower dividing streamlines will have to be computed.

The locations of streamlines can be determined by integrating the velocity and computing a displacement. By repeating the process, one can map a fluid path line from an arbitrarily selected initial upstream location (x_0, y_0) until the path line either enters the slot or passes it. The process uses a forward time step, such that the location of the path line at the end of the time step (δt) is based on the air velocity at the beginning of the time step, that is, $U_x(t)$ and $U_y(t)$,

$$x(t + \delta t) = x(t) + (\delta t)U_x(t) \qquad (7\text{-}169)$$

$$y(t + \delta t) = y(t) + (\delta t)U_y(t) \qquad (7\text{-}170)$$

The magnitude of the time step (δt) should be small, and progressively smaller values should be used in the close proximity of the inlet.

To compute U_x and U_y, the user must be able to compute Φ_w and Ψ_w at arbitrary values of x and y. Since these parameters are not explicit functions of x and y, a Newton–Raphson method for two simultaneous equations can be used. Details of this method can be found in Appendix A-13.

The upper dividing streamline can be determined by a trial-and-error procedure. The trial-and-error procedure consists of the following:

(1) Choose an upstream location x_0 to begin the analysis. This coordinate will not change.

(2) Guess a distance y_i through which the dividing streamline is believed to pass.

(3) Compute $\Phi_w(x_0, y_i)$ and $\Psi_w(x_0, y_i)$ using the Newton–Raphson method (Appendix A-13).

(4) Compute $\Psi = \Psi_w(x_0, y_i) + \Psi_f(x_0, y_i)$ and compare it with the value of Ψ (dividing streamline) which must be equal to $Q/2L$. If the values do not agree, select another value of y_i, and repeat step (3) until satisfactory agreement is obtained.

(5) Once the correct value of y_i is obtained repeat step (3). Once the correct values of Ψ_w and Φ_w have been found, compute $U_x(x_0, y_i)$ and $U_y(x_0, y_i)$ from Eqs. 7-166 and 7-167.

(6) Compute the location of the air parcel after an interval of time (δt) using Eqs. Eqs. 7-169 and 7-170.

$$x(t_0 + \delta t) = x_0 + (\delta t)U_x(x_0, y_i)$$

$$y(t_0 + \delta t) = y_i + (\delta t)U_y(x_0, y_i)$$

(7) Using the new location found in step 6, return to step 4 and repeat the process. In doing so repeatedly, users will "march" toward the inlet after each increment of time until the streamline passes through an arbitrarily small region surrounding the lip $(-w/\pi, w)$.

To accomplish this procedure, users will have to understand the relative location of streamlines and the general relationship between the initial location and the point where the streamline intersects the x axis.

If a streamline other than the dividing streamline is studied, one only needs to know its numerical value to use the procedure. If the numerical value of the stream function is known, the above process may be used intact. For example, consider the streamline passing through the center of the inlet. If the numerical value is not known but the location $(x_0, 0)$ of the streamline as it passes through the plane of the inlet is known, the process given above must be preceded by a step to compute the numerical value of Ψ.

(1) Compute $\Psi_w(x_0, 0)$ and $\Phi_w(x_0, 0)$ using the Newton–Raphson method and compute Ψ from Eq. 7-133.
(2) Repeat steps 1–7 above.

To compute the velocity at an arbitrary point (x, y) upwind of the inlet the Newton–Raphson method can be used in a manner similar to that described for the flanged slot (Eqs. 7-140 to 7-150).

7.7 MULTIPLE FLANGED RECTANGULAR INLETS

Inlets are generally designed with flanges because it increases the reach. To a first approximation the reach of an inlet is equal to the smaller of its characteristic dimensions. Figures 6.8 and 6.9 show that at upstream distances equal to the smaller of the inlet characteristic dimensions, the air velocity is between 10 and 20% of the face value. Thus for the same volumetric flow rate, a single large inlet influences contaminants farther from the inlet plane than a series of small inlets of the same total cross sectional area. However, the cross sectional area of the region of influence is larger for the multiple inlets. Esmen et al. (107, 122) have expanded the use of potential functions to analyze multiple inlets.

The analysis presumes that each inlet is a quadrilateral opening lying in the y-z plane. Figure 7.22 shows the dimensions and orientation of the surfaces used in the analysis. The coordinates x and z are measured from the center of the inlet. Each inlet is presumed to possess flanking surfaces parallel to the x-z and x-y planes. If a flanking surface is absent the dimension L_y or L_z is set equal to infinity.

Consider an arbitrary point P located at coordinates (x, y, z) in front of a single quadrilateral opening of dimensions A by B through which air is withdrawn at a volumetric flow rate Q. The absolute value of the velocity, $|U(x, y, z)|$, is

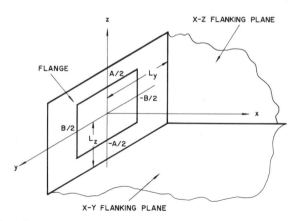

Fig. 7.22 Coordinate system for a flanged quadrilateral inlet with flanking planes.

$$|U(x, y, z)| = \frac{\left(\dfrac{Q}{AB}\right)}{f(R)} \tag{7-171}$$

where $f(R)$ is the following function:

$$f(R) = \frac{1}{2}\left[\left(\frac{R}{B}\right)[\pi + \arcsin k_1] + \left(\frac{R}{A}\right)[\pi + \arcsin k_2]\right]$$

$$+ \left[1 + \frac{4(|xy| + |zx|)}{(A^2 + B^2)}\right]^{1/2}$$

$$+ \left(\frac{\pi}{AB}\right)(k_1 + k_2)R^2 \tag{7-172}$$

The terms $|xy|$ and $|zx|$ denote the absolute value of the product. The parameter R is called a space parameter and defined as

$$R^2 = x^2 + k_3\left(|y| - \frac{B}{2}\right)^2 + k_4\left(|z| - \frac{A}{2}\right)^2 \tag{7-173}$$

where

$$k_3 = \begin{bmatrix} 0 & \text{if} & |y| < \text{or} = B/2 \\ 1 & \text{if} & |y| > B/2 \end{bmatrix} \tag{7-174}$$

$$k_4 = \begin{bmatrix} 0 & \text{if} & |z| < \text{or} = A/2 \\ 1 & \text{if} & |z| > A/2 \end{bmatrix} \tag{7-175}$$

$$k_1 = \begin{bmatrix} L_y/R & \text{if} & R > \text{or} = L_y \\ 1 & \text{if} & R < L_y \end{bmatrix} \tag{7-176}$$

$$k_2 = \begin{bmatrix} L_z/R & \text{if} & R > \text{or} = L_z \\ 1 & \text{if} & R < L_y \end{bmatrix} \tag{7-177}$$

The velocity at $P(x, y, z)$ has a directional (vector) defined as

$$D = \left\{\frac{1}{k_5}\right\}\left[\hat{i}4x + \hat{j}[2y(k_3 + 1) - Bk_3] + \hat{k}[2z(k_4 + 1) - Ak_4]\right] \tag{7-178}$$

$$k_5 = \sqrt{16x^2 + [2y(k_3 + 1) - Bk_3]^2 + [2z(k_4 + 1) - Ak_4]^2} \tag{7-179}$$

where $\hat{i}$, $\hat{j}$, $\hat{k}$ are unit vectors in the x, y and z directions.

When several quadrilateral inlets of different dimensions are located in the same plane, the velocity at any point $P(x, y, z)$ in front of the plane can be found by adding vectorially each of the velocities predicted by Eqs. 7-171 through 7-179. As shown in earlier sections, a useful property of ideal flows is that the net velocity field produced by a composite of sinks is the vector sum of the velocities each sink produces. The computations are tedious because the origin of the coordinate system used in the equations above is at the center of each opening, thus the coordinates (x, y, z) are unique for each opening. One computes the velocity that would exist at point P for each inlet, defining x, y, z for that inlet.

7.8 FLANGED INLETS OF ARBITRARY SHAPE

Up to this time, the discussion has involved inlets of conventional geometry, that is, circular or rectangular openings of finite or infinite aspect ratio. Occasions may arise when a flanged opening of face area A_i of unusual shape is to be used. In this case, the following analysis offers users the ability to compute the velocity potential $\Phi(x, y, z)$ at arbitrary points upwind of the inlet. Figure 7.23 illustrates the geometrical features one must take into account. Let point $P(x, y, z)$ be an arbitrary point upwind of and inlet of area A_i lying in an infinite x-y plane.

The essential feature of the analysis is to assume that the inlet is composed of elemental point-sinks lying in the x-y plane through which air is withdrawn at a volumetric flow rate equal to $(Q/A_i)\, dx\, dy$, where Q is the overall volumetric flow rate. The contribution to the overall velocity potential due to this elemental point-sink is

$$d\Phi = -\left(\frac{Q}{(2\pi A_i s)}\right) dx\, dy \tag{7-180}$$

where s is the distance between point $P(x, y, z)$ and the elemental area $dx\, dy$. The overall velocity potential $\Phi(x, y, z)$ is obtained by integrating over the inlet area.

$$\Phi(x, y, z) = -\int\int_{A_i} \left(\frac{Q}{2\pi A_i s}\right) dx\, dy \tag{7-181}$$

The crucial term in the integrand is the distance s, which is the straight line distance between the elemental area $dx\, dy$ and point P. As one integrates over the inlet area the distance s changes. Thus this seemingly benign integral is all but impossible to integrate in closed form except for elementary inlets that have been analyzed by other means in earlier sections. However, the integral can be solved numerically when repeated for an array of points P, and provides tabulated data of the velocity potential for inlets that cannot be analyzed by other means.

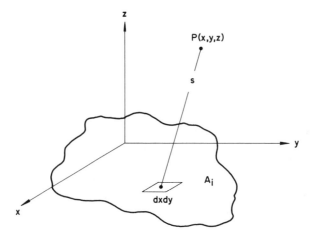

Fig. 7.23 Point $P(x, y, z)$ upwind of a flanged inlet of arbitrary shape.

Anastas and Hughes (347) obtained partial solutions for Eq. 7-181 at points along the centerline for a circular opening and along the midplane of slots. With these expressions, they differentiated with respect to z to obtain centerline velocities. The results agreed well with measurements.

To illustrate the technique, apply Eq. 7-181 to find the centerline velocity upwind of the flanged circular inlet of radius R shown in Fig. 7.24. Because of symmetry, Eq. 7-181 can be replaced by the following:

$$\Phi(0, 0, z) = -\int_0^R \int_0^{2\pi} \left(\frac{Q}{2\pi^2 R^2}\right) \frac{r \, d\theta \, dr}{\sqrt{(z^2 + r^2)}}$$

$$\Phi(0, 0, z) = -\left(\frac{Q}{\pi R^2}\right) \int_0^R \frac{r \, dr}{\sqrt{(z^2 + r^2)}} \tag{7-182}$$

$$\Phi(0, 0, z) = -\left(\frac{Q}{\pi R^2}\right)[\sqrt{(z^2 + R^2)} - z]$$

The air velocity along the centerline can be found by differentiating the velocity potential with respect to z. Thus

$$U_z = -\frac{d\Phi}{dz}$$

$$= -\left(\frac{Q}{\pi R^2}\right)\left[1 - \frac{\left(\dfrac{z}{R}\right)}{\sqrt{\left(\dfrac{z}{R}\right)^2 + 1}}\right] \tag{7-183}$$

At large z, it can be seen that the U_z approaches zero and that as z approaches zero, U_z approaches the face velocity $Q/\pi R^2$.

For a slot, one can repeat the above analysis but now use an infinite number of line-sinks lying along the slot. Using this approach, Anastas and Hughes (347)

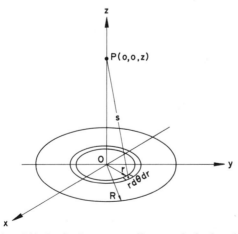

Fig. 7.24 Point $P(0, 0, z)$ along centerline upwind of a circular flanged inlet.

found that the velocity along the center plane upwind of a slot of width $2w$, can be written as

$$U_z = \left(\frac{2Q}{\pi 2wL}\right) \tan^{-1}\left(\frac{1}{2z}\right) \qquad (7\text{-}184)$$

For a rectangular slot of width A and height B, (aspect ratio $A_r = A/B$), they found that the centerline velocity can be written as

$$U_z = \left(\frac{2Q}{\pi A_r}\right) \tan^{-1}\left(\frac{1}{C}\right) \qquad (7\text{-}185)$$

where C is

$$\frac{2z}{A_r + 1} \sqrt{1 + A_r^2 + \frac{4z^2 A_r^2}{(A_r + 1)^2}} \qquad (7\text{-}186)$$

Equations 6-2 to 6-5 are analytical expressions for the centerline velocities for plain and flanged inlets abstracted from an extensive literature review by Braconnier (290,297). The use of centerline velocities is limited since contaminants are apt to be at any location upwind of an inlet, and to determine the reach of inlets and to assess their effectiveness, users need to be able to compute the velocity at arbitrary points upwind of an inlet.

NOMENCLATURE

Symbol	Description	Dimensions *
a	Radius of a sphere, cylinder	L
a, b	One-half major and minor diameter of an elliptical opening	L
a_1, a_2, a_3, a_4	Constants defined by equation	
A	Width of finite rectangular opening, constant defined by equation	L
A_i	inlet area	L^2
A_r	aspect ratio A/B	
B	Height of finite rectangular opening	L
c	Radius of circular inlet	L
c_v, c_p	specific heat at constant volume and constant pressure	Q/MT
C	Constant	
C_P	pressure coefficient defined by equation	
$\boldsymbol{D}$	Directional vector defined by equation	
D	Diameter	L
$f(R)$	function defined by equation	

* Q – energy, F – force, L – length, M – mass, N – mols, t – time, T – temperature

g	Acceleration of gravity	L/t^2
h_d	Height of dividing streamline at arbitrary x	L
h_0	Height of dividing streamline at $x = 0$ for axisymmetric flow	L
h_∞	Height of dividing streamline at x = infinity for axisymmetric flow	L
$h_{0,200}, h_{\infty,200}$	Height of streamline where velocity is 200 FPM at $x = 0$ and x = infinity	L
J	Jacobian	
k	Thermal conductivity	Q/tLT
k, K	Constant defined by equation	
$k_1, k_2, k_3, k_4,$ k_5, K_1, K_2	Constants defined by equation	
L	Length of slot perpendicular to x-y plane	
L_z, L_y	Distance between flanking planes and center of rectangular inlet	L
m	Coefficient defined by equation	
M	Doublet strength	L^4/t
n	Constant defined by equation	
P	Pressure	F/L^2
P_0	Far-field pressure	F/L^2
q_v	Viscous dissipation	Q/Mt
Q	Volumetric flow rate into inlet	L^3/t
R	Constant defined by equation	
R	Radius of disk	L
R_i	Gas constant for molecular species i	Q/NT
r	Radial distance	L
r_0	Radial distance to stagnation point	L
s	Distance from elemental area to point in space	L
t	Time	t
T	Temperature	T
U_f	Face velocity	L/t
$U, U_i, U_x, U_y,$ U_z, U_r, U_θ	Air velocity and its vector components	L/t
U_0	Free stream or far-field velocity	L/t
w	Half width of rectangular slot	L
x, y, z	Spatial coordinates	L

Greek	*Descriptions*	*Dimensions*
α	Angle defined in figure	
β	Corner angle	
δt	Increment of time	t
θ	Angular displacement	
π	3.14159	
ρ	Density	M/L^3
μ	Viscosity	Ft/L^2

ν	(μ/ρ) kinematic viscosity	L^2/t
τ_{ij}	Shear stress	F/L^2
Φ	Velocity potential	L^2/t
Ψ	Stream function	L^2/t
Ω	vorticity, angular velocity	t^{-1}

Subscripts   *Description*

$()_d$	Dividing streamline
$()_0$	Far field value, initial value
$()_w$	Relevant to inlet slot of width 2w, wall condition
$()_f$	Relevant to stream flow field, value at inlet face
$()_i$	Value at i-th iteration
$()_r$	r coordinate
$()_x, ()_y, ()_z$	x, y, z Coordinates
$()_\infty$	Value at $x = \infty$
$()_1, ()_2$	Locations 1 and 2
$()_{200}$	Value at velocity = 200 FPM
$()_\theta$	Theta coordiante

Abbreviations

$$\frac{D}{Dt} = \frac{\partial}{\partial t} + U_i \frac{\partial}{\partial x_i}$$ Total or substantial derivative

$\cdot$	Dot product
$\times$	Cross product
Curl	curl of a vector
∇	Gradient of a scalar
Grad	Gradient of a scalar
Div	Divergence of a vector
∇^2	Laplacian

PROBLEMS

1. Each of the questions below has a single correct answer.

(a) Which of the following defines an ideal fluid?
   (1) Constant density
   (2) Frictionless
   (3) Negligible viscous effects
   (4) All of the above

(b) Which of the following defines irrotational flow?
   (1) Incompressible
   (2) Divergence of the velocity is zero
   (3) Vorticity is zero
   (4) Frictionless

(c) What is the necessary and sufficient condition in steady incompressible flow without work or heat transfer to insure that the quantity $(P/\rho + U^2/2 + gz)$ is constant along a streamline?

(1) Frictionless

(2) Irrotational

(3) Vorticity is zero

(4) The streamlines are not curved

(d) What is the necessary and sufficient condition in steady incompressible flow without work or heat transfer to insure that the quantity $(P/\rho + U^2/2 + gz)$ is constant everywhere in the flow field?

(1) Vorticity is zero

(2) Frictionless

(3) Irrotational

(4) All of the above

(e) Irrotational flow requires streamlines to be straight lines. True or false?

(f) The vorticity is nonzero only if the streamlines are curved. True or false?

(g) The vorticity is nonzero only if the curl of the velocity is nonzero. True or false?

(h) Eulers equation is the restricted form of the Navier–Stokes equation when in addition to steady flow which of the following is true?

(1) The flow is incompressible.

(2) The flow is inviscid.

(3) The flow is irrotational.

(4) All of the above.

(i) The steady flow equation,

$$\left(\frac{P}{\rho}\right)_2 - \left(\frac{P}{\rho}\right)_1 + \frac{U_2^2 - U_1^2}{2} + g(z_2 - z_1) = \text{(head loss due to friction)}$$

$$+ \text{(head loss due to shaft work)}$$

has which restrictions?

(1) No restriction

(2) Perfect gas

(3) Incompressible flow

(4) Irrotational flow

(j) Which of the following conditions is <u>necessary</u> in order to define the stream function Ψ for a steady incompressible flow?

(1) Two-dimensional flow

(2) Irrotational flow

(3) Inviscid flow

(4) All of the above

(k) Which of the following conditions is <u>necessary</u> in order to define the velocity potential Φ?

(1) Two-dimensional flow

(2) Irrotational flow

(3) Inviscid flow

(4) All of the above

(l) What is a vortex? Do all vortices have vorticity?

(m) Answer true or false to the following:

Vorticity is a measure of the angular deformation of a fluid element.

Vorticity is always zero in a rotational flow field.

All curvilinear flow has vorticity.

Vorticity is a measure of the viscosity of the fluid.

Vorticity is always zero for incompressible flow.

Vorticity is a measure of rotation of a fluid element

2. Prove that a flow is steady and the divergence of the velocity is zero, then the flow is incompressible irrespective of whether the density is constant or not.

3. Derive analytical expressions for the appropriate potential functions in Illustrated Example 7.1

4. Consider a stationary right circular cylinder of radius (a) immersed in an ideal liquid that is moving in the negative x direction with a speed U_0. The stream function for the flow around the cylinder is given by

$$\Psi = U_0 y \left[1 - \left(\frac{a}{r} \right)^2 \right]$$

where r is the radial distance from the center of the cylinder.

(a) Show that the velocity potential is

$$\Phi = U_0 \left[r + \left(\frac{1}{r} \right) a^2 \right] \cos \theta$$

(b) Prepare a graph showing lines of constant stream function and velocity potential.

(c) Show that the pressure coefficient C_P (defined by Eq. 7-101) is equal to

$$C_P = \frac{1}{4(\sin \theta)^2}$$

5. A cylinder of radius a moves through a stationary ideal fluid with a velocity U_0 normal to its axis. The stream function is,

$$\Psi = U_0 \left(\frac{a^2}{r} \right) \sin \theta$$

(a) Derive an expression for the velocity potential (Φ).

(b) Show that the speed of the fluid adjacent to the surface of the cylinder is equal to U_0 for all values of the angle θ.

6. The flow of a fluid through a vertical row of identical right circular cylinders of radius (b) can be modeled as an ideal fluid if the Reynolds numbers are low. The cylinders are aligned in the y-direction, separated by a distance (a) between their centers. The axes of the cylinders is normal to the x-y plane. The upwind velocity (U_0) is uniform and in the negative x direction. The stream function for such a flow is

$$\Psi = U_0 b \left(\frac{y}{b}\right) - \left[\frac{\left(\frac{b}{a}\right)\pi \sin\left\{\frac{y}{m}\right\}}{\left\{\cosh\left(\frac{x}{m}\right) - \cos\left(\frac{y}{m}\right)\right\}}\right]$$

where $m = a/(2\pi)$, $x = 0$, $y = 0$ corresponds to the center of one of the cylinders. If $a/b = (3.14159)$, find the velocity components (U_x and U_y) midway between the cylinders, that is, find U_x and U_y at $(x = 0, y = a/2)$

7. The motion of a sphere of radius (a) traveling with a velocity U_0 in a stationary infinite ideal fluid is given by

$$\Phi = -a\left(\frac{U_0}{2}\right)\left(\frac{a}{r}\right)^2 \cos\theta$$

(a) Show that the stream function is given by

$$\Psi = \left(\frac{U_0}{2r}\right)a^3(\sin\theta)^2$$

(b) Are the velocity and pressure distributions on the sphere's surface different than the distributions when the roles are reversed and the fluid passes over a stationary sphere? Are the far-field velocity and pressure distributions different?

8. Examine the validity of a statement that will appear in the chapter on Particle Dynamics concerning circumstances under which spherical particles A and B (of radii a and b respectively) influence each other. It will be asserted that particle B will not affect the flow field around an adjacent particle A when the distance between the particles is greater than approximately 10 times the particle diameter.

Case I – Particle A is to the rear of particle B and both move along their line of centers with a velocity U_0 through quiescent air. If the particle centers are separated by a distance c, the velocity potential in the vicinity of particle A is,

$$\Phi = U_0\left(\frac{a}{2}\right)\left(\frac{a}{r}\right)^2 \cos\theta + \left(\frac{ab}{c}\right)^3 \left(\frac{U_0}{2c^2}\right)\left[1 + \left(\frac{2r}{c}\right)\cos\theta\right] + \left(\frac{U_0}{2}\right)\left(\frac{a}{c}\right)^6 b^3 \frac{\cos\theta}{r^2}$$

Show that the velocities on the surface of particle A are unaffected by particle B if c/b is of the order of 5.

Case II – The centers of particles A and B are a distance c apart and both particles move through quiescent air at a velocity U_0 at right angles to their line of centers. If the velocity potential in the vicinity of *A* is

$$\Phi = \left(\frac{aU_0}{2}\right)\left(\frac{a}{r}\right)^2\left[1 + \frac{3}{2}\left(\frac{b}{c}\right)^3\right]\cos\theta$$

show that the velocities on the surface of particle A are unaffected by particle B if c/b is of the order of 5.

9. The stream function

$$\Psi = -\left[\left(\frac{a}{m}\right)U_w\right]\left[\left(\frac{r}{a}\right)^m\right]\sin(m\theta)$$

$$m = \left(\frac{\pi}{\alpha}\right)$$

describes the flow of fluid entering a corner (from right to left) of angle α. When the angle α is greater than $\pi/2$, the stream function describes flow over a corner. The velocity (U_w) is the velocity adjacent to either wall at a radius a from the origin. Derive the expression for the velocity potential Φ.

10. Show that the conservation equations for mass, momentum and energy reduce to Eqs. 7-13 to 7-15 for an inviscid, constant temperature–constant density fluid.

11. In cylindrical coordinates, show that if the vorticity is constant, liquid rotates about its axis at an angular speed N as if it was a solid body (hence the phrase solid body rotation) and as a result,

$$\frac{U_\theta}{r} = N$$

If the fluid motion is irrotational, show that

$$U_\theta r = \text{Constant}$$

12. Prove the vector identity Eq. 7-12.

13. A vessel of circular cross section, outer radius R, is filled with a liquid to a uniform depth H. The vessel is rotated about its vertical axis at a constant angular speed N and the surface of the liquid is no longer horizontal. The pressure on the upper surface of the liquid is constant and equal to the atmospheric pressure. Write an equation that describes the shape of the liquid surface,

$$z = f(r, N, R, H, \text{fluid properties})$$

assuming (a) the motion is rotational and (b) the motion is irrotational.

14. Fully established flow in a circular duct of outer radius R has the following (axial) velocity profile:

$$U_x = \left[\frac{2Q}{\pi R^2}\right](R^2 - r^2); \qquad U_r = 0; \qquad U_\theta = 0$$

Is this irrotational flow? Is it rotational flow of constant vorticity?

15. Derive the stream function Ψ that corresponds to the following velocity potentials for two dimensional flows.

$$\Phi = Ar \cos \theta$$

$$\Phi = \frac{A \cos \theta}{r}$$

$$\Phi = A \ln r$$

where A is a constant. Describe these flows in physical terms.

16. For a two dimensional flow field show that values of Ψ and Φ are orthogonal.

17. Show that the velocity potential Φ satisfies both the continuity and Navier–Stokes equations.

18. If a two dimensional flow field has velocity components $U_x = Kx$, $U_y = -Ky$ where K is a constant, derive an analytical expression for the stream function and describe the flow field physically.

19. A line sink Q_1 lies in a plane located at $(c, 0)$ and a second line sink Q_2 $(Q_1 = 2Q_2)$ lies in the plane at $(-c, 0)$. Air streams above the surface from right to left at a uniform velocity U_0. Write expressions for the potential functions Ψ and Φ for the composite flow field. Find the velocity components at $(0, c)$

20. Consider the flow field formed by a line source lying at $(0, c, z)$ in a vertical surface $(y - z$ plane) and a line sink located at the origin $(0, 0, z)$. The vertical surface intersects a horizontal surface coincident with the z axis. If the input volumetric flow rate of air per unit length of the source is equal to the exhaust volumetric flow rate per unit length of the sink, write the potential functions for the composite flow field and describe the flow field physically. What are the velocity components at $(c, c/2, z)$

21. What physical flow field can be described by the potential functions

$$\Phi = Ar^2 \cos 2\theta \qquad \Psi = Ar^2 \sin 2\theta$$

What does A represent in physical terms?

22. What physical flow field is described by

$$\Psi = A \sinh\left(\frac{\pi x}{a}\right) \sin\left(\frac{\pi y}{a}\right)$$

$$\Phi = A \cosh\left(\frac{\pi x}{a}\right) \cos\left(\frac{\pi y}{a}\right)$$

Describe the flow field in the region $x > 0$, $0 < y < a$.

23. A process produces an unusual ideal, two dimensional flow described by the following:

$$\Psi_s = \left(\frac{K}{2\pi}\right) \ln r \qquad \Phi_s = -\left(\frac{K}{2\pi}\right)\theta$$

where $K = 2\pi R_1 U_0$ and R_1 is a constant. The process is placed in a streaming flow in the negative x direction whose potential functions are

$$\Psi_f = U_0 \sin \theta \qquad \Phi_f = U_0 \cos \theta$$

(a) Write general equations that express the velocity components U_r and U_θ at arbitrary points in the flow field.

(b) Describe the flow field in physical terms.

(c) Write an equation that will enable users to plot the values of r and θ at which the air speed is constant and equal to the magnitude of U_0.

(d) If the far-field pressure is P_0 and the velocity is U_0, find the pressure at the following points,

$$r = R_1, \qquad \theta = \text{zero}$$

$$r = R_1, \qquad \theta = \frac{\pi}{2}$$

24. It is claimed that the stream function

$$\psi = U_0 \left[y - \frac{\left(\frac{c}{4}\right) \sin\left(\frac{2\pi y}{c}\right)}{A} \right]$$

$$A = \left[\cosh\left(\frac{2\pi x}{c}\right) - \cos\left(\frac{2\pi y}{c}\right) \right]$$

describes the flow around a right circular cylinder (diameter $c/2$) located at the $(0,0)$ lying midway between parallel plates a distance $c/2$ above and below the x axis. The velocity U_0 is the approach velocity. Verify that the statement is true and find the air velocity $(-c, c/2)$.

25. The following stream functions describe different two-dimensional flow fields. Compute the x and y velocity components. Which of these flow fields are irrotational?

$$\Psi = Ar \sin \theta$$

$$\Psi = Ar(1 - B/r^2) \sin \theta$$

$$\Psi = -Ay + B e^{-y} \sin x$$

26. Consider the velocity field produced by a uniform flow streaming over an infinite slot inlet (width $2w$) in an infinite flanged surface. The uniform velocity of the streaming flow (U_0) is equal to one half of the average velocity through the inlet [$U_0 = Q/(L4w)$]. The values of w and U_0 are $1\,cm$ and $1\,m/s$ respectively.

   (a) Find the coordinates (x, y) of points along the dividing streamline.

   (b) Draw a graph that locates the streamline that passes through the center of the inlet, i.e. that passes through $(0, 0)$.

   (c) Find the numerical value of U_x, and U_y at $x = y = w/2$

27. Overburden from a quarry is piled in a large mound that has a semicircular cross section (radius, $a = 30\,m$) and a length (L) $200\,m$. During the winter months, freezing and thawing makes the pile's surface susceptible to erosion by the wind. For a first approximation assume that the material in the pile has a uniform diameter of $100\,\mu$ ($10^{-4}\,m$). Assume that air is an ideal fluid and approaches the pile with a uniform velocity U_0 ($10\,km/hr$) perpendicular to the long dimension. Assume that soil is made airborne at a rate (g_p, $gm/m^2\,s$) proportional to the local wind speed (m/s) in excess of a critical value (U_c) cubed,

$$g_p(gm/m^2\,s) = 12.8[U(a, \theta) - U_c]^3$$
$$U_c(m/s) = 116\sqrt{D_p(m)}$$

$U(a, \theta)$ is the tangential velocity adjacent to the surface of the pile. Neglect gravimetric settling of the dust after it is made airborne. Estimate the fugitive emission from the overburden (g/s).

28. For irrotational flow of a constant density fluid, show why the Laplacian of the velocity potential is zero, that is.

$$\nabla^2 \Phi = 0$$

29. For two-dimensional flow of a constant density fluid, show why the Laplacian of the stream function is equal to zero, that is,

$$\nabla^2 \Psi = 0$$

30. The flow of an ideal fluid around a stationary right circular cylinder of radius a is given by the stream function

$$\Psi = U_0 \left[r - \left(\frac{a^2}{r} \right) \right] \sin \theta - U_0 a \ln \left(\frac{r}{a} \right)$$

where U_0 is the far-field air velocity. Compute the velocity potential Φ, the radial and tangential velocity components. Describe the physical attributes of the flow including the location of the stagnation points and the location of the streamlines $\Psi = 0$ and $\Psi = \pm Ua/2$. Derive an expression that predicts the pressure distribution of the surface of the cylinder.

31. Prove that the stream function given below describes the flow of an ideal fluid through a semiinfinite aperture of width $2w$ by defining the location of several key streamlines.

$$\left[\frac{(x/w)}{\cos \Psi} \right]^2 - \left[\frac{(y/w)}{\sin \Psi} \right]^2 = 1$$

Derive an expression for the velocity component U_y at an arbitrary point in the aperture inlet plane, that is, $-w < x < w$ and $y = 0$.

32. A prairie dog is a small animal that lives in the western United States. Colonies of prairie dogs live in underground burrows (tunnels). One end of the tunnel has an opening level with the ground several meters from the other opening, which is in the center of a hemispherical mound (assume 2 m in diameter) above the ground. An interesting question arises (263) as to how the tunnel is ventilated since several prairie dogs may remain in the burrow for many hours. Whether by accident or instinct, prairie dogs use the venturi effect to draw fresh air into the ground level opening and exhaust air through the elevated opening (the pressure at the ground-level opening is atmospheric while the pressure at the top of the hemispherical mound is less than atmospheric because the air velocity reaches a maximum). Using the potential flow function for flow over a stationary sphere and neglecting boundary layers, estimate the ventilation volumetric flow rate (m³/sec) through a single tunnel. When applying the energy equation, define beginning and end points carefully and be sure to include friction and entry and exit losses properly.

(a) Each tunnel has a circular cross section (diameter 0.01 m) and two 90-degree elbows, each of which produces a pressure drop equivalent to 3 m of tunnel.

(b) Total tunnel length (exclusive of elbows) is 20 meters.

(c) Far-field wind speed is 4 m/s.

(d) The pressure drop due to the viscous wall shear stress can be expressed by Fig. 6.22 or by the following:

$$f = \frac{64}{Re_D} \quad \text{if } Re_D < 2000$$

$$\sqrt{\frac{1}{f}} = -2.0 \log_{10}\left[\frac{(\varepsilon/D)}{3.7}\right]$$

$$\varepsilon = 0.5 \text{ mm} \quad \text{if Re}_D > 2000$$

33. Compute the air velocity (speed and direction) along a line $(x, 0.75 \text{ m}, 0.3125 \text{ m})$ perpendicular to the plane containing two rectangular openings and intersecting the plane at a point midway between the two rectangular openings. The openings are located near two flanking planes (see Fig. 7.22). If a computer-graphics program is available, plot the surface over which the air speed is 200 FPM (typical capture velocity).

Opening Number 1

$$A_1 = 0.5 \text{ m}; \qquad B_1 = 0.25 \text{ m}$$

$$\text{Slot velocity} = 10 \text{ m/s}$$

$$L_{Y1} = 0.75 \text{ m}; \qquad L_{Z1} = 0.5 \text{ m}$$

Opening Number 2

$$A_2 = 1 \text{ m}; \qquad B_2 = 0.125 \text{ m}$$

$$\text{Slot velocity} = 10 \text{ m/s}$$

$$L_{Y2} = 0.75 \text{ m}; \qquad L_{Z2} = 0.125 \text{ m}$$

34. Consider a point sink and a flanged circular opening of diameter 0.01 m. The volumetric flow rate in both cases is $0.04 \text{ m}^3/\text{min}$. Compare the air speed along two radial lines (from the center of the opening) inclined 30 and 60 degrees to the flange. Also compare the results to Fig. 6.9.

35. Compute the location of the bounding streamline for streaming flow over a circular opening of diameter 0.01 m in a flanged surface. The volumetric flow rate is $0.04 \text{ m}^3/\text{min}$ and the far-field velocity U_0 is 5 m/s. Compare the dimensions of this dividing streamline with what would be predicted if the opening was a point sink with the same volumetric flow rate.

36. Consider an infinite surface lying in the $x - z$ plane that intersects an infinite surface lying in the $y - z$ plane. Air approaches the $y - z$ plane with a far-field velocity $(-U_0)$. Assume the air flow is incompressible, inviscid, and irrotational. The air velocity on the $x - z$ plane a distance a upwind of the y-z plane is measured and found to be $-3U_0$, that is,

$$U_x(a, 0, z) = -3U_0 \qquad U_y(a, 0, z) = 0$$

The y-z plane has a small hole (point sink) in it at $(0, a/z, 0)$ and air is withdrawn through the hole at a volumetric flow rate Q. Estimate the air velocity at the point $x = a/2$, $y = a/2$, $z = 0$.

37. A right circular cylinder moves through a stationary inviscid and incompressible fluid with a velocity U_0 in a direction perpendicular to its axis of symmetry. Show that the speed of the air is equal to $|U_0|$ at all points on the surface of the cylinder. Find the pressure at an arbitrary point on the cylinder surface.

38. Compute the pressure distribution on the surface of a stationary cylinder past which an inviscid incompressible fluid passes in a direction perpendicular to its axis of symmetry with a velocity U_0. Compare the results with the pressure distribution found in the previous problem and reconcile any differences.

39. Write a computer program that computes the coordinates of an arbitrary streamline for the following inlets in streaming flow.

(a) Flanged slot
(b) Plain slot
(c) Flanged circular inlet
(d) Flanged elliptical inlet

Apply the program and plot the location of the dividing streamlines where $U_0 = Q/A$.

40. Write a computer program that allows you to compute the velocity components at arbitrary points upwind of the following inlets in streaming flow.

(a) Flanged slot
(b) Plain slot
(c) Flanged circular inlet
(d) Flanged elliptical inlet

Apply the program and compute the velocity at a point one opening distance directly in front of the center of the inlet when $U_0 = Q/A$.

41. Using the methods discussed in Section 7.8, compute the centerline velocity along the centerplane for an infinite slot of width $2w$.

CHAPTER 8

Particle Dynamics

Goal: Students will be expected to learn,
  - dynamics of particles in a moving viscous fluid
  - statistical descriptions of aerosols
  - to model the performance of particle removal systems

The motion of gaseous and vapor contaminants is affected by the velocity field of the air through which they move. The motion of particle contaminants on the other hand is also affected by the inertia of the particle and aerodynamic drag. The purpose of this chapter is to provide an analytical basis to describe the motion of particles in a moving air stream so that engineers can design collection systems and predict their performance.

8.1 DRAG

When a particle moves through air the viscosity of the air produces a force on the particle in the opposite direction to the relative velocity. The *drag force* on a sphere (F) is written as,

$$\mathbf{F} = -c_D \left(\frac{\rho}{2}\right) \left[\frac{\pi D_p^2}{4}\right](\mathbf{v} - \mathbf{U})|\mathbf{v} - \mathbf{U}| \tag{8-1}$$

where

$\quad \mathbf{U}$ = fluid velocity
$\quad \mathbf{v}$ = particle velocity
$\mathbf{v} - \mathbf{U}$ = *relative velocity* = $\mathbf{v}_r$
$$\qquad = \mathbf{i}(v_x - U_x) + \mathbf{j}(v_y - U_y) + \mathbf{k}(v_z - U_z) \tag{8-2}$$
$$\qquad = \mathbf{i}\, v_{rx} + \mathbf{j}v_{ry} + \mathbf{k}v_{rz}$$
$$|\mathbf{v} - \mathbf{U}| = [(v_x - U_x)^2 + (v_y - U_y)^2 + (v_z - U_z)^2)]^{0.5}$$
$$\qquad = [v_{rx}^2 + v_{ry}^2 + v_{rz}^2]^{0.5} \tag{8-3}$$
$\quad c_D$ = drag coefficient

Figure 8.1 shows the relationship between the *drag coefficient* for a sphere and the Reynolds number. Similar curves exist for other geometrical shapes. The drag

429

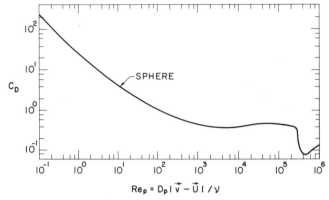

Fig. 8.1 Drag coefficient of a sphere, redrawn from ref. 335.

coefficient includes both drag caused by an unsymmetrical pressure distribution (*form drag*) and drag caused by shear stresses acting on the sphere's surface (*viscous drag*). The marked reduction in drag on a sphere at a Reynolds numbers of 200,000 is caused by a turbulent boundary layer that reduces the separation of flow on the downstream side of the sphere.

It is common practice to divide Fig. 8.1 into flow regimes:

Stokes flow regime: $Re < 1$, $c_D = 24/Re$
Transition flow regime: $1 < Re < 10^3$, $c_D = $ variable
Newtonian flow regime: $10^3 < Re < 10^5$, $c_D = 0.4$

The relationship between the drag coefficient and Reynolds number for $Re < 10^5$ can be expressed by the following empirical equation with an accuracy of 10%.

$$c_D = 0.4 + \frac{24}{Re} + \frac{6}{(1 + Re^{0.5})} \tag{8-4}$$

$$Re = \frac{\rho D_p |\mathbf{v} - \mathbf{U}|}{\mu}$$

It must be emphasized that the velocity used in the Reynolds number and in Eq. 8-4 is based on the *relative velocity* (v_r) between the particle and the fluid, that is, $|\mathbf{v} - \mathbf{U}|$. For particles moving in motionless air, the relative velocity is the particle velocity. For particles traveling through a moving fluid great care must be taken to evaluate the relative velocity. The viscosity is a function of temperature. Shown in Fig. 8.2 is the *dynamic viscosity* (μ) and *kinematic viscosity* (ν, where $\nu = \mu/\rho$) for air as a function of temperature. The dynamic viscosity of air can be expressed empirically as a function of temperature (in degrees K),

$$\mu(\text{kg/m s}) = 1.3554 \times 10^{-6} + 0.6738 \times 10^{-7}(T)$$

$$- 3.808 \times 10^{-11}(T^2) + 1.183 \times 10^{-14}(T^3) \tag{8-5}$$

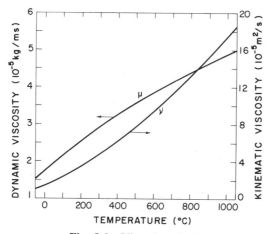

Fig. 8.2   Viscosity of air.

8.2 PHYSICAL PROPERTIES OF AEROSOLS

A suspension of particles in air is called an *aerosol*. The particles may be solid or liquid and the aerosol may be stationary or moving. Properties of the aerosol depend on the properties of the particles and the air. The air consists of molecules which travel at very high speed. The average distance traveled by molecules between collisions with each other (λ) is called the *mean free path*. The mean free path can be expressed (274) in terms of the viscosity by the following,

$$\lambda = \frac{\mu}{I} \sqrt{\frac{\pi}{8\rho P}} \tag{8-6}$$

where I is a constant equal to 0.499. For air at STP, $\lambda = 0.0667$ μm. Assuming air to be a perfect gas,

$$P = \rho RT = \rho(R_u/M)T \tag{8-7}$$

the mean free path at other temperatures and pressures can be expressed as

$$\lambda(T, P) = \lambda_0 \left(\frac{\mu}{\mu_0}\right)\left(\frac{P_0}{P}\right)\left(\frac{T}{T_0}\right)^{1/2} \tag{8-8}$$

where the subscript 0 refers to conditions at STP.

8.2.1 Flow Regime

A fluid is considered a *continuum* if the moving particle is considerably larger than the mean free path. The particle is unaware of individual molecules and distances between them and moves through the fluid as if it were a continuous medium. If on the other hand, the particle is comparable in size (or smaller) than the mean free path, the particle is affected by collisions with individual molecules and at times *slips* between molecules. Such motion is called *free molecular flow* or more generally *slip flow*.

The parameter used to distinguish a continuum regime from free molecular regime is the *Knudsen number* (K_n), which for spherical particles is

$$K_n = \frac{\lambda}{D_p} = \begin{cases} K_n > 10 & \text{Free molecular flow} \\ 10 > K_n > 0.1 & \text{Transition flow} \\ K_n < 0.1 & \text{Continuum regime} \end{cases} \qquad (8\text{-}9)$$

Engineers are familiar with the equations of continuum fluid mechanics but many contaminant particles are in the transition and free molecular flow regime, for example, smoke, fume, or fine dust. When dealing with these particles, one must modify Eq. 8-1. It is expedient and accurate to insert a *slip factor* (C) also called the *Cunningham correction factor* in the denominator of the equation for the viscous drag. The slip factor can be expressed as (274)

$$C = 1 + K_n\left[2.514 + 0.8 \exp\left(-\frac{0.55}{K_n}\right)\right] \qquad (8\text{-}10)$$

For spherical particles, Eq. 8-1 becomes

$$\mathbf{F} = -\left(\frac{c_D \rho}{2C}\right)\left(\frac{\pi D_p^2}{4}\right)(\mathbf{v} - \mathbf{U})|\mathbf{v} - \mathbf{U}| \qquad (8\text{-}11)$$

The relationship between the slip factor and the diameter of spherical particles in air at STP is shown in Table 8.1. It is clear that for large particles, $D_p > \lambda$, the slip factor (C) is essentially unity and can be omitted from Eq. 8-11. The slip factor is particularly significant for submicron particles. It should be noted that the slip factor is independent of the particle velocity and depends only on the particle size and physical properties of the fluid.

TABLE 8.1   Cunningham Correction Factors or Slip Factors for Spherical Particles in Air at STP

Diameter (μm)	Slip Factor (C)
0.01	22.218
0.02	11.415
0.05	4.969
0.10	2.867
0.30	1.561
0.50	1.330
1.00	1.164
5.00	1.033
10.00	1.016
20.00	1.008
50.00	1.003

8.2.2 Particle Shape

Particles vary in geometry, for example, perfect spheres such as condensed vapors, cylindrical or flat filaments such as cotton fibers or asbestos in which the ratio of length to width is large, platelets such as silica or mica, feathery agglomerates such as soot, or irregularly shaped fragments such as coal dust, foundry sand or metal grinding particles. If particles are not spheres the drag may be quite different than for spheres of the same mass.

To accommodate nonspherical particles, Fuchs (41) suggested introducing a unitless *dynamic shape factor*, χ, defined as the ratio of the drag forces between a nonspherical particle and a sphere having the same volume and velocity. Thus the drag force (Eq. 8-1) is written as

$$\mathbf{F} = -\left(\frac{\chi c_D \rho}{2C}\right)\left(\frac{\pi D_{e,p}^2}{4}\right)(\mathbf{v} - \mathbf{U})|\mathbf{v} - \mathbf{U}| \tag{8-12}$$

where $D_{e,p}$, is a *volume equivalent diameter* defined in terms of the actual particle volume (V_p)

$$D_{e,p} = \left[\frac{6V_p}{\pi}\right]^{0.33} \tag{8-13}$$

and the drag coefficient is expressed in terms of the Reynolds number based on the equivalent volume diameter. The Cunningham correction factor should be computed on the basis of the equivalent volume diameter. The equivalent volume diameter is a the diameter an actual particle would have if it were a sphere. For flow beyond the Stokes regime, one may use Eq. 8-12 to describe the drag in lieu of a better expression, but readers are urged to consult other sources (40–44, 301, 406) for information. Table 8.2 shows values of χ for a variety of common shapes. Dynamic shape factors for particles that have a length, width, and height of comparable value are close to unity and may be omitted for purposes of industrial ventilation. For these cases, $D_{e,p}$ may be replaced by either the length, width or height.

Leith (216) suggests that the dynamic shape factor χ can be given by the following (216):

$$\chi = \left[0.33 + 0.67 \frac{D_{s,p}}{D_{p,p}}\right]\left[\frac{D_{e,p}}{D_{p,p}}\right] \tag{8-14}$$

where $D_{p,p}$ and $D_{s,p}$ are called the *equivalent projected area diameter* and *equivalent surface area diameter*, respectively.

$D_{p,p}$ = diameter of a sphere with the same projected area, i.e. cross-sectional area of the actual particle normal to the direction of flow.

$D_{s,p}$ = diameter of a sphere with the same surface area as the actual particle.

The *Sauter mean diameter* is the ratio of the total volume of all particles in an aerosol divided by the total surface area of all the particles. For this reason the Sauter diameter can be thought of as a mean volume-surface diameter.

TABLE 8.2 Dynamic Shape Factors (χ) Averaged over all Orientations

Shape	χ
Sphere	1.00
Cube	1.08
Cylinder ($L/D = 4$)	
Axis horizontal	1.32
Axis vertical	1.07
Ellipsoid, across polar axis	
Major/minor diameters = 4	1.20
Parallelepiped, square base	
Height to base	
0.25	1.15
0.50	1.07
2.00	1.16
3.00	1.22
4.00	1.31
Clusters of spheres	
Chain of 2	1.12
Chain of 3	1.27
3 compact	1.15
Chain of 4	1.32
4 compact	1.17

Source: Abstracted from refs. 41 and 44.

Spherical particles whose density is equal to the density of water are called *unit density spheres*. For the most part, the density of a particle is the density of the compound of which it is composed. In the event the particle contains voids, is a composite material or is a loose feathery agglomerate, the density is more difficult to define. Details of how to cope with these circumstances can be found in ref. 42.

8.2.3 Size Distribution

Another physical property to distinguish one group of particles from another is their size distribution. For example if two aerosols have the same mass concentration (mg/m^3) but different size distributions, the dynamic behavior of one group of particles taken as an entity may be quite different than the other. If an aerosol is composed of particles of a single size it is called *monodisperse*. Monodisperse are rarely encountered in engineering practice. If particles vary in size, the aerosol is called *polydisperse*. What follows is a synopsis of a commonly encountered statistical description, called the *log-normal size distribution*. For a systematic study of particle size distributions, readers should consult texts devoted to the subject (60, 239).

If the physical properties of the particles in two polydisperse aerosols are the same and the mass of particles per unit volume of gas is the same, one way to distinguish one aerosol from the other is to compare the statistical parameters that describe the size distributions of the two aerosols. Suppose that an aerosol is analyzed on the basis of size and found to have the distribution shown in Table

8.3, the *arithmetic mean* diameter based on number $[D_{p,am}]$, *variance* (σ^2), and *standard deviation* (σ) can be computed from the following:

$$D_{p,am} = D_p \text{ (number arithmetic mean)}$$

$$= \left(\frac{1}{n_t}\right) \int_0^\infty D_p n(D_p)\, dD_p \tag{8-15}$$

$$= \sum_{i=1} n_i \frac{D_{p,i}}{n_t} \tag{8-16}$$

TABLE 8.3   Particle Size Distribution (51)

Frequency Distribution of 1000 Particles Classified According to Classes of Equal Linear Size.

Class Boundary (micron)	Class Width (micron)	Class Midpoint (micron)	Particles per Class	Percent Particles in Class	Percent of Particles Less Than Upper Interval Size
0 -0.5	0.5	0.25	16	1.6	1.6
0.5-1.0	0.5	0.75	159	15.9	17.5
1.0-1.5	0.5	1.25	235	23.5	41.0
1.5-2.0	0.5	1.75	200	20.0	61.0
2.0-2.5	0.5	2.25	133	13.3	74.3
2.5-3.0	0.5	2.75	97	9.7	84.0
3.0-3.5	0.5	3.25	55	5.5	89.5
3.5-4.0	0.5	3.75	36	3.6	93.1
4.0-4.5	0.5	4.25	24	2.4	95.5
4.5-5.0	0.5	4.75	15	1.5	97.0
5.0-5.5	0.5	5.25	10	1.0	98.0
5.5-6.0	0.5	5.75	6	0.6	98.6
6.0-6.5	0.5	6.25	5	0.5	99.1
6.5-7.0	0.5	6.75	3	0.3	99.4
7.0-7.5	0.5	7.25	2	0.2	99.6
7.5-8.0	0.5	7.75	1	0.1	99.7
8.0-8.5	0.5	8.25	1	0.1	99.8
8.5-9.0	0.5	8.75	1	0.1	99.9
9.0-9.5	0.5	9.25	1	0.1	100.0

Frequency Distribution of 1000 Particles Classified According to Size Classes of Equal Logarithmic Diameter Size Width.

Class Boundary (micron)	Class Width (micron)	Class Midpoint (micron)	Particles per Class	Percent Particles in Class	Percent of Particles Less Than Upper Interval Size
0.3-0.4	0.1	0.35	6	0.6	0.6
0.4-0.53	0.13	0.46	15	1.5	2.1
0.53-0.71	0.18	0.62	41	4.1	6.2
0.71-0.94	0.24	0.82	88	8.8	15.0
0.94-1.28	0.31	1.1	150	15.0	30.0
1.28-1.68	0.42	1.5	180	18.0	48.0
1.68-2.23	0.57	2.0	205	20.5	68.5
2.23-3.00	0.75	2.6	156	15.6	84.1
3.0-4.0	1.0	3.5	91	9.1	93.2
4.0-5.3	1.32	4.6	44	4.4	97.6
5.3-7.1	1.63	6.2	18	1.8	99.4
7.1-9.4	2.32	8.2	5	0.5	99.9
9.4-12.8	3.13	11.0	1	0.1	100.0

The variable $n(D_p)$ is the *number distribution function* such that the quantity (ndD_p) represents the number of particles between sizes D_p and $(D_p + dD_p)$, n_i is the number of particles in the size range "i" where the midrange particle size is $D_{p,i}$, and the symbol n_t is the total number of particles.

$$n_t = \int_0^\infty n \, dD_p = \sum_{i=1} n_i \tag{8-17}$$

$$\sigma^2 = \sum_{i=1} \left(\frac{n_i}{n_t}\right)(D_{p,am} - D_{p,i})^2 \tag{8-18}$$

The reader must be very careful to distinguish between the discrete number concentration n_i which has the units number of particles per unit volume and the continuous size distribution function $n(D_p)$ which has the units "number of particles/μm m^3". [In the material that follows the function $n(D_p)$ will be written as n to simplify equations.]

Figure 8.3 is a graph of the discrete number fraction n_i/n_t versus particle diameter D_p and shows that the distribution is not a Gaussian, bell-shaped (normal) distribution. Many particles encountered in industrial ventilation can be described as *log-normal*, that is, if the discrete number fraction n_i/n_t is plotted against $[\ln D_p]$ as shown in Fig. 8.4, the resulting distribution has the behavior associated with a normal distribution. It will be useful to generate an analytical function to describe a log-normal distribution since it can be incorporated in equations and computer programs to calculate parameters needed in industrial ventilation. Log-normal distributions can be characterized by two properties:

Geometric mean particle diameter based on number $(D_{p,gm})$
The geometric standard deviation (σ_g)

If an aerosol has a log-normal size distribution, the normalized *number distribution function* can be expressed as

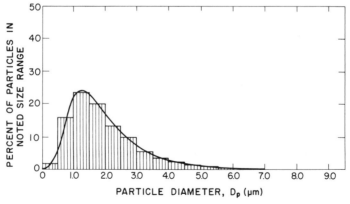

Fig. 8.3 Number distribution versus particle diameter for data in Table 8.3 (51).

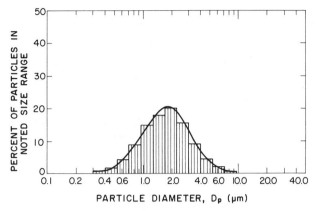

Fig. 8.4 Number distribution versus logarithm of particle diameter in Table 8.3 (51).

$$\frac{n}{n_t} = K \exp\left\{ -\frac{1}{2}\left[\frac{\ln\left(\frac{D_p}{D_{p,gm}}\right)}{\ln \sigma_g} \right]^2 \right\} \tag{8-19}$$

where

$$K = \frac{1}{\sqrt{2\pi}\, D_p \ln \sigma_g} \tag{8-20}$$

and σ_g is the *geometric standard deviation* which is related to the variance

$$\sigma_g = \exp[\sqrt{\ln(1+b)}]$$

$$b = \left[\frac{\sigma}{D_{p,am}} \right]^2 \tag{8-21}$$

The *geometric mean diameter* based on number $(D_{p,gm})$ is related to the arithmetic mean diameter based on number

$$D_{p,gm} = D_{p,am} \exp\left[-\frac{(\ln \sigma_g)^2}{2} \right] \tag{8-22}$$

The *cumulative number distribution* function $N(D_p)$ based on number is the total number of particles smaller than D_p. The cumulative distribution function will be written as N in the material that follows to simplify equations. The normalized cumulative distribution function for a log-normal distribution can be expressed in terms of the error function (erf)

$$\frac{N}{n_t} = \frac{1}{2}\left\{ 1 + \text{erf}\left[\frac{\ln(D_p/D_{p,gm})}{1.414 \ln \sigma_g} \right] \right\} \tag{8-23}$$

The fraction of particles between any two particles sizes $D_{p,1}$ and $D_{p,2}$ can be found by subtracting the values of the respective normalized cumulative distribution functions, viz.,

$$\frac{N(D_{p,2})}{n_t} - \frac{N(D_{p,1})}{n_t} \qquad (8\text{-}24)$$

There exists a unique particle diameter $(D_{p,50})$ called the *median particle diameter* such that when it is substituted in Eq. 8-23 the value of the normalized cumulative distribution function is one half, that is, $N/n_t = 0.5$,

$$\text{erf}\left[\frac{\ln\left(\dfrac{D_p}{D_{p,gm}}\right)}{1.414 \ln \sigma_g}\right] = 0 \qquad (8\text{-}25)$$

After simplification the equation above reduces to:

$$D_{p,gm} = D_{p,50} \qquad (8\text{-}26)$$

Thus for a log-normal size distribution, the median particle diameter based on number is equal to the geometric mean particle diameter based on number and the terms can be used interchangeably.

There also exists another unique particle diameter $(D_{p,84.1})$ defined by

$$\ln\left[\frac{D_{p,84.1}}{D_{p,gm}}\right] = \ln \sigma_g$$

$$\sigma_g = \frac{D_{p,84.1}}{D_{p,gm}} \qquad (8\text{-}27)$$

When $D_{p,84.1}$ is substituted into Eq. 8-23,

$$\frac{N}{n_t} = \frac{1}{2}\{1 + \text{erf}(0.707)\} = 0.841 \qquad (8\text{-}28)$$

The term $D_{p,84.1}$ corresponds to a particle whose normalized cumulative distribution function is 84.1%, for example, 84.1% of the particles are smaller than $D_{p,84.1}$. The geometric standard deviation can also be found from a cumulative distribution graph in which

$$\sigma_g = \frac{D_{p,84.1}}{D_{p,50}} \qquad (8\text{-}29)$$

The choice of subscripts in Eq. 8-29 is deliberate and indicates the particle diameters at which the normalized cumulative distributions are 84.1% and 50%. A similar analysis can be performed to show that

$$\sigma_g = \frac{D_{p,50}}{D_{p,15.9}} \qquad (8\text{-}30)$$

In the work to follow the subscripts will be rounded off to 16% and 84%. On the basis of number, 34% of the total number of particles are larger than $D_{p,gm}$ and 34% are smaller than $D_{p,gm}$.

The size distribution is determined from measurements of air samples withdrawn into instruments that measure the number, surface area, or mass of particles within various intervals of diameter. Computing the cumulative distribution may mask whether the aerosol has one or several modes. For example, if an aerosol consists of two slightly overlapping log-normal distributions, the cumulative distribution of the entire aerosol will not be log-normal and users may not be able to discern the separate modes.

To reveal the existence of different modes and their modal diameters, it is common practice to plot the data as

$$\frac{(N_i - N_{i-1})}{(\ln D_{p,i} - \ln D_{p,i-1})} \tag{8-31}$$

versus the mean diameter in the interval, where N_i is the cumulative number distribution. Why these variables reveal different modes in a distribution can be seen by differentiating the normalized cumulative distribution function with respect to $(\ln D_p)$

$$\frac{N}{n_t} = \int_0^{\ln D_p} \left(\frac{n}{n_t}\right) d(\ln D_p)$$

$$\frac{n}{n_t} = \frac{d(N/n_t)}{d(\ln D_p)} \tag{8-32}$$

Thus peaks in the value of the function of

$$\frac{d\left(\dfrac{N}{n_t}\right)}{d(\ln D_p)}$$

correspond to the existence of modes.

Figure 8.5 (169) depicts the origin and fate of particles in the atmosphere. Secondary submicron particles are formed from atmospheric photochemical reactions involving gases and vapors. *Secondary particles* begin as short-lived *nuclei* (particles less than 0.1 μm) that grow by processes of *agglomeration and coagulation* to sizes in the accumulation range (particles between 0.1 and 1.0 μm). *Primary particles* (particles larger than 1 μm) are emitted directly by man as solids or vapors that condense. Figure 8.6 (190) is the size distribution for an urban aerosol and confirms the existence of accumulation and primary mode. Plotting the data as a single cumulative distribution (191) masks the existence of the these two modes.

The discussion to this point has concerned only the number of particles of particular sizes. Many instruments used to analyze aerosols are based on light scattered by particles which depend on the square of the diameter. Other instruments discriminate particles on the basis of their mass which depends on the

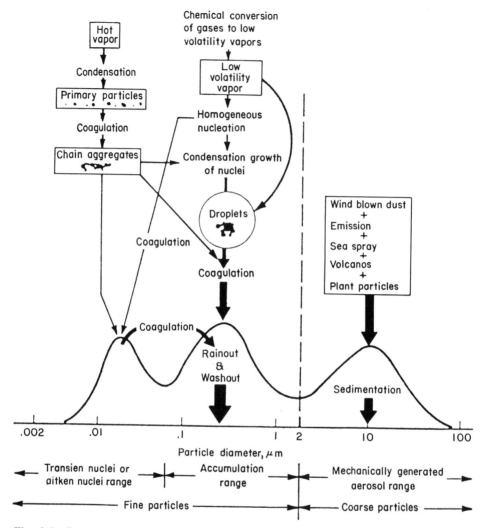

Fig. 8.5 Formation and removal mechanisms of atmospheric particles (with permission from ref. 169).

cube of the particle diameter. Consequently, there are particle area distributions and particle mass distributions. The expressions above have counterparts based on mass in which the normalized number distribution function (n/n_t) is replaced by a normalized *mass distribution function* (m/m_t) where m_t is the total mass of the aerosol, and the normalized cumulative number distribution (N/n_t) is replaced by the normalized *cumulative mass distribution function* (G/m_t). Again the mass distribution function (m/m_t) versus particle diameter can be drawn that is similar to Fig. 8.3. If the size distribution is log-normal, it can be shown (239) that the geometric standard deviation for all three distributions (number, area and mass) are the same and that the geometric mean diameter based on number and the geometric mean diameter based on mass are related by

$$\ln D_{p,gm} \text{ (mass)} = \ln D_{p,gm}(\text{number}) + 3[\ln \sigma_g]^2 \qquad (8\text{-}33)$$

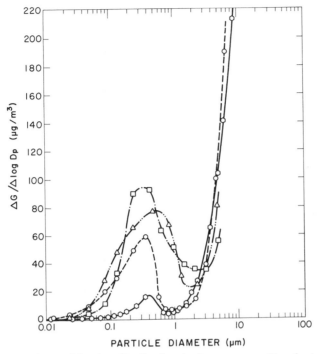

Fig. 8.6 Atmospheric particle size distribution in four communities (redrawn from ref. 190).

Special graph paper is printed for log-normal distributions called *log-probability paper* in which the cumulative distribution and particle diameter can be displayed. Conventional probability paper (for normal distributions) can also be used if log D_p is chosen as the variable. The paper is printed with the cumulative distribution scaled in such a way that graphs of log-normal distributions appear as straight lines. With such paper $D_{p,50}$ and σ_g can be computed easily using Eq. 8-27.

Illustrated Example 8.1—Particle Size Distribution

The concentration of foundry dust is particularly high in an area where castings are removed from their molds. Samples of air are taken and the particles are sized optically. Table 8.4 is a compilation of the measurements. Determine whether the particles have a log-normal size distribution.

The arithmetic mean diameter based on number, variance and standard deviation can be determined directly from Table 8.4 and are found to be

$$D_{p,am} = \sum_{i=1} n_i \frac{D_{p,i}}{n_t} = \frac{7781.5}{240} = 32.4 \ \mu m$$

$$\sigma^2 = \sum_{i=1} n_i \frac{(D_{p,am} - D_{p,i})^2}{n_t}$$

$$= \frac{42519.2}{240} = 177.2 \ \mu m^2$$

$$\sigma = 13.3 \ \mu m$$

TABLE 8.4   Foundry Dust Size Distribution for Illustrated Example 8.1

Class	Range	$D_{p,i}$	n_i	n_i/N_t	N/N_t	$n_i D_{p,i}$	$n_i(D_{p,am} - D_{p,i})^2$
1	5–6	5.5	0	0	0	0	0
2	6–9	7.5	0	0	0	0	0
3	9–13	11.0	2	0.008	0.008	22	915.9
4	13–18	15.5	29	0.121	0.129	449.5	8282.7
5	18–26	22.0	54	0.225	0.354	1188	5840.6
6	26–37	32.0	84	0.350	0.704	2688	13.4
7	37–52	42.5	54	0.225	0.929	2295	5508.5
8	52–73	62.5	14	0.058	0.987	875	12684.1
9	73–103	88.0	3	0.012	0.999	264	9274.1
Totals			240			7781.5	42519.3

The geometric standard deviation and geometric mean diameter based on number are found to be

$$\sigma_g = \exp\left[\left[\ln\left\{1 + \left[\frac{177.2}{(32.4)^2}\right]\right\}\right]^{0.5}\right] = 1.484$$

$$D_{p,gm} = 32.4 \exp\left\{-\frac{[\ln 1.484]^2}{2}\right\} = 30.0 \ \mu m$$

The slope of a log-normal distribution plotted on log-probability paper is found in terms of $D_{p,84}$ and $D_{p,16}$

$$D_{p,84} = (1.484)(30) = 44.5 \ \mu m$$

$$D_{p,16} = \frac{30}{1.484} = 20.2 \ \mu m$$

The geometric mean diameter based on mass is found to be

$$\ln D_{p,gm} \text{ (mass)} = \ln 30 + 3 \ [\ln 1.484]^2$$

$$D_{p,gm} \text{ (mass)} = 47.8 \ \mu m$$

Figures 8.7 and 8.8 show the results of these computations. The assumption of a log-normal size distribution is seen to be poor. Figures 8.7 and 8.8 show that the distribution is clearly not a Gaussian distribution but appears to be log-normal. The most accurate way to determine log-normality is to plot the cumulative distribution on log-probability paper. Figure 8.9 shows that the assumption of log-normality is satisfied in only an approximate fashion. Additional samples should be taken with instruments that divide particles into classes with smaller ranges than in Table 8.4.

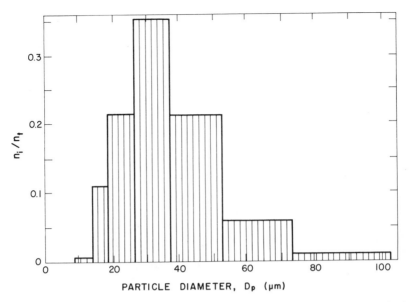

Fig. 8.7 Number distribution versus particle diameter for Illustrated Example 8.1.

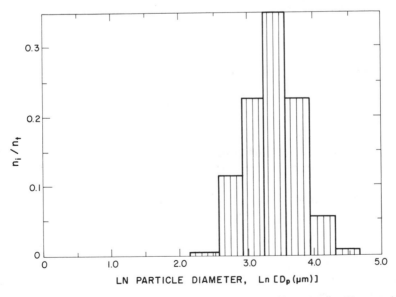

Fig. 8.8 Number distribution versus logarithm of particle diameter for illustrated Example 8.1.

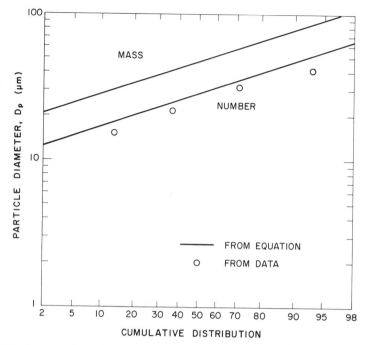

Fig. 8.9 Mass and number cumulative distributions for Illustrated Example 8.1.

8.3 OVERALL COLLECTION EFFICIENCY

Particles are removed from a gas stream by either settling or impacting other bodies and adhering to their surface. The removal process is traditionally expressed as a *grade*, or *fractional efficiency* in which the percent of particles of a particular size that are removed $\eta(D_p)$ is expressed as function of the particle size D_p and other operating parameters.

$$\eta(D_p) = 1 - \left[\frac{c(D_p)_f}{c(D_p)_i} \right] \tag{8-34}$$

where the subscripts i and f mean *initial* and *final* states. Fractional efficiency curves are obtained from experiment or predicted from first principles. Deriving these curves is a large subject in itself and will be developed later. What is important is how such data and knowledge of the aerosol size distribution can be used to predict the overall collection efficiency.

If an aerosol enters a device or process with a size distribution that can be expressed statistically, the aerosol will leave with another size distribution in which there is a higher percentage of small particles. The overall mass concentration is lower than when it entered, but the size distribution is skewed toward the smaller particles.

The *overall removal efficiency* by the process is equal to

$$\eta(\text{overall}) = 1 - \frac{c_f}{c_i} \tag{8-35}$$

where c_f and c_i are the overall mass concentration at the outlet and inlet of the device. The size distribution can be described by the number fraction or mass fraction of particles of a given size. For this analysis it is preferable to use the mass fraction. Define $g(D_{p,j})_i$, as the concentration of particles of a particular size j at the inlet $c(D_{p,j})_i$ divided by the total mass concentration at the inlet c_i,

$$g(D_{p,j})_i = \frac{c(D_{p,j})_i}{c_i} \tag{8-36}$$

The quantity $g(D_{p,j})_i$ is also called the inlet mass fraction. The overall collection efficiency $\eta(\text{overall})$, can now be expressed as

$$\eta = 1 - \sum_{j=1} \frac{c(D_{p,j})_f}{c_i} \tag{8-37}$$

where $c(D_{p,j})_f$ refers to the concentration of particles within size range j that appear at the outlet of the device. The fractional efficiency defined in Eq. 8-34 can be rearranged as follows:

$$c(D_{p,j})_f = c(D_{p,j})_i [1 - \eta(D_{p,j})] \tag{8-38}$$

Combine Eqs. 8-37 and 8-38 and simplify,

$$\eta = 1 - \sum_{j=1} [1 - \eta(D_{p,j})] \frac{c(D_{p,j})_i}{c_i}$$

$$\eta = 1 - \sum_{j=1} \frac{c(D_{p,j})_i}{c_i} + \sum_{j=1} \eta(D_{p,j}) \frac{c(D_{p,j})}{c_i} \tag{8-39}$$

$$\eta = 1 - 1 + \sum_{j=1} \eta(D_{p,j}) g(D_{p,j})_i$$

$$\eta = \sum_{j=1} \eta(D_{p,j}) g(D_{p,j})_i$$

Thus the overall collection efficiency is equal to the mass fraction for a certain range of particle sizes times the fractional efficiency for that size range summed over all the particle size ranges.

To compare collectors whose efficiencies are very close to 100%, it is useful to use the term penetration. The *penetration* (P) is defined as $(1 - \eta)$. Thus to compare two collectors whose efficiencies are 99.5% and 99.8%, emphasis can be gained by saying the penetration of one collector is 0.5% while the penetration of the other is 0.2%.

8.3.1 Collectors in Series

Consider a collection system consisting of several (n) collectors arranged in series, as shown in Fig. 8.10. The overall collection efficiency is

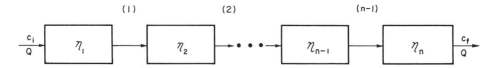

Fig. 8.10 A group of collectors in series.

$$\eta(\text{overall}) = 1 - \left(\frac{c_f}{c_i}\right) \tag{8-40}$$

$$= 1 - \left(\frac{c_1}{c_i}\right)\left(\frac{c_2}{c_1}\right)\left(\frac{c_3}{c_2}\right) \cdots \left(\frac{c_{n-1}}{c_{n-2}}\right)\left(\frac{c_f}{c_{n-1}}\right)$$

The efficiency of each collector can be expressed as

$$\eta_1 = 1 - \frac{c_1}{c_i}; \quad \frac{c_1}{c_i} = 1 - \eta_1 \tag{8-41}$$

$$\eta_2 = 1 - \left(\frac{c_2}{c_1}\right); \quad \left(\frac{c_2}{c_1}\right) = 1 - \eta_2 \tag{8-42}$$

$$\eta_n = 1 - \left(\frac{c_f}{c_{n-1}}\right); \quad \left(\frac{c_f}{c_{n-1}}\right) = 1 - \eta_n \tag{8-43}$$

Combining,

$$\eta(\text{overall}) = 1 - (1 - \eta_1)(1 - \eta_2) \cdots (1 - \eta_n) \tag{8-44}$$

Thus the overall collection efficiency is not the arithmetic average of the separate collection efficiencies, but unity minus the product of the penetration of each collector in series.

$$\eta(\text{overall}) = 1 - (P_1)(P_2) \cdots (P_n) \tag{8-45}$$

8.3.2 Collectors In Parallel

Consider two collectors arranged in parallel as shown in Fig. 8.11. Assume that the efficiencies and the volumetric flow rates through each collector are not in

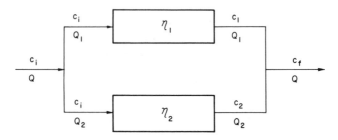

Fig. 8.11 Two collectors in parallel.

general the same although the concentration entering each collector is the same. The contaminant mass flow rate leaving each collector is

$$c_1 Q_1 = (1 - \eta_1) Q_1 c_i$$
$$c_2 Q_2 = (1 - \eta_2) Q_2 c_i$$

(8-46)

The overall collection efficiency of the parallel configuration is

$$\eta(\text{overall}) = 1 - \frac{[c_1 Q_1 + c_2 Q_2]}{[Q c_i]}$$

(8-47)

$$= 1 - \frac{[(1 - \eta_1) Q_1 c_i + (1 - \eta_2) Q_2 c_i]}{[Q c_i]}$$

$$= 1 - \left[(1 - \eta_1)\left(\frac{Q_1}{Q}\right) + (1 - \eta_2)\left(\frac{Q_2}{Q}\right) \right]$$

Thus the overall collection efficiency is not the arithmetic average of each collector's efficiency but depends on the manner in which the volumetric flow rates are split as well as the magnitudes of η_1 and η_2.

8.4 EQUATIONS OF PARTICLE MOTION

If the number of particles per unit volume of gas is below certain values (42), it will be assumed that particles move through the carrier gas independently of each other and that the particles do not influence the gas velocity field. The assumption is valid if particles do not collide with one another or do not pass through each other's wakes.

A useful rule of thumb that quantifies these two assumptions for a monodisperse aerosol is that the average distance between particles is at least 10 times the particle diameter. Assuming that 8 particles are located at the corners of a cube $L \times L \times L$, one finds that $L/D_p > 10$ when

$$\left(\frac{4\pi\rho_p}{3c}\right) = \left(\frac{8}{n_t D_p^3}\right) < 1000$$

(8-48)

where c is the mass of particles per volume of gas and n_t is the number of particles per volume of gas. Table 8.5 illustrates these upper limits. For water droplets, Eq. 8-48 corresponds to particle concentrations less than 4.2 kg/m^3 of air. For most

TABLE 8.5 **Particle Concentrations Beyond Which Flow Field is Dependent On Particle Concentration**

D_p (μm)	n (particles/m^3)
1.0	8×10^{15}
10.0	8×10^{12}
100.0	8×10^9

problems in air pollution and industrial ventilation the particle concentration is hundreds of times lower than this. Consequently, the motion of the carrier gas is independent of the motion of the particles.

To describe the motion of an aerosol, one must first compute the velocity field of the carrier gas and then compute the motion of the particles. The motion of the carrier gas can be expressed analytically if one is fortunate enough to have a system of simple geometry, or it may be established experimentally and the data stored numerically as a data file. If an analytical expression of the velocity field exists, one can compute the particle trajectories explicitly. For most industrial applications only experimentally measured velocity data are available and the computer will be needed to compute the particle trajectories.

The density of a particle is approximately 1000 times greater than the density of air. Thus the force of buoyancy on a particle is negligible with respect to its weight and can be neglected. The motion of a *single spherical particle* is given by,

$$\left(\frac{\pi D_p^3}{6}\right)\rho_p \frac{d\mathbf{v}}{dt} = -\left(\frac{c_D \rho}{2C}\right)\left(\frac{\pi D_p^2}{4}\right)(\mathbf{v} - \mathbf{U})|\mathbf{v} - \mathbf{U}|$$

$$-\left(\frac{\pi D_p^3}{6}\right)\rho_p \mathbf{g} \tag{8-49}$$

where $\mathbf{g}$ is the acceleration of gravity. To illustrate the use of this equation, examples of increasing complexity will be undertaken. Readers should try to cultivate physical intuition from these calculations so that they may have some way to check the appropriateness of solutions to more complex problems. Cases to be studied are:

Freely falling particles in quiescent air (gravimetric settling).
Horizontal motion in quiescent air (penetration distance).
Particles traveling through a moving air stream.

8.5 FREELY FALLING PARTICLES IN QUIESCENT AIR

Consider the motion of the spherical particle in quiescent air ($U = 0$) shown in Fig. 8.12. Since the only motion is downward, the instantaneous downward velocity $v_s(t)$ is,

$$v_s = -v_y \tag{8-50}$$

To illustrate a number of important concepts, consider only the motion of small particles in which the Reynolds numbers are always less than 1.0, that is, the condition called the *Stokes flow regime*. Thus,

$$c_D = \frac{24}{\text{Re}} = \frac{24\mu}{\rho D_p v_s} \tag{8-51}$$

Only the vertical component of Eq. 8-49 is to be solved. After substitution and simplification

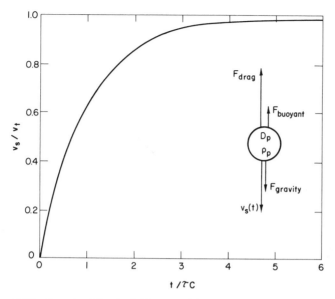

Fig. 8.12   Speed of freely falling particle versus time, Stokes flow.

$$\frac{dv_s}{dt} = g - \frac{v_s}{\tau C} \tag{8-52}$$

where

$$\tau = \frac{\rho_p D_p^2}{18\mu} \tag{8-53}$$

The term τ is called the *relaxation time* since it possesses the units of time and because it is customary to use this name when it appears in first-order differential equations like Eq. 8-52. If the particle starts from rest, the downward velocity is

$$v_s(t) = Cg\tau\left[1 - \exp\left(-\frac{t}{\tau C}\right)\right] \tag{8-54}$$

When $t \gg \tau$, a steady state condition occurs, which is equivalent to setting the left-hand side of Eq. 8-52 to zero. Under these conditions the downward velocity is called the *settling, fall,* or *terminal velocity* (v_t). Thus,

$$v_t = \tau g C \quad \text{for Re} < 1 \tag{8-55}$$

The variation of v_s with time is shown in Fig. 8.12. It is seen that it increases rapidly with time and achieves 63.2% of v_t after an elapsed time $t = \tau$ and 99.9% of its final value in $t = 7\tau$.

Since the particle starts from rest, the maximum velocity is the terminal velocity. If the Reynolds number is always to be less than unity, the results above pertain to particles and fluid such that

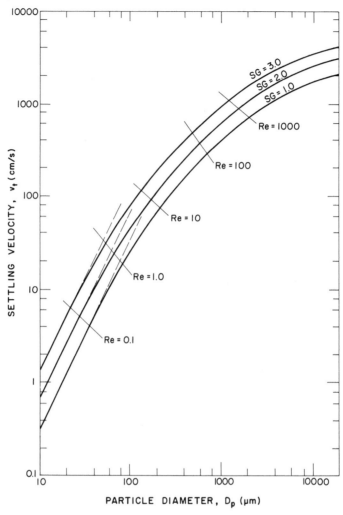

Fig. 8.13 Gravimetric settling velocity of spheres in air at STP; specific gravities (SG) of 1.0, 2.0, and 3.0.

$$\text{Re} = \rho D_{\text{p}} \frac{v_{\text{t}}}{\mu} \leq 1$$

$$= \frac{\rho \rho_{\text{p}} D_{\text{p}}^3 g C}{18 \mu^2} \leq 1 \tag{8-56}$$

The variation of the settling velocity v_{t} for spherical particles of different size and density in air at STP is shown in Fig. 8.13. It can be seen that the settling velocity varies with the square of the particle diameter in the Stokes regime. Table 8.6 shows values of the settling velocity (v_{t}) for several particles of specific gravity 1 i.e. unit density spheres in air at STP. It should be noted that the values are very small and that particles smaller than 100 μm have Reynolds numbers based on their settling velocity less than 1.0 and are therefore within the Stokes

TABLE 8.6 Gravimetric Settling Of Unit Density Spheres In Air at STP

D_p (μm)	τ (s)	C	v_t (m/s)	Re_t
0.01	3.069×10^{-10}	22.2	6.5×10^{-8}	4.3×10^{-11}
0.10	3.069×10^{-8}	2.9	8.5×10^{-7}	5.7×10^{-9}
1.00	3.069×10^{-6}	1.2	3.6×10^{-5}	2.4×10^{-6}
10.00	3.069×10^{-4}	1.0	3.1×10^{-3}	2.1×10^{-3}
100.00	3.069×10^{-2}	1.0	0.3	2
1000.00	3.069	1.0	4.0	267

regime. In industrial ventilation problems, characteristic times are of the order of seconds or minutes; consequently, the transient portion of vertical fall is of negligible importance. From a practical point of view the vertical velocity for these small particles (relative to air) can always be set equal to the settling velocity.

Suppose the particle is very large and the Reynolds number always above 1000, that is, *Newtonian flow regime*. The slip factor C is unity and will be omitted. The drag coefficient (c_D) is constant and will be set equal to 0.4. The steady-state vertical velocity (i.e., settling velocity), can be found by setting the left-hand side of Eq. 8-49 equal to zero. Thus for $Re > 1000$,

$$v_t = \sqrt{\left(\frac{4}{3}\right)\left(\frac{\rho_p D_p g}{0.4\rho}\right)} \tag{8-57}$$

A general expression for the settling velocity for particles of any diameter and Reynolds number can also be found by setting the left-hand side of Eq. 8-49 equal to zero. Thus one is obligated to solve the simultaneous equations

$$v_t = \left[\left(\frac{4C}{3c_D}\right)\left(\frac{\rho_p D_p g}{\rho}\right)\right]^{1/2} \tag{8-58}$$

$$c_D = 0.4 + \frac{24}{Re} + \frac{6}{(1 + Re^{1/2})} \tag{8-59}$$

$$Re = \rho \frac{D_p v_t}{\mu} \tag{8-60}$$

Illustrated Example 8.2—Settling Velocity

Write a general program to compute the settling velocity of a particle of any diameter and density falling in quiescent air at an arbitrary temperature and pressure. Compute the settling velocity for particles of density of 3000 kg/m^3 falling through air at 1 atm and 300 K. The particle diameter varies from 0.01 μm to 0.1 m.

The density of air can be found from the ideal gas law where R(air) = 0.287 kJ/kg K. The viscosity of air is given as a function of temperature by the power law expression in Appendix A-11.

```
  1 rem filename-settling velocity
  2 rem program computes settling velocity for spheres in
air
  3 rem user input: dp-diameter (micrometers)
          rp-particle density (kg/cu m)
          p-pressure (atm)
          T-temperature (K)
  4 rem physical properties: cd-drag coefficient
          vt-settling velocity (cm/s)
          vl-low velocity (cm/s)
          vh-high velocity (cm/s)
          va-avg velocity (cm/s)
          mu-air viscosity (kg/m s)
          ra-air density (kg/cu m)
  5 print ''dp (micrometers)          vt (cm/s)''
  6 print''_____    _____''
 10 dim d(8)
 15 for i=1 to 8
 20 read d(i)
 25 next i
 30 read rp, p,T
 35 rem mean free path (lam) is proportional to p/T,
    lam(STP)=0.066 micrometers
    lam (p,T)=0.0667 * (T/T0) * (p0/p)
 45 rem ra=p(atm) * 100 (kPa) / R(air) * T(K)
 50 ra=p * 100 / (0.287 * T)
 55 mu=(13.554+0.6738 * T-3.808 * T * T/10000
        +1.183 * T * T * T/10000000)/10000000
 60 for i=1 to 8
 65 dp=d(i)
 70 rem compute cunningham correction (slip) factor
 75 c=1+(lam/dp) * (2.514+0.8 * exp(-0.55 * dp / lam))
 80 vl=0.0000001
 85 vh=1000000
 90 va=(vl+vh)/2
 95 re=ra * dp * va / (mu * 100000000)
100 cd=0.4+(24 / re)+6 / (1+sqr(re))
105 a=4 * c * rp * dp * 9.8/(3 * cd * ra * 1000000)
110 v=100 * sqr(a)
115 r=v / va
120 if r>1 then vl=va
125 if r<1 then vh=va
130 if abs (r-1)<=0.1 then goto 140
135 goto 90
140 print dp,v
145 next i
150 data 0.01, 0.1, 1, 10, 100, 1000, 10000, 100000
155 data 3000, 1, 300
160 end
```

dp (micrometers)	vt (cm/s)
0.01	1.01999E-06
0.1	8.86383E-05
1	8.12171E-03
10	0.91789
100	82.36541
1000	685.06508
10000	2753.46481
100000	9098.76989

Consider two particles A and B that have the same settling velocity,

Particle A: perfect sphere of density 1000 kg/m^3
Particle B: nonspherical particle of unknown density

The *aerodynamic diameter* (D_a) is the diameter assigned to a particle of unknown density and shape that possesses the same settling velocity as a sphere of water. If the unknown particle has a settling velocity v_t(observed) within the Stokes regime,

$$D_a = \left[\frac{18 v_t(\text{observed})\mu}{\rho_w g C(D_a)} \right]^{1/2} \tag{8-61}$$

where $C(D_a)$ is the Cunningham factor based on the aerodynamic diameter and ρ_w is the density of water. If the unknown particle happens to be a sphere of diameter (D_p),

$$D_a = D_p \left[\frac{C(D_p)\rho_p}{C(D_a)\rho_w} \right]^{1/2} \tag{8-62}$$

where $C(D_p)$ and $C(D_a)$ are the Cunningham factors based on the actual and aerodynamic diameters and ρ_p is the actual particle density. For particles larger than 1 μm, the Cunningham slip factors are unity and

$$D_a = D_p \left[\frac{\rho_p}{\rho_w} \right]^{1/2} \tag{8-63}$$

For particles in a liquid medium settling is often called *sedimentation*. Sedimentation is described by the same model used for settling in a gaseous medium except that buoyancy cannot be neglected. As a result the equations above can be used, but the particle density should be replaced by the difference between the particle density and the density of water.

8.6   HORIZONTAL MOTION IN QUIESCENT AIR

To illustrate the dominating effect of viscosity to reduce the relative motion between the particle and carrier gas, consider the instantaneous horizontal velocity $v_x(t)$ of a sphere in quiescent air where the initial Reynolds number is less

TABLE 8.7   Penetration Distance Of Unit Density Spheres In Quiescent Air at STP, Initial Particle Velocity = 1 m/s

D_p (μm)	Penetration Distance (m)
0.01	68.18×10^{-10}
0.10	8.79×10^{-8}
1.00	3.57×10^{-6}
10.00	3.12×10^{-4}

than 1.0. Since subsequent Reynolds numbers will be smaller than the initial value, the particle's motion will be entirely within the Stokes regime. The differential equation for the horizontal velocity component is

$$\frac{dv_x}{dt} = -\frac{v_x}{\tau C} \tag{8-64}$$

If a particle is projected into quiescent air with an initial velocity $v_x(0)$, the horizontal velocity component at subsequent time is

$$v_x(t) = v_x(0) \exp\left[-\frac{t}{\tau C}\right] \tag{8-65}$$

The horizontal displacement can be found from

$$\int_0^{x(t)} dx = \int_0^t v_x \, dt = \tau C v_x(0)\left[1 - \exp\left(-\frac{t}{\tau C}\right)\right] \tag{8-66}$$

The maximum horizontal displacement is called the *penetration or pulvatation distance or stopping distance* ℓ, and is found by allowing $t \gg \tau$ Thus

$$\ell = v_x(0)\tau C \tag{8-67}$$

Table 8.7 is a compilation of penetration distances for several different sized water drops possessing initial Reynolds numbers of 1.0. It can be seen that penetration distances are small, indicating that viscosity damps relative motion very quickly. Illustrated Example 8.3 contains a program to compute the horizontal velocity component of spherical particles in quiescent air for arbitrary diameters and initial velocities.

8.7   PARTICLES TRAVELING IN A MOVING GAS STREAM

Consider the general motion of a spherical particle traveling through a moving two dimensional gas stream in which the air velocity (**U**) varies. Let the particle have an initial velocity [**v**(0)]. Figure 8.14 depicts such motion. The equation describing a particle's acceleration reduces to

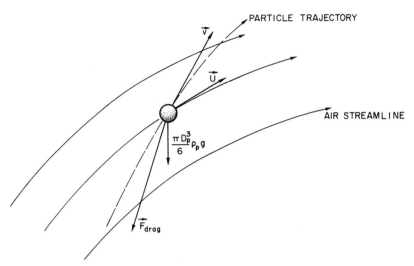

Fig. 8.14 Particle trajectory in moving air.

$$\frac{d\boldsymbol{v}}{dt} = -\mathbf{j}g - \left(\frac{3c_D}{4C}\right)\left(\frac{\rho}{\rho_p D_p}\right)(\mathbf{v} - \mathbf{U})|\mathbf{v} - \mathbf{U}| \qquad (8\text{-}68)$$

where the *relative velocity* $(\mathbf{v} - \mathbf{U})$ is

$$(\mathbf{v} - \mathbf{U}) = \mathbf{i}(v_x - U_x) + \mathbf{j}(v_y - U_y)$$

$$= \mathbf{i}v_{rx} + \mathbf{j}v_{ry} \qquad (8\text{-}69)$$

$$|\mathbf{v} - \mathbf{U}| = [v_{rx}^2 + v_{ry}^2]^{1/2} \qquad (8\text{-}70)$$

which reduce to a pair of coupled differential equations,

$$\frac{dv_x}{dt} = \left(\frac{-3c_D}{4C}\right)\left(\frac{\rho}{\rho_p D_p}\right)v_{rx}(v_{rx}^2 + v_{ry}^2)^{1/2} \qquad (8\text{-}71)$$

$$\frac{dv_y}{dt} = -g - \left(\frac{3c_D}{4C}\right)\left(\frac{\rho}{\rho_p D_p}\right)v_{ry}(v_{rx}^2 + v_{ry}^2)^{1/2} \qquad (8\text{-}72)$$

$$c_D = 0.4 + \frac{24}{\text{Re}} + \frac{6}{(1 + \text{Re}^{1/2})} \qquad (8\text{-}73)$$

$$\text{Re} = \rho D_p \frac{[v_{rx}^2 + v_{ry}^2]^{1/2}}{\mu} \qquad (8\text{-}74)$$

If the particle's motion is entirely within the Stokes regime, replacement of c_D by Eqs. 8-73 and 8-74 *uncouples the equations* and the above reduces to

$$\frac{dv_x}{dt} = -\frac{(v_x - U_x)}{\tau C} \qquad (8\text{-}75)$$

$$\frac{dv_y}{dt} = -g - \frac{(v_y - U_y)}{\tau C} \qquad (8\text{-}76)$$

To illustrate this kind of Stokes flow, consider the motion shown in Fig. 8.15 in which particles enter an air stream that is moving to right with a constant velocity $U_x(0)$. The initial particle velocity is $v_x(0)$ and $v_y(0)$. By direct integration, Eqs 8-75 and 8-76 become

$$v_x(t) = v_x(0) \exp\left[\frac{-t}{\tau C}\right] + U_x(0)\left[1 - \exp\left[\frac{-t}{\tau C}\right]\right] \qquad (8\text{-}77)$$

$$v_y(t) = v_y(0) \exp\left[\frac{-t}{\tau C}\right] - (g\tau C)\left[1 - \exp\left[\frac{-t}{\tau C}\right]\right] \qquad (8\text{-}78)$$

If the transients are neglected, $t \gg \tau$, the steady-state particle velocity becomes,

$$v(\text{steady state}) = \mathbf{i}U_x(0) - \mathbf{j}(g\tau C)$$
$$= \mathbf{i}U_x(0) - \mathbf{j}v_t \qquad (8\text{-}79)$$

This analysis justifies the commonplace assumption that small particles move horizontally at a speed equal to the carrier gas and simultaneously drift downward relative to the gas at a speed equal to the settling velocity. One must keep in mind that this is only true for Stokes flow and is only an approximation for other flow regimes. This behavior is often assumed as a general proposition, which may be adequate providing the accuracy of the analysis allows it.

If the Reynolds numbers are unknown and the flow regime is apt to be beyond the Stokes regime. The differential equations (Eqs. 8-71 to 8-74) remain coupled and numerical methods are required to compute the particle velocity and trajectory. For brevity, only two dimensional motion will be described. To solve these equations numerically, rewrite them as

$$\frac{dv_x}{dt} = Av_x + B \qquad (8\text{-}80)$$

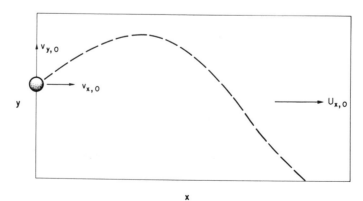

Fig. 8.15 Particle trajectory in a horizontal streaming flow.

$$\frac{dv_y}{dt} = Av_y + D \tag{8-81}$$

where

$$A = -\left(\frac{3c_D}{4C}\right)\left(\frac{\rho}{\rho_p D_p}\right)[(v_x - U_x)^2 + (v_y - U_y)^2]^{1/2} \tag{8-82}$$

$$B = -AU_x \tag{8-83}$$

$$D = -AU_y - g \tag{8-84}$$

The equations will be solved by numerical means in which the particle velocity and location will be computed at the end of a small interval of time δt. Knowing the initial location and velocity of the particle, the velocity and location will be computed at the end of the first interval, $t_1 = \delta t$. Using the velocity and location at t_1 as initial values, the process will be repeated and the velocity and location will be computed at the end of the second time interval $t_2 = t_1 + \delta t$. The process is then repeated until the particle travels some prescribed distance or for some prescribed elapsed time. The *Runge Kutta Method* (Appendix A-12) is recommended.

Illustrated Example 8.3—Trajectory Of Particle In A Moving Gas

Write a general program that predicts the velocity and location of particles traveling in a moving air stream. The program must be general and accommodate particles of arbitrary diameter, density, and initial location and velocity. The temperature and pressure of the air stream are to be input variables. Assume that the velocity field of the air stream can be expressed as an analytical function, $U(x, y)$. The program must be written in a manner that can be used as a subprogram in computer routines in which the air velocity field is obtained numerically. Apply the program to the following problem.

Analyze the motion of an aerosol consisting of drops of cutting oil used for lubrication and cooling on a high-speed punch. Air passes over a horizontal surface in the positive x direction with a velocity profile

$$U_x\left(\frac{m}{s}\right) = 1.2\left(\frac{y}{0.1}\right)^2$$

where y is measured in meters. The air has no vertical velocity component. The drops of cutting oil are injected in the negative x direction at an angle 30 degrees above the horizontal. Spherical particles of varying diameter leave the jet at 25 m/s. The jet is 0.1 m above the surface. The density of the drops is 3000 kg/m^3.

Workers claim that large drops are carried far downwind of the jet while small particles are not. The plant engineer claims that this is wrong, that the large particles will settle close to the jet and that no particles will be transported beyond 1 m from the jet. Compute the trajectories of the particles and decide who is right.

```
1 rem filename-particle trajectory
2 rem program computes the trajectories of spherical
  particles traveling through a moving stream of air
3 rem user input: rp-particle density (kg/cu m)
             dp-particle diameter (micrometers)
             U(x,y)-air velocity field
             P-air pressure (atm)
             T-temperature (K)
4 rem variables: ra-air density (kg/cu m)
             mu-air viscosity (kg/m s)
             re-Reynolds number
             uf, vf-particle velocity at end of
               time step (m/s)
             x,y-particle location at end of
               time step (m)
             ti,dt-elapsed time and time step (s)
5 print ''dp (micrometer)  time (s)  x (m)  y (m)''
6 print ''_____  _____  _____  _____''
10 read P,T,rp
15 k=0
20 dt=0.001
25 ra=p * 100 / (0.287 * T)
30 mu=(13.554+0.6738 * T-(3.808 * T * T/10000)
 +(1.183 * T * T * T/10000000)) / 10000000
35 dim d(4)
40 for i=1 to 4
45 read d(i)
50 next i
55 for i=1 to 4
60 dp=d(i)
65 ti=0
70 rem input initial particle location and velocity
75 x=0
80 y=0.1
85 ag=3.1415/6
90 v=25 * sin(ag)
95 u=-25 * cos(ag)
100 rem input expression for air velocity
105 ua=1.2 * (y/0.1) * (y/0.1)
110 va=0
115 ti=ti+dt
120 rem calculate relative velocity, Reynolds number and
    cd
125 vr=sqr((u-ua)(u-ua)+(v-va)(v-va))
126 re=ra * dp * vr/(mu * 1000000)
130 cd=0.4+(24 / re)+6 / (1+sqr(re))
135 rem calculate particle velocity and position at end of
    time step, calculate A, B and D in Runge-Kutta method
```

```
140 A=-0.75 * cd * ra * vr / (rp * dp / 1000000)
145 B=-A * ua
150 g=9.8
155 D=-(A * va)-g
160 uf=(u * (1+A * dt/2)+B * dt) / (1-A * dt/2)
165 vf=(v * (1+A * dt/2)+D * dt) / (1-A * dt/2)
170 rem print x and y every 25 iterations
175 k=k+1
180 if k=25 goto 215
185 rem calculate new values of x and y and reset initial
    values of u and v
190 x=x+dt * (uf+u) / 2
195 y=y+dt * (vf+v) / 2
200 u=uf
205 v=vf
210 goto 105
215 print dp,x,y
220 k=0
225 if y<=0.001 goto 235
230 goto 105
235 print
240 next i
245 data 1,298,3000
250 data 200, 100,50,25
255 end
```

Trajectories for these particles are shown in Fig. 8.16. The results show that the workers are right and the plant engineer is wrong. Unlike trajectories in air moving uniformly in the x direction, the calculations show that large particles

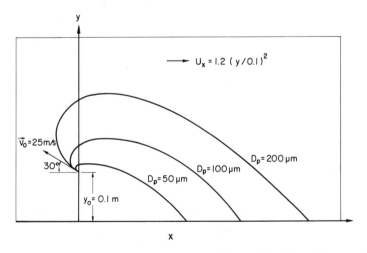

Fig. 8.16 Particle trajectories for Illustrated Example 8.3: initial particle speed = 25 m/s.

TABLE 8.8   Trajectories For Illustrated Example 8.3

D_p (μm)	Apogee			Ground		
	t(s)	x(m)	y(m)	t(s)	x(m)	y(m)
200	0.145	1.32	0.54	0.850	13.08	0
100	0.100	0.20	0.32	0.775	3.94	0
50	0.050	−0.03	0.19	1.050	1.56	0
25	0.025	−0.02	0.13	2.500	1.64	0

travel the farthest. The reason for this behavior is that large particles rise to greater values of y before their upward motion is damped by viscosity and gravity. At these large values of y, the larger particles encounter higher air velocities that sweep them farther downwind than the smaller particles.

8.8   GRAVIMETRIC SETTLING IN CHAMBERS

If one is interested in estimating the particle concentration in a chamber as a function of time, a crude model of the *sedimentation* process can be used that assumes that the particles have only two velocity components,

   Vertical component equal to the settling velocity
   Horizontal component equal to the carrier gas velocity

The assumption ignores transient behavior and assumes that the particles are in equilibrium with the carrier gas everywhere inside the chamber. It is often difficult to predict or measure carrier gas velocities in a chamber. Thus it is useful to consider an upper and lower limit to the particle concentration, knowing that reality lies somewhere in between. Define the upper limit as one where mixing is a maximum (well-mixed, turbulent model) and the lower limit where there is no mixing (laminar model).

   (a) *Laminar Model* (no mixing)—all particles of the same size fall at a uniform rate equal to the settling velocity and there is no mixing mechanism to redistribute the particles as they fall.
   (b) *Well-Mixed Model* (turbulent model)—all particles of the same size fall at a uniform rate equal to the settling velocity, but there is an idealized mixing mechanism that completely redistributes the remaining particles so that even though the concentration decreases with time, it is always uniform within the volume.

   Figure 8.17 illustrates the laminar and well-mixed models. Consider a chamber of volume V and horizontal cross-sectional area A, as shown in Fig. 8.17. In the laminar model all particles of the same size fall uniformly such that if $c(D_p)_0$ is the initial concentration of particles of diameter D_p in the chamber, then at any

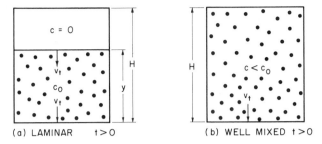

Fig. 8.17 Gravimetric settling in a chamber.

subsequent time there will be no particles of that size above a certain height $y(D_p)$ and a concentration $c(D_p)_0$ below this height. The same argument applies to particles of other sizes except the value of $y(D_p)$ will be different because they settle at a different velocity. The rate of change of mass of particles of this size within the chamber is equal to the rate of deposition.

8.8.1 Laminar Settling Model

$$\frac{d[Ayc(D_p)_0]}{dt} = Ac(D_p)_0 \frac{dy}{dt} = -v_t Ac(D_p)_0$$

$$\int_H^y dy = -\int_0^t v_t \, dt$$

$$y = H - (tv_t) \tag{8-85}$$

Let the average concentration of particles be $c(D_p)_a$.

$$c(D_p)_a = \frac{[Ayc(D_p)_0]}{V} = Ac(D_p)_0 \frac{(H - tv_t)}{V}$$

$$= c(D_p)_0 \left[1 - \frac{tv_t}{H}\right] \tag{8-86}$$

The average concentration of these particles will decrease linearly with time until a time t_c elapses,

$$t_c = \frac{H}{v_t} \tag{8-87}$$

and all of particles of the size whose settling velocity is v_t have settled to the floor.

8.8.2 Well-Mixed Settling Model

On the other hand, suppose the *well-mixed model* is valid. As particles of a particular size fall to the floor with a velocity v_t, a mixing mechanism redistributes the remaining particles throughout the room. The rate of change of mass of particles of this size within the room is equal to the rate of deposition,

$$\frac{d[c(D_p)V]}{dt} = V \frac{dc(D_p)}{dt} = -v_t c(D_p)A$$

$$\frac{c(D_p)}{c(D_p)_0} = \frac{c(D_p)_a}{c(D_p)_0} = \exp\left[\frac{-tv_t}{H}\right] \qquad (8\text{-}88)$$

Since the well-mixed model presumes that the concentration is the same throughout the enclosure at any instant of time, $c(D_p)$ in Eq. 8-88 also expresses the average concentration, $c(D_p)_a$.

The well-mixed model predicts that the average concentration decreases exponentially and the laminar model predicts that it decreases linearly. In the well-mixed model $c(D_p)_a/c(D_p)_0 = 0.368$ at $t = t_c$, while in the laminar model it is zero. Only after $t = 6.9t_c$ does the average concentration decrease to 0.001 of its initial value in the well-mixed model.

The laminar model leads one to believe that the dust will be removed too quickly because it ignores unavoidable thermal currents, drafts, and diffusion, that redistribute particles. The laminar model also predicts an infinite concentration gradient at an interface (see Fig. 8.17) which cannot exist in nature. The well-mixed model overestimates the time to clean the air because it exaggerates the mixing mechanisms. For small, inhalable particles it is, however, the more realistic model to use. Certainly it is a more conservative model.

8.9 GRAVIMETRIC SETTLING IN DUCTS

Gravimetric settling in ducts, also called sedimentation can be analyzed using the same concepts of a laminar model and the well-mixed model. Figure 8.18 illustrates the laminar and well-mixed models for flow in a horizontal duct of rectangular cross section ($A = WH$).

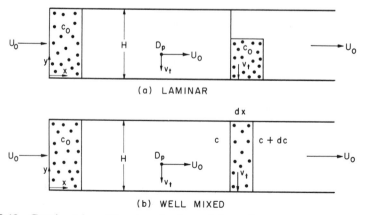

Fig. 8.18 Gravimetric settling in a horizontal duct of rectangular cross section.

8.9.1 Laminar Settling Model

Assume that the average duct velocity (U_0) is constant (i.e., *plug flow*). In the laminar model all particles of the same size fall at their settling velocity v_t and each possesses the horizontal velocity of the carrier gas (i.e., $v_x = U_0$). Thus at a downstream distance x, all particles of the same diameter have fallen the same distance and the uppermost particles have fallen a distance $(H - y)$,

$$H - y = tv_t = \left(\frac{x}{U_0}\right)v_t \tag{8-89}$$

The average concentration of particles of a certain size $c(D_p)_a$ distance x from the inlet can be written as

$$c(D_p)_a = \frac{yc(D_p)_0}{H} \tag{8-90}$$

Combining

$$\frac{c(D_p)_a}{c(D_p)_0} = 1 - \left(\frac{x}{H}\right)\left(\frac{v_t}{U_0}\right) \tag{8-91}$$

If a collection efficiency η is defined as

$$\eta = 1 - \frac{c(D_p)_a}{c(D_p)_0}$$

$$= \left(\frac{x}{H}\right)\left(\frac{v_t}{U_0}\right) \tag{8-92}$$

At a *critical distance* (downstream) $L_c = HU_0/v_t$, the collection efficiency will be 100% and the duct will contain no particles of the size defined by the settling velocity.

It can be shown (264) that the laminar collection efficiency given by Eq. 8-92 is valid for all laminar flow irrespective of the velocity profile so long as the profile does not change in the flow-wise direction, that is, so long as the flow is *fully established*, $\partial(\ )/\partial x = 0$.

8.9.2 Well-Mixed Settling Model

Following the assumptions of the well-mixed model used in Section 8.8, it can be assumed that the concentration $c(D_p)$ of particles of a particular diameter D_p is uniform in an element $(A\,dx)$. The difference in the mass concentration of particles entering and leaving the elemental volume is equal to the rate of deposition within the volume. Thus,

$$c(D_p)AU_0 = [c(D_p) + dc(D_p)]AU_0 + c(D_p)v_t W\,dx$$

$$\int_{c(D_p)_0}^{c(D_p)} \frac{dc(D_p)}{c(D_p)} = -\int_0^x \left(\frac{v_t}{U_0}\right)\frac{dx}{H}$$

$$\frac{c(D_p)}{c(D_p)_0} = \exp\left[-\left(\frac{v_t}{U_0}\right)\left(\frac{x}{H}\right)\right] \tag{8-93}$$

Using the definition above of collection efficiency, and realizing that for the well-mixed model $c(D_p)_a = c(D_p)$,

$$\eta = 1 - \exp\left[-\left(\frac{v_t}{U_0}\right)\left(\frac{x}{H}\right)\right] \qquad (8\text{-}94)$$

$$\eta = 1 - \exp\left[-\left(\frac{x}{L_c}\right)\right]$$

A comparison of the collection efficiencies for laminar and well-mixed models shows differences similar to those concluded for settling chambers.

(a)   The laminar model overestimates deposition because it ignores turbulence that mixes and redistributes particles. The well-mixed model exaggerates mixing but nevertheless provides a more accurate and conservative design estimate.

(b)   At a critical downstream distance L_c the well mixed model predicts a collection efficiency of 63.2% while the laminar model predicts 100%. At a downstream distance of $6.9L_c$ the well-mixed model predicts a collection efficiency of 99.9%.

It is common practice in industrial ventilation to design for duct velocities of 3500–4500 FPM (17.8–22.9 m/s) to minimize gravimetric settling. These expressions can be used to examine the settling one can expect at these velocities. Table 8.9 is a compilation of the duct lengths (expressed as x/H) at which 1% and 10% of unit density spheres settle.

It can be seen that 3500–4500 FPM is not an exaggerated recommendation. For particles less than 100 μm, settling is minimal in duct lengths up to 30 H; however, deposition is serious for larger particles. Thus it is a wise practice to provide a particle collector (cyclone or gravity drop-out chamber) whenever a ventilation system is connected to a hood and a significant concentration of large particles is expected. If this practice is not followed, a ventilation duct may unknowingly become a particle collector, which after a period of time reduces the duct cross sectional area and upsets the air distribution for the which the system

TABLE 8.9   Gravimetric Settling of Unit Density Spheres in Ducts; Duct Velocity = 22.86 m/s (4500 FPM)—Air at STP

D_p (μm)	v_t (m/s)	Efficiency = 1% $(x/H)_{lam} = (x/H)_{turb}$	Efficiency = 10% $(x/H)_{lam}$	$(x/H)_{turb}$
10	3.1×10^{-3}	74.6	745.8	785.8
50	7.3×10^{-2}	3.12	31.2	32.9
100	2.5×10^{-1}	0.92	9.2	9.7
200	0.7	0.33	3.3	3.4
600	2.5	0.09	0.9	1.0
1000	4.0	0.06	0.6	0.6

of ducts was designed. The volumetric flow rate in the duct will decrease and in the extreme the flow may cease.

If the flow in the rectangular duct is fully established, it can be shown that the expressions for the collection efficiency do not change for either model. Anand and McFarland (412) analyzed particle deposition for turbulent flow in transport lines of circular cross sectional area inclined at arbitrary angles to the horizontal. They found that there is an optimum inside diameter at which deposition is a minimum that is independent of the tube length but depends on particle diameter, volumetric flow rate and angle of inclination.

8.10 STOKES NUMBER

The *Stokes number* is an important dimensionless parameter in particle dynamics just as the *Reynolds* number is important in fluid dynamics. Rather than use the *Buckingham Pi theorem* (18) to show the uniqueness of the Stokes number, it is easier to transform the equation for particle motion into a dimensionless equation and allow the Stokes number to emerge through the derivation. This process is called *inspectional analysis*, as contrasted with *dimensional analysis*. To begin inspectional analysis, it is necessary to define a characteristic velocity U_0 and a characteristic length L. The characteristic time is equal to L/U_0. The characteristic velocity and length are merely two terms relevant to the problem that the engineer can easily identify and assign numerical values to. Typical values might be the velocity of an aerosol entering a device, the width or diameter of the inlet of a device. The choice of U_0 and L is generally dictated by the phenomena being modeled.

Consider a spherical particle traveling with a velocity v through a carrier gas that has a velocity U, see Fig. 8.14. The equation of motion for the particle is Eq. 8-49,

$$\left(\frac{\pi D_p^3}{6}\right)\rho_p \frac{d\mathbf{v}}{dt} = -\left(\frac{c_D \rho}{2C}\right)\left(\frac{\pi D_p^2}{4}\right)(\mathbf{v}-\mathbf{U})|\mathbf{v}-\mathbf{U}| - g\left(\frac{\pi D_p^3 \rho_p}{6}\right) \quad (8\text{-}95)$$

Gravity will be neglected since inertial and viscous effects are much larger. Define the Reynolds number in terms of the relative velocity as shown in Eq. 8-4.

$$\mathrm{Re} = \frac{\rho D_p |\mathbf{v}-\mathbf{U}|}{\mu}$$

Substitute the Reynolds number into Eq. 8-49 and simplify,

$$\frac{d\mathbf{v}}{dt} = -\frac{3}{4}\left(\frac{c_D \mu\, \mathrm{Re}}{C D_p^2 \rho_p}\right)(\mathbf{v}-\mathbf{U}) \quad (8\text{-}96)$$

To reduce Eq. 8-96 to a dimensionless equation, multiply and divide the velocity, length, and time by the known characteristic velocity U_0, length L, and time (L/U_0).

$$\frac{d\left[\frac{\mathbf{v}}{U_0}\right]U_0}{d\left[t\Big/\left(\frac{L}{U_0}\right)\right]\frac{L}{U_0}} = -\frac{3}{4}\left(\frac{c_D\mu\,\mathrm{Re}}{CD_p^2\,\rho_p}\right)\left[\left(\frac{\mathbf{v}}{U_0}\right) - \left(\frac{\mathbf{U}}{U_0}\right)\right]U_0$$

$$\frac{d\mathbf{v}^*}{dt^*} = -\left(\frac{18\mu}{D_p^2\,\rho_p}\right)\left(\frac{L}{U_0}\right)\left[\frac{c_D\,\mathrm{Re}}{24C}\right](\mathbf{v}^* - \mathbf{U}^*) \tag{8-97}$$

where the dimensionless particle and carrier gas velocties are $\mathbf{v}^* = \mathbf{v}/U_0$ and $\mathbf{U}^* = \mathbf{U}/U_0$ and the dimensionless time is $t^* = t/(L/U_0)$.

Define a dimensionless parameter called the *Stokes number* (Stk)

$$\mathrm{Stk} = \left[\frac{D_p^2\rho_p}{18\mu}\right]\left[\frac{U_0}{L}\right] = \tau\frac{U_0}{L} \tag{8-98}$$

The Stokes number is the ratio of the penetration distance (τU_0) (see Section 8.6) to the characteristic distance (L). Equation 8-97 can now rewritten as

$$\frac{d\mathbf{v}^*}{dt^*} = -\left(\frac{c_D\,\mathrm{Re}}{24\,C\,\mathrm{Stk}}\right)(\mathbf{v}^* - \mathbf{U}^*) \tag{8-99}$$

Since the differential equation is dimensionless, solutions to widely different problems are the same, providing the initial values of the dimensionless variables and the scaling parameter

$$\text{scaling parameter} = \frac{c_D\,\mathrm{Re}}{24\,C\,\mathrm{Stk}}$$

are the same. For example, providing the scaling parameters of the particles are the same, the dynamics of dust entering a personnel dust sampler will be similar to the dynamics of rain entering a jet engine during take-off. The dynamics of dust generated by trucks traveling on unpaved roads will be similar to the movement of welding fume from a workbench. In general, if two different physical phenomena (A and B) are such that the following is true

$$\left[\frac{c_D\,\mathrm{Re}}{24\,C\,\mathrm{Stk}}\right]_A = \left[\frac{c_D\,\mathrm{Re}}{24\,C\,\mathrm{Stk}}\right]_B \tag{8-100}$$

the motion of the particles written in dimensionless terms will be equivalent.

If in addition the Reynolds numbers of the two phenomena are less than unity (i.e., Stokes flow), the slip factors (C) are unity for both flows and the drag coefficients and Reynolds numbers are related by $c_D = 24/\mathrm{Re}$. In this case, the scaling parameter is the Stokes number.

8.11 INERTIAL DEPOSITION IN CURVED DUCTS

As an aerosol flows through a curved duct one can expect a particle's inertia to cause it to move radially outward. The velocity of the carrier gas can be obtained from the solution of the Navier–Stokes equations. For purposes of illustration, assume that the cross-sectional area is rectangular (width W) and that the tangential velocity is given by

$$U_\theta \, r^n = \text{constant} = C_n \qquad (8\text{-}101)$$

where the value of n lies between (or equal to) $+1$ and -1. When $n = 1$ the flow is irrotational (**Curl U** is zero) and when $n = -1$ the flow is called solid body rotation (constant vorticity). For lack of a better phrase, flow fields that can be described by Eq. 8-101 where $-1 < n < 1$, will be called n-degree rotational since irrotational implies the specific condition (**Curl U** $= 0$), while rotational is more broad (solid body rotation being a specific form of rotational flow).

$$\text{Irrotational: } n = 1, \text{ constant} = rU_\theta \, . \qquad (8\text{-}102)$$

$$\text{Solid body rotation: } n = -1, \text{ constant} = \frac{U_\theta}{r} \qquad (8\text{-}103)$$

$$n\text{-Degree rotational: } -1 < n < 1, \text{ constant} = U_\theta \, r^n \qquad (8\text{-}104)$$

Figure 8.19 shows the velocity profiles for rotational and irrotational flow. If the tangential velocity is given by Eq. 8-101, application of the continuity equation requires that $\dfrac{\partial U_r}{\partial r} = 0$. If in addition the flow is irrotational then U_r is zero.

The volumetric flow rate through a curved duct of constant rectangular cross section and constant radius is

$$Q = \int_{r_1}^{r_2} W U_\theta \, dr \qquad (8\text{-}105)$$

where W is the depth of the bend normal to the dimension r. Upon substituting Eq. 8-102 into 8-104, the constant in Eqs. 8-102 to 8-104 can be evaluated and the tangential velocity expressed as

Irrotational Flow $n = 1$:

$$C_n = \frac{Q}{W \ln \left(\dfrac{r_2}{r_1} \right)} \qquad (8\text{-}106)$$

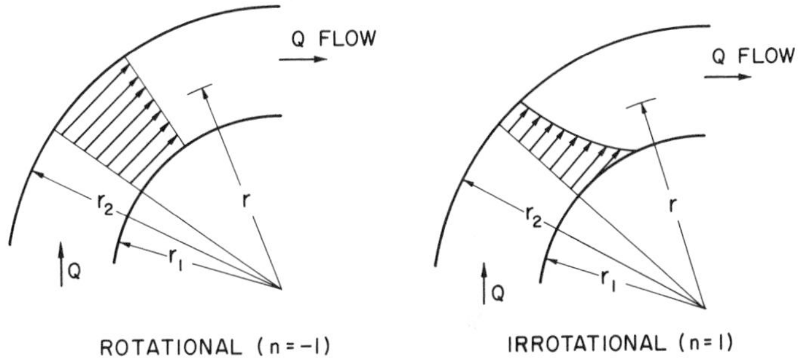

ROTATIONAL (n = -1)   IRROTATIONAL (n = 1)

Fig. 8.19 Velocity profiles for rotational and irrotational flow in a curved duct.

$$U_\theta = \left(\frac{Q}{W \ln\left(\frac{r_2}{r_1}\right)} \right) \frac{1}{r} \tag{8-107}$$

Solid Body Rotation $n = -1$:

$$C_n = \frac{2Q}{W(r_2^2 - r_1^2)} \tag{8-108}$$

$$U_\theta = \left(\frac{2Q}{W(r_2^2 - r_1^2)} \right) r \tag{8-109}$$

n-Degree rotational $-1 < n < 1$:

$$C_n = \frac{(1-n)Q}{W(r_2^{1-n} - r_1^{1-n})} \tag{8-110}$$

$$U_\theta = \left[\frac{(1-n)Q}{W(r_2^{1-n} - r_1^{1-n})} \right]\left(\frac{1}{r}\right)^n \tag{8-111}$$

Rather than solving the full set of equations describing the motion of particles in a moving flow field (Eqs. 8-80 and 8-81), a useful approximation can be had by assuming

(a)   Quasi-static equilibrium
(b)   Stokes flow
(c)   Gravimetric settling is insignificant

Consistent with the assumption of quasi-static equilibrium is the assumption that the tangential velocity of the particle and the gas are equal. In the tangential direction, there will be no relative motion between the particle and the carrier gas Fig. 8.20. In the radial direction, it will be assumed that the *centrifugal force* of the particle is balanced by the viscous drag. Since the gas velocity has no radial component ($U_r = 0$), the particle's relative velocity is its absolute velocity in the radial direction (v_r). Thus,

Tangential direction (θ):

$$v_\theta = U_\theta = \frac{C_n}{r^n} \tag{8-112}$$

Radial direction (r):

$$\rho_P\left[\frac{\pi D_P^3}{6} \right]\left[\frac{U_\theta^2}{r} \right] = c_D\left[\frac{\pi D_P^2}{4} \right]\left(\rho\, \frac{v_r^2}{2} \right)$$

$$v_r = U_\theta \sqrt{ \left(\frac{\rho_P}{\rho}\right)\left(\frac{4}{3c_D}\right)\left(\frac{D_P}{r}\right) } \tag{8-113}$$

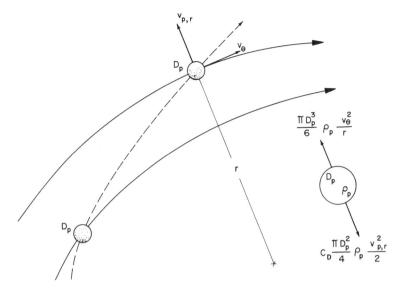

Fig. 8.20 Quasi-static equilibrium in curvilinear flow.

Lastly, it will be assumed that the particle's Reynold's number based on the relative velocity is less than unity such that Stokes flow can be assumed.

$$c_D = \frac{24}{Re} \tag{8-114}$$

$$Re = \rho \, \frac{D_p v_r}{\mu} \tag{8-115}$$

At any point in the flow field, the particle radial and tangential velocity components are

$$\text{Tangential } (\theta): \quad v_\theta = U_\theta = \frac{C_n}{r^n} \tag{8-116}$$

$$\text{Radial } (r): \quad v_r = \tau \, \frac{U_\theta^2}{r} \tag{8-117}$$

Deposition of particles on the outer wall of the bend will be modeled in a fashion similar to deposition in horizontal ducts (Section 8.9):

Laminar model
Well-mixed model (turbulent model)

8.11.1 Laminar Model

Figure 8.21 depicts the particle concentration downwind from the inlet of the bend. It will be useful to find the angle θ at which a particle entering the bend at a radius r impacts the outer wall. Such an expression can be found as follows:

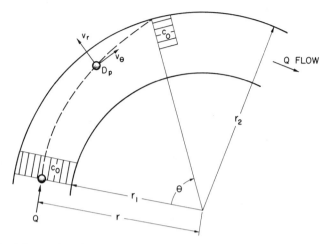

Fig. 8.21 Particle motion in a curved duct of rectangular cross section, laminar flow model.

$$\frac{v_\theta}{v_r} = \frac{\left[\dfrac{r\, d\theta}{dt}\right]}{\left[\dfrac{dr}{dt}\right]}$$

$$= r\frac{d\theta}{dr} = \frac{r}{\tau U_\theta}$$

$$= \frac{r^{1+n}}{\tau C_n} \tag{8-118}$$

where C_n is the constant in Eq. 8-101 and must be evaluated in terms of other flow parameters. For irrotational and n-degree rotational flow, (i.e., n not equal to -1),

$$\int_0^\theta d\theta = \int_r^{r_2} r^n \frac{dr}{\tau C_n}$$

$$r = [r_2^{n+1} - (n+1)\theta\tau C_n]^a \tag{8-119}$$

$$a = \frac{1}{1+n}$$

For solid body rotational flow ($n = -1$)

$$\int_0^\theta d\theta = \int_r^{r_2} \frac{dr}{r\tau C_n}$$

$$r = r_2 \exp[-\tau\theta C_n] \tag{8-120}$$

Define collection efficiency (η) as

$$\eta = \frac{(r_2 - r)}{(r_2 - r_1)} = \frac{\left[1 - \left(\dfrac{r}{r_2}\right)\right]}{\left[1 - \left(\dfrac{r_1}{r_2}\right)\right]} \qquad (8\text{-}121)$$

For irrotational flow ($n = 1$) the efficiency is

$$\eta_{n=1} = \frac{1 - \sqrt{1 - A\theta}}{1 - \left(\dfrac{r_1}{r_2}\right)}$$

$$A = \frac{2\tau Q}{Wr_2^2 \ln\left(\dfrac{r_2}{r_1}\right)} \qquad (8\text{-}122)$$

For solid body rotation ($n = -1$) the efficiency is

$$\eta_{n=-1} = \frac{1 - \exp(-A\theta)}{1 - \left(\dfrac{r_1}{r_2}\right)}$$

$$A = \frac{2\tau Q}{W(r_2^2 - r_1^2)} \qquad (8\text{-}123)$$

Equations 8-122 and 8-123 can be rearranged in terms of a Stokes number defined in terms of an average velocity,

$$U_{ave} = \frac{Q}{W(r_2 - r_1)}$$

$$Stk_a = \frac{\tau U_{ave}}{r_2} = \frac{\tau Q}{Wr_2(r_2 - r_1)} \qquad (8\text{-}124)$$

For irrotational flow ($n = 1$), the efficiency becomes

$$\eta_{n=1} = \frac{1 - \sqrt{1 - 2K\theta\, Stk_a}}{1 - \left(\dfrac{r_1}{r_2}\right)}$$

$$K = \frac{r_2 - r_1}{r_2 \ln\left(\dfrac{r_2}{r_1}\right)} \qquad (8\text{-}125)$$

and for solid body rotation ($\eta = -1$)

$$\eta_{n=-1} = \frac{1 - \exp(-2K\theta\, Stk_a)}{1 - \left(\dfrac{r_1}{r_2}\right)}$$

$$K = \frac{r_2}{r_1 + r_2} \qquad (8\text{-}126)$$

A useful parameter to consider is the critical angle (θ_c) at which all particles entering the bend impact on the outer wall. For the laminar model and for irrotational flow, the critical angle is

$$\theta_c = \frac{(r_2 + r_1) \ln\left(\frac{r_2}{r_1}\right)}{2r_2 \, \text{Stk}_a} \tag{8-127}$$

8.11.2 Well-Mixed Model

The well-mixed model assumes that within the flow field there exists a mixing mechanism that redistributes particles remaining in the duct after those along the outer radius impact the outer wall. For this reason, the particle concentration varies with angle θ but not the radius. To analyze the flow, begin by constructing a mass balance for an element of the flow as shown in Fig. 8.22.

$$cQ = Q(c + dc) + cv_r(r_2)Wr_2 \, d\theta$$

$$\int_{c_0}^{c_\theta} \frac{dc}{c} = \int_0^\theta \frac{v_r(r_2)Wr_2}{Q} \, d\theta \tag{8-128}$$

where $v_r(r_2)$ is the particle radial velocity (Eq. 8-117) evaluated at the outer radius (r_2). Define the collection efficiency as

$$\eta = \frac{1 - c_\theta}{c_0} \tag{8-129}$$

The right hand side of Eq. 8-129 can be found by integrating Eq. 8-128, replacing $v_r(r_2)$ by Eq. 8-117, simplifying and introducing the Stokes number based on the

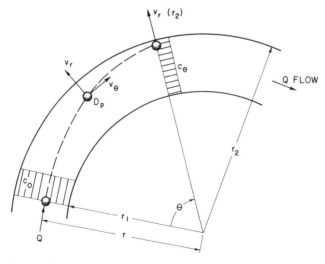

Fig. 8.22 Particle motion in a curved duct of rectangular cross section, well-mixed model.

average velocity (Eq. 8-124). As a consequence of these substitutions and simplifications the efficiency can be written as

$$\eta = 1 - \exp[-K\theta \, \mathrm{Stk_a}] \qquad (8\text{-}130)$$

where

$$K(\text{irrotational}) = \frac{r_2 - r_1}{r_2 \left[\ln\left(\dfrac{r_2}{r_1}\right) \right]^2} \qquad (8\text{-}131)$$

$$K(\text{solid body rotation}) = \frac{4r_2^3}{(r_2 + r_1)(r_2^2 - r_1^2)} \qquad (8\text{-}132)$$

$$K(n\text{-degree rotational}) = \frac{(1-n)^2(r_2 - r_1)r_2^{1-2n}}{(r_2^{1-n} - r_1^{1-n})^2} \qquad (8\text{-}133)$$

Table 8-10 summarizes the equations for the inertial impaction efficiency for curvilinear flow.

The assumption of quasi static equilibrium is valid for a large class of flows when one recalls the analysis of Section 8.5. The time required for small particles to achieve equilibrium is very small and only several times the relaxation time. Thus it is reasonable to assume viscosity acts rapidly on particles flowing in a curved duct. The assumption of Stokes flow is also not restrictive since it is only the particle's radial velocity component that produces the relative velocity. The radial velocity is small since it depends on the relaxation.

TABLE 8.10   Impaction of Particles on Outer Walls of Curved Duct with Rectangular Cross Section

$$\mathrm{Stk_a} = \frac{\tau Q}{W r_2 (r_2 - r_1)} \qquad\qquad \theta_c = (r_2 + r_1)\ln\left(\frac{r_2}{r_1}\right) \Big/ 2 r_2 \mathrm{Stk_a}$$

Flow	$U_\theta = \text{const } r^n$	Impaction Efficiency	
		Laminar Model	Well-Mixed Model
Irrotational $(n = 1)$	$\left[\dfrac{Q}{W\ln\left(\frac{r_2}{r_1}\right)}\right]\dfrac{1}{r}$	$\dfrac{1 - \left[1 - \dfrac{2\theta(r_2 - r_1)\mathrm{Stk_a}}{r_2 k \frac{r_2}{r_1}}\right]^{1/2}}{1 - \frac{r_1}{r_2}}$	$1 - \exp\left[-\dfrac{\theta(r_2 - r_1)\mathrm{Stk_a}}{r_2 (\ln\frac{r_2}{r_1})^2}\right]$
Solid Body Rotation $(n = -1)$	$\left[\dfrac{2Q}{W(r_2^2 - r_1^2)}\right] r$	$\dfrac{1 - \exp\left[-\dfrac{2\theta r_2 \mathrm{Stk_a}}{(r_2 + r_1)}\right]}{1 - \frac{r_1}{r_2}}$	$1 - \exp\left[-\dfrac{4\theta r_2^3 \mathrm{Stk_a}}{(r_2 + r_1)(r_2^2 - r_1^2)}\right]$
n-Rotational $-1 < n < 1$	$\left[\dfrac{(1-n)Q}{W(r_2^{1-n} - r_1^{1-n})r^n}\right]\left(\dfrac{1}{r}\right)^n$	$\dfrac{1 - \left[1 - \dfrac{(1-n^2)\,\theta r_2(r_2 - r_1)\mathrm{Stk_a}}{r_2^{1+n}(r_2^{1-n} - r_1^{1-n})}\right]^{\frac{1}{1-n}}}{1 - \frac{r_1}{r_2}}$	$1 - \exp\left[-\dfrac{(1-n)^2\theta(r_2 - r_1)r_2^{1-2n}\mathrm{Stk_a}}{(r_2^{1-n} - r_1^{1-n})^2}\right]$

The restrictive assumption in the analysis above is that the gas velocity is given by

$$U_\theta r^n = \text{constant} \tag{8-134}$$

The velocities entering the bend depend on conditions upstream of the inlet. If air enters the bend from the atmosphere, the inlet velocity profile will be flat. If a long rectangular duct precedes the bend, the velocity profile entering the bend will be fully established, or well on its way to becoming so. In either case, the velocity profile in the bend would not be given by the above. The velocity profile is also dictated by the pressure distribution in the bend. The radial and tangential pressure gradients for irrotational and rotational flow are quite different. A thorough analysis of particle motion depends on solving the equations of motion of the air in the bend.

8.11.3 Cyclone Collectors

The expressions above can be used to estimate the particle collection efficiency of a large class of particle collection devices called cyclones. Shown in Figs 8.23 and 8.24 are schematic diagrams of *straight-through* and *reverse-flow* cyclone collectors. Cyclone collectors can be made of inexpensive materials to capture high-temperature aerosols and liquid particles. Cyclones have no moving parts and may have a modest pressure drop. As a rule cyclones do not remove very small particles although they can be designed to do so. Traditionally, cyclones are used to remove coarse particles and find use inside coal-fired boilers, inlets to military gas turbines for helicopters, air inlet cleaners for diesel engines in construction vehicles and so on. Cyclones are also used in powder technology, in which case one is interested in increasing the concentration of particles in an air stream. Cyclones used for this purpose are called *concentrators*.

It is necessary to bleed air from straight through cyclones to remove the particles that migrate radially outward. If used as a concentrator, bleed air

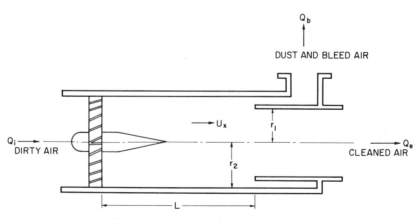

Fig. 8.23 Straight-through cyclone.

transports particles downstream to the next step in the process. If used as an air cleaning device, bleed air reduces the amount of clean air available for use. If a reverse flow cyclone is mounted with its conical hopper beneath the outlet, it is not necessary to bleed air to remove particles. Particles migrating radially outward strike the cyclone walls and fall downward into the hopper where they can be removed at a later time as a batch process or they can be removed continually by hopper valves. In either event, only a minimal amount of bleed air is required.

Consider the straight-through cyclone shown in Fig. 8.23. The helical turning vanes (angle α) impart rotation to the air such that

$$\tan \alpha = \frac{U_\theta}{U_x} \tag{8-135}$$

Flow through the cyclone can be conceived as a wrapped-around bend of angle θ (in radians) where

$$\theta = \left[\frac{U_\theta(r_2)}{r_2} \right]\left[\frac{L}{U_x} \right]$$

$$= \left(\frac{L}{r_2} \right) \tan \alpha \tag{8-136}$$

and r_2 is the outer radius of the straight-through cyclone. The *number of turns* is equal to $(\theta/2\pi)$. The width of the bend (W) can be scaled as follows:

$$\frac{W}{L} = \frac{2\pi}{\theta}$$

$$W = \frac{2L\pi}{\theta} \tag{8-137}$$

Flow through cyclones is turbulent, but is neither irrotational nor of constant vorticity. A realistic estimate is difficult to make, but assuming the flow to be n-degree rotational with n equal to 0.5 has been assumed by Strauss (44) for the reverse-flow cyclone. Assuming that $n = 0.5$ can also be used for straight-through cyclones an overall collection efficiency for straight through cyclones can be given by Eq. 8-130

$$n = 1 - \exp\left\{ -\frac{L(\tan \alpha)^2 \tau Q}{8\pi r_2^3(r_1 + r_2 - \sqrt{r_1 r_2})} \right\}$$

A reverse-flow cyclone in Fig. 8.24 can also be conceived as a wrapped-around bend of angle θ where θ is obtained from Eq. 8-137. The width (W) of the bend can be approximated as the width of the inlet duct. The overall collection efficiency can be expressed by Eq. 8-130 with K given by Eq. 8-131 to 8-133 with r_1 replaced by the radius of the exit duct ($D_e/2$).

The value of cyclone collectors is underestimated. Reverse cyclones have been used for decades and while their ability to remove small particles is limited, it is suspected that this is basically a reflection of the limited ability to model the flow

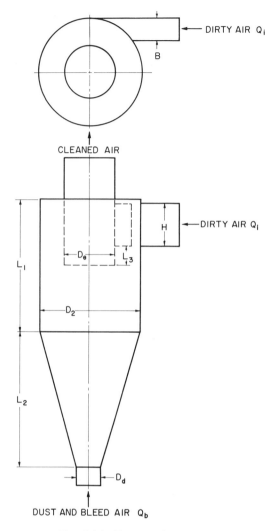

Fig. 8.24 Reverse-flow cyclone.

field. With the advent of numerical modeling programs to predict the flow field inside cyclones it is believed that superior performance may be achieved. Figure 8.25 (197) shows the predicted velocity field in a reverse-flow cyclone and illustrates the capabilities of contemporary fluid-modeling techniques. Since the numerical methods improve steadily with time, there is reason to be optimistic that the performance of cyclones will be improved so that they can be made to capture small particles with a tolerable pressure drop.

Shown in Fig. 8.26 are the dimensions of a *standard cyclone* which have received a great deal of attention for many years, see for example (19, 41, 44, 45, 142, 146, 197, 416). Standard cyclones are also called *Stairmand-type* cyclones. Geometrically similar *standard reverse-flow cyclones* are those whose dimensions scale as shown in the table in Fig. 8.26. The attraction for standard cyclones is that their performance can be described by a single *grade efficiency* or *fractional*

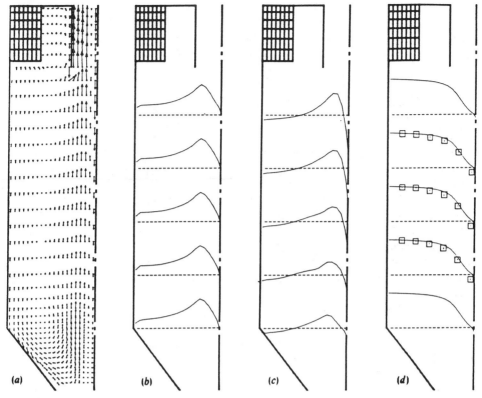

Fig. 8.25 Predicted velocity and pressure distribution in a reverse-flow cyclone; (*a*) maximum component of velocity; (*b*) tangential velocity; (*c*) axial velocity; (*d*) pressure distribution with experimental data (with permission, Transaction Institution of Chemical Engineering, Vol. 60, 1982, pp. 222–230.

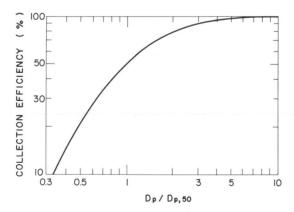

Fig. 8.26 Dimensions and performance of a standard reverse-flow cyclone (redrawn from ref. 146). Cut diameter $= D_{p,50} = \sqrt{\left(\dfrac{3}{4\pi}\right)\left(\dfrac{\mu B^2 H}{Q \rho_p}\right)}^{0.5}$. $L_1 = L_2 = 2D_2$. $D_e = H = \frac{1}{2}D_2$. $B = D_d = \frac{1}{4}D_2$. $L_s = \frac{1}{8}D_2$.

efficiency curve shown in Fig. 8.26. The abscissa of the graph is the particle diameter divided by the *cut diameter*, which is defined as the diameter of a particle that can be removed with a collection efficiency of 50%. The value of the cut-diameter is given by the equation shown in Fig. 8.26.

8.12 IMPACTION BETWEEN MOVING PARTICLES

Impaction is the general name given to the collision of a particle with a collecting surface as the air carrying the particle passes around the collecting surface. When used in this text, *impaction* will include the additional effects of interception and diffusion. The reader should consult standard texts on particle dynamics (19, 40–45, 142) to obtain a full understanding of impaction, interception and diffusion. *Interception* accounts for the fact that even though the particle's center of gravity may not collide with a collecting surface, collision still occurs if the distance between the path of the center of gravity and surface is less than the particle's radius. *Diffusion* accounts for the fact that in addition to inertial effects, particles exhibit Brownian movement and very small particles will migrate toward a collecting surface and be removed from the gas stream. A spherical particle's diffusion coefficient (Eq. 4-59) is inversely proportional to its diameter such that above 1 μm, diffusion is negligible. For convenience, all three processes (impaction, interception, and diffusion) occur simultaneously and hereafter will be subsumed under the phrase *impaction*.

Impaction occurs between small particles and large bodies so long as there is relative motion between the two, for example, rain drops falling through quiescent dusty air or a high-speed aerosol passing through slower moving water droplets in a venturi scrubber. For generality, consider the impaction between a small particle (D_p) possessing the velocity of the carrier gas and a large collecting body (D_c) moving through the gas stream at a different speed (Fig. 8.27). Assume both bodies (D_p and D_c) are spheres. Throughout this section the term

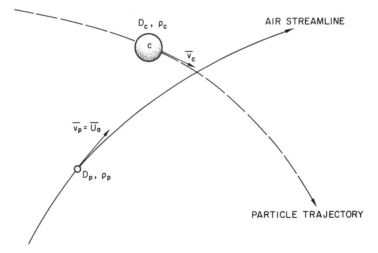

Fig. 8.27 Inpaction between a small particle and a collector particle.

"small particle" refers to the small contaminant particle (D_p) that is to be removed by the larger "collecting particle" of scrubbing liquid denoted by (D_c).

It will be assumed that if a small particle impacts a large particle, it will stick and be removed from the gas stream. The consequent growth in size and mass of the collecting particle will be neglected. The effectiveness of the removal process is expressed as the *single drop or particle collection efficiency* (η_d) defined as the rate (by mass) with which particles are removed by impacting with the collecting drop or sphere (D_c), divided by the mass rate of flow of particles in a stream tube of cross sectional area equal to the collecting sphere.

$$\eta_d = \frac{\text{mass removal rate}}{c_0 v_r \left[\dfrac{\pi D_c^2}{4} \right]} \tag{8-139}$$

where v_r is the speed of the gas stream carrying the particles relative to the collector and c_0 is the concentration of contaminant particles in the gas stream approaching the collector (Fig. 8.28). In terms of the average particle concentration in the stream tube (of cross sectional area ($\pi D_c^2/4$) approaching the collector (c_0) and the average concentration downstream of the collector (c_f), the single particle collection efficiency can also be written as

$$\eta_d = 1 - \frac{c_f}{c_0} \tag{8-140}$$

Similar expressions can be defined for impaction of spheres on cylinders or any pair of impacting bodies.

Calvert and Englund (142) recommend that for potential flow and viscous flow in which the Stokes numbers exceed 0.2, the single particle collection efficiency of *spheres impacting on spheres* can be approximated by

$$\eta_d = \left[\frac{\text{Stk}}{\text{Stk} + 0.7} \right]^2 \tag{8-141}$$

where

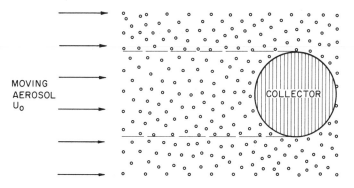

Fig. 8.28 Single-particle, single-fiber collection efficiency.

$$\text{Stk} = \frac{\tau_p v_r}{D_c} \qquad (8\text{-}142)$$

$$\tau_p = \frac{\rho_p D_p^2}{18\mu} \qquad (8\text{-}143)$$

v_r = velocity of particle relative to the collector

$$v_r = v_p - v_c \qquad (8\text{-}144)$$

v_p and v_c = velocity of particle and collector $\qquad (8\text{-}145)$

A graph of the single particle collection efficiency for spheres impacting on spheres shows that the η_d is zero at Stk = 0.083 and asymptotically approaches unity for large values of Stk.

8.12.1 Spray Chambers

To illustrate impaction as a method to control small particles, consider the impaction between a descending stream of collecting water drops (D_c) and an ascending stream of small particles (D_p). In nature this occurs when falling rain drops impact dust suspended in the air. In this case the process is called *scavenging* or *washout*. Two industrial processes in which scavenging is used are

  spray chamber used to remove small particles of cutting fluid generated by a high speed punch

  water spray used control dust generated by a conveyor that discharges material to a stockpile

The overall effectiveness of such a *counterflow spray chamber* (Fig. 8.29) can be modeled using an approach suggested by Crawford (19). Assume the following:

(1) The height of the control region is L. The number of liquid drops per unit volume (n_c), the diameter of the drops (D_c) and the absolute velocity (v_c) of the falling drops are constant.

(2) The diameter of the small particles (D_p) is constant and the concentration varies only with height.

(3) The volumetric flow rate of air (Q_a) and spray liquid (Q_s) are constant.

(4) Neglect gravimetric settling of the small particles; assume that the velocity of the small particles (v_p) is equal to the air velocity (U_a) and that it is constant.

$$v_p = U_a \qquad (8\text{-}146)$$

(5) The velocity of impaction is the velocity of the particle relative to the collector,

$$v(\text{impaction}) = v_p - v_c = U_a - v_c \qquad (8\text{-}147)$$

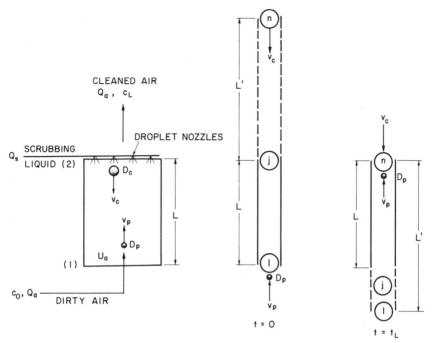

Fig. 8.29 Schematic diagram of a spray chamber illustrating the number of encounters between the particles and the falling drops.

Because the air and particles flow upward and the drops fall downward, the impaction velocity is the sum of the air speed and the drop speed, which in turn is equal to the settling velocity ($v_{t,c}$) of the collector particle in quiescent air

$$v_{t,c} = U_a + v_c \tag{8-148}$$

The number of collecting drops per unit volume of air (n_c) can be found from the following:

$$Q_s = n_c \left(\frac{\pi D_c^3}{6} \right) v_c A$$

$$n_c = \frac{6Q_s}{v_c A \pi D_c^3} \tag{8-149}$$

Multiplying and dividing by the volumetric flow rate of air (Q_a) where $Q_a = U_a A$,

$$n_c = \left(\frac{6Q_s}{v_c A \pi D_c^3} \right) \left(\frac{U_a A}{Q_a} \right)$$

$$n_c = \left(\frac{Q_s}{Q_a} \right) \left(\frac{6}{\pi D_c^3} \right) \left(\frac{U_a}{v_{t,c} - U_a} \right) \tag{8-150}$$

Since the only spatial variable is the height, collecting drops can be imagined to travel one behind the other in a column of height $L + L'$. Impaction occurs as small particles rise through a height L and encounter n drops. The single drop impaction efficiency (η_d) is constant since the velocities are constant and can be calculated from Eq. 8-141. The parameter n is the number of times small particles encounter collector drops as the small particles travel upward through a height L. The overall removal of small particles can be imagined as the flow of an aerosol through a stream tube containing n collectors in series, each with a single particle collection efficiency (η_i). Thus the overall collection efficiency for the column can be constructed as follows:

$$\eta_L = 1 - \frac{c_L}{c_0} = 1 - \left(\frac{c_L}{c_n}\right)\left(\frac{c_n}{c_{n-1}}\right) \cdots \left(\frac{c_3}{c_2}\right)\left(\frac{c_2}{c_1}\right)\left(\frac{c_1}{c_0}\right) \tag{8-151}$$

where n is number of collectors (number of encounters). Each of the ratios (e.g., c_2/c_1) can be expressed in terms of the single particle collection efficiency η_d. Thus

$$\eta_L = 1 - \frac{c_L}{c_0} = 1 - (1 - \eta_d)^n \tag{8-152}$$

where c_L and c_0 are the small particle concentrations at the exit and entrance of the chamber.

The total number of encounters can be deduced using Fig. 8.29. It must be kept in mind that as the small particles travel upward, n drops travel downward such that at the instant a particle has risen through the height L, drop number n is being encountered. The total number of encounters is

$$n = n_c(L + L')\frac{\pi D_c^2}{4} \tag{8-153}$$

$$= \left(\frac{Q_s}{Q_a}\right)\left(\frac{6}{\pi D_c^3}\right)\left(\frac{U_a}{v_c}\right)\left(\frac{\pi D_c^2}{4}\right)(v_c t_L + U_a t_L)$$

where t_L is the time it takes for the small particles to rise a height L, which is also the time it takes for drop number n to fall through a height L'.

$$t_L = \frac{L}{U_a} = \frac{L'}{v_c} \tag{8-154}$$

Thus

$$n = \left(\frac{3}{2}\right)\left(\frac{Q_s}{Q_a}\right)\left(\frac{U_a}{v_c}\right)\left(\frac{t_L}{D_c}\right)(v_c + U_a)$$

$$n = \left(\frac{3}{2}\right)\left(\frac{Q_s}{Q_a}\right)\left(\frac{v_{t,c}}{v_c}\right)\left(\frac{L}{D_c}\right) \tag{8-155}$$

The overall collection efficiency of spray chambers is inherently low. Since the collector drops fall by gravity alone, the relative velocity between falling collector drops and rising small particles is low, which in turn produces a low single particle collection efficiency.

An alternative model of a vertical (counter-flow) spray chamber can be formulated. A schematic diagram of a vertical spray chamber of height L is shown in Fig. 8.30. It will be assumed that the particle concentration varies only in the vertical direction (z direction) and that there are no variations in the transverse direction.

Consider a differential volume $A\,dz$. The conservation of mass for particles can be written as

$$cAU_a = (c + dc)AU_a + \eta_d c(v_c + U_a)n_c\left(\frac{\pi D_c^2}{4}\right)A\,dz \qquad (8\text{-}156)$$

where n_c is the number of collection drops per volume of the carrier gas. The quantity n_c is given by Eq. 8-149. Combine Eqs. 8-149 and 8-156 and simplify,

$$\frac{dc}{c} = -\eta_d\left(\frac{v_c + U_a}{v_c}\right)\left(\frac{3}{2}\right)\left(\frac{Q_s}{Q_a}\right)\frac{dz}{D_c} \qquad (8\text{-}157)$$

where Q_s and Q_a are the volumetric flow rates of the scrubbing fluid and carrier gas. The settling velocity of the collecting drop is sufficiently large so as to overcome the upcoming gas velocity. Thus the absolute velocity of the collecting drop (v_c) plus the carrier gas velocity (U_a) is equal to the settling velocity ($v_{t,c}$) of the collecting drop,

$$v_{t,c} = v_c + U_a \qquad (8\text{-}158)$$

Integrating Eq. 8-157 over the height of the spray tower (L), one obtains the

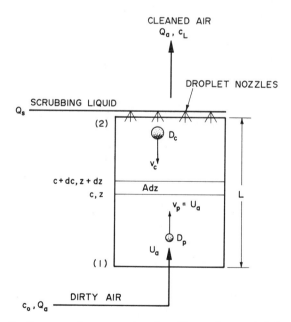

CLEANED AIR
Q_a, c_L

DROPLET NOZZLES

SCRUBBING LIQUID
Q_s

(2)

D_c

v_c

$c + dc$, $z + dz$
c, z

$A\,dz$

$v_p = U_a$

L

D_p

U_a

(1)

DIRTY AIR
c_o, Q_a

Fig. 8.30 Schematic diagram of a spray chamber illustrating an infinitesimal well-mixed volume.

following expression for the overall collection efficiency of a vertical, counter-flow spray chamber.

$$\eta = 1 - \exp\left[\eta_d\left(\frac{v_{t,c}}{v_c}\right)\left(\frac{3}{2}\right)\left(\frac{Q_s}{Q_a}\right)\left(\frac{L}{D_c}\right)\right] \tag{8-159}$$

Equations 8-159 and 8-152 predict the overall collection efficiency of a vertical spray chamber. Differences between the two expressions can be resolved if both the exponential term in Eq. 8-159 and the power term in Eq. 8-152 are expanded as Taylor series. Since the number of collecting drops (n) over the height of the tower (L) is large, it can be shown that both expressions yield similar numerical results.

In industrial plants there is often an insufficient vertical distance to accommodate a tall vertical spray chamber. Under these conditions, engineers often install a *cross-flow (transverse) spray chamber*. In a cross-flow spray chamber, the gas and particles flow in the horizontal direction and the scrubbing liquid is sprayed downward, That is, transverse to the carrier gas flow. By modeling the device in a manner similar to the above, and assuming that well-mixed conditions exist in a unit volume Adx, where A is the chamber cross-sectional area and x is in the direction of flow, the overall collection efficiency of the device can be expressed by an equation similar to Eq. 8-159.

8.12.2 Venturi Scrubbers

A second example in which impaction is used to remove particles from a gas stream is the venturi scrubber. Figure 8.31 is a sketch of a venturi scrubber and the downstream cyclone separator used to remove particles of the collecting liquid after they leave the scrubber. Figure 8.32 depicts the throat of a venturi scrubber that produces a large relative velocity and high overall collection efficiency. The operation of a venturi scrubber is based on the following:

(1) Air containing the undesired small particles is brought to a high velocity by passing it through a nozzle. Velocities of 50–100 m/s are common.

(2) Drops of the collecting liquid are injected into the aerosol at the inlet of the throat. There are a variety of methods to accomplish this injection that will be omitted for the moment. The drops have a low velocity upon entering the gas stream but accelerate as they travel downstream and approach (if not achieve) the gas velocity.

(3) The high-speed small particles impact on the slower moving collector drops and are removed from the gas steam. The single drop collection efficiency decreases as the collector drops travel downstream and the relative velocity between drop and collector decreases.

(4) The collector drops containing the impacted small particles are later removed from the gas stream by a cyclone or some other conventional particle removal system.

The *overall collection efficiency* of a venturi scrubber can be modeled using equations developed earlier if the following assumptions are made:

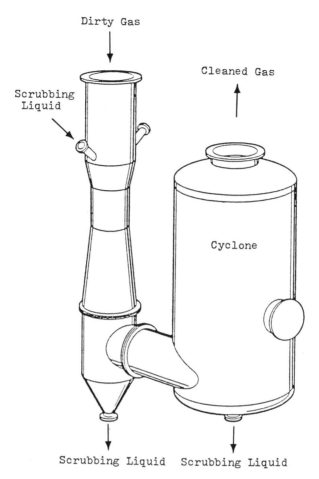

Fig. 8.31 Venturi scrubber with cyclone collector to remove drops of a scrubbing liquid (with the permission of Heil Process Equipment, Xerxes Coperation).

Steady-state, well mixed conditions at any value of y.

Only spatial variations in the direction of flow will be considered, variations transverse to this direction are assumed to be zero.

The velocity of the small particles (v_p) equals the air velocity (U_a) and is constant throughout the throat.

The evaporation of the scrubbing liquid is neglected and the pressure and temperature of the gas in the throat are constant.

Writing a mass balance for an elemental volume,

$$cAU_a = (c + dc)AU_a + v_r c\left[\frac{\pi D_c^2}{4}\right]\eta_d n_c A \, dy$$

$$\frac{dc}{c} = -\eta_d\left(\frac{v_r}{v_c}\right)\left(\frac{3}{2}\right)\left(\frac{Q_s}{Q_a}\right)\frac{dy}{D_c} \qquad (8\text{-}160)$$

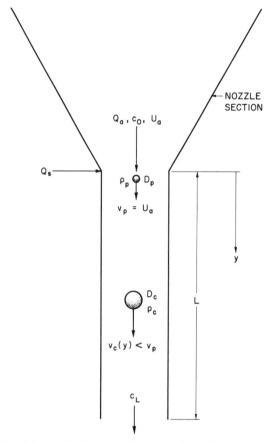

Fig. 8.32 Schematic diagram of a venturi scrubber throat section.

where η_d is the single-drop collection efficiency for an *accelerating drop* given by Eq. 8-141 and v_r is the relative velocity given by Eq. 8-147. The collector drop velocity (v_c) increases from $v_c(0)$ at the throat inlet and may ultimately reaches the gas velocity at some point well downstream of the inlet. At any point downstream of the inlet the collector velocity $v_c(y)$ can be found by solving the equation of motion (Eqs. 8-71 to 8-74), which reduces to

$$\frac{dv_c}{dt} = g - \left(\frac{3c_D}{4}\right)\left(\frac{\rho}{\rho_c D_c}\right)[v_c(y) - U_a]^2 \tag{8-161}$$

$$c_D = 0.4 + \frac{24}{Re_c} + \frac{6}{(1 + Re_c^{1/2})} \tag{8-162}$$

$$Re_c = \rho D_c \frac{[U_a - v_c(y)]}{\mu} \tag{8-163}$$

It must be emphasized that only the collector drop accelerates. Thus the equation of motion for a particle is written for the drop. The small particles that one wishes to remove travel at a constant velocity equal to the gas velocity. The equation for the collector velocity is a nonlinear, second-order differential equation. Numerical methods will be required to solve this equation and the equation to predict the overall collection efficiency. An elementary algorithm for this purpose consists of the following:

Step 1: Using the flow condition at the throat inlet, compute the single particle collection efficiency. Next select a small increment of distance δy in the direction of flow and compute a new concentration using Eq. 8-160.

Step 2: At this new location $y = \delta y$, compute the elapsed time $(\delta t = \delta y / U_a)$ and using the computer program in Illustrated Example 8.3 compute the new velocity of the collector particle. Be careful to use a small value of δy such that the time increment is sufficiently small for the program. With this new collector velocity, compute the new relative velocity, Stokes number and single drop collection efficiency. Because the collector drop accelerates rapidly, very small time steps will be needed near the throat inlet, but progressively larger increments can be used as the collector particle approaches the throat exit.

Step 3: Using the values computed in Step 2, return to Step 1.

Step 4: Repeat the process until the overall displacement of the collector particle is equal to the length of the throat.

In both the analysis of the spray chamber and the venturi scrubber it was assumed that the spray drops (D_c) were of the same size. Considering the ways sprays are formed, it is highly unlikely that this is achieved. It was also assumed that the particles to be collected were of constant diameter (D_p) and it is highly unlikely that this is true either. Modeling a system in which both the particles and spray drops have their own size distributions is complex. A parameter often used to characterize the spray is the *Sauter diameter*. The Sauter diameter is equal to the total volume of the spray particles in a volume of the carrier gas divided by the total surface area of these particles. Since gas absorption depends on absorber surface area the magnitude of the Sauter diameter for a spray is a measure of the absorption rate of the spray. For a monodisperse spray, the Sauter diameter is equal to the drop diameter divided by six.

8.13 FILTRATION

Filtration is the name given to the removal of particles as they pass through a bed of collectors. The collectors may be *woven fibers* (*filter bags*) as seen in Fig. 8.33, or a bed of tightly packed matted fibers (*felt*). Alternatively, the bed may consist of a layer of individual collecting particles through which the aerosol passes. Particles are removed as they impact on the collectors. Industrial filters often combine both phenomena since the dust cake on the upstream (dirty) side of a filter acts as a filter bed, while the filter removes particles and supports the dust cake.

Collection is provided by impaction, interception, and diffusion. Because the aerosol velocity is low, diffusion is of more importance than it was in scrubbing.

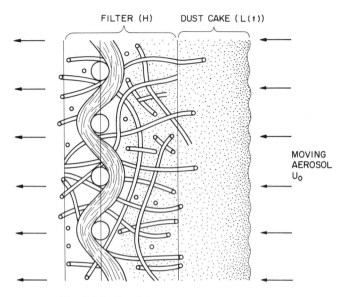

FILTER (H)   DUST CAKE (L(t))

MOVING
AEROSOL
U_0

Fig. 8.33   Particle removal by filter media.

The *single fiber collection efficiency for spheres impacting on cylinders* (η_f) can be given (142) by

$$\eta_f = \left[\frac{Stk}{Stk + 0.85} \right]^{2.2} \qquad (8\text{-}164)$$

When the Stokes number (Stk) is 0.05 or less, the single fiber impaction collection efficiency is negligible, but diffusion becomes important. Figure 8.34 illustrates these mechanisms. Hinds (42) suggests that the following *overall single-fiber collection efficiency*, η_f(overall), can be used to include impaction (η_I), interception (η_R), diffusion (η_D) and an enhanced collection due to the interception of a diffusing particle (η_{DR}) for large Peclet numbers,

$$\eta_f(\text{overall}) = \eta_R + \eta_I + \eta_D + \eta_{DR} \qquad (8\text{-}165)$$

where

$$\eta_R = \frac{\left[2(R+1)\ln(R+1) - (R+1) + \dfrac{1}{(R+1)} \right]}{2\,\mathrm{Ku}} \qquad (8\text{-}166)$$

$$R = \frac{D_p}{D_f} \qquad (8\text{-}167)$$

where Ku is the *Kuwabara hydrodynamic factor* that accounts for the alteration of the flow field as air flows around fibers in close proximity of each other,

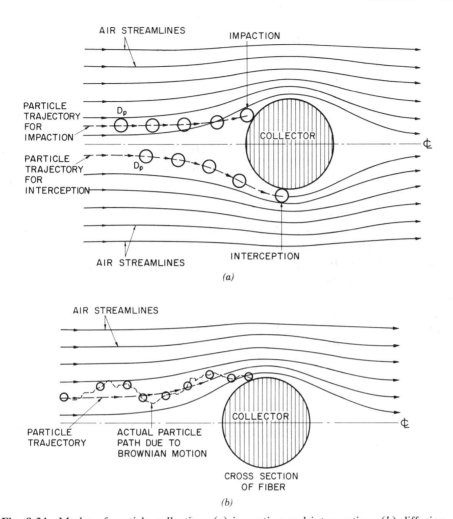

Fig. 8.34 Modes of particle collection: (*a*) impaction and interception; (*b*) diffusion.

$$\mathrm{Ku} = -\left(\frac{1}{2}\right)\ln f_{\mathrm f} - 0.75 + f_{\mathrm f} - \frac{(f_{\mathrm f})^2}{4} \tag{8-168}$$

where $f_{\mathrm f}$ is the *fiber solids fraction* in the filter.

$$\eta_{\mathrm I} = \mathrm{J}\,\frac{\mathrm{Stk}}{2(\mathrm{Ku})^2} \tag{8-169}$$

$$J = (29.6 - 28f_{\mathrm f}^{0.62})\mathrm R^2 - 27.5\mathrm R^{2.8} \quad \text{for } R < 0.4 \tag{8-170}$$

$$\eta_{\mathrm D} = \frac{2}{\mathrm{Pe}^{0.67}} \tag{8-171}$$

$$\mathrm{Pe} = \text{Peclet number} = D_{\mathrm f}\frac{U_0}{\mathscr{D}} \tag{8-172}$$

$$\eta_{DR} = \frac{1.24R^{0.67}}{\sqrt{(Ku\,Pe)}} \quad \text{for Pe} > 100 \tag{8-173}$$

where $\mathscr{D}$ is the particle diffusion coefficient,

$$\mathscr{D} = \frac{kTC}{3\pi D_p \mu} \tag{8-174}$$

where k is Boltzmann's constant and C is the Cunningham correction factor. Figures 8.35 and 8.36 show the single fiber collection efficiency (η_f) for different flow conditions. Note how sensitive η_f is for particles (D_p) in the range 0.1–0.5 μm.

To estimate the overall collection efficiency of a filter, consider the one dimensional flow of an aerosol through a filter as shown in Fig. 8.37. The velocity inside the filter (U) is larger than the velocity approaching the filter (U_0) owing to the blockage produced by fibers within the filter. To model the overall collection efficiency, it is necessary to define the following parameters:

(1) *Fiber solid fraction* (f_f): The *porosity* of the filter is defined as the fraction of the overall filter volume that is open. The porosity is also called the *voidage* or *void fraction*. Thus f_f, which is the fraction of the overall filter that is composed of solids (fibers), is related to the porosity by

$$f_f = 1 - \text{porosity} \tag{8-175}$$

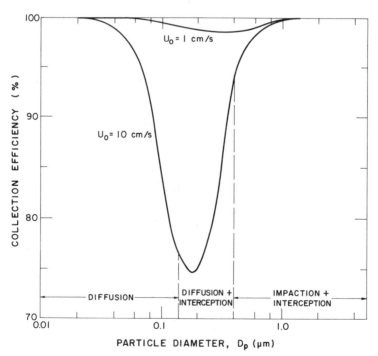

Fig. 8.35 Grade efficiency for two face velocites. Fiber thickness = 1 mm, solids fraction = 0.05, single fiber diameter = 2 μm (redrawn from ref. 42).

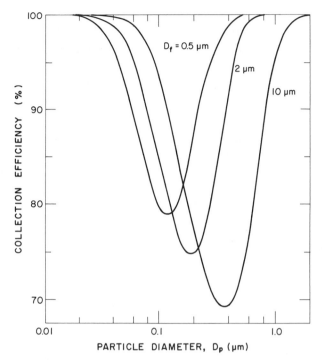

Fig. 8.36 Grade efficiency for three single-fiber thicknesses. Face velocity = 0.01 m/s, solid fraction = 0.05, filter thickness adjusted to produce a constant pressure drop (redrawn from ref. 42).

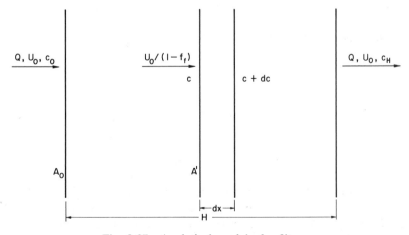

Fig. 8.37 Analytical model of a filter.

By similar reasoning, f_f is the ratio of the bulk density of the filter to the density of the fiber material. Similarly, f_f is equal to the *blockage*, the fraction of the cross sectional area of the filter that is composed of solid matter material, that is,

$$1 - f_f = \frac{A'}{A_0} \tag{8-176}$$

where A' is the cross sectional area for air moving inside the filter and A_0 is the frontal area of the filter. The velocity of the aerosol inside the filter (U) is related to the approach velocity (U_0) by

$$U = \frac{U_0}{1 - f_f} \tag{8-177}$$

   (2) *Length of fiber per unit volume of filter* (L_f): For simplicity imagine that the filter is composed of a single length of fiber of diameter D_f. Thus the solids fraction (f_f) can be written as

$$f_f = \frac{\pi D_f^2 L_f}{4} \tag{8-178}$$

   Consider the passage of air carrying particles as it passes through the filter. A mass balance for particles in air, can be written as

$$cUA' = (c + dc)UA' + UD_f L_f c \eta_f A_0 \, dx \tag{8-179}$$

where A' is the open area through which the aerosol passes and A_0 is the frontal area of the filter. Using Eq. 8-176, the above reduces to

$$\frac{dc}{c} = -\frac{D_f \eta_f L_f \, dx}{1 - f_f}$$

$$\int_{c_0}^{C_f} \frac{dc}{c} = -\eta_f \left[\frac{f_f}{(1 - f_f)} \right] \left(\frac{4}{\pi} \right) \int_0^H \frac{dx}{D_f} \tag{8-180}$$

The overall collection efficiency of the filter is

$$\eta(\text{overall}) = 1 - \frac{c_f}{c_0} \tag{8-181}$$

$$= 1 - \exp\left[-\eta_f \left(\frac{f_f}{1 - f_f} \right) \left(\frac{4}{\pi} \right) \left(\frac{H}{D_f} \right) \right]$$

Upon examination, each term in the expression above is of the order of unity except (H/D_f), which is considerably larger than unity. For this reason the overall efficiency is virtually 100%. The only exception occurs when the single particle collection efficiency (η_f) may be very low, which occurs for a narrow band of particles in the vicinity of 0.1 μm (Figs. 8.35 and 8.36). In this range collection by both diffusion and impaction may be small.

One of the attractive features of filters is the relative independence of the overall collection efficiency to the volumetric flow rate. Thus filters are well suited for batch processes where the volumetric flow rate is not constant. It is customary to refer to the volumetric flow rate in terms of the air to cloth ratio. The *air to cloth ratio* is defined as

$$\text{Air-to-cloth ratio} = \frac{Q_a}{\text{total filter area}} \tag{8-182}$$

If the air entered the filter normal to its surface, the air-to-cloth ratio is equal to the average air velocity entering the filter (U_0). With cyclones, scrubbers, and other inertial collectors, the overall collection efficiency is strongly dependent on the volumetric flow rate. Thus for industrial batch process or process in which the volumetric flow rate is subject to change, inertial collectors have an inherent disadvantage. With filters, the overall collection efficiency is virtually 100% for all volumetric flow rates. The principal sensitivity of filters to the volumetric flow rate occurs in the pressure drop across the filter. The *superficial velocity* is defined as the velocity of the carrier gas approaching the filter face. Consequently the air-to-cloth ratio and the superficial velocity are one and the same.

The pressure drop through filters has an unusual dependence on time, velocity (air to cloth ratio, U_0), and particle concentration (c_0). Assume that the dust cake seen in Fig. 8.33 is of *thickness $L(t)$*. The thickness can be expressed as

$$L(t) = \frac{w(t)}{\rho_p f_d} \tag{8-183}$$

where f_d is the solid fraction of the dust cake, ρ_p is the density of the particulate matter, and $w(t)$ is mass of dust per unit area of filter after a period of time t. The mass of dust that has accumulated after a period of time t is called the *dust cake* or *loading $w(t)$* can be written as

$$w(t) = c_0 U_0 t \eta(\text{overall}) = c_0 U_0 t \tag{8-184}$$

because the overall collection efficiency is virtually 100%. Combining Eqs. 8-183 and 8-184, the thickness of the dust cake becomes

$$L(t) = \frac{c_0 U_0 t}{\rho_p f_d} \tag{8-185}$$

The overall pressure drop (δP_d) across the filter and dust cake can be written as

$$\delta P_d = \delta P_f + \delta P_c \tag{8-186}$$

Since the velocities are low, the pressure drop can be expressed by *Darcy's law* which is applicable to flow through a porous media,

$$\frac{dP}{dx} = -C_1 \mu U \tag{8-187}$$

where C_1 is a constant. Thus the overall pressure drop is

$$\delta P_d = C_1 \mu U_0 H + C_2 \mu U_0 L(t)$$

$$= C_1 \mu U_0 H + \frac{C_2 \mu U_0 c_0 t U_0}{\rho_p f_d}$$

$$\delta P_d = K_1 U_0 + K_2 c_0 U_0^2 t \tag{8-188}$$

The constant K_1 is called the *residual drag* and the constant K_2 is called the dust cake *specific resistance*. Values of K_1 can be obtained from manufacturers of filter material. It must be remembered that the pressure drop for a new, unused filter will be slightly different than that of a filter that has undergone several cleaning cycles. In reverse-flow and shaker filter units (called *baghouses*) that use woven fabric filter, the largest pressure drop is through the dust cake. With pulse jet filter units which use felt material, the dust cake is thin and the pressure drop through the filter is of more importance. Calvert and Englund (142) tabulated values of K_2 for different combinations of filter material, powders, dusts, aerosols, and so on. The range of values is very large and users should measure the specific resistance of their own dusts.

There are occasions when Eq. 8-188 is shortened to

$$P_d = S U_0 \tag{8-189}$$

The parameter S is called the *filter drag*. It can be seen that the filter drag is not a property of the filter and the particles to be collected since it depends on the thickness of the dust cake, which in turn varies linearly with time and the dust concentration. For an entire baghouse in which the separate modules are taken off-line and cleaned in a prescribed fashion, the overall pressure drop of the baghouse may vary linearly with the air-to-cloth ratio (U_0). Under these conditions it is possible to measure the filter drag for the entire baghouse.

Figure 8.38 shows that the pressure drop increases linearly with time and that

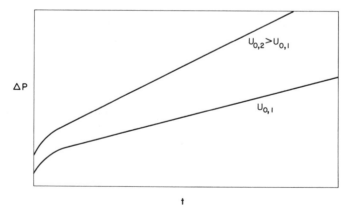

Fig. 8.38 Pressure drop across a filter versus time for an aerosol with different face velocities and constant particle concentration.

the slope varies with the square of the velocity (air-to-cloth ratio). Thus in baghouses consisting of several modular units on a staggered cleaning schedule, users must decide what maximum pressure drop they are willing to tolerate and adjust the cleaning cycle to clean each unit after this predetermined period of time.

NOMENCLATURE

Symbol	Description	Dimensions*
A	Area, constant defined by equation	L^2
A'	Open cross sectional area within filter	L^2
A_0	Frontal area of filter	L^2
b	Constant defined by equation	
B	Constant defined by equation	
c_i, c_0	Inlet or free-stream mass concentration	M/L^3
c_L, c_H, c_f	Final or exit mass concentrations	M/L^3
$c(D_p), c(D_p)_0, c(D_p)_a$	Local, initial and average mass concentration of particles of diameter D_p	M/L^3
$c(D_{p,j})$	Concentration of particles in the j-th interval	
C	Slip factor (Cunningham correction factor)	
c_D	Particle drag coefficient	
C_n	Constant used to describe curvilinear flow	
C_1, C_2	Constants defined by equation	
d_n	Collision diameter	L
D	Constant defined by equation	
$\mathscr{D}$	Particle diffusion coefficient constant	m^2/s
D_a	Aerodynamic diameter defined by equation	L
D_c	Diameter of the collecting particle	L
D_f	Fiber diameter	L
D_p	Particle diameter	L
$D_{p,j}$	Mean particle diameter in the j-th interval	L
$D_{e,p}$	Equivalent volume diameter, defined by equation	L
$D_{s,p}$	Equivalent surface area diameter, defined by equation	L
$D_{p,p}$	Equivalent projected area diameter, defined by equation	L
$D_{p,16}, D_{p,50}, D_{p,84}$	Particle diameter at which the cumulative distributions are 16%, 50% and 84%	L
$D_{p,am}$	Arithmetic mean diameter	L
$D_{p,gm}$	Geometric mean diameter	L

*Q – energy, F – force, L – length, M – mass, N – mols, t – time,

D_f	Characteristic diameter of a single fiber	L
f_f	Solids fraction of filter	
f_d	Solids fraction of dust cake	
$\mathbf{F}$	Drag force on a particle	F
g	Acceleration of gravity	L/t^2
$g(D_p)_j$	Mass fraction of particles in size range j	
G	Cumulative mass distribution function	
H	Height of a duct or enclosure, filter thickness	L
J	Constant defined by equation	
k	Boltzmann constant	Q/T
K	Constant defined by equation	
K_1	Residual drag	M/L^2
K_2	Dust cake specific resistance	t^{-1}
L_c	Critical length	L
$L(t)$	Dust cake thickness	L
ℓ	Penetration (pulvatation) distance	L
L, L'	Height of spray chamber, hypothetical height	L
L_f	Length of filter fiber per unit volume of filter	L^{-2}
m	Mass distribution function	L^{-1}
m_t	Total mass of particles	M
n_c	Number of drops per volume of air	L^{-3}
$n(D_p)$	Number distribution function	L^{-1}
n	Number of encounters between particle and falling drop	
n	Number of molecules per volume of gas	L^{-3}
n	Exponent for curvilinear flow	
n_i	Number of particles of diameter $D_{p,i}$	
n_t	Total number of particles	
n_t	Total number of mols	N
$N(D_p)$	Cumulative number distribution function	
N_i	Cumulative distribution for particles less than $D_{p,i}$	
P	Pressure	F/L^2
P	Penetration ($P = 1 - \eta$)	
δP_d	overall pressure drop across dust-caked filter	F/L^2
δP_c	Pressure drop across dust cake	F/L^2
δP_f	Pressure drop across the filter	F/L^2
Q	Volumetric flow rate	L^3/t
Q_a, Q_s	Volumetric flow rates of air and scrubbing fluid	L^3/t
r	Radius	L
R	D_p/D_f	
R_u, R_j	Universal gas constant, gas constant for a specific molecular species j	Q/NT

S	Filter drag	F/L^2
t	Time	t
δt	Incremental time step	t
t_c	Critical time	t
t^*	Dimensionless time	
T	Temperature	T
v_c	Velocity of the collecting particle	L/t
$v_s(t)$	instantaneous settling velocity	L/t
v_t	Terminal (gravimetric, steady state) settling velocity	L/t
$\mathbf{v}_{t,c}$	Settling velocity of collecting particle	L/t
$\mathbf{v}, v_i$	Particle velocity	L/t
$\mathbf{v}_r, v_{r,i}$	Relative velocity, $(\mathbf{v} - \mathbf{U})$	L/t
$\mathbf{U}, U_i$	Gas velocity	L/t
U_a	Air velocity	L/t
$\mathbf{U}^*, \mathbf{v}^*$	Dimensionless gas and particle velocities	
$\mathbf{U}_0$	Characteristic velocity	L/t
V	Volume	L^3
V_p	Particle volume	L^3
W	Width of a duct	L
$w(t)$	Dust cake mass per area of filter	M/L^2
x, y, z	Coordinates	L

Greek	*Description*	*Dimension*
α	Turning angle	
η	Efficiency (mass removed/mass entering)	
$\eta(D_p)$	Fractional or grade efficiency	
η_d	Single drop (particle) collection efficiency	
η_f	Single-fiber collection efficiency	
$\eta_I, \eta_R, \eta_D,$ η_{DR}	Single-fiber collection efficiencies of impaction, interception, diffusion and interception	
λ	Carrier gas mean free-path	L
μ	Viscosity	Ft/L^2
π	3.14159	
ρ, ρ_p, ρ_w	Density of carrier gas, particle and water	M/L^3
σ, σ_g	Standard deviation, geometric standard deviation	
τ	Relaxation time	t
ν	kinematic viscosity μ/ρ	L^2/t
X	Particle dynamic shape factor	

Subscripts		
$()_a$	Property of air, average property	
$()_{ave}$	Average	
$()_b$	Bleed air	
$()_c$	Collecting particle, critical property	

$()_d$	Drop property
$()_e$	Exit condition
$()_f$	Final condition, filter property, streaming flow property
$()_g$	Geometric
$()_i$	Initial condition
$()_L$	Condition at distance L
$()_p$	Particle property
$()_{p,j}$	Particle in size range j
$()_r$	Radial component
$()_s$	Scrubbing liquid property
$()_w$	Water, values relevant to inlet, width 2w
$()_0$	Far field, initial or incoming properties, STP
$()_x, ()_y, ()_z, ()_i$	Vector components
$()_\theta$	Component in angular direction

Abbreviations	*Description*
Kn	Knudsen number
Ku	Kuwabara number
Pe	Peclet number
Re	Reynolds number
Stk	Stokes number
STP	Standard temperature and pressure

PROBLEMS

1. High-speed stamping machines used to manufacture small electrical connectors require a fluid to be sprayed (as a mist) on the dies for lubrication and cooling. Unfortunately, an aerosol is formed around the machine that irritates the surface of the nasal cavity. From time to time the aerosol is analyzed and found to be log-normal. The results of a recent analysis are shown below (data taken from Crawford, p. 139). Compute the median particle size based on number and mass. The total particle concentration is 10,000 particles per cm^3 and the density of the fluid is 1.2 gm/cm^3, estimate the aerosol mass concentration (mg/m^3).

Range of D_p (μm)	(Number of Particles/Range)/10,000
0–2	0
2–8	0.0359
8–14	0.1999
14–20	0.2642
20–30	0.2940
30–45	0.1556
45–60	0.0365
60–80	0.0139

2. An aerosol produced during a bagging operation is log-normal with a mass median diameter of 10 μm and a geometric standard deviation of 2.5. What is the median particle size based on number? If the specific gravity of the dust is 2.5 and the number concentration is 100,000 particles/cm^3, what is the mass concentration (mg/m^3)?

3. In the grinding room of a foundry the aerosol is log-normal with a mass median diameter of 18 μm and a geometric standard deviation of 1.5. The mass concentration is 100 mg/m^3. The specific gravity of the particles is 4.5. What is the median diameter based on number and what is the particle concentration based on number (number of particles/cm^3)?

4. A thin layer of finely ground corn starch is placed on the surface of newly printed pages leaving a four-roll, color offset printing press to keep the pages from sticking together. Unfortunately some of the corn starch becomes airborne, settles on the rolls, ink reservoir, and fountain (wetting) solution and is transferred to the rolls and produces what printers humorously call "hickeys" on the printed page. A sample of the air above the press is passed through a filter and analyzed under the microscope. The number of particles (by number) within size intervals is as follows:

Particle Interval D_p (μm)	Number of Particles/Interval
2–3	10
3–4	0
4–5	0
5–6	30
6–7	100
7–8	210
8–9	450
9–10	910
10–11	850
11–12	800
12–13	440
13–14	230
14–15	90
15–16	30
16–17	20

Is the size distribution log-normal? What is the median size with respect to number and with respect to mass?

5. A cubic meter of air (at STP) in an urban environment is analyzed and found to have particles whose mass lies within the following limits.

Size interval D_p (μm)	Mass (mg)
0 – 0.5	2.5×0^{-6}
0.5– 1.0	0.0007
1.0– 1.5	0.0046
1.5– 2.0	0.0107
2.0– 2.5	0.0151
2.5– 3.0	0.0223
3.0– 3.5	0.0292
3.5– 4.0	0.0454
4.0– 4.5	0.0645
4.5– 5.0	0.1020
5.0– 5.5	0.1590
5.5– 6.0	0.2970
6.0– 6.5	0.4520
6.5– 7.0	0.6240
7.0– 7.5	0.9220
7.5– 8.0	1.3000
8.0– 8.5	1.6900
8.5– 9.0	2.0200
9.0– 9.5	2.3000
9.5–10.0	2.2200
10.0–10.5	2.1500
10.5–11.0	2.2400
11.0–11.5	2.1400
11.5–12.0	1.7800
12.0–12.5	1.6200
12.5–13.0	0.9740
13.0–13.5	0.6750
13.5–14.0	0.3380
14.0–14.5	0.1450
14.5–15.0	0.0321

(a) Plot the cumulative mass and cumulative number distribution on log-probability paper. Compute the mass median and number median diameters. Compute the geometric standard deviation for each curve. Are these values the same and if not why.

(b) The particle specific gravity is 1.15. Compute the overall mass concentration (mg/m^3) and number concentration (total number of particles/m^3).

(c) Plot $[dG/d \log D_p]$ versus $[\log D_p]$ where G is the cumulative mass distribution and estimate the modal diameter(s).

6. Paint particles from a spray booth in an auto body repair shop are sampled. The mass of particles per unit volume of air is found to be 2 g/m^3 at STP. The particle size distribution is found to be:

(a) Draw a histogram.

(b) Compute the arithmetic mean particle diameter and standard deviation.

Particle size range (μm)	Number of particles
2–3	1
3–4	1
4–5	1
5–6	3
6–7	10
7–8	21
8–9	45
9–10	91
10–11	85
11–12	80
12–13	44
13–14	23
14–15	9
15–16	3
16–17	2

(c) Compute the geometric standard deviation and geometric mean particle diameter.

(d) Plot the actual cumulative size distribution based on diameter.

(e) Using the data in part c, plot the cumulative size distribution assuming a log-normal size distribution.

(f) Compute the mass median diameter.

(g) If the particle density is 1800 kg/m^3, how many particles are there per cubic meter of air?

7. A widely used experimental technique to study the velocity, temperature and, concentration in a moving gas stream is by exciting small particles injected into the stream with a laser. By choosing unique particles and lasers, velocities can be studied (laser doppler velometry) or temperatures and concentrations studied by analyzing the emission spectra. It has been suggested that large particles can be used but others believe that gravimetric settling will provide bogus information. You have been asked to study particles motion the following conditions.

Duct diameter $(D) = 0.10$ m
Volumetric flow rate $(Q) = 0.07$ m^3/min
Initial particle velocity $[v_x(0)] = 10\ U(\text{avg})$
$T = 25°C$, $P = 1$ atm
Particle density $(\rho_p) = 1200$ kg/m^3

(a) If the air velocity is uniform and constant and $U_r = 0$, $U_\theta = 0$, plot a graph of D_p versus L/D for $D_p = 10$, 100 and 1000 μm, where L is the horizontal distance a particle travels before it encounters the duct wall.

(b) Repeat (a) if the flow is fully established, i.e. the axial air velocity is

$$\frac{U_x(r)}{U_{avg}} = 2\left[1 - \left(\frac{r}{R}\right)^2\right], \qquad U_{avg} = \frac{Q}{A}$$

(c) A polydisperse aerosol is injected along the axis at a mass flow rate 10 mg/min, $v_x(0) = 10 U_{avg}$. The mass median diameter is 15 μm and the geometric standard deviation is 1.2. If the velocity air flow is fully established as in (b) above, plot the mass fraction of the deposited particles as a function of L/D.

8. A nozzle produces a log-normal distribution of water particles in which the median particle diameter based on number is 0.5 μm and the geometric standard deviation is 1.5. At a particular point in the spray, the particle number density is 10^6 particles per cm^3. At this point, what is the mass concentration (mg/m^3)?

9. A size distribution of particles from a painter's spray gun is log-normal. The concentration of unit density spheres is 500 mg/m^3. The number median diameter is 10 μm and the geometric standard deviation is 1.5. What is the mass concentration (mg/m^3) of particles in the respirable range, that is, particles less than 10 μm? What is the mass concentration of particles between 10 and 20 μm?

10. A remote sensor monitors an aerosol in a foundry using electric arc furnaces and determines that

   (a) the particle size distribution is log-normal
   (b) the geometric mean diameter based on number is 20 μm
   (c) the geometric standard deviation is 1.8
   (d) the overall concentration is 1500 particles per cm^3

   On the basis of mass ($\rho_p = 8000$ kg/m^3) what percent of the particles are greater than 40 μm in diameter?

11. The concentration of particles in a quarry is 25 mg/m^3. The density of the particles is 1200 kg/m^3. The size distribution is log-normal. The geometric standard deviation is 2.0 and the mass median diameter is 10 μm. What is the concentration of particles (mg/m^3) between 5 and 25 μm?

12. Contaminant collectors 1 and 2 are arranged in series and remove smoke and fume from a stream of air (Q_t) taken from the workplace. The efficiencies of each collector are the same ($\eta_1 = \eta_2$). What is the overall efficiency (η_0) of the pair of units? The company wishes to increase the overall efficiency by 25% and plans to buy an additional collector (A) with an efficiency (η_a). Which configuration below requires the smallest amount of air to be diverted through the new unit (Q_A) and what is the minimum Q_a/Q_t.

   (a) Units A and 1 in parallel followed by unit 2.
   (b) Unit 1 followed by units 2 and A arranged in parallel.
   (c) Unit A in parallel with units 1 and 2 arranged in series.

13. Air in a welding shop contains one million particles per cubic meter. The particle size distribution is log-normal. Based on number, the log-mean

diameter is 0.5 μm and the geometric standard deviation is 1.8. If welding fume particles have a density of 1800 kg/m^3, what is the overall particle mass concentration at STP?

14. Two pollution control devices were installed in series many years ago. The collection efficiencies of each device are 75% and independent of the volumetric flow rate. Company management wishes to increase the overall collection efficiency to 95% by purchasing a new device with a collection of efficiency of 95% in parallel with the second device. What fraction of the flow should be diverted through the new device? Would it make any difference if the new device was placed in parallel with the first device?

15. A research laboratory is equipped with a ventilation system similar to Fig. 5.6:

$$V = 600 \text{ m}^3 \qquad Q_m = 10 \text{ m}^3/\text{hr} \qquad Q_r = 100 \text{ m}^3/\text{hr}$$

$$\text{air cleaner efficiency} = 80\%$$

The initial particle concentration in the room is 100 particles/m^3 but as soon as the work begins, particles are generated at a constant rate of 10^5 particles/hr. On a particular day the wind directs the exhaust from a nearby building toward the makeup air inlet so that the ambient particle concentration (c_a) becomes 10^5 particles/m^3. How long will it take before the particle concentration in the laboratory reaches 5000 particles m^3? What is the eventual steady-state concentration?

16. A particle collection unit consists of a particle concentrator A and a particle collector B. See Fig. 5.18 where collectors 1 and 2 are now called A and B. The volumetric flow rate entering the unit is Q_0. The concentrator A removes particles with an efficiency η_A. The collected particles from A and a fraction $(f_B = Q_B/Q_A)$ of the gas stream are then sent to a collector B which removes η_B of the particles and sends the remaining particles and the gas stream $(f_B Q_0)$ back to a point upstream of concentrator A. Write an expression for the overall collection efficiency of the unit. Corrosion produces a leak between the point where B returns gas and the inlet to A and a fraction f_L $(f_L = Q_L/Q_0)$ of the gas stream and its particles escape to the atmosphere. Write an expression for the overall collection efficiency of the unit including the leak fraction f_L.

17. A baghouse contains N_b bags arranged in parallel. The collection efficiency of each bag is η_b. The volumetric flow rate through each bag is constant Q_b. If k bags break, allowing all the dusty air to pass to the clean side of the baghouse, show that the rate of change of the overall collection efficiency (η) with respect to the number of broken bags (k) is given by

$$\frac{dE}{dk} = -\frac{\eta_b}{N_b}$$

18. Compute the terminal velocity of the following in air at STP:

   (a) Baseball ($D_p = 9$ cm, $\rho_p = 950$ kg/m^3).
   (b) Ping-pong ball ($D_p = 3$ cm, overall density $= 2$ kg/m^3).
   (c) agricultural dust ($D_p = 5$ μm, $\rho_p = 2$ kg/m^3).

19. The motion of a large spherical particle is entirely in the Newtonian regime where the drag coefficient is constant ($c_D = 0.4$). Derive an expression for the gravimetric settling velocity.

20. Compute the terminal settling velocity of a 50-μm spherical particle falling freely from rest in air (25 C, 1 atm) if the particle specific gravity is (a) 0.8, (b) 1.3, (c) 3.0 and (d) 13.6.

21. Compute the terminal settling velocity of an agglomerated particle consisting of four, spherical particles (5 μm each and density 1100 kg/m^3) connected to each other in a loose chain-like structure. The carrier gas is an auto exhaust at 1.15 atm, 200 C.

22. (a) How long will it take the following particles to achieve 34% of their steady state terminal settling velocity if they have an initial velocity (downward) of 100 m/s in quiescent air at -10 C, 0.9 atm.

   (1) $D_p = 1$ μm, $\rho_p = 980$ kg/m^3
   (2) $D_p = 10$ μm, $\rho_p = 980$ kg/m^3
   (3) $D_p = 100$ μm, $\rho_p = 98$ kg/m^3
   (4) $D_p = 25$ μm, $\rho_p = 3000$ kg/m^3
   (b) If the particles in the problem above have an initial velocity in the horizontal direction of 100 m/s in quiescent air at -10 C, 0.9 atm, what is the displacement and elapsed time before the horizontal velocity component has decreased to 0.001 m/s?

23. An unventilated enclosure surrounding a foundry shake-out operation is 5 m $\times$ 5 m and 3 m high. Initially the dust ($\rho_p = 2000$ kg/m^3) concentration is uniform and equal to 10 g/m^3. Your supervisor has asked you to estimate how the concentration varies with time.
   Part I—Assuming the particle size distribution is log–normal, the mass median diameter is 25 μm and the geometric standard deviation is 2.0, plot the overall concentration at point P, 1.5 m above the floor versus time assuming (a) settling is laminar and (b) settling is well mixed.
   Part II—Plot the mass median diameter versus time at point P assuming (a) settling is laminar and (b) settling is well mixed.

24. Compute the penetration distance of a unit density sphere, $D_p = 500$ μm, injected into still air at STP with a velocity of 100 m/s.

25. A "poor-man's" anemometer consists of a metal sphere hung by a string and suspended in air. When the wind blows at a velocity U_0, the sphere is displaced and the string makes an angle θ with the vertical. Derive an expression for the wind speed as a function of angle θ, the diameter and density of the sphere, and the density and viscosity of air. Neglect the drag on the string. Assume that the size of the sphere is chosen such that the Reynolds numbers are always between 1000 and 100,000.

26. A 1-mm sphere (density 8000 kg/m^3) is injected into still air (1 atm, 25 C) with an initial velocity of 50 m/s inclined at 60 degrees to the horizontal. How far will it travel in the horizontal direction? What maximum height will it attain?

27. A spherical particle ($D_p = 10$ μm, 1800 kg/m^3) is injected into still air at STP in a horizontal direction with a velocity of 5 m/s. After 10 sec, find the x and y displacement from the injection point.

28. A 6-mm water drop is given an initial downward velocity of 30 m/s into an air stream which is traveling uniformly to the right at 5 m/s. What is the steady-state vertical velocity component of the particle? How long does take for the drop to acquire a downward velocity of 15 m/s.

29. A water drop 20 μm in diameter is injected horizontally into motionless air with an initial velocity 15 m/s. Which of the following particles will have the same initial drag coefficient and experience the same initial drag force.

Particle	ρ_p (kg/m^3)	D_p (μm)	v (m/s)
a	1000	20.0	1
b	13,600	15.0	20
c	8000	2.5	15
d	2000	15.0	10

30. An aerosol consists of glass spheres ($\rho_p = 2724$ kg/m^3) in an upward moving stream of hot air. The air temperature and pressure are 1000 K and 0.8 atm. For particles of the following diameters: 1.0, 10.0 and 100.0 μm:

(a) What is the gas viscosity (kg/m s)?
(b) What is the Cunningham correction factor (C)?
(c) What are the aerodynamic diameters of the particles?
(d) What gas velocity will levitate the particles?

31. A water drop 20 μm in diameter is traveling in an irrotational flow field where the irrotational constant C ($C = rU_\theta$, Eq. 8-102) is equal to 10 m^2/s. If the particle tangential velocity is equal to the tangential air velocity, what is the radial velocity of the particle at a radius of 1.0 m?

32. Analyze the trajectories of particles traveling over an infinitely long mound having a semicircular cross sectional area of radius a. Assume that the velocity field around the mound can be described by the stream function $\psi = U_0 r[1 - (a/r)^2] \sin \theta$, where the radius r is measured from the center of the mound, the positive x axis corresponds to a value of θ equal to zero, and air with a uniform velocity (U_o) approaches the mound in the negative x direction. Begin the analysis with particles at points along a vertical line $x = 1.5\ a$, $y = h$, where the particle velocity $v_x = -1.5\ U_0$. Some particles will impact the mound and others will pass over it to settle to the ground downwind of the mound. Analyze the particle trajectories and determine where each particle impacts either the mound or the ground. Carry out the computations for particles originating at values of h from 0 to 1.5a.

(a) Plot the trajectory of a particle whose diameter is 50 μm.
(b) Plot a graph of h/a versus the normalized displacement x/a at which the particle impacts a solid surface. Repeat the calculation for several particles sizes. Assume the following physical conditions:

$\rho_p = 1000\ \text{kg/m}^3$
$U_0 = 1\ \text{m/s}$
$D_p = 20,\ 30,\ 50,\ \text{and}\ 100\ \mu\text{m}$
$a = 10\ \text{m}$

(c) Repeat b but omit the mound. Discuss the significance of the mound to enhance the collection of particles.
(d) A polydisperse particle stream having a log-normal size distribution (mass median diameter 30 μm, geometric standard deviation $= 1.2$) enters the air stream at $x = 1.5a$, $y = a$ with an initial velocity $v_p = -i\ 1.5U_0$. Plot the mass fraction versus y at the point $x = -1.5a$.

33. Air at STP passes over a horizontal flat plate in the positive x direction and a boundary layer develops on the plate. The air approaches the plate with a uniform velocity (U_0) of 1.0 m/s. Small particles are injected vertically upward from a point on the plate 10 m from the leading edge, are slowed by the air, travel to the right, pass back through the boundary layer, and strike the plate. Calculate the initial particle velocity if 50-μm unit density spheres strike the plate in a band 1 cm wide, 11 m downwind from the leading edge. The coordinate system is at the leading edge of the plate. The velocity profile in the boundary layer is given by $U/U_0 = (2y/\delta) - (y/\delta)^2$ and $V = 0$ where δ is the boundary layer thickness given by $\delta = 5x/\sqrt{(\text{Re}_x)}$ and $\text{Re}_x = \rho U_0\ x/\mu$.

34. Consider the gravimetric settling of particles on the floor of a long horizontal duct of square cross section through which an aerosol flows at a uniform velocity U_0. Assume that settling is governed by the well-mixed model. Define R as

$$R = \frac{\text{Local rate of settling at a distance } x \text{ from the inlet}}{\text{Total inlet mass flow rate/duct cross sectional area}}$$

Show that a graph of R versus (v_t/U_0) for constant values of (x/H) have maximum values and that the maximum values of R occur when $(v_t/U_0) = H/X$.

35. An aerosol enters a horizontal duct of width W and height H with a concentration c_0. The volumetric flow rate is Q. Show that the rate of

deposition (D, kg/sec) over the entire length of duct varies with respect to the length L as follows:

$$\frac{dD}{dL} = v_t c_0 W \exp\left[-\frac{LWv_t}{Q}\right]$$

36. Air from an enclosure containing a foundry shake-out process is to be ducted to a baghouse to remove particles of molding sand. The dust concentration c_0 entering the 0.1 m $\times$ 0.1 m duct is 10 g/m^3. The duct is 50 m long and lies in the horizontal direction.

   (a) Foundry dust settles to the bottom of the duct and collects in a layer in which the bulk density (ρ_b) is 5000 kg/m^3. Assuming that the volumetric flow rate Q is constant and that settling proceeds according to the well-mixed model, derive an expression for the thickness (h) of the deposited dust layer as functions of the time (t), distance downstream of the inlet (x), volumetric flow rate (Q), dust settling velocity (v_t), and duct width (W).

   (b) Assuming that the fan maintains a constant volumetric flow rate ($Q = 0.1$ m^3/s), how long will it take to block one half of the cross sectional area? Where will this occur?

   (c) Assuming that the fan maintains a constant volumetric flow rate ($Q = 0.1$ m^3/s), draw a graph of the friction head loss (h_f) versus time where $h_f = f\left(\frac{L}{W}\right)\frac{U^2}{2g} = f\left(\frac{L}{W}\right)\left(\frac{Q^2}{2gA^2}\right)$. The term f is the friction factor which for a first approximation can be given by its laminar value, 64/Re, where Re is the Reynolds number: $\text{Re} = \frac{(\rho WU)}{\mu} = \frac{(4\rho Q)}{(\pi W \mu)}$.

   (d) Repeat (c) above but assume that the friction factor has to be taken from the Moody chart, which can be described by the empirical equation

$$\frac{1}{\sqrt{f}} = -2.0 \log_{10}\left[\frac{\left(\frac{\varepsilon}{W}\right)}{3.7} + \frac{2.51}{\text{Re}\sqrt{f}}\right]$$

   where ε is the duct roughness factor (0.2 mm).

   (e) Repeat (d) above but take into account that as dust fills with duct, the volumetric flow rate decreases in accordance with the fan operating curve,

$$h_f(m) = \int_0^L \frac{\left(\frac{f}{W}\right)U^2 \, dx}{2g} = 75 - 52Q^2$$

   where Q has the units (m^3/s). Draw a graph of volumetric flow rate versus time.

37. A settling chamber is to be constructed in the form of N horizontal circular trays. The inner radius is R_1 and the outer radius is R_2. Air enters the center

of the apparatus, divides equally between the N trays and flows radially outward.

(a) Assuming the well-mixed model, derive an expression for the fractional gravimetric collection efficiency of the device.

(b) Assume that the flow is reversed an travels radially inward. Compute the fractional gravimetric collection efficiency.

38. A horizontal duct $(H \times H)$ transports a dusty air stream at very low velocity such that laminar flow exists everywhere. Particle motion satisfies Stokes flow. At a location x_0, the dust concentration (c_0) is uniform. The air velocity profile is fully developed, viz

$$\frac{U(y)}{U_{max}} = \frac{(2yH - y^2)}{H^2} \qquad U_{max} = \frac{3Q}{8H^2}$$

where y is measured vertically. Derive an expression for the particle deposition rate (kg/sec per m^2) as a function of x (in the direction of flow). Derive an expression for the laminar gravimetric settling efficiency as a function of x.

39. A monodisperse aerosol flows through a horizontal duct of circular cross section. The Reynolds number based on diameter is 5000. Derive an expression to predict the percent of particles that will settle over a distance L from the inlet.

40. An aerosol flows upward between parallel plates that are a distance H apart. The length of the plates is L. The air velocity (U_a) is constant. The plates are inclined to the horizontal at an angle θ. Assuming well-mixed conditions, show the gravimetric fraction is,

$$\eta(D_p) = 1 - \exp\left[\frac{-\rho_p D_p^2 L \cos \theta}{18 \mu U_a H} \right]$$

41. Derive an expression for the gravimetric settling efficiency of a monodisperse aerosol traveling through a horizontal duct of circular cross section in which the flow is laminar and fully established. If the duct is inclined upward at an angle α, what is the collection efficiency?

42. A horizontal duct of length, width, and height L, W, H is used as a gravimetric settling device.

(a) Using the well-mixed model, draw curves of the fractional efficiency $\eta(D_p)$ versus particle diameter D_p of unit density spheres for three volumetric flow rates:

$$\eta(D_p) = 1 - \frac{c(D_p)_f}{c(D_p)_0}$$

$H = 0.5 \, \text{m}; \ W = 1.0 \, \text{m}, \ L = 10 \, \text{m}, \ Q = 1.0, \ 1.25 \text{ and } 1.5 \, \text{m}^3/\text{s}.$

(b) An aerosol of unit density spheres enters the duct at an overall concentration of 100 mg/m^3 and volumetric flow rate of 1.25 m^3/s. The entering aerosol has a log-normal size distribution in which the mass median diameter is 30 μm and the geometric standard deviation is 1.3. What is the overall mass concentration (mg/m^3) leaving the duct and what is the mass median particle size (μm).

43. An aerosol passes through a duct of constant width W. The top of the duct slopes downward at an angle α while the bottom surface is horizontal.

(a) Assuming well-mixed conditions and constant air density, show that the collection efficiency η is

$$\eta = 1 - \exp\left[\frac{-LWv_t}{Q}\right]$$

where L is the length of the duct in the direction of flow, Q is the volumetric flow rate, and v_t is the settling velocity.

(b) Suppose the gravimetric settling is modeled as laminar, show that the collection efficiency η is

$$\eta = \left[\left(\frac{v_t}{H - x\tan\alpha}\right)\left(\frac{Wx}{Q}\right)\left(H - \frac{x\tan\alpha}{2}\right)\right]$$

where H is the height of the duct at the inlet ($x = 0$).

44. A duct of square cross section ($W \times W$) carries a dusty air stream upward at an angle α inclined to the horizontal. The airstream enters the duct with an average velocity U_0, volumetric flow rate Q and the Reynolds number based on W is 50,000. The mass flow rate of dust entering the duct is m_0. Gravity causes dust to settle on the duct floor. Show that the rate of dust deposition [$m(x)$ in kg/m^2 s] at a downstream distance x can be expressed as

$$m(x) = \left[\frac{m_0 v_t \cos\alpha}{U_0 W^2}\right]\exp\left(-\frac{x v_t W \cos\alpha}{Q}\right)$$

45. Air carrying dust particles flows between two horizontal parallel plates separated by a distance H. The velocity profile is given by

$$\frac{U}{U_0} = 1 - 4\left[\frac{2z - H}{2H}\right]^2,$$

where z is measured from the bottom plate and U_0 is the velocity along the midplane. At $x = 0$ the particle concentration at all points between the plates is equal to c_0 and the particle velocity in the flow-wise direction is equal to the local gas velocity. If gravimetric settling can be described as laminar and if at any downstream location the particle velocity in the flow-wise direction is equal to the local air velocity, write a general expression for the collection efficiency and show that all the particles will be removed within a critical downstream distance x_c given by $x_c = HU_0/v_t$.

46. A duct (0.3 m high by 0.6 m wide) is placed on the roof of a building for make-up air to a newly renovated optics laboratory. Unfortunately, the installers ignored an unpaved access road directly beneath the side of the building where the duct inlet was located. Vehicles using the access road produce fugitive dust that enters the make-up air duct. The dust drawn into the duct has a concentration of 10 mg/m^3, a density 1500 kg/m^3, and a log-normal size distribution (mass median diameter of 20 μm and geometric standard deviation of 2.0). The duct is 50 m long and has a 90-degree elbow at the inlet ($r_1 = 1$ m). What is the concentration and size distribution of particles leaving the duct if the average velocity in the duct is 10 m/s?

47. The flow through a 90-degree elbow is irrotational and well-mixed. At the exit, show that the concentration of particles of a particular diameter $c(D_p)$ varies with particle size as follows:

$$\frac{dc}{dD_p} = -\left[\frac{2\tau_p c(D_p)_0}{D_p}\right]\left[\frac{9\mu Q\pi}{A}\right]\exp\left[\frac{-\rho_p Q D_p^2 \pi}{2A}\right]$$

where $A = 18\mu R_2 W (R_2 - R_1)\ln\left(\frac{R_2}{R_1}\right)$

48. Hot exhaust (assume air) from a fluidized bed combustor contains large particles:

D_p, ρ_p: 50 μm, 2500 kg/m^3
Gas temperature and pressure: 1000°C, 1000 kPa

What is the particle settling velocity in such a gas? It has been suggested that 80% of these particles can be removed by a bank of reverse-flow cyclones. Assuming collection occurs at the above temperature and pressure, how many cyclones of the dimensions below should be installed in parallel? The total gas mass flow rate is 1 kg/sec.

Outer diameter = 10 cm
Exit diameter = 5 cm
Height and width of inlet = 5 cm and 2.5 cm

49. Predict the fractional efficiency curve for an air cleaning unit consisting of a straight through cyclone and a reverse flow cyclone. Bleed flow (Q_b) of the straight-through cyclone returns air to the air exiting the straight-through cyclone. Assume that a fan draws a total volumetric flow rate (Q_0) into the unit and produces a bleed flow of 10% ($Q_b/Q_0 = 0.2$). Assume also that the flow through both devices is n-degree rotational with $n = 0.5$. Assume the particles are unit-density spheres, that air is at STP and the following dimensions of the cyclones

Straight-Through Cyclone	Reverse-Flow cyclone
$D_2 = 5\,\text{cm}$	$D_2 = 10\,\text{cm}$
$D_1 = 4.5\,\text{cm}$	$D_1 = 5\,\text{cm}$
$L = 50\,\text{cm}$	$L_1 = L_2 = 20\,\text{cm}$
Vane angle $= \pi/4$	$W = 2.5\,\text{cm}$
	$H = 5\,\text{cm}$

Draw the fractional efficiency curve for $Q_0 = 0.05$, 0.075, and $0.10\,\text{m}^3/\text{s}$.

50. Derive an expression for the gravimetric settling efficiency of a horizontal duct of rectangular cross section in which the base is horizontal but the roof slopes downward at an angle α in the flow-wise direction. The duct width b is constant. The flow enters the duct at a Reynolds number of 100,000 and the particles are large such that the drag coefficient is constant. Show the collection efficiency as a function of some (but not necessarily all) of the following:

Volumetric flow rate (Q)

Flow-wise distance (x)

Angle α

Local average velocity $[U(x)]$

Duct width (b) and initial height (h_0)

Dust diameter and density

Air properties

51. A 90-degree elbow of square cross section, inner radius 0.9 m, and outer radius 1.0 m will be used to remove particles of coal ($\rho_p = 2000\,\text{kg/m}^3$). What volumetric flow rate (at STP) will be needed to remove 95% of particles 50 μm in diameter through a slit 1 mm wide lying along the outer radius at the exit of the elbow. Assume that the flow is irrotational throughout and that the particle concentration is uniform at the inlet.

52. A spray chamber uses water to capture small particles produced in grinding marble. Draw the fractional efficiency curves for Q_a/Q_s of 1000 and 5000 if the spray chamber has the following properties.

$D_c = 1\,\text{mm}$	$L = 5\,\text{m}$
$U_a = 3\,\text{m/s}$	$c_0 = 1000\,\text{mg/m}^3$
$\rho_p = 4000\,\text{kg/m}^3$	$\rho_c = 1000\,\text{kg/m}^3$
$D_p = 2, 5\ 10, 20, 40\ \mu\text{m}$	

53. Show that the single drop collection efficiency (η_d) varies with the relative velocity as follows:

$$\frac{d\eta_d}{dv_r} = \left(\frac{\text{Stk}}{v_r}\right)\left[\frac{1.4\,\text{Stk}}{(\text{Stk} + 0.7)^3}\right].$$

54. Spherical water drops ($D_c = 1\text{mm}$) are sprayed on dust generated when a conveyor discharges coal ($\rho = 1500\,\text{kg/m}^3$) to a stock pile. Assume the air

and coal dust velocities are negligible. The coal dust concentration is 100 mg/m³. Estimate the single drop collection efficiency assuming the drops to fall at their settling velocity. If there are 1000 drops/m³ of air, what is the initial rate of removal of coal dust (mg/m³ sec)?

55. Consider a transverse flow scrubber through which an aerosol passes in a horizontal direction. Assume that there is only one spatial variable (in the direction of flow) and that the following are known:

    Duct dimensions: height H, length L, depth W
    Scrubbing liquid volumetric flow rate: Q_s that produces a uniform stream of drops of diameter (D_c)
    Uniform air velocity: $U_0 = Q_a/HW$
    Aerosol: D_p, ρ_p, inlet and outlet concentration $c(0)$ and $c(L)$
    Drop velocity: $\mathbf{v}_d = \mathbf{i}U_0 - \mathbf{j}v_t$, where v_t is the constant settling velocity and $v_t \ll U_0$
    Drop diameter: D_c

    Assume that the single-drop collection efficiency (η_d) between drop and aerosol particle is known and constant. Show that the overall collection efficiency (η_0) is equal to

    $$\eta_0 = 1 - \exp\left[-\eta_d\left(\frac{3}{2}\right)\left(\frac{Q_s}{Q_a}\right)\left(\frac{v_t L}{v_t L + U_a H}\right)\left(\frac{H}{D_c}\right)\right]$$

56. Spherical rain drops 4 mm in diameter fall through still air. The air contains dust $(D_p = 10\ \mu m,\ \rho_p = 2000\ kg/m^3)$. What is the single particle collection efficiency to impaction? If the drop size is larger than 4 mm, will the impaction efficiency increase, stay the same, or decrease? Justify your answer.

57. It is a well-known fact that rain impacts dust particles and removes them from the atmosphere. Draw a graph of the single particle collection efficiency for spheres impacting spheres as a function of drop diameter (D_c) over the range 100–2000 μm and for several size dust particles $(\rho_p = 1000\ kg/m^3,\ D_p = 2, 5,$ and 10 $\mu m)$. Assume that the dust is suspended in air and the rain drops fall at their gravimetric settling velocity and impact the dust.

58. You are employed by a firm that makes bituminous concrete (roadway asphalt). The process produces a sticky aerosol of concentration (c_0) 2.0 g/m³ at STP and it has been decided to design a venturi scrubber to bring the process into compliance with state air-pollution regulations. You have been asked to conduct a preliminary analysis of the overall collection process and to produce a series of graphs showing how the overall collection efficiency varies with the operating parameters:

    Drop diameter (D_c)
    Throat length (L_t)
    Ratio of the volumetric flow rates of gas to liquid (G/S)

Process parameters:

Particle density $(\rho_p) = 1500 \, \text{kg/m}^3$
Throat diameter $= 0.6 \, \text{m}$
Throat gas velocity $(U_a) = 15 \, \text{m/s}$
Gas temperature and pressure 25 C, 1 atm
Scrubbing liquid is water and enters the throat with a negligible velocity
Once particles impact water drops they are removed, and liquid particles
do not impact one another

(a) Develop an analytical model that predicts the overall collection efficiency
by impaction. State all assumptions carefully and fully. Develop a finite
difference technique to integrate the differential equations so that the
overall collection efficiency can be computed for a venturi throat of
arbitrary length. Compute the velocity of a drop $(D_c = 0.25, 1.0, 2.0$ and
4.0 mm) versus throat length.

(b) Using $\eta_d = [\text{Stk}/(\text{Stk} + 0.7)]^2$, compute the single drop collection efficiency as a function of throat length for $D_c = 1$ and 2 mm and $D_p = 2 \, \mu\text{m}$.

(c) Plot the overall collection efficiency as a function of length assuming
$D_c = 1 \, \text{mm}$ and $D_p = 2 \, \mu\text{m}$ for $G/S = 3{,}000$, 9,000 and 15,000, for $L_t =$
0.1 to 10.0 m.

(d) Plot the overall collection efficiency as a function of the drop diameter
(D_c) assuming $D_p = 2 \, \mu\text{m}$, $L_t = 3.0 \, \text{m}$, $G/S = 9{,}000$, and $D_c = 0.25$, 1.0,
2.0, 3.0, and 4.00 mm.

59. Fog particles $(D_p = 10 \, \mu\text{m})$ in air pass through a window screen $(D_c =$
1000 $\mu\text{m})$ with a velocity 1.0 m/s. Some of the particles are collected by the
screen. Which of the following fog–air aerosols will be collected with the
same efficiency?

Particle	D_p (μm)	D_c (μm)	U (m/s)
a	20	1000	0.5
b	20	2000	1.0
c	10	2000	4.0
d	10	2000	2.0

60. Show (264) that the overall collection efficiency (η_0) of a filter bed of
thickness H consisting of spherical particles of diameter D_c is the following:

$$\eta_0(\text{overall}) = 1 - \exp\left[-\eta_d\left(\frac{f_s}{(1 - f_s)}\right)\left(\frac{3}{2}\right)\left(\frac{H}{D_c}\right)\right]$$

The bed has the following properties:

U_0 = approach velocity $U_0 = Q/A$, where Q is the volumetric flow rate
through a bed of frontal area A

η_d = single-particle collection efficiency, collection efficiency of each bed
particle D_c to collect particles of diameter D_p

f_s = Solids fraction, fraction of bed composed of solid material, the solids fraction is equal to $(1 - \text{Porosity})$

(a) Derive an expression for the overall collection efficiency in terms of the above parameters.

(b) Derive an expression for the rate of change of the overall collection efficiency η_0 with respect to the face velocity $d\eta_0/dU_0$, if the single particle collection efficiency is governed by $\eta_d = [\text{Stk}/(\text{Stk} + 0.7)]^2$, where $\text{Stk} = U_0\tau/D_c$ and $\tau = \rho_p D_p^2/(18\ \mu)$.

61. You wish to install a baghouse to capture particles removed from the air in a foundry. The baghouse manufacturer claims that the filter resistance (K_1) is 10,000 N sec/m^3 and estimates that the dust cake resistance (K_2) will be 80,000 sec^{-1}. The dust concentration entering the unit is 5 gm/m^3. The recommended air-to-cloth ratio is 1.0 m/min. If the baghouse is to operate at a maximum pressure drop of 500 N/m^2, how often should the bags be cleaned?

62. A horizontal duct of square cross section $W \times W$ and length (L) carries an aerosol at a volumetric flow rate Q. The inlet dust concentration is c_0. The duct is only mounted at the inlet end and has no other supports. Dust settles to the floor of the duct. After a while a sufficient amount of dust accumulates to produce a critical bending moment M_c that fractures the duct mounting. Write an expression to predict the time t_c when the mounting fails.

CHAPTER 9

Control of Particles

Goal: Students will be expected to learn to
- predict the effectiveness of inlets to capture
  particle
- use finite difference techniques to
  predict the effectiveness of inlets to capture
  particles

The purpose of this chapter is to combine techniques learned separately in previous chapters, introduce new concepts and propose computational methods to evaluate the effectiveness of devices to capture particles.

9.1 REACH OF INLETS

To prevent contaminants from entering the workplace it is necessary to withdraw air from a region in the vicinity of the source of contamination. Air is withdrawn by inlets of various shapes: openings in a planar surface (flanged inlets) (Fig. 9.1) and straight-sided openings (unflanged, plain inlets) (Fig. 9.2). The openings may be circular, square, or rectangular in cross section. Selecting the dimensions of the opening, its orientation to the source, the distance between the opening and the source, and the volumetric flow rate of the withdrawn air are decisions that have to be made by engineering designers. The technology that describes the velocity field upstream of the inlet is presently limited. The technology that relates the designer's selections and the effectiveness of the opening to capture contaminants does not exist. A design procedure is needed which has the following capabilities:

(a) Give designers the ability to predict the location of the stream tube of air entering the inlet. The boundaries of this stream tube are called the *dividing streamlines*.

(b) Give designers the ability to predict the location of the stream tube of contaminant entering the inlet. The boundaries of this stream tube are called the *bounding trajectories*. Bounding trajectories are called *limiting trajectories* by some authors (58, 376).

The location of the dividing streamlines and bounding trajectories are functions of the following:

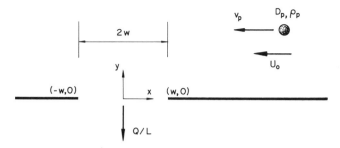

Fig. 9.1   Flanged inlet in streaming flow.

Volumetric flow rate of withdrawn air

Inlet dimensions

Orientation of the inlet to the source of contamination

Velocity of surrounding air

Characteristics of the source

 Gas or vapor: mass emission rate per unit area, chemical composition

 Particles: mass emission rate per unit area, initial particle velocity, initial size distribution

How to define an *inlet's effectiveness* in quantitative terms is not obvious nor is there general agreement on how effectiveness should be defined. The effectiveness should be a ratio in which the numerator is equal to what is captured and the denominator should be what one would like to capture. One would like to capture all the contaminant, but upon reflection one soon realizes the inability to capture everything may not be a limitation in the inlet design but a consequence of the environment in which the inlet is used.

There are several ways to describe the effectiveness of an inlet to capture contaminants that exist at some point(s) (x_0, y_0) upstream of the inlet. The most obvious measure of effectiveness would be an efficiency which typically is the ratio of what is captured to what is generated. Such a definition has been used by Ellenbecker et al. (121, 129, 189) and is useful if the mass of generated contaminant is finite. Unfortunately, the efficiency of an inlet immersed in an infinite flow of contaminant and air would be zero simply because the captured mass is finite and the generated mass is infinite. The efficiency would not reflect how well the inlet captured contaminant. Consequently, the denominator of any

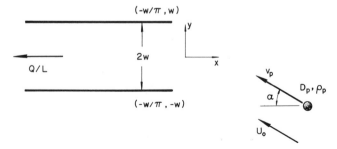

Fig. 9.2   Unflanged inlet in oblique streaming flow.

effectiveness should not be the total mass of generated contaminant. The clearest way to arrive at a definition of effectiveness is to begin by tracking the paths of contaminants from the point(s) of generation to the inlet. Figures 9.1–9.6 illustrate two different industrial environments in which contaminants are captured by inlets.

(a) Infinite source
      Uniform streaming flow of gaseous or particle contaminants
(b) Finite source
      Line source of particles in a streaming flow of air
      Line source of gaseous contaminants in streaming flow of air

In case b an inlet could conceivably capture all the generated contaminant, in case a all the contaminant could not be captured owing to the enormity of the polluted air mass streaming past the inlet.

The quantitative measure of inlet effectiveness used in this text will be the *reach*. The reach denotes the relative size of the region in front of the inlet that is influenced by the inlet. The reach defines the boundaries of the region from which the inlet *reaches out and captures* contaminants. Admittedly such a phrase is simplistic and anthropomorphic, but nevertheless it is a vivid and useful phrase. The *reach* will be defined as the ratio of the cross sectional area of the stream tube of contaminants entering the inlet to the cross sectional area of the steam tube of air entering the inlet,

$$\text{Reach} = \frac{\text{Cross-sectional area of contaminant stream tube}}{\text{Cross-sectional area of air stream tube}} \qquad (9\text{-}1)$$

The reach is not always unity because, owing to their inertia, not all particles in a volume of air at some point in front of the inlet actually enter the inlet even though the air may do so. The inertia of some particles prevents them from entering the inlet even though the air with which they were associated enters the inlet. With respect to contaminants that are gases or vapors, the reach is unity.

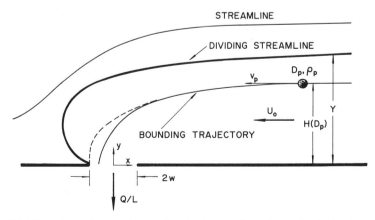

Fig. 9.3 Dividing streamline and bounding trajectory for a flanged inlet in streaming flow.

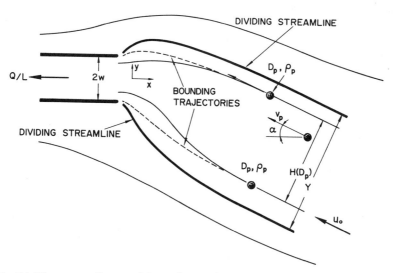

Fig. 9.4 Dividing streamlines and bounding trajectories for an unflanged inlet in oblique streaming flow.

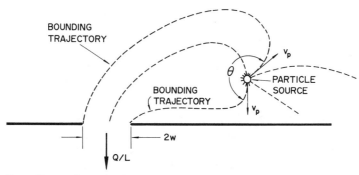

Fig. 9.5 Bounding trajectories for a flanged inlet and a line source in quiescent air.

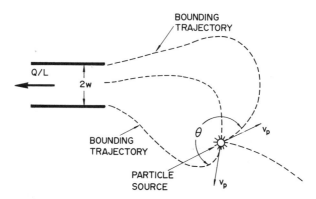

Fig. 9.6 Bounding trajectories for an unflanged inlet and a line source in quiescent air.

For finite sources, the reach and efficiency have the same value. For infinite source, the reach of different inlets have unique but different values while all values of the efficiency are zero.

To illustrate this definition as it applies to particles, consider the two types of particle motion shown in Figs. 9.1–9.6. For simplicity, consider only a two dimensional flow field, that is, rectangular inlet of large aspect ratio (length/width).

9.1.1 Uniform Streaming Flow of Particles

When an inlet removes particles from an air stream in which the concentration is uniform, the reach is defined as

$$\text{Reach } (x_0, y_0) = \frac{H(D_p)}{Y} \qquad (9\text{-}2)$$

The coordinates (x_0, y_0) define a point upstream of the inlet at which the reach is to be found, Y is the width of the stream tube of air entering the inlet, and $H(D_p)$ is the width of the stream tube of particles entering the inlet. The reach is thus the ratio of the cross sectional area defined by the bounding trajectories divided by the cross sectional area defined by the dividing streamlines. Figures 9.3 and 9.4 illustrate the reach of a flanged inlet and unflanged inlet in streaming flow.

9.1.2 Line Source of Particles in Quiescent Air

When a rectangular inlet removes particles generated by an isotropic line source located at (x_0, y_0) upstream of the inlet, the reach is defined as

$$\text{Reach } (x_0, y_0) = \frac{\theta}{2\pi} \qquad (9\text{-}3)$$

where θ is the included angle between the bounding trajectories. Figures 9.5 and 9.6 illustrate the reach of an unflanged and flanged inlet for an isotropic line source in quiescent air. Similar definitions of reach can be written for a stream of particles moving through quiescent air or a line source located in a stream of flowing air.

The reach is an unambiguous quantity that has a unique value based on the physical parameters of the problem. Values of the reach vary from 100% to zero. In no case is it undefined or infinite. A value of 100% indicates that the inlet removes all the contaminant from the air drawn into the inlet and is performing at its maximum. The reach has a meaning similar to the single-particle collection efficiency introduced in Chapter 8. Like the single-particle collection efficiency, the reach addresses only the ability to capture contaminants in a well defined stream tube influence by the collecting device (impacting particle or inlet).

9.2 PARTICLE SAMPLING

There is a large body of professional literature on stack sampling (65) and ambient sampling in the outside and workplace environment (42, 58, 123, 124,

328, 329, 376, 409, 417) relevant to the design of contaminant control systems. Whether engineering designers can use the technology directly is another question since the objective of sampling is to obtain particle concentrations and size distributions that replicate the sampled stream. Thus, while designers in industrial ventilation can gain a great deal of guidance and qualitative information, and can use basic analytic formulations, they will not be able to obtain information about how to estimate the effectiveness of an inlet.

Figure 9.7 shows the streamlines and particle trajectories upstream of a probe used to obtain a sample of an aerosol in process gas streams (58). The streamlines can be computed assuming the air to be an ideal fluid and the particle trajectories computed from the equations described in Chapter 8. The figures illustrate the disparity between the dividing streamline and bounding trajectory.

Consider a probe of internal radius R (Fig. 9.8), used to withdraw a sample of

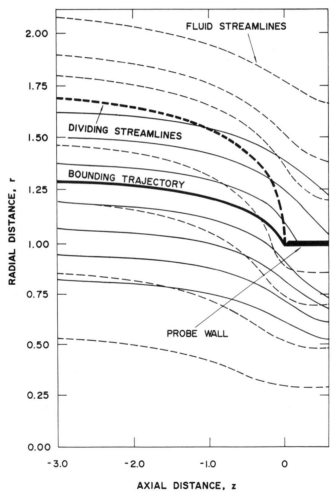

Fig. 9.7 Particle trajectories and streamlines of a sampling probe in streaming flow; $U(\text{probe})/U_0 = U_i/U_0 = 3.0$; $\text{Re}_p = 23.2$; Stk(based on nozzle radius) = 1.0 (redrawn from ref. 58).

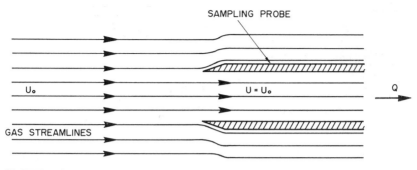

(a) ISOKINETIC SAMPLING, $U_o = Q/A$, $\theta = 0$

Fig. 9.8a Particle trajectories and streamlines for isokinetic sampling (redrawn from ref. 42).

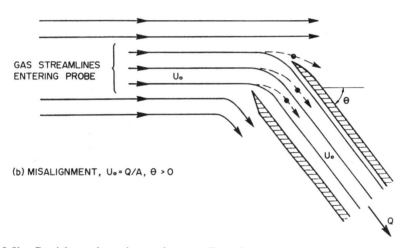

(b) MISALIGNMENT, $U_o = Q/A$, $\theta > 0$

Fig. 9.8b Particle trajectories and streamlines for a sampling nozzle misaligned to direction of flow (redrawn from ref. 42).

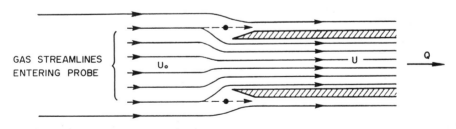

(c) SUPER-ISOKINETIC SAMPLING, $Q/A > U_o$, $\theta = 0$

Fig. 9.8c Particles trajectories and streamlines for nonisokinetic sampling (redrawn from ref. 42).

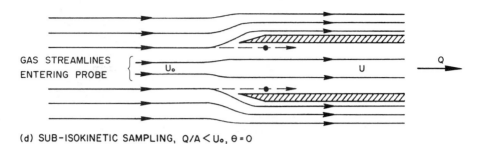

(d) SUB-ISOKINETIC SAMPLING, Q/A < U₀, θ = 0

Fig. 9.8d   Particles trajectories and streamlines for nonisokinetic sampling (redrawn from ref. 42).

an aerosol approaching from the right with a velocity of U_0. The axis of the probe is parallel to the direction of flow. The free stream contains particles of known density and diameter that travel to the right with an initial velocity equal to the air velocity U_0. The particle concentration in the free stream is c_0. Air is withdrawn through the probe at a volumetric flow rate Q that can be varied. Of concern to an experimentalist are the values of Q at which the particle concentration in the sample (c) are no longer equal to the concentration in the free stream (c_0). The values will be the same ($c = c_0$) if the gas velocity entering the probe (called the sampling velocity, $U = Q/\pi R^2$) is equal to the free-stream velocity U_0. *Isokinetic sampling* is the name given to test conditions in which $U/U_0 = 1.0$ and *anisokinetic sampling* is when the ratio is not unity. Figures 9.7 and 9.8 show the streamlines for U/U_0 equal to three. On the basis of the continuity equation, the radius of the dividing streamline well upstream of the probe should be equal to R times the square root of 3. Figure 9.7 shows this to be the case. The figures also shows that the most rapid change in the direction of streamlines near the dividing streamline occurs within a distance one radius upstream of the inlet.

Figure 9.7 shows the pathlines for particles that have an initial Stokes number of unity. It is clear that the bounding trajectory and dividing streamline are not coincident. The ratio of the particle concentration in the sample stream to the free stream value (c/c_0) can be calculated. The ratio is called the *aspiration coefficient* (*efficiency*). Rader and Marple (328) and Liu et al. (376) studied the aspiration coefficient as functions of

$$\frac{U}{U_0}$$

$$\text{Reynolds number} = \frac{\rho R U_0}{\mu}$$

$$\text{Stokes number} = \frac{\rho_p D_p^2 U_0 C}{18 \mu R}$$

where R is the radius of the sampling probe and C is the Cunningham correction factor. They found that the principal parameters affecting the aspiration coefficient are the Stokes number and U/U_0. The Reynolds numbers were of only minor significance at Stokes numbers above unity. For Stokes numbers interception phenomena became important and has to be included in the analysis. Figure 9.9 shows the aspiration coefficient at a sampling probe Reynolds number of 300 as a function of the Stokes number and U/U_0 for a typical aerosol in which the ratio of air density to particle density times the Cunningham coefficient ($\rho/C\rho_p$) is 1.2×10^{-3}. When U/U_0 exceeds unity, some of the particles that were originally in the air entering the probe pass outside the probe, that is, cross the dividing streamline. As a result, the air entering the probe does not contain all the particles it formerly had and the measured concentration (c) is less than c_0. On the other hand, if U/U_0 is less than unity, particles from outside the dividing streamline enter the air withdrawn by the probe and c/c_0 exceeds unity, see Fig. 9.7. For small particles with low Stokes numbers, the aspiration coefficient approaches unity irrespective of U/U_0. Thus,

$$\text{Superisokinetic: } \frac{U}{U_0} > 1, \quad \text{one can expect } \frac{c}{c_0} < 1$$

$$\text{Isokinetic: } \frac{U}{U_0} = 1, \quad \text{one can expect } \frac{c}{c_0} = 1 \qquad (9\text{-}4)$$

$$\text{Subisokinetic: } \frac{U}{U_0} < 1, \quad \text{one can expect } \frac{c}{c_0} > 1$$

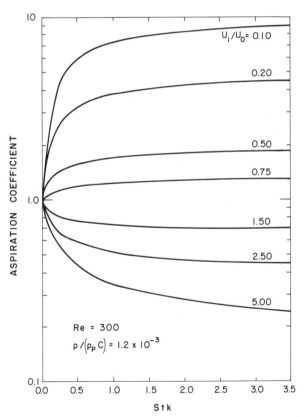

Fig. 9.9 Aspiration coefficient versus Stokes number and velocity ratio (U_i/U_0) (redrawn from ref. 42).

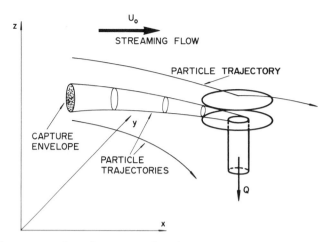

Fig. 9.10 Capture envelope for an annular slot-sampling probe immersed in streaming flow (redrawn from ref. 123).

Strict adherence to isokinetic conditions can be relaxed as U_0 and or D_p decrease.

Dust samplers using an annular slot have been used for years to measure the size distribution and concentration of an aerosol in the outdoor and workplace environment. Armbruster and Zebel (123) examined the influence of the air speed on the particle size distribution in the sample. Figures 9.10 and 9.11 show the particle trajectories predicted by the equations of motion described in Chapter 8 assuming the air was an ideal fluid whose stream functions and velocities could be predicted by the equations in Chapter 7. The authors define a sampling efficiency (η) to reflect the disparities in the sample and ambient particle size distributions.

$$\eta = \frac{\left(\dfrac{dm}{dD_p}\right)(\text{sample})}{\left(\dfrac{dm}{dD_p}\right)(\text{ambient})} \tag{9-5}$$

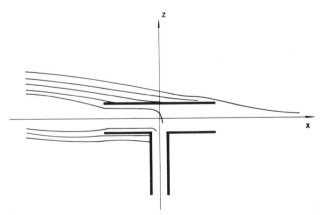

Fig. 9.11 Particle trajectories for an annular slot sampling probe immersed in streaming flow (redrawn from ref. 123).

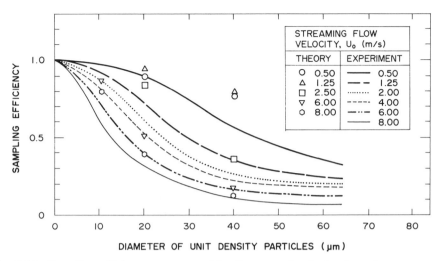

Fig. 9.12 Sampling efficiency versus particle diameter of unit density spheres at several streaming flow velocities (redrawn from ref. 123).

where dm/dD_p is the mass of dust per unit size interval in the diameter (mg/μm). Figure 9.12 shows the predicted and experimentally measured efficiencies as functions of the wind speed U_0 and particle size. The figure suggests that the size distribution of the sample is an accurate representation of the source for particle less than 10 μm for a wide range of air velocities but that the disparity becomes increasingly worse for larger particles and higher wind speeds. In short, annular slots are appropriate devices to sample for respirable particles.

The aspiration coefficient associated the human head and mouth breathing has been analyzed by Dunnett (405) assuming ideal flow. Unless the free stream velocity is unusually large, the aspiration coefficient is virtually unity and unaffected by the breathing frequency and volumetric flow rate.

9.3 PARTICLE REMOVAL BY LINE SINK IN STREAMING FLOW

To improve an understanding of the dividing streamline and bounding trajectory, begin by analyzing the reach of a line sink lying in a plane over which air and particles pass in the negative x direction. The mathematics of a line sink are considerably simpler than the mathematics of rectangular inlets of finite width. In Chapter 7 it was shown that the air streamlines for the two cases are similar several slot widths upstream of the inlet. The anticipated particle trajectories for a line sink in a horizontal plane are shown in Fig. 9.13. The air travels uniformly over the surface in the negative x direction with a speed U_0. The particles possess an initial velocity $[v(0)]$ a distance x_0 upwind of the line sink. Captured particles will be defined as those that strike the horizontal surface anywhere upstream of the stagnation point. The analytical method must determine which particles strike the surface at $x > x_1$ and which particles escape capture, $x < x_1$. The value x_1 defines the location of the stagnation point. For the configuration in Fig. 9.13, the value of x_1 is negative.

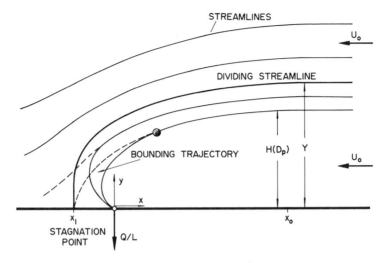

Fig. 9.13 Dividing streamline and bounding trajectory for a line sink in streaming flow, Illustrated Example 9.1.

The dividing streamline defines the region of flow above the surface within which all the air is withdrawn. Far upwind of the line sink (i.e., $x_0 \gg 0$), the distance between the surface and the dividing streamline is

$$Y = \frac{Q}{LU_0} \tag{9-6}$$

For smaller positive values of x_0, the vertical distance between the surface and dividing streamline can be found from Eq. 7-54 by solving the equation for the value of ψ_d corresponding to the dividing streamline of a line sink in streaming flow for y when

$$\psi_d = \frac{Q}{L} = U_0 r \sin\theta + \left(\frac{Q}{L\pi}\right)\theta \tag{9-7}$$

$$r \sin\theta = y \tag{9-8}$$

$$\theta = \arctan\left(\frac{y}{x}\right) \tag{9-9}$$

In the absence of gravity, no particles in the air above the dividing streamline will enter the line sink. If gravity is included, large particles above the dividing streamline may be captured, depending on the value of U_0, D_p, and x_0. Particles close to the horizontal surface are removed by the line sink, but there is a unique height $H(D_p)$ above the surface at which the particle's inertia will carry it through the dividing streamline. The height $H(D_p)$ defines the height of the bounding trajectory at x_0.

A procedure to determine $H(D_p)$ involves determining particle trajectories using techniques developed in Chapter 8. A value of y is selected and the trajectory of a particle beginning at (x_0, y) calculated. If the particle strikes the surface at $x < x_1$, it will be assumed that it is not captured. If the particle is not captured, a lower value of y is selected and the process repeated until the particle's trajectory causes it to strike the surface at $x > x_1$. The maximum value of y which satisfies this criterion is defined as $H(D_p)$.

Illustrated Example 9.1—Reach of a Line Sink in Streaming Flow
Consider the flow depicted in Fig. 9.13. Compute the height of the dividing streamline (Y), bounding trajectory $H(D_p)$, and reach for the following conditions:

$Q/L = 0.5 \, \text{m}^3/\text{s}$ per meter of inlet (322.7 CFM per foot of inlet)
$U_0 = 0.5 \, \text{m/s}$
Particle density $= 1000 \, \text{kg/m}^3$
Initial particle velocity $v_p(0) = 10 \, \text{m/s}$
Initial particle location $x_0 = 1.0 \, \text{m}$
Include the effects of gravity

From Chapter 7, the following quantities in Fig. 9.12 can be found,

$$Y(x_0 \gg 0) = \frac{Q}{LU_0} = \frac{0.5}{0.5} = 1.0 \, \text{m}$$

$$x_1 = -\frac{Q}{LU_0 \pi} = -\frac{0.5}{(0.5)(3.14)} = -0.3183 \, \text{m}$$

Since $x_0 = 1.0 \, \text{m}$ is close to the sink, the height of the dividing streamline (Y) at $x_0 = 1.0 \, \text{m}$ has to be computed from the equation for the stream function. The value of Y is found to be $0.7877 \, \text{m}$. The components of the air velocity (U_x, U_y) are computed from the velocity potential

$$\phi = \left(\frac{Q}{L\pi} \right) \ln r + U_0 r \cos \theta$$

$$= \left(\frac{Q}{L\pi} \right) \ln \left[\sqrt{x^2 + y^2} \right] + U_0 x$$

$$U_x = -\frac{\partial \phi}{\partial x} = -U_0 - \left(\frac{Q}{L\pi} \right) \left(\frac{x}{x^2 + y^2} \right)$$

$$U_y = -\frac{\partial \phi}{\partial y} = -\left(\frac{Q}{L\pi} \right) \left(\frac{y}{x^2 + y^2} \right)$$

The algorithm below computes the initial location $[x_0, H(D_p)]$ on the uppermost trajectory of the particle removed from the approaching air by the line sink. By printing the subsequent values of x and y that are computed, the trajectory itself can be computed. The algorithm uses the particle trajectory program in Illustrated Example 8.3 and computes values of the air velocity from the program in Illustrated Example 7.2.

```
  1 rem filename-reach of line sink for particles in
streaming flow
  2 rem this program computes a point (x0,y0) on the
bounding trajectory
  3 rem variables
    k=Q/(L*Pi) (sq m/s)
    dt=time step (s)
    rp, ra=particle and air density (kg/cu m)
    dp=particle diameter (m)
    vpi=initial particle velocity (m/s)
    y0 (m)=height of bounding trajectory at x0
    hds=height of dividing streamline at x0=1.0 m
    P,T=air pressure (atm) and temperature (K)
    cd=drag coefficient
  8 print ''        Dp (m)        H(Dp) (m) at x0=1.0 m''
  9 print ''_____    _____'',''
 10 rem input initial values
 20 k=0.159
 30 dt=0.05
 40 rp=1000
 50 g=9.8
 60 x0=1.0
 70 hds=0.7877
 80 vpi=-10.0
 90 p=1
100 T=298
110 dim d(5)
120 for i=1 to 5
130 read d(i)
140 next i
150 for i=1 to 5
160 dp=d(i)
170 mu=(13.554+0.6738*T-(3.808*T*T/10000)+
        (1.183*T*T*T/10000000))/10000000
180 ra=p*100/(0.287*T)
190 yl=0.01*hds
200 yh=1.2*hds
210 y0=(yl+yh)/2
220 ui=vpi
230 vi=0
240 x=x0
250 y=y0
260 rem calculate air velocity at x,y
270 ua=-(0.5+(k*x / (x*x+y*y))
280 va=-k*y / (x*x+y*y)
290 rem calculate particle velocity at x,y
300 vr=sqr(( ui-ua)*(ui-ua)+(vi-va)*(vi-va))
310 re=ra*dp*vr / mu
```

```
320 cd=0.4 + 24 / re + 6/( 1+sqr(re))
330 a=-0.75*cd*ra*vr / (rp*dp)
340 b=-a*ua
350 d=-( a*va)-g
360 uf=(ui*(1+a*dt/2)+b*dt) / ( 1-a*dt/2)
370 vf=( vi*(1+a*dt/2)+d*dt) / ( 1-a*dt/2)
380 rem compute new location of particle
390 x=x+(uf+ui)*dt / 2
400 y=y+(vf+vi)*dt / 2
410 vi=vf
420 ui=uf
430 rem logic to select the uppermost trajectory
440 r=(x+.3183) / .3183
450 if y<-.001 goto 470
460 goto 260
470 if abs(r)<0.05 then print dp,y0 : next i
480 if x>-.3183 then yl=y0 : goto 210
490 if x<-.3183 then yh=y0 : goto 210
500 data 0.000080, 0.000060, 0.000040, 0.000020, 0.000010
510 end
```

Figure 9.13 illustrates the trajectory of a particle that escapes capture and the air streamline it is coincident with at $[x_0, H(D_p)]$. The table below gives the reach $(H(D_p)/Y)$ at $x_0 = 1.0$ m. Large particles within (and even some beyond) the dividing streamline are captured. For smaller particles, the dividing streamline is larger than the capture region. Gravity has a stronger influence than inertia on the large particles and aids in their capture while the reverse is true for smaller particles. Many of the smaller particles have sufficient inertia to escape capture as the fluid streamline bends toward the sink to the left of the origin. The results below also indicate that the higher the initial particle velocity, the less likely the particle will be captured. From an overall intuitive basis, one can conclude that the dividing streamline is larger than the capture region for the majority of particles of interest in industrial ventilation.

v_i (m/s)	D_p (μm)	Reach ($x_0 = 1.0$ m)
10	80	1.394
10	60	1.229
10	40	1.087
10	20	1.005
10	10	0.971
100	80	1.252
100	60	1.111
100	40	1.016
100	20	0.971
100	10	0.971

The program shown above can be altered and statements added to print x, y along particle trajectories anywhere below the dividing streamline. Better accuracy can be obtained by reducing the time step (line 30) and reducing the tolerance (line 470).

9.4 FLANGED AND UNFLANGED RECTANGULAR INLETS IN STREAMING FLOW

Consider the capture of particles by flanged and unflanged inlets in which the particles are either generated by a line source in quiescent air or contained in a moving air stream. Figures 9.3–9.6 illustrate the streamlines and particle trajectories that can be expected. The location of the dividing streamlines and bounding trajectories and the computation of the reach can be accomplished by a single rather basic computer program. Different boundary conditions need to be introduced, depending on which flow configuration (Figs. 9.3–9.6) applies. Of particular importance will be the need to use appropriate values for the initial guesses for ϕ and ψ in the *Newton–Raphson* subroutine.

In the absence of wakes, jets, and temperature gradients, the flow field is irrotational:

$$\nabla^2 \phi = 0 \tag{9-10}$$

$$\nabla^2 \psi = 0 \tag{9-11}$$

and the air velocities can be expressed as

$$U_x = -\frac{\partial \phi}{\partial x} = -\frac{\partial \psi}{\partial y} \tag{9-12}$$

$$U_y = -\frac{\partial \phi}{\partial y} = \frac{\partial \psi}{\partial x} \tag{9-13}$$

If the flow field consists of an inlet immersed in a streaming flow,

$$\phi + \phi_w + \phi_f \tag{9-14}$$

$$\psi = \psi_w + \psi_f \tag{9-15}$$

where the subscript w denotes the inlet and the subscript f denotes streaming flow. The velocities of the flow field can be found by differentiation,

$$U_x = -\frac{\partial \phi_w}{\partial x} - \frac{\partial \phi_f}{\partial x} \tag{9-16}$$

$$U_y = -\frac{\partial \phi_w}{\partial y} - \frac{\partial \phi_f}{\partial y} \tag{9-17}$$

Comparable equations can be written with respect to the stream function. If air is withdrawn from a quiescent air mass, the potential functions for streaming flow are set equal to zero.

9.4.1 Flanged Slot

Consider a slot of width $2w$ located in an infinite horizontal surface shown in Fig. 9.3. Through the slot air will be withdrawn uniformly at a volumetric flow rate per unit length of slot Q/L. In addition, assume that air passes over the surface at a uniform speed U_0 in the negative x direction. By neglecting the boundary layer along the surface, the flow field can be assumed to be conservative. The potential functions for the slot and streaming flow are:

Inlet:

$$\frac{x}{w} = \cosh\left(\frac{\phi_w}{k}\right)\cos\left(\frac{\psi_w}{k}\right) \tag{9-18}$$

$$\frac{y}{w} = \sinh\left(\frac{\phi_w}{k}\right)\sin\left(\frac{\psi_w}{k}\right) \tag{9-19}$$

$$k = \frac{Q}{L\pi} \tag{9-20}$$

Streaming flow:

$$\psi_f = U_0 x \tag{9-21}$$

$$\phi_f = U_0 y \tag{9-22}$$

The potential functions for the slot are not given as functions of x and y. Thus it will not be possible to define overall potential functions for the composite flow field as functions of x and y as was possible when analyzing a line sink in streaming flow. Alternatively, the velocity components of the composite flow field will be computed by adding the velocity components for a flanged inlet and streaming flow,

$$U_x = -U_0 - \frac{k}{Aw}\left[\sinh\left(\frac{\phi_w}{k}\right)\cos\left(\frac{\psi_w}{k}\right)\right] \tag{9-23}$$

$$U_y = -\frac{k}{Aw}\left[\cosh\left(\frac{\phi_w}{k}\right)\sin\left(\frac{\psi_w}{k}\right)\right] \tag{9-24}$$

$$A = \left[\sinh\left(\frac{\phi_w}{k}\right)\cos\left(\frac{\psi_w}{k}\right)\right]^2 + \left[\cosh\left(\frac{\phi_w}{k}\right)\sin\left(\frac{\psi_w}{k}\right)\right]^2 \tag{9-25}$$

9.4.2 Unflanged Slot

Consider the horizontal, straight-sided opening (unflanged inlet) of width $2w$ shown in Fig. 9.4. Through the inlet, air will be withdrawn at a uniform rate per unit length of slot Q/L. The inlet will be immersed in an air mass moving at a uniform speed U_0 inclined at an angle α to the walls of the unflanged inlet. The potential functions of the unflanged inlet and streaming flow can be found in Table 7.2.

Inlet:

$$\frac{x}{w} = \frac{1}{\pi}\left[\frac{2\phi_w}{k} + \exp\left(\frac{2\phi_w}{k}\right)\cos\left(\frac{2\psi_w}{k}\right)\right] \tag{9-26}$$

$$\frac{y}{w} = \frac{1}{\pi}\left[\frac{2\psi_w}{k} + \exp\left(\frac{2\phi_w}{k}\right)\sin\left(\frac{2\psi_w}{k}\right)\right] \tag{9-27}$$

$$k = \frac{Q}{L\pi} \tag{9-28}$$

Oblique streaming flow:

$$\phi_f = U_0 x \cos\alpha - U_0 y \sin\alpha \tag{9-29}$$

$$\psi_f = U_0 y \cos\alpha + U_0 x \sin\alpha \tag{9-30}$$

It is not possible to define overall potential functions as explicit functions of x and y. Thus the velocity field will be computed by adding the velocity components for unflanged inlet and streaming flow,

$$U_x = -U_0\cos\alpha - \frac{k\pi}{2wA}\left[1 + \exp\left(\frac{2\phi_w}{k}\right)\cos\left(\frac{2\psi_w}{k}\right)\right] \tag{9-31}$$

$$U_y = U_0\sin\alpha - \frac{k\pi}{2wA}\left[\exp\left(\frac{2\phi_w}{k}\right)\sin\left(\frac{2\psi_w}{k}\right)\right] \tag{9-32}$$

$$A = \left[\exp\left(\frac{2\phi_w}{k}\right)\sin\left(\frac{2\psi_w}{k}\right)\right]^2 + \left[1 + \exp\left(\frac{2\phi_w}{k}\right)\cos\left(\frac{2\psi_w}{k}\right)\right]^2 \tag{9-33}$$

9.4.3 Upper Dividing Streamline For An Flanged Slot

In Section 7.6.1 a method was described to compute the coordinates (x, y) of the dividing streamline for a flanged slot. The method was explicit and relatively simple owing to Eqs. 7-106 and 7-107 which expressed x and y as functions of ψ and ϕ. Described below is a general (but more complicated) trial-and-error method to compute the coordinates (x, y) of the dividing streamline for a variety of inlets when equations such as 7-106 and 7-107 do not exist. In Section 9.4.4 the method will be used for an unflanged slot.

To compute the location of the dividing streamline, a *marching process* can be used in which the user locates a point on the dividing streamline and marches forward in time and computes the new location of the air parcel. The process is repeated until the parcel passes into the inlet.

(1) Choose an upstream location (x_0) at which you wish to compute the reach. This coordinate will not change.

(2) Guess a distance (y_i) at which together with (x_0) defines the coordinate (x_0, y_i) through which the upper dividing streamline is believed to pass.

(3) Compute $\phi_w(x_0, y_i)$ and $\psi_w(x_0, y_i)$ using the Newton–Raphson method.

(4) Compute $\psi = \psi_w(x_0, y_i) + \psi_f(x_0, y_i)$ and compare it with the value of ψ that is known to exist on the upper dividing streamline (i.e., $\psi = Q/L$. If the

values do not agree, return to step 2, select another value of y_i and repeat step 3 until satisfactory agreement is obtained. Call the correct value of (y_i) on the dividing streamline (y_0).

(5) Using the correct coordinates (x_0, y_0), repeat step (3) to obtain the values of ψ_w and ϕ_w and compute the air velocity at the point (x_0, y_0) from Eqs. 9-23 to 9-25.

(6) Select an increment of time δt. Compute the location of the air parcel after an interval of time δt.

$$x(t_0 + \delta t) = x_0 + \delta t U_x(x_0, y_0) \tag{9-34}$$

$$y(t_0 + \delta t) = y_0 + \delta t U_y(x_0, y_0) \tag{9-35}$$

where (t_0) is zero for the first time step and the elapsed time for subsequent time steps.

(7) Using the new location found by step 6, repeat steps 3–6. The value of δt should be small so that the displacement of the air parcel during an increment of time is several orders of magnitude less than the value of (x_0). By repeating the process, one marches toward the inlet until the streamline passes through an arbitrarily small region surrounding the inlet lip. Defining such a region is necessary because the lip itself is a singular point where derivatives of the potential functions are infinite.

The algorithm for finding the coordinates of the dividing streamlines for an unflanged inlet is the same except that the values of the upper and lower dividing streamlines are $Q/2L$ and $-Q/2L$, respectively, and Eqs. 9-31 to 9-33 are used to compute the air velocity.

9.4.4 Bounding Trajectory

To compute the coordinates (x, y) of a bounding trajectory, a marching process can be used in which the user begins at some arbitrary point (x_0, y_i) upstream of the inlet, computes the displacement of the particle after an increment of time, and marches forward in time to find the new location of the particle. The process is repeated until the particle either enters the inlet or passes near it. If the particle does not strike a small region (defined by the user) near the lip, the user adjusts the initial location (y_i) and repeats the marching process until the particle strikes a region reasonably close to the inlet. The algorithm can be used for either flanged or unflanged inlets by using the appropriate potential functions for the inlet.

(1) Choose an upstream location (x_0) for which you wish to find the bounding trajectory. This coordinate will not change.

(2) Guess a value y_i which together with x_0 defines the coordinate (x_0, y_i) that hopefully constitutes a point on the bounding trajectory; it will be presumed that the particle velocity is known at (x_0, y_i).

(3) At the point (x_0, y_i), compute the air velocity $U_x(x_0, y_i)$ and $U_y(x_0, y_i)$

using steps 3 and 4 from the algorithm to find the dividing streamline (Section 9.4.3).

(4) Find the particle velocity, displacement and new location (x_j, y_j) at the end of the first interval of time using Eqs. 8-80 to 8-84. The location (x_j, y_j) of the particle at the end of an interval of time can be found from

$$x_j = x_0 + \delta t v_x(x_0, y_i) \tag{9-36}$$

$$y_j = y_i + \delta t v_y(x_0, y_i) \tag{9-37}$$

(5) Using the location (x_j, y_j) as the initial location, repeat steps 3 and 4 above until the particle passes through or suitably near the inlet.

(6) If the particle does not pass satisfactorily close to the inlet lip, select another initial particle location (y_i) and repeat steps 2 and 5 above until the particle passes satisfactorily close to the lip.

Figure 9.14 shows the measured (125) coordinates of the capture envelope (bounding trajectory) when smoke is removed from air moving over a flanged square inlet at a variety of free stream velocities (U_0) and volumetric flow rates (Q). Because smoke particles are very small, the dividing streamline and bounding trajectory are coincident. Figure 9.14 suggests that the coordinates for the different capture envelopes can be represented by a single curve when expressed by the dimensionless variables

$$x\left(\frac{U_0}{Q}\right)^{1/2} \quad \text{and} \quad y\left(\frac{U_0}{Q}\right)^{1/2} \tag{9-38}$$

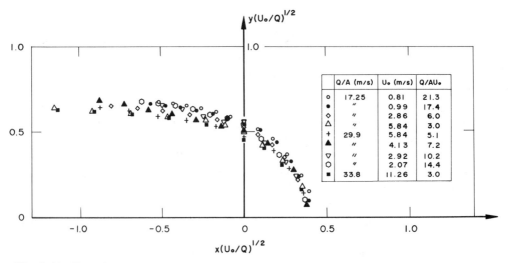

Fig. 9.14 Experimental values of the capture envelope for smoke moving over a flanged square inlet of area A at various velocities (U_0) flowing in the positive x direction, (redrawn from ref. 125).

TABLE 9.1 A Comparison of Points on the Capture Envelope

Location	Potential Functions for a point sink	Experiment (125) for a square inlet
Stagnation point $x = r_0$, $y = 0$	$r_0\sqrt{U_0/Q} = 0.398$	$r_0\sqrt{U_0/Q} = 0.4$
above inlet $x = 0$, $y = h_0$	$h_0\sqrt{U_0/Q} = 0.562$	$h_0\sqrt{U_0/Q} = 0.5$
Downstream of inlet $y = h_d$	$x = $ infinity $h_d\sqrt{U_0/Q} = 0.798$	$x\sqrt{U_0/Q} = 1$ $h_d\sqrt{U_0/Q} = 0.6$

The potential functions for streaming flow over a flanged *square inlet* can be constructed by adding the potential functions of streaming flow to those in Section 7.6, more specifically by adding the streaming flow velocity to the velocities predicted in Section 7.6. Alternatively, an approximation can be made by using the potential functions for streaming flow over a point sink. The equations in Section 7.4 can be reformulated for the nondimensional variables in Eq. 9-38. Table 9.1 is a comparison of three points on the capture envelope for a square inlet and a point sink, (Section 7.3.3) confirming the accuracy of approximating an inlet of finite dimensions as a point sink for points far from the inlet face.

Illustrated Example 9.2—Reach of an Unflanged Inlet in Streaming Flow

Consider an unflanged inlet of width 0.1 m ($c = 0.05$ m) that withdraws air at a volumetric flow rate Q/L equal to 0.762 m^3/s per meter of length. The inlet is immersed in a moving air stream that travels in the negative x direction at a speed (U_0) of 3.81 m/s. With respect to Fig. 9.4 angle α is zero. Particles are generated at x_0 upstream of the inlet and that have an initial velocity of $v_x(0) = -10$ m/s, $v_y(0) = 0$. Analyze the flow field of the air and particles and compute:

(a) Location of the dividing streamlines

(b) Location of the bounding trajectories

(c) Compute the reach

```
  1 rem filename-capture of particles by an unflanged slot
in streaming flow
  2 rem this program computes the bounding trajectory and
locates a point x0,y0 on this trajectory
  3 rem variables: psi=s , phi=f ; air velocity=u,v parti-
cle velocity=up,vp
  5 rem input initial values
  6 pi=3.14159 : e=2.71828 : ga=9.8
 10 QL=0.762 : c=0.05 : rp=1000 : x0=4*c : xl=x0
 18 u=-3.81 : v=0 : upi=-10 : vpi=0
 21 k=QL / (2*pi)
 24 p=1 : t=298
 29 rem input particles, compute viscosity, air density
 30 dim par(5)
```

```
31 for i=1 to 5
32 read par(i)
33 next i
40 ra=p*100 / (0.287*t)
50 for i=1 to 5
51 dp=par(i)
60 mu=(13.554+0.6738*t-(3.808*t*t / 10**4)+
       (1.183*t*t*t/ 10** 7)) / 10** 7
62 rem input upper and lower limits on initial particle
location
63 yh=1.5*QL / (2*abs(u))
64 yl=0.5*c
70 y0=(yl+yh) / 2
71 dt=0.002 : ui=upi : vi=vpi
72 yg=y0
74 x=x0 : y=y0
76 print '' dp (m) ''
77 print ''_____''
78 print dp
79 print
80 print ''        x/c           y/c ''
81 print ''_____        _____''
84 gosub 600
90 a1=(exp ((f2/k))*sin ((s2/k)))**2 +
       (1+exp ((f2/k))*cos ((s2/k)))**2
95 ua=u-(k*pi/(2*c))*(1+exp ((f2/k))*cos ((s2/k))) / a1
100 va=v-(k*pi/(2*c))*exp ((f2/k))*sin ((s2/k)) / a1
120 vr=sqr ((ua-ui)**2+(va-vi)**2)
140 re=ra*dp*vr / mu
160 cd=0.4+24 / re+(6 / (1+sqr(re)))
180 a=-0.75*cd*re*vr / (rp*dp)
200 b=-a*ua
210 d=-(a*va)-ga
211 rem compute particle velocity at x,y
215 if x<0 then dt=0.0001
220 uf=(ui*(1+a*dt/2)+b*dt) / (1-a*dt/2)
240 vf=(vi*(1+a*dt/2)+d*dt) / (1-a*dt/2)
241 rem compute new location of particle
260 x=x+(uf+ui)*dt/2
280 y=y+(vf+vi)*dt/2
281 vi=vf : ui=uf
290 print x/c, y/c
300 r=(c-y) / c
315 if x<- 0.5*c / pi and abs (r)<0.02 then goto 350
320 if x<- 0.5*c / pi and y>c then yh=y0 : goto 70
322 if x<- 0.5*c / pi and (x-xl)>=0 then yh=y0 : goto 70
325 if x<- 0.5*c / pi and y<c then yl=y0 : goto 70
326 xl=x
```

```
330 goto 84
350 print ''upper bounding trajectory (dp , x0/c , y0/c)''
351 print ''_____''
355 print dp , x0/c , yg/c : next i
550 data 0.000025 , 0.000100 , 0.000500 , 0.000010 ,
0.000050
551 rem
552 rem
600 rem newton raphson subroutine to compute psi and phi in
        the vicinity of (x, y)
605 rem input approximate values of psi and phi in the
        vicinity of (x, y)
610 f2=0.1 : s2=0.1
685 rem compute values of f, g, ff, fs, gf, gs, df, dg, ja and
        f1 and s1
703 g=(s2 / k)-(pi*y / c)+exp ((f2 / k))*sin ((s2 / k))
705 f=(f2 / k)-(pi*x / c)+exp ((f2 / k))*cos ((s2 / k))
710 ff=(1+exp((f2 / k))*cos ((s2 / k))) / k
720 fs=-(exp ((f2 / k))*sin ((s2 / k))) / k
730 gs=ff : gf=-fs
740 ja=(ff*gs)-(fs*gf)
750 df=(g*fs-f*gs) / ja
760 ds=(f*gf-g*ff) / ja
770 if abs (ds)<0.00001 and abs (df)<0.00001 then goto
        910
780 s2=s2+ds : f2=f2+df
900 goto 685
910 return
930 end
```

Figure 9.15 shows the dividing streamlines and bounding trajectories for several particle sizes.

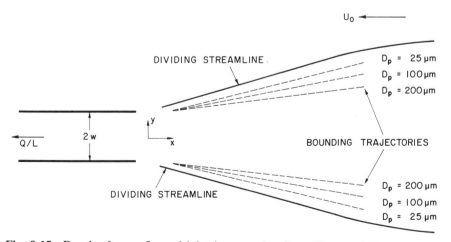

Fig. 9.15 Reach of an unflanged inlet in streaming flow, Illustrated Example 9.2.

9.5 FINITE-DIFFERENCE METHODS

When the flow field of an ideal fluid involves boundaries that are irregular or for which the potential functions are unknown or cannot be created by superposition, numerical finite-difference methods can be used to obtain values of the stream function at specified nodes within the flow field. Once values of the stream function (ψ) have been computed at nodes throughout the flow domain, the computer programs described previously can acquire the air velocity at arbitrary points (x, y) by computing the following derivatives as finite differences:

$$U_x = -\frac{\partial \psi}{\partial y} \cong -\frac{\delta \psi}{\delta y} \tag{9-39}$$

$$U_y = \frac{\partial \psi}{\partial x} \cong \frac{\delta \psi}{\delta x} \tag{9-40}$$

What follows is a brief summary of how an elementary finite-difference technique can be applied to analyze aerosol impaction in a rectangular inertial impactor. Readers should consult texts (195, 196) devoted to numerical methods in fluid mechanics to learn the subject thoroughly and more sophisticated techniques. Figure 9.16 is a schematic diagram of a rectangular inertial impactor. Air and particles pass downward through a nozzle of width $2w$ with a uniform velocity U_0 into a region between two parallel planes separated by a distance s. The flow leaves the region with a uniform velocity. The parallel planes are of width $2b$. Inertial impactors are used commonly in air pollution and industrial hygiene to collect airborne particles. *Cascade impactors* are a series of inertial impactors arranged in series in which each stage removes a particular fraction of the particles and are commonly used to determine the aerosol size distribution. Marple et al. (198, 199) analyzed the velocity field in impactors by solving the Navier–Stokes equations by finite difference techniques and computed the trajectory of particles in order to determine the particle collection efficiency. These predictions were then confirmed by experiments.

The variable to be analyzed will be the stream function which in two-dimensional Cartesian coordinates satisfies Laplace's equation

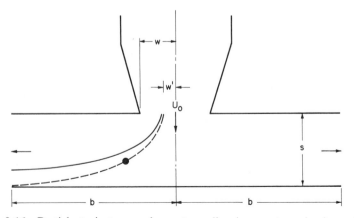

Fig. 9.16 Particle trajectory and gas streamline in a rectangular impactor.

$$\frac{\partial^2 \psi}{\partial x^2} + \frac{\partial^2 \psi}{\partial y^2} = 0 \tag{9-41}$$

The finite-difference technique begins by dividing the flow field into nodes, as shown in Fig. 9.17. The nodes do not have to be equally spaced, although equally spaced nodes will be used in the discussion that follows. The first derivative $\partial \psi / \partial x$ will be approximated by the finite difference,

$$\frac{\partial \psi}{\partial x} = \frac{\psi(x + \delta x, y) - \psi(x, y)}{\delta x} \tag{9-42}$$

and similarly for $\partial \psi / \partial y$. It should be noted that the difference is a *forward difference* in that the derivative at a point uses values of the function evaluated at the "downstream" (direction of increasing x or y) nodal location. The second derivative $\partial^2 \psi / \partial x^2$ will be approximated by

$$\frac{[\psi(x + \delta x, y) - \psi(x, y)] - [\psi(x, y) - \psi(x - \delta x, y)]}{\delta x^2} \tag{9-43}$$

The differences above can be written more succinctly as

$$\frac{\partial \psi}{\partial x} = \frac{\psi_{i,j+1} - \psi_{i,j}}{\delta x} \tag{9-44}$$

$$\frac{\partial^2 \psi}{\partial x^2} = \frac{\psi_{i+1,j} - 2\psi_{i,j} + \psi_{i-1,j}}{\delta x^2} \tag{9-45}$$

Derivatives in y are written in the same manner

$$\frac{\partial \psi}{\partial y} = \frac{\psi_{i,j+1} - \psi_{i,j}}{\delta y} \tag{9-46}$$

$$\frac{\partial^2 \psi}{\partial y^2} = \frac{\psi_{i,j+1} - 2\psi_{i,j} + \psi_{i,j-1}}{\delta y^2} \tag{9-47}$$

When the finite differences replace the second derivatives, Laplace's equation becomes a series of algebraic equations for each node in the flow field.

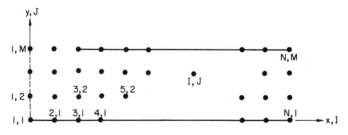

Fig. 9.17 Location of nodes in a rectangular impactor.

$$2(1 + B)\psi_{i,j} = \psi_{i,j+1} + \psi_{i+1,j} + B(\psi_{i,j-1} + \psi_{i-1,j}) \tag{9-48}$$

where

$$B = \left(\frac{\delta x}{\delta y}\right)^2 \tag{9-49}$$

The task of solving an elliptic, partial differential equation has been replaced by the task of solving a set of simultaneous linear algebraic equations for every node in the flow field.

Very often the grid mesh is square, such that $\delta x = \delta y$ and $B = 1$ and Eq. 9-48 becomes

$$\psi_{i,j} = \frac{\psi_{i,j+1} + \psi_{i,j-1} + \psi_{i+1,j} + \psi_{i-1,j}}{4} \tag{9-50}$$

Thus for a square mesh, the value of the potential function at an interior point (i, j) is the algebraic average of the stream functions at each of the adjacent (four) nodes. In BASIC or FORTRAN language the above can be written as

$$P(I, J) = 0.25 * (P(I, J + 1) + P(I, J - 1) + P(I + 1, J) + P(I - 1, J)) \tag{9-51}$$

Along solid boundaries one can assume that the stream function possesses a constant value users assign depending on the volumetric flow rate. For the rectangular impactor in Fig. 9.16, symmetry can be employed such that to the right of the plane of symmetry (coincident with the y axis) the volumetric flow rate per unit length of slot is $Q/2L$, where L is a dimension perpendicular to the X-Y plane.

$$\frac{Q}{2L} = \int_{\psi(\text{lower})}^{\psi(\text{upper})} d\psi \tag{9-52}$$

One value of ψ, such as $\psi(\text{upper})$ can be assigned the value zero, thus making the value of psi(lower) equal to $Q/2L$. All values of ψ lying along the upper surface $[P(3, M)$ to $P(N, M)]$ will have the value zero and all values of ψ lying along the lower surface $[P(1, 1)$ to $P(N, 1)]$ will have the value $Q/2L$.

If the inlet velocities are uniform, the value of the stream function at each node in the inlet plane will have values determined by

$$U_y = \frac{\partial \psi}{\partial x} = \left(\frac{Q}{2L}\right)\left(\frac{1}{w}\right) \tag{9-53}$$

where $Q/2L$ is the volumetric flow rate per unit length normal to the X-Y plane and w is one-half the width of the inlet. Integrating,

$$P(I + 1, M) = P(I, M) + \left(\frac{Q}{2L}\right)\left(\frac{\delta x}{w}\right) \tag{9-54}$$

Thus, beginning at the node on the boundary for which the stream function is zero, one merely assigns values to the next node according to the above equation.

If the velocities at the outlet plane are uniform, a similar strategy can be followed with the result that

$$P(N, J + 1) = P(N, J) - \left(\frac{Q}{2L}\right)\left(\frac{\delta y}{s}\right) \tag{9-55}$$

For nodes lying along the plane of symmetry, one may assume that the gradients normal to the plane of symmetry are zero. Values of ψ for nodes lying on the plane of symmetry can be computed using the equation for interior nodes assuming the existence of an imaginary node on the far side of the plane of symmetry. Assume that the imaginary node on the opposite side of the plane of symmetry has a value equal to the value of ψ the same as the node 180 degrees from it.

Figures 9.18–9.21 show the geometry and equations for interior nodes, nodes

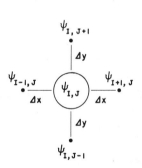

Fig. 9.18 Interior node Eq. 9-51.

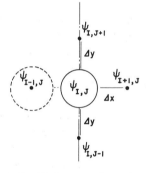

Fig. 9.19 Node on a line of symmetry. Eq. 9-51 with $\psi_{I-1, J} = \psi_{I+1, J}$.

Fig. 9.20 Node at a uniform outlet. Eq. 9-55.

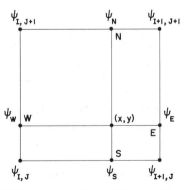

Fig. 9.21 Computation of U_x and U_y at an arbitrary point (x, y) between nodes. Eqs. 9-58 to 9-63.

lying on planes of symmetry, and nodes in the inlet and outlet plane. Suffice it to say that great care must be exercised to use correct values of ψ on the boundaries because the final solution is no more accurate than the accuracy of the input data.

To begin the solution, one assigns numerical values to each node. It is suggested that each interior node be given the same value, equal to the average of the values on the two boundaries. The iteration process begins by computing a new value of ψ at a node and then replacing the value before it is stored. The computation is repeated by "sweeping" over each node in the flow field replacing each value of ψ by its updated version. Such iteration is called *Gauss Sidel iteration*. Convergence can be achieved with fewer iterations by a process called *successive over-relaxation* in which the value of new $P(I, J)$ (left-hand side) of the equation is obtained from

$$P(I, J) = P(I, J) + A[P(I, J)^* - P(I, J)] \tag{9-56}$$

where $P(I, J)^*$ is the value computed on the current iteration and the values of $P(I, J)$ on the right hand side of Eq. 9-56 are the values for the previous iteration. The constant A is called the *over-relaxation coefficient* and is in the range $1 < A < 2$. Over-relaxation is used to accelerate convergence. The optimum value of A is chosen by trial and error. A value of $A = 1.7$ is typical.

Because the derivatives were approximated in terms of values of ψ from adjacent nodes, the matrix describing the coefficients for each variable contains only the major diagonal and diagonals above and below it. The remaining coefficients in the matrix are zeros. Such a matrix can be solved quickly by *tridiagonal matrix* techniques available as standard library computer programs.

The iteration is stopped when successive values of the variables converge to suitably constant values. *Convergence* can be quantified by computing a residual (R) defined as the sum of the difference in successive values of ψ at each node between any two iterations n and $n + 1$, viz,

$$R = \sum \left| \frac{[P(I, J)^{n+1} - P(I, J)^n]}{P(I, J)^n} \right| \tag{9-57}$$

is below some prescribed tolerance, such as 10^{-3} or 10^{-9}.

Once satisfactory convergence is obtained, numerical values of ψ are available at each node for the computer to use to compute the air velocity at any point (x, y) in the flow field. The program to compute particle trajectories begins with the particle at the inlet boundary. The location of the particle after one time step is calculated by the previous programs and one marches ahead in time and space until the particle either strikes the wall or passes out the outlet. Unfortunately, after one time step the location of the particle will probably be somewhere between a set of four nodes, as depicted in Fig. 9.21. To compute the air velocity at this point one should use values of ψ at points NESW, as shown in Fig. 9.21, and an interpolation technique needs to be developed to compute ψ at points NESW.

A simple *interpolation* scheme that will be satisfactory if the nodes are reasonably close is to compute ψ at the points NESW by averaging. Thus,

$$U_x = -\frac{\psi_N - \psi_S}{\delta y} \tag{9-58}$$

$$U_y = \frac{\psi_E - \psi_W}{\delta x} \tag{9-59}$$

where ψ_N, and so on are approximated by

$$\psi_N = \frac{[P(I+1, J+1) + P(I, J+1)]}{2} \tag{9-60}$$

$$\psi_S = \frac{[P(I+1, J) + P(I, J)]}{2} \tag{9-61}$$

$$\psi_E = \frac{[P(I+1, J+1) + P(I+1, J)]}{2} \tag{9-62}$$

$$\psi_W = \frac{[P(I, J) + P(I, J+1)]}{2} \tag{9-63}$$

The ability of a rectangular impactor to collect particles is expressed as its collection efficiency. If the particle concentration and air velocity (assumed to be equal to the particle velocity) are uniform at the inlet plane, the collection efficiency can be expressed as

$$E = \frac{w'}{w} \tag{9-64}$$

where w is the half width of the inlet and w' is the distance between the plane of symmetry and the initial location of the particle that impacts the lower surface a distance $x = b$, that is, the lower right hand corner of the flow domain.

The particle trajectory program is used to compute the particle's trajectory for a particle at a preselected location on the inlet plane, that is, $y = s$ and x, until a value of x is selected in which the particle impacts the lower plate at the lower right hand corner of the flow domain. The value of x for this condition will then be equal to w'. The process should be repeated for several volumetric flow rates, and dimensions s and w and several particle diameters and densities. The dimensionless equation of motion for particles

$$\text{Stk}\, \frac{d\mathbf{v}^*}{dt^*} = \mathbf{U}^* - \mathbf{v}^* \tag{9-65}$$

where $\mathbf{U}^*$ and $\mathbf{v}^*$ are the *dimensionless velocities* of air (see Section 8.10) and particle and Stk is the Stokes number,

$$\text{Stk} = \frac{\tau_p U_0}{D_0} \tag{9-66}$$

where U_0 and D_0 are a characteristic velocity and distance. The characteristic velocity (U_0) is taken to be the nozzle exit velocity

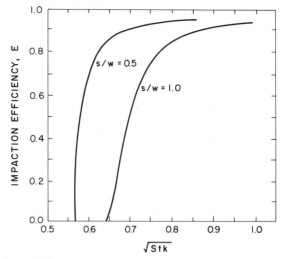

Fig. 9.22 Impaction efficiency versus Stokes number for two rectangular impactor ($s/w = 0.5$ and 1.0) at $Re = 1000$ (redrawn from ref. 199).

$$U_0 = \frac{Q}{L2w} \qquad (9\text{-}67)$$

but the *characteristic distance* D_0 can be either the plate separation distance s or the nozzle width $2w$. Figure 9.22 shows results obtained by Marple et al. (198, 199) in which the characteristic distance (D_0) is the nozzle width ($2w$).

9.6 GRINDING BOOTH

Hand-held, portable grinders (Fig. 1.5) and sanders (Fig. 1.10) ordinarily produce copious small particles, including respirable particles (203) which if not controlled, travel throughout the workplace. In the case of gray iron foundries, portable grinders used to remove molding sand adhering to the casting surface may produce airborne silica concentrations in excess of TLV.

When the castings are small, pedestal (or stand) grinders can be used to remove gates, risers, parting fins, or surface imperfections from the castings. The grinding produces particles, many of which are respirable. Particles leave the wheel with a velocity equal to the tangential velocity of the wheel. Small particles become trapped in the *boundary layer* on the wheel and are stripped from the wheel at the point where the wheel encounters the workpiece. These particles may enter the workers breathing zone. O'Brien et al. (203) suggest that a baffle or air jet should be placed inside the housing surrounding the wheel to strip the particles in the *boundary layer* so that they can be carried away with the air exhausted from the grinder housing.

When the castings are large and require the worker to move around the casting, lean over (Fig. 1.23) and perhaps inside the casting it is very difficult to control the grinding particles. For such large castings, grinding booths shown in

Fig. 6.10 are often used. Present design guidelines suggest values for the average air velocity (face velocity) entering the booth, slot velocity, and booth geometry, but do not provide insight regarding,

Air velocities inside the booth

Article trajectories

Particle concentrations in the operators breathing zone

Techniques described in previous chapters can be used to obtain quantitative data for the above (66–68).

The grinding booth can be modeled as a rectangular parallelepiped (Fig. 9.23). Room air enters the booth through the open face, circulates around the worker (who is modeled as a second parallelepiped) and work piece, and finally exits through one or more slots in the wall opposite the entry face. The dimensions of the booth and the slot, and the size and location of the worker are selected by the user. The work piece should be included, but will be omitted in the discussion that follows. The grinding wheel can be modeled as a right circular cylinder.

Inside the booth, the air velocities are small and viscous effects occur only along solid surfaces and in the region downstream of the worker and work piece. The viscous boundary layers along the walls have little influence on particle trajectories and can be ignored. By contrast, the low velocity wake region adjacent to the worker's body and face has a significant effect on the trajectories

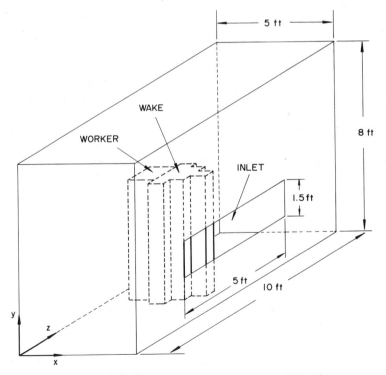

Fig. 9.23   Grinding booth for large castings (67, 68).

because particles are apt to be generated in, or in close proximity to the wake. Air flow in the main portion of the booth can be modeled as an ideal fluid. The *wake* can be modeled analytically as a low-velocity, recirculating flow field downstream of the worker and included as a boundary condition for analyzing the main flow field. The aerodynamic effects of the wheel can be included by superimposing a rotating boundary layer on the main flow field.

Because the main flow field is assumed to be ideal and three dimensional, the velocity will be computed from the velocity potential and

$$\nabla^2 \phi = 0 \tag{9-68}$$

$$U_x = -\frac{\partial \phi}{\partial x} \tag{9-69}$$

$$U_y = -\frac{\partial \phi}{\partial y} \tag{9-70}$$

$$U_z = -\frac{\partial \phi}{\partial z} \tag{9-71}$$

Three boundary conditions are required.

(1) The inlet velocity distribution must be specified across the open face of the booth.
(2) The outlet velocity distribution must be specified across the exit slot.
(3) No-flow conditions (zero normal gradients of ϕ) must be specified on the walls of the booth, on the blockage (the worker), and along the boundaries of the wake region.

The ideal flow field can be calculated from the velocity potential equation and the boundary conditions by standard finite difference procedures. The main flow field is divided into a three-dimensional grid system. For interior nodes, central differences on an unequal grid can be used. These finite difference relations are determined by performing a mass balance on each small nodal volume in the grid. Finite difference equations for the three types of boundary nodes are written in a similar fashion. The resulting linear system of finite difference equations can be solved (70) iteratively using a point-over-relaxation method (Eq. 9-56) which can be rewritten as,

$$P(I, J, K)(n) = A[P(I, J, K)(n + 1)] + (1 - A)[P(I, J, K)(n)] \tag{9-72}$$

where

$P(I, J, K)(n)$ = velocity potential that will be assigned to node i, j, k for iteration number n

$(n + 1)$ = represents the current iteration

$P(I, J, K)(n + 1)$ = is the value of the unknown computed by the Gauss–Sidel method on iteration $(n + 1)$ iteration

A = over-relaxation factor [an optimum value of 1.7 was found (70) by trial and error]

The notation $(1 - A)[P(I, J, K)(n)]$ in Eq. 9-72 denotes the operation of multiplying the value of $(1 - A)$ times the value of the velocity potential during iteration number n. Once the velocity potential is found at every node, the components of air velocity are obtained by central differences.

The worker and wake can be modeled as regions through which no air passes. The wake region contains air that recirculates because of viscous forces; for a first approximation it can be assumed that no air crosses the boundaries of the wake region. Rather than include all details of this localized flow phenomena in the computational model, the wake can be modeled in a phenomenological sense as an isolated region around which the external air flow passes and within which air circulates in a describable fashion. Since the size and shape of the wake is a weak function of the incoming velocity, the geometry of the wake can be specified prior to the calculations. The inviscid flow around the worker and the wake can be calculated as if the worker and wake are a single "blockage" element. To fit the wake and worker into the rectangular grid, the characteristic shape shown in Fig. 9.24 can be assumed.

To calculate particle trajectories, the velocities inside the wake must be known. After the main flow field has been determined, a simple analytical velocity field in the wake region can be matched with the external flow. The following third-order polynomial can be used to represent the x-component velocity $[U_x(x, z)]$ inside the wake (Fig. 9.24). The other velocity components (U_y and U_z) can be neglected.

$$U_x = \frac{U_b}{51} \left[108 \left(\frac{z_w}{b} \right)^3 - 45 \left(\frac{z_w}{b} \right)^2 - 12 \right] \tag{9-73}$$

where

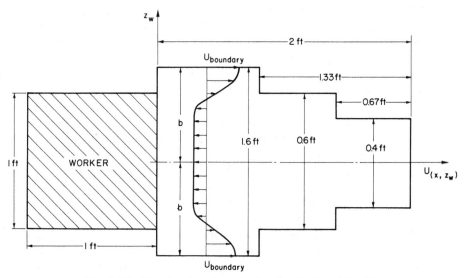

Fig. 9.24 Simulated wake velocity distribution (67, 68).

$U_b = x$ component of the free-stream velocity of air at wake boundary

b = distance between wake centerline and boundary

z_w = distance between wake centerline and point of interest inside the wake

The following conditions are used to define the polynomial coefficients in Eq. 9-73:

1. The x component of the wake velocity must be equal to the value in the main flow.

$$z_w = b; \qquad U(z, x) \text{ wake} = U_x \text{ main flow} \qquad (9\text{-}74)$$

2. The wake must be symmetrical.

$$z_w = 0; \qquad \frac{dU_x}{dz_w} = 0 \qquad (9\text{-}75)$$

3. There must be no net flow into (or out of the) wake

$$\int_0^b U_x \, dz = 0 \qquad (9\text{-}76)$$

4. Demarcation between the reverse and forward flow is specified to occur two-thirds of the way from the centerline to the outer edge of the wake. Along with condition 3, this insures low velocities in the reverse flow region.

$$z_w = \frac{2b}{3}; \qquad U_x = 0 \qquad (9\text{-}77)$$

Grinding wheels for hand-held grinders rotate at speeds between 3000 and 20,000 RPM, depending on their diameter, and all possess a high-speed *boundary layer* into which particles are ejected. Large particles pass through the boundary layer quickly, smaller particles may be trapped in the boundary layer. Particles trapped in the boundary layer are stripped from the boundary layer after each revolution of the wheel. The air velocity inside the boundary layer has been measured experimentally (69) and can be expressed empirically as a polynomial which can be included in the computer program for the particle trajectory. Alternatively, one may assume a boundary layer thickness that increases around the wheel and express the velocity profile by equations describing the developing boundary layer on flat plate.

The particle trajectory equations of Chapter 8 have to be expanded to three dimensions. Particle trajectories are calculated by integrating the equations of motion over short increments of time. The time step should be sufficiently small so that the displacement of each particle is considerably smaller than the dimension of the nodal volume through which it passes. Thus the program must adjust the size of the time step automatically. The integration begins when the particle is located on the surface of the wheel and its velocity is equal to the tangential velocity of the wheel. The trajectory is computed by a time marching

procedure identical to that used in the previous examples. At each point in space
the program determines whether the particle is inside the grinding wheel boundary layer, the wake region, or the main flow field and provides the appropriate air
velocity to the equations of particle motion. A family of trajectories are computed
for particles generated over the entire circumference of the grinding wheel.
Equations 4-8 and 4-9 describe how the mass generation rate varies with angular
position on the grinding wheel.

To determine the particle concentration at a certain point in the booth, a small
volume defined as the concentration volume V_c can be chosen to surround the
point in question. Typically V_c is a portion (or all) of the operators breathing
zone, because this is the region in which OSHA standards must be satisfied,
although users are free to choose any point in space they wish. The particle
concentration in V_c is found by identifying the number of particle trajectories that
pass through V_c. A particle's trajectory is a function of its aerodynamic diameter

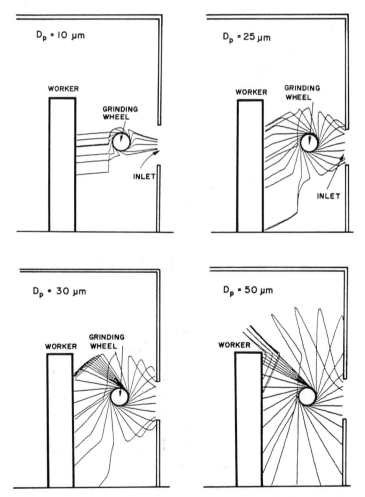

Fig. 9.25 Side view of computed particle trajectories in a grinding booth, worker with
back to incoming air, booth face velocity of 100 FPM (67, 68).

and the angular position on the wheel from which it was ejected. The concentration is also a function of the booth configuration, suction flow rate, position of worker, and location of the grinding wheel. For a given particle size, a family of trajectories can be computed as a function of the angular position on the wheel. Figure 9.25 shows particle trajectories produced by workers positioned with their backs to the incoming air in a booth where the face velocity is 100 FPM. It can be seen that many particles are trapped in the wake region. Figure 9.26 shows particle trajectories produced by a worker facing the side of the booth. It is clear that it is unwise for workers to grind while facing the slot, that is, to perform grinding in the body's wake.

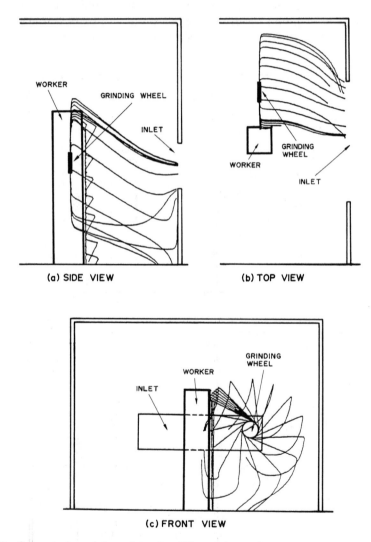

Fig. 9.26   Computed particle trajectories (35 μm) in a grinding booth, worker facing the side of the booth, booth face velocity 100 FPM (67, 68).

Once the trajectories are determined, the particle mass concentration (c) in V_c can be found by computing the mass flow rate associated with each trajectory generated on the wheel, computing the amount of time the trajectory spends inside V_c and summing over the number of trajectories (N) passing through V_c,

$$c = \sum_{i=1}^{N} \frac{(m_{gi})(t_i)}{V_c} \qquad (9\text{-}78)$$

where

m_{gi} = local (mass) generation rate for the trajectory i
t_i = time particles in trajectory i are inside V_c
N = the total number of the trajectories that enter V_c
V_c = volume of the concentration volume

The time particles in each trajectory spend inside V_c is determined as part of the particle trajectory calculation. Figure 9.27 shows how the concentration varies with ventilation volumetric flow rate and location in the worker's wake if all the particles are of the same diameter. The entire process can be used for a polydisperse aerosol by repeating the process for each size range.

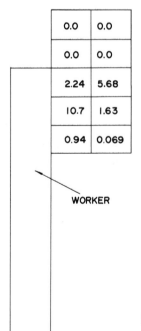

Fig. 9.27 Concentration (g/m^3) of 30-μm particles in the vicinity of a worker in a grinding booth, worker with back to incoming air, grinding rate 1 kg/hr, booth face velocity 100 FPM.

9.7   WINDBREAKS

The prevention of soil erosion in agriculture, the control of material in stock piles (200–202, 297, 355), the control of dust generated by vehicles traveling on unpaved roads (75), or the control of wind-blown snow (300) are not issues generally associated with industrial ventilation. Nevertheless, the analysis of how windbreaks can be used in these situations is an excellent application of the principles of industrial ventilation. *Windbreaks* are natural or man-made barriers that produce regions of low air velocity. If placed upwind of a stock pile or farm field (Fig. 9.28) windbreaks may actually reduce the generation of dust. If placed downwind of a stock pile (Fig. 9.29), windbreaks capture particles by providing a low-velocity sheltered region on the lee side and allowing gravimetric settling to occur. Whether placed upwind or downwind of a stock pile or unpaved road, windbreaks provide regions of low velocity in which the removal of particles by gravimetric settling is enhanced. The porosity of the windbreak can vary from 0 to 100%, solid barrier to no windbreak at all.

The objective of the analysis is to develop an analytical method to predict the fractional efficiency of a windbreak to control fugitive dust generated by vehicles on paved and unpaved roads as functions of

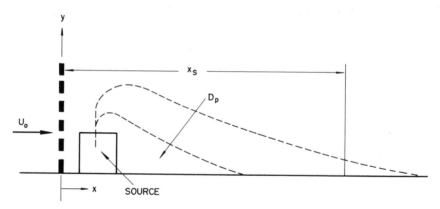

Fig. 9.28   Particle trajectories from a source downwind of a porous windbreak.

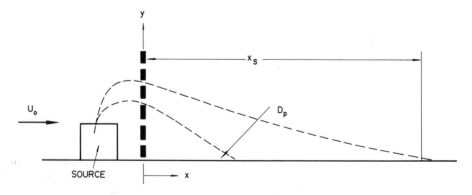

Fig. 9.29   Particle trajectories from a source upwind of a porous windbreak.

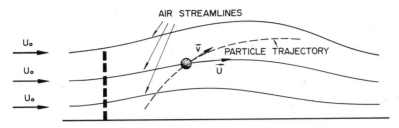

Fig. 9.30 Region downwind of a porous windbreak.

(a) Distance downwind of the windbreak

(b) Location of the source

(c) Wind speed, height and porosity of the windbreak

Figures 9.30 and 9.31 are schematic diagrams of the particle motion that will be modeled. The following assumptions will be made:

(a) The air velocity has only a horizontal component that varies with x and y. (The vertical component is very important but since no data is available it will be omitted for the moment.)

$$U_x = f(x, y); \qquad U_y(x, y) = 0 \qquad (9\text{-}79)$$

(b) The particle velocity has x and y components.

$$v_x = f(x, y) \qquad v_y = g(x, y) \qquad (9\text{-}80)$$

(c) The source will be divided into 100 nodes, each of which emits particles of a particular diameter in the vertical direction with an initial velocity $v(0)$ selected by the user.

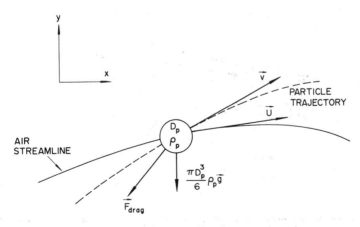

Fig. 9.31 Force on particle traveling in a moving air stram.

(d) The windbreak is of height H and porosity (ϵ) and is placed upwind of the source as per Fig. 9.28.

(e) The distance x_s is the distance within which the user wishes particles to settle.

(f) The particles are spherical and do not collide with one another, agglomerate or interact in anyway.

(g) Only gravity and viscous forces act on the particles.

The velocity profile of air on the lee side of the windbreak has been studied by Jensen (71), Segman (72), and Hagen et al. (73, 74). Figure 9.32 shows the ratio of the mean leeward to windward wind speed for neutral and stable atmospheric conditions and wind speeds (U_0) between 3 and 11 m/s for a windbreak of 40% porosity. The authors report (73) comparable profiles for porosities of 0, 20% and 60%. The vertical velocity components were not given in a manner or detail that permits their use in the analysis. For each porosity, the experimental data can be expressed by an empirical nondimensional power law of the form

$$\frac{U_x}{U_0} = \sum_{n=1} a_n \left(\frac{x}{H}\right)^p \left(\frac{y}{H}\right)^q \tag{9-81}$$

where the coefficients a_n are constants and the exponents p and q are integers.

Assume the source of fugitive dust is the wake region of automotive vehicles traveling on unpaved (or paved) roads. The source strength (mass of particles emitted per unit time) can be estimated by the emission factors or independent measurement. For convenience, it will be assumed that the wake region has a square cross section ($1.5 \text{ m} \times 1.5 \text{ m}$) in which dust is generated by 100 equally spaced nodes for which the mass generation rate and size distribution are the same.

The velocity and displacement of the particles can be computed by the procedures in Chapter 8. The computer program calculates the trajectory of the particles from each of the 100 nodes from the time the particle leaves the node until it strikes the ground. The computational algorithm is organized as follows:

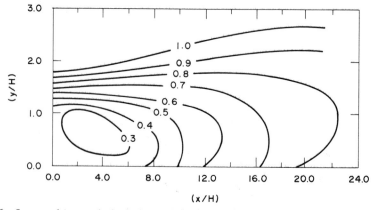

Fig. 9.32 Leeward to upwind wind speed (U_x/U_0) profiles in the wake region of a porous (40%) windbreak (redrawn from ref. 73).

(a) Particles leave each node with known initial velocities

$$v_x(0), \qquad v_y(0)$$

(b) At each particle location (x, y) the air velocity U_x is obtained from Eq. 9-81

(c) For each computation, the relative velocity, Reynolds number, and drag coefficient, are computed using initial values.

(d) The particle velocity and location at the end of each time interval (δt)

$$v_x(t + \delta t), \quad v_y(t + \delta t), \quad x(t + \delta t), \quad y(t + \delta t)$$

are computed using equations in Chapter 8.

(e) Steps a–d are repeated for the next time interval (a value of 0.01 s can be used) until the particle strikes the ground.

Figure 9.33 is a sketch of particle trajectories for a tier of nodes at $y = 0.75$ m and $y = 0.15$ m. At $y = 0.15$ m the trajectory of 50 μm and 100 μm particles are virtually the same. The ability of windbreaks to enhance settling can be quantified by using a settling fractional efficiency $\eta(D_p)$ that will be defined as the fraction of

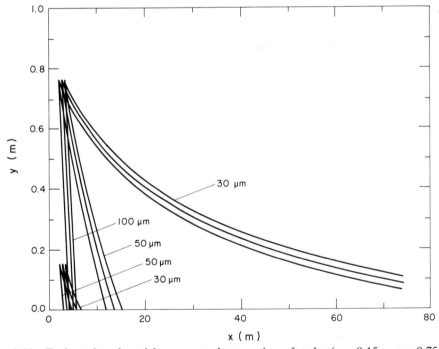

Fig. 9.33   Trajectories of particles generated at two tiers of nodes ($y = 0.15$ m, $y = 0.75$ m) in the wake region. At $y = 0.15$ m, the trajectories of 50 and 100 μm particles are virtually the same. $U_0 = 5$ m/s, $H = 2$ m, porosity = 40%, $\rho_p = 1000$ kg/m^3 (redrawn from ref. 75).

particles of a particular size that fall to the ground within a selected downwind distance x_s. If the source is characterized by a series of N nodes the settling fractional efficiency, $\eta(D_p)$, can be computed by summing the parameter $F(D_{p,j})$ over all the nodes $(j = 1, 2 \ldots, N)$, viz

$$\eta(D_p) = \frac{1}{N} \sum_{j=1} F(D_{p,j}) \tag{9-82}$$

where $F(D_{p,j})$ has the value unity if the particles of size D_p generated from node j, fall to the ground within x_s or a value of zero if they do not. The overall windbreak collection efficiency can be computed by multiplying the fraction of particles emitted by each node $g(D_{p,i})$ by $F(D_{p,j})$ and summing over all the particle sizes $(i = 1, 2, \ldots, M)$ and then summing over all the nodes,

$$\eta(\text{overall}) = \frac{1}{N} \sum_{i=1} \sum_{j=1} F(D_{p,j}) g(D_{p,i}) \tag{9-83}$$

Analyzed in this way the user can repeat the calculation for any source in which the following parameters can be selected:

(a) Size and location downwind of the windbreak
(b) Number and location of nodes
(c) particle size distribution at each node (for vehicular traffic on unpaved roads (see ref. 299)
(d) particle density and initial velocity

Figures 9.34 and 9.35 show the results of this procedure for a vehicular wake

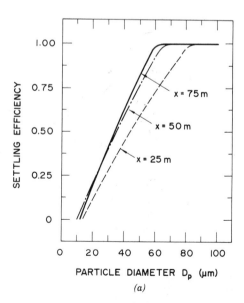

Fig. 9.34 Fractional efficiency of a porous (40%) windbreak at three downwind locations, $H = 2$ m, $U_0 = 5$ m/s, $\rho_p = 1000$ kg/m^3.

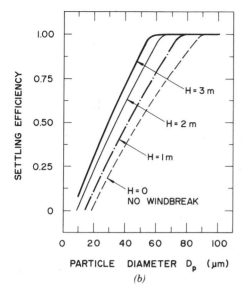

Fig. 9.35 Fractional efficiency of a porous (40%) windbreak at $x = 50$ m, and three different heights (H), $U_0 = 5$ m/s, $\rho_p = 1000$ kg/m^3.

consisting of 100 equally spaced nodes in a region 1.5 m $\times$ 1.5 m. The figures are generated by applying Eq. 9-82 for monodisperse aerosol over the range 10 μm $\leq$ $D_p \leq 100$ μm. The figures can be interpreted as the fractional (grade) efficiency of a windbreak for the specific values of U_0, H, ϵ and ρ_p. In the absence of a windbreak (100% porosity), particles fall to the ground naturally owing to gravity. Thus the fractional efficiency for 100% porosity is a baseline reference to judge the effectiveness of a windbreak. The trajectories in Fig. 9.33 show that upon leaving the source, particles reach their apogee very quickly. The vertical penetration for even large particles is so small that for a first approximation, it can be neglected. The figures also show that particles fall to earth with nonlinear trajectories. Nonlinearity implies that the particle's vertical and horizontal velocity components change as the particle moves downwind and is a consequence of the unique velocity profile downwind of the windbreak (Fig. 9.32). The wind velocity profiles show that this conclusion is to be expected. Computations show that the fractional efficiencies at $x = 50$ m are significantly larger than they are at $x = 25$ m but they are virtually the same at 50 m as they are at 75 m.

The equations used to describe the air velocity downwind of the wind break are crude since they neglect the vertical components. Deposition of particles is controlled by gravimetric settling, and since settling velocities are small, fractional efficiencies will be affected significantly by small vertical components of the air velocity. It is clear that a more accurate description of the air velocity field is needed.

The usefulness of windbreaks seems obvious. While there may be a large first cost to install a natural or man-made windbreak, they bring about a significant reduction in fugitive dust emission. If the cost of the material in the stock pile is significant, the savings in material lost can be large.

NOMENCLATURE

Symbol	Description	Dimensions*
A	Constant defined by equation, over-relaxation factor	
A	Inlet area	L^2
a_n	Constant defined by equation	
b	Distance between wake centerline and boundary, Half-width of impactor plates	L
B	Constant defined by equation	
c	Mass concentrations of particles entering inlet	M/L^3
c_0	Mass concentration of particles in free stream	M/L^3
C	Cunningham correction factor	
D_0	Characteristic distance	L
D_p	Particle diameter	L
$F(D_p)$	Unity or zero, depending whether particle settles to the ground	
$g(D_p)$	Fraction of particles generated at each node	
H	Height of windbreak	L
$H(D_p)$	Width of stream tube of particles entering an inlet	L
I	Nodal coordinate	
J	Nodal coordinate	
k	Constant defined by equation	
L	Length (value depends on problem)	L
m	Mass of dust	M
m_{gi}	Mass generation rate for trajectory i	M/t
M	Iteration number, node coordinate	
n	Iteration number	
N	Total number of trajectories that enter V_c	
$P(I, J)$	value of ψ at node (i, j)	
Q	Volumetric flow rate of air	L^3/t
r	Radius	L
R	Residual defined by equation	
R	Universal gas constant	Q/NT
R	Sampling probe inlet radius	L
s	Distance between impactor surfaces	L
t_i	Time particles in trajectory i are inside V_c	t
$t*$	Dimensionless time	
U	Velocity of air entering a sampling probe	L/t
$\mathbf{U}*$	Dimensionless air velocity	
U_b	x-component of air velocity at air-wake boundary	L/t

* Q – energy, F – force, L – length, M – mass, N – mols, t – time, T – temperature

U_x, U_y, U_z	x, y and z components of air velocity	L/t
U_0	Free stream air velocity	L/t
$\boldsymbol{v}_x, \boldsymbol{v}_y$	x and y components of particle velocity	L/t
$\mathbf{v}^*$	Dimensionless particle velocity	
V_c	Volume of concentration volume	L^3
w	Inlet half-width	L
w'	Dimension that defines trajectory of impacting particle	L
x, y	Arbitrary coordinates in space	L
x_0, y_0	Unique position in space	L
x_s	Distance within which particles settle to ground	L
Y	Width of stream tube of air entering an inlet	L
z	Axial distance	L
z_w	Distance between wake centerline and point of interest inside wake	L

Greek	*Description*	*Dimensions*
α	Angle of oblique streaming flow	
π	3.14159	
$\delta x, \delta y$	Incremental distances in x and y directions	L
δt	Increment of time	t
η	Collection efficiency	
$\eta(D_p)$	Fractional or grade efficiency	
θ	Angle defined in each problem	
ϕ	Velocity potential	L^2/t
ψ	Stream function	L^2/t
τ_p	Particle relaxation time	t
μ	Viscosity	Ft/L^2
ρ	Density	M/L^3

Subscripts	*Description*
$()_0$	far field or upstream values
$()_p$	particles
$()_w$	inlet flow, width $2w$
$()_f$	streaming flow
$()_i$	initial guess for iteration in the Newton–Raphson method
$()_i$	face conditions at plane of inlet
$()_j$	iteration index, for Newton–Raphson
$()_i, ()_j$	coordinates for finite difference techniques
$()_N$	properties of node north of node P, similarly for nodes NESW

Abbreviations	*Description*
Re	Reynolds number
Stk	Stokes number

PROBLEMS

1. Air passes over a flat surface with a uniform velocity U_0. A particle of diameter D_p and density ρ_p is ejected at an angle θ to the horizontal with an initial velocity $v_p(0)$. The particle rises and then falls by gravity, all the while being transported in the direction of the moving air. Write a computer program to compute the angle θ that results in the maximum downwind displacement L. Neglect boundary layers. Write the program for arbitrary particle diameters and initial velocities.

2. Analyze the trajectories of particles traveling over an infinitely long mound having a semicircular cross sectional area of radius a. Assume that the velocity field around the mound can be described by the stream function $\psi = U_0[1 - (a/r)^2]\sin\theta$ where r is measured from the center of the mound, the positive x axis corresponds to a value of θ equal to zero and that air with a uniform velocity (U_0) approaches the mound in the negative x direction. Neglect the boundary layer on the mound. Begin the analysis with particles at points along a vertical line at $x = 1.5a$, $y = h$ where the particle velocity is equal to the air velocity. Some particles will impact the mound and others will pass over it to settle to the ground downwind of the mound. Analyze the particle trajectories and determine where each particle impacts either the mound or the ground. Carry out the computations for particles originating at values of h from zero to $1.5a$.

   Draw a graph of (h/a) versus the normalized displacement distance (x/a) at which the particle impacts a solid surface. Repeat the calculation for several particle sizes. Assume the following physical conditions:

   $\rho_p = 1000 \text{ kg/m}^3$
   $U_0 = 1 \text{ m/s}$
   $D_p = 5, 10, 20, 50 \ \mu\text{m}$
   $a = 10 \text{ m}$

3. Air at STP passes over a horizontal flat plate in the positive x direction and a boundary layer develops on the plate. The air approaches the plate with a velocity (U_0) of 1 m/s. Small particles injected vertically from the leading edge of the plate, are slowed by the air, travel to the right, pass through the boundary layer and strike the plate. Calculate the initial particle velocity if $50 \ \mu\text{m}$ (D_p) unit density spheres strike the plate 1 m downwind from the leading edge. The velocity profile in the boundary layer is given by,

$$\frac{u}{U_0} = \left(\frac{2y}{\delta}\right) - \left(\frac{y}{\delta}\right)^2 \quad \text{and} \quad v = 0$$

where δ is the boundary layer thickness,

$$\delta = \frac{5x}{\sqrt{\text{Re}_x}}; \qquad \text{Re}_x = \frac{x\rho U_0}{\mu}$$

4. Consider a horizontal plane containing a line source that emits spherical particles in all directions above the plane into quiescent air at STP. The plane also contains a slot of negligible width (line sink) through which air is withdrawn at a constant rate per unit length (Q/L). Compute the particle trajectories between the time they leave the source and strike the plane. The particles have a constant diameter and density of 50 μm and 3000 kg/m^3. Draw the trajectories of particles leaving the plane at angels equal to 30, 60, 90, 120, and 150 degrees. The initial speed of the particle $|v_p(0)|$ is 50 m/s. The sink is located 10 cm to the right of the source and withdraws air at a rate (Q/L) of 0.5 m^3/sec per meter length of sink. Find the unique angle (θ') corresponding to the trajectory that strikes the plane at a point A midway between the source and sink. It is logical to expect that θ' is a function of the particle diameter and initial velocity. Keeping the initial speed constant (50 m/s) vary the particle diameter, compute θ' and plot θ' versus the particle diameter.

5. A duct transporting glass frit has a hole in its top surface that inadvertently discharges particles into the workplace at a rate of 10 g/s. The hole is 0.5 m $(y = 0.5)$ above the floor and all the particles are discharged vertically upward with a velocity of 10 m/s. The particles are glass spheres with a density 3000 kg/m^3, and a size distribution that is log-normal with a mass median diameter of 50 μm and a geometric standard deviation equal to 1.3. Room air at STP moves horizontally to the right with a velocity,

$$U = U_0(y/H)^2; \qquad U_0 = 1 \text{ m/s}; \qquad H = 0.5 \text{ m}$$

Compute and plot the deposition rate on the floor versus distance downwind of the duct.

6. A spray bar injects a monodisperse aerosol of kerosene $(D_p = 10 \ \mu$m, $\rho_p = 865$ kg/m$_3$) into a stream of air passing between two horizontal parallel plates that are a distance $2H$ apart. Particles are injected across the entire region between the plates and have a uniform initial velocity $v_x(0) = 2U_0$, where U_0 is the velocity along the mid-plane between the parallel plates. The velocity profile of the air is fully established. The initial aerosol concentration at the plane of injection is constant. Derive an analytical expression (or prepare a computer program) to compute the overall collection efficiency by gravimetric settling if the following different conditions exist.

(a) Well-mixed conditions prevail and the fully developed profile is turbulent,

$$\frac{U_x}{U_0} = \left[1 - \left(\frac{y}{H}\right)^{0.2}\right]$$

where y is measured from the mid-plane.

(b) Laminar mixing conditions prevail and fully developed profile is laminar,

$$\frac{U_x}{U_0} = \left[1 - \left(\frac{y}{H} \right)^2 \right]$$

where y is measured from the mid-plane.

7. Analyze the reach and aspiration efficiency (average concentration of inhaled aerosol divided by the far-field concentration) for a long distance runner in the New York City marathon. For simplicity, assume the head is an isolated sphere moving at a velocity U_0 in stationary air containing particles usually found in an urban aerosol. Assume the nose is a circular opening in the sphere orientated such that the plane of the opening is perpendicular to the velocity U_0. For simplicity neglect the boundary layer on the sphere and assume the velocity field in the vicinity of the sphere can be described by the potential flow functions for an ideal fluid. Assume the urban aerosol is a monodisperse aerosol with a constant Cunningham correction factor greater than unity (see ref. 329 for details).

8. Air at STP is withdrawn through a slot 2 cm wide ($2w$) in an infinite plane at a rate (Q/L) of $0.03 \, \text{m}^3/\text{m}$ s. Air passes over the plane in the negative x direction at a uniform speed (U_0) of $0.5 \, \text{m/s}$. At $x_0 = 30 \, \text{cm}$ and $y = y$ upwind of the center of the slot, unit density spheres are injected into the air in the negative x direction at an initial speed ($v_p(0)$) of $30 \, \text{m/s}$. Compute the reach of the slot for particles 10, 20, 40, 60 and 80 μm in diameter. Plot the bounding trajectories. See Fig. 9.1 for a sketch.

9. A grinding table consists of a side draft hood similar to Fig. 7.13b. The opening is 6 cm high. Surface imperfections on gray iron castings are removed by grinding performed on this table. Grinding injects particles ($D_p = 100 \, \mu$m, $8000 \, \text{kg/m}^3$) into the air in a vertical direction from a point $x = 0.5 \, \text{m}$, $y = 0.01 \, \text{m}$ upwind of the inlet at a velocity of $200 \, \text{m/s}$. With what volumetric flow rate (Q/L) should air be withdrawn through the slot to prevent the particles from achieving an elevation 20 cm above the table.

CHAPTER 10

Control of Gases and Vapors

Goal: Students will be expected to learn to,
- write and solve the Navier-Stokes equations in finite difference form to predict velocity fields
- use commercially available Navier-Stokes solvers to predict velocity fields
- predict the concentrations of contaminant vapors and gases in moving viscous turbulent flow fields

There are many industrial operations in which undesirable vapors and gases will be generated and enter the workplace unless preventive steps are taken. If the process can be changed to cease producing contaminants or enclosed to prevent the contaminants from entering the workplace, it is wise to do so. If state and federal air pollution regulations permit, the contaminants can be discharged to the atmosphere if not, a pollution control system will have to be installed. In rare cases the vapor concentration may approach the lower combustion limit and additional precautions must be taken. For brevity, contaminant gases and vapors will simply be called vapors throughout this chapter.

In some industrial processes it is obvious how vapors are generated, in others their generation may escape attention. To illustrate the point, consider vapors generated by the following processes:

Open vessels containing volatile liquids

Drying operations in which liquids evaporate rapidly, for example, painted surface coatings on metal, wood, and paper, and textile dying and finishing.

Curing operations in which vapors are generated slowly such as food processing, manufacturing of wood products, textiles, and plastics, bonding, gluing, and sealing agents

Air vents in liquid storage containers

Foreign materials on the surface of metals which vaporize when the metal is heated

Liquids, mops, and rags used in cleaning operations

Vapors that leak from valve packing and pump seals

Vapors evaporating from liquid spills

Exhausts from vehicles within a plant

Fugitive emissions of vapor during filling and bottling operations

Vapors evaporating from liquids used for cooling and lubrication

It is not suggested that vapors from these operations are necessarily hazards to health or require controls. The list is given only to arouse an awareness of diverse processes that emit vapors.

If a process cannot be changed to reduce the emission of vapors, then whether ventilation measures will be required depends on two considerations:

What contaminants are workers exposed to?

Are the concentrations in excess of TLV (or more stringent industrial standards)?

Identifying the contaminant is a matter of locating the process generating the vapors and identifying the constituents in the liquid producing the vapors. Difficulties arise because the composition of many commercial liquids are trade secrets and will not be disclosed by their vendors. Material safety data sheets require vendors to identify hazardous agents that may be produced when the liquid is used and should be used to alert engineers which contaminant vapors can be expected.

This chapter is concerned with ways to compute the concentration at arbitrarily selected points in the vicinity of the source. Such computation is necessary for two reasons:

(a) If no ventilation is needed, it will be necessary that engineers calculate concentrations and show that they are well below TLV (or more stringent company standards if they exist) at locations where workers perform duties.

(b) If general or local ventilation measures are needed, it will be necessary to show that the proposed system reduces concentrations below some agreed upon level before the actual system is constructed, installed, and tested.

In both cases engineers must be able to predict the concentration with precision. Simply building and operating a ventilation system in some predetermined way is no guarantee that these standards will be met. Measuring concentrations after the system is built without adequately analyzing its performance beforehand can be costly. Reliable methods are needed to predict performance before the system is built and field tests are made.

The problem is to compute $c(x, y, z, t)$ in a control volume selected by the designer. The phrase *control volume* will be used throughout the chapter to designate the volume (generally containing the source) selected by designers. The volume may be an entire room or building in which air enters and leaves through known points on the volume's surface. The control volume could be the enclosure surrounding the source. The control volume could also involve an imaginary surface in the vicinity of the source through which air enters and leaves. The selection of the control volume is of vital importance. It should be chosen carefully and include the points at which one wants to compute $c(x, y, z, t)$. It should also be selected because of the ease with which the designer can define boundary conditions. The *control surface* or *system boundaries* is the surface of the control volume.

The rate at which contaminants enter the control volume is handled in two ways.

(a) Experimentally measured emission rates, emission factors, or other empirical relationships can be used to specify the generation rate as a boundary condition. This procedure would be logical if readers want to compute $c(x, y, z, t)$ in a large volume in which the surface area over which generation occurs is small or if general ventilation is appropriate.

(b) The concentration at an air–liquid interface can be specified and the actual generation rate computed using equations of mass transfer presented in earlier chapters.

In contrast to the macro-models developed in the chapter on general ventilation, a micro-model approach will now be employed. *Micro-models* are mathematical models based on describing the conservation of mass, momentum, and energy for a very large number of elemental volumes. The approach may also be referred to as *distributed parameter model* in contrast to a *lumped parameter model*. Details of the flow field will not be assumed as they are in macro-models but are unknowns to be computed along with temperatures and concentrations throughout the space under study. Micro-models may be either two- or three-dimensional, depending on the physical programs and inventive numerical solution techniques.

10.1 TUNNELS

Tunnels for automotive and railway traffic, underground mining, wastewater transport and underground passageways for communications conductors or piping are places where unhealthy air is apt to occur because ventilation is inadequate and, or emissions are greater than expected. Assuming that steady-state conditions occur and that the only spatial coordinate is in the longitudinal direction, tunnels can be modeled easily (402) and the concentration of gaseous contaminants predicted as a function of location in the tunnel. Figure 10.1 depicts four classes of automotive tunnels that will be modeled.

Natural ventilation refers to air motion without fans or plenum chambers to add, withdraw and distribute air. Natural ventilation occurs by moving autos that draw air into the tunnel by viscous forces. Automotive and railway vehicles traveling through the tunnel in a single direction draw air through the tunnel, as if a such vehicles were "pistons". Natural ventilation also occurs if the tunnel is constructed on a slant and thermal buoyancy causes air to travel in a single direction. Natural ventilation is typically used for tunnels less than 300 m.

Local make-up air ventilation refers to fresh air introduced into the tunnel at a single point (usually at the inlet) at a volumetric flow rate Q_m. Local make-up air ventilation is typically used for tunnels less than 600 m.

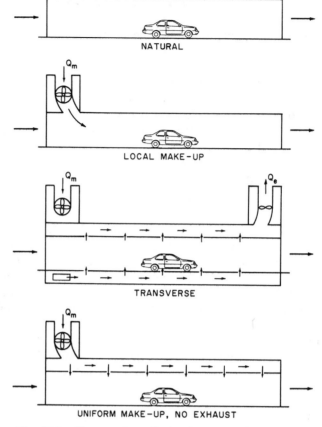

Fig. 10.1 Classes of ventilation in automotive tunnels.

Transverse ventilation involves plenums and fans to introduce outside air (Q_m) uniformly along the length of the tunnel and plenums and fans to withdraw contaminated air (Q_e) uniformly along the tunnel length. If $Q_e = Q_m$, the ventilation is *balanced*. It is common for $Q_m > Q_e$, but it is rare that the reverse is true. Transverse ventilation is typically used for long tunnels in excess of 1500 m.

Uniform make-up air, no exhaust ventilation refers to plenums and fans that only introduce outside air $(Q_m > 0)$ but withdraw no contaminated air $(Q_e = 0)$. Uniform make-up air with no exhaust ventilation is typically used for tunnels less than 1500 m.

Automobile emissions are often given in terms of an emission factor, e_c (mg/vehicle-km). The number of vehicles per mile (n_c, vehicles/km) is called the *traffic density*. If automobiles travel through a tunnel of diameter D (m), length L (km) and cross sectional area A (m^2) at a uniform speed v_c (km/hr), the total rate at which pollutants are emitted S (mg/hr) is,

$$S(\text{mg/hr}) = e_c n_c v_c L \qquad (10\text{-}1)$$

Figure 10.2 is a schematic diagram representing air and contaminant entering and leaving an elemental volume ($A\,dx$) within a tunnel of uniform make-up air and uniform exhaust ventilation. The model also applies to all of the above classes of tunnel. Within an elemental volume ($A\,dx$), the conservation of mass for air results in the following,

$$\frac{dU}{dx} = (q_m - q_e) \tag{10-2}$$

where q_m and q_e are defined as,

$$q_m = \frac{Q_m}{LA} \qquad q_e = \frac{Q_e}{LA} \tag{10-3}$$

If q_m and q_e are constant, or vary in a known fashion with x, Eq. 10-2 can be integrated. When q_m and q_e are constant, the air velocity in the tunnel at any location x in the tunnel can be expressed as,

$$\frac{U(x)}{U_0} = 1 + \frac{x(q_m - q_e)}{U_0} \tag{10-4}$$

where U_0 is the air velocity entering the tunnel. So long as $q_m > q_e$, $U(x)$ increases linearly in the direction of travel. If a tunnel has only an exhaust system or if $q_e > q_m$, $U(x)$ decreases in the direction of travel. If the tunnel is long enough, a stagnation point will occur at x', $U(x') = 0$. Such a situation is undesireable because the contaminant concentration at x' is infinite, $c(x') = \infty$. It is rare that tunnels are designed with only an exhaust system but tunnels are often designed with only make-up air systems.

Within an elemental volume ($A\,dx$), the conservation of mass for contaminant is,

$$\frac{c\,dU}{dx} + \frac{U\,dc}{dx} = s + q_m c_m - cq_e - kc \tag{10-5}$$

where

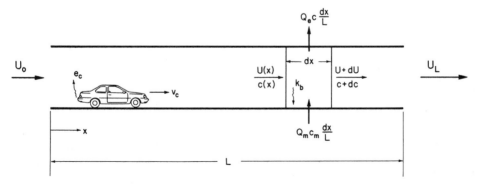

Fig. 10.2 Air and contaminant transport in a tunnel.

$$s = \frac{S}{LA} \tag{10-6}$$

$$k = \frac{4k_d}{D} \tag{10-7}$$

D is the tunnel diameter, c_m is the contaminant concentration in the make-up air and k_d (m/s) is a constant representing the rate at which the contaminant adsorbs or settles on the tunnel walls. Combining Eqs. 10-2 and 10-5, yields the following equation describing the variation of contaminant concentration in the tunnel,

$$\frac{U \, dc}{dx} = s + q_m c_m - c(k + q_m) \tag{10-8}$$

The maximum contaminant concentration is,

$$c(\text{max}) = \frac{s + q_m c_m}{k + q_m} \tag{10-9}$$

Combining Eqs. 10-4 and 10-8,

$$\frac{dc}{(q_m c_m + s) - (k + q_m)c} = \frac{dx}{U_0 + (q_m - q_e)x} \tag{10-10}$$

If q_m and q_e are constant, unequal and nonzero, Eq. 10-10 can be integrated in closed form.

$$c(x) = \frac{s + q_m c_m}{k + q_m} + \left[c_0 - \left(\frac{s + q_m c_m}{k + q_m} \right) \right] \left[\frac{U(x)}{U_0} \right]^{-b} \tag{10-11}$$

where

$$b = \frac{k + q_m}{q_m - q_e} \tag{10-12}$$

and $U(x)/U_0$ can be replaced using Eq. 10-4. If q_m and q_e are equal or identically zero, Eq. 10-11 can not be used and one must return to Eq. 10-8 to obtain a solution. If q_m and q_e and other parameters in the above equations vary with x, Runge-Kutta techniques will be needed to integrate the expressions.

Illustrated Example 10.1—Formaldehyde From Methanol-fueled Vehicles In Tunnels

Methanol-fueled autos receive increasing attention because of their potential to reduce ozone levels in urban areas. While methanol combustion produces fewer unburned hydrocarbons that ultimately produce ozone, the combustion of methanol produces more formaldehyde than the combustion of gasoline. The TLV for formaldehyde is currently 3 PPM ($3670 \ \mu m/m^3$) but the EPA is concerned (402) that outdoor concentrations of $500 \ \mu m/m^3$ may cause irritation for many individuals. Estimate the formaldehyde concentration in several classes of tunnel assuming that the following conditions reflect (402) roadway tunnels under severe conditions of traffic congestion, poor tunnel ventilation and engines with high rates if formaldehyde emission.

Parameter	Natural	Uniform make-up, no exhaust	transverse
L (m)	300	1000	2000
D (m)	7.57	7.57	7.57
U_0 (m/min)	60	60	60
q_m (min^{-1})	0	0.2	0.2
q_e (min^{-1})	0	0	0.18 and 0.2
c_m (μg/m^3)	5.0	5.0	5.0
c_0 (μg/m^3)	7.4	7.4	7.4
n_c (autos/km)	100	100	100
v_c (km/hr)	8	8	8
e_c (mg/km)	100	100	100
k (min^{-1})	0.02	0.02	0.02

The source term s is common to several of the equations above.

$$s = \frac{S}{LA} = e_c n_c v_c \left(\frac{4}{\pi D^2} \right)$$

$$= (100 \text{ mg/km})(100 \text{ autos/km})(8 \text{ km/hr})(\text{hr}/60 \text{ min})4/(\pi(7.57 \text{ m})^2)$$

$$= 29.64 \ \mu g/(m^3 \ min)$$

Natural Ventilation

For natural ventilation $q_m = q_e = 0$, Eq. 10-11 can not be used and one must integrate Eq. 10-8 directly. Equation 10-4 shows that $U(x) = U_0$. Thus

$$U_0 \frac{dc}{dx} = s - kc$$

and integration yields

$$c(L) = \left(\frac{s}{k} \right) - \left[\left(\frac{s}{k} \right) - c_0 \right] \exp\left(\frac{-kL}{U_0} \right)$$

The value of the maximum concentration is equal to (s/k)

$$c(\text{max}) = \frac{s}{k} = \frac{(29.64 \ \mu g/m^3 \ min)}{(0.02 \ min^{-1})} = 1482 \ \mu g/m^3$$

The concentration at any location L is,

$$c(L)(\mu g/m^3) = 1482 - [1482 - 7.4] \exp(-0.02L/60)$$

$$= 1482 - 1474.6 \exp(-0.00033L)$$

Uniform Make-up, No Exhaust
When $q_m = 0.2$ min^{-1} and $q_e = 0$, Eq 10-11 can be used.

$$\frac{s + q_m c_m}{k + q_m} = \frac{[29.64 + (0.2)(5)]}{[0.02 + 0.2]}$$

$$= 139.27 \ \mu g/m^3$$

the exponent b is

$$b = \frac{0.02 + 0.2}{0.2} = 1.1$$

and the concentration at any location L is,

$$c(L)(\mu g/m^3) = 139.27 + (7.4 - 139.27)\left[1 + \left(\frac{0.2L}{60}\right)\right]^{-1.1}$$

$$c(L) = 139.27 - 131.87(1 + 0.0033L)^{-1.1}$$

The maximum value of the concentration occurs when dc/dx is zero. Thus

$$c(\text{max}) = \frac{s + q_m c_m}{k + q_m} = 139.27 \ \mu g/m^3$$

Transverse Ventilation
When the system is balanced $q_m = q_e = 0.2$ min^{-1}. Equation 10-4 reduces to $U(x) = U_0$ but Eq. 10-11 can not be used. Returning to Eq. 10-8 and integrating one obtains,

$$c(L)(\mu g/m^3) = 139.27 - (7.4 - 139.27)\exp\left[-(0.02 + 0.2)\left(\frac{L}{60}\right)\right]$$

$$c(L) = 139.27 - 131.87 \exp(-0.003667L)$$

The maximum concentration is the same as above $c(\text{max}) = 139.27 \ \mu g/m^3$.
   If $q_e = 0.18$ min^{-1} and $q_m = 0.2$ min^{-1}, the system is unbalanced and Eq. 10-11 can be used. The exponent b is now,

$$b = \frac{k + q_m}{q_m - q_e} = \frac{(0.02 + 0.2)}{(0.2 - 0.18)} = 11$$

and the concentration at any location L is,

$$c(L)(\mu g/m^3) = 139.27 - (7.4 - 139.27)\left[1 + (0.2 - 0.18)\left(\frac{L}{60}\right)\right]^{-11}$$

$$c(L) = 139.27 - 131.87[1 + 0.00033L]^{-11}$$

A summary of formaldehyde concentrations ($\mu g/m^3$) for the different classes of tunnel is as follows,

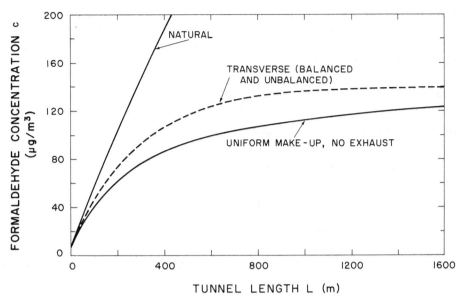

Fig. 10.3 Formaldehyde concentration in an automotive tunnel, illustrated in Example 10.1.

L (m)	Natural	Make-up, but no exhaust	Transverse balanced	unbalanced
100	55	43	47	47
200	102	64	75	74
400	190	87	108	106
600	272	100	124	122
800	350	107	132	129
1000	422	113	136	134
1200	490	117	138	135
1400	552	120	138	137
1600	612	122	139	138
2000	720	125	139	139

Figure 10-3 shows the results of these computations. It is clear that as the tunnel length increases, natural ventilation is inadequate and it is necessary to use uniform make-up air ventilation and perhaps to also withdraw contaminated air.

10.2 LAMINAR FLOW MODEL

Jets of air, wakes, and regions of recirculation produce turbulence that may affect the transport of contaminants in a significant way. The analysis of turbulent flow fields is complex and will be presented in Section 10.6. On the other hand, many other flow fields are either totally laminar or predominately laminar, turbulence is isolated in small regions of little importance to the rest of the flow field. This section will present the analysis of laminar flow fields. Once readers understand how laminar flow computations are made they will be in a better position to understand the more complicated analysis of turbulent flow fields. The assumptions of the laminar flow model are as follows:

(a) Steady-state
(b) Transport properties are independent of temperature
(c) Viscous energy dissipation can be neglected
(d) No chemical reaction
(e) No external energy addition

The steady-state equations for a nonreacting flow have been given in earlier chapters, but they will be restated for ease of reference. If the continuity equation is added to the momentum equation, it can be stated in the alternative form shown below. For brevity, the equations will be written for two dimensional flow, although the techniques to be discussed in Chapter 10 do not require such a restriction.

State

$$P = \rho RT \tag{10-13}$$

Continuity:

$$\frac{\partial(\rho U_x)}{\partial x} + \frac{\partial(\rho U_y)}{\partial y} = 0 \tag{10-14}$$

Momentum (x direction)

$$\frac{\partial(\rho U_x^2)}{\partial x} + \frac{\partial(\rho U_x U_y)}{\partial y} = -\frac{\partial P}{\partial x} + \frac{\partial\left(\mu_{eff}\frac{\partial U_x}{\partial x}\right)}{\partial x} + \frac{\partial\left(\mu_{eff}\frac{\partial U_x}{\partial y}\right)}{\partial y} \tag{10-15}$$

Momentum (y direction)

$$\frac{\partial(\rho U_y^2)}{\partial y} + \frac{\partial(\rho U_x U_y)}{\partial x} = -\frac{\partial P}{\partial y} + \frac{\partial\left(\mu_{eff}\frac{\partial U_y}{\partial x}\right)}{\partial x} + \frac{\partial\left(\mu_{eff}\frac{\partial U_y}{\partial y}\right)}{\partial y} + g\rho\beta\theta \tag{10-16}$$

where ($g\rho\beta\theta$) accounts for *buoyancy* and

$$\theta = (T - T_0) \tag{10-17}$$

and β is the *coefficient of thermal expansion*. For an ideal gas,

$$\beta = -\frac{1}{T} \tag{10-18}$$

Energy:

$$\frac{\partial(\rho U_x T)}{\partial x} + \frac{\partial(\rho U_y T)}{\partial y} = \frac{\partial\left[\left(\frac{k}{c_p}\right)_{\text{eff}} \frac{\partial T}{\partial x}\right]}{\partial x} + \frac{\partial\left[\left(\frac{k}{c_p}\right)_{\text{eff}} \frac{\partial T}{\partial y}\right]}{\partial y} \qquad (10\text{-}19)$$

Species:

$$\frac{\partial(\rho U_x Y_s)}{\partial x} + \frac{\partial(\rho U_y Y_s)}{\partial y} = \frac{\partial\left(\rho \mathscr{D}_{\text{eff}} \frac{\partial Y_s}{\partial x}\right)}{\partial x} + \frac{\partial\left(\rho \mathscr{D}_{\text{eff}} \frac{\partial Y_s}{\partial y}\right)}{\partial y} \qquad (10\text{-}20)$$

where Y_s is the *mass fraction* of chemical species s. By definition

$$\sum_{i=1} Y_i = 1 \qquad (10\text{-}21)$$

The pressure will be denoted by P. It is useful to rearrange the equations above (10-14 to 10-20) into a single *generalized Reynolds average transport equation* having three parts,

$$\frac{\partial(\rho U_j \phi)}{\partial x_j} - \frac{\partial\left(G_\phi \frac{\partial \phi}{\partial x_j}\right)}{\partial x_j} - S_\phi = 0 \qquad (10\text{-}22)$$

where ϕ, G_ϕ and S_ϕ are parameters defined in Table 10.1. (The parameter ϕ *IS NOT* the velocity potential.) The derivatives with respect to x and y shown in Eqs. 10-14 to 10-20 are accounted for by the subscript j. The three terms in Eq. 10-22 are also called the convection, diffusion, and source terms, viz,

Convection term: $\rho U_j \phi$

Diffusion term: $G_\phi \dfrac{\partial \phi}{\partial x_j}$

Source term: S_ϕ

The discussion that follows is compressed from material in the field called *computational fluid dynamics* (CFD). Readers should study the substantial literature in this field. Of particular value to the techniques recommended in this chapter are references 55, 57, and 93.

The generalized equation (Eq. 10-22) will not be solved as a differential equation; it will be transformed into an algebraic equation and solved by numerical means. To solve equations by numerical means the user must select a computational mesh or grid. It is convenient to use an orthogonal grid. Intersections of grid lines define nodes at which the dependent properties will be calculated. The process of transforming differential equations into algebraic equations is called *discretization*. The resulting equations relate properties at a node ϕ_P to properties at the nearest neighbors, that is, nodes NESW

TABLE 10.1 General Conservation Equation (312)

$$\text{Div}(\phi\rho\mathbf{U} - G_\phi \, \text{grad } \phi) = S_\phi + S_B$$

ϕ	G_ϕ	S_ϕ	S_B
1	zero	zero	zero
U_i	μ_{eff}	$\dfrac{\partial\left[\mu_{eff}\left(\dfrac{\partial U_j}{\partial x_i} + \dfrac{\partial U_i}{\partial x_j}\right)\right]}{\partial x_i}$	$-\rho\beta g_i\theta$
h	μ_{eff}/σ_h	zero	zero
k	μ_{eff}/σ_k	$G - \rho\epsilon$	G_B
ϵ	$\mu_{eff}/\sigma_\varepsilon$	$\epsilon(C_1 G - C_2\rho\epsilon)/\kappa$	$C_3\epsilon G_B/K$
c	μ_{eff}/σ_c	zero	zero

$\mu_{eff} = \rho(\nu + \nu_t)$

$\nu_t = C_\mu k^2/\epsilon$

$\theta = T - T_0$; T_0 is a reference value

$$G = \mu_t\left[\frac{\partial U_i}{\partial x_j} + \frac{\partial U_j}{\partial x_i}\right]\left(\frac{\partial U_j}{\partial x_i}\right)$$

$$G_B = \rho\beta g\left(\frac{\nu_t}{\sigma_h}\right)\frac{\partial\theta}{\partial x_i}$$

$C_1 = 1.44 \qquad C_2 = 1.92 \qquad C_\mu = 0.09 \qquad C_3 = 1.44$

$\sigma_K = 1.0 \qquad \sigma_\varepsilon = 1.3 \qquad \sigma_h = 0.9 \qquad \sigma_c = 1.14$

boundary conditions; use wall function method (93)

$$a_P\phi_P = \sum_{j=1} a_j\phi_j + b \qquad\qquad (10\text{-}23)$$

where the a's and b's include the combined effects of geometry, transport, sources, and accumulation. There are four nearest neighbors in two dimensional flow and six in three dimensional flow. Discretized equations for energy and species will not be developed since they follow procedures identical to those shown below.

Each node in the region is indexed as shown in Fig. 10.4; the internodal distances and dimensions of each node are computed and stored. Around each node a control volume is defined by surfaces lying (generally midway) between adjacent nodes. The normal velocity components are assumed on the surface of each control volume. Nodal regions do not have to be squares or cubes or be of equal volume, but the three principal dimensions of each node should not differ by 20% from their neighbors. The capitalized subscripts NESW signifies properties at the nodes to the north, east, south, and west of node P, whereas the lowercase letters nesw signify properties that exist on the north, east, south, or west surface of the control surface surrounding node P.

The general conservation equation is integrated over the control volume and transformed by Green's Theorem

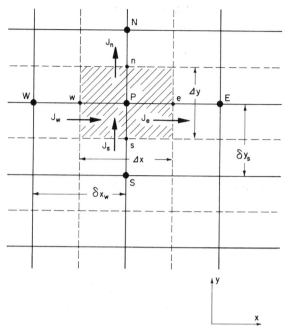

Fig. 10.4 Grid notation, control volume, nodal locations.

$$0 = \iiint_V \left(\text{Div}(\rho \mathbf{U}\phi - G_\phi \, \text{Grad } \phi - S_\phi) \right) dV \qquad (10\text{-}24)$$

$$= \iint_S (\rho \mathbf{U}\phi - G_\phi \, \text{Grad } \phi) \cdot d\mathbf{S} - \iiint_V S_\phi \, dV \qquad (10\text{-}25)$$

where $\mathbf{U}$ is the velocity vector, $d\mathbf{S}$ is the differential area taken as a vector.

The source term S_ϕ will be linearized by assuming it can be divided into a constant portion S_C and a variable portion S_P which varies linearly with the property (ϕ_P)

$$S_\phi = S_C + \phi_P S_P \qquad (10\text{-}26)$$

(Note the subscripts C and P are capitalized, but the subscript P does not denote the value of S at point P.) How one chooses S_C and S_P is important, see reference 55 for details. For the flows to be analyzed, in this chapter both S_C and S_P are zero since the pressure gradient in the momentum equation will be treated in a special fashion.

There are a variety of ways to evaluate the surface integral given above and readers are encouraged to study reference 55 for a full discussion of the subject. A versatile and general formulation that is valid for a variety of two dimensional control volumes of different aspect ratios, is the power law scheme in which Eq. 10-25 is written in the form of Eq. 10-23,

$$a_P \phi_P = a_E \phi_E + a_W \phi_W + a_S \phi_S + a_N \phi_N + b \qquad (10\text{-}27)$$

where the subscripts on the right hand side are NESW and the subscript on the left is uppercase P. Other terms are defined as follows,

$$a_E = D_e A(|P_e|) + [-F_e, 0] \tag{10-28}$$

$$a_W = D_w A(|P_w|) + [F_w, 0] \tag{10-29}$$

$$a_N = D_n A(|P_n|) + [-F_n, 0] \tag{10-30}$$

$$a_S = D_s A(|P_s|) + [F_s, 0] \tag{10-31}$$

The lowercase subscripts on the terms on the right hand side of the equations 10-28 to 10-31 signify that the properties or operations are conducted along the boundaries of the control volume. The second symbol on the right in Eqs. 10-28 to 10-31, viz

$$[A, B] \tag{10-32}$$

denotes the operation in which the user selects the larger of A and B. In the computer language FORTRAN, the symbol is written as $AMAX(A, B)$. The function

$$A(|P_e|) \tag{10-33}$$

in Eq. 10-28 is defined as

$$A(|P_e|) = [0, (1 - 0.1|P_e|)^5] \tag{10-34}$$

where $|P_e|$ is the absolute value of the *Peclet number* on the east boundary of the control volume (Fig. 10.4). (The symbols P_e, P_w, etc. in Eqs. 10-28 to 10-31) should not be confused with the pressure.) The Peclet number is the product of the Reynolds and Prandtl numbers.

$$Pe = Peclet = \frac{LU\rho c_p}{k} = Re_L \, Pr \tag{10-35}$$

The remaining quantities in Eq. 10-27 are

$$b = S_C \Delta x \, \Delta y + (a_P^0 \phi_P^0) \tag{10-36}$$

$$a_P = a_E + a_S + a_W + a_N + a_P^0 - S_P \Delta x \, \Delta y \tag{10-37}$$

$$a_P^0 = \frac{\rho_P^0 \Delta x \, \Delta y}{\delta t} \tag{10-38}$$

where the subscripts on the right are NESW, Δx and Δy which are the widths of the nodal region (see Fig. 10.4). The terms ϕ_P^0 and ρ_P^0 refer to values known at time t, while all the other values are the unknown values at time $(t + \delta t)$. The mass flow rate F_e on the eastern boundary is,

$$F_e = (\rho U)_e \Delta y \qquad (10\text{-}39)$$

The term D_e is called the *conductance* and is defined in terms of the generalized diffusion term (see Table 10.1) on the eastern boundary:

$$D_e = \frac{G_e \Delta y}{\delta x_e} \qquad (10\text{-}40)$$

where δx_e is the distance between node P and the nearest eastern node (Fig. 10.4). The Peclet number P_e can be expressed as

$$P_e = \frac{F_e}{D_e} \qquad (10\text{-}41)$$

Similar values are defined for the south, west, and north surfaces of the control volume.

10.3 NUMERICAL SOLUTION—SIMPLER ALGORITHM

In industrial ventilation the geometry and volumetric flow rates are known at the inlet and outlet, but the pressure and the velocity fields within the control volume are unknown. When both the pressure and the velocity fields are unknown, special provision must be made to solve the equations. The pressure gradient comprises a part of the source term in the momentum equation and solutions of the momentum equations for the velocities cannot be obtained unless the pressures are known. Patankar (55) created a numerical computational method called <u>SIMPLE</u>, and its revised version called <u>SIMPLER</u> (semi-implicit method for pressure-linked equations), to solve Eqs. 10-13 to 10-21. The methods have been applied to a variety of compressible and incompressible flows of interest to engineers and found to be accurate. The methods are respected and are widely used in engineering. The algorithms make use of a *staggered grid* in which control volumes for the x-momentum equation are slightly displaced from the control volumes for the y-momentum equation. Figure 10.5 shows the actual locations of the staggered control volumes for the momentum equations and the normal control volume for the continuity equation.

The solution to be discussed is two dimensional. To simplify terminology, it will be useful to dispense with some of the subscript notation and designate the velocity components as follows,

$$U_x = U \quad \text{and} \quad U_y = V \qquad (10\text{-}42)$$

In the discussion to follow an iterative procedure will be developed in which estimated values of P, U and V will approach the correct values in a progressive manner. It is useful to imagine that P, U and V are composed of an estimated value (designated by $*$) consisting of the last iterated value and a correction (designated by a prime), viz,

$$P = P^* + P'; \quad U = U^* + U' \quad V = V^* + V' \qquad (10\text{-}43)$$

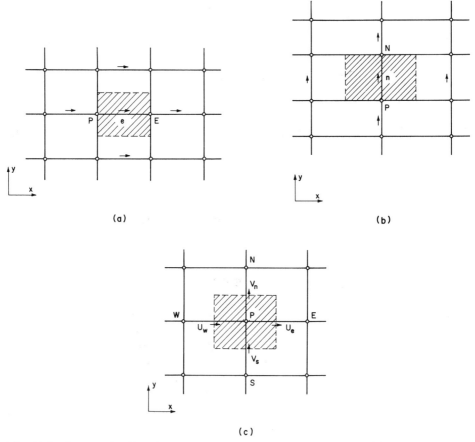

Fig. 10.5 Staggered grid and control volumes for different variables: (a) velocity in the x direction; (b) velocity in the y direction; (c) pressure, pressure correction, temperature, and concentration.

When discretized, the x-momentum equation becomes,

$$a_e U_e = \sum_{j=1} a_j U_j + b + (P_P - P_E)A_e \qquad (10\text{-}44)$$

The boundaries over which the sum is obtained are shown in Fig. 10.5. The coefficients a_j in Eq. 10-44 account for the convection and diffusion occurring at the boundaries of the control volume. The parameter b in Eq. 10-44 is defined by Eq. 10-36 but the pressure gradient *is not* to be included in the source term quantities S_C and S_P because it is included explicitly in the equation itself. It is convenient to rearrange Eq. 10-44.

$$U_e = \sum_{j=1} \frac{a_j U_j + b}{a_e} + d_e(P_P - P_E) \qquad (10\text{-}45)$$

where

$$d_e = \frac{A_e}{a_e} \tag{10-46}$$

A similar derivation for the y-momentum equation results in Eqs. 10-47 to 10-49. Figure 10.5 shows that the dimensions of the staggered control volume for the V velocities is not the same as it is for the U velocities.

$$a_n V_n = \sum_{j=1} a_j V_j + b + (P_P - P_N) A_n \tag{10-47}$$

$$V_n = \sum_{j=1} \frac{(a_j V_j + b)}{a_n} + d_n (P_P - P_N) \tag{10-48}$$

$$d_n = \frac{A_n}{a_n} \tag{10-49}$$

As with the x-momentum equation, the parameter b is not to include the pressure gradient. Define a *pseudovelocity*,

$$\hat{U}_e = \sum_{j=1} \frac{a_j U_j + b}{a_e} \tag{10-50}$$

(The symbols $\hat{U}_e$ $\hat{V}_n$ and so on in Eqs. 10-38 to 10-40 and 10-49, should not be interpreted as vectors, as was the custom in previous chapters.) Equation 10-45 becomes

$$U_e = \hat{U}_e + d_e (P_P - P_E) \tag{10-51}$$

In a similar fashion,

$$V_n = \hat{V}_n + d_n (P_P - P_N) \tag{10-52}$$

The pressure field can be expressed as a function of the velocity field by integrating the continuity equation (Eq. 10-14) over the control volume in which the node P is the center (see Fig. 10.5c)

$$O = \int_V \int \int \nabla \cdot (\rho \mathbf{U}) \, dV = \int_S \int \rho \mathbf{U} \cdot d\mathbf{S} \tag{10-53}$$

and obtain

$$[(\rho V)_n - (\rho V)_s] \Delta x + [(\rho U)_e - (\rho U)_w] \Delta y = 0 \tag{10-54}$$

The subscripts in the equation above are lowercase values nesw. Replace U and V by substituting Eqs. 10-51 and 10-52 and obtain

$$a_P P_P = a_E P_E + a_S P_S + a_W P_W + a_N P_N + b$$

$$P_P = \frac{a_E P_E + a_S P_S + a_W P_W + a_N P_N + b}{a_P} \tag{10-55}$$

where all the subscripts are the upper case P and NESW and

$$a_E = \rho_e d_e \, \Delta y \qquad\qquad (10\text{-}56)$$

$$a_W = \rho_w d_w \, \Delta y \qquad\qquad (10\text{-}57)$$

$$a_S = \rho_s d_s \, \Delta x \qquad\qquad (10\text{-}58)$$

$$a_N = \rho_n d_n \, \Delta x \qquad\qquad (10\text{-}59)$$

$$a_P = a_E + a_S + a_N + a_W \qquad\qquad (10\text{-}60)$$

$$b = [(\rho \hat{U})_w - (\rho \hat{U})_e] \, \Delta y + [(\rho \hat{V})_s - (\rho \hat{V})_n] \, \Delta x \qquad\qquad (10\text{-}61)$$

Reconsider the momentum equations. When written in terms of the correct values of P, U and V one obtains Eqs. 10-44 and 10-47. When written in terms of the approximate values (designated by *), the momentum equations become

$$a_e U^*_e = \sum_{j=1} a_j U^*_j + b + (P^*_P - P^*_E) A_e$$

$$U^*_e = \sum_{j=1} \frac{a_j U^*_j + b}{a_e} + d_e (P^*_P - P^*_E) \qquad\qquad (10\text{-}62)$$

$$a_n V^*_n = \sum_{j=1} a_j V^*_j + b + (P^*_P - P^*_N) A_n$$

$$V^*_n = \sum_{j=1} \frac{a_j V^*_j + b}{a_n} + d_n (P^*_P - P^*_N) \qquad\qquad (10\text{-}63)$$

Subtract Eqs. 10-62 and 10-63 from Eq. 10-44 and obtain

$$a_e U'_e = \sum_{j=1} a_j U'_j + (P'_P - P'_E) A_e \qquad\qquad (10\text{-}64)$$

Disregard $\sum_{j=1} a_j U'_j$ (see ref. 55 for details) and obtain

$$U'_e = d_e (P'_P - P'_E) \qquad\qquad (10\text{-}65)$$

Equation 10-65 is called the *pressure correction*. Using Eq. 10-43 the new improved value of the velocity U_e becomes

$$U_e = U^*_e + d_e (P'_P - P'_E) \qquad\qquad (10\text{-}66)$$

and similarly,

$$V_n = V^*_n + d_n (P'_P - P'_N) \qquad\qquad (10\text{-}67)$$

Return to the continuity equation, (Eq. 10-54), and evaluate it in terms of the velocity correction U'_e and V'_n. After simplification the following equations are obtained

$$P'_P = \frac{a_E P'_E + a_W P'_W + a_S P'_S + a_N P'_N + b}{a_P} \tag{10-68}$$

where P' refers to corrected pressure, a_E, a_S, a_W and a_N are identical to Eqs. 10-56 to 10-60 but b is now equal to

$$b = [(\rho U^*)_w - (\rho U^*)_e] \Delta y + [(\partial V^*)_s - (\rho V^*)_n] \Delta x \tag{10-69}$$

The following is an outline of the steps of the SIMPLER algorithm. The procedure makes use of the following sets of equations:

Set I: momentum equations in terms of U, V, and P (Eqs. 10-45 and 10-48)
Set II: momentum equations in terms of U^*, V^*, and P^* (Eqs. 10-62 and 10-63)
Set III: corrected velocities (Eqs. 10-66 and 10-67)
Set IV: pressure in terms of pseudovelocities (Eq. 10-55)
Set V: pressure correction (Eq. 10-68)

Algorithm

1. Guess a velocity field U^* and V^*.
2. Calculate $\hat{U}$ Eq. 10-50 (and calculate $\hat{V}$ from a comparable equation).
3. Calculate the pressure field from Eq. 10-55.
4. Using the pressures in step 3 as values for P^*, obtain the approximate velocities U^* and V^* from Eqs. 10-62 and 10-63.
5. Using the velocity corrections U'_e and V'_n, but not correcting the pressure, compute the correct velocity field using Eqs. 10-66 and 10-67.
6. Compute P'_P from Eq. 10-68.
7. If the density, concentration and temperature change, compute values for these variables using the velocity field in step 6.
8. Return to step 2 and repeat until satisfactory convergence is obtained.

10.4 RIM EXHAUSTERS FOR OPEN VESSELS

There are many industrial operations in which vapors and gases are generated by workpieces immersed in liquids contained in open vessels see for example Fig. 1.25. Examples include metal plating, metal cleaning, electroplating, acid dipping, stripping, degreasing, and etching. If the open vessel can be enclosed, it is obviously wise to do so since both emissions and the amount of air to be moved will be minimal. Unfortunately, some processes require workers to raise and lower workpieces into vessels or visually monitor the process and enclosures are impractical.

When enclosures are impractical, a popular method to control vapors is a lateral ventilation system also called a rim exhauster. Figure 10.6 illustrates the essential features of rim exhausters in an environment in which air streams over

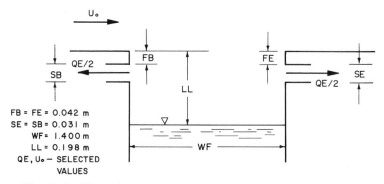

Fig. 10.6 Dimensions of a rim exhauster in streaming flow (77).

the vessel in the positive x direction. While the designs are popular, there is no way to estimate vapor concentrations in the vicinity of the vessel where workers may be stationed or how concentrations vary with ventilation volumetric flow rate, the dimensions and location of the inlet, distance between liquid and rim, liquid thermodynamic properties, and room air currents. Ventilation volumetric flow rates per unit area of liquid surface (QE/A) recommended by Hemeon (47) are 0.762–0.889 m^3/sec per m^2 of surface area (150–175 CFM/ft^2). For a vessel containing trichlorethylene, the ACGIH procedures (27) discussed in Chapter 6 result in the same values. Silverman (178,179) measured the concentration above open vessels equipped with lateral exhausts and recommended a series of empirical equations relating the volumetric flow rate to the geometry of the tank. Exhaust volumetric flow rates of 150–175 CFM/ft^2 are the lower end of values recommended by Silverman.

Rim exhausters can also be designed to totally surround the opening of drums, crucibles, and so on (Fig. 6.1) to prevent the escape of contaminants into the workplace. Klein (294) reports on experiments in which volumetric flow rates per unit drum area of 260–434 CFM/ft^2 and slot widths of 0.75–2 in protected workers needing to reach into the drum.

Rim exhausters can be modeled by neglecting viscous effects (78) using potential flow functions. Alternatively, they can be modeled to include viscous effects (77, 79) using the algorithm described in Section 10.3. In both cases, a rectangular vessel will be assumed in which the side containing the rim exhausters is large with respect to the width of the vessel. With such a configuration, spatial variations in the long dimension can be neglected and the only spatial variables are the distance across the vessel (x) and distance above the surface of the liquid (y).

If viscosity is neglected and the liquid and air are the same temperature, the velocity field for a rim exhauster located in a room in which air streams over the top of the open vessel in the x direction can be modeled using potential functions ϕ and ψ.

$$\nabla^2 \phi = 0 \tag{10-70}$$

$$U = -\frac{\partial \phi}{\partial x} \tag{10-71}$$

$$V = -\frac{\partial \phi}{\partial y} \tag{10-72}$$

Since air and vapor velocities are the same, the upper case symbols U and V will be used to express velocities. If the vapor concentration in air is small, so that the air density can be assumed constant, the velocity field can be computed independently of the concentration field. Once the velocity field is determined, the diffusion equation (Eq. 10-20) can be solved to obtain the vapor concentration. A flow domain is defined by the user and values of the velocities (U, V) and concentration (c), or their derivatives, are specified on the surfaces of the domain. Values of the velocities can be computed by adding the velocity of the streaming flow to the air velocities produced by the rim openings modeled as line sinks. The same finite numerical techniques can be used to solve Eqs. 10-70 to 10-72 that were used to compute the velocities in the main flow field in the grinding booth. Using these techniques it was found (78) that the volumetric flow rate is the dominant operating parameter affecting performance. The dimensions and location of the inlet slot are not of major importance, since at distances more than one slot width from the slot, the air velocity field can be described accurately by assuming the slots are line sinks. The distance between the liquid surface and rim (free board) is not a sensitive parameter. In the operator's breathing zone, the concentration is unaffected by the free board distance.

To predict the velocity and concentration fields above the vessel accurately, the effects of viscosity should be included and consequently Eqs. 10-13 to 10-21 will have to be solved. Once again it will be assumed that ambient air passes over the vessel in the positive x direction with a velocity U_0. As a first approximation to including the effects of turbulence, the viscosity can be increased by a scalar multiple. This procedure has the effect of forcing turbulence to be uniform throughout the flow domain, which of course is not true, but by varying the scalar the technique affords the opportunity to examine the effect of varying degrees of turbulence.

If the liquid is hot and the effects of heat transfer are to be included, Eqs. 10-13 to 10-21 must be solved. To solve these equations by the techniques of Section 10.3, a control volume shown in Fig. 10.7 must be defined. The sides of the control volume should extend above the vessel to insure that velocities (or their derivatives) can be defined as boundary conditions. To compute conditions on the right hand edge (downwind edge), the horizontal surface of the vessel will have to extended to the right. Mathematically the extension corresponds to a solid surface. Equations 10-13 to 10-21 can be solved numerically subject to the following boundary conditions:

Liquid–air interface:

$$\text{Concentration } (c) = \text{contaminant partial pressure} \tag{10-73}$$

$$V = 0 \tag{10-74}$$

$$T = T(\text{liquid}) \tag{10-75}$$

$$\frac{\partial U}{\partial y} = 0 \tag{10-76}$$

Vessel walls:

$$U = 0 \tag{10-77}$$

$$V = 0 \tag{10-78}$$

$$\frac{\partial [c, T, P]}{\partial x} = 0 \tag{10-79}$$

Inlet openings:

$$U = \text{volumetric flow rate/area} = \text{constant} \tag{10-80}$$

$$\frac{\partial [V, T, P, c]}{\partial x} = 0 \tag{10-81}$$

Top boundary of control volume:

$$V = 0 \tag{10-82}$$

$$\frac{\partial [U, T, P, c]}{\partial y} = 0 \tag{10-83}$$

Left hand boundary of control volume above vessel:

$$U = U_0 \tag{10-84}$$

$$T = T_0 \tag{10-85}$$

$$c = 0 \tag{10-86}$$

$$\frac{\partial [V, P]}{\partial x} = 0 \tag{10-87}$$

Right hand boundary of control volume above vessel:

$$\frac{\partial [V, T, P, c]}{\partial x} = 0 \tag{10-88}$$

At the end of each iteration an overall mass balance must be computed for the entire control volume. The relative difference in mass flow rate is computed and the velocity components (U) at each grid point on the outflow boundary adjusted by this factor. These modified velocities are then used for the next iteration.

For the rim exhauster shown in Fig. 10.7 a grid consisting of 14 horizontal nodes by 47 vertical nodes was used (77, 79). The algorithm described in Section 10.3 has been used. The procedure begins by assuming an initial velocity field and then computing a pressure field. Using these pressures, the momentum equations are solved for new values of the velocities (U, V). The new values of (U, V) are then used to calculate pressure corrections which in turn are used to correct the velocity field. When the vessel liquid is hot and produces a buoyant plume, the

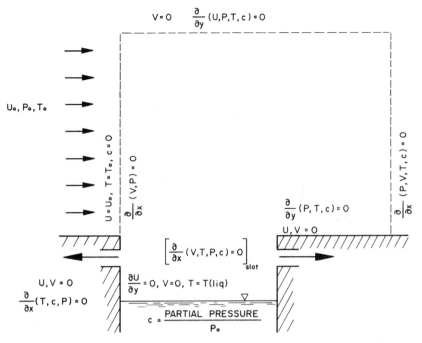

Fig. 10.7 Boundary conditions for rim exhauster in streaming flow (77).

energy equation is coupled to the continuity and momentum equations and must be solved before going back to compute the pressure. Once the final velocity field is computed, the contaminant concentration is determined from the diffusion equation in the same fashion as the temperature is computed from the energy equation. Because the equations are nonlinear and coupled, under relaxation techniques (relaxation factors less than unity) should be employed. For the most part convergence will occur in 200–250 iterations.

Figures 10.8 and 10.9 show the velocity field for an open vessel and rim exhauster containing trichlorethylene at an elevated temperature. Room air is assumed to pass over the vessel in the x direction at a velocity of 1 m/s. The ventilation flow rates are 0.025 and 0.0025 m^3/sec per m^2 of liquid surface (4.9 and 0.49 CFM per ft^2). The low value is chosen to represent an open vessel without a rim exhauster. The velocity field shows the presence of a large, low velocity eddy in the lower left hand corner inside the vessel. Since a ventilation rate of 0.0025 corresponds to an uncontrolled open vessel, it can be seen that the draft velocity of 1.0 m/s has a significant effect on the concentration profile. Figures 10.10 and 10.11 show the concentration isopleths for these two flow fields. The TLV for trichlorethylene is 100 PPM (mol fraction 0.0001), thus in the workers' breathing zone, marginally safe working conditions exist on the windward (left) side of the vessel, but unhealthy conditions exist on the downwind (right) side of the vessel. Computations show that the vapor concentrations are considerably higher when the liquid in the vessel is hot. The high concentrations are not due to a buoyant plume, but rather because the vapor concentration at the air–liquid interface is high which in turn is because the vapor pressure of the

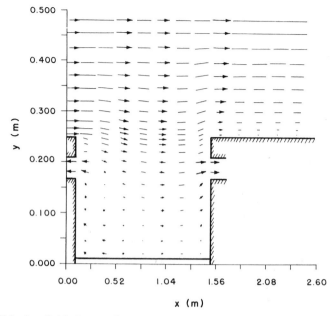

Fig. 10.8 Velocity field for a rim exhauster in streaming flow, $QE/A = 0.025$ m/s, $T(\text{liquid}) = 67$ C, $U_0 = 1$ m/s (77).

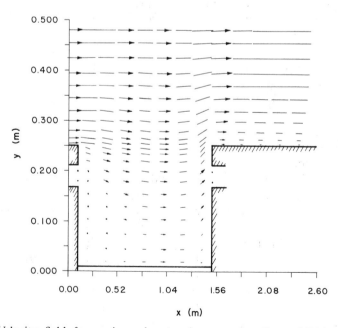

Fig. 10.9 Velocity field for a rim exhauster in streaming flow, $QE/A = 0.0025$ m/s, $T(\text{liquid}) = 67$ C, $U_0 = 1$ m/s (77).

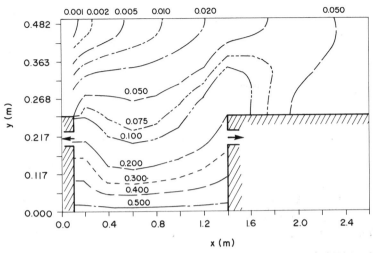

Fig. 10.10 Concentration field (mol fraction) for a rim exhauster, $QE/A = 0.025 \, \text{m/s}$, $T(\text{liquid}) = 67 \, \text{C}$, $U_0 = 1 \, \text{m/s}$ (77).

liquid is high. Thus heated liquids pose hazards not because of buoyancy but because their vapor pressures are strongly dependent on temperature. When these computations are repeated for QE/A of 0.0025 but a draft velocity of 0.1 m/s, the concentrations are in excess of TLV on both sides of the vessel for both hot and room temperature liquids. Thus room air currents are significant factors in the transport of vapor.

The SIMPLER algorithm will not converge if one of the boundaries passes through a recirculation eddy (55), but eddies inside the control volume pose no such problem. Thus the user must choose the control volume carefully and anticipate the presence of recirculation eddies. For example, if the ventilation

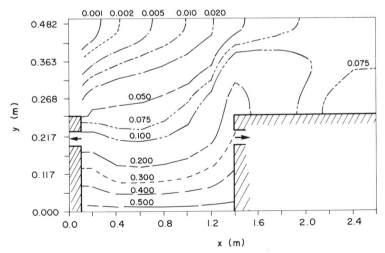

Fig. 10.11 Concentratin field (mol fraction) for a rim exhauster, $QE/A = 0.0025 \, \text{m/s}$, $T(\text{liquid}) = 67 \, \text{C}$, $U_0 = 1 \, \text{m/s}$ (77).

flow rate exceeds 0.025 m^3/second per m^2 when the draft velocity of 1.0 m/s, a large recirculation eddy will form above the right-hand rim of the vessel in Fig. 10.7 and users will have to extend the control volume to the right.

10.5   TURBULENT FLOW MODEL

Flow fields encountered in industry are generally unsteady and often turbulent. A turbulent flow is one in which all flow properties possess fluctuations or perturbations over and above time averaged values. In addition, the fluctuations can be described statistically. Workers often ascribe to turbulence, features and properties it does not possess. Thus turbulence becomes a straw man, which is conveniently blamed for behavior that cannot be explained any other way. This is unfortunate, because turbulence may not be the reason why predictions and observations do not agree. In the final analysis turbulence merely increases mixing, which may be described by increases in the local transport properties (viscosity, thermal conductivity, diffusion coefficient, etc.) above their molecular values. If turbulence affected each transport property by a constant amount, the conventional conservation equations written in terms of the average velocities and pressures could be used and molecular transport properties replaced by new, higher values enhanced by turbulence. Unfortunately, larger transport properties is a local event and the enhancement varies throughout the flow field. Thus engineers are forced to analyze turbulence as a spatially varying phenomena requiring new and special partial differential equations to be solved.

The initial step in analyzing turbulent flow fields is to conceive that each variable can be represented as the sum of an average value and a fluctuation (noted by a prime) (Fig. 10.12). Two types of averaging can be used to model turbulent flows. The oldest and most common are unweighted time averages called *Reynolds averages*. The second is a density-averaged method called *Favre average* (318).

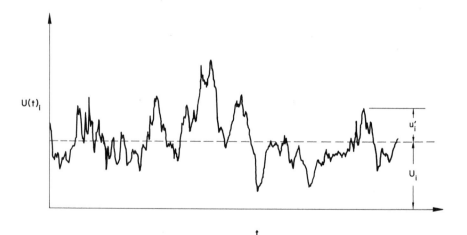

Fig. 10.12   Turbulent velocities, $U_i(t)$, U_i, and U_i'.

For unweighted averages, the instantaneous value of any variable is decomposed into a time-averaged and a fluctuating component:

$$U_i(t) = U_i + u_i'$$
$$P(t) = P + p'$$
$$T(t) = T + T' \tag{10-89}$$

The fluctuation u_i' is also called the *turbulent velocity scale*. By definition, the *average* velocity (or any other property denoted by an overbar) is defined by the following limit, where t_0 is an averaging time that is considerably longer than the period of the fluctuations,

$$\overline{U_i} = \frac{1}{t_0} \int_0^{t_0} U_i(t)\, dt = U_i \tag{10-90}$$

Thus the average of a fluctuation is zero, viz,

$$\overline{u_i'} = \frac{1}{t_0} \int_0^{t_0} u_i'\, dt = 0 \tag{10-91}$$

Using this logic, it can be shown that

$$\overline{T'} = 0, \qquad \overline{p'} = 0 \tag{10-92}$$

Let f and g be any two dependent variables and m an independent variable such as time (t) or position (x, y, z). It can be shown that

$$\overline{\overline{f}} = \overline{f}$$

$$\overline{f + g} = \overline{f} + \overline{g}$$

$$\overline{\overline{f}g} = (\overline{f})(\overline{g})$$

$$\overline{\frac{df}{dm}} = \frac{d\overline{f}}{dm} \tag{10-93}$$

$$\overline{\int f\, dm} = \int \overline{f}\, dm$$

$$\overline{fg} = (\overline{f})(\overline{g}) + \overline{f'g'}$$

Equation 10-93 is called *Reynolds averaging*. The average of the product of fluctuations such as,

$$\overline{u_i'u_i'}, \qquad \overline{u_i'u_j'u_k'}, \qquad \overline{u_i'T'} \tag{10-94}$$

are called correlations and are not necessarily zero. The *intensity, degree, or level of turbulence* are synonyms used to describe turbulence and can be defined as the root-mean-square value of the sum of the fluctuations normalized by the air speed $|\mathbf{U}|$

$$\frac{\sqrt{\dfrac{\overline{u_i' u_i'}}{3}}}{U(\text{ave})} \tag{10-95}$$

The Favre of density-weighted average for a property ϕ is defined as follows,

$$\tilde{\phi} = \frac{\overline{\rho \phi}}{\overline{\rho}} \tag{10-96}$$

The density-weighted mean and fluctuating components become

$$\rho \phi = \overline{\rho \tilde{\phi}} + \rho \phi'' \tag{10-97}$$

In this case taking the time average as in Eqs. 10-92 and 10-93 above results in

$$\overline{\rho \phi''} = 0 \tag{10-98}$$

but

$$\overline{\phi''} = -\frac{\overline{\rho' \phi''}}{\overline{\rho}} \tag{10-99}$$

is not zero in general. Reynolds and Favre averages are identical in constant density flows, but can be appreciably different in variable density flow fields such as occurs in exothermic chemical reactions, flames, and natural convection. In most cases of industrial ventilation the density is essentially constant, the distinction between Favre and Reynolds averages is negligible, and conventional unweighted averages will be used throughout this chapter.

When the dependent variables in the conservation equations are expressed as the sum of an average and its fluctuation, a new and surprising set of equations is created that becomes the operational equations that describe turbulent flow fields. To illustrate this, consider a steady-state turbulent flow of a constant density fluid,

$$\rho = \overline{\rho} \tag{10-100}$$

The continuity equations become

$$\frac{\partial (U_i + u_i')}{\partial x_i} = 0 \tag{10-101}$$

Integrate the equation with time to obtain the average,

$$\overline{\frac{\partial (U_i + u_i')}{\partial x_i}} = \frac{1}{t_0} \int_0^{t_0} \left\{ \frac{\partial (U_i + u_i')}{\partial x_i} \right\} dt \tag{10-102}$$

Since the velocities and other properties are continuous, the order of integration and differentiation can be reversed and the equation above becomes

$$\frac{\partial}{\partial x_i} \left\{ \frac{1}{t_0} \int_0^{t_0} (U_i + u_i') \, dt \right\} = \frac{\partial U_i}{\partial x_i} \qquad (10\text{-}103)$$

Thus the continuity equation for a turbulent incompressible flow reduces to the familiar equation written for average (or mean) values. An expression relating the velocity fluctuations can be obtained by returning to Eq. 10-102 and differentiating before taking the average. If this is done and Eq. 10-103 is subtracted from the result, one obtains

$$\frac{\partial u_i'}{\partial x_i} = 0 \qquad (10\text{-}104)$$

Thus the fluctuations satisfy a continuity equation which is the same as if the mean values were replaced by their fluctuations.

Consider the momentum equation written with respect to the x direction. Begin by replacing the single dependent variable by the sum of the average and fluctuating values. Assuming steady-state and neglecting gravity, the momentum equation in the x-direction is

$$\rho \left[(U_i + u_i') \frac{\partial (U_x + u_x')}{\partial x_i} \right] = \frac{-\partial (P + p')}{\partial x} + \mu \, \nabla^2 (U_x + u_x') \qquad (10\text{-}105)$$

Differentiate and take the average of each term. Making use of the logic used above, the momentum equation in the x direction becomes

$$\rho \left[U_i \frac{\partial U_x}{\partial x_i} \right] = -\frac{\partial P}{\partial x} + \mu \nabla^2 U_x - \rho \overline{\left[u_i' \frac{\partial u_x'}{\partial x_i} \right]} \qquad (10\text{-}106)$$

Using the continuity equation involving the velocity fluctuations, (i.e., Eq. 10-104), the last term on the right hand side can be rewritten

$$\rho \overline{u_i' \frac{\partial u_x'}{\partial x_i}} = -\rho \frac{\overline{\partial (u_x' u_i')}}{\partial x_i} \qquad (10\text{-}107)$$

The terms

$$-\rho \overline{u_x' u_x'} \, ; \qquad -\rho \overline{u_x' u_y'} \, ; \qquad -\rho \overline{u_x' u_z'} \qquad (10\text{-}108)$$

in Eq. 10-107 are called *turbulent stresses* because they affect the flow in the same manner as the viscous stress ($\mu \partial U/\partial y$). Actually, turbulent stresses are convective accelerations, nonetheless, they are called turbulent stresses. The quantity

$$-\rho \overline{u_i' u_j'} \qquad (10\text{-}109)$$

is called the *Reynolds stress* or *eddy stress*. Unlike the continuity equation in which the expressions for average and fluctuating components resemble their laminar counterparts, the momentum equation written in terms of the mean velocity is unlike its laminar counterpart. The momentum equation cannot be written solely in terms of the average velocities. A turbulent flow field has turbulent stesses to convect momentum and hence can be thought to possess an *effective viscosity* larger than the laminar (i.e. molecular) viscosity.

It is convenient to express the Reynolds stress as a function of an *eddy viscosity* (ν_t)

$$-\rho \overline{u_i' u_j'} = \nu_t \left[\frac{\partial U_i}{\partial x_j} + \frac{\partial U_j}{\partial x_i} \right] - \frac{2}{3} \delta_{ij} \kappa \qquad (10\text{-}110)$$

where δ_{ij} is the Kronecker delta and κ is the *turbulent kinetic energy*

$$\kappa = \frac{1}{2} \overline{(u_i' u_i')} \qquad (10\text{-}111)$$

Consider the transport of energy in a constant density fluid. Let the instantaneous value of the temperature be written as Eq. 10-89. If the specific heat is constant, the transport of energy satisfies the equation

$$\frac{\partial(\rho T)}{\partial t} + \frac{\partial(\rho T U_j)}{\partial x_j} = \frac{\partial \left[\left(\frac{\rho k}{c_p} \right) \frac{\partial T}{\partial x_i} \right]}{\partial x_i} + S_T \qquad (10\text{-}112)$$

where k is the thermal conductivity and S_T is a source term whose value depends on chemical reactions in the flow field and can be composed of an average value and a fluctuation. If the variable T is replaced by the concentration of species r (c_r in the units mols of species r per unit volume) and the thermal conductivity (k) is replaced by the diffusion coefficient ($\mathscr{D}$), an equation for the transport of mass is obtained.

Replace the dependent variable by it's average and fluctuating value and take the average of each term in the equation. Following the logic used above one obtains,

$$\rho U_i \frac{\partial T}{\partial x_i} + \overline{\rho u_i' \frac{\partial T'}{\partial x_i}} = \frac{\partial \left[\left(\frac{\rho k}{c_p} \right) \frac{\partial T}{\partial x_i} \right]}{\partial x_i} + S_T \qquad (10\text{-}113)$$

Since the density is constant, the continuity equation written for the fluctuation u_i' can be used to establish the following

$$\frac{\partial(u_i' T')}{\partial x_i} = u_i' \frac{\partial T'}{\partial x_i} + T' \frac{\partial u_i'}{\partial x_i} \qquad (10\text{-}114)$$

If an average is now taken, Eq. 10-104 shows that the second term on the right is zero. Thus the energy equation becomes

$$\rho U_i \frac{\partial T}{\partial x_i} = \frac{\partial}{\partial x_i} \left[\left(\frac{\mu}{\text{Pr}} \right) \frac{\partial T}{\partial x_i} - \overline{\rho u_i' T'} \right] + S_T \qquad (10\text{-}115)$$

The quantity $\rho\overline{u_i'T'}$ represents an additional way energy is transferred in a turbulent flow and is called the *turbulent heat flux*.

It is convenient to express the turbulent transport of a property ϕ in terms of a gradient of the *average* value of ϕ (the overbar will be omitted for brevity) by introducing the concept of an *eddy transport property* (G_t)

$$\rho\overline{u_i'\phi'} = \rho G_t \frac{\partial\phi}{\partial x_i} \tag{10-116}$$

and the gradient of the average value of the scalar property. In the case of the energy equation the eddy transport property is equal to the eddy viscosity (ν_t) divided by an empirical constant called the *eddy Prandtl number* (σ_t). Thus

$$\rho\overline{u_i'T'} = -\left(\frac{\mu_t}{\sigma_t}\right)\frac{\partial T}{\partial x_i} \tag{10-117}$$

and the energy equation becomes,

$$\rho U_i \frac{\partial T}{\partial x_i} = \frac{\partial\left[\left\{\left(\dfrac{\mu}{\mathrm{Pr}}\right)+\left(\dfrac{\mu_t}{\sigma_t}\right)\right\}\dfrac{\partial T}{\partial x_i}\right]}{\partial x_i} + S_T \tag{10-118}$$

By similar reasoning the species conservation equation can be written for the molar concentration of species r $(c_r$ in mols/m$^3)$

$$\rho U_i \frac{\partial c_r}{\partial x_i} = \frac{\partial}{\partial x_i}\left[\left\{\left(\frac{\mu}{\mathrm{Sc}}\right)+\left(\frac{\mu_t}{\sigma_c}\right)\right\}\left(\frac{\partial c_r}{\partial x_i}\right)\right] \tag{10-119}$$

where Sc is the *Schmidt number* and σ_c is an empirical constant.

In order to solve the conservation equations, the Reynolds stress in the momentum equations has to be specified by means of a turbulence model. A model that has received considerable attention because of its simplicity and ease of solution and because its results have been verified for a large number of physical applications is the κ-ϵ *model* of Launder and Spalding (93). The parameter κ is the turbulent kinetic energy Eq. 10-111 and ϵ is the turbulent energy dissipation

$$\epsilon = \nu \overline{\frac{\partial u_i'}{\partial x_j}\frac{\partial u_i'}{\partial x_j}} \tag{10-120}$$

The turbulent kinematic viscosity ν_t is related to κ and ϵ by

$$\nu_t = C_\mu \frac{\kappa^2}{\epsilon} \tag{10-121}$$

where C_μ is an empirical constant to be determined by experiment. The distribution of κ and ϵ over the flow field can be shown (93) to be given by the differential equations

Turbulent, kinetic energy (κ):

$$\frac{D\kappa}{Dt} = \frac{\partial\left[\left(\frac{\nu_t}{\sigma_\kappa}\right)\frac{\partial\kappa}{\partial x_j}\right]}{\partial x_j} + \nu_t\left[\frac{\partial U_i}{\partial x_j}\left(\frac{\partial U_i}{\partial x_j} + \frac{\partial U_j}{\partial x_i}\right)\right] - \epsilon \qquad (10\text{-}122)$$

Turbulent energy dissipation (ϵ):

$$\frac{D\epsilon}{Dt} = \frac{1}{\rho}\frac{\partial\left[\left(\frac{\nu_t}{\sigma_\epsilon}\right)\frac{\partial\epsilon}{\partial x_j}\right]}{\partial x_j} + C_1\left(\frac{\mu_t}{\rho}\right)\left(\frac{\epsilon}{\kappa}\right)\left[\frac{\partial U_i}{\partial x_j} + \frac{\partial U_j}{\partial x_i}\right]\frac{\partial U_i}{\partial x_j} - C_2\frac{\epsilon^2}{\kappa}$$
$$(10\text{-}123)$$

The constants C_1, C_2, σ_ϵ, and σ_κ are empirical constants that need to be determined by independent means. The notation $D[\ ]/Dt$ is the *total or substantial derivative*,

$$\frac{D[\ ]}{Dt} = \frac{\partial[\ ]}{\partial t} + U_i\frac{\partial[\ ]}{\partial x_i} \qquad (10\text{-}124)$$

For ease of numerical computation, the conservation equations of mass, momentum energy, and species and κ and ϵ can be rearranged in a general form,

$$\text{Div}[\rho\mathbf{U}\phi - G_\phi\ \mathbf{grad}\ \phi] = S_\phi + S_B \qquad (10\text{-}125)$$

where the ϕ is a property of the flow field and source terms are on the right hand side and U is the vector velocity. Table 10.1 defines the values of these parameters.

To compute the velocities, temperatures and concentrations in flow fields of interest to industrial ventilation users solve the equations in Table 10.1 using the algorithm described in the previous section. There are a large number of publications in which users have applied the κ-ϵ model to different flow fields. In some cases users have used empirical constants slightly different than those in Table 10.1 and the literature should be consulted.

10.6 PUSH–PULL VENTILATION SYSTEMS

To improve the control of vapors from liquids in open vessels, ambient air can be induced to flow across the liquid surface by blowing air through an inlet on one side of the vessel and removing air through an opening on the other side of the vessel, as shown in Fig. 10.13. Such a configuration is called a push–pull ventilation system. The results of research in the last 40 years (27, 47–49, 63, 64, 83–92, 138, 139, 219) can be summarized as follows:

(1) Recommended values of the volumetric flow rate of the exhaust air and blowing jet are given in terms of the width and surface area of the open vessel. Figure 6.21 summarizes these recommendations.

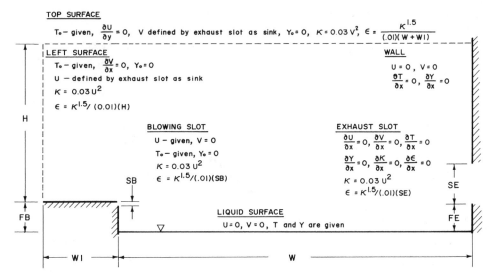

TOP SURFACE

T_0 − given, $\frac{\partial U}{\partial y} = 0$, V defined by exhaust slot as sink, $Y_0 = 0$, $K = 0.03\, V^2$, $\epsilon = \dfrac{K^{1.5}}{(.01)(W + WI)}$

LEFT SURFACE

T_0 − given, $\frac{\partial V}{\partial x} = 0$, $Y_0 = 0$

U − defined by exhaust slot as sink

$K = 0.03\, U^2$

$\epsilon = K^{1.5}/(0.01)(H)$

WALL

$U = 0$, $V = 0$

$\frac{\partial T}{\partial x} = 0$, $\frac{\partial Y}{\partial x} = 0$

BLOWING SLOT

U − given, $V = 0$

T_0 − given, $Y_0 = 0$

$K = 0.03\, U^2$

$\epsilon = K^{1.5}/(.01)(SB)$

EXHAUST SLOT

$\frac{\partial U}{\partial x} = 0$, $\frac{\partial V}{\partial x} = 0$, $\frac{\partial T}{\partial x} = 0$

$\frac{\partial Y}{\partial x} = 0$, $\frac{\partial K}{\partial x} = 0$, $\frac{\partial \epsilon}{\partial x} = 0$

$K = 0.03\, U^2$

$\epsilon = K^{1.5}/(.01)(SE)$

LIQUID SURFACE

$U = 0$, $V = 0$, T and Y are given

Fig. 10.13 Dimensions and boundary conditions for push–pull ventilation system in a quiescent environment, $H = 0.781$ m, $FB = FE = 0.163$ m, $SB = 0.015$ m, $SE = 0.195$ m, $W1 = 0.375$ m, $W = 1.8$ m (137).

(2) The width of the exhaust opening can be determined by assuming the expansion angle between the upper edge of the jet and the horizontal is 10–20 degrees.

(3) The width of the jet inlet is selected so that the jet velocity (U_B) is 5–10 m/s. The base of the exhaust opening and blowing jet are the same height above the level of the liquid.

While useful, these guidelines do not enable designers to predict the vapor concentration at points in the vicinity of the open vessel where workers may be stationed or understand how the concentration varies with volumetric flow rates of exhaust and blowing air, geometry of the open vessel, width of exhaust or blowing openings, thermodynamic properties of the liquid, or room air currents. Such understanding is only acquired by modeling push-pull ventilation systems. Since the flow field has both viscous and turbulent properties, the model must include the effects of viscosity and turbulence.

The flow domain is defined by the control volume shown in Fig. 10.13. The analysis in this section presumes that the open vessel is in a quiescent environment. The analysis must be modified to accommodate room air currents that pass over the vessel. The equations in Table 10.1 are elliptic differential equations, thus all the velocity components (or their derivatives) must be specified on all boundaries to insure the existence of a solution. In addition, the mass of air in the control volume must be conserved. The mass flux associated with the velocity components on the free boundaries must be equal to the difference between the mass through the exhaust slot minus the mass entering the jet slot. Four types of boundaries are encountered, walls, fluid surface, jet and exhaust slots, and free-stream boundaries. Boundary conditions for concentration and temperature are similar to those in Section 10.4 for rim exhausters; zero gradients normal to all solid surfaces and the plane of the suction slot, and specified values at the fluid surface, jet slot, and free stream boundaries. The velocity boundary conditions at walls, fluid surfaces, and exhauster slot are also similar to those in Section 10.4; no slip condition at walls and the fluid surface, and specified values at the suction and jet slot but the velocities on the free stream boundaries are not the same as they were in Section 10.4.

Since turbulence diminishes rapidly close to a wall, gradients of velocity, temperature, and concentration are steep near a wall. Consequently, the effective viscosity and other effective transport properties approach their laminar values near walls. It is possible to compute these variations in two ways.

(1) Nodes can be packed densely in regions of steep gradients to produce numerically accurate results.

(2) A special function called *wall function* can bridge the gap between the highly turbulent flow region and the laminar region adjacent to the wall.

Adoption of the first method consumes considerable computer time. Adoption of the second method requires empirical knowledge and insight to select a useful wall function. Considerable research in turbulence is available to provide this insight.

The nearest grid point to the wall (hereafter denoted by the subscript P) must be located far enough from the wall (denoted by the subscript w) so that the value of the ratio of effective viscosity to laminar viscosity is much greater than unity. The distance from the wall to the point P is denoted by y_P in Fig. 10.14. The transport processes occurring between the first grid point (P) and the wall (w) will be described by a *wall function* that relates the flux of a transport property ϕ to the wall with the local difference in the property ϕ, and other relevant quantities. A well known "wall function" is the logarithmic *law of the wall*, which is used to describe the velocity profile in the absence of a pressure gradient.

$$U^+ = \frac{1}{X} \ln(Ey^+) \qquad (10\text{-}126)$$

The term X is known as the *von Karman constant* and is taken to be 0.4 and the constant E is a *roughness parameter* equal to 9 for smooth walls (82). The term U^+ is a nondimensional velocity defined as

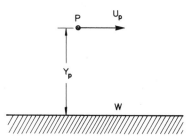

Fig. 10.14 Node for wall function.

$$U^+ = \frac{U}{U_t} \tag{10-127}$$

where U_t is the *friction velocity*

$$U_t = \sqrt{\frac{\tau_w}{\rho}} = C_\mu^{0.25} \kappa_P^{0.5} \tag{10-128}$$

The term y^+ is a nondimensional distance measured from the wall

$$y^+ = \frac{U_t y \rho}{\mu} \tag{10-129}$$

and κ_P is the turbulence kinetic energy at point P. The constant C_μ has been found (214) to be equal to 0.09. The law of the wall can be used to derive a relationship between the wall shear stress (τ_w) and velocity at point P. Substituting Eqs. 10-127 to 10-129 into Eq. 10-126 results in

$$\left(\frac{U_P}{\left(\frac{\tau}{\rho} \right)_w} \right) B = \frac{1}{X} \ln \left[BEy_P \left(\frac{\rho}{\mu} \right) \right] \tag{10-130}$$

$$B = C_\mu^{0.25} \kappa_P^{0.5} \tag{10-131}$$

The wall function for the energy equation can be obtained from the Reynolds analogy and experimental measurements. The following form is used by Yu (311)

$$(T_P - T_w) c_p \rho \left(\frac{B}{q_w''} \right) = \frac{1}{X} \ln \left[BEy_P \left(\frac{\rho}{\mu} \right) \right] \tag{10-132}$$

where T_P, T_w are time-averaged temperatures and q_w'' is the average heat flux at the wall. If it is assumed that the turbulent Lewis number is unity, a wall function for the species equation can be written as

$$(Y_P - Y_w) \rho \left(\frac{B}{m_w''} \right) = \frac{1}{X} \ln \left[BEy_P \left(\frac{\rho}{\mu} \right) \right] \tag{10-133}$$

where Y_P, Y_w are time-averaged concentrations and m_w'' is the mass flux at the wall. There is no wall function for κ and ϵ because of the steep velocity gradients near the wall, the source term for the turbulence kinetic energy equation can be expressed as a function of the wall shear stress (τ_w)

$$S_\kappa = G_\kappa - \rho \epsilon_P + S_B$$
$$= \left| \tau_w \frac{\partial U}{\partial y} \right| - \left[\left(\frac{C_\mu}{\tau_w} \right) (\rho^2 \kappa^2) \frac{\partial U}{\partial y} \right] + S_B \tag{10-134}$$

where S_B is the source term (Table 10.1). Near the wall the length scale is proportional to y_P and flow conditions can be approximated by Couette flow. Thus

$$\left(\frac{\tau}{\rho}\right)_{\mathrm{w}} = \frac{U_{\mathrm{P}}}{y_{\mathrm{P}}} \tag{10-135}$$

$$q''_{\mathrm{w}} = \frac{k(T_{\mathrm{P}} - T_{\mathrm{w}})}{y_{\mathrm{P}}} \tag{10-136}$$

$$m''_{\mathrm{w}} = \frac{\rho\mathcal{D}(Y_{\mathrm{P}} - Y_{\mathrm{w}})}{y_{\mathrm{P}}} \tag{10-137}$$

where k in Eq. 10-136 is the thermal conductivity. Substituting Eqs. 10-135 to 10-137 into Eqs. 10-130 to 10-133, one can obtain the equation for ϵ_{P} near the wall.

$$\epsilon_{\mathrm{P}} = \frac{C_{\mu}\kappa_{\mathrm{P}}^{1.5}}{y_{\mathrm{P}}} \tag{10-138}$$

The SIMPLER algorithm described in Section 10.4 can be used to solve the finite difference equations for turbulent flow. *Under-relaxation coefficients* in the analysis are:

(a) 0.5 for U and V equations
(b) 1.0 for the pressure equation
(c) 0.7 for the κ and ϵ equations

Satisfactory convergence is obtained in 1600 CPU seconds of IBM-370 mainframe computer time to compute 800 iterations. About 20% of CPU time can be saved by ignoring buoyancy.

The velocity and concentration fields have been predicted (82) for an open vessel 1.8 m wide containing trichlorethylene. Graphs of velocity, concentration, temperature and viscosity ratio are shown in Figs 10.15 and 10.16. Schlerin photographs of push–pull flow fields for three combinations of exhaust and

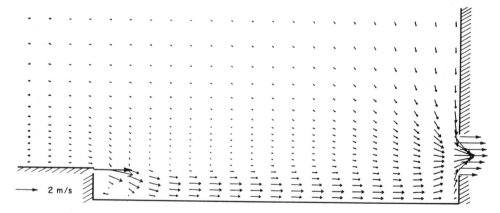

Fig. 10.15 Velocity field for push–pull ventilation system in a quiescent environment, $QB/A = 0.0571$ m/s, $SB = 0.015$ m, $UB = 3.81$ m/s, $QE/QB = 8.67$, T(liquid) $= 67$ C (137).

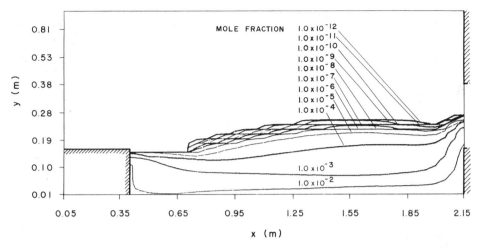

Fig. 10.16 Concentration field (mol fraction) for a push–pull ventilation system in a quiescent environment, $QB/A = 0.0571$ m/s, $SB = 0.015$ m, $UB = 3.81$ m/s, $QE/QB = 8.67$, T(liquid) = 67 C (137).

blowing volumetric flow rates are shown in Fig. 10.17 (311). Figure 10.18 shows experimentally measured velocities for a push-pull system for an electroplating tank (64).

The results show that the jet sweeps over the liquid surface like a wall jet entraining room air as the jet approaches the exhaust slot. Energy from the hot

Fig. 10.17 Schlieren Photographs: (a) $QE = QB = 0$, (b) $QB = 0$, $QE = 0.51$ m^3/s per m of slot; (c) $QE = 0.051$, $QB = 0.006$ m^3/s per m of slot; vessel contains water at 56 C, T(air) = 24 C vessel width = 0.52 m (with permission (311)).

Fig. 10.17 (*Continued*)

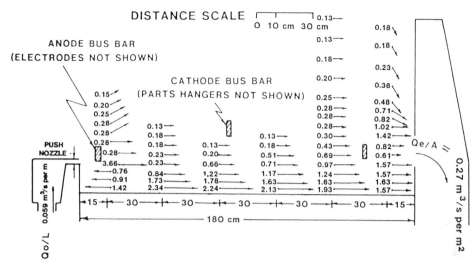

Fig. 10.18 Measured midplane velocities (m/s) of a push–pull ventilation system for an electroplating tank, $QE = 0.486$, $QB = 0.016 \, \text{m}^3/\text{s}$ per m of slot (with permission (64)).

liquid is not transferred to the workplace but is carried to the exhaust slot. In all cases a recirculation eddy forms directly below the push jet. Near the exhaust slot, the jet bends upward and flows into the slot. Concentrations of vapor above the vessel and on the left hand side are well below TLV. Examination of the ratio of turbulent to laminar viscosity shows that turbulence intensity decays quickly but gradually increase as the air approaches the exhaust slot. Convection is the dominant transport process near the liquid surface.

The principal asset of computer-aided design is that it enables designers the opportunity to examine the performance of different geometric configurations and operating parameters quickly. Results from such studies improve their understanding of their designs as effectively as if physical experiments had been run. Once a computer program has been written such "computer experiments" can be run more cheaply and rapidly that physical experiments. Once the most attractive design configuration is selected, experiments should be made to confirm that performance specifications have been satisfied. The exhaust volumetric flow rate (QE) and vessel width (W) were constant in all four cases in Table 10.2, thus only effects due to jet velocity, jet momentum, and QB/QE can be revealed. Nevertheless, there is reason to believe that these relationships are valid for push-pull systems with different exhaust volumetric flow rates.

The entrainment of air in a jet is directly proportional to its momentum. While large entrainment is desirable, the exhaust volumetric flow rate must be sufficient, otherwise contaminated air will spill over and exhaust and enter the workplace. The velocity field in Case 1 did not produce "spill," although a recirculation eddy developed above the jet and a dead air region occurred over the center of the vessel. Both occurrences signify that "spill" is imminent and will occur if the jet momentum is larger. The jet momentum for Case 3 is the lowest and produces the most gentle sweep of air across the liquid surface. Examination of the concen-

TABLE 10.2 Configurations of Push-Pull Ventilation Systems

Case	QB/A m^3/s per m of slot	SB (m)	UB (m/s)	QE/QB	Liquid Temperature (K)
1	0.0571	0.00725	7.88	8.67	340
2	0.0285	0.00725	3.93	17.34	340
3	0.0571	0.01500	3.81	8.67	340
4	0.0571	0.01500	3.81	8.67	293

Exhaust volumetric flow rate (QE) per m^2
  of liquid surface − 0.283 m^3/s per m^2

Exhaust volumetric flow rate (QE) per m
  of slot—0.495 m^3/s per m

Vessel width (W)—1.8 m

Freeboard ($FE = FB$)—0.163 m

Exhaust slot width (SB)—0.195 m

$$T_0 = 293 \text{ K} \qquad Y_0 = 0 \qquad U_0 = 0$$

tration fields shows that at any level above the liquid surface, Case 3 produces lower concentrations than Case 2, which in turn produces lower concentrations than Case 1.

On the basis of the four cases studied, designers can conclude the following for an open vessel, 1.8 m wide, in a quiescent environment:

(1) An exhaust volumetric flow rate (QE/A) of 0.238 m^3/sec m^{-2} of surface area is adequate, even though it is considerably less than the recommended ACGIH value.

(2) A jet velocity (UB) of 3.81 m/s reduces concentrations more effectively than 7.88 m/s.

(3) If UB is 3.81 m/s, satisfactory control can be obtained for QE/QB between 8.67 and 17.34

(4) The jet sweeping over the liquid surface deflects hot gas and vapor into the exhaust slot.

(5) A buoyant plum does not rise from the hot liquid and perturb the velocity field or enhance the transfer of vapor to the workplace. Heating the liquid affects the concentration field because the liquid vapor pressure is strongly dependent on temperature.

10.7 FUTURE DEVELOPMENTS

Rapid progress is being made in the field of computational fluid mechanics and it will not be long before specialists will apply their talents to problems in industrial ventilation. Since 1967 many computer codes have been developed for solving the conservation equations. The codes are known by both commercial and generic names such as TEACH, GENMIX, PHOENICS, CHAMPION 2/E/FIX, FLUENT, FIDAP, FLOTRAN, FLOW3D, NEKTON, and NISA/3D-Fluid (187, 401, 404) and are used to solve the conservation equation written in the form seen in Table 10.1. The codes have been applied to a wide variety of physical problems and validated by experimental measurements. Nielsen (134) predicted the velocity fields in ventilated rooms of different geometries; Qingyan and Van Der Kooi (312,314) predicted the velocity and temperature fields in climate rooms; Hjertager and Magnussen (315, 316) predicted the velocity field of a turbulent three-dimensional jet in a rectangular enclosure; Yang and Lloyd (317) predicted the concentration field of turbulent buoyant flow resulting from fires in aircraft cabins. Dellagi and colleagues at the INRS in France used such codes to predict the velocity and concentration fields inside paint spray booths for automobiles (135, 136). Using finite element techniques, Garrison and his students (291, 292, 396) modeled the velocity fields upstream of flanged and unflanged inlets (Fig. 10.19). Kurabuchi and Kusuda (254) report using κ-ϵ

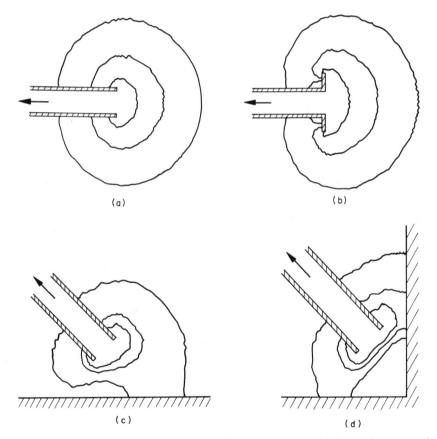

(a)   (b)

(c)   (d)

Fig. 10.19 Predicted velocity contours for various inlet geometries in a quiescent environment (redrawn from 291)
(a), (b) – 5, 10, 25% face velocity
(c), (d) – 20, 30, 40% face velocity.

models to model velocity fields inside rooms of unusual geometry. Using SIM-PLE, Kuehn (352) analyzed the flow field and particle transport mechanisms in tunnel clean rooms and Ye and Pui (415) analyzed particle deposition in a tube containing an abrupt contradiction. In the research programs conducted by R J Heinsohn and his colleagues (66–68, 70, 77–79, 82, and 107), computer programs to solve the Navier–Stokes equations were written from scratch. This is no longer necessary.

For most of the predictions that need to be made, commercially available *Navier–Stokes Solvers* can be used. Several commercial firms presently advertise the availability of these programs in engineering journals. There is still need for research in this area, the accuracy, speed, and "user-friendly" features of the programs need improvement. Given the imagination and vigor of computer science engineers and software specialists, one can be confident that these improvements will occur in the near future. For the bulk of work in designing new ventilation systems and improving old systems, even the present programs are adequate. Consider just two issues of current popular interest in which predictions can now be made using these programs.

(1) There is considerable controversy about rules concerning cigarette smoking on commercial aircraft and in office buildings and restaurants. It is not known if ventilation can provide areas with tolerably low smoke concentrations or whether there should be a prohibition on smoking.

(2) Industry is making increasing use of plastics, laminates, glues, finishes, epoxies, and resins to find new, cheaper, and lighter materials as substitutes for metals. The health effects of these materials, taken singularly or in combination, is a cause for concern. What is not known is the workplace concentration and the exposure of workers.

In both cases, commercially available Navier–Stokes Solvers can be used to predict the concentration at arbitrary points as functions of the materials generating the vapors, geometry of inlet and outlet ducts, and ventilation volumetric flow rates. Using these programs, engineers together with respiratory physiologists and toxicologists can estimate the health risks. Users of Navier–Stokes Solvers must be very sensitive to the limitations in these programs and careful not to assume a false sense of reality. Nonetheless, these predictions cannot be made any other way.

The most serious limitation to the κ-ϵ model is its assumption of *isotropic turbulence*, that is, turbulence intensity that is the same is each of the coordinate directions. Buoyant plumes are not isotropic and the κ-ϵ model may predict results that cannot be verifed fully by measurement. To analyze flow fields containing large buoyant sources, careful consideration must be given to the *Richardson number* (362),

$$\text{Ri} = \frac{-(g\overline{u_z'\theta'})}{-(\overline{u_x'u_z'})T_\infty \dfrac{\partial U_x}{\partial z}} \tag{10-139}$$

where θ' is the temperature fluctuation associated with the buoyant plume

$$\theta' = T'(z) - T_\infty \qquad (10\text{-}140)$$

T_∞ is the temperature of the surrounding air and z is in the vertical direction. The *Richardson number* is the ratio of the turbulent kinetic energy produced by buoyancy to the turbulent kinetic energy produced by shear stress. The term $\partial U_x / \partial z$ is normally positive for air above a heat source, thus the sign of the Richardson number is governed by the sign of the θ'. The Richardson number describes whether an eddy is amplified or damped by buoyancy.

$\mathrm{Ri} = 0$: The temperature fluctuation $T'(z)$ is equal to the local average temperature and there is no contribution to turbulent kinetic energy from temperature fluctuations.

$\mathrm{Ri} = {>}0$, $\overline{u_z'\theta'} < 0$: Positive values of u_z' occur with negative values of θ' and turbulent fluctuations are damped by buoyancy.

$\mathrm{Ri} < 0$, $\overline{u_z'\theta'} > 0$: Positive values of u_z' occur with positive values of θ' and turbulent fluctuations are magnified by buoyancy.

In the field of architectural engineering there is a growing interest in using CFD to predict the temperature, humidity, and contaminant concentrations in buildings prior to their construction. A sampling of CFD research readers should be familiar with include Nielsen (370) in Denmark, Murakami et al (362) and Kurabuchi et al (364) in Japan, Kusuda (364), Axley (365), Ferizger (366), Schiller (369), Albright et al (367), amd Baker and Kelso (368) in the United States.

Improving the ability to predict concentrations is the single most important need facing engineers in industrial ventilation. The need to control contaminants remains the objective of their work, but the key to accomplishing it quickly, economically, and precisely, lies in their ability to predict. How else will they know if a process produces unhealthy conditions and warrants a control device? How else will they know whether a design they propose satisfies design specifications? How else can designers persuade their superiors that their designs will reduce concentrations to tolerable levels? Physical measurements have to be made, but modern industry cannot always afford the time and money to perform as many as they would like and many decisions will continue to be made on the basis of predictions. The capacity to predict is the engineers's stock in trade. Advances in other fields of technology attest to the economic value of prediction. Technological accomplishments improve, as does profitability, when this mentality flourishes.

There is nothing sacrosanct about the numerical methods contained in this book. The methods are only a means to an end, and when better means come along inferior ones should give way to superior ones. Superior means are easier to understand and apply, more accurate, cheaper, and faster. The rate at which improved software appears suggests that superior numerical methods will appear on a regular basis and engineers must be ready to replace old methods with better ones.

NOMENCLATURE

Symbol	Description	Dimensions*
a_P	Parameter evaluated at point P (subscripts NESW refer to values of the constant evaluated at adjacent nodes)	
a_e	Constant evaluated along eastern face of nodal volume (subscripts nesw refer to values evaluated along adjacent faces)	
a_E	Parameter defined by equation for eastern node (subscripts NESW refer to values of adjacent nodes)	
A_e	Area of eastern face of nodal volume (subscripts nesw refer to areas of adjacent faces)	L^2
$A(\|Pe\|)$	Defined function of the eastern face Peclet number evaluated on the eastern face of the nodal volume	
A	Cross sectional area	L^2
b	Constant uniquely defined in each equation	
B	Constant defined by equation	
c	Concentration	M/L^3
c_i	Concentration of molecular species "i"	M/L^3
c_p	Specific heat at constant pressure	Q/MT
C_μ	Constant	
C_ε	Constant	
C_1, C_2	Constants	
d_e	Constant A_e/a_e	ML^3t
$\mathscr{D}$	Diffusion coefficient	L^2/t
D	Tunnel diameter	
D_e	Conductance evaluated along east face of the nodal volume (subscripts nesw refer to values of the conductance evaluated on nesw faces of the nodal volume)	
$\mathscr{D}_{eff}$	Effective diffusion coefficient	L^2/t
e_c	Vehicle emission rate	M/L
E	Constant defined by equation, roughness parameter	
F_e	Mass flow rate through eastern face of nodal volume	M/t
g	Acceleration of gravity	L/t^2
G	Function defined by equation	
G_B	Function defined by equation	
G_t	Eddy transport property	
G_e	Generalized transport property along eastern face of nodal volume	

*Q – energy, F – force, L – length, M – mass, N – mols, t – time, T – temperature

G_ϕ	Generalized transport property for physical property ϕ	
h	Enthalpy	Q/M
J_e	Flux of a quantity crossing the east face of a nodal region (subscripts nesw refer to fluxes crossing the nesw faces)	
k	Thermal conductivity	Q/tLT
k	Parameter defined by equation	
k_d	Wall loss coefficient	L/t
L	Characteristic length	L
L_m	Mean free path	L
ℓ	Turbulent mixing length	L
M_i	Molecular weight of species i	M/N
m_w''	Mass flux at wall	$M/L^2 t$
n_c	Traffic density	L^{-1}
$P(t), P, p'$	Instantaneous, average and fluctuating pressure	F/L^2
P_P	Pressure at node P (subscripts NESW refer to values of the pressure evaluated at adjacent nodes)	F/L^2
P^*	Estimated pressure used in SIMPLER algorithm	F/L^2
P_E'	Corrected value of the pressure at the east node	F/L^2
Q_m, Q_e	Makeup and exhaust volumetric flow rate	L^3/t
q_m, q_e	Q_m/LA, Q_e/LA	t^{-1}
q_w''	Heat flux at wall	$Q/L^2 t$
R_i	Gas constant for molecular species i	Q/MT
R_u	Universal gas constant	Q/NT
s	S/LA	
S	Surface area of control volume	L^2
S	Source term defined by equation	
S_T	Energy source due to a chemical reaction	$Q/L^2 t$
S_ϕ	Source term for physical property ϕ, S_ϕ uniquely defined in each equation	
S_C, S_P	Constant and variable portion of the source term	
S_κ, S_B	Source terms for $\kappa\text{-}\epsilon$ equations	
t_0	Characteristic time	t
$T(t), T, T'$	Instantaneous, average, and fluctuating temperature	T
T_P, T_W	Temperature at point P and at wall	T
T_0	Far-field temperature, a reference temperature	T
U_0	Far field velocity	L/t
U	Velocity, $U = U_x$	L/t
U_e	x component of the velocity on the east face of the nodal region (subscripts nesw refer to values on the nesw faces of the nodal volume)	L/t
U_e^*	Approximate value of the x component of the velocity on the east face of the nodal volume	L/t
U_e'	Pressure corrections on the east face of the nodal region	F/L^2

$U(t)_i, U_i, u'_i$	Instantaneous, average and fluctuating velocity for i-th vector component	L/t
U^+	Nondimensional velocity, defined by the equation	
U_t	Friction velocity defined by equation	L/t
U_m	Mean molecular speed	L/t
$\hat{U}_e$	Pseudovelocity, defined by equation	L/t
V	Volume of control volume	L^3
V	Velocity, $V = U_y$	L/t
v_c	Velocity of a car	L/t
Y_i	Mass fraction of molecular species i	
Y_P, Y_w	Mass fraction at point P and at the wall	
y_P	Distance between wall and point P	L
y^+	Nondimensional distance from wall	

Greek	Description	Dimension
β	Thermal expansion coefficient	T^{-1}
Δx	Width of the nodal volume in the x direction	L
δx_e	Distance between nodes P and E	L
δ_{ij}	Kronecker delta	
ϵ	Turbulent energy dissipation	FL/tM
θ	$(T - T_0)$	T
κ	Turbulent kinetic energy	L^2/t^2
μ_{eff}	Effective viscosity in a turbulent flow field	Ft/L^2
μ_t	Eddy viscosity	Ft/L^2
ρ	Density	M/L^3
ν	Kinematic viscosity	FtL/M
ν_t	Turbulent kinematic viscosity	FtL/M
$\rho\phi''$	Fluctuating component in Favre density-weighted average	M/L^3
σ_k	Empirical constant	
σ_h	Empirical constant	
σ_c	Empirical constant, eddy Schmidt number	
σ_ε	Empirical constant	
σ_t	Empirical constant, eddy Prandtl number	
ϕ	Generalized property	
τ_w	Wall shear stress	F/L^2
X	Von Karman constant (0.4)	

Subscripts	Description
$()_{eff}$	Effective property in a turbulent flow field
$()_i$	i-th vector component
$()_j$	j refers to running index
$()_N$	Value of the property at the north node, similarly for nodes NESW
$()_n$	Value of the property evaluated along the northern face of the nodal volume, similarly for faces nesw

$()_P$	Value of the property at point P
$()_i$	Molecular species i
$()_t$	Eddy property
$()_w$	Value of the property at the wall
$()_0$	Far-field value
$()_\infty$	Far-field value

Superscripts	Description
$()'$	Corrected value of a property in SIMPLER algorithm
$()'$	Fluctuating property in turbulent flow
$()^0$	Values of a variable at time t
$()^*$	Estimated value of a property in SIMPLER algorithm
$\overline{()}$	Time-averaged mean quantity
$\widetilde{()}$	Favre density-weighted average

Abbreviations	Description
$\dfrac{D[\ ]}{Dt} = \dfrac{\partial[\ ]}{\partial t} + U_i \dfrac{\partial[\ ]}{\partial x_i}$	Total derivative
CFD	Computational fluid dynamics
CFM	Cubic feet per minute
DIV	Divergence (vector operator)
FB, FE	Distance between the tank top and the opening for lateral exhausters
Grad	Gradient (vector operator)
H	Height of control volume
LL	Distance between tank top and liquid surface
Pe	Peclect number
Pr	Prandtl number
QB, QE	Blowing and exhaust volumetric flow rates
PPM	Parts per million
Re	Reynolds number
Ri	Richardson number
SIMPLE, SIMPLER	Algorithm name
Sc	Schmidt number
SB,SE	Width of openings for lateral exhauster
UB	Velocity of air leaving blowing port
W	Width of vessel
$W1$	Distance to the left of the blowing jet in a push-pull system

PROBLEMS

1. Derive Eq. 10-1.

2. Consider the transverse ventilated tunnel described in Illustrated Example 10.1. If $q_m = 0.18$ min^{-1} and $q_e = 0.2$ min^{-1}, find the location of the stagnation point and show that the concentration approaches infinity at this point.

3. Predict the concentration of carbon monoxide (CO) in the tunnels described in Illustrated Example 10.1. Assume that the tunnel parameters are the same but that the emission factor for CO is 12.5 g/vehicle-km and that the auto speed is 40 km/hr. Find the maximum CO concentration and the CO concentration at 1000 m for the four tunnels.

4. Using a commercial CFD program (such as FLUENT), analyze the trajectories of small particles passing through a two dimensional slot-impactor discussed in Section 8.5. If the slot width (2w) is 2 mm, the impactor plates are separated (s) by 3 mm and the plates are 4 cm wide (in the direction of flow), plot the impaction efficiency versus the Stokes number where

$$\text{Stk} = \left(\frac{\rho_p D_p^2}{18\,\mu}\right)\left(\frac{U_0}{s}\right)$$

and U_0 is the slot exit velocity of air and particles. Compare these results with those obtained in Chapter 8 where the flow was assumed to be an ideal fluid. Discuss why assuming the fluid to be ideal produces no recirculation zones while the solution of the Navier Stokes equations (by FLUENT) predicts a recirculation zone near the nozzle exit. Comment on the significance of the second (but smaller) recirculation zone lying along the bottom surface near the exit of the impactor. Compare the results obtained by using FLUENT, with the analytical and experimental results obtained by Marple et al (198,199)

References

1. Formica, P. N., Controlled and Uncontrolled Emission Rates and Applicable Limitations for Eighty Processes, EPA-340/1-78-004, April 1978.

2. Bond, R.G. and Straub, C. P. (Ed.), *Handbook of Environmental Control*, Chemical Rubber Company, Cleveland, OH, 1972.

3. Danielson, J.A., *Air Pollution Engineering Manual*, 2nd ed. EPA, Office of Air and Water Programs, Office of Air Quality Planning and Standards, Research Triangle Park, NC, May 1973.

4. Hoogheem, T. J., Horn, D. A., Hughes, T. W., and Marn, P. J. Source Assessment: Solvent Evaporation–Degreasing Operations, EPA-600/2-79-019f, August 1979.

5. ———, *Volatile Organic Compound (VOC) Species Manual*, 2nd ed., EPA 450/4-80-015, July 1980.

6. ———, *Compilation of Air Pollutant Emission Factors (Including Supplements 1–7)*, 3rd ed., Supplement 9, NTIS PB81-244097, July 1979.

7. ———, *Compilation of Air Pollutant Emission Factors (Including Supplements 1–7)*, 3rd ed., Supplement 12, NTIS, PB82-101213, April 1981.

8. ———, *Compilation of Air Pollutant Emission Factors (Including Supplements 1–7)*, 3rd ed., Supplement 10, NTIS PB80-199045, February 1980.

9. ———, *Compilation of Air Pollutant Emission Factors (Including Supplements 1–7)*, 3rd ed., Supplement 11, NTIS, PB81-178014, October 1980.

10. ———, *Compilation of Air Pollutant Emission Factors*, EPA Air Programs Publications, No. AP-42, February 1972.

11. Hahn, R. S. and Lindsay, R. P., Principles of Grinding, *Machinery*, New York, Part I, July 1971, pp. 55–62; Part II, August 1971; pp. 33–39, Part III, September 1971, pp. 33–39, Part IV, October 1971; pp. 57–67; Part V, November 1971, pp. 48–53.

12. Heinsohn, R. J., Johnson, D., and Davis, J. W., Grinding Booth for Large Castings, *Am. Ind. Hygiene Assoc. J.* **43**(8), 587–595 (1982).

13. Bastress, E. K., Niedzwecki, J. M., and Nugent, A. E., Ventilation Requirements for Grinding, Buffing and Polishing Operations, NTIS, PB-277332, 1974.

14. Shen, T. T., Estimation of Organic Compound Emissions from Waste Lagoons, *J. Air Pollution Control Assoc.*, **32**(1), 79–82, Jan. 1982.

15. Mackay, D., Bobra, A., Chan, D. W., and Shiu, W. Y., Vapor Pressure Correlations For Low-Volatility Environmental Chemicals, *Environ. Sci. Tech.* **16**(10), 645–649 (1982).

16. Vargaftik, N. B., *Tables on the Thermophysical Properties of Liquids and Gases*, 2nd ed., Hemisphere Publishing Corp., Washington, DC, 1975.

17. Sherwood, T. K., Pigford, R. L., and Wilke, C. R., *Mass Transfer*, McGraw Hill, New York, 1975.

18. Perry, R. H. and Chilton, C. H., *Chemical Engineers' Handbook*, McGraw Hill, New York, 1973.

19. Crawford, M., *Air Pollution Control Theory*, McGraw Hill, New York, 1976.

20. McCabe, W. L. and Smith, J. L., *Unit Operations of Chemical Engineering*, McGraw Hill, New York, 1967.

21. Treybal, R. E., *Mass-Transfer Operations*, McGraw Hill, New York, 1968.

22. Hirschfleder, J. O., Curtiss, C. F., and Bird, R. B., *Molecular Theory of Gases and Liquids*, Wiley, New York, 1954.

23. Bird, R. B., Stewart, W. E., and Lightfoot, E. N., *Transport Phenomena*, Wiley, New York, 1960.

24. Repace, J. L. and Lowery A. H., Indoor Air Pollution, Tobacco Smoke, and Public Health, *Science*, **208,** 464–472, May 1980.

25. Ishizu Y., General Equation for the Estimation of Indoor Pollution, *Environ. Sci. Technol.* **14**(10), 1254–1257, Oct. 1980.

26. Esmen, N. A., Characterization of Contaminant Concentrations in Enclosed Spaces, *Environ. Sci. Technol.* **12**(3), 337–339, March 1978.

27. Committee on Industrial Ventilation, American Conference of Governmental Industrial Hygienists, *Industrial Ventilation, A Manual of Recommended Practice*, 20th ed., Committee on Industrial Ventilation, PO Box 16153, Lansing MI, 48901, 1988.

28. American Society of Heating, Refrigerating and Air-Conditioning Engineers, *ASHRAE Handbook & Products Directory*, *1984 Systems*, Chapter 22, Local Exhaust Systems, Atlanta, GA, 1984, pp 22.1–22.14.

29. Sterling, T. P., Dimich, H., and Kobaryashi, D., Indoor Byproduct Levels of Tobacco Smoke: A Critical Review of the Literature, *J. Air Pollut. Control Assoc.*, **32**(3) 250–259, March 1982.

30. Leaderer, B. P., Air Pollutant Emissions From Kerosene Space Heaters, *Science*, **218**, 1113–1115, 10 Dec. 1982.

31. Dixon, J. R., *Design Engineering: Inventiveness, Analysis, and Decision Making*, McGraw Hill, New York, 1966.

32. Woodson, T. T., *Introduction To Engineering Design*, McGraw Hill, New York, 1966.

33. Office of the Federal Register, Code of Federal Regulations, Title 29, Parts 1900 to 1910, (commonly called "OSHA General Industry Standards"), revised as of 1 July 1988.

34. Milne-Thomson, L. M., *Theoretical Hydrodynamics*, MacMillan Co., New York, 1950.

35. Streeter, V. L., *Fluid Dynamics*, McGraw Hill, New York, 1948.

36. US Department of Health, Education, and Welfare, Public Health Service, Center for Disease Control, National Institute of Occupational Safety and Health, The Industrial Environment—It's Evaluation and Control, Superintendent of Documents, US Government Printing Office, Washington DC, 20402, 1973.

37. Molburg, J. C., and Rubin, E. S., Air Pollution Costs for Coal–To–Electricity Systems, *J. Air Pollut. Control. Assoc.*, **33**(5), 523–530 (1983).

38. Grant, E. L. and Ireson, W. G., *Principles of Engineering Economy*, 5th ed., Ronald Press, New York, 1970.

39. Neveril, R. B., Price, J. U., and Engdahl, K. L., Capital and Operating Costs of Selected Air Pollution Control Systems, *J. Air Pollut. Control Assoc.*, Part I, **28**(8), 829–836, Aug. 1978; Part II, **28**(9) 963–968, Sept. 1978; Part III, **28**(10), 1069–1072, Oct. 1978; Part IV, **28**(11), 1171–1172, Nov. 1978; Part V, **28**(12), 1253–1256, Dec. 1978.

40. Hidy, G. M. and Brock, J. R., *The Dynamics of Aerocolloidal Systems*, Pergamon Press, New York, 1970.

41. Fuchs, N. A., *The Mechanics of Aerosols*, Pergamon Press, New York, 1964.

42. Hinds, W. C., *Aerosol Technology*, Wiley-Interscience Publications, New York, 1982.

43. Davies, C. N. (Ed.), *Aerosol Science*, Academic Press, New York, 1966.

44. Strauss, W., *Industrial Gas Cleaning*, Pergamon Press, New York, 1966.

45. Licht, W., *Air Pollution Control Engineering*, Marcel Dekker, New York, 1980.

46. Leaderer, B. P., Cain, W. S., Isseroff, R., and Berglund, L. G., Ventilation Requirements in Buildings—II Particulate Matter and Carbon Monoxide From Cigarette Smoking, *Atmos. Environ.* **18**(1), 99–106 (1984).

47. Hemeon, W. C. L., *Plant and Process Ventilation*, Industrial Press, New York, 1963.

48. Baturin, V. V., *Fundamentals of Industrial Ventilation*, Pergamon Press, New York, 1972.

49. McDermott, H. J., *Handbook of Ventilation for Contaminant Control*, Ann Arbor Science Publishers Inc., Ann Arbor, MI, 1976.

50. Constance, J. D., *Controlling In-Plant Airborne Contaminants*, Marcel Dekker, Inc., New York, 1983.

51. American Conference of Governmental Industrial Hygienists, *Air Sampling Instruments*, 4th ed., PO Box 1937, Cincinnati, OH, 45201, 1972.

52 Cain, W. S., Leaderer, B. P., Isserhoff, R., Berglund, L. G., Huey, R. J., Lipsitt, E. D., and Perlman D., Ventilation Requirements in Buildings—I. Control of Occupancy Odor and Tobacco Smoke Odor, *Atmos. Environ*, **17**(6), 1183–1197 (1983).

53. Goodfellow H. D., and Smith J., Industrial Ventilation—A Review and Update, *Am. Ind. Hyg. Assoc. J.* **43**, 175–184, March 1982.

54. Billings, C. E., and Vanderslice, S. F., Methods for Control of Indoor Air Quality, *Environ. Int.* **8**, 497–504, 1982.

55. Patankar, S. V., *Numerical Heat Transfer and Fluid Flow*, Hemisphere Publishing Corp., New York, 1981.

56. Pipes, L. A., and Harvill, L. R., *Applied Mathematics for Engineers and Physicists*, 3rd ed., McGraw Hill, New York, 1970.

57. Gosman, A. D., Pun, W. M., Runchal, A. K., Spalding, D. B., and Wolfshtein, M., *Heat and Mass Transfer in Recirculating Flows*, Academic Press, New York, 1969.

58. Vitols, V, Theoretical Limits of Errors Due to Anisokinetic Sampling of Particulate Matter, *J. Air Pollut. Control Assoc.* **16**(2), 79–83 (1966).

59. Kunkel, B. A., A Comparison of Evaporative Source Strength Models for Toxic Chemical Spills, Report AFGL-TR-0307, Air Force Geophysics Laboratory, Air Force Systems Command, USAF, 50 pages, 16 Nov. 1983.

60. Cadle, R. D., *Particle Size*, Reinhold, New York, 1965.

61. Schmidt, F. W., Henderson, R. E., and Wolgemuth, C. H., *Introduction to Thermal Sciences*, John Wiley and Sons, New York, 1984.

62. Kirchhoff, R. H., *Potential Flows*, Marcel Dekker, Inc., New York, 1985.

63. Hayashi, T., Howell, R. H., Shibata, M., and Tsuji, K., *Industrial Ventilation and Air Conditioning*, CRC Press, Boca Raton, FL, 1985.

64. Huebener, D. J., and Hughes, R. T., Development of Push–Pull Ventilation, *Am. Ind. Hyg. Assoc. J.* **46**(5), 262–267 (1985).

65. Cooper, H. B. H., Jr. and Rassano, A. T., Jr., *Source Testing for Air Pollution Control*, McGraw Hill, New York, 1971.

66. Bennett, A. K., Heinsohn, R. J., and Merkle, C. L., Predicting Trajectories and

Concentrations of Particles in a Grinding Booth, ASME Paper 8-FE-23, Presented Joint ASME/ASCE Bioengineering, Fluids Engineering and Applied Mechanics Conference, June 22–24, 1981, Boulder, CO.

67. Zarouri, M. D., Heinsohn, R. J., and Merkle, C. L., Numerical Computation of Trajectories and Concentrations of Particles in a Grinding Booth, Paper No. 2768, *ASHRAE Trans.*, Part 2A, 1983, pp. 119–135.

68. Zarouri, M. D., Heinsohn, R. J., and Merkle, C. L., Computer-Aided Design of a Grinding Booth for Large Castings, Paper No. 2767, ASHRAE Transactions, Part 2A, pp. 95–118, 1983.

69. Bastress, K. E., Niedzwecki, J. M., and Nugent, A. E., Jr., Ventilation Requirements for Grinding, Buffing and Polishing Operations, IKOR Inc., Contract No. HSM 99-72-126, HEW Publication No. (NIOSH) 75-107.

70. Zarouri, M. D., Computer-Aided Design of Grinding Booths, MS Thesis, Department of Mechanical Engineering, The Pennsylvania State University, University Park, PA 1981.

71. Jensen, M., *Shelter Effect Investigation into the Aerodynamic of Shelter and its Effect on Climate and Crop*, The Danish Press, Copenhagen, 1954.

72. Segman, I., and Sagi, R., Drag On A Windbreak in Two-Dimensional Flow, *Agric. Meteorol.*, **9**, 323–333 (1971–1972).

73. Hagen, L. J., and Skidmore, E. L., Turbulent Velocity Fluctuations and Vertical Flow as Affected by Windbreak Porosity, *Trans. Soc. Agric. Eng.*, **14**(4), 634–637 (1971).

74. Hagen, L. J., Skidmore, E. L., Miller, P. L., and Kipp, J. E., Simulation of Effect of Wind Barriers on Airflow, *Trans. Soc. Agric. Eng.*, **24**(4), 1002–1008 (1981).

75. Heinsohn, R. J., and Megherhi, M. H., Control of Fugitive Dust by Windbreaks, *Particulate Sci. Tech.*, **4**, 87–112, (1986).

76. Hagapian, J. H., and Bastress, K. E., *Recommended Industrial Ventilation Guidelines*, A. D. Little Co., Cambridge, MA, NIOSH Technical Report, Contract No. CDC-99-74-33, January 1976.

77. Heinsohn, R. J., Hsieh, K. C., and Merkle, C. L., Lateral Ventilation Systems for Open Vessels, Paper CH-85-08 No. 1, ASHRAE Transactions, Part 1B, pp 361–382, 1985.

78. McCall, R. D., Computer-Aided Design of Open Vessel Ventilation Hoods, MS Thesis, Department of Mechanical Engineering, The Pennsylvania State University, University Park, PA, 1982.

79. Hsieh, K. C., Computer-Aided Design of Open Vessels, MS Thesis, Department of Mechanical Engineering, The Pennsylvania State University, University Park, PA, 1983.

80. Jones, W. P., and Launder, B. E., The Production of Laminarisation with 2-Equation Model of Turbulence, *Int. J. Heat Mass Transfer*, Vol. 15, p. 301, 1972.

81. Tennekes, H., and Lumley, J. L., *A First Course in Turbulence*, The MIT Press, Cambridge, MA, 1972.

82. Heinsohn, R. J., and Merkle, C. L., Control of Contaminants from Open Vessels, Final Report—Grant R18-OH01296-02, CAES Report 741-85 Center for Air Environment Studies, The Pennsylvania State University, University Park, PA, March 1985.

83. Battista, W. P., Semi-Lateral Tank Ventilation Hood Controls Contamination, Cuts Costs, *Heat. Piping Air Cond.*, **19**(1), 85–89 (1947).

84. Ege, J. E., and Silverman, L. S., Design of Push–Pull Exhaust System, *Heating Ventilating*, **10**, 73–78 (1950).

85. Hama, G. M., Supply and Exhaust Ventilation for Metal Acid Pickling Tanks, *Air Cond. Heat Vent.* 61–63 (1957).

86. Hama, G. M., Push–Pull Hoods, Duct Design and Lab Hood Ventilation, *Air Cond. Heat Vent.*, 51, 1964.

87. Hama, G. M., Jet Stream Replaces Vertical Duct in Push–Pull Ventilation of Pickling Tank, *Heat. Piping Air Cond.*, **45**, 121–123 (1973).

88. Parker, J. H., Cost Advantage of Push–Pull Ventilation, *Air Eng.*, **9**, 26–27 (1967).

89. Institute of Heat and Ventilating Engineering, *I.H.V.E. Guide Book B* 1970, B3–8, 1972.

90. Van, N. Q., and Howell, R. H., Influence of Initial Turbulent Intensity on the Development of Plane Air Curtain Jets, *ASHRAE Trans.*, **85**, Part 1, 208–228 (1976).

91. Shibata, M., Howell, R. H., and Hayashi, T., Characteristics and design Method for Push–Pull Hoods: Part 1—Cooperation Theory of Airflow, *ASHRAE Trans.*, Part 1, **88**, 535 (1982).

92. Shibata, M., Howell, R. H., and Hayashi, T., Characteristics and Design Method for Push–Pull Hoods: Part 2—Streamline Analyses of Push–Pull Flow, *ASHRAE Trans.*, Part 1, **88**, 557 (1982).

93. Launder, B. E., and Spalding, D. B., The Numerical Computation of Turbulence flows, *Comput. Methods Appl. Mech. Eng.*, **3**, 269–289 (1974).

94. American Conference of Governmental and Industrial Hygienists, Threshold Limit Values for Chemical Substances and Biological Exposure Indices from 1989–1990, 6550 Glenway Ave, Bldg. D-7, Cincinnati, Ohio 45211–9438

95. National Institute for Occupational Safety and Health, Occupational Diseases, A Guide to Their Recognition, NIOSH Pub. No. 77–181, Superintendent of Documents, US Government Printing Office, Revised June 1977.

96. Zimmerman, N. J., Principles of Occupational Safety and Health Engineering, Instructors Guide published by NIOSH, PO No. 81-3030, March 1984.

97. American Society of Heating, Refrigerating and Air-Conditioning Engineers, *ASHRAE Handbook 1981 Fundamentals Volume*, Chapter 22, Atlanta, GA, 1981, pp. 22.1–22.20.

98. Wadden, R. A., and Scheff, P. A., *Indoor Air Pollution*, Wiley-Interscience Publication, John Wiley & Sons, New York, 1983.

99. Skaret, E., Ventilation By Displacement—Characterization and Design Implications, *Ventilation '85*, H. D. Goodfellow (Ed.), Elsevier Science Publishers, Amsterdam, pp. 827–842, 1986.

100. Skaret, E. and Mathisen, H. M., Ventilation Efficiency–A Guide to Efficient Ventilation, *Trans. ASHRAE*, Part 2B, 480–495 (1983).

101. Niemela, R., Toppila, E., and Rolin, I., Characterization of Supply Air Distribution in Large Industrial Premises by the Tracer Gas Technique, *Ventilation '85*, H. D. Goodfellow (Ed.), Elsevier Science Publishers, Amsterdam, pp. 797–805, 1986.

102. Perera, M. D. A. E. S., Walker, R. R., and Oglesby, O. D., Infiltration Measurements in Naturally Ventilated, Large Multicelled Buildings, *Ventilation '85*, H. D. Goodfellow (Ed.), Elsevier Science Publishers, Amsterdam, pp. 807–827, 1986.

103. American Society of Heating, Refrigerating and Air-Conditioning Engineers, *ASHRAE Handbook & Products Directory, 1984 Systems*, Chapter 20, Local Exhaust Systems, Atlanta, GA, 1984.

104. Burton, D. J., *Industrial Ventilation—A Self Study Companion to the ACGIH Ventilation Manual*, 2nd ed., ACGIH, 650 Glenway Ave., Bldg D-5, Cincinnati, OH, 45211, 1984.

105. Hanna, S. R., Briggs, G. A., and Hosker, R. P., Jr., *Handbook on Atmospheric Diffusion*, DOE/TIC-11223 (DE82002045), US Dept. of Energy, NTIS, 1982.

106. Cooper, D. W., and Horowitz, M., Exposures From Indoor Powder Releases: Models and Experiments, *Am. Ind. Hyg. Assoc. J.*, **47**(4), 214–218 (April 1986).

107. Esmen, N. A., Weyel, D. A., and McGuigan, F. P., Aerodynamic Properties of Exhaust Hoods, *Am. Ind. Hyg. Assoc. J.*, **47**(8), 448–454 (1986).

108. Pickrell, J. A., Griffis, L. C., Mokler, B. V., Kanapilly, G. M., and Hobbs, C. H., Formaldehyde Release from Selected Consumer Products: Influence of Chamber Loading, Multiple Products, Relative Humidity and Temperature, *Environ. Sci. Tech.*, **18**, 682–686 (1984).

109. Silver, W., and Mackay, D., Evaporation Rate of Spills of Hydrocarbons and Petroleum Mixtures, *Environ. Sci. Technol.*, **18**, 834–840 (1984).

110. Drivas, P. J., Calculation of Evaporative Emissions from Multicomponent Liquid Spills, *Environ. Sci. Technol.* **16**, 726–728 (1982).

111. Roberts, P. V., and Dandilker, P. G., Mass Transfer of Volatile Organic Contaminants from Aqueous Solution to the Atmosphere During Surface Aeration, *Environ. Sci. Technol.* **17**(8), 484–489 (1983).

112. Cain, W. S., Leaderer, B. P., Isseroff, R., Berglund, L. G., Huey, R. J., Lipsitt, E. D., and Perlman, D., Ventilation Requirements in Buildings—I. Control of Occupancy Odor and Tobacco Smoke Odor, *Atmos. Environ.*, **17**(6), 1183–1197 (1983).

113. Ryan, P. B., Spengler, J. D., and Letz, R., The Effects of Kerosene Heaters on Indoor Pollutant Concentrations: A Monitoring and Modeling Study, *Atmos. Environ.*, **17**(7), 1339–1345 (1983).

114. Sandberg, M., The Use of Moments for Assessing Air Quality in Ventilated Rooms, American Society of Heating, Refrigerating and Air-Conditioning Eng., Building and Environment, **18**(4), 9–25 (1983).

115. Mackay, D., and Yeun, A. T. K., Mass Transfer Coefficient Correlations for Volatilization of Organic Solutes from Water, *Environ. Sci. Technol.*, **17**(4), 211–217 (1983).

116. Traynor, G. W., Girman, J. R., Apte, M. G., Dillworth, J. F., and White, P. D., Indoor Air Pollution due to Emissions from Unvented Gas-Fired Space Heaters, *Air Pollut. Control Assoc. J.*, **35**(3), 231–237.

117. Meyer, B., and Hermanns, K., Reducing Indoor Air Formaldehyde Concentrations, *Air Pollut. Control Assoc. J.*, **35**, 816–821 (1985).

118. Traynor, G. W., Allen, J. R., Apte, M. G., Girman, J. R., and Hollowell, C. D., Pollutant Emissions from Portable Kerosene-Fired Space Heaters, *Environ. Sci. Technol.*, **17**(6), 369–371 (1983).

119. Traynor, G. W., Anthon, D. W., and Hollowell, C. D., Technique for Determining Pollutant Emissions from Gas-Fired Range, *Atmos. Environ.*, **16**(12) 2979–2987 (1982).

120. Hall, R. E., and DeAngelis, D. G., EPA's Research Program for Controlling Residential Wood Combustion Emissions, *J. Air Pollut. Control Assoc.*, **30**(8), 862–867 (1980).

121. Flynn, M. R., and Ellenbecker, M. J., The Potential Flow Solution for Air Flow Into a Flanged Circular Hood, *Am. Ind. Hyg. Assoc. J.*, **46**(6), 318–322 (1985).

122. Esmen, N. A., and Weyel, D. A., Aerodynamics of Multiple Orifice Hoods, H. D. Goodfellow (Ed.), *Ventilation '85*, Elsevier Science Publishers, Amsterdam, pp. 735–741, 1986.

123. Armbruster, L., and Zebel, G., Theoretical and Experimental Studies for Determin-

ing the Aerosol Sampling Efficiency of Annular Slot Probes, *J. Aerosol Sci.*, **16**(4), 335–341 (1985).

124. Gallily, I., Schiby, D., Cohen, A. H., Hollander, W., Schless, D., and Stober, W., On the Inertial Separation of Nonspherical Aerosol Particles from Laminar Flows I. The Cylindrical Case, *Aerosol Sci. Technol.*, **5**, 267–286 (1986).

125. Fletcher, B., and Johnson, A. E., The Capture Efficiency of Local Exhaust Ventilation Hoods and the Role of Capture Velocity, *Ventilation '85*, H. D. Goodfellow (Ed.), Elsevier Science Publishers, Amsterdam, 1986, pp. 369–379.

126. Kranzberg, M. The Disunity of Science-Technology, *Am. Sci.*, **56**(1), 21–34 (1986).

127. Kranzberg, M. The Unity of Science-Technology, *Am. Sci.*, **55**(1), 60 (1967).

128. Masood, T., The Capture of Dust by Hoods; A New Design Method, *Ventilation '85*, H. D. Goodfellow (Ed.), Elsevier Science Publishers, Amsterdam, 1986, pp. 703–720.

129. Ellenbecker, M. J., Gempel, R. F., and Burgess, W. A., Capture Efficiency of Local Exhaust Ventilation Systems, *Am. Ind. Hyg. J.*, **44**(10) 752–755 (1983).

130. Heinsohn, R. J., and Choi, M. S., Advanced Design Methods in Industrial Ventilation, *Ventilation '85*, H. D. Goodfellow (Ed.), Elsevier Science Publishers, Amsterdam, 1986, pp. 81–109.

131. Braconnier, R., Thiebaut, D., Aubertin, G., Serieys, J. C., and Muller, J. P., Double-Flow Devices for the Capture of Contaminants, *Ventilation '85* H. D. Goodfellow, (Ed.), Elsevier Science Publishers, Amsterdam, 1986, pp. 743–753.

132. Hooper, F. C., and Hooper, J. T., The Relationship between Observed Pollutant Concentrations and Building Ventilation System Design, *Ventilation '85*, H. D. Goodfellow, (Ed.), Elsevier Science Publishers, Amsterdam, 1986, pp. 309–322.

133. Siurna, D. L., and Bragg, G. M., Stochastic Modeling of Room Air Diffusion, *Ventilation '85*, H. D. Goodfellow, (Ed.), Elsevier Science Publishers, Amsterdam, 1986, pp. 121–135.

134. Nielsen, P. V., Restivo, A., and Whitelaw, J. H., The Velocity Characteristics of Ventilated Rooms, *J. Fluids Eng.*, *Trans. ASME*, **100**, 291–298, Sept. 1978.

135. Dellagi, F., Dumaine, J. Y., and Aubertin, G., Numerical Simulation of Air Flows—Application to the Ventilation of a Paint-Booth, *Ventilation '85* H. D. Goodfellow, (Ed.), Elsevier Science Publishers, Amsterdam, 1986, pp. 391–403.

136. Cornu, J. C., Leleu, J., Gerber, J. M., Vincent R., Voirin, D., and Aubertin, G., Determination of Ventilation Criteria for Closed Paint Booths—Validation on 14 Boots in Operation, *Ventilation '85* H. D. Goodfellow, (Ed.), Elsevier Science Publishers, Amsterdam, 1986, pp. 193–204.

137. Heinsohn, R. J., Yu, S. T., Merkle, C. L., Settles, G. S., and Huitema, B. C., Viscous Turbulent Flow in Push–Pull Ventilation Systems, *Ventilation '85* H. D. Goodfellow, (Ed.), Elsevier Science Publishers, Amsterdam, 1986, pp. 529–566.

138. Hampl, V., and Hughes, R. T., Improved Local Exhaust Control by Directed Push–Pull Ventilation System, *Am. Ind. Hyg. Assoc. J.*, **47**(1) 59–65 (1986).

139. Hughes, R. T., Design Criteria for Plating Tank Push–Pull Ventilation, *Ventilation '85* H. D. Goodfellow, (Ed.), Elsevier Science Publishers, Amsterdam, 1986, pp. 521–528.

140. First, M., Billings, C. E., Harris, R. L., Peterson, J. E., and Leamon, T. B., Engineering Control of Occupational Safety and Health Hazards: Recommendations for Improving Engineering Practice, Education and Research, prepared for US Department of Health and Human Services, Centers for Disease Control, NIOSH, 1983.

141. Association of Schools of Public Health, Proposed National Strategies for the Prevention of Leading Work-Related Diseases and Injuries, Part 1, National Institute for Occupational Safety and Health, 1986.

142. Calvert, S., and Englund, H. M., *Handbook of Air Pollution Technology*, John Wiley and Sons, NY, 1984.

143. Alden, J. L., and Kane, J. M., *Design of Industrial Ventilation Systems*, 5th ed., Industrial Press, Inc., NY, 1982.

144. Starr, C., Social Benefit Versus Technological Risk, *Science*, **165**, 1232–1238, 19 Sept. 1969.

145. Envirex Company, An Evaluation of Occupational Health Hazard Control Technology for the Foundry Industry, NIOSH Publication 79–114, 1978.

146. Danielson, J. A., *Air Pollution Engineering Manual*, EPA Publication AP-40, 1973.

147. Cralley, L. J., and Cralley, L. V., (Ed.), *Patty's Industrial Hygiene And Toxicology*, Volume I—*General Principles*, Volume II—*Toxicology*, and Volume III—*Theory and Rationale of Industrial Hygiene Practice*, John Wiley and Sons, New York, 1979.

148. Guyton, A. C., *Textbook of Medical Physiology*, 7th ed., W. B. Saunders Co., Philadelphia, PA, 1986.

149. *Casarett and Doull's Toxicology*, 3rd ed., Klaassen, C. D., Amdur, M. O., and Doull, J. (Eds.), Macmillan Publishing Company, New York, 1986.

150. Murray, J. F., *The Normal Lung*, 2nd ed., W. B. Saunders Company, Philadelphia, PA, 1986.

151. Chang, H. K., *Mechanisms of Gas Transport During Ventilation by High-Frequency Oscillation*, Journal of Applied Physiology, **56**(3), pp. 553–563, 1984.

152. Haselton, F. R., and Scherer, P. W., Bronchial Bifurcations and Respiratory Mass Transport, *Science*, **208**, 69–71 (1980).

153. Hanna, L. M., and Scherer, P. W., A Theoretical Model of Localized Heat and Water Vapor Transport in the Human Respiratory Tract, *Trans. ASME, J. Biomech. Eng.*, **108**, 19–27 (1986).

154. Hanna, L. M., and Scherer, P. W., Regional Control of Local Airway Heat and Water Vapor Losses, *Journal of Applied Physiology*, **61**(2), 624–632 (1986).

155. Hanna, L. M., and Scherer, P. W., Measurement of Local Mass Transfer Coefficients in a Cast Model of the Human Upper Respiratory Tract, *Trans. ASME, J. Biomech. Eng.*, **108**, 12–18 (1986).

156. Scherer, P. W., Neufeld, G. R., Aukburg, S. J., and Hess, G. D., Measurement of Effective Peripheral Bronchial Cross Section from Single-Breath Washout, *Trans. ASME*, **105**, 290–293 (1983).

157. Colucci, A. V., and Strieter, R. P., Dose Considerations in the Sulfur Dioxide Exposed Exercising Asthmatic, *Environ. Health Perspect.*, **52**, 221–232 (1983).

158. Menkes, H., Cohen, B., Permutt, S., Beatty, T., and Shelhamer, J., Characterization and Interpretation of Forced Expiration, *Ann. Biomed. Eng.*, **9**, 501–511 (1981).

159. Permutt, S., and Menkes, H. A., Spirometry, Analysis of Forced Expiration within the Time Domain, contained in *The Lung in the Transition Between Health and Disease*, P. T. Macklen and S. Permutt, (Eds.), Marcel Dekker Inc., New York, pp. 113–152, 1979.

160. US Department of Health and Human Services, Public Health Service, Centers for Disease Control, National Institute for Occupational Safety and Health, *NIOSH Pocket Guide to Chemical Hazards*, US Government Printing Office, 1985.

161. Nichols, A., and Zeckhauser, R., *The Dangers of Caution: Conservatism in Assess-*

ment and the Mismanagement of Risk, Harvard University, November 1985, #E-85-11.

162. Brain, J. D., and Beck, B. D., Bioassays for Mineral Dusts and Other Particulates, *In Vitro Effects of Mineral Dusts*, E. G. Beck and J. Bignon (Eds.), NATO ASI Series Vol. G3, Springer Verlag, Berlin, pp. 323–335, 1985.

163. Rowe, W. B., *An Anatomy of Risk*, John Wiley and Sons, New York, 1977.

164. National Institute for Occupational Safety and health, *Registry of Toxic Effects of Chemical Substances*, Vols. I, II, and III, Tatken, R. L. and Lewis, R. J. (Eds.), 1981–1982.

165. Steering Committee on Identification of Toxic and Potentially Toxic Chemicals for Consideration by the National Toxicology Program, Board on Toxicology and Environmental Health Hazards, Commission on Life Sciences, National Research Council, *Toxicity Testing, Strategies to Determine Needs and Priorities*, National Academy Press, Washington DC, 1984.

166. Office of Technology Assessment, *Preventing Illness and Injury in the Workplace*, 1985.

167. Goodfellow, H. D., *Advanced Design of Ventilation Systems for Contaminant Control*, Elsevier Science Publishers, Amsterdam, 1986.

168. Hatch, T. F., and Gross, P., *Pulmonary Deposition and Retention of Inhaled Aerosols*, Academic Press, NY, 1964.

169. Perera, F. P., and Ahmed, A. K., *Respirable Particles*, Ballinger Publishing Company, Cambridge, MA, 1979.

170. Rothenberg, S. J., and Swift, D. L., Aerosol Deposition in the Human Lung at Variable Tidal Volumes: Calclation of Fractional Deposition, *Aerosol Sci. Technol.*, **3**, 215–226 (1984).

171. Kasper, G, Nida, T, and Yang, M., Measurements of Viscous Drag on Cylinders and Chains of Spheres with Aspect Ratios Between 2 and 50, *J. Aerosol Sci.*, **16**(6) 535–556 (1985).

172. Gallily, I., Schiby, D., Cohen, A. H., Hollander, W., Schless, D., and Stober., W., On the Inertial Separation of Nonspherical Aerosol Particles from Laminar Flows. I. The Cylindrical Case, *Aerosol Sci. Technol.*, **5**, 267–286 (1986).

173. Garrison, R. P., and Byers, D. H., Noise Characteristics of Circular Nozzles for High Velocity/Low Volume Exhaust Ventilation, *Am. Ind. Hyg. Assoc. J.*, **41**(10), 713–720 (1980).

174. Garrison, R. P., and Byers, D. H., Static Pressure, Velocity, and Noise Characteristics of Rectangular Nozzles for High Velocity/Low Volume Exhaust Ventilation, *Am. Ind. Hyg. Assoc. J.*, **41**(12), 855–863 (1980).

175. American Welding Society, Characterization of Arc Welding Fumes, Publication CAWF, 1983; Fumes and Gases in the Welding Environment, Publication FGW, 1979, PO Box 351040, Miami, FL, 33135.

176. Silverman, L. Velocity Characteristics of Narrow Exhaust Slots, *J. Ind. Hyg. Toxicol.*, **24**(9), 267–276 (1942).

177. Silverman, L., Centerline Velocity Characteristics of Round Openings Under Suction, *J. Ind. Hyg. Toxicol.*, **24**(9), 259–266 (1942).

178. Silverman, L., Fundamental Factors in the Design of Lateral Exhaust Hoods for Industrial Tanks, *J. Ind. Hyg. Toxicol.*, **23**(5) 187–195 (1941).

179. Silverman, L., Ventilation Requirements for Solvents in Industrial Tanks, *J. Ind. Hyg. Toxicol.*, **25**(7), 306–319 (1943).

180. Perkins, H. C., *Air Pollution*, McGraw Hill, NY, 1974.

181. Wark, K., and Warner, C. F., *Air Pollution*, 2nd ed., Harper and Row, NY, 1981.

182. Seinfeld, J. H., *Air Pollution*, McGraw Hill, NY, 1975.

183. Hosker, R. P., Jr., Methods for Estimating Wake Flow and Effluent Dispersion Near Simple Block-Like Buildings, National Oceanic and Atmospheric Administration, Report NUREG/CR-2521, ERL-ARL-108, 1982.

184. Chemical Engineering Branch, Economics and Technology Division, Office of Toxic Substances, Environmental Protection Agency, *A Manual for the Preparation of Engineering Assessments*, Environmental Protection Agency, Washington DC, 20460, 1986.

185. American Society of Heating, Refrigerating and Air-Conditioning Engineers, *Fundamentals Handbook*, American Society of Heating, Refrigerating Engineers, Inc., 1791 Tullie Circle, N.E., Atlanta, GA, 1985.

186. Cooper, C. D., and Alley, F. C., *Air Pollution Control: A Design Approach*, PWS Publishers, Boston, MA, 1986.

187. Howell, R. H., and Sauer, H. J., *Bibliography on Available Computer Programs in the General Area of Heating, Refrigerating, Air Conditioning and Ventilating*, 2nd ed., prepared by the University of Missouri at Rolla for ASHRAE, sponsored by DOE, July 1980.

188. Consolidated Environmental Services, Inc., *Ventilation System Design Manual Using the Apple II Microcomputer System*, prepared for the Division of Training and Manpower Development, National Institute for Occupational Safety and Health, Cincinnati, OH, 45226.

189. Flynn, M. R., and Ellenbecker, M. J., Capture Efficiency of Flanged Circular Local Exhaust Hoods, *Ann. Occup. Hyg.*, **30**(4), 497–513 (1986).

190. Whitby, K. T., Charlson, R. E., Wilson, W. E., and Stevens, R. K., The Size of Suspended Particle Matter in Air, *Science*, **183**, 1098–1100, 15 March 1974.

191. Lee, R. E., Jr., The Size of Suspended Particle Matter in Air, *Science*, **178**, 567–575, 10 Nov. 1972.

192. Leonardos, G., Kendall, D., and Barnard, N., Odor Threshold Determinations of 53 Odorant Chemicals, *J. Air Pollut. Control Assoc.*, **19**(2), 91–95, January 1969.

193. Hellman, T. M., and Small, F. H., Characterization of the Odor Properties of 101 Petrochemicals Using Sensory Methods, *J. Air Pollut. Control Assoc.*, **24**(10), 979–982, October 1974.

194. Dravnieks, A., Schmidtsdorff, W., and Meilgaard, M., Odor Thresholds by Forced-Choice Dynamic Triangle Olfactometry: Reproducibility and Methods of Calculation, *J. Air Pollut. Control Assoc.*, **36**(8), 900–905, August 1986.

195. Lapidus, L., and Pinder, G. F., *Numerical Solution of Partial Differential Equations in Science and Engineering*, Wiley–Interscience, New York, 1982.

196. Hornbeck, R. W., *Numerical Methods*, Prentice Hall, Englewood Cliffs, New Jersey, 1975.

197. Boysan, F., Ayers, W. H., and Swithenbank, J., A Fundamental Mathematical Modeling Approach to Cyclone Design, *Trans. Inst. Chem. Eng.*, **60**, 222–230 (1982).

198. Marple, V. A., Liu, B. Y. H., and Whitby,, K. T., Fluid Mechanics of the Laminar Flow Aerosol Impactor, *Aerosol Sci.*, **5**, 1–16 (1974).

199. Marple, V. A., and Liu, B. Y. H., Characteristics of Laminar Jet Impactors, *Environ. Sci. Technol.*, **8**(7), 648–654 (1974).

200. Cai, S., Chen, F. F., and Soo, S. S., Wind Penetration into a Porous Storage Pile and Use of Barriers, *Environ. Sci. Technol.*, **17**(5), 298–305 (1983).

201. Bilman, B. J., Windbreak Effectiveness for the Control of Fugitive-Dust Emissions from Storage Piles—A Wind Tunnel Study, Proceedings 5th Symposium on the Transfer and Utilization of Particulate Control Technology, Sponsored by EPA and EPRI, Aug. 27–30, 1984.

202. Zimmer, R. A., Axetell, K., and Ponder, T. C., Field Evaluation of Windscreens as a Fugitive Dust Control Measure for Material Storage Piles, NTIS Publ. No. PB86-231289/WEP, 126 pp., July 1986.

203. O'Brien, D., Baron, P., and Willeke, K., Respirable Dust Control in Grinding Gray Iron Castings, *Am. Ind. Hyg. Assoc. J.*, **48**(2), 181–187 (1987).

204. Klien, M. K., A Demonstration of NIOSH Push–Pull Ventilation Criteria, *Am. Ind. Hyg. Assoc. J* **48**(3), 238–246 (1987).

205. Launder, B. E., Reynolds, W. C., Rodi, W, Mathieu, J., and Jeandel, D., Turbulence Models and Their Applications, Direction Des Etudes Et Recherches D'Electricite De France, 1984.

206. Hinze, J. O., *Turbulence*, McGraw Hill, New York, 1959.

207. Stahl, W. H., (Ed.), Compilation of Odor and Taste Threshold Value Data, American Society of Testing and Materials, ASTM DS 48A, 1978.

208. Cebeci, T., and Smith, A. M. O., *Analysis of Turbulence Boundary Layers*, Academic Press, New York, 1974.

209. Baldwin, B. S., and Lomax, H., Thin Layer Approximation and Algebraic Model For Separated Flows, AIAA Paper, 78–257, 1987.

210. Prandtl, L., Uber ein neues Formelsystem fur die ausgebildete Turbulenz, Nachrichten von der Akad. der Wissenschaft in Gottingen, 1945.

211. Kolmogorov, A. N., Equations of Turbulent Motion of an Incompressible Turbulent Fluid, *Izv. Akad. Nauk SSSR Ser. Phys.*, **VI**, (1–2) (1956).

212. Jones, W. P., and Launder, B. E., The Production of Laminarisation with 2-Equation Model of Turbulence, *Int. J. Heat Mass Transfer*, **15**, 301 (1972).

213. Tennekes, H., and Lumley, J. L., *A First Course in Turbulence*, The MIT Press, Cambridge, MA 1972.

214. Jones, W. P., and Launder, B. E., A Reynolds Stress Model of Turbulence and Its Applications to Thin Shear Flows, *J. Fluid Mech.* **68**, 537–566 (1972).

215. Burns, A. D., and Jones, I. P., Numerical Prediction of Turbulent Three-Dimensional Jet Flows in Rectangular Enclosures, Report AERE–R11892, Computer Science and Systems Division, Harwell Laboratory, Oxfordshire, OX11 ORA, HL86/1268(C7), England, May 1986.

216. Leith, D, Drag on Nonspherical Objects, *Aerosol Sci. Technol.*, **6**, 153–161 (1987).

217. Paustenbach, D., and Langner, R., Corporate Occupational Exposure Limits: The Current State of Affairs, *Am. Ind. Hyg. Assoc. J.*, **47**(12), 809–818 (1986).

218. Butcher, S. S., and Ellenbecker, M. J., Particulate Emission Factors for Small Wood Stoves and Coal Stoves, *J. Air Pollut. Control Assoc.*, **32**(4), 380–384 (1982).

219. Klein, M. K., A Demonstration of NIOSH Push–Pull Ventilation Criteria, *Am. Ind. Hyg. Assoc. J.*, **48**(3), 238–246 (1987).

220. Flynn, M. R., and Ellenbecker, M. J., Empirical Validation Of Theoretical Velocity Fields into Flanged Circular Hoods, *Am. Ind. Hyg. Assoc. J.*, **48**(4), 380–389 (1987).

221. Klein, M. K., An Introductory Study in Center Push–Pull Ventilation, *Am. Ind. Hyg. Assoc. J.*, **47**(6), 369–373 (1986).

222. Verschueren, K., *Handbook of Environmental Data on Organic Chemicals*, 2nd ed., Van Nostrand Reinhold Co., New York, 1983.

223. Committee on Airliner Cabin Air Quality, National Research Council, *The Airliner Cabin Environment*, National Academy Press, Washington DC, 1986.

224. Ryan, P. B., Spengler, J. D., and Halfpenny, P. F., Sequential Box Models for Indoor Air Quality: Application to Airliner Clean Cabin Air Quality, Paper 86-7.4, 79th Annual Meeting of the Air Pollution Control Association, 22–27 June 1986.

225. Committee on Indoor Pollution, *Indoor Pollution*, National Academy Press, Washington DC, 1981.

226. Meyer, B., *Indoor Air Quality*, Addison-Wesley Publishing Co., Reading, MA, 1983.

227. Matthews, T. G., Fung, K. W., Tromberg, B. J., and Hawthorne, A. R., Impact of Indoor Environmental Parameters on Formaldehyde Concentrations in Unoccupied Research House, *J. Air Pollut. Control Assoc.*, **36**(11), 1244–1249 (1986).

228. Kashdan, E. R., Coy, D. W., Spivey, J. J., and Cesta, T., Technical Manual: Hood System Capture of Process Fugitive Particulate Emissions, EPA/600/7-86-016, PB86-190444, National Technical Information Service, Springfield, VA, 22161, April 1986.

229. Dennis, R., and Bubenick, D. V., Fugitive Emissions Control for Solid Materials Handling Operations, *J. Air Pollut. Control Assoc.*, **33**(12), 1156–1161 (1983).

230. Bender, M, Fume Hoods, Open Canopy Type—Their Ability to Capture Pollutants in Various Environments, *Am. Hyg. Assoc. J.*, **40**, 118–127 (1979).

231. Morton, B. R., Taylor, G. I., and Turner, J. S., Turbulent Gravitational Convention from Maintained and Transient Sources, *Proc. R. Soc.*, **A234**, 1–23 (1956).

232. Turner, J. S., *Buoyancy Effects in Fluids*, Cambridge University Press, 1973.

233. Morton, B. R., Forced Plumes, *J. Fluid Mech.*, **5**, 151–163 (1959).

234. Kashdan, E. R., Coy, D. W., Spivey, J. J., Cesta, T., and Harmon, D. L., Highlights from Technical Manual on Hood System Capture of Process Fugitive Particulate Emissions, *Ventilation '85*, H. D. Goodfellow, (Ed.), Elsevier Science Publishers, Amsterdam, pp. 497–520, 1986.

235. Clapp, D. E., Groh, D. J., and Nenadic, C. M., Ventilation Design by Microcomputer, *Am. Ind. Hyg. Assoc. J.*, **43**, 212–217 (1982).

236. Traynor, G. W., Apte, M. G., Carruthers, A. R., Dillworth, J. F., Grimsrud, D. T., and Gundel, L. A., Indoor Air Pollution due to Emissions from Wood-Burning Stoves, *Environ. Sci. Technol.*, **21**(7), 691–697 (1987).

237. Prugh, R. W., Plant Safety, in *Kirk–Othmer Encyclopedia of Chemical Technology*, 3rd ed., **18**, 60 (1982).

238. Sax, N I., *Dangerous Properties of Industrial Materials*, 5th ed., Van Nostrand Reinhold, New York, 1979.

239. Seinfeld, J. H., *Atmospheric Chemistry and Physics of Air Pollution*, Wiley Interscience Publications, New York, 1986.

240. Kleinman, M. T., Sulfur Dioxide and Exercise; Relationships between Response and Absorption in Upper Airways, *J. Air Pollut. Control Assoc.*, **34**(1), 32–37 (1984).

241. American Society of Heating, Refrigerating and Air-Conditioning Engineers, *ASHRAE Handbook 1985 Fundamentals*, Chapter 8, Physiological Principles, Comfort and Health, American Society of Heating, Refrigerating and Air-Conditioning Engineers, 1791 Tullie Circle, NE, Atlanta, GA, 30329, 1981.

242. Wadden, R. A., and Scheff, P. A., Engineering Design for the Control of Workplace Hazards, McGraw-Hill, New York, 1987.

243. Hayashi, T., Howell, R. H., Shibata, M., and Tsuji, K., *Industrial Ventilation and Air Conditioning*, CRC Press, Boca Raton, FL, 1985.

244. Herrigel, E., *Zen in the Art of Archery*, Pantheon Books, Inc., New York, 1953.

245. Stock, T. H., Formaldehyde Concentrations Inside Conventional Housing, *J. Am. Air Pollut. Control Assoc.*, **37**(8) 913–918 (1987).

246. Kulle, T. K., Sauder, L. R., Hebel, J. R., Green, D. J., and Chatham, M. D., Formaldehyde Dose–Response in Healthy Nonsmokers, *J. Air Pollut. Control Assoc.*, **37**(8) 919–924 (1987).

247. Hattis, D., Wasson, J. M., Page, G. S., Stern, B., and Franklin, C. A., Acid Particles and the Tracheobronchial Region of the Respiratory System—An Irritation-Signaling Model for Possible Health Effects, *J. Air Pollut. Control Assoc.*, **37**(9), 1060–1066 (1987).

248. Donovan, R. P., Locke, B. R., and Ensor, D. S., Measuring Particle Emissions from Cleanroom Equipment, *Microcontamination*, **5**(10), 36–39, 60–63, Oct. 1987.

249. Whitby, K. T., Anderson, G. R., and Rubow, K. L., Dynamic Method for Evaluating Room-Size Air Purifiers, *Trans. ASHRAE* Paper No. 2771, Part 2A, pp. 172–184, 1983.

250. Siu, R. G. H., *The Tao of Science*, MIT Press, Cambridge, MA, 1957.

251. Matthews, T. G., Wilson, D. L., Thompson, A. J., Mason, M. A., Basiley, S. N., and Nelms, L. H., Interlaboratory Comparison of Formaldehyde Emissions from Particleboard Underlayment in Small-Scale Environmental Chambers, *J. Air Pollut. Control Assoc.*, **37**(11), 1320–1326 (1987).

252. McKone, T. E., Human Exposure to Volatile Organic Compounds in Household Tap Water: The Indoor Inhalation Pathway, *Environ. Sci. Technol.*, **21**(12), 1194–1201 (1987).

253. Thurow, L. C., A Weakness in Process Technology, *Science*, **238**, 1659–1663, 18 December 1987.

254. Kurabuchi, T., and Kusuda, T., Numerical Prediction for Indoor Air Movement, *ASHRAE J.*, **29**(12) 26–30, December 1987.

255. American Society of Heating, Refrigerating and Air-Conditioning Engineers, Chapter 35, Sound and Vibration Control, *ASHRAE Handbook*, 1980 Systems Volume, American Society of Heating, Refrigerating and Air-Conditioning Engineers, Inc., 1791 Tullie Circle, NE, Atlanta, GA, 1980.

256. Wilson, R., and Crouch, E. A. C., Risk Assessment and Comparisons: An Introduction, *Science*, **236**, 267–270, 17 April 1987.

257. Ames, B. N., Magaw, R., and Gold, L. S., Ranking Possible Carcinogenic Hazards, *Science*, **236**, 271–280, 17 April 1987.

258. Slovic, P., *Perception of Risk*, *Science*, **236**, 280–285, 17 April 1987.

259. Russell, M., and Gruber, M., Risk Assessment in Environmental Policy-Making, *Science*, **236**, 286–290, 17 April 1987.

260. Lave, L. B., Health and Safety Risk Analyses: Information for Better Decisions, *Science*, **236**, 291–295, 17 April 1987.

261. Okrent, D., The Safety Goals of the Nuclear Regulatory Commission, *Science*, **236**, 296–300, 17 April 1987.

262. Dewees, D. N., Does the Danger From Asbestos in Buildings Warrant the Cost of Taking It Out?, *Am. Sci.*, **75**, 285–288, May–June 1987.

263. Kabel, R. L., *Prairie Dog Problem*, *Chem. Eng. Educ.*, **14**(4), 70–71, 198–199 (1980).

264. Flagan, R. C., and Seinfeld, J. H., *Fundamentals of Air Pollution*, Prentice Hall, Englewood Cliffs, NJ, 1988.

265. Lewis, B., and von Elbe, G., *Combustion, Flames and Explosions of Gases*, Academic Press, New York, 1951.

266. Chigier, N, *Energy, Combustion and Environment*, McGraw Hill, New York, 1981.

267. Penner, S. S., and Mullins, B. P., *Explosions, Detonations, Flammability and Ignition*, Pergamon Press, New York, 1959.

268. Strehlow, R. A., *Fundamentals of Combustion*, International Textbook Co., Scranton, PA, 1968.

269. Williams, F. A., *Combustion Theory*, Addison-Wesley, Reading, MA, 1965.

270. Bare, J. C., Indoor Air Pollution Source Database, *J. Air Pollut. Control Assoc.*, **38**(5), 670–671 (1988).

271. Microcomputer Users Committee, J. C. Clary, Chairman, Microcomputer Software Reviews, *J. Air Pollut. Control Assoc.*, **38**(1), 70 (1988).

272. Hertzberg, M., Cashdollar, K. L., and Zlochower, I. A., Flammability Limit Measurements for Dusts and Gases: Ignition Energy Requirements and Pressure Dependencies, 21st Symposium (International) on Combustion, The Combustion Institute, Pittsburgh, PA, pp. 303–313, 1986.

273. Hertzberg, M., Cashdollar, K. L., Ng, D. L., and Conti, R. S., Domains of Flammability and Thermal Ignitability for Pulverized Coals and Other Dusts: Particle Size Dependencies and Microscopic Residue Analysis, 19th Symposium (International) on Combustion, The Combustion Institute, Pittsburgh, PA, pp. 1169–1180, 1982.

274. Jenning, S. G., The Mean Free Path, *J. Aerosol Sci.*, **19**(2), 159–166 (1988).

275. Leaderer, B. P., Zagraniski, R. T., Berwick, M., and Stolwijk, J. A. J., Assessment of Exposure to Indoor Air Contaminants from Combustion Sources: Methodology and Application, *Am. J. Epidemiol.*, **24**(2), pp. 275–289, 1986.

276. Guerin, M. R., Higgins, C. E., and Jenkins, R. A., Measuring Environmental Emissions from Tobacco Combustion: Sidestream Cigarette Smoke Literature Review, *Atmos. Environ.*, **21**(2), 291–297 (1987).

277. Crawford, W. A., On the Health Effects of Environmental Tobacco Smoke, *Arch. Environ. Health*, **43**(1), 34–37 (1988).

278. Ingebrethsen, B. J., Heavner, D. L., Angel, A. L., Corner, J. M., Steichen, T. J., and Green, C. R., A Comparative Study of Environmental Tobacco Smoke Particulate Mass Measurements in an Environmental Chamber, *J. Air Pollut. Control Assoc.*, **38**(4), 413–417 (1988).

279. Oldakar, III, G. B., and Conrad, F. C., Estimation of Effect of Environmental Tobacco Smoke on Air Quality within Passenger Cabins of Commercial Aircraft, *Environ. Sci. Technol.*, **21**(10), 994–999, 1987.

280. Repace, J. L., and Lowery, A. H., Tobacco Smoke, Ventilation, and Indoor Air Quality, *Trans. ASHRAE, Soc.*, 88 Part 1, Paper HO-82-6 No. 2, 895–914 (1982).

281. Traynor, G. W., Apte, M. G., Carruthers, A. R., Dillworth, J. F., Grimsrud, D. T., and Gundel, L. A., Indoor Air Pollution due to emissions from Wood-Burning Stoves, *Environ. Sci. Technol.*, **21**(7) 691–697 (1987).

282. Girman, J. R., and Hodgson, A. T., Source Characterization and Personal Exposure to Methylene Chloride from Consumer Products, Paper No. 86-52.7, Proceedings 79th Annual Meeting of the Air Pollution Control Association, 22–27 June 1986.

283. Tichenor, B. A., and Mason, M. A., Organic Emissions from Consumer Products and Building Materials in the Indoor Environment, *J. Air Pollut. Control Assoc.*, **38**(3) 264–268 (1988).

284. Hawthorne, A. R., and Matthews, T. G., Models for Estimating Organic Emissions from Building Materials: Formaldehyde Example, *Atmos. Environ.*, **21**(2), 419–424 (1987).

285. Matthews, T. G., Hawthorne, A. R., and Thompson, C. V., Formaldehyde Sorption and Desorption Characteristics of Gypsum Wallboard, *Environ. Sci. Technol.*, **21**(7), 629–634 (1987).

286. Nixon, W., and Egan, M. J., Modeling Study of Regional Deposition of Inhaled Aerosols with Special Reference to Effects of Ventilation Asymmetry, *J. Aerosol Sci.*, **18**(5), 563–579 (1987).

287. Yu, C. P., and Xu, G. B., Predicted Deposition of Diesel Particles in Young Humans, *J. Aerosol Sci.*, **18**(4), 419–429 (1987).

288. Xu, G. B., and Yu, C. P., Deposition of Diesel Exhaust Particles in Mammalian Lungs, *J. Aerosol Sci. Technol.*, **7**, 117–123 (1987).

289. Cox, G. V., and Strickland, G. D., Risk is Normal to Life Itself, *Am. Ind. Hyg. Assoc. J.*, **49**, 223–227, 1988.

290. Braconnier, R., Bibliographic Review of Velocity Fields in the Vicinity of Local Exhaust Hood Openings, *Am. Ind. Hyg. Assoc. J.*, **49**(4), 185–198 (1988); Errata, **49**, 475–478 (1988).

291. Garrison, R. P., and Wang, Y., Finite Element Application for Velocity Characteristics of Local Exhaust Inlets, *Am. Ind. Hyg. Assoc. J.*, **48**(12), 983–988 (1987).

292. Garrison, R. P., and Park, C., Graphical Approximation Model for Velocity Characteristics of Local Exhaust Inlets, *Am. Ind. Hyg. Assoc. J.*, **49**(2), 45–57 (1988).

293. Garrison, R. P., and Erig, M., Velocity Characteristics of Local Exhaust Inlets Facing an External Boundary Surface, *Am. Ind. Hyg. Assoc. J.*, **49**(4), 176–184 (1988).

294. Klein, M. K., The Air Flow Characteristics of an Annular Exhaust Hood, *Appl. Ind. Hyg*, **3**(4), 105–109 (1988).

295. Hampl, V., Johnston, O. E., and Murdock, Jr., D. L., Application of an Air Curtain–Exhaust System at a Milling Process, *Am. Ind. Hyg. Assoc. J.*, **49**(4), 167–175 (1988).

296. Conroy, L. M., Ellenbecker, M. J., and Flynn, M. R., Prediction and Measurement of Velocity into Flanged Slot Hoods, *Am. Ind. Hyg. Assoc. J.*, **49**(5), 226–234 (1988).

297. Braconnier, R., Champs de vitesse au voisinage de l'entre'e des dispositifs d'aspiration localise'e (Velocity Fields Near the Inlet of Local Exhaust Hoods), Cahiers Notes Docum, ND-1586-124-86 (1986) (in French).

298. Stunder, B. J. B., and Aya, S. P. S., Windbreak Effectiveness for Storage Pile Fugitive Dust Control: A Wind Tunnel Study, *J. Air Pollut. Control Assoc.*, **38**(2), 135–143 (1988).

299. Pinnick, R. G., Fernandez, G., Hinds, B. D., Bruce, C. W., Schafer, R. W., and Pendleton, J. D., Dust Generated by Vehicular Traffic on Unpaved Roadways: Sizes and Infrared Extinction Characteristics, *Aerosol Sci. Technol.*, **4**, 99–121, 1985.

300. Tabler, R., Geometry and Density of Drifts Formed by Snow Fences, *J. Glaciol.*, **26**(94), 405–419 (1980).

301. Cheng, Y-S., Yeh, H-C., and Allen, M. D., Dynamic Shape Factor of a Plate-Like Particle, *Aerosol Sci. Technol.*, **8**, 109–123 (1988).

302. Weibel, E., *Morphology of the Lung*, Academic Press, New York, 1963.

303. Ultman, J. S., Gas Transport in the Conducting Airways, Chapter 3, *Gas Mixing in the Lung*, Engle, L. A., and Paiva, M. (Eds.), Marcel Dekker, New York, pp. 63–136, 1985.

304. Slutsky, A. S., Kamm, R. D., and Drazen, J. M., Alveolar Ventilation at High

Frequencies Using Tidal Volumes Smaller than the Anatomical Dead Space, Chapter 4, *Gas Mixing in the Lung*, Engle, L. A., and Paiva, M. (Eds.), Marcel Dekker, New York, pp. 137–176, 1985.

305. Paiva, M., Theoretical Studies of Gas Mixing in the Lung, Chapter 6, *Gas Mixing in the Lung*, Engle, L. A., and Paiva, M. (Eds.), Marcel Dekker, New York, pp. 221–286, 1985.

306. Engle, L., Intraregional Gas Mixing and Distribution, Chapter 7, *Gas Mixing in the Lung*, Engle, L. A., and Paiva, M. (Eds.), Marcel Dekker, New York, pp. 287–358, 1985.

307. Ultman, J. S., Transport and Uptake of Inhaled Gases, *Air Pollution, The Automobile, and Public Health*, A. Y. Watson, R. R. Bates, and D. Kennedy (Eds.), Health Effects Institute, National Academy Press, Washington, DC, pp. 323–366, 1988.

308. Sargent, E. V., and Kirk, G. D., Establishing Airborne Exposure Control Limits in the Pharmaceutical Industry, *Am. Ind. Hyg. Assoc. J.*, **49**(6), 309–313 (1988).

309. Folinsbee, L. J., McDonnell, W. F., and Horstman, D. H., Pulmonary Function and Symptom Responses after 6.6-Hour Exposure to 0.12 PPM Ozone with Moderate Exercise, *J. Air Pollut. Control Assoc.*, **38**(1), 28–35 (1988).

310. Miller, F. J., Overton, J. H., Jaskot, R. H., and Menzel, D. B., A Model of the Regional Uptake of Gaseous Pollutants in the Lung, 1. The Sensitivity of the Uptake of Ozone in the Human Lung to Lower Respiratory Tract Secretions and to Exercise, *J. Toxicol. Environ. Health*, **79**, 11–27 (1985).

311. Yu, S. T., Push–Pull Ventilation System for Open Vessels, PhD Thesis, The Pennsylvania State University, May 1989.

312. Qingyan, C., and Van Der Kooi, J., Experiments and 2D Approximated Computations of 3D Air Movement, Heat and Concentration Transfer in a Room, Proceedings of the International Conference on Air Distribution in Ventilated Spaces, ROOM-VENT-87, Stockholm, 10–12 June 1987.

313. Qingyan, C., and Van Der Kooi, J., Measurements and Computations on Air Movement and Temperature Distribution in a Climate Room, Paper No. 54, Committee E.1.1, Proceedings of the XVIIth International Congress of Refrigeration, Vienna, 23–29 Aug. 1987.

314. Van Der Kooi, J., and Qingyan, C., Improvement Of Cooling Load Programs By Combination With An Air Flow Program, Paper No. 55, Committee E.1.1, Proceedings of the XVIIth International Congress of Refrigeration, Vienna, 23–29 Aug. 1987.

315. Hjertager, B. H., and Magnussen, B. F., Calculation of Turbulent Three-Dimensional Jet Induced Flow in Rectangular Enclosures, *Comput Fluids*, **9**(4), 395–407 (1981).

316. Hjertager, B. H., and Magnussen, B. F., Laser Doppler Velocity Measurements in the Three-Dimensional Flow of a Jet in a Square Enclosure, *Lett. Heat Mass Transfer*, **8**, 171–186 (1981).

317. Yang, K. T., and Lloyd, J. R., Turbulent Buoyant Flow in Vented Simple and Complex Enclosures, in *Natural Convection, Fundamentals and Applications*, edited by Kakac, S., Aung, W., and Viskanta, R. (Eds.), Hemisphere Publishing Corp., New York, pp. 303–329, 1985.

318. Bilger, R. W., Turbulent Jet Diffusion Flames, *Prog. Energy Combust. Sci.*, **1**, 87–109 (1976).

319. Ultman, J. S., Exercise and Regional Dosimetry: Factors Governing Gas Transport in the Lower Airways, paper given at the Susceptibility To Inhaled Pollutants

conference, Williamsburg, VA, 30 September 1987, and to be published as an ASTM monograph.

320. Sittig, M., Environmental Sources and Emissions Handbook, Noyes Data Corp, Park Ridge, NJ, 1985.

321. Bond, R., and Staub, C. P. (Eds.), *Prober R coordinating editor, Handbook Of Environmental Control*, Vol. I, Air Pollution, CRC Press, Cleveland, OH, 1972.

322. Lyman, W. J., Reehl, W. F., and Rosenblatt, D. H., *Handbook Of Chemical Property Estimation Methods*, McGraw Hill, 1982.

323. Reid, R. C., Prausnitz, J. M., and Sherwood, T. K., *The Properties of Gases And Liquids*, 3rd ed., McGraw Hill, New York, 1977.

324. Mackay, D., and Shiu, W. Y., A Critical Review of Henry's Law Constants for Chemicals of Environmental Interest, Reprint No. 193, *J. Phys. Chem. Ref. Data*, **10**(4), 1175–1199 (1981).

325. American Society of Heating, Refrigerating and Air-Conditioning Engineers, Chapter 19, *Ventilation of the Industrial Environment*, System Volume, 1984.

326. Ricci, P. F., and Molton, L. S., Regulating Cancer Risks, *Environ. Sci. Technol.*, **19**(6), 473–479 (1985).

327. Ryan, P. B., Spengler, J. D., and Halfpenny, P. F., Sequential Box Models for Indoor Air Quality: Application to Airliner Cabin Air Quality, *Atmos. Environ.*, **22**(6), 1031–1038 (1988).

328. Rader, D. J., and Marple, V. A., A Study of the Effects of Anisokinetic Sampling, *Aerosol Sci. Technol.*, **8**, 283–299 (1988).

329. Dunnett, S. J., and Ingham, D. B., An Empirical Model for the Aspiration Efficiencies of Blunt Aerosol Samplers Orientated at an Angle to the Oncoming Flow, *Aerosol Sci. Technol.*, **8**, 245–264 (1988).

330. Licht, W., *Air Pollution Control Engineering*, Marcel Dekker, New York, 1980.

331. Kuo, K., *Principles of Combustion*, Wiley-Interscience, New York, 1986.

332. National Safety Council, *Accident Facts 1987 Edition*, 444 N Michigan Ave, Chicago, IL, 1987.

333. Breum, N., Air Exchange Efficiency of Displacement Ventilation in a Printing Plant, *Ann. Occup. Hyg.*, **32**(4), 481–488 (1988).

334. Holvey, D. N., (Ed.), *The Merck Manual of Diagnosis and Therapy*, 12th ed., Merck Sharp & Dohme Research Laboratories, Rahway, NJ, 1972.

335. Incropera, F. P., and DeWitt, D. P., *Fundamentals of Heat Transfer*, John Wiley & Sons, New York, 1981.

336. Kreith, F., *Principles of Heat Transfer*, 3rd ed., Harper & Row, New York, 1973.

337. Stiver, W., Shiu, W. Y., and Mackay, D., Evaporation Times and Rates of Specific Hydrocarbons in Oil Spills, *Environ. Sci. Technol.*, **23**(1), 101–105 (1989).

338. Nirmalakhandan, N. N., and Speece, R. E, QSAR Model for Predicting Henry's Constant, *Environ. Sci. Technol.*, **22**(11), 1349–1357 (188).

339. Silberstein, S, Grot, R. A., Ishiguro, K., and Mulligan, J. L., Validation of Models for Predicting Formaldehyde Concentrations in Residences due to Pressed-Wood Products, *J. Air Pollut. Control Assoc.*, **38**(11), 1403–1411 (1988).

340. Matthews, T. G., Hawthorne, A. R., Thompson, C. V., Formaldehyde Sorption and Desorption Characteristics of Gypsum Wallboard, *Environ. Sci. Technol.*, **21**(7), 629–634 (1987).

341. Pruppacher, H. R., and Klett, J. D., Microphysics of Clouds and Precipitation, D. Reidel Publishing Co., Dordrecht, Holland, 1978.

342. Nazaroff, W. W., and Cass, G. R., Mathematical Modeling of Indoor Aerosol Dynamics, *Environ. Sci. Technol.*, **23**(2), 157–166 (1989).

343. Tilton, B. E., Health Effects of Tropospheric Ozone, *Environ. Sci. Technol.*, **23**(3), 257–263 (1989).

344. Alenandersson, R., and Hedenstierna, G., Pulmonary Function in Wood Workers Exposed to Formaldehyde: A Prospective Study, *Arch. Environ. Health*, **44**(1), 5–11, Jan/Feb 1989.

345. Cowherd, C., Jr., Grelinger, M. A., Englehart, P. J., Kent, R. F., and Wong, K. F., An Apparatus and Methodology for Predicting the Dustiness of Materials, *Am. Ind. Hyg. Assoc. J.*, **50**(3), 123–130 (1989).

346. Cowherd, C., Jr., Grelinger, M. A., and Wong, K. F., Dust Inhalation Exposures from the Handling of Small Volumes of Powders, *Am. Ind. Hyg. Assoc. J.*, **50**(3), 131–138 1989.

347. Anastas, M. Y., and Hughes, R. T., Center-line Velocity Models for Flanged Local Exhaust Openings, *Appl. Ind. Hyg.*, **3**(12), 342–347 (1988).

348. Rothstein, M. A., *West's Handbook Series, Occupational Safety and Health Law*, 2nd ed., West Publishing Co., St. Paul, MN, 1983.

349. Lofroth, G., Burton, R. M., Forehand, L., Hammond, S. K., Seila, R. L., Zweidlnger, R. B., and Lewtas, J., Characterization of Environmental Tobacco Smoke, *Environ. Sci. Technol.*, **23**(5), 610–614 (1989).

350. US Department of Labor, Air Contaminants—Permissible Exposure Limits (Title 29 Code of Federal Regulations Part 1910.1000), 1989.

351. CRC Press, *Handbook of Chemistry and Physics*, 56th ed. Cleveland, OH, 1975.

352. Kuehn, T. H., Predicting Air Flow Patterns and Particle Concentration in Clean Rooms, *J. Aerosol Sci.*, **19**(7), 1405–1408 (1988).

353. Davies, C. N., Cigarette Smoke: Generation and Properties of The Aerosol, *J. Aerosol Sci.*, **19**(4), 463–469 (1988).

354. Leung, H. W., and Paustenbach, D. J., Application of Pharamacokinetics to Derive Bilogical Exposure Indexes from Threshold Limit Values, *Am. Ind. Hyg. Assoc. J.*, **49**(9), 445–450 (1988).

355. Nicholson, K. W., A review of Particle Resuspension, *Atmos. Environ.*, **22**(12), 2639–2651 (1988).

356. Flynn, M. R., and Miller, C. T., Comparison of Models for flow through Flanged and Plain circular Hoods, *Ann. Occup. Hyg.*, **32**(2), 373–384 (1988).

357. Dunn, J. E., and Tichenor, B. A., Compensating for Sink Effects in Emissions Test Chambers by Mathematical Modeling, *Atmos. Environ.*, **22**(5), 885–894 (1988).

358. Traynor, G. W., Apte, M. G., Carruthers, A. R., Dillworth, J. F., Prill, R. J., Grimsrud, D. T., and Turk, B. H., The Effects of Infiltration and Insulation on the Source Strengths and Indoor Air Pollution from Combustion Space Heating Appliances, *J. Air Pollut. Control Assoc.*, **38**(8), 1011–1015 (1988).

359. Fleischer, R. L., Basement Ventilation Needed to Lower Indoor Radon to Acceptable Levels, *J. Air Pollut. Control Assoc.*, **38**(7), 914–916 (1988).

360. Heinsohn, R. J., Macro-Modeling Using Sequential Box Models, Proceedings of Buildings Systems: Room Air and Air Contaminant Distribution, University of Illinios, Dec. 5–8, 1989, L. L. Christianson (Ed.), American Society of Heating, Refrigerating and Air-Conditioning Engineers, pp. 200–205, 1989.

361. O'Brien, D. M., and Hurley, D. E., An Evaluation of Engineering Control Technology for Spray Painting, DHHS(NIOSH) Publication No. 81–121, 1981.

362. Murakami, S., and Kato, S., Current Status of Numerical and Experimental Method for Analyzing Flow Field and Diffusion Field In a Room, Proceedings of Building Systems: Room Air and Air Contaminant Distribution, L. L. Christianson (Ed.), ASHRAE, pp. 39–56.

363. Kurabuchi, T., Sakamoto, Y., and Kaizuka, M., Numerical Predictions of Indoor Airflows by Means of the K-E Turbulence Model, Proceedings of Building Systems: Room Air and Air Contaminant Distribution, L. L. Christianson (Ed.), ASHRAE, pp. 57–67.

364. Kusuda T., Status of Mathematical Modeling for Room Convection Analysis, Proceedings of Building Systems: Room Air and Air Contaminant Distribution, L. L. Christianson (Ed.), ASHRAE, pp. 111–115.

365. Axley, J. W., Integrating Microscopic and Macroscopic Models of Air Movement and Contaminant dispersal in Buildings, Proceedings of Building Systems: Room Air and Air Contaminant Distribution, L. L. Christianson (Ed.), ASHRAE, pp. 116–122.

366. Ferziger, J. H., Application of Zonal Modeling to Ventilation Simulation, Proceedings of Building Systems: Room Air and Air Contaminant Distribution, L. L. Christianson (Ed.), ASHRAE, pp. 123–124.

367. Choi, H. L., and Albright, L. D., Modeling the Effect of Observations in Slot-Ventilated Enclosures, Proceedings of Building Systems: Room Air and Air Contaminant Distribution, L. L. Christianson (Ed.), ASHRAE, pp. 127–130.

368. Baker, A. J., Kelso, R. M., Noronha, W. P., and Woods, J. B., On the Maturing of CFD in Design of Room Air Ventilation Systems, Proceedings of Building Systems: Room Air and Air Contaminant Distribution, L. L. Christianson (Ed.), ASHRAE, pp. 149–152.

369. Schiller, G. E., A Numerical Model of Indoor Radon Decay Product Deposition, Proceedings of Building Systems: Room Air and Air Contaminant Distribution, L. L. Christianson (Ed.), ASHRAE, pp. 169–176.

370. Nielsen, P. V., Numerical Prediction of Air Distribution in Rooms—Status And Potentials, Proceedings of Building Systems: Room Air and Air Contaminant Distribution, L. L. Christianson (Ed.), ASHRAE, pp. 31–38.

371. Flynn, M. R., and Miller, C. T., The Boundary Integral Equation Method (BIEM) for Modeling Local Exhaust Hood Flow Fields, *Am. Ind. Hyg. Assoc. J.*, **50**(5), 281–288 (1989).

372. Petzold, C., and Hummel, R. L., BOOTREC.PCM New Boot Record, a section in *PC Magazine DOS Power Tools Techniques*, *Tricks and Utilities*, P. Somerson (Ed.), Bantam Computer Books, pp. 901–908.

373. Lippmann, M., Health Effects of Ozone, A Critical Review, *J. Air Pollut. Control Assoc.*, **39**(5) 672–695 (1989).

374. Eatough, D. J., Benner, C. L., Bayona, J. M., Richards, G., Lamb, J. D., Lee, M. L., Lewis, E. A., and Hansen, L. D., Chemical Composition of Environmental Tobacco Smoke. 1. Gas-Phase Acids and Bases, *Environ. Sci. Technol.*, **23**(6), 679–687 (1989).

375. Benner, C. L., Bayona, J. M., Caka, F. M., Hongmao, T., Lewis, L., Crawford, J., Lamb, J. D., Lee, M. L., Lewis, E. A., Hansen, L. D., and Eatough, D. J., Chemical Composition of Environmental Tobacco Smoke. 2. Particulate-Phase Compounds, *Environ. Sci. Technol.*, **23**(6), 688–699 (1989).

376. Liu, B. Y. H., Zhang, Z. Q., and Kuehn, T. H., A Numerical Study of Inertial Errors in Anisokinetic Sampling, *J. Aerosol Sci.*, **20**(3), 367–380 (1989).

377. West, J. B., *Respiratory Physiology—The Essentials*, Williams and Wilkins Co., Baltimore, MD, 1974.

378. Slonim, N. B., and Hamilton, L. H., *Respiratory Physiology*, 5th ed., C. V. Mosby Co., St Louis, MO, 1987.

379. Levitzky, M. G., *Pulmonary Physiology*, 2nd ed., McGraw-Hill, New York, 1986.

380. Halitsky, J., A Jet Plume Model for Short Stacks, *J. Air Pollut. Control Assoc.*, **39**(6), 856–858 (1989).

381. National Institute for Occupational Safety and Health, Control of Emissions from Seals and Fittings in Chemcial Process Industries, technical report DHHS Publication No. 81-118, Harold Van Wagenen (author), April 1981.

382. Slutsky, A. S., Drazen, J. M., Ingram, R. H., Kamm, R. D., Shapiro, A. H., Fredberg, J. J., Loring, S. H., and Lehr, J., Effective Pulmonary Ventilation with Small-Volume Oscillations at High Frequency, *Science*, **209**, 609–610 (1980).

383. Talty, J. T., *Industrial Hygiene Engineering, Recognition Measurement, Evaluation and Control*, (2nd ed.), Noyes Publications, Noyes Data Corporation, Park Ridge, NJ, 07656, 1988.

384. Hama, G. M., Supply and Exhaust Ventilation for Metal Acid Pickling Tanks, *Air Cond. Heating and Vent.*, 61–63 (1957).

385. Hama, G. M., Push–Pull Hoods, Duct Design and Lab Hood Ventilation, *Air Cond. Vent.*, 51 (1964).

386. Hama, G. M., Jet Stream Replaces Vertical Duct in Push–Pull Ventilation of Pickling Tanks, *Heating Piping Air Cond.*, **45**, 121–123 (1973).

387. Institution of Heating and Ventilating Engineers, *IHVE Guide Book B* 1970, B3–8, 1972.

388. Burgess, W. A., Ellenbecker, M. J., and Treitman, R. T., *Ventilation for Control of the Work Environment*, John Wiley and Sons, New York, 1989.

389. American Society of Heating, Refrigerating and Air-Conditioning Engineers, Inc. *Ventilation for Acceptable Indoor Air Quality*, ASHRAE Standard 62-1989, 1791 Tullie Circle NE, Atlanta Ga, 30329.

390. Berger, R. S., The Carcinogenicity of Radon, *Enviro. Sci. Technol.*, **24**(1), 30–31 (1990).

391. Bossard, F. C., LeFever J. J., LeFever, J. B., and Stout, K. S., *A Manual Of Mine Ventilation Design Practices*, Floyd C Bossard and Associates, Inc, Box 3837, Butte, MT, 59702, 1983.

392. Mossman, B. T., Bignon, J., Corn, M., Seaton, A., and Gee, J. B. L., Asbestos: Scientific Developments and Implications for Public Policy, *Science*, **247**, 294–301, 19 January 1990.

393. American Society of Heating, Refrigerating and Air-Conditioning Engineers, *ASHRAE Journal HVAC & R Software Directory*, American Society of Heating, Refrigerating and Air-Conditioning Engineers, 1791 Tullie Circle NE, Atlanta, GA 30329, 1990.

394. Garrison, R. P., Nabar, R., and Erig, M., Ventilation to Eliminate Oxygen Deficiency in the Confined Space—Part One: Cubical Model, *Am. Ind. Hyg. Assoc. J.*, **4**, 1–11 (1989).

395. Park, C., and Garrison, R. P., Multicellular Model for Contaminant Dispersion and Ventilation Effectiveness with Application for Oxygen Deficiency in a Confined Space, *Am. Ind. Hyg. Assoc. J.*, **51**(2), 70–78, February 1990.

396. Park C., and Garrison, R. P., Evaluation of Models for Local Exhaust Velocity

Characteristics—Part One and Two: Velocity Contours For Local Free Standing And Bounded Inlets, *Am. Ind. Hyg. Assoc. J.*, **50**(4), 196–203, April 1989.

397. Tim Suden, K. D., Flynn, M. R., Goodman, R., Computer Simulation in the Design of Local Exhaust Hoods for Shielded Metal Arc Welding, *Am. Ind. Hyg. Assoc. J.*, **51**(3), 115–126, March 1990.

398. The Center for Chemical Process Safety of the American Institute of Chemical Engineers, Safety, Health, and Loss Prevention in Chemical Processes, Problems for Undergraduate Engineering Curricula and the Instructor's Guide, *Am. Inst. Chem. Eng.*, 345 East 47th Street, New York, NY 10017, 1990.

399. McCartney, M. L., Sensitivity Analysis Applied to Coburn-Foster-Kane Models of Carboxyhemoglobin Formation, *Am. Ind. Hyg. Assoc. J.*, **51**(3), 169–177 (1990).

400. National Institute for Occupational Safety and Health, Proceedings of the Symposium on Occupational Health Hazard Control Technology in the Foundry and Secondary Non-Ferrous Smelting Industries, DHHS (NIOSH) Publication Number 81–114, August 1981.

401. Lueptow, R. M., Software for Computational Fluid Flow and Heat Transfer Analysis, *Computers in Mechanical Engineering*, **6**(5), 10–17, March/April 1988.

402. Chang, T. Y., and Rudy, S. J., Roadway Tunnel Air Quality Models, *Environ. Sci. Technol.*, **24**(5), 672–676 (1990).

403. Nazaroff, W. W., Teichman, K, *Indoor Radon*, *Environ. Sci. Technol.*, **24**(6), 774–7822 (1990).

404. Jameson, A., Computational Aerodynamics For Aircraft Design, *Science*, 245, 361–371, 28 July 1989.

405. Dunnett, S. J., An Investigation into the Effects of Breathing Upon the Aspiration Efficiency of the Head, *Annals Occupational Hygiene*, **33**(2), 209–218 (1989).

406. Lee, C. T., Leith, D, Drag Force on Agglomerated Spheres in Creeping Flow, *J. Aerosol Sci.*, **20**(5), 503–513 (1989).

407. Wadden, R. A., Scheff, P. A., Franke, J. E., Emission Factors for Trichlorethylene Vapor Degreasers, *Am. Ind. Assoc. J.*, **50**(9), 496–500 (1989).

408. Eisner, A. D., Martonen, T. B., Simulation of Heat and Mass Transfer Processes in a Surrogate Bronchial System Developed for Hygroscopic Aerosol Studies, *Aerosol Science and Technology*, **11**, 39–57 (1989).

409. Zhang, Z. Q., Liu, B. Y. H., On The Empirical Fitting Equations for Aspiration Coefficients for Thin-Walled Sampling Probes, *J. Aerosol Sci.*, **20**(6), 713–720 (1989).

410. Gradon, L., Yu, C. P., Diffusional Particle Deposition in the Human Nose and Mouth, *Aerosol Science and Technology*, **11**, 213–220 (1989).

411. Kukkonen, J., Vesala, T, Kulmala, M., The Interdependence of Evaporation and Settling for Airborne Freely Falling Droplets, *J. Aerosol Sci.*, **20**(7), 749–763 (1989).

412. Anand, N. K., McFarland, A. R., Particle Deposition in Aerosol Sampling Lines Caused by Turbulent Diffusion and Gravitational Settling, *Am. Ind. Hyg. Assoc. J.*, **50**(6), 307–312 (1989).

413. Gradon, L., Orlicki, D., Deposition of Inhaled Aerosol Particles in a Generation of the Tracheobronchial Tree, *J. Aerosol Sci.*, **21**(1), 3–19 (1990).

414. Yu, C. P., Asgharian, B., A Kinetic Model of Alveolar Clearance of Amosite Asbestos Fibers from the Rat Lung at High Lung, Burdens, *J. Aerosol Sci.*, **21**(1), 21–27 (1990).

415. Ye, Y., Pui, D. Y. H., Particle Deposition in a Tube with an Abrupt Contraction, *J. Aerosol Sci.*, **21**(1), 29–40 (1990).

416. Moore, M. E., McFarland, A. R., Design of Stairmand-Type Sampling Cyclones, *Am. Ind. Hyg. Assoc. J.*, **51**(3), 151–159 (1990).

417. Hangal, S., Willeke, K., Aspiration Efficiency: Unified Model for all Forward Sampling Angles, *Enviro. Sci. Technol.*, **24**(5), 688–691 (1990).

418. Muller, W. J., Hess, G. D., Scherer, P. W., A Model of Cigarette Smoke Particle Deposition, *Am. Ind. Hyg. Assoc. J.*, **51**(5), 245–256 (1990).

419. Weschler, C. J., Shields, H. C., and Rainer, D., Concentrations of Volatile Organic Compounds at a Building with Health and Comfort Complaints, *Am. Ind. Hyg. Assoc. J.*, **51**(5), 261–268 (1990).

420. Kriebel, D., Smith, T. J., A Nonlinear Pharmacologic Model of the Acute Effects of Ozone on the Human Lungs, *Environmental Research*, **51**, 120–146 (1990).

421. Tichenor, B. A., Sparks, L. E., Jackson, M. D., Emission of Perchlorothylene from Dry Cleaned Fabrics, *Atmospheric Environment*, **24A**(5), 1219–1229 (1990).

422. Joyner, W. M., Compilation of Air Pollutant Emission Factors, AP-42. Volume 1. Stationary and Area Sources, 4th Edition. Supplement A, PB87-150959, USEPA Office of Air Quality Planning Standards, Research Triangle Park, NC, 27711, October 1986.

OSHA Permissible Exposure Limits for Common Industrial Materials as of 1989[a]

(abstracted from 350)

Substance (Cas No)	M	TWA-PEL (PPM)
Acetaldehyde (75-07-0)	44	100
Acetic acid (64-19-7)	60	10
Acetic anhydride (108-24-1)	102	5(Ceiling)
Acetone (67-64-1)	58	750
Allyl alcohol (107-18-6)	58	2[d]
Ammonia (7664-41-7)	17	35(STEL)
Aniline (62-53-3)	93	2[d]
Benzene (71-43-2)	78	10[c]
Biphenyl (92-52-4)	154	0.2
Boron trifluoride (7637-07-2)	68	1(Ceiling)
Butyl acetate (123-86-4)	116	150
Butyl alcohol (71-36-3)	74	50(Ceiling)
Butyl amine (109-73-9)	73	5(Ceiling)[d]
Butyl mercaptan	90	0.5
Carbon disulfide (75-15-0)	76	4[c]
Carbon monoxide (630-08-0)	28	35
Carbon tetrachloride (56-23-5)	154	2[c]
Chlorine (7782-50-5)	71	0.5
Chlorobenzene (108-90-7)	113	75
Chloroform (67-66-3)	119	2
Cyclohexane (110-82-7)	84	300
Decaborane (17702-41-9)	122	0.05[d]
Diazomethane (334-88-3)	42	0.2
Diborane (19287-45-7)	28	0.1
Diethylamine (109-89-7)	73	10
Dichloroethyl ether (111-44-4)	143	5[d]
Diisopropylamine (108-18-9)	101	5[d]
Dimethylamine (124-40-3)	45	10
Dioxane (123-91-1)	88	25[d]
Ethyl acetate (141-78-6)	88	400
Ethyl acrylate (140-88-5)	100	5[d]
Ethyl alcohol (64-17-5)	46	1000
Ethylamine (75-04-7)	45	10
Ethyl benzene (100-41-4)	106	100

Substance (Cas No)	M	TWA-PEL (PPM)
Ethyl chloride (75-00-3)	65	1000
Ethyl formate (109-94-4)	74	100
Ethylene dibromide (106-93-4)	188	#
Ethylene dichloride (107-06-2)	99	1
Ethyl mercaptan (75-08-1)	65	0.5
Ethylene oxide (75-21-8)	44	1
Fluorine (7782-41-4)	38	0.1
Formaldehyde (50-00-0)	30	3[c]
Formic acid (64-18-6)	46	5
Furfural (98-01-1)	96	2[d]
Heptane (142-82-5)	100	400
Hexachloroethane (67-72-1)	237	1[d]
Hexane (110-54-3)	86	50
Hydrazine (302-01-2)	32	0.1[d]
Hydrogen bromide (10035-10-6)	81	3(Ceiling)
Hydrogen chloride (7647-01-0)	37	5(Ceiling)
Hydrogen cyanide (74-90-8)	27	4.7(STEL)[d]
Hydrogen fluoride (7664-39-3)	20	3
Hydrogen peroxide (7722-84-1)	34	1
Hydrogen sulfide (7783-06-4)	34	10
Iodine (7553-56-2)	254	0.1(Ceiling)
Isobutyl acetate (110-19-0)	116	150
Isobutyl alcohol (78-83-1)	74	50
Isopropyl acetate (108-21-4)	102	250
Isopropyl alcohol (67-63-0)	60	400
Isopropyl ether (108-21-3)	60	500
Mercury, vapor (7439-97-6)	201	0.05 mg/m^3
Methacrylic acid (79-41-4)	86	20
Methyl acetate (79-20-9)	74	200
Methyl acrylate (96-33-3)	86	10[d]
Methyl alcohol (67-56-1)	32	200
Methylamine (74-89-5)	31	10
Methyl bromide (74-83-9)	95	5[d]
Methyl chloride (74-87-3)	51	100*
Methyl ethyl ketone (78-93-3)	72	200
Methyl formate (107-31-3)	60	100
Methyl iodide ((74-88-4)	142	2[d]
Methyl isocyanate (624-83-9)	102	0.02[d]
Methyl mercaptan (74-93-1)	48	0.5
Methylene chloride (75-09-2)	85	#
Naphthalene (91-20-3)	128	10
Nickel carbonyl (13463-39-3)	171	0.001
Nitric acid (7697-37-2)	63	2
Nitric oxide (10102-43-9)	30	25
Nitrobenzene (98-95-3)	123	1[d]
Nitroethane (79-24-3)	75	100
Nitrogen dioxide (10102-44-0)	46	1(STEL)
Nitromethane (75-52-5)	61	100
Nitrotoluene (88-72-2)	137	2[d]
Octane (111-65-9)	114	300
Ozone (10028-15-6)	48	0.1

Substance (Cas No)	M	TWA-PEL (PPM)
Perchloroethylene (127-18-4)	166	25
Pentaborane (19624-22-7)	63	0.005
Phenol (108-95-2)	94	5^d
Phosgene (75-44-5)	99	0.1
Phosphoric acid (7664-38-2)	98	$1 \, mg/m^3$
Picric acid (88-89-1)	229	$0.1^d \, mg/m^3$
n-Propyl acetate (109-50-4)	102	200
n-Propyl alcohol (71-23-8)	60	200
Propylene oxide (75-56-9)	58	20
Pryidine (110-86-1)	79	5
Styrene (100-42-5)	104	50
Sulfur dioxide (7446-09-5)	64	2
Sulfuric acid (7664-93-9)	98	$1 \, mg/m^3$
Toluene (108-88-3)	92	100
Trichlorothylene (79-01-6)	131	50
1, 2, 3-Trichloropropane (96-18-4)	147	10
Turpentine (8006-64-2)	136	100
Xylene (1330-20-7)	100	100

INDUSTRIAL DUSTS AND FUMES

Substance (Cas No)	TWA-PEL (mg/m^3)
Chromium metal (7440-47-3)	1
Coal dust, respirable fraction	
$\quad SiO^2 < 5\%$	2
$\quad SiO^2 > 5\%$	0.1
Cobalt metal dust and fume (7440-48-4)	0.05
Copper	
$\quad$ dust (7440-50-8)	1
$\quad$ fume (7440-50-8)	0.1
Grain dust (oat, wheat, barley)	10
Graphite, (7782-42-5)	
$\quad$ respirable dust	2.5
Gypsum, (7778-18-9)	
$\quad$ total dust	15
$\quad$ respirable dust	5
Iron oxide, total dust (1309-37-1)	10
Kaolin,	
$\quad$ total dust	10
$\quad$ respirable dust	5
Limestone, (1317-67-3)	
$\quad$ total dust	15
Magnesium oxide, (1309-48-4)	
$\quad$ total dust	10
$\quad$ respirable dust	5
Manganese fume (7439-96-5)	1
Marble	
$\quad$ total dust	15
$\quad$ respirable dust	5

Substance (Cas No)	TWA-PEL (mg/m^3)
Molybdenum, insoluble compounds (7439-98-7)	
total dust	10
respirable dust	5
Nickel, metal dust (7440-02-0)	
insoluble compounds	1
soluble compounds	0.1
Oil mist, mineral	5
Particulates, not otherwise regulated	
total dust	15
respirable dust	5
Platinum (7440-06-4)	
metal	1
soluble salts	0.002
Portland cement, (65997-15-1)	
total dust	10
respirable dust	5
Rhodium (7440-16-6)	
insoluble compounds	0 .1
soluble compounds	0 .001
Selenium (7782-49-2)	0.2
Silica	
amorphous, crystalline silica < 1% (68855-54-9)	6
crystalline cristobalite, respirable (14464-46-1)	0.05
crystalline quartz, respirable (14808-60-7)	0.1
crystalline tripoli, respirable (1317-95-9)	0.1
crystalline tridymite, respirable (15468-32-3)	0.05
fused (60676-86-0)	0.1
Silicon, (7440-21-3)	
total dust	10
respirable dust	5
Silver, metal (7440-22-4)	0.01
Tin	
oxide (7440-31-5)	2
organic compounds (7440-31-5)	0.1
Tungsten, soluble compounds (7440-33-1)	1
Vanadium, respirable dust (1314-62-1)	0.05
Vegetable oil mist	
total dust	15
respirable dust	5
Wood dust	
general	5
western red cedar	2.5
Zinc oxide fume, (1314-13-2)	
total dust	10
respirable dust	5
Zirconium compounds (7440-67-7)	5

Source:
* See (350) for ceiling and maximum transitory values
# In process of rulemaking
[a] See Appendices A-8 and A-10 for PEL of additional materials. Note PEL and TLV are reviewed on a regular basis and readers should always use currently approved values.
[d] Denotes additional entry through the skin.

Emission Factors for Particles from Uncontrolled Metallurgical Processes*

Process	Emissions (kg particles/ Mg raw material)
Gray iron fugitive emissions	
Scrap and charge handling	0.30
Magnesium treatment	2.50
Pouring	2.50
Cooling	5.00
Inoculation	1.5–2.5
Shakeout	16.00
Cleaning, finishing	8.50
Sand handling, preparation, mulling	20.00
Core making, baking	0.60
Gray iron furnaces	
Electric arc	5.00
Electric induction	0.75
Reverberatory	1.00
Cupolas (average)	8.55
Less than 48 in ID	6.45
48–60 in ID	9.75
greater than 60 in ID	9.45
Iron and steel mills	
blast furnaces	
Slips (kg/slip)	39.5
Cast house (kg/Mg hot metal)	
Tap hole and trough	0.15
Sintering (kg/Mg finished sinter)	
Windbox leaving grate	5.56
Discharge (breaker and hot screens)	3.40
Basic oxygen furnaces (kg/Mg steel)	
Top blown melting and refining	14.25

*Taken from references 2 and 6-10.

Process	Emissions (kg particles/ Mg raw material)
Iron and steel mills–*Continued*	
Charging, at source	0.3
Tapping, at source	0.46
Hot metal transfer, at source	0.095
BOF monitor (all sources)	0.25
Electric arc furnaces (kg/Mg steel)	
Melting and refining (carbon steel)	19.0
Charging, tapping, slagging	0.7
Melting, refining, charging, tapping and slagging	
Alloy steel	5.65
Carbon steel	25.0
Open hearth furnaces (kg/Mg steel)	
Melting and refining	10.55
Teeming (kg/Mg steel)	
Leaded steel at source	0.405
Unleaded steel at source	0.035
Machine scarfing (kg/Mg steel)	0.05
Miscellaneous (boilers, soaking pits, Slab reheat furnaces, blast furnace gas, Coke oven gas) (kg/1000 mJ)	0.015
Melting of red brass (<7% zinc)	
Crucible or pot furnaces	1.65
Rotary furnace	10.60
Reverberatory furnaces	8.40
Electric furnaces	1.50
Melting of bronze	
Crucible furnace	1.90
Rotary furnace	15.30
Aluminum production	
Bauxite grinding	3.00
Aluminum hydroxide	100.00
Prebake cell	47.00
Crucible furnace	0.95
Reverberatory furnace	2.60
Secondary aluminum	
Sweating furnace	7.20
Smelting	
Crucible furnace	0.95
Reverberatory furnace	2.15

Process	Emissions (kg particles/ Mg raw material)
Secondary zinc smelting	
Reverberatory	
General metallic scrap	6.50
Residual scrap	16.00
Rotary sweating	5.5–12.5
Muffle sweating	5.4–16.0
Kettle sweating	
General metallic scrap	5.50
Residual scrap	12.50
Electric resistance sweating	<5
Crushing, screening	0.5–3.8
Sodium carbonate leaching	
Crushing, screening	0.5–3.8
Calcining	44.50
Retort and muffle distillation	
Pouring	0.2–0.4
Casting	0.1–0.2
Muffle distillation	22.50
Retort distillation, oxidation	10–20
Muffle distillation, oxidation	10-20
Retort reduction	23.50
Galvanizing	2.50
Fugitive emissions	
Reverberatory sweating	0.63
Rotary sweating	0.45
Muffle sweating	0.54
Kettle (pot) sweating	0.28
Electric resistance sweating	0.25
Crushing, screening	2.13
Kettle (pot) melting	0.0025
Crucible melting furnace	0.0025
Reverberatory melting furnace	0.0025
Electric induction melting	0.0025
Retort and muffle distillation	1.18
Casting	0.0075
Secondary copper smelting	
Cupola	
Insulated copper wire	115
Scrap copper and brass	35
Reverberatory	
Brass and bronze	18
Copper	2.55
Rotary, brass and bronze	150

Process	Emissions (kg particles/ Mg raw material)
Secondary copper smelting–*Continued*	
Crucible, pot for brass and bronze	10.50
Electric arc	
Copper	2.50
Brass and bronze	5.50
Electric induction	
Copper	3.50
Brass and bronze	10.00
Primary lead smelting	
Ore crushing	1.00
Sintering (updraft)	106.50
Blast furnace	180.50
Dross reverberatory furnace	10.00
Fugitive emissions	
Ore mixing and pelletizing	1.13
Conveyor loading, car charging (sinter)	0.25
Sinter machine leakage	0.34
Sinter crushing, screening, discharge	0.75
Sinter transfer to dump	0.10
Ladle operation	0.46
Slag cooling	0.24
Zinc fuming furnace vents	2.30
Dross kettle	0.24
Reverberatory furnace leakage	1.50
Silver retort building	0.90
Lead casting	0.44
Secondary lead	
Sweating	16–35
Smelting	
Reverberatory	73.50
Blast (cupola)	96.50
Fugitive emissions	
Sweating	0.8–1.8
Smelting	1.4–7.9
Casting	0.44
Lead products	
Type metal production	0.35
Cable covering	0.30

Emission Factors for Volatile Hydrocarbons from Uncontrolled Sources*

Source	Emission Factor
Surface coating	
Paint	560 kg/Mg
Varnish and shellac	500 kg/Mg
Lacquer	770 kg/Mg
Enamel	420 kg/Mg
Primer (zinc chromate)	660 kg/Mg
Plywood veneer dryers	
Douglas fir, sapwood	
Steam fired	2.3 kg/10,000 m^2
Gas fired	38.6 kg/10,000 m^2
Douglas fir, heartwood	6.7 kg/10,000 m^2
Larch	1.0 kg/10,000 m^2
Southern pine	15.1 kg/10,000 m^2
Interior printed panels	
Filler	
Water borne	0.3 kg/100 m^2
Conventional paint	3.0 kg/100 m^2
Ultraviolet coating	Negligible
Sealer	
Water borne	0.2 kg/100 m^2
Conventional paint	0.5 kg/100 m^2
Ultraviolet coating	0
Basecoat	
Water borne	0.2 kg/100 m^2
Conventional paint	2.4 kg/100 m^2
Ultraviolet coating	0.24 kg/100 m^2
Ink	
Water borne	0.1 kg/100 m^2
conventional paint	0.3 kg/100 m^2
ultraviolet coating	0.1 kg/100 m^2

* Compiled from references 2, 4–10, and 381.

Source	Emission Factor
Ink–*Continued*	
Topcoat	
Water borne	$0.4\,\text{kg}/100\,\text{m}^2$
Conventional paint	$1.8\,\text{kg}/100\,\text{m}^2$
Ultraviolet coating	Negligible
Rotogravure printing	$840\,\text{kg}/\text{Mg}$
Dryer exhaust	$130\,\text{kg}/\text{Mg}$
Fugitive	$30\,\text{kg}/\text{Mg}$
Printed product	
Nonmethane hydrocarbon vapor emissions (lbm/hr)	
Valves	
Gas–vapor streams	0.047
Light liquid/two-phase streams	0.023
Heavy liquid streams	0.0007
Pump seals	
Light liquid streams	0.26
Heavy liquid streams	0.045
Flanges (all)	0.00058
Gas–vapor	0.0005
Light liquid/two-phase streams	0.0005
Heavy liquid streams	0.0007
Compressor seals	
Hydrocarbon service	0.98
Hydrogen service	0.10
Drains (all)	0.070
Light liquid/two-phase streams	0.085
Heavy liquid streams	0.029
Relief valves (all)	0.19
Gas–vapor streams	0.36
Light liquid/two-phase streams	0.013
Heavy liquid streams	0.019
Transfer of hydrocarbons by tank cars and trucks	
Submerged loading—normal service	
Gasoline	$0.6\,\text{kg}/\text{m}^3$
Crude oil	$0.4\,\text{kg}/\text{m}^3$
JP-4	$0.18\,\text{kg}/\text{m}^3$
Kerosene	$0.002\,\text{kg}/\text{m}^3$
No 2	$0.001\,\text{kg}/\text{m}^3$
No 6	$0.00001\,\text{kg}/\text{m}^3$

Source	Emission Factor
Transfer of hydrocarbons by tank cars and trucks–*Continued*	
Splash loading—normal service	
Gasoline	$1.4\,kg/m^3$
Crude oil	$0.8\,kg/m^3$
JP-4	$0.5\,kg/m^3$
Kerosene	$0.005\,kg/m^3$
No 2	$0.004\,kg/m^3$
No 6	$0.00004\,kg/m^3$
Submerged loading—balance service	
Gasoline	$1.0\,kg/m^3$
Crude oil	$0.6\,kg/m^3$
JP-4	$0.3\,kg/m^3$
Splash loading—balance service	
Gasoline	$1.0\,kg/m^3$
Crude oil	$0.6\,kg/m^3$
JP-4	$0.3\,kg/m^3$
Gasoline in transit (fully loaded)	$0.001–0.009\,kg/m^3$
Degreasers	
All	$1000\,kg/Mg$
Cold cleaner ($0.39\,kg/hr\,m^2$)	$430\,kg/Mg$
Open top vapor ($0.73\,kg/hr\,m^2$)	$775\,kg/Mg$
Conveyorized	$850\,kg/Mg$
Sabric scouring	$500\,kg/Mg$

SOLVENT EMISSION (RATE RELATIVE TO EVAPORATION OF CARBON TETRACHLORIDE)

Hydrocarbon (HC)	$\left(\dfrac{\text{HC evaporation rate}}{CCl_4\ \text{evaporation rate}}\right)$
Carbon tetrachloride	1.00
Toluene	0.12
Methyl ethyl ketone	0.45
Acetone	0.91
n-Butanol	0.035
Sec-Butanol	0.094
Naphtha, coal tar	0.015–0.12
Naphtha, safety (standard)	0.015–0.12
Mineral spirits	0.0063
Ethers (petroleum)	1.00
Benzene	0.49
o-Xylene	0.055
Cyclohexane	0.02

Hydrocarbon (HC)	$\left(\dfrac{\text{HC evaporation rate}}{\text{CCl}_4 \text{ evaporation rate}} \right)$
Hexane	1.13
Trichlorotrifluroethane	2.80
Methylene chloride	1.47
Perchloroethylene	0.27
Trichloroethylene	0.69
1, 1, 1-trichloroethane	1.39

Emission Factors for Uncontrolled Mineral Processes*

Process	Emissions (kg/Mg raw material)
Brick	
Raw material handling (particles)	
Drying	35
Grinding	38
Storage	17
Curing and firing (fluorides)	
Gas, oil, or coal-fired kiln	0.4
Portland cement (particles)	
Dry process	
Kilns	123
Dryers, grinders	48
Wet process	
Kilns	100
Dryers, grinders	16
Ceramic clay (particles)	
Drying	35
Grinding	38
Storage	17
Clay and fly-ash sintering (particles)	
Fly ash	55
Clay mixed with coke	20
Natural clay	6
Concrete batching (particles)	
Transfer of sand and aggregate to bins	0.02
Transfer of cement to silos	0.12
Loading mixer with raw materials	0.01
Loading mix truck	0.01
Loading dry-batch truck	0.02

* Compiled from references 6–10.

Process	Emissions (kg/Mg raw material)
Glass fiber (particles)	
Unloading and conveying	1.5
Storage bins	0.1
Mixing and weighing	0.3
Glass furnace—wool	
Electric	0.25
Gas-regenerative	11
Gas-recuperative	13-15
Gas-unit melter	4.5
Glass furnace—textile	
Recuperative	1
Regenerative	8
Unit melter	3
Forming—wool	
Rotary spun	29
Flame attenuation	1
Oven curing—wool	
Rotary spun	4.5
Flame attenuation	3
Rotary spun	
Cooling—wool	0.65
Oven curing and cooling—textile	0.6
Glass frit smelters	
Rotary furnace	
Particles	8
Fluorides	2.5
Glass manufacture (soda—lime)	
Particles	1
Gypsum (particles)	
Raw material dryer	20
Primary grinder	0.5
Calciner	45
Conveying	0.35
Lime (particles)	
Crushing, primary	15.5
Calcining	
Vertical kiln	4
Rotary kiln	100

Process	Emissions (kg/Mg raw material)
Mineral wool (particles)	
Cupola	11
Reverberatory furnace	2.5
Blow furnace	8.5
Curing oven	2
Cooler	1
Stone quarrying and processing (particles)	
Crushing	
Primary	0.25
Secondary crushing and screening	0.75
Tertiary crushing and screening	3
Recrushing and screening	2.5
Fines mill	3
Screening, conveying, and handling	1
Storage pile losses	5
Phosphate rock processing (particles)	
Drying	2.9
Calcining	7.7
Grinding	0.8
Transfer and storage	1
Open storage piles	20
Asphalt concrete	
Hot mix dryer drum (particles)	2.45
Stack gas	
Particles	0.137
Sulfur dioxide	0.146 times % sulfur
Volatile organic hydrocarbons	0.1
Carbon monoxide	0.019
Polycyclic organic compounds	0.000013
Aldehydes	
Formaldehyde	0.000075
2-Methylpropanal	0.00065
1-Butanal	0.0012
3-Methylbutanal	0.008
Fugitive particles	
Unloading aggregate to bins	0.05
Cold and dried aggregate elevator	0.10
Screening hot aggregate	0.013

Emission Factors for Uncontrolled Chemical Processes*

Process	Emissions (kg/Mg raw material)
Carbon Black	
Oil furnace	
Main process vent	
Particles	3.27
Carbon monoxide	1.4
Hydrocarbons	50
Hydrogen sulfide	30
Flare	
Particles	1.35
Carbon monoxide	122
Hydrocarbons	1.85
Sulfur oxides	25
Hydrogen sulfide	1
CO boiler and incinerator	
Particles	1.04
Carbon monoxide	0.88
Hydrocarbons	0.99
Sulfur oxides	17.5
Hydrogen sulfide	0.11
Charcoal (without chemical recovery plant)	
Particles	200
Carbon monoxide	160
Hydrocarbons (as methane)	50
Crude methanol	76
Acetic acid	116
Paint and varnish manufacture	
Paint	
Particles	1 of pigment
Undefined hydrocarbons	15 of pigment

*Compiled from references 6–10.

Process	Emissions (kg/Mg raw material)
Paint and varnish manufacture–*Continued*	
Varnish (undefined hydrocarbons)	
Bodying oil	20 of pigment
Oleoresinous	75 of pigment
Alkyd	80 of pigment
Acrylic	10 of pigment
Plastics manufacturing	
Polyvinyl chloride	
Particles	17.5
Vinyl chloride gas	8.5
Polypropylene	
Particles	1.5
Propylene gas	0.35
Printing ink (condensed organics)	
Vehicle cooking	
General	60
Oils	20
Oleoresinous	75
Alkyds	80
Synthetic fibers	
Viscose rayon	
Carbon disulfide	27.5
Hydrogen sulfide	3
Nylon	
Hydrocarbons	3.5
Oil mist	7.5
Dacron, oil mist	3.5
Synthetic rubber	
Alkenes	
Butadiene	20
Methyl propene	7.5
Butyene	1.5
Pentadiene	0.5
Alkanes	
Dethylheptane	0.5
Pentane	1
Ethanenitrile	0.5
Carbonyls	
Acrylonitrile	8.5
Acrolein	1.5

Process	Emissions (kg/Mg raw material)
Ammonium sulfate	
Rotary dryers	
Particles	23
Volatile organic hydrocarbons	0.74
Fluidized bed dryers	
Particles	109
Volatile organic hydrocarbons	0.11

Emission Factors for Uncontrolled Food Processes*

Process	Emissions	
Coffee roasting		
Direct-fired		
Particles	3.8	kg/Mg
Aldehydes	0.1	kg/Mg
Organic acids	0.45	kg/Mg
Indirect-fired		
Particles	2.1	kg/Mg
Aldehydes	0.1	kg/Mg
Organic acids	0.45	kg/Mg
Stoner and cooler, particles	0.7	kg/Mg
Instant coffee spray dryer, particles	0.7	kg/Mg
Cotton gin, particles		
Unloading fan	2.27	kg/bale
Cleaner	0.45	kg/bale
Stick and burr machine	1.36	kg/bale
Miscellaneous	1.36	kg/bale
Feed and grain mills and elevators (particles)		
terminal elevators		
Shipping or receiving	0.5	kg/Mg
Transferring, conveying,etc	1	kg/Mg
Screening and cleaning	2.5	kg/Mg
Drying	3	kg/Mg
Grain processing		
Corn meal	2.5	kg/Mg
Soybean	3.5	kg/Mg
Barley or wheat cleaner	0.1	kg/Mg
Milo cleaner	0.2	kg/Mg
Barley flour milling	1.5	kg/Mg

*Compiled from references 6–10.

Process	Emissions
Fermentation, beer and whiskey	
Particles	
Grain handling	1.5 kg/Mg
Drying spent grains	2.5 kg/Mg
Hydrocarbons (whiskey)	0.024 kg/Mg
Meat smoking	
Particles	0.15 kg/Mg
Carbon monoxide	0.3 kg/Mg
Hydrocarbons (methane)	0.035 kg/Mg
Aldehydes (HCOH)	0.04 kg/Mg
Organic acids (acetic)	0.1 kg/Mg

Emission Factors for Indoor Processes and Activities

CIGARETTE (ABSTRACTED FROM REF. 225 AND 349)

General	Mainstream (inhaled)	Sidestream (smouldering)
Duration (sec)	20	550
Tobacco burned (mg)	347	441
Particles no/cigarette	1.05×10^{12}	3.5×10^{12}

Emissions mg/cigarette

	Mainstream (inhaled)	Sidestream (smouldering)
Particles		
Unfiltered cigarettes		
Tar	20.8	44.1
Nicotine	0.92	1.69
Filtered cigarettes		
Tar	10.2	34.5
Nicotine	0.46	1.27
CO	18.3	86.3
NH_3	0.16	7.4
NO_x	0.014	0.051
HCN	0.24	0.16
Acrolein	0.084	0.825

GAS RANGE AND KEROSENE SPACE HEATERS (MICROGRAMS/KCAL)
(ABSTRACTED FROM REF. 226)

Compound	Gas Range (2500 kcal/hr)	Space Heater (2800 kcal/hr)
CO	890	632
CO_2	209,000	200,000
NO	31	76
NO_2	85	46
SO_2	0.8	
CH_2O	7.1	

FORMALDEHYDE (ABSTRACTED FROM REF. 226)

Material	Emission rate (mg/m^2 day)
UF-foam insulation	1–50
Plywood (UF–bonded)	1–34
Hardwood paneling (UF)	1–34
Particleboard (std, UF)	2–34
Fiberglass ceiling panel	2.8
100% cotton drapery fabric	0.2–0.7
Paper cups and plates	0.33–0.7
Fiberglass insulation	0.45
Latex-backed fabric	0.19
Foam backed carpet	0.12
Nylon upholstery fabric	0.018

COMMON HOUSEHOLD AEROSOLS IN 1970 (ABSTRACTED FROM REF. 226)

Material	Emission Rate (g/month)
Deodorant spray	112–140
Hairspray	84–112
Shaving foam	84–112
Air fresheners	28–56
Disinfectant sprays	112
Furniture polish	56
Dust sprays	28–56
Oven cleaners	84

SMALL STOVES (ABSTRACTED FROM REF. 218)

Material	Emission Factor (g/kg fuel)
Wood	
Particles	1.6–6.4
Carbon monoxide	100
Bituminous Coal	
Particles	10.4
Carbon Monoxide	116
Anthracite Coal	
Particles	0.5
Carbon Monoxide	21

Vapor Pressures and OSHA PEL for Industrial Volatile Liquids

Substance	TWA-PEL (PPM)	M	Temperature (°C)[a]									
			1 mm	5 mm	10 mm	20 mm	40 mm	60 mm	100 mm	200 mm	400 mm	760 mm
Acetaldehyde	100	44	−81.5	−65.1	−56.8	−47.8	−37.8	−31.4	−22.6	−10.0	+4.9	20.2
Acetic acid	10	60	−17.2	+6.3	17.5	29.9	43.0	51.7	63.0	80.0	99.0	118.1
Acetic anhydride	5C	102	1.7	24.8	36.0	48.3	62.1	70.8	82.2	100.0	119.8	139.6
Acetone	750	58	−59.4	−40.5	−31.1	−20.8	−9.4	−2.0	+7.7	22.7	39.5	56.5
Acrolein	0.1	56	−64.5	−46.0	−36.7	−26.3	−15.0	−7.5	+2.5	17.5	34.5	52.5
Allyl alcohol[d]	2	58	−20.0	+0.2	10.5	21.7	33.4	40.3	50.0	64.5	80.2	96.6
Aniline[d]	2	93	34.8	57.9	69.4	82.0	96.7	106.0	119.9	140.1	161.9	184.4
Benzene[c]	10	78	−36.7	−19.6	−11.5	−2.6	+7.6	15.4	26.1	42.2	60.6	80.1
Biphenyl[c]	0.2	154	70.6	101.8	117.0	134.2	152.5	165.2	180.7	204.2	229.2	254.9
Carbon disulfide[c]	4	76	−73.8	−54.3	−44.7	−34.3	−22.5	−15.3	−5.1	+10.4	28.0	46.5
Carbon tetrachloride[c]	2	154	−50.0	−30.0	−19.6	−8.2	+4.3	12.3	23.0	38.3	57.8	76.7
Chlorobenzene	75	113	−13.0	+10.6	22.2	35.3	49.7	58.3	70.7	89.4	110.0	132.2
Chloroform	2	119	−58.0	−39.1	−29.7	−19.0	−7.1	+0.5	10.4	25.9	47.7	61.3
1-Chloroprene	10	89	−81.3	−63.4	−54.1	−44.0	−32.7	−25.1	−15.1	+1.3	18.0	37.0
Cyclohexane	300	84	−45.3	−25.4	−15.9	−5.0	+6.7	14.7	25.5	42.0	60.8	80.7
Dimethylamine	5	45	−87.7	−72.2	−64.6	−56.0	−46.7	−40.7	−32.6	−20.4	−7.1	+7.4
Dimethylaniline	10	121	29.5	56.3	70.0	84.8	101.6	111.9	125.8	146.5	169.2	193.1
Ethyl acetate	400	88	−43.4	−23.5	−13.5	−3.0	+9.1	16.6	27.0	42.0	59.3	77.1
Ethyl acrylate[d]	5	100	−29.5	−8.7	+2.0	13.0	26.0	33.5	44.5	61.5	80.0	99.5
Ethyl alcohol	1000	46	−31.3	−12.0	−2.3	+8.0	19.0	26.0	34.9	48.4	63.5	78.4
Ethylamine	10	45	−82.3	−66.4	−58.3	−48.6	−39.8	−33.4	−25.1	−12.3	+2.0	16.6
Ethyl benzene	100	106	−9.8	+13.9	25.9	38.6	52.8	61.8	74.1	92.7	113.8	136.2
Ethyl chloride	1000	65	−89.8	−73.9	−65.8	−56.8	−47.0	−40.6	−32.0	−18.6	−3.9	+12.3
Ethylene dibromide[d]		188	−27.0	+4.7	18.6	32.7	48.0	57.9	70.4	89.8	110.1	131.5
Ethylene dichloride	1	99	−44.5	−24.0	−13.6	−2.6	+10.0	18.1	29.4	45.7	64.0	82.4
Ethyl formate	100	74	−60.5	−42.2	−33.0	−22.7	−11.5	−4.3	+5.4	20.0	37.1	54.3
Ethyl mercaptan	0.5	62	−76.7	−59.1	−50.2	−40.7	−29.8	−22.4	−13.0	+1.5	17.7	35.0
Formaldehyde[c]	3	30	—	—	−88.0	−79.6	−70.6	−65.0	−57.3	−46.0	−33.0	−19.5
Formic acid	5	46	−20.0	−5.0	+2.1	10.3	24.0	32.4	43.8	61.4	80.3	100.6
Furfural[d]	2	96	18.5	42.6	54.8	67.8	82.1	91.5	103.4	121.8	141.8	161.8
Heptane	400	100	−34.0	−12.7	−2.1	+9.5	22.3	30.6	41.8	58.7	78.0	98.4
Hexachloroethane[d]	1	237	32.7	49.8	73.5	87.6	102.3	112.0	124.2	143.1	163.8	185.6
Hexane	50	86	−53.9	−34.5	−25.0	−14.1	−2.3	+5.4	15.8	31.6	49.6	68.7

Compound	PEL	MW	1	2	3	4	5	6	7	8	9	10
Isobutyl acetate	150	116		+1.4	12.8	25.5	39.2	48.0	59.7	77.6	97.5	118.0
Isobutyl alcohol	50	74		−9.0	21.7	32.4	44.1	51.7	61.5	75.9	91.4	108.0
Isopropyl acetate	250	102		−38.3	−7.2	+4.2	17.0	25.1	35.7	51.7	69.8	89.0
Isopropyl alcohol	400	60		−26.1	+2.4	12.7	23.8	30.5	39.5	53.0	67.8	82.5
Methacrylic acid	20	86		25.5	60.0	72.7	86.4	95.3	106.6	123.9	142.5	161.0
Methyl acetate	200	74		−57.2	−29.3	−19.1	−7.9	−0.5	+9.4	24.0	40.0	57.8
Methyl acrylate[d]	10	86		−43.7	−13.5	−2.7	+9.2	17.3	28.0	43.9	61.0	80.2
Methyl alcohol	200	32		−44.0	−16.2	−6.0	+5.0	12.1	21.2	34.8	49.9	64.7
Methylamine	10	31		−95.8	−73.8	−65.9	−56.9	−51.3	−43.7	−32.4	−19.7	−6.3
Methylene chloride		85		−70.0	−52.1	−43.3	−22.3	−15.7	−6.3	+8.0	24.1	40.7
Methyl chloride	50	51		—	−99.5	−92.4	−76.0	−70.4	−63.0	−51.2	−38.0	−24.0
Methyl formate	100	60		−74.2	−57.0	−48.6	−39.2	−28.7	−21.9	−12.9	16.0	32.0
Methyl iodide[d]	2	142	—	−55.0	−45.8	−35.6	−24.2	−16.9	−7.0	+8.0	25.3	42.4
Naphthalene	10	128	52.6	74.2	85.8	101.7	119.3	130.2	145.5	167.7	193.2	217.9
Nitrobenzene	1	123	44.4	71.6	84.9	99.3	115.4	125.8	139.9	161.2	185.8	210.6
Nitroethane	100	75	−21.0	+1.5	12.5	24.8	38.0	46.5	57.8	74.8	94.0	114.0
Nitromethane	100	61	−29.0	−7.9	+2.8	14.1	27.5	35.5	46.6	63.5	82.0	101.2
2-Nitrotoluene[d]	2	137	50.0	79.1	93.8	109.6	126.3	137.6	151.5	173.7	197.7	222.3
Octane	300	114	−14.0	+8.3	19.2	31.5	45.1	53.8	65.7	83.6	104.0	125.6
Perchloroethylene	25	166										
Phenol[d]	5	94	40.1	62.5	73.8	86.0	100.1	108.4	121.4	139.0	160.0	181.9
n-Propyl acetate	200	102	−26.7	−5.4	+5.0	16.0	28.8	37.0	47.8	64.0	82.0	101.8
n-Propyl alcohol[d]	200	60	−15.0	+5.0	14.7	25.3	36.4	43.5	52.8	66.8	82.0	97.8
Propylene oxide	20	58	−75.0	−57.8	−49.0	−39.3	−28.4	−21.3	−12.0	+2.1	17.8	34.5
Pyridine	5	79	−18.9	2.5	13.2	24.8	38.0	46.8	57.8	75.0	95.6	115.4
Styrene, monomer	50	104	−7.0	+18.0	30.8	44.6	59.8	69.5	82.0	101.3	122.5	145.2
Toluene	100	92	−26.7	−4.4	+6.4	18.4	31.8	40.3	51.9	69.5	89.5	110.6
Trichloroethylene	50	131	−43.8	−22.8	−12.4	−1.0	+11.9	20.0	31.4	48.0	67.0	86.7
1, 2, 3-Trichloropropane	10	147	9.0	33.7	46.0	59.3	74.0	83.6	96.1	115.6	137.0	158.0
Xylene	100	106	−3.8	+20.2	32.1	45.1	59.5	68.8	81.3	100.2	121.7	144.4

Source:

c See (350) for ceiling and maximum transitory values

in process of rulemaking

a See Appendices A-8 and A-10 for PEL of additional materials. Note PEL and TLV are reviewed on a regular basis and readers should always use currently approved values.

d denotes additional entry through the skin

Henry's Law Constant and Diffusion Coefficients of Contaminants in Air and Water

Substance	M	Henry's Law Constant (10^7 N/m^2)	Diffusion Coefficient Air (10^{-5} m^2/s)	Water (10^{-9} m^2/s)
Acetic acid	60		1.06	1.19
Acetone	56		0.83	1.16
Acetonitrile	41			1.26
Acetylene	26	13.5	1.7	2.0
Ammonia	17	0.03	2.2	2.0
Aniline	93		0.75	0.92
Benzene	78	3.05^S	0.77	1.02
Benzoic acid	122			1.00
Benzyl alcohol	94			0.82
Biphenyl	154	0.03^S		
Bromine	160	0.747	1.0	1.3
n-Butane	58		0.96^V	0.89
n-Butanol	74		0.89	0.77
Carbon dioxide	44	16.5	1.5	2.0
Carbon monoxide	28	587.0	2.0	2.0
Carbon disulfide	76		0.89^P	
Carbon tetrachloride	154	11.1^S	0.62	0.82
Carbonyl sulfide	60	26.3	1.3	1.5
Chloroform	119	2.66^L	0.87	0.92
Chlorine	71	6.82	1.2	1.5
Chlorobenzene	113	2.0^S	0.62	0.86^M
Cyclohexane	84	18.0^S	0.86^P	
Dibromochloropropane		0.021	0.69	0.72
Diethylamine			0.88	0.97
Ethane	30	281^S	1.5	1.4
Ethyl alcohol	46		1.02	0.84
Ethyl acetate	88		0.72	1.00
Ethyl benzene	106	4.44^S	0.66^P	0.81^R
Ethylene	28	116.0	1.6	1.5
Ethylene dibromide	188	85.66	0.81	0.89
Ethylene dichloride	99	0.61^L		
Ethyl formate	74		0.84^P	

Substance	M	Henry's Law Constant (10^7 N/m^2)	Diffusion Coefficient Air $(10^{-5} \text{ m}^2/\text{s})$	Water $(10^{-9} \text{ m}^2/\text{s})$
Ethylene glycol				1.16
Formaldehyde	30			
Formic acid	46		1.31[P]	0.69[P]
Furfural	96			1.04[R]
Glycerol				0.82
Glycine				1.06
Heptane	100		0.71[R]	
Hexane	86	944[S]	0.8[R]	
Hydrogen cyanide	27	0.064	1.5	1.8
Hydrogen sulfide	34	5.52	1.7	1.6
Isobutyl acetate	116	0.61[C]		
Isopropyl alcohol	60		1.07[V]	0.87[R]
Methane	16	374	2.2	1.8
Methyl alcohol	32		1.33	0.84
Methyl acetate	74		0.84[P]	
Methyl chloroform	133	0.346	0.78	0.81
Methyl chloride	51	13.3[L]	1.3	1.5
Methylene chloride	85	1.67		
Methyl formate	60		0.87[P]	
Napthalene	128	0.043[S]	0.51[P]	
Nitric oxide	30	291.0	2.0	2.4
Nitrous oxide	44	22.7	1.5	1.8
Nitrobenzene	123		0.86[V]	
Octane	114	1667[S]		
Oxalic acid	90			1.53
Ozone	48	46.4		2.0
Perchloroethylene	166	2.42	0.74	0.76
Phosgene	99		0.80	
Phosphine	34	398[S]	1.6	
Propane	44		0.88	0.97
Propylene	42	57.3		1.1
n-Propyl acetate	102		0.67[P]	
Propyl alcohol	60		0.85[P]	1.1[P]
Pyridine	79			0.58
Sulfur dioxide	64	0.485	1.3	1.7
Toluene	92	3.72[S]	0.71	0.844[M]
Trichloroethylene	131	0.922	0.78	0.81
Urethane				1.06
o-Xylene	106	2.78[S]		

Source: Taken from reference 19 except where noted.
[a] Superscripts: R, reference 323; P, reference 18; S, converted from data in reference 324; V, reference Vargaftik; M, reference 115; C, critical tables see Crawford for details.

Critical Temperatures, Pressures and PEL for Common Toxicants

Name	Formula	T_c(C)	P_c(atm)	TWA-PEL* (PPM)
Acetaldehyde	C_4H_4O	187.8	54.7	100(Ceiling)
Acetic acid	$C_2H_4O_2$	321.6	57.1	10
Acetone	C_3H_6O	235.5	47	750
Acetonitrile	C_2H_3N	274.7	47.7	40
Aniline	C_6H_7N	425.6	52.3	2[d]
Benzene	C_6H_6	288.9	48.6	10[c]
Benzyl chloride	C_6H_5Cl	359.2	44.6	1
Boron triflouride	BF_3	−12.3	49.2	1(Ceiling)
Carbon disulfide	CS_2	279	78	4[c]
Carbon tetrachloride	CCl_4	283.4	45.6	2[c]
Diethylamine	$C_4H_{11}N$	223.3	36.6	10
Diemethylamine	C_2H_7N	164.6	52.4	10
Ethylene Oxide	C_2H_5O	195.8	71	1
Hydrogen chloride	HCl	51.4	82.1	5(Ceiling)
Hydrogen cyanide	HCN	183.5	48.9	4.7(STEL)
Hydrogen sulfide	H_2S	100.4	88.9	10
Methyl alcohol	CH_4O	240	78.5	200
Methylamine	CH_5N	156.9	40.2	10
Methylene chloride	CH_2Cl_2	237	60	#
Methyl mercaptan	CH_4S	196.8	71.4	0.5
Naphthalene	$C_{10}H_8$	474.8	40.6	10
Nitric oxide	NO	−93	64	25
Ozone	O_3	−5.2	67	0.1
Phenol	C_6H_6O	421.1	60.5	5[d]
Propylene oxide	C_3H_6O	209	48.6	20
Styrene	C_8H_8	374.4	39.4	50
Triethylamine	$C_6H_{15}N$	258.9	30	10
Toluene	C_7H_8	320.8	41.6	100

Source:

Abstracted from reference 27, 350 and 351.

* in process of rulemaking

Notes

[d] denotes additional entry through skin

[c] See (350) for ceiling and maximum transitory values

Thermophysical Properties of Air (61)

T (K)	ρ (kg/m³)	C_p (kJ/kg·K)	μ (10^{-7} N·s/m²)	ν (10^{-6} m²/s)	κ (10^{-3} W/m·K)	α (10^{-6} m²/s)	Pr
200	1.7458	1.007	132.5	7.590	18.1	10.3	0.737
250	1.3947	1.006	159.6	11.44	22.3	15.9	0.720
300	1.1614	1.007	184.6	15.89	26.3	22.5	0.707
350	0.9950	1.009	208.2	20.92	30.0	29.9	0.700
400	0.8711	1.014	230.1	26.41	33.8	38.3	0.690
450	0.7740	1.021	250.7	32.39	37.3	47.2	0.686
500	0.6964	1.030	270.1	38.79	40.7	56.7	0.684
550	0.6329	1.040	288.4	45.57	43.9	66.7	0.683
600	0.5804	1.051	305.8	52.69	46.9	76.9	0.685
650	0.5356	1.063	322.5	60.21	49.7	87.3	0.690
700	0.4975	1.075	338.8	68.10	52.4	98.0	0.695
750	0.4643	1.087	354.6	76.37	54.9	109.0	0.702
800	0.4354	1.099	369.8	84.93	57.3	120.0	0.709
850	0.4097	1.110	384.3	93.80	59.6	131.0	0.716
900	0.3868	1.121	398.1	102.9	62.0	143.0	0.720
950	0.3666	1.131	411.3	112.2	64.3	155.0	0.723
1000	0.3482	1.141	424.4	121.9	66.7	168.0	0.726

Formulas for interpolation (T = absolute temperature)

$$\rho = \frac{348.59}{T} \qquad (\sigma = 9 \times 10^{-4})$$

$$f(T) = A + BT + CT^2 + DT^3$$

$f(T)$	A	B	C	D	Standard deviation (σ)
c_p	1.0507	-3.645×10^{-4}	8.388×10^{-7}	-3.848×10^{-10}	4×10^{-4}
$\mu \times 10^7$	13.554	0.6738	-3.808×10^{-4}	1.183×10^{-7}	0.4192
$\kappa \times 10^3$	-2.450	0.1130	-6.287×10^{-5}	1.891×10^{-8}	0.1198
$\alpha \times 10^6$	-11.064	7.04×10^{-2}	1.528×10^{-4}	-4.476×10^{-8}	0.4417
Pr	0.8650	-8.488×10^{-4}	-1.234×10^{-6}	-5.232×10^{-10}	1.623×10^{-3}

Runge–Kutta Methods for Solving First-order Differential Equations

Suppose one wants to find the value of the continuous function $U(t)$ at arbitrary values of t where the function satisfied the nonlinear differential equation,

$$\frac{dU}{dt} = UA(t) + B(t) \tag{A-12.1}$$

and the quantities A and B are known analytical functions of t. Assume that the initial value of $U(0)$ is known, i.e.

$$U(t_0) = U(0) = \text{known} \tag{A-12.2}$$

Express the function $U(t)$ as a Taylor series,

$$U(t + \delta t) = U(t) + U'(t)\,\delta t + U''(t)\,\frac{\delta t^2}{2} + \ldots \tag{A-12.3}$$

Where the function $U(t)$ and its derivatives $U'(t)$ and $U''(t)$ are evaluated at time t and δt is a small increment of time. Eliminate the second, third, and all higher order derivatives and replace the first derivative by Eq. A-12.1

$$U(t + \delta t) = U(t) + \delta t[U(t)A(t) + B(t)] \tag{A-12.4}$$

Since dropping the higher-order derivatives introduces error, replace $U(t)$ inside the brackets on the right-hand side by the average value of the function between (t) and $(t + \delta t)$,

$$U(t + \delta t) = U(t) + \delta t\left[A(t)\frac{[U(t) + U(t + \delta t)]}{2} + B(t) \right] \tag{A-12.5}$$

Rearrange and obtain

$$U(t + \delta t) = \frac{\left\{ U(t)\left[1 + \left(\frac{\delta t}{2}\right)A(t) \right] + \delta t\, B(t) \right\}}{M(t)}$$

$$M(t) = 1 - \left(\frac{\delta t}{2}\right)A(t) \tag{A-12.6}$$

To obtain the value of U at the end of a time period t_n, that is, $U(t_n)$, begin by evaluating Eq. A-12.6 at t_1:

$$t_1 = t_0 + \delta t$$

$$U(t_1) = \frac{\left\{ U(t_0)\left[1 + \left(\frac{\delta t}{2} \right) A(t_0) \right] + (\delta t) B(t_0) \right\}}{M(t_0)}$$

$$M(t_0) = 1 - \left(\frac{\delta t}{2} \right) A(t_0) \tag{A-12.7}$$

Now evaluate Eq. A-12.6 at t_2 but use $U(t_1)$, $A(t_1)$, and $B(t_1)$ on the right hand-side of Eq. A-12.6,

$$t_2 = t_1 + \delta t$$

$$U(t_2) = \frac{\left\{ U(t_1)\left[1 + \left(\frac{\delta t}{2} \right) A(t_1) \right] + (\delta t) B(t_1) \right\}}{M(t_1)}$$

$$M(t_1) = 1 - \left(\frac{\delta t}{2} \right) A(t_1) \tag{A-12.8}$$

Repeat the process until you have computed $U(t_n)$

$$t_n = t_{n-1} + \delta t \tag{A-12.9}$$

The quantity δt is the time step; its magnitude is selected by the user. The value should be small such that

$$\left(\frac{\delta t}{2} \right) A(t) < 1 \tag{A-12.10}$$

Fourth-order Runge–Kutta Method

More accuracy can be obtained using Runge–Kutta methods of higher order (196). In such methods estimates are made at intermediate values between t and $t + \delta t$. A commonly used expression is the Runge-Kutta fourth-order method. Begin by expressing the differential equation as,

$$\frac{dU}{dt} = f(U, t) \tag{A-12.11}$$

It will be presumed that the value of U is known at the initial time t_i, that is

$$U_i = U(t_i) \tag{A-12.12}$$

The value of U at the end of the time step δt, that is,

$$U(t + \delta t) = U_{i+i} \tag{A-12.13}$$

is given by

$$U_{i+1} = U_i + \delta t \left[\frac{f_i}{6} + \frac{f_1}{3} + \frac{f_2}{3} + \frac{f_3}{6} \right] \qquad \text{(A-12.14)}$$

where the functions f_i, f_1, f_2, and f_3 are given by

$$f_i = f(t_i, U_i) \qquad \text{(A-12.15)}$$

$$t_1 = t_i + \frac{\delta t}{2} \qquad \text{(A-12.16)}$$

$$U_1 = U_i + \left(\frac{\delta t}{2} \right) f_i \qquad \text{(A-12.17)}$$

$$f_1 = f(t_1, U_1) \qquad \text{(A-12.18)}$$

$$U_2 = U_i + \left(\frac{\delta t}{2} \right) f_1 \qquad \text{(A-12.19)}$$

$$f_2 = f(t_1, U_2) \qquad \text{(A-12.20)}$$

$$U_3 = U_i + (\delta t) f_2 \qquad \text{(A-12.21)}$$

$$t_2 = t_i + \delta t \qquad \text{(A-12.22)}$$

$$f_3 = f(t_2, U_3) \qquad \text{(A-12.23)}$$

Simultaneous Ordinary Differential Equations

If one is asked to solve a simultaneous set of ordinary differential equations of the form

$$\frac{dp}{dt} = f(p, q, \ldots, z, t)$$

$$\frac{dq}{dt} = g(p, q, \ldots, z, t)$$

$$\vdots$$

$$\frac{dz}{dt} = h(p, q, \ldots, z, t) \qquad \text{(A-12.24)}$$

Equations A-12.14 to A-12.23 can also be used. Each intermediate functions f_1, f_2, f_3; g_1, g_2, g_3; and so on involves intermediate values of time t_1 and t_2 and intermediate values of the variables. Since there are several independent variables, each variable has intermediate values noted by subscripts 1, 2, and 3. Thus before intermediate function f_1, g_1, $\ldots$, h_1 are computed, users will have to compute p_1, q_1, r_1, $\ldots$, z_1. Similarly p_2, q_2, r_2, $\ldots$, z_2 will have to computed before the functions f_2, g_2, $\ldots$, h_2 are found. Lastly, p_3, q_3, r_3, $\ldots$, z_3 will have to be computed before functions f_3, q_3, $\ldots$, h_3 are found.

Newton Raphson Method for Solving Simultaneous Equations

Suppose the functions ϕ and ψ are implicit functions of x and y and you wish to evaluate $\phi(x_1, y_1)$ and $\psi(x_1, y_1)$ at arbitrary values of (x_1, y_1). For illustrative purposes, assume

$$\phi = -\exp(\phi + x) \cos(\psi + y)$$
$$\psi = -\exp(\phi + x) \sin(\psi + y) \tag{A-13.1}$$

If Eq. A-13.1 could be rearranged in the form,

$$\phi = f(x, y) \text{ and } \psi = g(x, y) \tag{A-13.2}$$

you could easily evaluate $\phi(x_1, y_1)$ and $\psi(x_1, y_1)$. Unfortunately there are many instances when Eq. A-13.1 can not be so rearranged. The Newton Raphson method is a numerical technique in which the user guesses values for $\phi(x_1, y_1)$ and $\psi(x_1, y_1)$ and through an iterative technique recomputes ϕ and ψ until satisfactory convergence is obtained. Convergence is rapid.

Define two functions F and G,

$$F = \phi + \exp(\phi + x) \cos(\psi + y)$$
$$G = \psi + \exp(\phi + x) \sin(\psi + y) \tag{A-13.3}$$

Differentiate the functions F and G,

$$\frac{\partial F}{\partial \phi} = 1 + \exp(\phi + x) \cos(\psi + y)$$

$$\frac{\partial G}{\partial \phi} = \exp(\phi + x) \sin(\psi + y)$$

$$\frac{\partial F}{\partial \psi} = -\exp(\phi + x) \sin(\psi + y)$$

$$\frac{\partial G}{\partial \psi} = 1 + \exp(\phi + x) \cos(\psi + y) \tag{A-13.4}$$

Note,

$$\frac{\partial F}{\partial \phi} = \frac{\partial G}{\partial \psi}$$

$$\frac{\partial F}{\partial \psi} = -\frac{\partial G}{\partial \phi} \qquad \text{(A-13.5)}$$

Since F and G are continuous functions the chain rule applies.

$$dF = \left(\frac{\partial F}{\partial \phi}\right) d\phi + \left(\frac{\partial F}{\partial \psi}\right) d\psi$$

$$dG = \left(\frac{\partial G}{\partial \phi}\right) d\phi + \left(\frac{\partial G}{\partial \phi}\right) d\psi \qquad \text{(A-13.6)}$$

Rewrite Eq. A-13.6 in finite difference form:

$$F_{i+1} - F_i = \frac{\partial F}{\partial \phi} [\phi_{i+1} - \phi_i] + \frac{\partial F}{\partial \psi} [\psi_{i+1} - \psi_i]$$

$$G_{i+1} - G_i = \frac{\partial G}{\partial \phi} [\phi_{i+1} - \phi_i] + \frac{\partial G}{\partial \psi} [\psi_{i+1} - \psi_i] \qquad \text{(A-13.7)}$$

If the partial derivatives on the right-hand side of Eq. A-13.7 are evaluated at (x_i, y_i), the equation is similar to the Taylor series expansion of F and G in which the higher-order derivatives have been dropped.

The Newton–Raphson method is an iterative technique in which the numerical value of ϕ and ψ at (x_1, y_1) will be obtained in an iterative fashion beginning with guessed values. Define F_{i+1} and G_{i+1} as the correct values of F and G at $\phi(x_1, y_1)$ and $\psi(x_1, y_1)$. The values of F_i and G_i are the approximate values of F and G. Solve the pair of simultaneous equations (Eq A-13.7) for the quantities $[\phi_{i+1} - \phi_i]$ and $[\psi_{i+1} - \psi_i]$

$$[\phi_{i+1} - \phi_i] = \frac{G_i \dfrac{\partial F}{\partial \psi} - F_i \dfrac{\partial G}{\partial \psi}}{J}$$

$$[\psi_{i+1} - \psi_i] = \frac{F_i \dfrac{\partial G}{\partial \phi} - G_i \dfrac{\partial F}{\partial \phi}}{J}$$

where J is the Jacobian,

$$J = \left(\frac{\partial F}{\partial \phi}\right)\left(\frac{\partial G}{\partial \psi}\right) - \left(\frac{\partial F}{\partial \psi}\right)\left(\frac{\partial G}{\partial \phi}\right) \qquad \text{(A-13.8)}$$

Algorithm

The algorithm to evaluate $\phi(x_1, y_1)$ and $\psi(x_1, y_1)$ is as follows:

(a) Select x_1, y_1 at which values of $\phi(x_1, y_1)$ and $\psi(x_1, y_1)$ are sought.

(b) Guess values ϕ_i and ψ_i.

(c) Evaluate F and G using Eq. A-13.3; evaluate the derivatives

$$\frac{\partial F}{\partial \phi}, \quad \frac{\partial F}{\partial \psi}, \quad \frac{\partial G}{\partial \phi}, \quad \frac{\partial G}{\partial \psi}$$

using Eq. A-13.4 at x_1, y_1, ϕ_i, ψ_i.

(d) Compute $[\phi_{i+1} - \phi_i]$ and $[\psi_{i+1} - \psi_i]$ using Eq. A-13.8.

(e) Repeat steps $b-d$ using ϕ_{i+1} and ψ_{i+1} as the guessed values until such time $|\phi_{i+1} - \phi_i|$ and $|\psi_{i+1} - \psi_i|$ are sufficiently small.

Newtons Method

Suppose engineers want to know the coordinates (x, y, ϕ) at which a known but complicated analytical function $G(\phi, x, y)$ is zero. In particular suppose x and y are known but the value ϕ at which the function is zero is not known and cannot be computed in closed form. Newton's method is an uncomplicated method to compute a numerical value in an iterative fashion. For illustrative purposes suppose engineers want to compute the value ϕ for arbitrary values x_1, y_1 for the following function:

$$1 = \left[\frac{\left(\frac{x}{c}\right)}{\cosh \frac{\phi}{k}} \right]^2 + \left[\frac{\left(\frac{y}{c}\right)}{\sinh \frac{\phi}{k}} \right]^2$$

where k is a constant. To begin, define the function $G(\phi, x.y)$

$$G(\phi, x, y) = \left[\frac{\left(\frac{x}{c}\right)}{\cosh \frac{\phi}{k}} \right]^2 - \left[\frac{\left(\frac{y}{c}\right)}{\sinh \frac{\phi}{k}} \right]^2 - 1$$

and its derivative, $\dfrac{\partial G}{\partial \phi}$,

$$\frac{\partial G}{\partial \phi} = - \frac{\left(\frac{2}{k}\right)\left(\frac{x}{c}\right)^2 \sinh \frac{\phi}{k}}{\left(\cosh \frac{\phi}{k}\right)^3} - \frac{\left(\frac{2}{k}\right)\left(\frac{y}{c}\right)^2 \cosh \frac{\phi}{k}}{\left(\sinh \frac{\phi}{k}\right)^3}$$

Algorithm

Step 1: At the value x_1, y_1, guess a value ϕ_i

Step 2: Calculate G and its derivative $\left(\dfrac{\partial G}{\partial \phi}\right)$ at the coordinates (ϕ_i, x_1, y_1)

Step 3: Compute an error E

$$E = |G(\phi_i, x_1, y_1)|$$

Step 4: Compute a new value of ϕ,

$$\phi_{i+i} = \phi_i - \frac{G(\phi_i, x_1, y_1)}{\left(\dfrac{\partial G}{\partial \phi}\right)_{\phi_i, x_1, y_1}}$$

Step 5: Repeat steps 2–4 until E is less than some desired value.

Basic Iteration Methods

BISECTION METHOD

1 Rem: Basic iteration program to find the remaining root
  of f(x, y)=c if either x or y is known
2 Rem: Assume f(x, y) is a monotonically increasing
  function in y. If f(x, y) is a monotonically decreasing
  function in y, reverse the instructions in line 60 to
  read,....yl=yg and in line 70 to read, ... yh=yg
3 Rem: input known values of c and x1
10 yl=0.001
11 yh=100
15 Rem: The function f(x, y) must be studied to
  determine whether it is monotonically increasing or
  decreasing and how to choose the appropriate values of
  yh and yl
20 yg=(yl+yh)/2
30 fg=f(x1,yg)
40 r=fg/c
50 if abs(r-1)<=0.001 then goto 90
60 if abs(r)>1 then yh=yg
70 if abs(r)<1 then yl=yg
80 goto 20
90 print x1,yg

NEWTON METHOD

1 Rem: Assume y=f(x) is a known differentiable
  function and you want to find the value of × at which
  y=C
2 Rem: Expand f(x) as a Taylor Series about x0, i.e.
  x0=xguess. Keep the first two terms and solve for x
  x=x0+(C-f(x0))/f'(x0)
3 Rem: Define error (ER) as x-x0. Thus ER=(C-f(x0))/
  f'(x0)
30 Rem: Input C, xguess and the functions f(x), f'(x)
40 x=xguess

```
45 C=C
50 A=f(x)
55 B=f'(x)
60 ER=(C-B)/A
65 if abs (ER)<=0.001 then goto 80
70 x=x+ER
75 goto 50
80 Print x
```

Outdoor Air Requirements for Ventilation*

	Estimated Maximum person/1000 ft^2	CFM per person	CFM per ft^2
Commercial facilities			
Commercial laundry	10	25	
Commercial dry cleaner	30	30	
Bar, cocktail lounge	100	30	
Cafeteria, fast food	100	20	
Auto repair room			1.5
Office space	7	20	
conference room	50	20	
corridors			0.05
smoking lounge	70	60	
Elevators			1.0
Warehouses	5		0.5
Shipping and receiving	10		1.5
Beauty shop	25	25	
Supermarket	8	15	
Clothier, furniture			0.3
Spectator areas	150	15	
Gymnasium playing floor	30	20	
Auditorium	150	15	
Meat processing	10	15	
Transportation vehicles	150	15	
Photography darkrooms	10		0.5
Duplicating, printing			0.5
Hospital, operating room	20	30	
patient room	10	25	
Education, laboratory	30	20	
classroom	50	15	
Prison cell	20	20	

Private Home
Living area: 0.35 air changes/hr but not less than 1.5 CFM/person
Kitchen: 100 CFM intermittent or 25 CFM continuous or operable windows
Separate garage: 100 CFM per car

* Abstracted from reference 389.

Ventilation Duct Design

This computer program is written for IBM PC/XT/AT/PS2 or compatible computers. It requries PC/MS DOS 3.00 or higher operating systems and an 80-column monochromatic or color monitor. Printed output is directed through LPT 1 and requires no specific printer command sequences.

To safeguard the original disk contained in the text (hereafter called "original disk") and to produce a bootable copy of the Ventilation Duct Design program, prepare a backup disk that will contain a DOS System and the material in the original disk. To make this bootable backup disk, perform the following.

(a) Put a DOS System disk containing the DOS Utility programs DISK-COPY and SYS in drive A. Turn on the computer.

(b) At the DOS prompt, type DISKCOPY A: B: ⟨return⟩, put the original disk in drive A and a blank disk in drive B, then press return (hereafter noted as⟨cr⟩).

(c) Put the DOS disk back in drive A and the backup disk you've just created in drive B and type SYS B: ⟨cr⟩. After the DOS system files are transferred, type COPY A: COMMAND.COM B: ⟨cr⟩. This step will put the command processor on the backup disk. Your backup disk will now boot DOS automatically, load, and run the Ventilation Duct Design program.

To use the program on a hard disk, users should create a subdirectory (example, MD VENT ⟨cr⟩), then type CD to change that directory and make it the default directory. Then copy the VENT.EXC and VENT.INF files. If this is not done there may be difficulty finding the VENT.INF file.

The file VENT.INF provides the option to change the program credits without recompiling the source code. Simply edit the second (and last) line in the file and the credit will appear on the screen of the program. If you have a color monitor and wish to select your own screen colors, type VENT/COLOR ⟨cr⟩at the DOS prompt. This switch allows users to change the foreground and background colors of both the main screen and the control line at the bottom of the display. Once these colors are selected, they will be the default colors provided in the file VENT.INF. This small file carries the color information as well as a brief credit field displayed on the program video interface. If this file is not present on the disk, VENT.EXE will produce a default file with preset colors and credit stored in executable code.

When the backup disk is booted, the program will request the user to specify the drive and/or path where the ventilation data files are located or, if they do not exist, what they will be called after they are created. Typing ⟨cr⟩will display the default disk/subdirectory where the VENT.EXE file is located. A list of all the ventilation (*.DAT) files in the selected disk location will be displayed and the user is then asked for the name of the file to use or create. If a mistake is made in naming the directory, simply press ⟨cr⟩and you will be asked for the disk/path. When the user is prompted to input numerical entries, the alphabetical keys are disabled to minimize improper keystrokes. Also, all the alphabetical entries are converted to upper case before being stored to the disk.

The disk has a modified boot track (372) to allow a screen to be displayed even if no DOS system files are present. This screen will be displayed in RED on a color monitor and explains how (shown above) to make a bootable backup disk containing the DOS system. Once the DOS system is installed the AUTOEXEC.BAT file will load VENT.EXE and run it on bootup. This is adequate for IBM AT computers and computers that keep the current time when they are booted, however IBM PC's and XT's should include a command to set the time on the clock in the AUTOEXEC.BAT file.

The original disk contains the following files:

AUTOEXEC.BAT Batch file used to run the program after the DOS system file had been installed

VENT.EXE Actual duct design program rewritten from the program described in reference 235

VENT.INF An information file used by VENT.EXE to set the screen colors and define credits displayed by the program

READ.ME Program description and last minute change explanation

VENT.DOC An electronic manual

Common Physical Constants and Conversions

Energy	1 joule = 1 kg m^2/s^2 1 BTU = 778.16 ft lbf = 1.055 × 10^{10} ergs = 252 cal 1 cal = 4.186 joules 1 erg = 1 g cm^2/s^2
Force	1 newton = 1 kg m/s^2 = 10^5 dynes 1 dyne = 1 g cm/s^2 1 lbf = 4.448 × 10^5 dynes = 4.448 newtons
Length	1 m = 100 cm = 1000 mm = 3.280 ft = 39.37 cm 1 μm = 10^{-6} m = 10^4 Å 1 mile = 5280 ft = 1609.344 m
Mass	1 kg = 1000 g = 2.2046 lbm = 6.8521 × 10^{-2} slugs 1 slug = 1 lbf s^2/ft = 32.174 lbm 1 ton = 2000 lbm 1 metric ton = 1000 kg 1 ounce (troy) = 3.110347 × 10^{-2} kg
Power	1 watt = 1 joule/s = 1 kg m^2/s^3 1 HP = 550 ft lbf/s = 746 watts = 2545 BTU/hr 1 Mw = 10^3 kw = 10^6 w
Pressure	1 atm = 14.696 lbf/in^2 = 760 torr = 101.325 kPa 1 kPa = 1000 Pa = 1000 newton/m^2 = kJ/m^3 1 mm Hg = 1 torr = 0.01934 lbf/in^2 1 bar = 10^6 dynes/cm^2 = 14.504 lbf/in^2
Temperature	T(R) = 1.8 T(K) T(K) = T(C) + 273.15 T(R) = T(F) + 459.67
Time	1 hr = 60 min = 3600 s 1 msec = 10^{-3} s 1 μs = 10^{-6} s = 1000 nsec

Universal Gas Constant	$8.314 \text{ J/gmol K} = 8.314 \text{ kJ/kmol K}$
	$0.082 \text{ li atm/gmol K}$
	$1.987 \text{ BTU/lbmol R} = 1.987 \text{ cal/gmol K}$
	$1545.33 \text{ ft lbf/lbmol R}$

Volume

1 gal (liquid) $= 0.13368 \text{ ft}^3 = 3.785$ liters
1 liter $= 10^{-3} \text{ m}^3 = 1000.028 \text{ cm}^3$
1 barrel (petroleum) $= 42$ gal
1 fluid ounce $= 2.957352 \times 10^{-5} \text{ m}^3$

Absorption efficiency, lung, Fig 2.27, 80
Acceleration:
  drop, 486
  factor, 341
  total derivative, 594
Accident statistics, Table 1.1, 4
Accumulation range, Fig 8.5, 440
ACGIH, 8, 44, 45, 46
  design plates, 297
  duct design, 333
  manual, 294
  organization, 46
  publications, 44, 294
  TLV, 8, 120
Acinus, 55
Acute response, 101
Activity, physical:
  airway ozone dose, Figs 2.29, 2.30, 83–84
  oxygen uptake, Table 2.3, 73
Activity coefficient, 203
Additives, coatings, 177
Adduct, 108
Administrative controls, 19
Adsorption on walls, 244
Aerodynamic diameter, 95, 453
Aerodynamics of airways, Fig 2.7, 57
Aerosols:
  accumulation range, Fig 8.5, 440
  coarse, 18, Fig 2.36, 94
  definition, 18, 431
  densities, 434
  emission factors, 178
  fine, 18, Fig 2.36, 94
  log-normal distribution functions, 434
  monodisperse, 434
  nuclei range, Fig 8.5, 440
  polydisperse, 434
  primary, 439
  secondary, 439
  size distribution, 434
Age of makeup air, 273, Fig 5.17, 271
Agglomerate, 18
Agglomeration, 439
AICHE, 44
AIHA, 46
Air changes per hour, 230, 237

Air cleaning devices:
  characteristics, 258
  efficiency, 260
Air-to-cloth ratio, 493
Air composition, lungs, Table 2.3, 73
Air curtain, 144, 319, Figs 6.16, 319, Fig 6.18,
  321, 322
  illustrated example, 333
Air density, 119
  moist air density, 119
Air inlet, 324
Airless spraying, 177
Air monitoring, 126
Air motion, lung, Fig 2.21, 72
Air Movement and Control Association, 46
Air pollution, periodicals, 46
Air Pollution Control Association, 46
Air properties, (A-11) 661
Airway:
  abnormalities, Figs 2.13a&b, 62, 63,
    Fig 2.34, 90
  aerodynamics, Fig 2.7, 57
  constriction, Fig 2.13a, 62
  generations, Fig 2.6, 56
  irritation, 96
  obstruction, Fig 2.13a, 62, Fig 2.34, 90
  particle deposition, Fig 2.35, Fig 2.36, 91
  resistance, 89, Fig 2.34, 90
  temperatures, Fig 2.31, 86
  water flux, Fig 2.33, 86
  Weibel model, Fig 2.6, 56
Airway model heat and mass transfer,
    Figs 2.31–2.33, 86
  distributed parameter, Fig 2.28, 82
  extended Bohr model, 76, Fig 2.25, 77
Allergen, 101
Allergic response, 96, 98
Alveoli, 55
  alveolar-capillary barrier, Fig 2.19, 71,
    Fig 2.26, 78
  alveolar clearance, 92
  alveolar duct, 55
  alveolar gas composition, Fig 2.23, 74,
    Fig 2.24, 75
  alveolar sacs, 56
  alveolar surface area, 57

Alveoli (*Continued*)
  alveolar ventilation, 61, 64
  alveolar wall, Fig 2.19, 71
Ambient air quality standards (EPA), 124
AMCA, 46
American Congress of Governmental Industrial
    Hygienists, 8, 44–46
American Industrial Hygiene Association, 46
American Institute of Chemical Engineers, 44
American National Standards Association, 46,
    120
American Society of Heating, Refrigerating
    and Air-Conditioning Engineers, 44
American Society of Mechanical Engineers, 46
Ames test, 11
Analytical models of lung:
  Bohr, 62
    distributed parameter, 81
    extended Bohr, 76
Anatomic dead space, 55, 60, Fig 2.11, 61
Anemic hypoxia, 92
Anisokinetic sampling, 522
Annular risk, 12, Table 1.2, 13
ANSI, 46
  MAC, 20
  organization, 20
Annual cost, 158
Annual operating cost, 155
Annular heating degree-days, 276
APCA, 46
Appliances, 172
Arithmetic mean:
  particle size, 435
  transient time, 91
Arteries, 57
Arterioles, 57
Asbestosis, 98
Asbestos removal, 12
ASHRAE, 44
  Fundamentals Handbook, 335, 338
  duct design, 333
ASME, 46
Asphyxiant, 101
Aspiration coefficient, 522
Assignment scheduling, 19
Asthma, 97
Asymmetric bronchial tree, 66
Asymmetric velocity profiles, 68
Atmospheric aerosol, Fig 8.5, 440
  accumulation range, 440
  coarse 18, 440
  fine 18, 440
  nucleation range, 440
  primary, 439
  secondary, 439
Atmospheric dispersion, 179
Audit, 7
Average density, 198

Average molar concentration, 198
Average molecular weight, 198
Axisymmetric three dimensional flow, 372, 381
  point sink, 382
  potential functions, Table 7.1, 372
  streaming flow, 383

Bagassosis, 99
Bag dumping, Fig 1.26, 40
Bag filling, Fig 1.21, 38
Bag filters, 487
Baghouses, 494
BAL, 11
Batch processes, 493
BEI, 126
Benefit–cost analysis, *see* Risk *vs.* benefit
    analysis
Bernoulli's constant, 365
Bernoulli's equation, 365
Bifurcated flow, Fig 2.15, 66
Binder, coatings, 177
Bioavailability, 102
Biohazard cabinet, 261, 309
Biological Exposure Index, 126
Biological monitoring, 102
Biological safety cabinet, 261, 309
Biot Number, 213
Biotransformation, 104
Bisection Methods, (A-15) 671
Black lung, 99
Blast gate, 334, 335
Blockage, 492
Blood flow rate, Table 2.2, 64
Body burden, 102
Body surface area, 147
Bohr model, 62, Fig 2.14, 63
  extended, Fig 2.25, 77
Boiling temperature, 183
Booths, 31
  grinding, 8–25, Fig 6.10, 308
  spray, Fig 1.24, 39
Bottom filling, 174
Boundary integral method, 405
Boundary layer, grinding wheel, 544, 548
Bounding trajectory, 515
  algorithm, 533
  capture envelope, 405, 524
  inlet:
    flanged, Fig 9.3, 517, Fig 9.5, 518
    line sink, Fig 9.10, 524
    unflanged, Fig 9.4, 518, Fig 9.6, 518
Box model, 228
Branch section, 338
Breathing zone, 125
Bronchial system:
  bronchi, 52, 54, 55
  bronchioles, 52, Fig 2.5, 54, 55
  walls, Fig 2.4, 54

Bronchitis, 98
Bronchoaveolar lavage fluid, 11
Brownian motion:
  coagulation, 439
  filtration, Fig 8.34, 489
Buckingham Pi theorem, 465
Building:
  air inlet, 324
  atmospheric dispersion, 328
  building eddy, Fig 6.19, 325, Fig 6.20, 326
  eddies, 327
  exhaust stack, 324
  material emissions, 255
  recirculation, 324
  roof eddy, Fig 6.19, 325, 327
  roof wake, Fig 6.20, 326, 327
  stack height (EPA), 325
  turbulent shear region, 324
  wakes, 325, 326
Bulk concentration, 256
Bulk convection of gases, pulmonary system, 65
Bulk density, 311
Bulk materials handling, 310
Buoyant sources, 314, 572
  canopy hoods, 314
  plumes, 328
Bureau of Mines, 46
Burning velocity, 132
Bursts, 179
Byssinosis, 99

Cancer, 108
Cancer suspect agents, Table 3.1, 124
Canopy hood, 36, Fig 1.13, 33
  ACGIH, Fig 6.13, 315
  buoyant plumes, Fig 1.13, 33, 314
  close-fitting, Figs 5.28, 5.29
  with curtains, Fig 6.3, 296
  Hayashi, Fig 6.13, 315
  hopper, 317
  pool, 316
Capillaries, 57
  alveolar walls, Fig 2.20, 72
  lymphatic, 59
Capital cost (TCC), 153
Capital recovery cost (CRC), 156
Capture efficiency:
  collectors, 444
  fractional efficiency, 444
  grade efficiency, 444
  impaction:
    curved ducts, 473
    particle–fiber, 488
    particle–particle, 479
  overall, 444
  single fiber, 487
  single particle, 478

Capture envelope, 405, Fig 9.10, 524, Fig 9.14, 534
  reach, 517
Capture hood, *see also* Hoods
  close capture hood, Fig 6.5, 297
  definition, 28
Capture velocity, 28, 297, Table 6.1, 298
Carbon monoxide, Fig 2.41, 107
  illustrated example, 253
Carboxyhemoglobin, 107
Carcinogens, 101
Cardiotoxin, 101
Carina, 53
CAS, 8
Cascade impactors, 538
Ceiling TLV and PEL, 121
Cellular damage, 97
Center for Chemical Process Safety, 44
Centrifugal acceleration, 468
Centrifugal concentrators, 474
Centrifugal cyclone collectors, 474
Centrifugal force, 468
CFD, 573
Chambers and enclosures, gravimetric settling, 460
Chapman–Jouguet points, 131
CHAMPION 2/E/FIX, 603
Chemical Abstract Service, 8
Chemical fume hood, 37, 261, 309
Chemical Information Service, 8
Chemical processes, emission factor, (A-5) 648
Chemicals of commerce, 7
Chipping booth, Fig 1.23, 39
Chronic response, 101
Cigarette emissions, (A-7) 653
Cilia, Fig 2.3, 53
Circular equivalent of rectangular duct, 336
Circular inlet:
  flanged surface, 407
  velocity contours, Fig 6.9, 303
Circulatory hypoxia, 92
CIS, 8
Classification:
  of metal surface process, Table 6.3, 306
  of ventilation systems, Table 1.4, 31
Clausius–Clapeyron equation, 183
  illustrated example, 183
Clean-out ports, 334
Cleanrooms, Fig 1.17, 35, 260
  classification, Fig 5.13, 262
Clearance mechanism, 93
  alveolar, 93
  mucociliary, 93
Close capture hood, 32, 360
Closed system, 229
Close-fitting hood, Fig 1.18, 37, Fig 1.19, 37, Fig 1.29, 42
Cloud of drops, 211

Coagulation, 246, 439
Coarse particles, 18
Coating, spray emission rate, 177
Coefficient of thermal expansion, 139, 572
Collagen, 58, 98
Collection efficiency, 444
  cyclones, 474
  filter, 487
  fractional, 444
  overall, 7–12
  parallel, 446
  series, 445
  single fiber, 488
  single particle, 479
  spray chamber, 480
  venturi scrubber, 484
Combustion wave, 127
Combustion zone, 127
Compartment model of respiratory system, 76
Compliance, 59, Fig 2.10, 60
Components of ventilation system, 24
Composition of respiratory gas, Table 2.3, 73
Compound interest, 158
Compressibility factor, 119, 184
Computational fluid dynamics (CFD), 573
Computer programs:
  air speed on streamline, illustrated example,
    397
  bounding trajectory, 573
  iteration programs:
    bisection, (A-15) 671
    Newton, (A-15) 671
  mapping ϕ and ψ, illustrated example, 396
  Newton's Method, (A-15) 671
  Newton–Raphson, (A-13) 666
  particle trajectory, 455
  reach, streaming flow:
    point sink, 383
    line sink, 377
    unflanged inlet, 402
  Runge–Kutta, (A-12) 663
  settling velocity, 448, Fig 8.3, 450
  time-varying source, 238
Concentration, 118
  average molar, 198
  bulk, 256
  conversion to PPM, 120
  mass gas per unit volume, 118
  mass particles per unit volume of gas, 118
  mole fraction, 120
  partial molar, 199
  total molar, 198
Concentrators, 474
Conditioning inspired air, 84, 88
Conductance, 577
Conducting airways, 52
Confined space, 309, 310
Conservation equations:
  energy, 363, 364, 572

generalized, Table 10.1, 571
  mass, 363, 572
  momentum, 363, 572
  state, 363, 364, 572
Conservative velocity field, 369
Constricted lung, 89, Fig 2.13a, 62
Contaminant:
  classification, 18
  concentration, 118
  definition, 17
  exposure levels, 120
Contaminant control strategy, 17
  administrative controls, 19
  engineering controls, 20
  personal protective devices, 22
Contaminant generation, *see also* Evaporation
  rate
  emission factors, 178
  evaporation, filling vessels, 173, 174
  grinding, 175
  liquids in vessels, 173
  pouring powders, 174
  puff, 179
  removal by walls, 3
  spills, 174
  spray finishing, 177
Continuity equation:
  molecular species, 179, 573
  single-species, 179, 363
Continuum regime, 431
Contours constant velocity, Figs 6.8, 6.9, 301–303
Control:
  gases and vapors, 563
  mass, 229
  particles, 515
  strategy, 19
  surface, 76, 228, 565
  volume, 76, 228, 564
Control velocity, definition, 304, Table 6.3, 305
Convection heat transfer:
  from body, 148
  coefficient, 192, Table 4.2, 195
Convection mass transfer coefficient, 192
Convection term, 573
Convective dispersion, bronchial system,
  Fig 2.21, 72
Convergence, 542
Conversion factors, (A-18) 626
Convertible coating, 177
Cooperation theory, 45
Corner, flow field, 392
  flow around, Fig 7.4, 392
  with inlet, Fig 7.13, 394
Cost, make-up air, 276
Cost analysis, *see* Engineering economics
Cost–benefit ratio, 6. *See also* Risk *vs.* benefit
  analysis
Cotton dust, 99
Couette flow, 597

Coughing, 55
Countercurrent flow, lungs, 68
Counterflow spray chamber, 483
CRC, 156
Cristobalite, 98
Critical temperature and pressure, 184, 233,
       (A-10) 660
Cross-flow spray chamber, 484
Cumulative mass distribution, 440
Cumulative number distribution, 437
Cunningham correction factor, Table 8.1, 432
Curved air motion, 466
Curved duct, 466
Curved particle motion, 468
Curvilinear flow, 362, Fig 8.20, 469
    irrotational, 467
    laminar, 469
    quasi static, Fig 8.2, 469
    rotational, 467
    solid body rotation, 467
    well-mixed, Fig 7.16
Cut diameter, cyclone, $D_{p,50}$, Fig 8.26, 477
Cut-off saw, Fig 6.4, 296
Cyanosis, 92
Cyclone, 474
    grade efficiency, 477
    reverse flow, Fig 8.24, 476
    standard, 476, 477
    standard reverse flow, 476
    straight through, 474
Cylinder, flow field:
    moving, Table 7.3, 391
    stationary, Table 7.3, 391

$D_{p,am}$, 435
$D_{p,gm}$, 437
$D_{p,50}$, 438
$D_{p,84}$, 438
Dalton's law, 119
Darcy's law, 493
Daughters of Radon, 99
Deaths due to disease, 2
Decelerating drop, 486
Decibels, 140. *See also* Sound
    sound pressure, Fig 3.12, 144
    threshold, Table 3.3, 143
Deep pocket, 16
Defense mechanisms, respiratory system, 55
Deflagration branch, 131
Deflagration wave, 127
Deformation rate, 363
Density:
    aerosol, 118
    apparent, bulk, 311
    constant, 362
    dry air, 119
    moist air, 119
    particles, 434
    unit density, 434

Department of Energy, 46
Deposition:
    alveolar, Fig 2.35, 94
    chambers and enclosures, 460
    gravimetric, 450
    horizontal ducts, 462
    inertial in curved ducts, 466
    laminar, 460
    particles:
        in airway, Fig 2.35, 94
        in respiratory tract, 94
    settling velocity, 448, Fig 8.13, 450
    turbulent, 460
    well-mixed, 460
Derivative, substantial, total, 594
Design:
    ACGIH design plates, 297
    criteria, 116, 117
    defect, 17
    engineering, 116
    essential steps, 116
    method, 116
    recommendations of ACGIH, 297
    standards, 117
Design plates, ACGIH, 297
Desorbtion, 246
Detonation branch, 131
Detonation wave, 128
Deviation:
    geometric standard, 437
    standard, 436
Diameter:
    aerodynamic, 95, 453
    arithmetic mean, 435
    geometric mass mean, 437
    median, 438
Diaphragm, 55
Differential model, 229
Differential volume, 229
Diffusion:
    capillary membrane, 58
    coefficient, Table A-9, 658
        equation, gas, 184
            illustrated example, 184
    in filter, 488
    flame, 132
    moving air, 192
    particle, 180, 488
    respiratory system, 71
    single film theory, 203
    stagnant air, Fig 4.5, 185, 187
    turbulent diffusivity, 184
    two-film theory, 203
Diffusion term, 573
Diffusivity, *see* Diffusion, coefficient
Dilution:
    odors, 153
    ventilation, 31, 228
Dimensional analysis, 465

Dimensionless parameters, 465, 543
Direct cost, 155
Directivity of sound, 139
Discretization, 573
Diseases of the lung, Table 2.1, 50
   allergies, 98
   cell damage, 97
   fibrosis, 98
   irritation, 96
   oncogenesis, 96
Dispersion, atmosphere, 328
Dispersion coefficient, 179
   bronchioles, 70
   puff diffusion, 179
Dispersion modeling, 328
Displaced air, 311
Displacement ventilation, 228, 271
Dissipation, turbulent energy, 593
Distributed parameter respiratory model, 76, 81
   analytical model, 229, 565
Distribution functions, 434
   particle size:
      arithmetic mean, 435
      geometric mean, 437
      log-normal, 436
      mass, 440
      median, 438
      number, 436
      surface area, 440
Dividing streamline, 379, 515, Fig 9.3–9.6,
   517–518, Fig 9.13, 526
   algorithm, 528
DOE, 46
Dopants, 261
Dose, 101, 271
   total, 106
Dose rate, 107
   function of exercise, Fig 2.2, 80
Dose-response, 100
   class, 106
   extrapolated, Fig 2.40, 106
Doublets, point, 388
Downdraft bench, 40, Fig 1.23, 39, Fig 1.26, 40
Downwash, 328
Drafts, 29, 328
Drag, filter, 494
Drag, particles:
   coefficient, 429, Fig 8.1, 430
   force, 429
   form drag, 430
   irregular size, 433
   Newtonian regime, 430
   Stokes regime, 430, 448
   transition regime, 430
   viscous drag, 430
Droplet evaporation, 211
Drop-out boxes, 334
Dry etching, 261

Duct design, 333
   blast gates, 333
   branches, 338
   clean-out ports, 334
   computer programs, 342, (A-17) 674
   design procedure, 341
   drop-out boxes, 334
   fires, 334
   fittings, 336, Figs 6.23, 6.24, 339–340
   friction factor, 336
   friction loss, 337
   head loss, 335, 338
   hood entry loss, 338, Fig 6.23, 339
   hydraulic diameter, 336
   illustrated example, 334
   local loss, 336, 338
   roughness height, 336
   static pressure, 335
   total presure, 335
   transport velocities, Table 6.6, 334
Duct transport velocities, 334
Dust, 18
Dust–air mixtures, explosions, 136
Dust cake:
   loading, 493
   specific resistance, 494
   thickness, 493
Dust suppression, 19
Dynamic pressure, 335
Dynamic shape factor, 433, Table 8.2, 434
Dynamic velocity, 335
Dynamic viscosity, 430, (A-11) 662, Fig 8.2,
   431
Dyspnea, 92, 96

ECL, 125
Economics, *see* Engineering economics
Edema, 58, 97
Eddy:
   building, 324–326
   diffusivity, 593
   Prandtl number, 593
   recirculation, 324
   roof, Fig 6.20, 326
   stress, 591
   transport properties, 593, Table 10.1, 574
   viscosity, 592
Effectiveness coefficient, 272
   illustrated example, 274
Effectiveness of an inlet, 516
Effective stack height, 325
Efficiency, 444
   accelerating drop, 486
   air cleaning device, 260
   collection device, 444
   collectors:
      in parallel, 446
      in series, 445

cyclones:
    standard reverse flow, 476
    straight through, 474
experimental, 260
filters, overall, 492
fractional efficiency, 444
grade efficiency in curved duct laminar,
    Table 8.10, 473
on particles or fibers, 488
    diffusion, 488
    impaction, 488
    interception, 488
polydisperse aerosol, 434
settling in chambers, 460
well-mixed, Table 8.10, 473
Efficiency, 444
    filter, 492
    single fiber, 488
    single particle, 479
    spray chamber, 484
    venturi scrubbers, 484
Ehrlich Index, Fig 2.39, 105
Electrostatic spraying, 178
Elimination, 19
Elliptical opening, flanged, 408
Emission factors, 178
    chemical, (A-5) 648
    food, (A-6) 651
    indoor activities, (A-7) 653
    metallurgical, (A-2) 637
    mineral processes, (A-4) 645
    VOC, (A-3) 641, (A-8) 655
Emission from vessels:
    displacement, 174
    evaporation, 174, 190
    filling, 174
Emphysema, 81, 97
Enclosure, 31, Figs 1.18–1.20, 36–37, Fig 1.22, 38,
    Fig 6.2, 295, Fig 6.3, 296, 309, Fig 6.11,
    312, Fig 6.12, 313
    transfer point, 312
Energy:
    balance, body, 149
    conservation equation, 363
Engineering controls, 19
Engineering design, 116
Engineering economics, 153
    capital recovery factor, 156
    fixed cost factor, 156
    illustrated example, 156
    indirect cost factor, 155
    overhead expenses, 155
    total capital cost, 153
    total direct cost, 155
    total fixed cost, 155
    total indirect cost, 155
    total revenue requirement, 154
    total variable cost, 155

Enthalpy of vaporization, 183
Entrained air, 311
Entrainment of jets, 319
Entry loss, 338
Environmental monitoring, 19
Environmental Protection Agency (EPA), 46
    ambient air quality standards, 124
Epidemiology, 10
Epiglottis, 55
Equation of motion:
    carrier gas, 363, 364
    particle, 447
Equation of state, 363, 364
Equivalent diameter:
    projected area diameter, 433
    Sauter, 433, 487
    surface area diameter, 433
    volume diameter, 433
Equivalent duct length, 341
Error function, 437
Etching, wet and dry, 261
Eulerian method, frame of reference, 362
Eulers equation, 364
Evaporation rate, 182
    drops, 211
    empty barrels, 190
    human body, 149
    illustrated example, barrel, 190
    through moving air, 192, 197
    multicomponent liquids, 198
    open vessels, 196
    respiratory system, 84
    spills, 174
    through stagnant air, 187
Evaporative cooling, 211
Exchange coefficient, 266
    illustrated example, 269
Exercise and lung absorption efficiency,
    Fig 2.27, 80
Exfiltration, 262, Table 5.2, 264
Exhaust air, 24
Exhaust duct design, 333. See also Duct design
Exhaust stack height, 324
    ASHRAE, 327
    EPA, 325
Expiratory flow rate, Fig 2.13a, 62, Fig 2.13b,
    63
Explosion limits:
    dust–air mixtures, 136, 137
    gases and vapors, 127
Exposure control limits (ECL), 125
Exposure parameter:
    illustrated example, 123
    mixture of gases, 121
    puff diffusion, 181
Extended Bohr model, Fig 2.25, 71, 76
Exterior hoods, 31, 36
Extrathoracic airway, 52

Face:
  area, 298
  masks, 22
  noise, 44
  velocity, 298
Fall velocity, Fig 8.13, 450
Fan:
  inlet loss coefficient, 341
  laws, 342
Farmer's lung, 99
Fatalities, work, Table 1.1, 4
Fatigue, olfaction, 152
Favre average, 588
FCF, 156
Felt, 487
FEV, FEV1, 62
Fiber:
  filters, 488
  length per volume of filter, 492
  particles, 18
  solid fraction, 489, 490
Fibrosis, 96, 98
Fick's law, 186
FIDAP, 603
Filling, vessel, 313
  bottom, 174
  splash, 174
Film:
  coefficient, convection, 193, Table 4.2, 195
  theory:
    single-film, 198
    two-film, 200
Filter:
  bags, 487
  drag, 494
  residual drag, 494
Filtration, 487, Fig 8.33, 488
  air-to-cloth ratio, 493
  analytical model, 492
  length of fiber per volume of filter, 492
  overall efficiency, 492
  porosity, 490
  pressure drop, 490
  residual drag, 494
  single-fiber efficiency, 488
  solids fraction, 490
Fine particles, 18
Finishing, emissions, 177
Finite difference methods, 538
Fire, 127
First generation respiratory passage, 53
First law of thermodynamics, 146
First moment, 91, 272
Fitting loss coefficient, 336, Figs 6.23, 6.24,
    339–340
Fixed cost factor (FCF), 156
Flame, 127
  diffusion, 132

front, Fig 3.7, 136
  premixed, 132
  speed, 127
Flammability, 127
Flammability limits, common materials, Table
    3.2, 134
  dependence on hydrocarbon, Fig 3.6, 135
  ignition energy, Fig 3.8, 137
  lower explosion limits (LEL), 132
  oxygen and dilutents, Fig 3.4, 132
  particle size, Fig 3.9, 138
  pressure, Fig 3.5, 133
  upper limits, 132
Flange, definition, 302
Flanged inlet:
  arbitrary shape, 414
  circular inlet, 407
  elliptical opening, 408
  slot, 392, 401
    in streaming flow, 403
  velocity profiles:
    circular inlet, 300
    rectangular inlet, 302
Flash point temperature, Table 3.2, 134
Floor sweep, 231
Flow:
  airways, Fig 2.21, 72
  bifurcated, Fig 2.15, 66
  around bodies, 387
  over buildings, 325
  into corner, Fig 6.20, 392
  curvilinear, 362
  in cyclone, Fig 8.25, 477
  in ducts, 333
  flanged inlet, quiescent air:
    circular, 407
    eliptical, 408
    slot, Fig 7.12, 393
  flanged slot, Fig 7.12, 393
  friction, in ducts, Fig 6.22, 337
  ideal, 301, 362
  incompressible, 362
  inlet with streaming flow:
    flanged slot, 403
    unflanged slot, 409
  inlet velocity contours, 300
  inviscid, 361
  line sink, 377
  near, moving:
    cylinder, Table 7.3, 391
    rotating disk, 365
    sphere, Table 7.3, 391
  point sink, 383
  pulsatile, 65
  push–pull, 594
  rectilinear, 362
  regime, 431
  rim exhausters, Fig 10.7, 585

stationary:
  cylinder, Table 7.3, 391
  sphere, Fig 7.10, 390
  streaming flow, Table 7.2, 374
  unflanged slot, Table 7.2, 374, 402
  velocity contours for inlets, 300, 302
  windbreaks, 552
FLOTRAN, 603
FLOW3D, 603
FLUENT, 603
Flux plotting, 373
Fog, 18
Food processes, emission factors, (A-6) 651
Force:
  buoyant, 572
  on particle:
    centrifugal, Fig 8.20, 469
    drag, 429
    gravity, 448
Forced expiratory volume, 62
Forced vital capacity (FVC), 90, Fig 2.13, 63
  in one second (FVC1), 90, Fig 2.13, 63
Formaldehyde emissions:
  building materials, 255
  home products, 172
Form drag, 430
Forward difference, 539
Foundry:
  canopy hoods, Fig 1.13, 33
  electric arc furnace, Fig 1.19, 37
  grinding, Figs 1.17(a&b)
  side draft hoods, Fig 1.14, 34, Fig 1.16, 35
  vapors, Fig 5.1, 233
Four-layer barrier, Fig 2.26, 78
Fractional efficiency, 444
Freely falling particle, 448
Free-molecular flow regime, 432
Frequency functions, *see* Distribution functions
Fresh air:
  islands, 144
  requirements (ASHRAE), 230, (A-16) 673
Friction, *see also* Duct design
  factor, 336
  velocity, 208, 597
Fugitive emissions, 30, 310, Figs 1.9, 31
  roadways, 552
  stockpiles, 552
Full-face masks, Fig 1.4, 22
Fully alveolated alveolar ducts, 56
Fully established flow,, 68, 463
Fume:
  definition, 18, 118
  hoods, Fig 1.20, 37, 261, 309
Function:
  cumulative distribution 437
  size distribution, 436
Functional residual capcity, 60, Fig 2.11, 61
Future developments, 602

Future worth of annuity, 158
FVC, 90
FVC1, 90

Gametoxin, 102
Gas:
  composition, lungs, Table 2.3, 73
  contaminant concentration calculation, 119
  equation of state, 363, 572
  exchange, four layer, Fig 2.2, 78
  mixture, 119
  universal constant, (A-18) 676
Gas exchange in lung, Fig 2.19, 71, Fig 2.25, 77
Gas range emissions, 172
Gas transport, pulmonary system:
  asymmetric, 68
  bifurcated, Fig 2.15, 66
  bulk convection, 65
  modes, Fig 2.21, 72
  molecular diffusion, 71
  pendelluft, 66
  Taylor-type diffusion, 69
Gas uptake, 65
Gaussian plume equation, 179, 328
Gauss Sidel iteration, 542
General conservation equations:
  aerosol motion, 448
  gases, 363, 364, 572
  Tables, 574
Generalized conservation, Reynolds transport
  equations, 573, Table 10.1, 574
General ventilation, 31, 228
  adsorption on walls, 244
  air changes per unit time, 230, 237
  definition, 228
  dilution ventilation, 228
  exfiltration, 264
  fresh air requirements, 230, (A-16) 673
  infiltration, 264
  make-up air (100%), 236
    fraction, 246
  mixing factor, 250
  multicell well-mixed model, 264
  partially mixed, 249
  perfectly stirred, 228
  recirculation, 246
  sequential box model, 264
  well-mixed model, 228
  well-stirred, 228
GENMIX, 603
Geometric log-probability plot, illustrated
  example, 444
Geometric mean diameter, 439
Geometric standard deviation, 437
Global model, 229
Globe temperature, 148
Glove boxes, 261
Good engineering practice, 17

Grade efficiency curve, 444
Gravimetric settling, Fig 8.13, 450
  chambers and enclosures, 460
  in duct, 462
Grid notation, SIMPLER:
  finite difference, Figs 9.18–9.21, 541
  staggered, Fig 10.2, 578
Grinder:
  hand-held, Fig 1.5, 23, Fig 1.10, 32
  swing, Fig 1.28, 42
Grinding:
  booth, 544, Fig 6.10, 308, 545
  boundary layer, 544
  concentration, breathing zone, 551
  downdraft bench, Fig 1.23, 39
  local generation rate, 176
  metal removal parameter, Fig 4.1, 175
  metal removal rate, 175, 176
  particles, size distribution, Fig 4.2, 176
  particle trajectories, Fig 9.25, 549, Fig 9.26, 550

Haber's law, 104
Half-mask, Fig 1.4, 22
Hand-held grinder, Fig 1.5, 23
Harmonic functions, 373
Hay fever, 97
Hazardous materials, Table 2.1, 50
Hazard potential, Table 6.2, 304
Head loss, 335
Hearing impairment, 143
  age and dBA, Fig 3.13, 145
Heat balance, body, 146
  capacity, lumped, 213
  convection, 148
  evaporation, 149
  globe temperature, 148
  Heat Stress Index, Table 3.5, 149
  illustrated example, 150
  metabolic, 147
  radiation, 148
Heat flux, turbulent, 593
Heating degree days, annual definition, 276,
    Table, 276
Heat stress, 145
Heat Stress Index, Table 3.5, 149
Heat transfer, convection, 193, Table 4.2, 195
Heat of vaporization, 183
Hedonic tone, 153
Hematopoietic toxin, 101
Hemoglobin, 107
Henry's law, 205
  constants, 206, Table, (A-9) 658
HEPA filter, 262
Hepatotoxin, 101
HERP, 13
High frequency ventilation, 65
High velocity–low volume hood,
    Figs 1.10–1.11, 32, Fig 1.12, 33

Histamine, 99
Histotoxic hypoxia, 92
Homeostasis, thermal, 146
Hoods, *see also* Ventilation
  ACGIH design plates, 307
  bag filling and dumping, Fig 1.21, 38, Fig
    1.26, 40
  canopy, Fig 1.13, 33, 294, 315–317
  classification, Table 1.4, 31
  definition, 25, 294, 360
  down-draft, Fig 1.23, 39, Fig 1.26, 40, 294
  drum filling, Fig 6.2, 295
  enclosures, Fig 1.18, 36, Fig 1.22, 38,
    Fig 1.19, 37, Figs 6.2, 6.3, 295, 296
  entry loss coefficient, 338
  exterior, Table 1.4, 31
  flanged, 302
  fume, Fig 1.20, 37
  grinding booth, Fig 6.10, 308
  hood entry loss, 338
  hopper hood, Fig 6.14, 317
  inlet velocity contours, 301–303
  laboratory or fume, Fig 1.20, 37
  lateral, Fig 1.25, 40, 294, Fig 6.1, 295, 581
  local, Table 1.4, 31
  open vessel, Fig 1.25, 40
  painting booth, Fig 1.22, 38, Fig 1.24, 39
  partial enclosure, 294
  plane openings, 294
  pool hood, Fig 6.14, 316
  push–pull, 333, 594
  receiving, Fig 1.28, 42, 294, Fig 6.4, 296
  rim exhauster, Fig 6.1, 295
  side draft, Figs 1.14–1.15, 34, Fig 1.16, 35
  snorkel, Fig 1.27, 41
  surface sander, Fig 1.10, 32
  welding gun, Fig 1.11, 32
Hopper hoods, 317
Household contaminant emissions:
  appliances, 172
  formaldehyde, 255
  products, 172
Housekeeping, 19
HSI, Table 3.5, 149
Hugoniot and Rankine curve, Fig 3.3, 130
Human Exposure Dose/Rodent Potency,
    Table 1.3, 14
  definition, 13
Humidity, relative, 149
Hydraulic diameter, 336
Hydrocarbon:
  emission factors, (A-3) 641, (A-8) 655
  LEL, 132, Table 3.2, 134
  PEL, (A-1) 633
  properties, (A-8) 655
  uptake, 76
  vapor pressure, (A-8) 655
Hydrolase, 95

Hypercapnia, 92
Hypoxemia, 98
Hypoxia, types, 92, 97
Hystereis, 60

ICF, 155
Ideal flow:
  definition, 361, 362
  line sink:
    quiescent air, 376
    streaming flow, 377
  multiple flanged rectangular inlets, 412
  particle capture by line sink, 525
  particle capture by plane and flanged slot, 532
  point sink:
    quiescent air, 382
    streaming flow, 383
  potential functions of inlets, Table 7.2, 374
  slots, 300, 302
  streaming flow, 377
Ideal fluids, definition, 361
Ideal gas, gas mixtures, 119
Ideal solution, 198, 200
Ignition, 127
  minimum energy, Fig 3.8, 137
  particle size, Fig 3.9, 138
  temperature, 134
Impaction, 478. *See also* Single fiber collection
    efficiency; Single particle collection
    efficiency
  in curved ducts, 473
  sphere on cylinder, 488
  sphere on sphere, 479
Impact noise, 143
Impactors, cascade, 538
  grade efficiency, Fig 9.22, 544
Implicit differentiation, 394
Inadequate makeup air, symptoms, 329
Incompressible flow, 362
Indirect cost factor (ICF), 155
Indoor activities, emission factors, (A-7) 653
Indoor air quality, ventilation requirements,
    230, (A-16) 673
Induced air, 311
Industrial accidents, Table 1.1, 4
Industrial spill, 174
Industrial toxicants, Table 2.1, 50
Industrial ventilation, 19
Inertial impaction, particles:
  accelerating drop, 486
  curved ducts, 473
  cyclones, 474
  on cylinders, 488
  in filters, 487
  in respiratory system, 93
  on spheres, 479
  in venturi scrubbers, 484
Inertial sources, 310

Infiltratioin–exfiltration, Table 5.2, 264
Inflammability, 127. *See also* Flammability
Inhalable particles, 18, 95
Inlet, *see also* Ideal flow; Reach
  buildings, 324
  effectiveness, 516
  fan inlet loss, 341
  flanged:
    arbitrary shape, 414
    circular, 300
    definition, 302
    elliptical, 408
    rectangular, 302
    velocity contours, 302, 303
  loss coefficients, 338, Fig 6.23, 339
  multiple inlets, 412
  noise, Fig 3.14, 146
  plane, definition, 303
  potential functions, 374
  unflanged:
    circular, 300
    rectangular, 300
    velocity contours, 301, 302
Inspectional analysis, 465
Inspirable particles, 18
Instantaneous source, 179
Intensity:
  sound, 139
  turbulence, 589
Interception, 483, Fig 8.34, 489
International system of units conversion,
    (A-18) 676
Interpolation, 542
Interstitial fluid, 58
Interstitium, Fig 2.8, 58
Inviscid flow, 361
Involuntary risk, 1, Fig 1.1, 2
Irregular particles:
  dynamic shape factor, 433, Table 8.2, 434
  volume equivalent diameter, 433
Irritant, 89, 101
Irritation, bronchi, 96
Irrotational flow, 361, 367, 370, 467
  illustrated example, 365
Isokinetic sampling, 522, Fig 9.8, 521
Isolation, 19
Isothermal expansion coefficient, 139
  saturation boundary, 85
Isotropic turbulence, 604
Iteration programs:
  Bisection, (A-15) 671
  Newton, (A-15) 671

Jacobian $J(\Phi, \psi)$, 396

κ-ε turbulence model, 593
Kerosene space heater, 172, (A-7) 653
Kinematic viscosity, Fig 8.2, 431, (A-11) 661

Kinetic energy, turbulent, 592
Kitchen gas range, 172
Knudsen number, 432
Kohn, pores, Fig 2.5, 54
Kuwabara hydrodynamic factor, 488

Labeling, 19
Laboratory fume hood, 36, Fig 1.2, 37, 261, 309
Laminar computer model, 571
Laminar deposition:
  chambers and enclosures, 461
  curved ducts, 466
  straight ducts, 463
    critical length, 363
Laminar flame, 127
Laminar flow, bronchioles, 68
Laminar Navier–Stokes equations, 572
Laplace equation, 370
Lapse rate, 328
Larynx, 52
Lateral exhauster, 40
  illustrated examples:
    suction, 332
    push–pull, 332
  minimum volumetric flow rate, Table 6.4, 305
  open vesel, Fig 1.25, 40, Fig 6.1, 295
  rim, Fig 9.3
Launder–Spalding model, 593
Law of the wall, 596
LEL, 132
Length of filter per unit volume of filter, 492
Level of turbulence, 589
Lewis number, 194
Liability, 16
  design defect, 17
  negligence, 16
  strict liability, 16
Limiting trajectory, 515
Limits of flammability, 132
Line sink, 376
  illustrated example, 380
  in streaming flow, 377
Lipids, 58
List, toxic and hazardous substances, (A-1) 633,
    Table 2.1, 50–51
LMCD, 189
Loading, dust cake, 493
Local convection film coefficient, 192
Local mean age, 272
Local toxicity, 101
Local ventilation, Table 1.4, 31. *See also* Hoods
  ACGIH design plates, 297
  categories, 294
  inlets, velocity contours, 300–303
  loss coefficients, 338, 339
Logarithmic law of the wall, 596
Log mean difference, 189
Log mean ratio, 189
  mol fraction, 189

partial pressure, 192
Log-normal size distribution, 434
  cumulative number distribution, 437
  geometric standard deviation, 437
  graphs, 441, Fig 8.3, 436, Fig 8.9, 444
  median, 438
  number distribution, 436, 437
  urban aerosol, Fig 8.5, 440
Log-probability paper plotting procedure, 441
Loss:
  adsorption to wall, 244
  coefficients, *see* Duct design
  major, minor, 336
Lower explosion limit, 132. *See also*
    Flammability
Low velocity–high volume hood, 32
Low volume–high velocity hood, 32, 332
Lumped heat capacity, 213
Lumped parameter, 228, 565
Lung, *see also* Respiratory system
  analytical models:
    Bohr, Fig 2.25, 77
    distributed parameter, 81
    extended Bohr, Fig 2.25, 77
    geometrical parameters, Fig 2.7, 57
  diseases:
    action sites, Table 1.1, 50
    classifications, 96
    constricted, 89, Fig 2.13a, 62
    normal, Fig 2.13a, 62
    obstructed, Fig 2.13a, 62, Fig 2.34, 90
    spirometric measurements, Fig 2.13a, 62
    toxicants, Table 2.1, 50
  spirometry, Fig 2.11, 61
    clearance mechanisms, 53
    forced vital capacity, 90
    hysteresis, Fig 2.10, 60
    mode of air motion, Fig 2.21, 72
    total capacity, 60
    volumes, Figs 2.11–2.12, 61
Lymphatic system, 59
  capillaries, Fig 2.9, 59
  nodes, 59
Lysosme, 95

MAC, 120
Macro-model, 229
Macrophage, 55, 95
Maintenance, 19
Major head loss, 336
Make-up air, 246
  costs, 276
  fraction, 246
  fresh air requirements, 230, (A-16) 673
  illustrated example, 277
  inadequate, symptoms, 328
Management, 19
Man coolers, 150
Mask, face, 22

Mass concentration, 118
  aerosol, 118
  conversioin to PPM, 120
  gases and vapors, 119
Mass distribution function, 444
Mass fraction, 199, 444, 573
Mass transfer, coefficient:
  moving air, 192
  open vessel, 196
  overall, 206
  respiratory system, *see* Bohr model;
    Distributed parameter respiratory model;
    Extended Bohr model
  Reynolds analogy, 194, 211
  spills, 174
Mast cells, 99
Material Safety Data Sheet, 9
Maximum acceptable concentration, 120
Mean age (local), 272
Mean free path, air, 431
Mean particle diameter:
  arithmetic number, 436
  geometric mass, 440
  geometric number, 437
Mean properties, turbulent flow, 598
Median particle diameter by number, 438
Medical suirveillance, 19
Membrane, alveolar, Fig 2.19, 71
Metabolic rate, Table 3.4, 147
Metallurgical processes, emission factors, (A-2)
  637
Metal removal parameters, Fig 4.1, 175
Metal surface treatment, classification, Table
  6.5, 306
Micro-model, 229, 565
Mineral processes, emission factors, (A-4) 645
Mine Safety and Health Administration, 46
Mine ventilation, 46
Minimum control velocities, Table 5.4
Minimum volumetric flow rate lateral
    exhauster, 305
  illustrated example, 307
Minor head loss, 336
Minute normal respiratory rate, 61
Mist, 18
Mixing factor, 250, Table 5.1, 251
Mixture, TLV's:
  gases, 122
  liquids, 122
Mode, atmospheric aerosol, Fig 8.5, 440
Moist air, density, 119
Molar average velocity, 186, 199
Molar concentration:
  average or total, 198
  partial, 199
Molecular diffusion of air in alveoli, 71
Molecular weight, average, 198, Tables, (A-8)
  655, (A-9) 658, (A-10) 660
Mol fraction, 199

definition, 119
  log mean ratio, 189
Moments, 90, 272
Momentum equation, 363, 572
Monitoring, 19
Monodisperse aerosol, 434
Moody chart, 336
Motion of lung air motion, Fig 2.21, 72
MSDS, 9
MSHA, 46
Mucociliary escalator, 53
  clearance, 53
  lung, 53
Mucosal layer, 53
Mucous membrane, 53
Mucus, 53
Multicell well-mixed model, 265. *See also*
    Box model
Multicomponent mixtures, 121, 200
  evaporation, 198
Multi-hit model, 108
Multiple flanged inlets, 412
Multi-stage model, 109
Mutagens, 102

Nasal turbinates, 56
Nasopharyngeal region, 52
National Institute for Occupational Safety and
    Health, 46
Natural background aerosol, Fig 8.5, 440
Navier–Stokes equation, 363
  generalized, Table 10.1, 574
  numerical solvers, 604
N-degree rotational flow, 467
Negligence, 16
Nephrotoxin, 101
Neurotoxin, 101
Neutrally-buoyant plume, 328
Newton convection heat transfer, 192
Newtonian fluid, 363
Newton iteration method:
  one variable, (A-15) 671
  two variables, (A-14) 669
Newton–Raphson method, 406, 530
  computer program, (A-13) 663
Newton regime, 430, 451
NIOSH, 46
NISA/3D-fluid, 603
Nodes, *see* Grid notation, SIMPLER
Noise, 138. *See also* Sound
Nondimensional velocity, distance, law of the
    wall, 596
Non-flammable, 127. *See also* Flammability
Nonideal solutions, 203
No observable effect limit (NOEL), 125
Normal boiling temperature, 183
Normal breathing, 60
Normal minute respiratory rate, 61
Nu, 193

Nucleation range, Fig 8.5, 440
Nuclei, 18, 439
Null point, 298
Number:
    of drops per volume, 481
    of effective turns, 475
    of room air changes, 230
Number distribution, 436
    arithmetic, mean diameter, 435
    concentration, 436
    cumulative number distribution, 437
    geometric mean diameter, 437
    median diameter, 438
    relation to mass, 440
    standard deviation, 435, 436
Number of heating degree-days, 276
Numerical methods:
    Newton method, (A-14) 669, (A-15) 671
    Newton–Raphson, (A-13) 660
    Runge–Kutta, (A-12) 663
Nusselt number, 193, Table 4.2, 195

Obstructed lung, Fig 2.13a, 62, Fig 2.34, 90
Occupational disease and injury, Table 1.1, 4,
        Table 2.1, 50
    ranking, 3
Occupational Safety and Health
        Administration, 46
    PEL, 120, (A-1) 633
Odors, 151
    acceptability, 152
    character, 153
    detection threshold, 153
    dilution factor, 153
    index, 153
    intensity, 153
    olfaction, 152
    quality, 151, 153
    quantify, 152
    recognition threshold, 153
    threshold values, Table 3.6, 152
Off-design performance, 28
Oncogenesis, 96
One-hit model, 109
Open system, 228
Open vessel, Fig. 1.25, 40, 581. *See also* Hoods
    evaporation, 196
    illustrated example, 332
    lateral exhaust, Fig 1.25, 40, Fig 6.1, 295
    pure suction system, 581
    push–pull system, Fig 6.21, 333, 594
OSHA, 46
    Cancer-Suspect Agents, Table 3.1, 124
    PEL, (A-1) 633
    Threshold Limit Values for Noise, Table 3.3,
        143
Overall efficiency, 444

collectors:
    in parallel, 446
    in series, 445
Overall mass transfer coefficient, 206
Overhead expenses, 155
Over-relaxation coefficient, 542
Overspray, 177
Oxygen uptake, Table 3.4, 147
    adsorption *vs.* alveolar ventilation Fig 2.27, 80
    physical activity, Table 2.2, 64
Ozone adsorption, lung, 83
Ozone dose, Fig 2.42, 108
Ozone mucus reaction, 83

Packing density, *see* Solid fraction
Paint spraying booth, Fig 1.24, 39
Paper, log-probability, 441, Fig 8.9, 444
Parallel collectors, 446
Partially-alveolated respiratory bronchiole, 56
Partially mixed, 249
Partial molar concentration, 199
Partial pressure, 119
    hydrocarbons, (A-8) 655
Particle(s):
    acceleration, 486
    capture:
        by flanged slot, 531
        by line sink, 527
        by unflanged slot, 531
    classification, 18
    deposition in airway, Figs 2.35–2.36, 94
    diameter:
        aerodynamic, 95, 453
        arithmetic mean, 435
        equivalent projected area, 433
        equivalent surface area, 433
        equivalent volume, 433
    dimensionless time, 466
    drag coefficient, 430
    drag force, 429
    dynamic shape factor, Table 8.2, 434
    fibers, 433
    flow regime, 431
    geometric mean, 437
    impaction:
        alveoli, 94
        curved duct, 473
        fibers, 488
        spheres, 479
    influence on carrier gas, 447
    mass concentration, 118
    median, 435
    motion:
        due to gravity, 448
        in moving gas, 454
        settling velocity, 449
    particle shape, 433

relaxation time, 449
sampling, 519
  errors, Fig 9.7, 520, Fig 9.8, 522
Sauter, mean diameter, 433, 487
shape factor, 434
size classification, 18
size distribution, 434
  grinding, Fig 4.2, 176
  illustrated example, 441
slip factor, 432
spray chamber, 480
Stokes flow, 448
unit density sphere, 434
urban atmosphere, 440
venturi, 484
Particle relative velocity, 429, 430, 448, 455
Particle trajectories:
  annular slot in streaming flow, Fig 9.10, 524
  computer program, 457
  curvilinear motion, 466
  flanged inlet, Fig 9.3, 517, Fig 9.5, 518
  grinding booth, Fig 9.25, 549, Fig 9.26, 550
  illustrated example, 457
  impacting on:
    curved ducts, 473
    fibers, 488
    particles, 479
  impactor, 538
  sampling nozzles, Fig 9.7, 520, Fig 9.9, 523
  unflanged inlet, Fig 9.4, 518, Fig 9.6, 518
  windbreaks, Fig 9.33, 555
Parts per million, 119
  conversion to mass concentration, 119
Path, mean free, 431
Pathline, 364
Pathway for toxins, Fig 2.38, 101
Peclet number, 576
PEL, 120
Pendelluft, 66, 68, Fig 2.6, 67
Penetration distance, 454, Table 8.7, 454. *See also* Collection efficiency
Perception of risk, 11
Perfect gas law, 119, 363
  mixtures, 119
Perfectly stirred or mixed, 228
Performance standards, 171
Perfusion, 58, 74
Perfusion–ventilation ratio, Fig 2.27, 80
Permeability through capillary membrane, 58, 72
Permissible exposure limits (PEL), 120, (A-1) 633, (A-8) 655
Personal protection devices, 19, Fig 1.3, 21, Fig 1.4, 22
Perspective, 11
Phagocytes, 55
Phagocytosis, 55, 95

Pharmacokinetics, 102
PHOENICS, 603
Photoresist, 261
Physical activity:
  airway ozone dose, Fig 2.29, 83, Fig 2.30, 84
    absorption efficiency, Fig 2.27, 80
  oxygen uptake, Table 2.3, 73
Physical constants and conversions, (A-18) 676
Physiology, 49
Picocuries, 100
Pigment, coating, 177
Pinocytosis, Fig 2.37, 96
Plane inlet:
  quiescent air, 392
  streaming flow, 409
Plate-out, 244
Pleura, 60
Plug flow, 271, 463
  in bronchioles, 68
Plume, buoyancy, 314
Plume rise, 328
PM_{10}, 18
PMN, 10
Pneumoconiosis, 98
Pocket Guide to Chemical Hazards, 8
Point doublet, 388
Point-mutation, 108
Point sink, 382
  illustrated example, 385
  in streaming flow, 383
Point source emissions, 30, 310
Pollutant properties, Table 2.1, 50, (A-1) 633
Polydisperse aerosol, 434
Pool hood, 316
Porosity, filter, 490
Positive pressure helmet, Fig 1.5, 23
Potential flow, 370. *See also* Ideal flow
  derivatives, Table 7.1, 372
  functions, Table 7.2, 374
  inlets, Table 7.2, 374
Powders, pouring, 176, 310
Power, acoustical, 140
Power law scheme, 576
PPM, 119
Pr, 193
Prandtl number (Pr), 193, (A-11) 661
Present value of an annuity, 158
Precipitation scavenging particles from air, 478
Premanufacture Notice, 10
Premixed flame, 132
Present worth, 158
Pressure, *see also* Duct design
  correction, 580
  critical, (A-10) 660
  dynamic, 335
  stagnation, 335
  standard, 119

Pressure (*Continued*)
  static, 335
  vapor pressures, Table, (A-8) 655
  velocity pressure, 335
Primary particles, 18, 439, Fig 8.5, 440
Probability distribution, 434. *See also*
    Distribution functions
Probability plot, 441
Process modification, 19
Product change, 19
Projected area equivalent diameter, 433
Propaganda, 11
Properties:
  of air, Table, (A-11) 661
  of hydrocarbons, (A-8) 655
Proteoglycan, 58
Pseudovelocity, 579
Puff diffusion, 179, Fig 4.3, 180, Fig 4.4, 181
  illustrated example, 181
Pulmonary fibrosis, 96, 98
Pulmonary region, Fig 2.1, 51, Fig 2.5, 54, 55, 56
Pulmonary toxins, 101
Pulmonary ventilation, Table 2.2, 64
  mass transfer, 72
  spirometry, 61
Pulsatile flow in lungs, 68
Pulse injection technique, 274
Pulvation distance, 454
Push–pull ventilation, 41, 594
  ACGIH recommendations, Fig 6.21, 333
  analysis of flow field, 594
  boundary conditions, Fig 10.13, 595
  concentrations, Fig 10.16, 599
    illustrated example, 333
  Schlieren, Fig 10.17, 599, 600
  velocities, Fig 10.15, 598, Fig 10.18, 601

Quiet breathing, 55

Radiation:
  nuclear, 99
  thermal, 148
Radioactivity, 100
Radon:
  carcinogenicity, 99
  "daughters of," 99
Rain, 18
Rankine curve, 129
Rankine–Hugoniot, 128
Re, 193
Reach, illustrated examples:
  flanged line sink, streaming flow, 380, 525
  flanged point sink, 385
  unflanged inlet, streaming flow, 531
Reach of inlets, 303, 361, 379
  definition, 517
  line source in quiescent air, 519

unflanged inlet in streaming flow, illustrated
    example, 535
  uniform streaming flow of particles, 519
Receiving hood, 31, 40, Fig 1.28, 42, Fig 6.4, 296
Recirculation, 246
Recirculation eddy, 324
Rectangular slot, 392
  streaming flow, 409
Rectilinear flow, 362
Regimes:
  drag:
    Newtonian, 430
    Stokes, 430
    transition, 430
  flow:
    continuum, 430, 432
    free-molecular, 432
    transition, 432
Regional deposition, respiratory system, Fig
    2.35, 93, Fig 2.36, 94
Registry of Toxic Effects of Chemical
    Substances, 8
Relative humidity, 149
Relative velocity, 429, 430, 455
Relaxation coefficient, 542
Relaxation time, 449
Reproductive toxin, 102
Residence time, 273
Residual drag, 494
Residual volume, 60
  functional, Fig 2.11, 61
Respirable particles, 18, 95
Respirators, Fig 1.4, 22
  program, 23
Respiratory system, *see also* Airway
  abnormalities, Fig 2.13a&b, 62, 63
  air space, Fig 2.1, 51, 55
  alveolar ventilation, Table 2.2, 64
  bronchioles, 52
  clearance, 93
  components, Fig 2.1, 57
  deposition, 94
  disease categories, 96–98
  fluid mechanics, 59
  forced expiratory volume, 62
  gas composition, Table 2.3, 73
  mass transfer membrane, Fig 2.19, 71, Fig
    2.20, 72, Fig 2.26, 78
  modes of gas motion, Fig 2.21, 72
  particle deposition, 94
  physiology, 49
  pulmonary region, 55
  quotient, 147
  respiratory rate, Table 2.2, 64
  toxicology, 92
Response, 101
  dose-response class, Fig 2.40, 106

Revenue requirements (TRR), 154
Reverse-flow cyclone, Fig 8.24, 476
    standard, 476, Fig 8.26, 477
Reynolds number (Re), 193
    accelerating drop, 486
    airway generations, Fig 2.7, 57
    analogy, 194, 211, 597
    averages, 588
    ducts, 336
    generalized transport equations, 573
    particles, 430
    stress, 591
    venturi scrubbers, 486
Richardson number, 604
Rim exhauster, Fig 1.25, 40
    ACGIH recommendations, Table 6.4, 305
    analysis of flow field, 581
    concentrations, Fig 10.11, 587
    sketch, Fig 10.7, 585
    velocities, Figs 10.8–10.9, 586
Risk assessment, 6
    voluntary and involuntary, 1
Risk vs. benefit analysis, Fig 1.1, 2
Risk perspectives, 11
Risks, Table 1.2, 13
Roof eddy, 327
Room air changes, 230, 237
Rotational flow, 365, 467
    illustrated example, 365
Roughness:
    height, 336
    parameter, 596
RTECS, 8
Runge–Kutta, 239, 457, (A-12) 663

Sampling, particles, 519
    efficiency, Fig 9.12, 524
    errors, 520, 521
    isokinetic sampling, Fig 9.8a, 521
Sanitation, 19
Saturation pressure and temperature, 149, 233,
    (A-8) 655. See also Clausius–Clapeyron
    equation
Sauter diameter, 433, 487
Sc, 193
Scavenging by rain, 480
Schmidt number (Sc), 193
Scientific method, 116
Scrubbers:
    spray chambers, 480
    venturi, 484
Secondary air, 311
Secondary particles, 18, 439
Sedimentation, 453
    bronchial tree, 93
    chambers and enclosures, 461
    ducts, 462

Selecting a ventilation system, 29
Self-contained breathing apparatus, 22
Sequential box model, 264, Fig 5.15, 266
    illustrated example, 269
Series collectors, 446
Settling:
    chamber and enclosures, 460
    ducts, 462
    laminar, 461, 463
    respiratory system, 94
    settling velocity, 448, Fig 8.13, 450
        illustrated example, 451
    terminal velocity, 449, Fig 8.13, 450
    unit density spheres, 434, Table 8.6, 451
    well-mixed 461, 463
Sh, 193
Shape factor, dynamic, 433, Table 8.2, 434
Shear stress, 363, 597, 598
Sheet Metal & Air Conditioning Contractors
    National Association, 46
Sherwood number (Sh), 193
Short-time exposure limit, 121
Side draft hood, Figs 1.14–1.16, 34–36
Silica oxide, 98
Silicosis, 98
Similarity, 466
SIMPLE, SIMPLER, 577
    algorithm, 581
Simultaneous differential equations,
    Newton–Raphson, (A-13) 666
Single fiber collection efficiency, 488
Single film theory, 198, 200
Single particle collection efficiency, 479, 488,
    Fig 8.34, 489
Singular points, 375
Sink line, 376
Sink point, 382
SI units, conversions, (A-18) 676
Size distribution, 434. See also Distribution
    functions; Number distribution,
    concentration
Slip factor, Table 8.1, 432
Slip flow, 432
Slot:
    flanged, 392
    plain, 392
    rectangular, 392
    sampling, Fig 9.10, 524
    unflanged, 392
SMACNA, 46
Smog, 18
Smoke, 18
Smoking emissions, (A-7) 653
Sneeze, 55
Snorkel, Fig 1.27, 41
Snowfence, 551
Soil erosion, 552

Solid body rotation, 366, 367, 467
Solid fraction, 490
Solubility of gases in mucus and blood, 78, 79
Sound:
  definition, 138
  directivity, 139
  ear frequency response, Fig 3.10, 142
  energy and dB, Fig 3.12, 144
  exposure index, 143
  fans, blowers, 144
  hearing, 142
  illustrated example, 141
  impact, 139
  impairment and age, Fig 3.13, 145
  inlets, Fig 3.14, 146
  intensity, 139
  level, decibels, 140
  occupational standards, Table 3.3, 143
  permissible noise exposure, Table 3.3, 143
  power, 139
  pressure and dB, 130
  from several sources, 141
  sound level scales, 143
  source characteristics, Fig 3.12, 144
  speed, 139
Source, contaminant, 30
  area source, 30
  characteristics, 259
  fugitive, Fig 1.9, 31, 310
  inertial, 310
  net source strength, 256
  point, 30, 310
  strength, 236
  virtual, 315
Source term in differential equations, 573,
    Table 10.1, 574
Spatial uniformity, 228
Species continuity equation, 187
Specification requirements, 117
Specific resistance of dust cake, 494
Sphere, flow field around moving sphere,
    Table 7.3, 391
  stationary sphere, Fig 7.10, 390
Spillage, 316, 601
Spills, 174
Spirometry, 60, Fig 2.11, 61, Fig 2.13a, 62
Splash, 311
  bulk materials handling, 310
  filling, 17, Fig 6.2, 295, Fig 6.12, 313
Spot cooling, 149
Spray:
  chamber, 480, Fig 8.29, 481
  counter-flow, 480
  cross-flow, 484
  finishing, 177
  particles, 18
Spurious air currents, drafts, 29, 328
Stack, 24
  height, 328

Staggered grid, 577, Fig 10.5, 578
Stagnant air, diffusion, 187
Stagnation:
  point, 379, 384, 386
  pressure, 335
Stairmand-type cyclone, 476
Standard contaminant exposure levels, 120,
    (A-1) 633, (A-8) 655, (A-10) 660
Standard deviation, particles:
  arithmetic, 435
  geometric, 437
Standard reverse flow cyclones:
  cut diameter, Fig 8.26, 477
  grade efficiency, 477
  performance, 477
  proportions, Fig 8.26, 477
Standard temperature and pressure, 119
Star states, 206
Static pressure, 335
Statistics, *see* Distribution functions
Steady flow, 364
Steady state, definition, 237
  concentration, 253
STEL, 121
Step-down, tracers, 274
Step-up, tracers, 274
Stirred, perfectly:
  duct flow, 463
  enclosures, 228, 460
Stockpiles, windbreaks, 552
Stokes:
  drag coefficient, 448
  flow regime, 431, 448
  number, 465, 466
  uncoupling equations of motion, 455
Stopping distance, 454
Stoves, wood, coal, kerosene, 172
STP, 119
Straight-through cyclone, Fig 8.23, 474
Stream function, 370, Table 7.2, 374, Table 7.3,
    391, Table 7.4, 392
Streaming flow, 377
  flanged inlet, 403
  unflanged inlet, 403
Streamline, 364, 372
Stream tube, 370
Stress:
  eddy, 591
  heat stress index, 149
  Reynolds, 591
  turbulent, 591
Strict liability, 16
Subisokinetic sampling, 523
Submerged filling, 174
Submicron particles, 18
Substantial derivative, 594
Substitution, 19
Successive over-relaxation, 542
Superficial velocity, 493

Superisokinetic sampling, 523, Fig 9.8, 521
Superposition, 374, 456
Supply-air hoods, 309
Surface area equivalent diameter, 433
Surface metal treatment classification,
    Table 6.5, 306
Surfactant, 57
Swing grinder, Fig 1.28, 42
Switching point, 56
Symmetrical Weibel model, 55
System boundary, 228, 565
Systemic toxicity, 101

Taylor-type dispersion, 69
TCC, 153
TD_{50}, 13
TDC, 155
Temperature:
    critical, (A-10) 660
    globe, 148
    standard, 119
Temperature and water flux in airway, 86
Temporal uniformity, 364
Teratogens, 102
Terminal bronchioles, 55
Terminal velocity, 449, Fig 8.13, 450
TFC, 155
Thermal conductivity, 363, 573, (A-11) 661
    eddy thermal conductivity, 592
    expansion coefficient, 139, 572
    homeostasis, 146
Thermodynamics, first law, 146
Thermophysical properties of air, (A-11) 661
Thermoplastic coating, 177
Threshold limit values:
    ACGIH, 120
    ceiling, 121
    gas mixtures, 121
    liquid mixtures, 121
    short-time exposure limits, 121
    time-weighted average, 122
Threshold values:
    contaminant, (A-1) 633, (A-8) 655, (A-10) 660
    noise, Table 3.3, 143
    odors, Table 3.6, 152
Throw, 361
Tidal volume, 60, Fig 2.11, 61
    normal tidal volume, 61
Time:
    constant, well-mixed model, 253
    local mean age, 272
    nondimensional, 466
    relaxation, 449
    residence, 273
    value of money, 158
Time-varying source, 238
    illustrated example, 238
    ventilation rate, 238
Time-weighted average, 121

Tissue gel, 58
TLV, *see* Threshold limit values
Tobacco emissions, 172, (A-7) 653
Tort, 16
Total costs, *see also* Engineer economics
    capital cost (TCC), 153
    direct cost (TDC), 155
    fixed cost (TFC), 155
    indirect cost (TIC), 155
    revenue required (TRR), 154
    variable cost (TVC), 155
Total derivative, 594
Total head loss, 335
Total lung capacity, 60, Fig 2.11, 61
Total molar concentration, 198
Total pressure, 335
Toxicants, Table 2.1, 50
Toxicity:
    local, 101
    systemic, 101
Toxicology, 49, 92
Toxic Substances Control Act:
    CAS, 8
    CIS, 8
    MSDS, 8
    PMN, 10
    RTECS, 8
Toxin:
    classificaiton, 101
    pathways, Fig 2.38, 101
Tracer techniques, step-up, step-down, 274
Trachea, 52
Tracheobronchial region, Fig 2.2, 52
Trajectory of particle:
    computer program, 457
    illustrated problem, 457
Transfer of gas across respiratory membrane,
    Fig 2.26, 78
Transfer point, 310
Transition flow regime, drag coefficient, 430
Translocation, Fig 2.38, 101
Transplacental carcinogen, 102
Transport property, eddy, 592
Transport velocity, ducts, 334
Transverse spray chamber, 484
Tridiagonal matrix, 542
Tridymite, 98
TRR, 154
Trupent lung model, 81
TSCA, 8
Tuberculosis, 97
Tunnels, 565, Fig 10.1, 566
    illustrated example, 568
    local make-up air, 565
    natural, 565
    transverse, 566
    uniform make-up, no exhaust, 566
Turbulence:
    degree, 589

Turbulence (*Continued*)
  diffusivity, 184, 593
  eddy transport properties, 593
  eddy viscosity, 592
  energy dissipation (ε), 593
  heat flux, 593
  intensity, 589
  kinetic energy (κ), 594
  level, 589
  Prandtl number, 593
  Reynolds stress, 591
  Schmidt number, 593
  velocity scale, 589
Turbulent deposition, *see also* Well-mixed model
  chambers and enclosures, 462
  curved ducts, 472
  straight ducts, 463
Turbulent flow model (κ-ε), 588
TVC, 155
TWA, 121
Two dimensional flow, 375
Two-film theory, 200
  illustrated example, 200

ULPA filter, 262
Uncoupling equations of motion, 455
Under-relaxation, 598
Unflanged inlets, 402. *See also* Flow
  slot:
    quiescent air, 402
    streaming flow, 409
  velocity contours, 300–303
Uniform flow, 271, 377
  plug flow, 271
Uniformity:
  spatial, 228
  temporal, steady-state, 237, 253
Uniform streaming flow, 383
Unit density spheres, 43
  settling velocity, Table 8.6, 451
Units, Table of Conversions, (A-18) 676
Universal gas constant, 119
Unsatisfactory performance, 328
Unventilated space, 232
  illustrated example, 234
Upper airways, 52
Upper explosion limit, 132
Uptake absorption efficiency, 76
Uvula, 55

Vapor pressure, (A-8) 655
Variance, 435
Veins, 58
  venules, 58
Velocity:
  average, 589
  conservative field, 369
  contours of inlets, 300–303
  dimensionless, 466, 543
  friction, 597
  head, 335
  irrotational flow, 367
  pressure (VP), 335
  relative, 429, 430, 486, 487
  rotational, 367
  settling, 449
  terminal, 449
Velocity potential, 369
  function, Table 7.2, 374, Table 7.3, 391,
    Table 7.4, 392
Ventilation, *see also* Hoods
  air, requirements, 230
    ASHRAE, (A-16) 673
  ACGIH design procedures, 304
  alveolar, 61
  classifications, 31
  components, 24
  design criteria, 116
  dilution, 31, 228
  fresh air requirements, 230, (A-16) 673
  general, 31, 228
  local, 31, 294
  manual, 44, 294
  method of selection, 29
  performance standards, 117
  rate, lung, 61, Table 2.2, 64
  systems, *see* Hoods
Ventilation–perfusion ratio, 65, Table 2.2, 64
Venturi scrubber, 484, Fig 8.31, 485
Venules, 58
Vesicles, Fig 2.8, 58
Vessel filling, Fig 6.2, 295, Fig 6.12, 313
Virtual diffusion, 70
Virtual source, 315
Viscosity:
  dynamic, 430
  eddy, 592
  effect of temperature, (A-11) 661
  kinematic, 430
Viscous dissipation, 363
Viscous drag, 430
Vital capacity, 60, Fig 2.11, 61
VOC, 177
Void fraction, voidage, 490
Volatile organic compounds, 177
  critical temperatures and pressures, (A-10) 660
  diffusion coefficients, (A-9) 658
    Henry's law constant, (A-9) 658
  vapor pressures, (A-8) 655
Volume equivalent diameter, 433
Volumetric concentration:
  gas or vapor:
    mass, 118
    PPM, 119
Voluntary risk, 1, Fig 1.1, 2
von Karman constant, 596

Vortex, 362, 366
Vorticity, 365
VP, 335

Wakes:
    buildings, 324, Fig 6.19, 325, Fig 6.20, 326
    grinding booth, 547
Wall:
    adsorption, 244
    function, 596
    law of the wall, 596
    losses, 244
    plate-out, 244
    settling, 244
    shear stress, 597
    temperature, 148
Warning devices, 19
Washout, 480
Waste disposal, 19
Weber–Fechner Law, 153
Weibel symmetric model, 55, Fig 2.6, 56
Weibull model, 109
Welding, Fig 1.15, 34
Welding in confined spaces, 310
Well-mixed model, 228
    adsorption on walls, 244
    basic equation, 252
    Bohr model, 62
    characteristics, 258

compartment model, respiratory system, 76
deposition:
    curved ducts, 466
    enclosures, 228, 461
    straight ducts, 463
extended Bohr model, 76
make-up air, 246
mixing factor, 250
multicell model, 264
plate-out, 244
sequential box model, 264
steady-state, 253
well-stirred model, 228
Wet etching, 261
Whole population fatalities, 2
Windbreaks, 552
    collection efficiency, Fig 9.34, 556, Fig 9.35, 557
    particle trajectories, Fig 9.33, 555
    velocity profiles, Fig 9.32, 554
Womersley Number, 70
Wood-burning stoves, 2, 172, (A-7) 653
Work accidents, Table 1.1, 4
Work practices, 19
Worth:
    future, 158
    present, 158
Woven fiber filter, 487